OXFORD MEDICAL PUBLICATIONS

Textbook of Fetal Physiology

Textbook of Fetal Physiology

Edited by

GEOFFREY D. THORBURN

and

RICHARD HARDING

Department of Physiology, Monash University,
Melbourne, Australia

Oxford   New York   Tokyo
OXFORD UNIVERSITY PRESS
1994

Oxford University Press, Walton Street, Oxford OX2 6DP

Oxford   New York   Toronto
Delhi   Bombay   Calcutta   Madras   Karachi
Kuala Lumpur   Singapore   Hong Kong   Tokyo
Nairobi   Dar es Salaam   Cape Town
Melbourne   Auckland   Madrid
and associated companies in
Berlin Ibadan

Oxford is a trade mark of Oxford University Press

Published in the United States
by Oxford University Press Inc., New York

A catalogue record for this book is available from the British Library

Library of Congress Cataloging in Publication Data
Textbook of fetal physiology/edited by Geoffrey D. Thorburn and Richard Harding.
Includes bibliographical references and index.
1. Fetus–physiology.   I. Thorburn, Geoffrey D.   II. Harding, Richard, Dr.
[DNLM: 1. Fetus–physiology.   2. Fetal Development.   WQ 210.5 T355 1994]
RG610.T48   1994   612.6'4–dc20   93–28855

ISBN 0 19 857748 6

Typeset by EXPO Holdings Sdn. Bhd., Malaysia
Printed in Great Britain by
Butter & Tanner Ltd., Frome, Avon

Dedication

This book is dedicated to the memory of the late John Elgin Patrick, MD, FRCS(C), 1943–1990 — a superb scientist and obstetrician, and a very good friend.

Preface

Fetal physiology is the physiology of, roughly, the last two-thirds of intra-uterine life. It concerns the development of the organism from the end of the period of embryonic development until parturition. This book is focused on fetal physiology in placental mammals, with considerable emphasis placed on the human fetus. The scope of fetal physiology is, of necessity, very broad; it is broader than the physiology of the adult, as it encompasses the physiology of the placenta, the mother, and the major changes that occur at birth. It is a relatively recent science, due largely to the development over the last two decades of techniques for studying the undisturbed fetus *in utero*. Much of our present knowledge has been derived from studies of the ovine and human fetus, although in recent years studies of non-human primates and rodent species, as well as *in vitro* techniques, have yielded invaluable information on physiological and pathophysiological processes during early development. The discipline of fetal physiology has now become an integral part of medical courses and undergraduate, as well as graduate, training programmes in the biomedical sciences. Partly, this is due to the recognition that the quality of fetal development, and of the transition to postnatal life, has a substantial bearing on the quality of postnatal life through to adulthood.

Over the last decade, it became apparent to many of us who teach undergraduate students, and who conduct graduate programmes in fetal physiology, that there was a need for an up-to-date, comprehensive textbook on the subject. Although there is an ever-increasing number of sources of information on specific aspects of prenatal development, such as reviews and monographs, there was no single, comprehensive text which focused on the fetus and its relationship to its mother. We saw the need for a textbook of fetal physiology which would act as a companion to classical texts on adult physiology, and which would suit the needs of teachers, students, clinicians, and the research community. To undertake this task, we sought the help of John Patrick who we believed represented the epitome of the scientist-obstetrician, and who had successfully married the clinical practise of obstetrics with the science of perinatal physiology. It was John's plan to add a section at the end of each chapter explaining the clinical relevance of the text. His untimely death in 1990 brought our ideas to a sudden halt. When we started to recover from the shock of his death we realised that John would have wanted us to continue with the book. At that time, we could not think of anyone who could replace John as co-editor and play the vital role he was to play in the writing and editing of this book. We decided, instead, to ask some of John's close colleagues who had been fortunate enough to work with him and had been imbued with the same philosophy, to write the chapters which he would have written. We thank them for their excellent contributions.

The two of us have entered the fascinating field of fetal physiology via different routes. Geoff Thorburn trained in medicine and subsequently completed his MD on the measurement of blood flow using the inert gas clearance technique. In 1963, after training with Paul Korner and A. Clifford Barger in cardiovascular physiology, he moved to the Commonwealth Scientific and Industrial Research Organization laboratories at Prospect, New South Wales, where his research on reproductive physiology commenced. He was instrumental in the development of the chronic fetal sheep preparation which led to his research in fetal and placental endocrinology and the initiation of parturition. In 1972 he moved to Oxford, UK, to head a Medical Research Council programme grant at the Nuffield Institute for Medical Research. He returned to Australia in 1978 where, with Richard Harding, he established a research group with broad interests in fetal development, first at the University of Queensland, and then at Monash University in Melbourne.

Richard Harding trained in physiology and pharmacology, and completed a PhD on the neural control of ruminant gastric motility. With an interest in sensory physiology, he undertook postdoctoral study at the Nuffield Institute for Medical Research, working on a project with Paul Johnson and Geoffrey Dawes concerned with the effects of laryngeal stimulation on respiratory control in the neonate. For a deeper understanding of the development of neonatal laryngeal and respiratory reflexes, studies were undertaken in the fetal sheep, and these studies led, inevitably, to a strong interest in broader aspects of fetal development. During this period at the Nuffield Institute, the editors, both of whom originated from Australia, met and developed a lasting friendship and collaboration. After Oxford, they both moved to Brisbane and subsequently to Monash University, where they still work on aspects of fetal and neonatal physiology.

We trust that the *Textbook of fetal physiology* will engender in the reader as much interest in fetal physiology as this field has for us.

Melbourne, Australia   G. D. T.
October 1993   R. H.

Acknowledgements

The production of this book would not have been possible without the support and encouragement of many of our colleagues. We are particularly indebted to the late John Patrick and to G.C. (Mont) Liggins for their invaluable advice in the planning stages. We are also indebted to the chapter authors for their diligence and patience throughout the production process. Each chapter was subjected to review by experts in fetal physiology. Comments from reviewers were of value for enhancing the clarity and overall quality of the chapters. Our reviewers included: L. Aitkin, S. Brennecke, C. Browne, I. Caple, A. Care, N. Fisk, A. Fowden, C. Gibbs, S. Hooper, G. Liggins, C. McMillen, M. Payne, U. Proske, M. Ralph, S. Rees, G. Rice, J. Robinson, J. Schwartz, R. Short, M. Simpson Morgan, G. Taylor, D. Walker, I. Wendt, and M. Wlodek.

We are also grateful to Judi Herschell, Lynne Hepburn, Jan Deayton, and Kerryn Billings for secretarial assistance, and to Jill Poynton, Dianne Clare, and Michelle Mulholland for their assistance with artwork and photography.

The support and professional guidance of the staff at Oxford University Press have been essential to the production of this book, and their help is gratefully acknowledged.

Finally, we are indebted to our wives, Alison Thorburn and Wendy Harding, and to our families, for their patience and understanding through this project.

Contents

Contributors

Philip L. Ballard, Division of Neonatology, Children's Hospital of Philadelphia, Philadelphia, Pennsylvania, USA.

Alan D. Bocking, The Lawson Research Institute, St Joseph's Health Centre, 208 Grosvenor Street, London, Ontario N6A 4V2, Canada.

Richard Boyd, Department of Human Anatomy, University of Oxford, South Parks Road, Oxford OX1 3QX, UK.

Robert A. Brace, Department of Reproductive Medicine, University Hospital, University of California, Medical Center, 225 Dickinson Street, San Diego, California 92103-8433, USA.

Iain C. Bruce, Department of Physiology, University of Hong Kong, Hong Kong.

Ross N.P. Cahill, Laboratory for Foetal and Neonatal Immunology, School of Veterinary Science, University of Melbourne, Parkville, Victoria 3052, Australia.

Barbara Cannon, The Wenner-Gren Institute, The Arrhenius Laboratories F3, Stockholm University, S-106 91 Stockholm, Sweden.

Harold A. Coleman, Department of Physiology, Monash University, Clayton, Victoria 3168, Australia.

Alan J. Conley, Cecil H. and Ida Green Center for Reproductive Biology Sciences, University of Texas Southwestern Medical Center, 5323 Harry Hines Boulevard, Dallas, Texas 75235-9051, USA.

Geoffrey S. Dawes, 8 Belbroughton Road, Oxford OX2 6UZ, UK.

Delbert A. Fisher, UCLA School of Medicine, Harbor-UCLA Medical Center, 1000 West Carson Street, RB-1, Torrance, California 90509, USA.

Maria Fitzgerald, Department of Anatomy and Developmental Biology, University College and Middlesex School of Medicine, Gower Street, London WC1E 6BT, UK.

Abigail L. Fowden, Physiological Laboratory, Downing Street, Cambridge CB2 3EG, UK.

Deborah K. Froh, The Cardiovascular Research Institute, School of Medicine, University of California, San Francisco, San Francisco, California 94143-0130, USA.

Robert Gagnon, The Lawson Research Institute, St Joseph's Health Centre, 208 Grosvenor Street, London, Ontario N6A 4V2, Canada.

Peter Gluckman, Faculty of Medicine, University of Auckland, Private Bag, Auckland, New Zealand.

Edward N. Guillery, Department of Pediatrics, University of Iowa, Hospitals and Clinics, Iowa City, Iowa 52242, USA.

Victor K.M. Han, The Lawson Research Institute, University of Western Ontario, 268 Grosvenor Street, London, Ontario N6A 4V2, Canada.

Richard Harding, Department of Physiology, Monash University, Clayton, Victoria 3168, Australia.

William W. Hay, Jr, Division of Perinatal Medicine and Research, University of Colorado School of Medicine, 4200 E 9th Avenue, Box B195, Denver, Colorado 80262, USA.

David J. Hill, The Lawson Research Institute, University of Western Ontario, 268 Grosvenor Street, London, Ontario N6A 4V2, Canada.

Oussama Itani, Division of Neonatology, University of Cincinnati Medical Center, 231 Bethesda Avenue, Cincinnati, Ohio 45267-0451, USA.

Glen Jeffery, Department of Visual Science, Institute of Ophthalmology, University of London, Bath St., London EC1V 9EL, UK.

Graham Jenkin, Department of Physiology, Monash University, Clayton, Victoria 3168, Australia.

Pedro A. Jose, Department of Pediatrics, University of Iowa, Hospitals and Clinics, Iowa City, Iowa 52242, USA.

Wayne G. Kimpton, Laboratory for Foetal and Neonatal Immunology, School of Veterinary Science, University of Melbourne, Parkville, Victoria 3052, Australia.

Yoshiki Kudo, Tokyo Women's Medical College, 8-1 Kawada-Cho, Shinjuku-ku, Tokyo 162, Japan.

Graham C. Liggins, Department of Obstetrics and Gynaecology, University of Auckland, National Women's Hospital, Claude Road, Auckland, New Zealand.

J. Ian Mason, Cecil H. and Ida Green Center for Reproductive Biology Sciences, University of Texas Southwestern Medical Center, 5323 Harry Hines Boulevard, Dallas, Texas 75235-9051, USA

I. Caroline McMillen, Department of Physiology, University of Adelaide, Adelaide, South Australia 5001, Australia.

Murray D. Mitchell, Department of Obstetrics and Gynecology, University of Utah School Medicine, Room 2B200, 50 North Medical Drive, Salt Lake City, Utah 84132, USA.

David R. Moore, University Laboratory of Physiology, Parks Road, Oxford OX1 3PT, UK.

Mark Morton, Department of Physiology, School of Medicine, Oregon Health Sciences University, 3181 S W Sam Jackson Park Road, Portland Oregon 97201-3098, USA.

Peter W. Nathanielsz, Laboratory for Pregnancy and Newborn Research, New York State College of Veterinary Medicine, Cornell University, Ithaca, New York 14853, USA.

Jan Nedergaard, The Wenner-Gren Institute, The Arrhenius Laboratories F3, Stockholm University, S-106 91 Stockholm, Sweden.

Julie A. Owens, Department of Obstetrics & Gynaecology, University of Adelaide, GPO Box 498, Adelaide, South Australia 5001, Australia.

Philip C. Owens, Division of Human Nutrition, CSIRO, Adelaide, South Australia 5001, Australia.

Helena C. Parkington, Department of Physiology, Monash University, Clayton, Victoria 3168, Australia.

Daniel H. Polk, UCLA School of Medicine, Harbor-UCLA Medical Center, 1000 West Carson Street, RB-1, Torrance, California 90509, USA.

Uwe Proske, Department of Physiology, Monash University, Clayton, Victoria 3168, Australia.

John A. Rawson, Department of Physiology, Monash University, Clayton, Victoria 3168, Australia.

Sandra M. Rees, Department of Anatomy and Cell Biology, University of Melbourne, Parkville, Victoria 3052, Australia.

Bryan S. Richardson, Lawson Research Institute, St Joseph's Health Centre, 268 Grosvenor Street, London, Ontario N6A 4V2, Canada.

Jean E. Robillard, Department of Pediatrics, University of Iowa Hospitals and Clinics, Iowa City, Iowa 52242, USA.

Jeffrey S. Robinson, Department of Obstetrics & Gynaecology, University of Adelaide, GPO Box 498, Adelaide, South Australia 5001, Australia.

Daniel W. Rurak, The Research Centre, Department of Obstetrics and Gynecology, Faculty of Medicine, University of British Columbia, 950 West 28th Street, Vancouver, British Columbia V52 4H4, Canada.

Jeffrey L. Segar, Department of Pediatrics, University of Iowa Hospitals and Clinics, Iowa City, Iowa 52242, USA.

Arthur Shulkes, Department of Surgery, University of Melbourne, Austin Hospital, Melbourne, Victoria 3084, Australia.

Francine G. Smith, Department of Obstetrics and Gynaecology, University of Calgary, 3330 Hospital Dr. N.W., Calgary, Alberta T2N 4N1, Canada.

Kent L. Thornburg, Department of Physiology, School of Medicine, Oregon Health Sciences University, 3181 S W Sam Jackson Park Road, Portland, Oregon 97201-3098, USA.

Jeffrey F. Trahair, Child Health Research Institute, Adelaide Medical Centre for Women and Children, 72 King William Road, North Adelaide, South Australia 5006, Australia.

Reginald C. Tsang, Perinatal Research Institute, University of Cincinnati Medical Center, 231 Bethesda Avenue, Cincinnati, Ohio 452878-0451, USA.

David W. Walker, Department of Physiology, Monash University, Clayton, Victoria 3168, Australia.

Elizabeth A. Washington, Laboratory for Foetal and Neonatal Immunology, School of Veterinary Science, University of Melbourne, Parkville, Victoria 3052, Australia.

Randall B. Wilkening, Division of Perinatal Research, University of Colorado School of Medicine, 4200 E 9th Avenue B195, Denver, Colorado 80262, USA.

Jeremy S.D. Winter, Section of Endocrinology and Metabolism, Department of Paediatrics, University of Alberta, 671 Heritage Medical Research Centre, Edmonton, Alberta T6G 2S2, Canada.

Charles E. Wood, Department of Physiology, J Hillis Miller Health Center, Box J-274, College of Medicine, University of Florida, Gainesville, Florida 32610, USA.

Abbreviations

AA	arachidonic acid
ACS	apical canalicular/tubule system (see AEC)
ACh	acetylcholine
ACTH	adreno-corticotrophic hormone
ADP	adenosine diphosphate
AEBP	auditory evoked brainstem potential
AEC	apical endocytic complex (see ACS)
aFGF	acidic fibroblast growth factor
AG	antigen
AGA	appropriate for gestational age
AMP	adenosine monophosphate
AN	auditory nerve
ANF, ANP	atrial natriuretic factor/peptide
APUD	amine precursor uptake and decarboxylating
ASD	atrial septal defect
ATP	adenosine triphosphate
ATPase	adenosine triphosphatase
AV	arterio-venous
AVP	arginine vasopressin
BALT	bronchial-associated lymphoid systems
BAT	brown adipose tissue
bFGF	basic fibroblast growth factor
BPD	bronchopulmonary dysplasia
BPS	biophysical profile score
BSV	bovine seminal vesicle
CaBP	calcium binding proteins
cal	calory
CAM	cell adhesion molecule
cAMP	cyclic adenosine monophosphate
CBG	cortisol binding globulin
CCK	cholecystokinin
cDNA	complementary deoxyribonucleic acid
CDP	cytidine diphosphate
CFC	capillary filtration coefficient
CG	chorionic gonadotrophin
CGRP	calcitonin gene-related peptide
cGMP	cyclic guanosine monophosphate
CHO	Chinese hamster ovary
CNS	central nervous system
CoA	coenzyme A

CRE	cAMP response elements
CRF, CRH	corticotrophin-releasing factor/hormone
CRL	crown rump length
CSF	cerebrospinal fluid
CT	calcitonin
CTNF	ciliary neuronotrophic growth factor
D	delivery
DA	dopamine
DAG	diacylglycerol
DCN	dorsal cochlear nucleus
DEAS	see DHAS
DEX	dexamethasone
DGD	Asp-Gly-Asp
DHA (DHEA)	dehydroepiandrosterone
DHAS (DEAS)	dehydroepiandrosterone sulfate
DHPR	dihydropyridine receptor
DHT	dihydrotestosterone
DIT	diiodotyrosine
DLF	dorsolateral funiculus
DNA	deoxyribonucleic acid
DOC	deoxycorticosterone
DPPC	dipalmitoyl phosphatidylcholine
DRG	dorsal root ganglion
E	embryonic (stage)
ECF	extracellular fluid
EAL	electronic artificial larynx
eCG	equine chorionic gonadotrophic
ECM	extracellular matrix
ECoG	electrocorticogram
EDRF	epithelium derived relaxing factor
EDTA	ethylene diamino tetra-acetic acid
EEG	electroencephalogram
EGF	epidermal growth factor
EIPS	endogenous inhibitor of prostaglandin synthase
eLH	equine luteinizing hormone
EMG	electromyogram
EPSP	excitatory postsynaptic potential
FBM	fetal breathing movements
Fc	crystallizable portion (of IgG)
FF	fast fatiguable (muscle fibre)
FFA	free fatty acid
FGF	fibroblast growth factor
FHR	fetal heart rate
FR	fatigue-resistant (muscle fibre)

FRC	functional residual capacity
FSH	follicle stimulating hormone
FVW	flow velocity waveform
GABA	gamma amino butyric acid
GAG	glycosaminoglycans
GALT	gastrointestinal-associated lymphoid system
GAP	growth associated protein
GBM	gross body movements (of the fetus)
GDP	guanosine diphosphate
GFR	glomerular filtration rate
GH	growth hormone
GI	gastrointestinal
GIP	gastric inhibitory peptide,
Glut 1-5	glucose transporters 1-5
GM-CSF	granulocyte-macrophage colony-stimulating factor
GnRH	gonadotrophin-releasing hormone
GRF	growth hormone releasing factor
GRP	gastrin releasing peptide
GTP	guanosine triphosphate
h	hour
haPL	hamster placental lactogen
hCG	human chorionic gonadotrophin
hCS	human placental chorionic somatomammotrophin
HDL	high density lipoprotein
HETE	hydroxyeicosatetraenoic acid
hGH	human growth hormone
hLH	human luteinizing hormone
HMGCoA	3-hydroxy-3-methylglutaryl co-enzyme A
HPETE	hydroperoxyeicosatetraenoic acid
hPL	human placental lactogen
hPRL	human prolactin
HSD	hydroxysteroid dehydrogenase
HSOR	hydroxysteroid oxidoreductase
iCa	serum ionized calcium
Ig	immunoglobulin
IGF	insulin-like growth factor
IGFBP	insulin-like growth factor binding protein
IP_3	inositol triphosphate/trisphosphate (IP_3)
IRMA	immuno-radiometric assay
IU	international units
IUGR	intrauterine growth retardation (restriction)
IUP	intrauterine pressure
J	joule

Kb	kilobase
Kbp	kilobase pairs
kDA	kiloDalton
K-FGF	Karposi sarcoma fibroblast growth factor
KGF	keratinocyte growth factor
kPA	kilopascal

LDL	low density lipoprotein
LGN	lateral geniculate nucleus
LH	luteinizing hormone
LHRH	luteinizing hormone-releasing hormone
L/S ratio	lecithin to sphingomyelin ratio
LT	leukotriene
LV	left ventricle

M cells	microfold cells
MDI	iodothyronine monodeiodinase
mGH	murine growth hormone
MGN	medial geniculate nucleus
MHC	major histocompatibility complex
MIS	Müllerian inhibitory substance
MIT	monoiodotyrosine
MLCK	myosin light-chain kinase
mPL	murine placental lactogen
M_r	molecular weight
mRNA	messenger ribonucleic acid
MSH	melanocyte-stimulating hormone

NCAM	neural cell adhesion molecule
NGF	nerve growth factor
NLL	nuclei of the lateral lemniscus
NMDA	N-methyl d-aspartate
NMR	nuclear magnetic resonance
non-REMS	non-rapid eye movement sleep
NPY	neuropeptide Y
NRM	nucleus raphe magnus
NRPG	nucleus reticularis paragigantocellularis
NST	non stress test (change in FHR during uterine contraction)

oPL	ovine placental lactogen

P_{50}	partial pressure at which haemoglobin is 50% saturated with oxygen
P	pressure
P	permeability (of capillary)
PAF	platelet-activating factor
PAGE	polyacrylamide gel electrophoresis

PAI	plasminogen activator inhibitor
PAPSI	pregnancy-associated prostaglandin synthase inhibitor
PC	phosphatidylcholine
PCA	posterior cricoarytenoid (muscle)
PCO_2	partial pressure of carbon dioxide
PDGF	platelet derived growth factor
PG	prostaglandin
PGD_2	prostaglandin D_2
PGDH	15-hydroxy-prostaglandin dehydrogenase
PGE_2	prostaglandin E_2
PGEM-II	11-deoxy-13,14-dihydro-15-keto-cyclo-prostaglandin E_2
$PGF_{2\alpha}$	prostaglandin $F_{2\alpha}$
PGFM	13,14-dihydro-15 keto-prostaglandin $F_{2\alpha}$
PGHS	prostaglandin H synthase (cyclo-oxygenase)
PGI_2	prostacyclin
PGT	placental glucose transport (capacity)
pI	isoelectric point
PLA_2	phospholipase A_2
PL	placental lactogen
PNMT	phenylethanolamine-methyl transferase
PNS	peripheral nervous system
PO_2	partial pressure of oxygen
POMC	pro-opiomelanocortin
PP12	placental protein 12
PRA	plasma renin activity
PRL	prolactin
PTH	parathyroid hormone
PTHRP	parathyroid hormone-related peptide
PTU	propylthiouracil
PUBS	percutaneous umbilical blood sampling
$\dot{Q}$	flow
RBF	renal blood flow
RDS	respiratory distress syndrome
REM	rapid eye movements
REMS	rapid eye movement sleep
Rf	rate of infusion
RGD	Arg-Gly-Asp
RNS	renal nerve stimulation
ROO	peroxide radical
rPL	rat placental lactogen
rT_3	reverse tri-iodothyronine
RT-PCR	reverse transcription – polymerase chain reaction
RV	right ventricle
RVR	renal vascular resistance

S	surface area (of capillary)
SCG	superior cervical ganglion
SCN	suprachiasmatic nucleus
S/D ratio	peak-systolic flow velocity/end-diastolic flow velocity
SDS	sodium dodecyl sulfate
SE(SEM)	standard error (of the mean)
SHBG	sex hormone-binding globulin
SIF	small intensely fluorescent (cells)
SNB	spinal nucleus of the bulbocavernosus muscle
SOC	superior olivary complex
SP	surfactant protein
SP	Substance P
SP_1	pregnancy specific β-glycoprotein
SP-A(-D)	surfactant proteins A-D
SR	sarcoplasmic reticulum
SRIF	somatostatin
SWS	slow-wave sleep
T_3	tri-iodothyronine
T_4	thyroxine (tetra-iodothyronine)
TA	thyroarytenoid (muscle)
TBG	thyroxine-binding globulin
TcR	T cell receptor
TDF	testis determining factor
TeBG	sex steroid-binding globulin
TG	triglycerides
TG	trigeminal ganglion
TGF	transforming growth factor
TIMP	tissue inhibitor of metalloproteinases
Tm	tubular maxima
TMP	thiamine monophosphatase
$TR\alpha,\beta$	thyroid hormone receptors α and β
TRF	thyrotrophin-releasing factor
TRH	thyrotrophin-releasing hormone
TSH	thyroid stimulating hormone (thyrotrophin)
TSH	thyrotrophin
$TxA_2(B_2)$	thromboxane A_2 (B_2)
VAS	vibroacoustic stimulation
VCN	ventral cochlear nucleus
VGF	pox virus genes (viral growth factor)
VIP	vasoactive intestinal peptide
VLDL	very low density lipoprotein
$\dot{V}O_2$	oxygen consumption
VSD	ventricular septal defect
W	watt

Z	impedance
6-keto-PGF$_{1\alpha}$	6-keto-prostaglandin F$_{1\alpha}$
25 OHD	25 hydroxyvitamin D
1,25(OH)$_2$D	1,25 dihydroxyvitamin D
24,25(OH)$_2$D	24,25-dihydroxyvitamin D
3-QNB	3-quinylclidinyl benzylate
5-HT	5-hydroxytryptamine
α-MG	α-methyl-D-glucopyranoside
α MSH	α melanocyte-stimulating hormone

1. Fetal physiology: historical perspectives

Geoffrey S. Dawes

Sir Joseph Barcroft and researches on prenatal life, 1926–46

The modern history of fetal physiology goes back to the experiments of A. St G. Huggett who, when a young lecturer in St Mary's Hospital, London, did the first successful experiments to measure the relation between the partial pressure of oxygen (PO_2) in the blood of the mother and that in the blood of the fetus. Huggett had been brought up in a farming district in north Wales and had seen fetal lambs delivered. He found a discarded bath-tub in the hospital grounds and filled it with saline solution that was kept warm by Bunsen gas burners beneath the bath. He immersed a pregnant sheep, under anaesthesia, in the bath and opened the uterus beneath the saline so that he could take blood samples from the fetus under conditions as natural as possible given the facilities available. He demonstrated that the PO_2 of the fetus was much below that of the maternal arterial blood. Barcroft in Cambridge had made an extensive study of the oxyhaemoglobin dissociation curve in mammalian blood under a wide variety of circumstances. In 1913 he had recorded the first measurement of the O_2 saturation of human arterial blood (in 180 words!). He repeated Huggett's observations and was able to show that the fetal oxyhaemoglobin dissociation curve was much to the left of the maternal, a fact that was validated and investigated extensively in a variety of species during the next 15 years. But, more importantly, it led Barcroft to recognize that Huggett had designed an experimental preparation in which it would be possible to investigate more directly both fetal development and the first elements of the 'architecture of physiological function', his elegant phrase for the integration of physiological systems in the adult.

Thus, during the years between 1930 and 1939, Barcroft and his colleagues in Cambridge laid the foundations of fetal physiology. It happened that in 1936 Barcroft was at a meeting of the Physiological Society in London and went on to Oxford by train in the company of K.J. Franklin, a fellow of Oriel College. Franklin had made his reputation in the physiology of veins. In 1935 Sir William Morris, later Lord Nuffield, gave a generous benefaction to the University of Oxford to found a postgraduate medical school and the Nuffield Institute of Medical Research housed in the former Radcliffe Observatory, a magnificent stone building, based on the Tower of the Winds in Athens. Franklin had engaged the services of A.E. Barclay, a radiologist, to develop cineangiography of the venous system at the Nuffield Institute. Barcroft had been present at Franklin's communication to the Physiology Society and saw the possibility of applying this novel method to solving the long-standing problem of the course of the circulation in the fetus. Between 1936 and 1940 Barcroft, Franklin, Barclay, and Prichard collaborated in joint studies in Oxford. They showed the division of inferior vena caval blood flow (by the crista dividens) into streams entering the left atrium through the foramen ovale and the right atrium directly. (It is regrettable that some diagrams of the fetal circulation still show the foramen ovale as lying between the atria.) And they also showed the joint contribution of blood flow from the left ventricle (via the aortic arch) and the right ventricle (via the ductus arteriosus) to descending aortic blood flow.

Barcroft and his colleagues in Cambridge also undertook a wide investigation of fetal physiology, measuring the growth of the fetus and its component parts, the relationship between fetal and placental weight, the development and composition of the body, and the fetal heart rate and its relation to blood pressure. They also began the first serious studies of central nervous development in the fetal lamb. All these investigations were carried out on exteriorized fetuses, i.e. withdrawn from the uterus in a saline bath under general, spinal, or local anaesthesia. Towards the end of these investigations a young man named Don Barron came to Cambridge from New England to study with Barcroft. This collaboration had long-term effects on the development of the subject. Barron's philosophy was to try and make observations under conditions that were as natural as possible. He was a determined seeker after truth and, when he returned to Yale, he set up a laboratory that attracted many young North American workers whose influence on the subsequent development of the subject was great. During the war (1939–45) all fetal research came to a stop.

Acute fetal preparations under anaesthesia, 1946–65

The most important event in the early 1950s for the history of fetal physiology was the meeting in 1954 at Cold Spring Harbor, New York. This brought the new group of basic scientists who had begun to work on fetal physiology together with the clinicians who were concerned with the future of perinatal medicine. There were several new contributions that had long-term consequences for the subject. First, there was the demonstration by Jost that decapitation of fetal rabbits at an early stage of gestation did not prevent the subsequent growth of their trunk and limbs. This demonstrated that fetal growth was not under the control of growth hormone secreted by the pituitary and that the subsequent development of the heart and circulation, the trunk, and the limbs could take place independently of the central nervous system for a limited period of time. These observations marked the beginning of fetal endocrinology. The experiments were simple, elegant, and of great importance.

Second, Kenneth Cross reported that the initial hyperventilation, on exposure of new-born human infants to 15 per cent oxygen, was followed by a decrease in ventilation to or below the initial level, an observation that continued to puzzle paediatricians and physiologists for many years to come.

I was asked to describe the changes in the circulation at birth. I had been appointed in 1948 to direct the Nuffield Institute for Medical Research on the retirement of Professor Gunn and was searching for a subject that might afford a closer connection with clinical medicine than the analysis of the cardiac and pulmonary (von Bezold) reflexes on which I was then engaged. I received a request from Sam Reynolds of the Carnegie Institute in Washington to spend a sabbatical year (1950) in Oxford to test an hypothesis of the mechanism of venous return from the placenta, using the unique cineangiographic equipment developed in the 1930s. This led me to look critically at the recent literature and to realize that no explanation had been offered of how ventilation of the lungs at birth caused a redirection of the fetal circulation. There were not even measurements of vascular pressures in the different parts of the circulation. So, after preliminary experiments in 1950, we measured the changes in pressures and blood flows through the fetal lungs before and after ventilation and across the ductus arteriosus. It became evident that this was a more complicated system than it had first appeared to be. Blood flow through the ductus arteriosus reversed in direction and a murmur appeared. Blood flow through the lungs increased greatly due to vasodilatation; the mechanisms that controlled this continued to provide food for thought and experimentation for many years.

During the next 10 years steady progress was made in a number of different fields using, for the most part, exteriorized fetal lambs. Experiments on fetal lambs from 1950 onwards were rarely carried out in a warm saline bath. An operating table was constructed to hold the ewe, on her side, with a smaller adjustable warmed table elevated a few inches and pressed against the maternal abdomen, to ease delivery of the fetus without tension on the umbilical cord. This gave direct access to the fetus. Experiments were then undertaken either under general anaesthesia (of both mother and fetus) or using maternal spinal, epidural, or local anaesthesia and local anaesthesia of the fetus. This permitted many types of experiments on physiological mechanisms and the development of function with gestational age, but precluded study of fetal behaviour or of normal blood gas values (for example) under wholly natural conditions. These preparations were used to establish the lower limit of oxygen delivery at which survival could be maintained with artificial ventilation of the lungs and to establish the mechanisms that controlled the fetal circulation near term and at earlier gestational ages. Kenneth Cross in London and my group in Oxford combined in experiments on resuscitation of new-born animals, which showed that the use of drugs or hyperbaric oxygen to establish regular breathing was less satisfactory than artificial positive-pressure ventilation. A start was also made at trying to unravel some of the mechanisms by which the young fetus was able to withstand prolonged periods of hypoxia and asphyxia, a study made on a variety of species including rats, rabbits, sheep, and rhesus monkeys. Prevention of the fall in pH was found to prolong breathing movements, to facilitate resuscitation, and prevent or reduce brain damage.

In 1955 Pattle working in a defence research establishment in Porton (in the south of England) was studying the behaviour of bubbles in lung fluid from animals subjected to war gases. His highly original observations on their stability ultimately led to the discovery that premature infants deficient in lung surfactant developed the respiratory distress syndrome. This new subject area of great clinical importance was subsequently opened up by John Clements in Edgewood and Mel Avery and Jerry Mead in Boston, Massachusetts. This aspect of perinatal physiology did not always require large animal experiments. It is strange to reflect that even now, 30 years after the initial discovery, there is as yet no international agreement on an effective form of replacement therapy for surfactant deficiency after premature birth. Although the bovine derivative looks very promising, its source is thought by some to be a cause for concern.

As a student I remember that the outcome of rhesus disease was described as three distinct clinical entities.

However, with the discovery of the rhesus factor our understanding of its pathophysiology was set on a firm foundation. The discovery by Clarke of a method of prevention, facilitated by his interest in the genetics of butterflies, has much reduced its incidence in developed countries. Nevertheless, some cases still occur. Bill Liley, in Auckland, New Zealand, recognized the importance of measuring the products of fetal red cell haemolysis in the amniotic fluid, as a test of the progress of the disease. In the later 1950s he showed that health could temporarily be restored by infusion of red cells into the fetal peritoneum, whence they were absorbed into the vascular system from the lymph drainage of the peritoneal surface of the diaphragm, as described by Florey.

This work of Liley marks the beginning of fetal medicine. Within the next few years amniotic fluid samples began to be used for the analysis of factors that might indicate the presence of other genetically determined diseases. Saling showed that it was possible to get direct information on the acid–base status of the fetus in labour from fetal scalp blood samples. Hence, even in man more direct physiological studies became possible. It was another 20 years (i.e. about 1985) before samples were taken directly from fetal blood vessels *in utero*, before the onset of labour.

By the 1960s several laboratories in Europe, Australia, and North America were busily engaged in studying different aspects of fetal physiology. As early as 1953 extensive surgery had been done on fetal lambs, in Australia with survival *in utero*, to study immunological tolerance. In the 1960s there were two further technological developments. Don Barron's laboratory in Yale developed the first reported chronic fetal sheep preparations in which catheters were implanted into fetal vessels *in utero* to measure fetal arterial blood gas tensions and pH under normal physiological conditions with the ewe unanaesthetized and unrestrained. This, and the work of Liggins (see the next section) opened a whole new area for studies of fetal physiology which are of great importance to perinatal medicine. The second technological advance was by Rudolph and his colleagues, who measured systemic blood flow and cardiac output in unanaesthetized fetuses *in utero* using isotope-labelled microspheres. In particular, they were able to show for the first time that the outputs of the two ventricles of the fetus were normally different, that of the right heart exceeding that of the left heart. At about the same time prostaglandin, which had been discovered by U.S. Von Euler in 1934, was found to cause uterine contractions. Thus within a few years the vast gap, which physiologists deplored, between the steroids and the contractile mechanism of uterine musculature was partly filled. The scene was set for an explosion in knowledge.

Chronic fetal preparations, 1966 onwards

Mont Liggins was a young obstetrician who was inspired by work with Linton Snaith in Newcastle-on-Tyne and Bill Lilley in Auckland, New Zealand to wonder about the future of fetal endocrinology and medicine. Several observations had pointed to the idea that the onset of parturition might be dependent on the fetus. In an elegant series of experiments Liggins was able to demonstrate that in singleton ovine pregnancies destruction of the fetal pituitary prevented the onset of labour. He also showed that the adrenals were involved, that infusion of adrenocorticotrophin (ACTH) in a hypophysectomized fetal lamb led to parturition, and that the active agent was secretion of cortisol rather than a mineral corticoid. It was the use of operative procedures on the fetus *in utero*, under general anaesthesia, and with full recovery of both mother and fetus that made these observations possible.

It is interesting to quote from a grant application of Liggins in 1964 in which he says: 'The field of fetal physiology has been a somewhat neglected one when compared with the work done on the neighbouring fields of life — embryonic and neonatal. This has been largely due to the lack of suitable techniques by which to come to grips with the fetus in its protected environment. Until recently, experiments have been usually of an acute nature and often under most unphysiological conditions. Techniques such as those described above allow a direct approach while at the same time preserving a physiological state as far as possible'. What he had just described was bilateral fetal adrenalectromy and hypophysectomy, with recovery and without interruption of pregnancy, and separation of the sources of amniotic fluid, with identification of a high concentration of surface-active material from a plastic bag covering the fetal head.

He got the grant, which took him to the University of California in Davis, and later found that pulmonary surfactant production could be enhanced by the administration of dexamethasone. The result was an explosion of interest in fetal endocrinology that is still not exhausted. For we still do not understand in sufficient detail the control of fetal growth by hormonal or other local factors that are the subject of so much present study. Similarly, although we know much about the control of parturition in ungulates, we are still uncertain of the ultimate control (the subject of so many feed-back and feed-forward mechanisms) that determines the time of their parturition so closely. In primates the mechanisms are yet more complex.

The new, improved chronic fetal sheep preparations also made possible investigations of fetal breathing movements and the rediscovery that such movements are normally present in man. They had been described by a German obstetrician in the 1880s, whose interpretation was rejected by his contemporaries. In 1970 it became clear that fetal breathing movements near term were associated in the sheep with rapid eye movement sleep characterized by low-voltage electrocortical activity. The way was also now open for studies of fetal behaviour, both in chronic sheep preparations and, since 1975, in the human infant *in utero*, using the new methods of ultrasound observation of fetal movements, of the fetal heart rate, and of blood velocity profiles in fetal blood vessels. This has been an interesting development because it has demonstrated the hazard of extrapolation from experiments undertaken when the fetus can change behavioural states and thereby alter physiological patterns. Without adequate control studies under such different states it is all too easy to draw the wrong conclusions, just as the pulsatile release of hormones or their seasonal variation, once appreciated, has led to problems in interpretation.

One of the interesting features of biological research is how commonly simultaneous discoveries of the same general phenomenon appear in widely separated laboratories across the world. For example, the appearance of fetal sleep states in late gestation was reported as a consequence of electrocortical and other measurements from Yale, Oxford, Toulouse, and Rochester, USA within a 2-year period, 1970–72. It was as if the common threads of knowledge had prepared the ground; the time was ripe. Yet the reasons for the experiments undertaken were different. Similarly in the early 1960s Dawkins and Hull, working at the Nuffield Institute for Medical Research in Oxford on the origin of the large heat production by the new-born rabbit upon exposure to cold, discovered the thermogenic property of brown adipose tissue, independently, but at the same time as workers in North America reached similar conclusions as to its importance in recovery from hibernation.

But it is not always so. Key observations by van Neergaard in 1929 on pulmonary surface tension, calculated from the pressure–volume curves of either air- or fluid-filled lungs, were ignored until Clements, who had been impressed by Pattle's (1955) report, used a dynamic method to show that compression of a surface film of lung liquid caused the measured surface tension to fall dramatically. This provided a quantitative basis for Pattle's expectation that 'absence of the lining substance may sometimes be one of the difficulties with which a premature baby has to contend; such a defect may possibly play a part in causing some cases of atelectasis neonatorum'. Similarly, it has been pointed out that the logical basis for continuous positive airway pressure (to prevent alveolar collapse) in ventilating infants with the respiratory distress syndrome was established in 1959, but it was not until 10 years later that Gregory and his colleagues in San Francisco demonstrated that it worked in practice.

The past 25 years have seen a general recognition of the complexity of physiology and development in fetal life. I, like some others, started with the idea that studying the fetus, where physiological systems might be expected to be simpler, might unravel some of the complexities which make interpretation difficult in the adult. To a certain extent this is true. A good example is the discovery that hypoxia causes an immediate arrest of fetal breathing movements. This has provided a satisfactory explanation for Kenneth Cross's discovery of the biphasic response to hypoxia in the new-born (hyperventilation followed by hypoventilation). And it has led to the observation that the basic neural mechanism by which hypoxia causes respiratory depression, or arrest, persists in the adult. But this is an isolated example. More importantly, we recognize that cellular interaction is the basis of systems physiology. Growth and the development of the organs and limbs segmentally in the embryo are determined by the expression at the right time and in the appropriate cell line of the appropriate determinants of differentiation and growth. Thus, in early fetal life the development of the organ systems and their behaviour is determined by cytokines rather than by systems mechanisms and cytokine control of cellular function continues to underlie all physiological systems.

Towards the future

In the last few years, studies in different laboratories across the world, using different methods, have shown that the mature fetal lamb adapts rapidly to prolonged hypoxia. The complex changes of heart rate and arterial pressure and the arrest of breathing and of other fetal movements, disappear within some hours, before the oxygen-carrying capacity of the fetal blood is increased by accelerated red cell formation. This unexpectedly rapid adaptation is yet to be explained. Similarly, we seek an explanation of the gradual adaptation of the systemic arterial chemoreceptors after birth, ultimately relating to postnatal rather than antenatal values. More than 20 years ago I gave up working on the mechanism(s) by which the pulmonary vasculature vasoconstricted in response to hypoxia (a change better suited to fetal than adult life and first discovered by U.S. von Euler), because it was already then evident that the tools for unravelling such cellular mechanisms were not yet to hand. It seems likely that mechanisms at the cellular level must be responsible for the other features of fetal adaptation to hypoxia. As more is learned about the general features of fetal physiology, the problems have become more complex, not simpler.

We are now entering a phase where the tools of molecular biology can help to unravel some of the aspects of fetal development, especially the development of fetal control mechanisms. Fetal physiology is still developing and showing every sign of vigour. There is now evidence of its future importance in adult medicine. Recently, David Barker has shown that the incidence of ischaemic heart disease in the adult is related directly to placental size and inversely to birth weight. We do not know whether this association is genetically determined or whether it might be a consequence of the unusual fetal: placental imbalance during the last trimester, for example, over the critical period when the systemic cardiovascular system is established. The long-term effects of fetal development on adult behaviour and health are still to be unravelled.

In the immediate future Barker's observations suggest that the factors that control placental growth deserve more consideration. There are hundreds of scientific papers, on many species, on the factors that may determine fetal growth, but few on placental growth. Perhaps too much weight has been given to the fetus. Current experimentation, in producing fetal growth retardation by a variety of means, suggests that the placenta gets the first grab at O_2, glucose, and other maternal metabolites, and may survive at the expense of the fetus.

The history of fetal physiology has so far been dominated by enquiry for its own sake, not primarily aimed at practical application. Indeed, applied research, notably into trying to deduce the health of the fetus from observations on heart rate, has not been outstandingly successful. Nevertheless, the causes of human premature birth (infection? or multiple pregnancy? the latter now so commonly associated with *in vitro* fertilization or gamete intra fallopian transfer (GIFT) and of perinatal cerebral damage are of such practical importance as to deserve the continued attention of physiologists. Historically, there has been a close and fruitful link between perinatal physiologists and clinicians. Though their primary interests may differ, there is a broad area of common knowledge and concern that deserves fostering.

Further reading

Barclay, A.E., Franklin, K.J., and Prichard, M.M.L. (1944). *The fetal circulation and cardiovascular system, and the changes that they undergo at birth.* Blackwells Scientific Publications, Oxford.

Barcroft, J. (1946). *Researches on prenatal life.* Blackwells Scientific Publications, Oxford.

Dawes, G.S. (1968). *Foetal and neonatal physiology.* Year Book Medical Publishers, Chicago.

Gluckman, P.D., Johnston, B.M., and Nathaniels, P.W. (ed.) (1989). *Advances in fetal physiology: reviews in honor of G.C. Liggins.* Perinatology Press, Ithaca, New York.

Wolstenholme, G.E.W. and O'Connor, M. (1969). *Foetal autonomy.* J. & A. Churchill Ltd, London.

2. Transport functions of the placenta and fetal membranes

Richard Boyd and Yoshiki Kudo

In the human, the maternal and fetal circulations are separated by the trophoblast, its underlying basal lamina, and the mesenchymal elements in the core of the chorionic villus. Amongst these elements there is the endothelium of the fetal capillary circulation. At term the trophoblast is made up of a single cellular barrier (the syncytiotrophoblast) which has the unique property of being a true syncytium lacking a paracellular route between adjacent cells. Earlier in gestation the trophoblast is made up of two cell layers, the syncytiotrophoblast lying external to the cytotrophoblast which provides a population of stem cells. The trophoblast (as shown in Fig. 2.1 (a),(b)) thus forms an epithelium, albeit an unusual one because of its syncytial structure, which at its external surface lies in direct contact with the maternal circulation; at its basal surface this epithelium is separated by a basal lamina and some loose connective tissue from the underlying fetal capillary circulation.

It is important when considering placental transport to remember the epithelial nature of the barrier and in particular to note that, as in all other epithelia, the trophoblast cell possesses two quite separate plasma membranes, the one facing the mother (the maternal or apical brush border) the other facing the fetus (the fetal or basal plasma membrane). The contrasting properties of the transport systems present in the two surfaces of the trophoblast underlie the ability of the placenta to produce overall net active transport either from mother to fetus (for example, for amino acids) or from fetus to mother (for example, for bile acids). This will be discussed later in this chapter.

In general terms, the factors that determine the rate of diffusion J of substances across an epithelium depend on the concentration gradient $(C_1 - C_2)$ across the epithelium; the permeability P of the epithelium to the substance; the surface area S of the epithelium, and the thickness x of the diffusion barrier. This is described by Fick's law of diffusion,

$$J = \frac{P \times (C_1 - C_2) \times S}{x}$$

For substances that are very permeable (e.g. CO_2) the rate of transfer across the epithelium becomes limited by the rate of delivery of the substance by the blood flowing to the epithelium: such substances are described as being 'flow-limited'. More often, however, membrane permeability is rate-limiting ('membrane-limited') and for such substances (e.g. glucose) the rate of transport across the epithelium will not be influenced by changes in blood flow. When substances are transported actively against an overall electrochemical gradient, the epithelium itself must be powering the transfer of this substance at the input or output membrane of the epithelium. Only under pathological conditions will such transport become flow-limited.

The epithelium that the trophoblast forms at the surface of the chorionic villus has the important property (poorly understood in structural terms) of possessing a slight leak permeability to macromolecules. This interesting topic has recently been studied in the human by intravenous infusion into the mother before Caesarean section of molecules such as inulin that are then found to be present after delivery in the urine of the baby. It is possible that normally occurring areas of unusual anatomy may provide the structural basis for such findings.

Another potential route for transfer between maternal and fetal circulations is via extra-placental fetal membranes, particularly the chorion and amnion. In humans, the quantitative importance of this route in the transfer of any solute or solvent is probably minimal, but in other species such as the sheep a substantial movement of water across the allantoic membrane has been described.

Cellular basis of transport

The asymmetry of membrane transport systems between the apical and basal surfaces of the placenta (indicated above) is matched by an asymmetry in the composition of the two plasma membranes with respect to all components including lipids and proteins. For example, electron microscopy reveals the predominant localization of the glycocalyx, composed of glycoproteins and glycolipids, to the apical surface of the trophoblast (it is this biochemical feature that allows ready biochemical separation of the apical from the basal plasma membrane of the trophoblast as will be described later). The distribution between the two faces of the placental epithelium of the transport proteins themselves (be they pumps, carriers, or channels) has, to date, been studied by measurement of their function; information

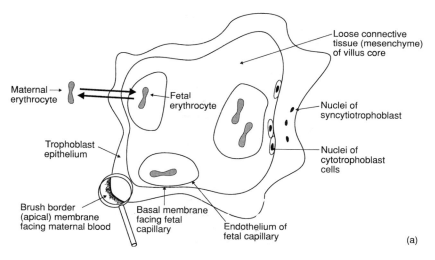

Maternal → erythrocyte

Fetal erythrocyte

Loose connective tissue (mesenchyme) of villus core

Nuclei of syncytiotrophoblast

Trophoblast epithelium

Nuclei of cytotrophoblast cells

Brush border (apical) membrane facing maternal blood

Basal membrane facing fetal capillary

Endothelium of fetal capillary

(a)

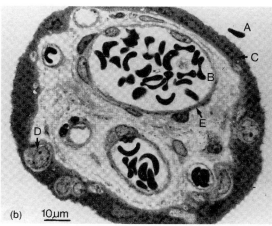

(b) 10 μm

Fig. 2.1 (a) Schematic illustration of the chorionic villus structure of human placenta. (b) Electron micrograph showing structural features separating maternal and fetal circulations in the term human placenta. Note the clearly separated apical and basal surfaces of the placenta and the relationship of the fetal blood vessels to the trophoblast. (A) Maternal erythrocyte in the intervillus space; (B) fetal erythrocyte in large fetal capillary; (C) syncytiotrophoblast; (D) cytotrophoblast (this layer of cells is nearly completely absent by this stage of gestation); (E) fetal capillary endothelium.

at the molecular level on the distribution of these transport proteins is only just emerging: doubtless our knowledge of this area will increase very substantially in the near future.

Pumps may be defined as transport proteins that are directly driven by ATP-hydrolysis. (Formally, active transport is divisible into primary active transport, which is ATP driven, and secondary active transport in which the translocation of one molecule against its electrochemical gradient is driven by coupling to the flux of another down its electrochemical gradient, as in sodium-coupled transport systems). ATPases, which are membrane-bound, are therefore good candidates as potential pumps (although obviously many other ATP-driven biochemical processes occur in association with the cell membrane other than the translocation of solutes across the plasma membrane). Thus, when the placental epithelium is fractionated into its constituent apical and basal membranes, it is the basal plasma membranes that are found to possess the $Na^+ K^+$ ATPase activi-

ty. This cellular location of the sodium pump is confirmed by the distribution of binding sites for [^{3}H] ouabain determined by autoradiography. This basal location is exactly that predicted by analogy with quite different epithelia, for example, those of the small intestine, an epithelium with which the placenta shares many common design features.

A large number of carrier (transporter) proteins have been identified in the plasma membranes of the trophoblast. In addition to functional studies described below, binding of ligands (e.g. cytochalasin B for glucose transport proteins) has also been used. Very recent work on this transporter in the placenta has used molecular methods to identify which isoform of the glucose transport family is expressed at the maternal-facing and the fetal-facing sides of the trophoblast. Placental glucose transport appears to be independent of insulin and is sodium-independent (in other words, it occurs by facilitated transport rather than by secondary active transport).

With regard to the third group of membrane proteins involved in transport, namely channels, the placenta is as yet largely unexplored. The presence of potassium channels (calcium-activated) and of chloride channels in the apical (brush border) surface of the placenta have recently been described; both are likely to be important for regulating the membrane potential across this surface of the trophoblast. Much remains to be done to characterize the single channel properties of these and of other channels; and it will be important in particular to identify mechanisms leading to their activation.

Experimental approaches to investigating placental transport

In vivo studies have been widely used with experimental animals but for obvious reasons must be of very limited scope in humans. Important work has been done recently using simultaneous umbilical and maternal blood sampling, for example, on amino acid transport. However, most investigations on human placental transfer have depended upon *in vitro* methods, ranging from studies using placental perfusion to experiments using plasma membranes isolated from the trophoblast *in vitro*. It is important in assessing such studies to consider the strengths and weaknesses of each experimental approach. It is also important to remember the distinction between one-way (unidirectional) and net transport since *in vitro* studies in which isotopically labelled solutes have been used will readily permit the determination of unidirectional

flux but will not directly allow one to assess the direction of overall net transport across the placenta, be it from mother to fetus or in the reverse direction. Bidirectional (two-way) flux determination, in which the same molecule is present on each side of the placenta but labelled with different isotopes (e.g. ^{22}Na and ^{24}Na), allows the net transport to be estimated as the difference between the two unidirectional isotope fluxes; and from this it may be possible to use the flux ratio (Ussing) equation to determine whether active transport is or is not involved.

It is also possible to measure the short-circuit current across the placenta *in vitro*, if the geometry of the tissue permits the transepithelial potential difference to be determined, and if current is applied which is of the right magnitude and polarity to abolish the spontaneous transepithelial potential. This short-circuit current is a measurement of the active transport processes driving ion movement across the tissue. Measurement of this, however, is limited to those placentae that are geometrically sufficiently simple (which the human placenta is not) to allow such electrical monitoring. When analysing the transport of any electrolyte (be it inorganic or organic) it is important to remember that electrical potentials (either across the placenta or across the individual cell membrane of the trophoblast) will be as important as are chemical gradients in influencing transport. This will apply to any organic molecule that is charged at physiological pH, or in considering the transport of any neutral (uncharged) species that is co-transported with an ion.

One *in vitro* method that has been widely used in the last decade is that of isolated plasma membrane vesicles

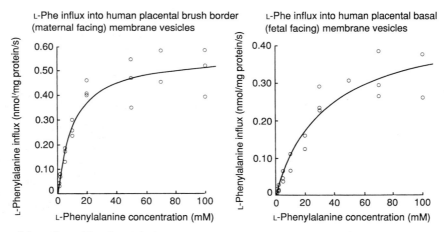

Fig. 2.2 Transport of the amino acid L-phenylalanine into isolated plasma membrane vesicles from, on the left, the apical surface (brush border), and on the right, from the basal surface of the term human placenta. Note that the kinetics of transport differ between the two membranes with the brush border having a higher affinity (lower K_t) than that of the basal surface.

prepared from either the maternal- or from the fetal-facing surface of the trophoblast (Fig. 2.2). Such isolated membranes re-seal to form closed structures (vesicles) that contain a compartment that is separated from the bulk phase of the experimental medium by a purified plasma membrane containing the transport proteins that are found in that membrane *in vivo*. It is therefore very straightforward to study each step of the transport pathway in isolation and, for example, to determine the driving forces responsible for solute translocation. Examples of this will be given in the next section.

Diffusive, facilitated, and active transport

Gas transfer

As with the lung, the transport of the respiratory gases across the placental epithelium is driven solely by diffusion; transfer will be optimized by the design of the blood flow to both surfaces of the placenta and by the attenuation of the trophoblast diffusion barrier. Gas transfer is considered in detail in Chapter 9 and will not be further described here other than to point out, with respect to carbon dioxide transport, that the handling of bicarbonate ions is dependent upon the rate of hydration of CO_2 and that this will depend on the activity of the carbonic anhydrase isoenzyme found in the trophoblast. Since bicarbonate is an excellent substrate for a variety of anion exchange proteins found in both brush border and basal surfaces of the placenta, the movement of HCO_3 through such proteins will indirectly contribute to the rate of CO_2 clearance from fetus to mother.

Water transfer

Rather little is known of the physical constants that will determine the rate at which water will flow across the placenta in any species; virtually nothing is known in the human. The possible forces driving water flow across the placenta are transplacental hydraulic pressure and osmotic pressure gradients. However, in the absence of knowledge of the reflection coefficients for macromolecules at the trophoblast or knowledge of the hydraulic conductivity, it is not at present possible to describe quantitatively which of these are more important determinants of the rate of flow of water from mother to baby. However, when the rate of water production by metabolism in the fetal compartment is accounted for, the net rate of water transport towards the baby must be sufficient to match fetal growth. The route of water transport across the placenta is also not

known and work is needed to characterize the proteins responsible for water permeability in the plasma membranes of the trophoblast.

Lipophilic transfer

Because of the chemical nature of the cell membrane it has been widely assumed that the high permeability of epithelium to lipophilic molecules (e.g. anaesthetics) reflects a non-specific permeability through the lipid bilayer. However, the evidence for this is indirect and the finding that certain inhibitors will slow transfer may make this conclusion less likely than was once thought. The absence of saturation in the rate of transfer as a function of concentration of the molecule does not preclude the possibility that a specific mechanism is involved, since linear kinetics only imply that cell membrane transport has not become rate-limiting under the conditions investigated. Certainly, with lipophilic molecules with a high maximum rate of transport it is very important to consider the possibility that diffusion to the membrane (unstirred layers) may be limiting access to the transport system involved. Thus the experimental procedure that is used to investigate transport will be very important since, in the absence of adequate mixing, unstirred layers may produce diffusion barriers that will be absent *in vivo*.

One particular group of lipophilic molecules of considerable biological importance are the fatty acid chains of lipids. Transport of free fatty acid across the placenta is quantitatively important in the human with its haemochorial placenta but probably much less important to the nutrition of the fetus in the ruminant with its epitheliochorial placental structure. In the human the proportion of fatty acid transferred from mother to fetus is much greater for short-chain (< 16 carbon atoms) than for longer-chain molecules, and the lipid composition of fetal tissues is changed very substantially by maternal diet. In addition to the widespread transport of such fatty acids across the placenta there is more specific transport to the fetus of essential fatty acids. These transporters may be powered (i.e. active) since transport, for example, for arachidonic acid to the fetus, remains high even when maternal concentrations are low.

In terms of fetal energy balance, placental lipid transport is probably of increasing importance as gestation proceeds since in early fetal life the fat stores are very low and may be accounted for by fatty acid synthesis by the fetal liver. The large increase in stores of fetal lipid later in gestation is probably accounted for by increased placental delivery from the mother. It seems probable that maternal circulating very-low-density-lipoproteins provide the source of this triglyceride, which is split by a placental lipoprotein lipase shown to be present only on the maternal surface of the placenta. The activity of this lipase appears to increase substantially towards the end of gestation leading to local release of free fatty acid as well as of monoglyceride and glycerol. The uptake of the released fatty acids is probably by mechanisms very similar to those found in other tissues such as the plasma membrane of the liver. In other words what determines the rate of transport is largely the concentration gradient between maternal circulation and trophoblast as well as the total area of the diffusion barrier. It seems probable that receptor-mediated uptake of cholesterol-rich low-density lipoproteins is receptor-mediated and involves endocytosis and trophoblast lysosomal degradation. The release in the trophoblast of cholesterol is probably important for the synthesis of placental steroid hormones (e.g. progesterone). Since cholesterol is readily synthesized by the fetal liver, it is uncertain how essential transfer of maternal cholesterol into the fetal compartment actually is.

Fatty acids, after entering the placenta, may be used for direct transfer to the fetus or as substrates for placental energy needs. Alternatively, they may be used for glycerolipid synthesis or cholesterol esterification in the placenta before being released into the fetal circulation. The nature of the regulation of those different pathways in the trophoblast is not known, but the existence of these various possible routes emphasizes that the placenta should not be viewed as being inert metabolically.

Solute transfer

Ions

Because of the importance of electrical factors in determining conductive transport a description of the pathways available for ion flow does not necessarily indicate the rate of ion flow through any particular pathway. For ion movement through channels, two factors will determine the ion flux: the channel open-state probability and the driving force acting on the ion. The latter is determined by the prevailing electrochemical gradient across the particular plasma membrane of the placenta (apical or basal) in question. Although a start has been made in identifying ion channels using patch-clamp techniques, no coherent overview relating physiological phenomena to molecular transport properties can yet be made. For organic solutes it is very important to remember that the charge may be influenced by the pH of the environment in which the particular molecule is present. Thus, for example, the pK of an organic molecule will determine whether, at a given intracellular pH, the transport of this solute across the basal membrane is wholly or only partially electrogenic (influenced by membrane potential).

Cations

Sodium transport is classically described in epithelia by the Ussing model of electrodiffusive sodium entry across the apical surface followed by active sodium efflux via the sodium pump located in the basal membrane. The evidence for such a system in placenta is not complete, but it is known that sodium entry at the apical brush border surface is in part electrogenic with a further substantial flux being used to power either Na-coupled co-transport systems (e.g. amino acids) or antiporters (e.g. the sodium proton exchanger).

Potassium transport across the brush border membrane has the interesting and unexpected property that its permeability is very strongly gated by the pH gradient across this membrane. External acidification increases potassium permeability and the pK for this activation is about 6.9. This implies that the physiological response of the placenta to maternal acidosis will be hyperpolarization of the brush border with a concomitant increase in electrogenic driving force for solutes whose transport is sodium-coupled. Potassium channels in the brush border of the placenta are also regulated by circulating maternal factors (e.g. arachidonic acid).

The transport of the divalent cations such as calcium and magnesium has not been investigated at the membrane level although there is a substantial literature on the development processes responsible for regulation, in particular, of calcium transport by vitamin D.

Anions

Chloride transport across the brush border membrane is by at least two pathways — one electrogenic and likely to be through chloride channels; the other electroneutral and inhibited by stilbene sulfonate derivatives such as DIDS. This latter pathway is shared with other anions such as sulfate. The oxides of trace elements such as selenate and chromate turn out to be substrates for this pathway, which they share with sulfate. The basal membrane, although possessing very similar transport properties to the brush border, has an interesting feature in that it is able to support trans-stimulation of sulfate. This means that net transport from mother to fetus will occur following periods of raised maternal sulfate concentration; however, efflux from the fetal compartment back to the mother will be very much slower if there is a fall in maternal sulfate concentration. This phenomenon, which depends upon the intrinsic kinetic properties of the two sulfate transporters found in the separate faces of the trophoblast, may be important in fetal nutritional homeostasis; its molecular basis is still unclear but it is similar to the asymmetries of transport, for example, of glucose in quite different cells.

Phosphate transport has been shown to occur via a sodium-coupled transporter in the brush border membrane. Efflux across the basal membrane has not yet been characterized.

Glucose

There is, at present, very intensive investigation of the distribution of glucose transporters in many tissues because of the discovery of the different properties of the members of this family of transport proteins. In the placenta no work has yet characterized at a molecular level the transporters that are present separately in the brush border and in the basal surfaces of the trophoblast. What is quite clear is that glucose transport across the placenta is sodium-independent and is therefore similar to the process, very well characterized, in red blood cells; it is not similar to the mechanism by which glucose is removed across the brush border membrane of either small intestinal or renal proximal tubule epithelial cells since in these two tissues transport of glucose is coupled to sodium (giving rise to secondary active transport). The flow of glucose between the maternal and fetal circulations is therefore dependent upon the concentration of glucose on each side of the trophoblast with net transport always being in the direction dictated by the prevailing transepithelial concentration gradient.

Glucose is, of course, a metabolic fuel for active tissues including the placenta, and the role of this tissue in glucose oxidation is probably substantial. Moreover, under certain circumstances, gluconeogenesis from fetal lactate will permit a functional Cori cycle to occur between fetal tissues and the placenta. The nature of the membrane transport systems responsible for glucose movement at each side of the syncytiotrophoblast must be important in determining the regulation of glucose delivery to the fetus. The sodium-independent glucose transporter family (Glut 1–5) differ in their kinetics, substrate specificity, and insulin sensitivity. There is now good evidence that insulin is not a major regulator of placental glucose delivery, although controversy persists as to whether there is any regulation of glucose uptake by maternal insulin. The family of glucose transporters all show marked stereospecificity (D-glucose transport being very much more rapid than that of the L-isomer) and they are also inhibited by phloretin and by cytochalasin B; glucose transport at both the apical and basal surfaces of human trophoblast shows these properties. It now appears that Glut 1 (the red blood-cell-like isoform of the glucose transporter) is probably the major molecule involved in placental glucose transport, although mRNA for the brain isoform Glut 3 has also been described.

Perhaps of greater significance to the overall physiology of the fetoplacental unit is the question of whether transepithelial glucose transport does or does not show functional rectification; is transplacental glucose transport equally as rapid in the feto-maternal as in the materno-fetal direction for the same chemical glucose concentration gradient? There is some indirect evidence that this may not be so.

Glucose transport across the placenta is particularly important with respect to the consequences for the fetus of raised maternal plasma glucose concentration, as in maternal diabetes mellitus. Because of the low circulating maternal insulin concentration, hyperglycaemia is found in the mother; transfer of glucose to the fetus is therefore augmented and the increased fetal plasma glucose concentration stimulates increased insulin secretion from the fetal endocrine pancreas. Increased growth of fetal tissues (associated with increased deposition of adipose tissue) is the consequence, and this process (macrosomia) is associated with increased risk of fetal abnormality and mobility both at birth and in the immediate postnatal period.

A fuller understanding of the placental role in carbohydrate delivery to the fetus will require a characterization of the transporters at both the brush border and the basal surface of the trophoblast and of the interplay between placental transport and metabolism. It is also likely that other carbohydrates (e.g. fructose) may have a role to play in placental homeostasis.

Although, at term, glycogen levels in the placenta are very low compared to earlier in gestation this macromolecule may provide an important source of placental fuel. A question that needs further investigation is to what extent is glycogenolysis regulated by local or systemic hormones. The adrenoceptor that is most likely to be coupled to cyclic AMP production (the β_1 adrenoceptor) is known to be present exclusively on the basal rather than on the apical surface of the trophoblast. It is of interest to know to what extent there is fetal regulation of placental metabolism through kinase-regulated mechanisms.

Amino acids

It has long been realized that the fetal environment must be appropriate for the cellular growth requirements of the growing conceptus. Delivery of the essential amino acids to the fetal tissues (and during fetal life those amino acids which are essential are more numerous than the 10 classical ones needed in the adult diet) must be via the placenta. Moreover, the fetal extracellular fluid differs markedly from that in the mother with regard to amino acid concentrations (see Table 2.1). Overall, the amino nitrogen concentration in the fetus is roughly twice that found in the mother.

Table 2.1 A survey of the concentrations of important solutes in the human maternal and fetal circulations. (Modified from Yudilevich D.L. and Boyd, C.A.R. (1987). *Amino acid transport in animal cells*. Manchester University Press.)

	Human maternal plasma (mmol/L)	Fetal/ maternal plasma (ratio)	Placenta tissue human (mmol/kg wet wt)
Leucine, Leu	0.06	1.98	0.38
Isoleucine, Ile	0.03	2.00	0.17
Valine, Val	0.10	2.04	0.30
Methionine, Met	0.02	1.28	0.06
Tryptophan, Trp	–	–	–
Tyrosine, Tyr	0.03	2.03	0.17
Phenylalanine, Phe	0.04	2.00	0.19
Glutamine, Gln	0.32	1.37	2.23
Asparagine, Asn	0.16	1.33	–
Histidine, His	0.07	3.95	0.17
Alanine, Ala	0.26	1.61	0.97
Serine, Ser	0.10	1.43	0.57
Cysteine, Cys	–	–	–
Threonine, Thr	0.06	1.63	0.54
Glycine, Gly	0.14	1.63	1.34
Proline, Pro	–	–	–
Aspartate, Asp	0.02	1.05	1.05
Glutamate, Glu	0.10	1.31	0.69
Lysine, Lys	0.10	2.95	0.62
Arginine, Arg	0.04	2.12	0.28
Ornithine, Orn	0.05	1.98	–
Citrulline, Cit	0.01	0.93	–
Taurine, Tau	0.06	2.25	–

Entry of amino acids into the trophoblast across the brush border membrane

Recent experimental work has concentrated on the membrane transport mechanisms underlying delivery of amino acids across the placenta to the fetus. Much work has been performed using membrane vesicles. A very wide variety of transporters have been described on the basis of their function, determined, for example, using kinetic, sodium-dependent, and cross-inhibition studies. There are present in the brush border of the term human placenta trans-porters that are sodium-dependent (e.g. system A); there are also transporters that are sodium-independent (e.g. systems L and y$^+$). Interestingly, there is also evidence for a class of transporters that have not been found in mature differentiated tissues, but which are known to exist in very early mammalian blastocysts. It may be that these are related to some of the relatively well-characterized amino acid transporters but they may represent a 'fetal' isoform with wider specificity and very high flux rates. Whatever the nature of these proteins it is important to realize that there is very substantial overlap in their specificity; thus, for example, an amino acid such as alanine may well be a substrate for perhaps four different transport systems. This 'redundancy' means that entry of amino acids into the trophoblast across the brush border membrane will be highly dynamic and the transporters that catalyse exchange (in particular system L) will play an important role in defining the blend of amino acids that are removed from the maternal circulation as it flows through the intervillus space.

Because of the sodium coupling to a number of neutral amino acid transporters (such as system A), the translocation of a substrate amino acid will be associated with a flow of positive ions across the brush border membrane. This will give rise to a depolarization of this membrane (the process of transport being described as electrogenic); more importantly for such electrogenic transport the membrane potential will be a determinant driving force for amino acid entry into the trophoblast. In other words, factors that influence the placental membrane potential will have a direct effect on the rate of amino acid uptake. As in most cells, membrane potential is largely determined by the passive permeabilities for a number of cations (particularly potassium) that are themselves kept out of electrochemical equilibrium by primary active transport processes (particularly the sodium pump, the Na$^+$ K$^+$ ATPase).

In addition to considering the importance of the physicochemical factors influencing placental amino acid transport there is important regulation of this process by intrinsic and by extrinsic mechanisms (Fig. 2.3). The intrinsic mechanisms relate to the control of amino acid transporter activity in the plasma membrane, for example, as a result of alteration of intracellular placental amino acid concentrations. This fascinating phenomenon, now known to be widely distributed in animal cells although originally discovered in placenta, probably involves both *de novo* synthesis of carriers and also regulation of the insertion into the plasma membrane of preformed transporters, probably located in intracellular membrane vesicles near the plasma membrane. The extrinsic control of placental amino acid transport has not been adequately established but numerous factors are known to have effects

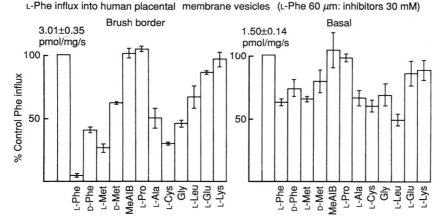

Fig. 2.3 Contrasting properties of the brush border (left) and basal (right) membrane for specificity for phenylalanine transport. Note that the pattern of inhibition by the amino acids indicated differs; for example, there is marked stereospecificity in the brush border for inhibition by phenylalanine which is not present for the transport systems located in the basal membrane.

on transport of amino acids, including classical hormones and growth factors. Thus insulin-like growth factors have been shown to stimulate amino acid intake through system A in human trophoblast cells.

Amino acid exit from the placenta into the fetal circulation

The asymmetry typical of epithelia is very pronounced in amino acid handling so that net transfer from maternal to fetal blood occurs. The concentrations of amino acids in the three compartments are shown in Table 2.1 and it is apparent that for virtually all amino acids the trophoblast concentration is higher than that on either the maternal or fetal side. However, the fetal concentration of a very wide variety of amino acids is approximately double the maternal concentration and this reflects the widespread (but not universal) distribution of sodium-coupled transport systems particularly in the apical surface. Very recent work has shown that the systems for amino acid transport in the basal membrane are numerous and show remarkable cross-reactivity between different substrates. This too may reflect the presence of embryonic amino acid transport mechanisms.

Placental amino acid metabolism

Two further features of amino acid handling by the placenta need to be emphasized. First, the placenta is a major

site of amino acid metabolism so that entry and efflux are not necessarily of the same molecular species. Second, the placenta can also remove both amino acids and their precursors from the fetal circulation and, for example, by transamination, generate other amino acids in the trophoblast. An example of this relates to fetal glutamic acid, which is removed by the placenta by a sodium/potassium-coupled transporter and is found in very high levels within the trophoblast. Alanine is generated by transamination and may readily enter the fetal compartment.

Other solutes

Many other organic solutes are known to be transported by the placenta, and recent work using vesicles has shown the presence of dicarboxylic acid transporters in the brush border membrane. Similarly, specific transport systems for individual water-soluble vitamins have been recently described in the brush border membrane (Fig. 2.4).

The direction of net transplacental transfer

It is easy to catalogue the concentrations of a very wide range of solutes in the maternal and in the fetal circulations. Differences in concentrations are nearly universal, but this does not always mean that the placenta is responsible for such differences. Thus, differences between the

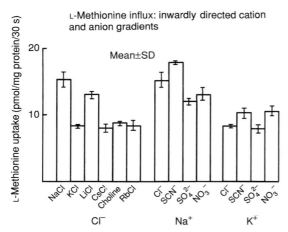

Fig. 2.4 Ionic dependence of amino acid transport in the brush border of human placenta. L-methionine influx was studied with inwardly directed gradients of sodium (centre), potassium (right), or chloride (left). Note that only sodium (and to a minor extent lithium) is able to stimulate uptake above basal levels and that this sodium-dependent transport of the neutral amino acid methionine is stimulated when a permeant anion such as SCN^- is present as compared to an impermeant anion such as SO_4^{2-}. In the absence of sodium no such effect of membrane potential is seen.

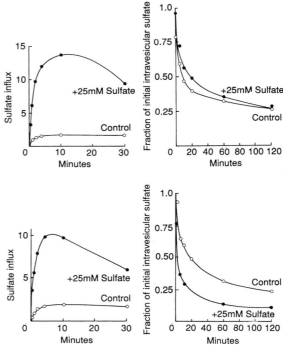

Fig 2.5 Sulfate transport by membrane vesicles prepared from both apical (brush border) top and basal (bottom) surfaces of the placenta. On the left data from influx experiments are shown, either in the absence (open circles) or in the presence (filled circles) of intravesicular sulfate (25 mM) to *trans*-stimulate entry of radioactive sulfate; on the right efflux experiments are shown again with and without 25 mM sulfate in the *trans*- (in this case external) medium. Note that, whereas *trans*-stimulation of influx is found in both apical and basal membranes, for efflux *trans*-stimulation is only found in the basal membrane preparation. The implication of this finding is discussed in the text.

maternal and fetal compartments in their metabolic clearance rates of the solute in question may equally be responsible for the observed concentration differences. In general terms, net transfer across the placenta is most readily measured as the difference between two unidirectional transplacental fluxes and only if the observed flux ratio is different from that predicted by the prevailing electrochemical gradients across the tissue do active transport systems need to be invoked.

Such active transport systems will have to be asymmetrically distributed in order to achieve net transport from mother to fetus or vice versa. Rather few molecular species will interact with primary active transport mechanisms in which ATP hydrolysis directly powers solute flux. Most are now known to be secondarily active and are, for example, powered by the prevailing electrochemical gradient of sodium ions. An increasing minority turn out to be driven by electrochemical gradients of H^+ ions. An important recent consideration relates to the systems by which solutes may interact with each other, as in the case of amino acids that are substrates for more than a single transporter (Fig. 2.5). Thus, for tyrosine, net transport across the trophoblast may result from the differing abilities of the two distinct transporters (both of which are sodium-independent) to interact with other neutral amino acids.

Transport in relation to blood supply

When looking at a section of term placenta, the intimate relationship between trophoblast and underlying fetal capillary is a most striking structural feature. For gas transport such minimizing of diffusion path length is clearly critical, although its significance for membrane-limited transporters is very much less. What are the factors regulating the proliferation of fetal capillaries during placenta development? The tortuosity factor for fetal capillary growth continues to rise until the end of gestation, although trophoblast development has matured by the end of the second trimester. It seems likely that the trophoblast itself is responsible for activating signalling pathways that finally produce capillary growth. This relationship must be important for trophoblast function both in relation to transport and to endocrine activity.

Regulation of placental transport and the matching of supply to demand

A critical issue in placental physiology is the extent to which placental transport is or is not controlled by external circulating factors. (Neural control of a classical sort is excluded by the absence of placental nerves, although the pathways for neurotransmitter biosynthesis, release, and receptor activation are all present within the trophoblast.) For at least calcium, phosphate, magnesium, and potassium there is good evidence of some form of hormonally regulated function. It may be that the intrinsic design of the placenta allows much of the regulation to be local, for example, through pathways that are regulated by intracellular pH, ATP levels, or membrane potentials.

Further reading

Bain, M.D., Copas, D.K., Taylor, A., Landon, M.J., and Stacey, T.E. (1990). *Journal of Physiology*, **431**, 505–13.

Barros, L.F. *et al.* (1992). In human placenta glucose transporter GLUT 1 is abundant in the brush border and basal membrane of the trophoblast. *Journal of Physiology*, **446**, 345P.

Battaglia, F.C. and Meschia, G. (1986). *An introduction to fetal physiology*. Academic Press, New York.

Boyd, C.A.R. (1991). The use of membrane vesicles to study placental transport. *British Journal of Nutrition*, **50**, 337–43.

Cetin, I. (1990). Amino acid in cordocentesis samples of fetal blood. *American Journal of Obstetrics and Gynecology*, **162**, 253–61.

Coleman, R.A. (1989). The role of the placenta in lipid metabolism and transport. *Seminars in Perinatology*, **13**, 180–91.

Faber, J.J. and Thornburg, K.L. (1983). *Placental physiology*. Raven Press, New York.

Furesz, T.C., Moe, A.J., and Smith, C.H. (1991). Two cationic amino acid transport systems in human placental basal plasma membrane. *American Journal of Physiology*, **261**, C246–C252.

Ganapathy, V. *et al.* (1988). Sodium-gradient-driven, high-affinity, uphill transport of succinate in human placental brush border membrane vesicles. *Biochemical Journal*, **249**, 179–84.

Greenwood, S.L., Boyd., R.D.H., and Sibley, C.P. (1991). Membrane potential of human trophoblast. *Journal of Physiology*, **438**, 267P.

Hoeltzli, S.D., and Smith, C.H. (1989). Alanine transport in isolated basal plasma membrane of human placenta. *American Journal of Physiology*, **256**, C630–C637.

Kudo, Y. and Boyd, C.A.R. (1989). Human placental L-tyrosine transport. *Journal of Physiology*, **426**, 381–95.

Ohlsson, R. (1989). Growth factors, protooncogenes and human placental development. *Cell Differentiation and Development*, **28**, 1–15.

Shennan, D.B. and Boyd, C.A.R. (1987). Ion transport by placenta. *Biochimica et Biophysica Acta*, **906**, 437–57.

Sibley, C.P. and Boyd, R.D.H. (1989). Control of transfer across the mature placenta. In *Oxford reviews of reproductive biology*, Vol. 10 (ed. J.R. Clarke), pp. 382–435. Oxford University Press, Oxford.

Smith, C.H. *et al.* (1992). Nutrient transport pathways across the epithelium of the placenta. *Annual Reviews of Nutrition*, **12**, 183–206.

Steven, D.H. (1975). *Comparative placentation*. Academic Press, New York.

Stulc, J. (1988). Is there control of solute transport at placental level? *Placenta*, **9**, 19–26.

Stulc, J. (1989). Extracellular transport pathways in the haemochorial placenta. *Placenta*, **10**, 113–19.

Yudilevich, D.L. and Sweiry, J. (1985). Transport of amino acids across the placenta. *Biochimica et Biophysica Acta*, **822**, 169–201.

3. Endocrine function of the placenta

Alan J. Conley and J. Ian Mason

From an endocrine perspective, the placenta of most mammals is virtually a universal gland. Notwithstanding species differences, the placenta synthesizes proteins, peptides and steroid hormones, biologically active amines, and other factors that are produced in similar form by various endocrine tissues of the fetus and adult. To date, however, it has been difficult to provide convincing evidence of definitive physiological functions for many of these factors. Therefore, this chapter will focus on chorionic gonadotrophins, placental lactogens, and steroids, the placental hormones for which physiological functions are more easily identified and which are released in significant quantities. In this chapter hormones are defined in the classical sense to mean substances released into the circulation that act to alter the function of another organ at a distal site. Therefore, little attention will be given to identified placental products that have been postulated to act in a paracrine or autocrine manner. This is not intended to suggest that such potential regulatory factors are not physiologically significant. Pregnancy is, however, a process of such fundamental importance that it is unlikely that its success would hinge on any single component. Instead, it is more reasonable to assume that successful pregnancy requires that multiple components, interacting in a complex way, provide several fail-safe mechanisms whereby the absence of any component can be compensated for by others. At the present time, we are far from understanding this complexity.

We have chosen to limit the scope of this chapter for another reason. Professor Medawar once stated, 'it is not hormones which have evolved but the uses to which they are put.' From this perspective it might be suggested that the presence or synthesis of a substance in the placenta does not provide evidence for a need or a function or that a function existing at one point in the evolution of an organism need necessarily be conserved. It is apparent that many different but equally successful strategies for maintaining pregnancy have evolved among animal species and what might represent a functional adaption in one could well be redundant in another. Placental hormones may also influence processes other than those involved in the maintenance of pregnancy. Mammals are distinguished by their ability to lactate. Therefore, additional functions of the placenta may relate equally well to the initiation of lactation as to pregnancy and the initiation of parturition. Although the function of many placental factors is not yet known, their discovery is an exciting first step toward understanding placental endocrinology.

The necessity for hormonal synthesis by the placenta is pre-empted by the endocrine contribution of the corpus luteum. In almost all mammals the establishment and/or the successful maintenance of pregnancy is dependent largely on luteal progesterone production. In some species, such as dogs, cats, and at least some mustelids, luteal function continues for the normal length of gestation, regardless of whether conception has taken place. Moreover, there may be no need for placental products to prepare the animal for motherhood. Pseudopregnancy in the bitch is typified by lactation and nesting behaviour sometimes indistinguishable from that of a pregnant female. In other species, those characterized by cyclical luteal regression, the blastocyst must bring about luteal maintenance for the entire gestational period or until placentation is sufficiently developed to be endocrinologically functional. In humans, chorionic gonadotrophin synthesized by the embryo is thought to maintain and extend the life span of the corpus luteum. In the elephant and equine species, pregnancy is maintained by the formation of additional accessory luteal structures which in mares has been related to the production of chorionic gonadotrophin by the endometrial cups (trophoblastic tissue embedded in the endometrium). The initial signal for luteal maintenance in cows, sheep, and goats is thought to be a trophoblastic protein of blastocyst origin, whereas, in the expanding pig blastocyst, oestrogen production is correlated with the time of maternal recognition of pregnancy. It is believed that trophoblastic proteins and oestrogen achieve luteal maintenance indirectly by averting the normal luteolytic mechanism originating in the uterus. The existence of antiluteolytic or luteotrophic activity of the early conceptus is the first obvious interaction between it and the maternal organism. It represents also the first identifiable attempt by the conceptus to influence and control its environment. Even so, in some species, such as dogs, cats, and mustelids, it is apparent that the conceptus need not play an active role in order to maintain successful pregnancy.

Chorionic gonadotrophins

Human chorionic gonadotrophin

As early as 1905 Halban postulated an endocrine role for the placenta based on his observations of continued gestation and subsequent lactation in ovariectomized women. It may be argued, from the foregoing observations as well as from ectopic pregnancies, hydatiform moles, and the continued function of placentae in the absence of a fetus, that influences of the ovaries, uterus, and the conceptus are not essential for placental (and therefore fetal) survival. Human chorionic gonadotrophin (hCG) was one of the first hormones identified as a placental product. It was demonstrated that placental explants induced luteinization in rabbits and later shown that tissue collected from early stages was more effective than tissue from later stages. Finally, it was demonstrated that hCG was excreted in urine from pregnant women, ultimately leading to the development of methods of pregnancy diagnosis.

It is now known that hCG belongs to the family of gonadotrophic hormones which includes luteinizing hormone (LH), follicle-stimulating hormone (FSH), and thyroid-stimulating hormone (TSH). These structurally related hormones are dimeric in form consisting of two dissimilar protein subunits designated α and β. The α subunit is common to all glycoprotein hormones. It is a product of a single gene expressed probably constitutively in both the pituitary and the placenta. The α subunit is expressed both in the cytotrophoblast and the syncytiotrophoblast. The β subunit of hCG and the other glycoprotein hormones confers the unique biological attributes upon each hormone of this group. While hCG has activities very similar to those of LH, these two hormones share only 80 per cent amino acid homology and hCG has an additional carboxy terminal extension not present in hLH. In fact, four separate genes code for hCG, although only two or three are expressed, while an additional gene encodes for hLH. In addition, hCG has a much higher carbohydrate content than its pituitary counterpart (hLH) and this may contribute to its much longer half-life in plasma. Certainly, removal of sialic acid residues appears to be associated with a decrease in half-life and therefore biological activity. Although biological specificity resides primarily in the β subunit, neither subunit is active on its own and binding of the β subunit to the LH receptor is dependent on its correct association with the α subunit. Production of each subunit occurs independently of the other in both the pituitary and the placenta and in each case the α subunit is synthesized in excess. Therefore secretion of the bioactive molecule is dependent on the rate of synthesis of the β subunit, which indeed mirrors the levels of circulating bio-

logically active hormone during pregnancy. This is true also because hCG appears not to be stored in membrane-bound granules as is pituitary LH; instead, synthesis and release occur as simultaneous processes at relatively low but constant rates from granules.

Levels of hCG in maternal plasma have been studied extensively and generally reflect levels of synthesis of the β subunit. Detectable in plasma at as early as the tenth day following the LH peak, levels rise to their maximum between weeks 8 and 12 of gestation after which they gradually decline to much lower but steady levels for the remainder of pregnancy. Levels of the free α subunit, however, continue to increase to week 36, and then plateau for the remainder of the pregnancy.

Synthesis of hCG occurs mainly in the syncytiotrophoblast. Despite the increasing percentage of syncytiotrophoblast relative to cytotrophoblast as pregnancy proceeds, relatively little hCG is localized to syncytiotrophoblast of term placentae compared with placentae obtained in the first trimester. These data suggest that hCG synthesis is not regulated solely by the differentiation of cytotrophoblastic cells. It has been possible to dissociate these two events in vitro such that prevention of syncytialization of cytotrophoblasts by culturing cells in serum-free medium does not prevent the increased, spontaneous secretion of hCG. Moreover, whereas hCG production is maintained at high levels for several days in both instances, it also spontaneously decreases with no corresponding decrease in cell viability. Whether or not the factor(s) regulating hCG synthesis under these in vitro conditions reflects in any way those occurring in vivo is unknown.

Studies utilizing various in vitro systems suggest that cyclicAMP (cAMP) increases hCG secretion independently of effects on differentiation. Endogenous agonists likely to increase intracellular cAMP in vivo are difficult to identify. β–Adrenergic agonists increase hCG secretion in vitro but the physiological relevance of these observations is difficult to assess. Epidermal growth factor (EGF), gonadotrophin-releasing hormone (GnRH), activin, inhibin, and glucocorticoids have all been shown to affect hCG production in vitro in some studies, while other studies have failed to duplicate some of these findings. hCG itself has been shown to increase intracellular cAMP levels and, therefore, it must be included as a possible autocrine regulator as well. In short, it would seem that experimental conditions can greatly alter observations relating to the short-term regulation of hCG release. Thus, at present, no clear picture has emerged as to the roles of these in the control of hCG secretion.

The secretion of hCG by the placenta, and indeed the implanting blastocyst, is thought to be important for several reasons. Most notable is that it is believed to play a role in bringing about luteal maintenance. The evidence supporting this notion includes: (1) hCG stimulates cAMP and progesterone secretion by luteal slices and cells in culture; (2) administration of hCG to non-pregnant women stimulates luteal progesterone production and delays the normal onset of menses; (3) hCG levels in plasma correlate with rising levels of progesterone during early pregnancy, and the corpus luteum of early pregnancy is more responsive to hCG than at later stages; and (4) passive immunization of monkeys against CG in early pregnancy results in abortion. However, pertinent questions still remain. Why do hCG levels continue to increase in early pregnancy, while luteal function apparently wanes? Why does exogenously administered hCG extend luteal function for a few days only? In defence of the second of these two questions, studies to date have failed to achieve high levels of plasma hCG in non-pregnant women equal to those achieved in early pregnancy. But this raises another question. Although the human blastocyst secretes hCG that can be detected in plasma as early as 8–10 days following the mid-cycle surge, do plasma levels of hCG reach minimally required luteotrophic levels at the time at which luteolysis would otherwise begin? If not, other, as yet unidentified, factors of trophoblastic origin might reasonably be expected to contribute, directly or indirectly, to luteal maintenance.

A second important postulated function of hCG relates to sexual differentiation of the male fetus. Although fetal plasma hCG levels are only about 3 per cent of those measured in maternal blood, it has been postulated that they are sufficient to influence fetal testicular testosterone secretion. Plasma hCG correlates well with fetal testicular testosterone concentrations which also reach a peak at about 12 weeks and decline thereafter. During this time Wolffian duct development takes place and differentiation of the external genitalia also begins. Anencephalic male fetuses undergo normal sexual development at this time further suggesting that an exogenous LH source may be influencing steroidogenesis. Finally, hCG has been implicated in the maintenance of the fetal zone of the human fetal adrenal. It has been suggested that fetal zonal integrity is dependent on hCG and both *in vivo* and *in vitro* administration of hCG can increase synthesis of dehydroepiandrosterone or dehydroepiandrosterone sulfate. The physiological relevance of these observations is unknown. Other purported effects of hCG on placental steroidogenesis, thyroid function, decidual relaxin and prolactin release, or immunosuppression have been hypothesized on the basis of *in vitro* studies. Although some interesting possibilities exist, no definitive roles have been ascribed to hCG in the regulation of these endocrine systems.

Equine chorionic gonadotrophin

The serum from a pregnant mare early in gestation possesses potent gonadotrophic activity, which can be attributed to the secretion of a chorionic gonadotrophin from unique structures called endometrial cups located in the endometrial wall. Initially referred to as pregnant mare serum gonadotrophin or, more recently, equine chorionic gonadotrophin (eCG), this hormone first appears in serum at 38 days of gestation; after peaking at 60–70 days it declines, reaching undetectable levels by mid-gestation. The endometrial cups develop from the fetal cells of the chorionic girdle of the equine placenta, the junction between the developing chorioallantois and the yolk sac. These trophoblastic cells migrate into the adjacent endometrium, multiply, and hypertrophy into discrete populations of binucleate cells forming concave disc-like structures 1–2 cm in diameter. Regression of the cups probably occurs by immune rejection and precedes the disappearance of eCG from the circulation.

Functionally, it has been proposed the eCG brings about the formation of secondary corpora lutea at day 42 when serum progesterone from the primary corpus luteum is declining. Serum progesterone levels rebound with secondary corpus luteum formation, an event believed to be important in the maintenance of pregnancy until the establishment of chorioallantoic placentation, which leads to placental progesterone production. Equine CG exhibits both FSH- and LH-like activity, although FSH-like activity often predominates. In equine tissues, however, eCG binds to LH receptors and has little or no FSH-like activity. When pregnant mares are hysterectomized prior to the development of endometrial cups, follicles develop at the time of expected secondary corpus luteum formation but luteinization does not ensue. From these data it has been suggested that, if eCG serves a physiological function in pregnancy, it is to cause ovulation and/or luteinization of follicles forming accessory luteal structures.

Studies at the molecular level support the data suggesting that eCG acts as a luteotrophin in the pregnant mare. eCG is a glycoprotein like hCG, composed of α and β chains that are joined by disulfide bonds and that are heavily, though variably, glycosylated (Fig. 3.1). Cloning of the α and β subunits of eCG suggests that both subunits are homologous with eLH at the amino acid level and that the major difference between eLH and eCG is the differential post translation modification by carbohydrate addition. In fact, eCG/eLH shares considerable homology with hCG, these hormones having longer carboxy terminal amino acid tails on the β chain than LH proteins of other species. Glycosylation of the carboxy terminus is thought to increase its half-life in serum but desialation also

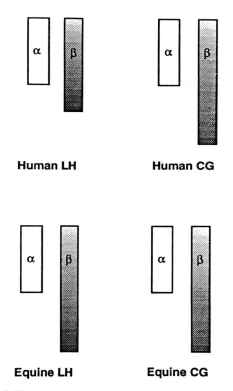

Fig. 3.1 Illustration of relative molecular sizes of the α- and β-subunits of human and equine LH and CG. Note that eLH and eCG have identical α- and β-peptide structure and are similar to hCG; hLH and LH of other species have smaller β-subunits structure and are less glycosylated.

increases receptor binding. Post translation modifications of eCG by the addition of carbohydrate prior to secretion result in differences between the secreted molecule and that extracted from tissues or isolated from media following tissue culture. How such differences relate to the physiology of pregnancy is not clear.

In summary, at the present time, hCG and eCG are the only well characterized chorionic gonadotrophins. Why equine and primate placentae synthesize and secrete such significant quantities of these compounds is not known. Studies characterizing the sequence in the 5'-regulatory region of both hormones suggest that higher primates and the human may have tandem cAMP-response elements (CRE) compared to Old World monkeys which have only one. This suggests that tandem CREs could provide those species with the ability to express highly the CG gene at basal levels of intracellular cAMP. However, identifiable CREs are absent from the equine gene, which suggests that alternate molecular mechanisms exist to confer high tissue-specific transcription of these genes in the placenta of mares.

Placental lactogens

Human placental lactogen (hPL)

Human placental lactogen is a major placental protein produced by the placenta throughout most of gestation. In fact at term hPL mRNA comprises an astounding 25 per cent of the total mRNA in the placenta. Even though it was first discovered through its ability to induce secretion of milk by mammary explants, much of the succeeding work on hPL has dealt with its effects on intermediary metabolism.

In the mother hPL has at least two major metabolic actions: (1) increasing lipolysis, and (2) inhibiting glucose uptake and gluconeogenesis. The effect of hPL is classically described as glucose-sparing, by increasing the supply of carbohydrate and amino acids to the developing fetus. Although in other animals hPL appears to have lactogenic properties, the evidence for such an effect in humans is poor. It has been suggested that hPL stimulates mammary gland epithelial proliferation but it appears that is incapable of maintaining lactation *per se*. The plasma concentration of hPL in the fetus is several orders of magnitude lower than that in the mother. Nevertheless, some of the proposed actions of hPL include fetal growth as well as metabolic partitioning in the maternal system. It has been proposed that the growth that occurs in anencephalic fetuses may continue in part because of the influence of hPL; however, direct evidence is lacking.

Despite the many suggested or implied functions of hPL, one important caveat should be noted. Cases have been reported of women having low or undetectable levels of hPL due to specific gene deletions. These pregnancies were successful and healthy babies were delivered at term suggesting that hPL was unnecessary for the mother or the fetus to maintain a normal pregnancy. Although these observations indicate that pregnancies can be sustained in the virtual absence of hPL, this does not necessarily negate a physiological role for this hormone that may be important under suboptimal conditions of pregnancy, such as during nutritional deprivation.

The mature polypeptide is a single-chain structure that shares 96 per cent amino acid homology with hGH. The gene encoding hPL is one of five in the growth hormone-placental lactogen (GH–PL) gene family, two of which encode identical hPL mature proteins that are produced in equivalent amounts in the placenta. The syncytiotrophoblast or fully differentiated cytotrophoblast is the major cellular source of the hormone (Fig. 3.2). Like hCG, it is packaged in granules and released by exocytosis without appreciable storage suggesting that the principal mode of regulating circulating levels resides in the control of the rate of transcription. No obvious increase in hPL mRNA occurs in

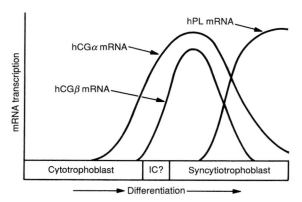

Fig. 3.2 Expression of mRNA for human chorionic gonadotrophins α and β and for human placental lactogen (hPL) during trophoblast differentiation. IC indicates intermediate cells. (Taken with permission from Mochizuka, M. and Hussa, R. (1988). In *Placental Protein Hormones*. Elsevier B.V., Amsterdam.)

individual syncytial units as pregnancy progresses. This suggests that the increases in plasma levels of hPL that occur up to 34 weeks of gestation may reflect an increase in the mass of syncytial tissue that is expressing the gene.

Evidence of regulation, over and above that determined by placental growth, has not been convincingly demonstrated. Calcium is apparently required for basal release *in vitro* but can paradoxically inhibit stimulated release. *In vitro* studies utilizing many potential agonists including hormones and neurotransmitters have failed to alter hPL release. Other studies have also failed to demonstrate an effect of cAMP on hPL secretion *in vitro*. However, arachidonic acid, protein kinase C activation, and inositol triphosphate (IP_3) all increase hPL release presumably by mobilizing intracellular Ca^{2+}. Even so, it would appear that basal release or long-term regulation by way of placental growth are likely to be more important in determining daily levels and therefore the potential biological effects of this hormone. In this regard, although concentration changes in amino acids, glucose, or lipids have little effect on plasma hPL, starvation induces an increase that is detectable within 60 h. Furthermore, hyperglycaemia has been associated with transient declines in hPL levels. How these effects are mediated is unknown, but they may depend on a combination of endocrine interactions. The biological role of hPL during pregnancy remains to be further defined.

Rodent placental lactogen

In recent years much progress has been made in the characterization of placental lactogens (PL) in rodents. PLs

have been identified and characterized in mice, rats, and hamsters. Rodent PLs (rPLs) possess lactogenic rather than somatogenic (i.e. growth-promoting) activity. In fact, some studies suggest that rodent PLs may bind prolactin receptors with greater affinity and, in some systems, are more potent than either GH or even prolactin itself. In the mouse, for example, mPLII demonstrates greater activity in terms of the stimulation of α-lactalbumin in mouse mammary tissues than does prolactin. Although this might seem anomalous at first, an examination of the genes provides a basis for understanding the relationships among these proteins. In the mouse it seems that the genes encoding four placental proteins and that for prolactin (PRL) coexists on chromosome 13 (Fig. 3.3). The gene for mGH, however, has been localized to chromosome 11. Therefore, the prospect of gene duplication within the GH–PRL–PL gene family in the mouse would seem to favour duplication of PL from prolactin rather than from GH, in contrast to the proposed situation in the human. Whether similar relationships exist in species other than the human and rodents, i.e. between chromosomal location and structure/function homology among members of the PRL–GH–PL gene family, remains to be investigated.

The cloning of several of the genes encoding placental lactogens, expression of the cDNA, and production of recombinant protein have greatly advanced our knowledge

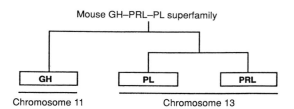

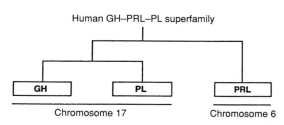

Fig. 3.3 Chromosomal localization of the genes encoding placental lactogens (PL), prolactin (PRL), and growth hormone (GH) in mouse and human suggesting the evolution of the PL genes from either PRL (mouse or GH (human). The proposed origin of PL from either GH or PRL correlates with the major physiological effects of PL during pregnancy in the mouse and human.

of these hormones. Two major forms exist in the mouse, rat, and hamster (mPLI and II; rPLI and II; haPLI and II, respectively). In each case PLI is the predominant circulating form at mid-gestation while PLII is the major form in later stages, declining following parturition. In the mouse, mPLI appears between days 8 and 12 reaching levels 10-fold higher than the peak levels of mPLII attained in late gestation. A similar pattern of secretion is described for rPLI and II and, in the hamster, haPLII also appears during the second half gestation. In rats, both rPLI and II are synthesized in the trophoblastic giant cells of the developing placenta and mPLII has also been localized to trophoblastic giant cells in mice.

As is the case in other species, the regulation of the secretion of rodent PLs is not well understood. Concentrations of mPLII in maternal serum are correlated with litter size and placental mass. It has also been suggested that circulating levels may be influenced by genotype, fasting, and possibly pituitary hormone concentrations, in particular, prolactin and GH. In mice, hypophysectomy at mid-pregnancy increased mPLII levels in part by increasing the half-life of the hormone. It also appears that GH and other pituitary factors may suppress mPLII production and provide another potential mechanism for the rise in mPLII levels seen after hypophysectomy. It is not known whether or not the apparent effects of the pituitary are direct. A similar rise after hypophysectomy is seen in rats, and it has been suggested that factors from the ovaries and the adrenals may negatively regulate these genes.

Earlier studies of rodent placental lactogens were complicated by the existence of several proteins of different apparent molecular weight. More recently, it has been recognized that some of these proteins differ only by virtue of their degree of glycosylation. In addition, haPLII forms of larger molecular weight represent aggregates of monomers bonded by disulfide bridges, or bonded to other circulating proteins such as α_2-macroglobulin. The way in which complex formation influences biological activity or function is presently unknown.

Ruminant placental lactogens

Placental lactogens have been identified in sheep, goats, and cattle by both radioreceptor and activity studies. Unlike rodent and human PL, which preferentially bind one receptor type and demonstrate one activity predominantly, ruminant placental lactogens compete with both growth hormone (GH) and prolactin for binding to somatotrophic and lactogenic receptors and exhibit both activities. The genes encoding ovine and bovine PL have been cloned and the availability of molecular probes, purified protein, and better antisera will, in future, facilitate the isolation of cDNAs encoding goat PL. At the amino acid level, ruminant PLs exhibit greater homology with prolactin than they do with GH. More recently, data suggest that, while ovine PL (oPL) can stimulate the release of insulin like growth factor II (IGFII) from fetal hepatocytes, it may do so through a specific interaction with a third, fetal-type lactogenic receptor. This receptor has a high affinity for oPL but appears not to bind either GH or prolactin. Furthermore, it has been suggested that neither GH nor prolactin exhibit somatotrophic activity in the fetus further supporting an exclusive role for oPL in fetal growth. If indeed a specific oPL receptor does exist, the concept of a family of PL/GH/prolactin genes must be expanded to include, perhaps, a family of receptors also. If so, the activity of the ruminant PLs may not reflect their origins by gene duplication from either the prolactin or GH genes, as appears to be the case in humans and in rodents, but by the evolution of the receptor instead.

In ewes, does, and cows PL appears to be synthesized exclusively in the binucleate cells of the placenta and can be identified immunocytochemically within the first 3 weeks of pregnancy (almost as soon as the binucleate cells themselves are recognizable). PL is contained in discrete membrane-bound vesicles that are released into the intracellular space by exocytosis. Histological evidence suggests that binucleate cells migrate from the fetus across the placental feto-maternal interface and release their contents in the maternal compartment.

Despite the early detection of PL in binucleate cells by immunocytochemical techniques, maternal plasma levels of PL are not detectable until at least the sixth week of gestation in either sheep or goats. They increase thereafter, almost doubling every couple of days, peak at over 500 ng/mL between days 120 and 140, then decline prior to parturition. In the ovine fetus, plasma levels of oPL remain relatively constant varying between 50 and 150 ng/mL throughout gestation. Therefore, fetal levels of PL exceed those in the maternal circulation during the first half of gestation even though maternal levels are considerably higher during the second half or later stages of gestation. Although binucleate cell numbers remain constant at about 20 per cent of trophoblastic cells in the placenta throughout pregnancy, the increase in total placental mass probably explains in large part the increases seen in PL as pregnancy progresses. During later stages of gestation PL levels are higher in ewes and does carrying twins and triplets than in animals with single fetuses. Since the placental mass is greater with multiple fetuses, this supports the concept that placental mass is an important determinant of PL production. In contrast to both ewes and does, cows have relatively low maternal concentrations

(<5 ng/mL) of PL throughout gestation, whereas fetal plasma concentrations are high (20–30 ng/mL), at least during the later stages of pregnancy, declining prior to parturition as they do in sheep and goats.

Although it is assumed that PL in ruminants functions to maintain glucose supply to the fetus and to support mammogenesis, direct supporting evidence in these species is scant. *In vivo* administration of protein or antisera to oPL, aimed at investigating metabolic effects, has so far yielded equivocal results. The difficulties of performing studies of this kind in large animal species is obvious. It is even more difficult to examine long-term effects, leaving many questions unanswered as to the physiological role of PL. In fact, oPL may even complicate certain metabolic conditions such as pregnancy toxaemia in late pregnancy in ewes. The incidence of this disorder is higher in ewes carrying multiple fetuses. So too are the levels of oPL in these animals. Although the pathophysiology of this disease is complex and poorly understood, it is possible that oPL may be involved in its pathogenesis.

At the cellular level, several factors have been examined for possible effects on PL release with inconsistent results. More recent studies that demonstrate that activation of phosphoinositol metabolism and alterations in cytoplasmic Ca^{2+} concentrations can influence oPL release *in vitro* still leave in question the physiological ligand that activates the system *in vivo*. Therefore, it remains a matter of speculation whether or not activation of phosphoinositol metabolism induces alterations in PL release of sufficient magnitude and duration to influence either the maternal or the fetal system.

Steroidogenesis

This section on placental steroidogenesis focuses on progesterone and the oestrogens. A general discussion of steroidogenic enzymes is followed by a consideration of how steroidogenesis might be regulated. Finally, progesterone and oestrogen production during pregnancy, and how it might relate to the physiology of gestation, is described. Although progesterone and oestrogens are regarded widely as the most biologically important steroids synthesized and secreted during pregnancy, other steroids, and their metabolites may be equally vital but are as yet less well characterized. For instance, catechol oestrogens are potent calcium channel blockers and vasodilators in the vascular beds of the reproductive tract. Also, certain 5α-reduced pregnanes are potent anaesthetic agents thought to act by modifying the activation of GABA receptors. Both the nature and physiological significance of these and other steroids produced by the placenta or gravid uterus remain to be defined.

Steroidogenesis is a series of catalytic reactions decreasing the size of the basic 27-carbon-unit structure (cholesterol) to C_{21} (progestogen), C_{19} (androgen), and C_{18} (oestrogen) molecules (Fig. 3.4). These redox reactions involve several enzymes of the cytochrome P450 family including cholesterol side-chain cytochrome P450 ($P450_{scc}$), 17α-hydroxylase cytochrome P450 ($P450_{17\alpha}$), and aromatase cytochrome P450 ($P450_{arom}$). These steroidogenic cytochrome P450 enzymes act by way of hydroxylation and subsequent cleavage of side-chain groups. They bring about conversion of cholesterol to the progestogens, to androgens, and finally to oestrogens by way of reactions that are essentially irreversible. Within each steroid family (progestogens, androgens, and oestrogens) changes in biopotency of member steroids are brought about by modifications made to the four-ring structure by oxidoreductase enzymes such as 3β-hydroxysteroid dehydrogenase (3β-HSD), 3α-hydroxysteroid oxidoreductase (3α-HSOR), 17β-hydroxysteroid oxidoreductase (17β-HSOR), $20\alpha/\beta$-hydroxysteroid oxidoreductase ($20\alpha/\beta$-HSOR), and steroid 5α- and 5β-reductases. For example, among other reactions, 3β-HSD catalyses oxidation at the 3 position of pregnenolone and isomerization of the Δ^5-double bond to yield progesterone, 5α-reductase catalyses the formation of 5α-dihydrotestosterone from testosterone, and 17β-HSOR is responsible for the formation of oestradiol-17β from oestrone. These reactions are not necessarily irreversible and may be influenced by product accumulation or substrate availability. Finally, conjugation of steroids, as for instance by sulfotransferase action or deconjugation by sulfatase, can dramatically affect tissue levels of active steroids as well as those in the circulation.

As complex as this series of reactions is in the placenta, the general picture can be somewhat simplified in two important ways: (1) regulation rarely occurs by major shifts in the activities of these enzymes; and (2) when expressed, the activities of the various enzymes are generally within an order of magnitude of one another, which allows regulation to occur by way of relatively small changes in levels of expression of selected enzymes. The notable exception to the first rule involves the induction and repression of a key enzyme, cytochrome $P450_{17\alpha}$, which occupies a central and crucial role in the process of steroidogenesis. This enzyme exhibits the relatively unusual property among placental cytochromes P450 of exhibiting marked and rapid changes in transcriptional levels and activity as well as showing marked differences in substrate specificity that are characteristic of a species. Therefore, with this exception, those enzymes present in the placenta at the time of its initial development and differentiation are present throughout gestation. Relative to the second general issue, steroidogenic enzymes are, for the most part, poised in an equilibrium

Steroidogenic pathways

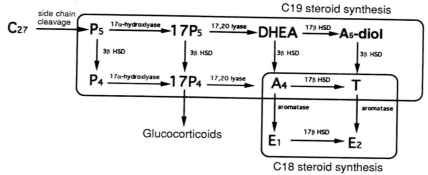

Rodent–porcine–equine model

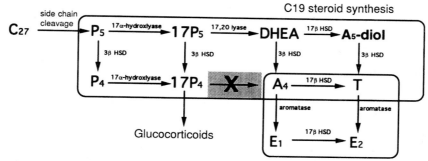

Human–bovine–ovine model

Fig 3.4. Steroidogenic pathways of cholesterol (C_{27}) metabolism to C_{21}, C_{19}, and C_{18} steroids in various species. Steroid names are abbreviated as follows: Pregnenolone (P_5), 17α-hydroxypregnenolone ($17P_5$), dehydroepiandrosterone (DHEA), 5-androstene-3β,17β-diol (A_5-diol), progesterone (P_4), 17α-hydroxyprogesterone ($17P_4$), 4-androstene-3,20-dione (A_4), testosterone (T), oestrone (E_1), oestradiol-17β (E_2). Note the inability of cattle, humans, and sheep to utilize progesterone for oestrogen synthesis, specifically because the 17α-hydroxylase/17,20 lyase of these species very inefficiently converts $17P_4$ to A_4.

such that no great differences exist among the activities of various enzymes of these pathways. The principles are best exemplified in domestic animal species where the balance between progestogen and oestrogen synthesis is changed dramatically following the induction of $P450_{17\alpha}$ at the time of parturition, a metabolic event unlikely if there were marked imbalances existing in the levels of other placental sterodogenic enzymes. The notable exception to this second rule is the human placenta which possesses exceptionally high levels of both 3β-HSD and steroid sulfatase such that few other enzymes can compete for substrate under any circumstances. Perhaps this is evidence for the greater autonomy of steroidogenesis in the human placenta in comparison to that of other species.

Regulation

Whether or not placental steroidogenesis is regulated remains a contentious issue. Although maternal plasma concentrations vary with stage of gestation, few studies have investigated ontogenic changes in enzyme levels. These data are important in order to assess placental steroidogenic potential, which is not necessarily reflected by plasma steroid levels for two major reasons. First of all, in some species, placental progesterone synthesis is foreshadowed by secretion from a continually active corpus luteum. Second, steroid output is a product of both synthesis and metabolism and, as will be discussed later, metabolism may be a major component of net steroid production by the placenta or its action on the uterus of some species. Consequently, without data relevant to changes in steroidogenic enzyme levels, it is difficult to assess the enzymatic background upon which regulation may take place.

With these limitations, placental steroid output is determined by several factors that can be discussed under the broad headings of placental mass, level of steroidogenic enzymes, and steroid metabolism. In the rest of this section each of these will be discussed in the context of progesterone and oestrogen secretion.

Placental steriodogenic capacity is dependent on mass and placental growth. Even in species exhibiting low levels of placental steroidogenesis, the initial rapid growth of the placenta is accompanied by corresponding increases in steroid synthesis. Oestrogen and PL secretion during human gestation are highly correlated with placental mass. In addition to an overall increase in cell number, a degree of differentiation accompanies the development of steroid synthesis. In humans, syncytialization of cytotrophoblast is thought to be associated with an increased steroidogenic capacity, and some evidence suggests that, in ruminants, fetal binucleate cells may be the steroidogenic elements in the placenta. Therefore, it would appear that development of steroidogenically active elements within the placenta requires both an increase in cell number and subsequent differentiation. Ultimately, growth and differentiation will determine the stage at which the placental contribution to the steroid pool becomes significant. In ovine and bovine species this may not occur until the second trimester. In the equine species, chorioallantoic placentation supersedes choriovitelline placentation after 50–60 days. In humans, this process is rapid, and significant progesterone production by developing trophoblast occurs by the end of the first month of gestation.

Both growth and differentiation are undoubtedly influenced by growth factors. Several studies have examined the effects of various factors on placental steroidogenesis. In cultures of human cytotrophoblastic cells, insulin and insulin-like growth factors I and II (IGFI and IGFII) all increase progesterone production possibly by effects on both 3β-HSD and P450$_{scc}$. At the same time, oestrogen production is decreased particularly in the presence of IGFII. The cellular effects of these growth factors are presumed to occur through cAMP or protein kinase C pathways and *in vitro* studies support this proposition. Despite these reports and although both proteins and peptide growth factors represent likely candidates for such a regulatory role, these factors are undoubtedly involved in cellular differentiation and possibly proliferation. The identification of inhibin in the placentae of humans and primates and the known effects and interactions of inhibins and activins on mesodermal differentiation certainly support the notion that differentiated cell function may be maintained in the placenta by local control mechanisms. Therefore, the specificity of the response relative to steroidogenesis is difficult to estimate as is the importance *in vivo*. However, these types of experiments have of necessity been performed *in vitro* and the expression of steroidogenic enzymes is difficult to maintain at *in vivo* levels while cells and tissues are in culture. More often,

stimulation by secretogogues represents a return of the level of steroidogenic enzymes, which have otherwise dropped in culture, to the range seen in the first day or two in culture. Those factors regarded as stimulators of steroidogenic function are perhaps only performing a maintenance function supporting our proposal that placental endocrine function is not acutely regulated. The challenge remains to determine the relative importance and potential interactions between and among the host of factors produced in and bathing the placenta.

The expression of the genes encoding enzymes involved in placental steroidogenesis, translation of the transcripts, and the stability of both the transcripts and the translated protein products determine, in large part, placental steroidogenic activity. This aspect of placental steroid production is very species-specific. We will attempt to give a brief overview of the major species differences by discussion of progesterone and oestrogen secretion.

The circulating levels of steroids vary considerably during the course of gestation just as they do among different species as illustrated in Fig. 3.5. In the sheep, luteal phase circulating progesterone concentrations of 1–3 ng/mL increase along with placental mass to values as high as 20 ng/mL in late gestation. Much less of a difference is seen between the luteal phase levels of progesterone and those attained during pregnancy which range from 5 to 10ng/mL in systemic plasma. Pigs are unusual in exhibiting a fall in systemic progesterone concentrations from 25–30 ng/mL during the oestrous cycle to 10–15 ng/mL throughout the remainder of gestation. Humans and guinea-pigs are also unusual because they have extremely high levels of circulating progesterone during pregnancy, almost 10-fold higher than those of many other animal species. Some of these variations are undoubtedly due to differences in the relative rates of synthesis of progestrone. For instance, progesterone production may be as much as 10-fold higher during pregnancy in women than in sheep. Less pronounced differences exist in the metabolic clearance rates of progesterone in these two species. However, the pregnant guinea-pig is exceptional in having a very low metabolic clearance rate of progesterone, in part due to unusually high levels of plasma-binding proteins. Therefore, despite having a relatively low rate of progesterone synthesis, the guinea-pig is able to maintain high levels of circulating steroid throughout most of gestation.

Progesterone synthesis

Many different functions have been proposed for progesterone in the physiology of the maintenance of pregnancy. The list includes its influence on the secretion of various uterine proteins and milk formation prior to placentation,

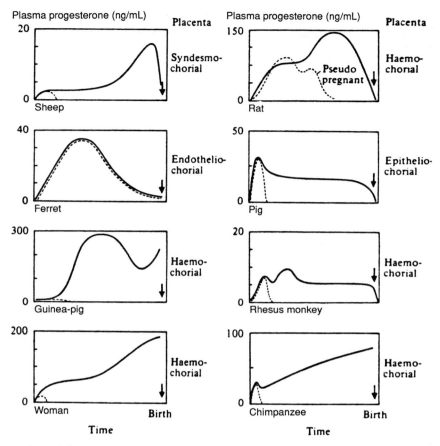

Fig. 3.5 The concentrations of plasma progesterone in different species, pregnant (solid lines) and non-pregnant (dotted lines), and the histological type of placenta. The arrows indicate the time of parturition. Note the differences in scale. (Reproduced with permission from *Reproduction in mammals*, Book 3, Hormonal control of reproduction, 2nd edition (ed. C.R. Austin and R.V. Short), Cambridge University Press, 1984).

the maintenance of myometrial quiescence, stimulation of mammary gland development, suppression of immune rejection of the conceptus, suppression of ovulation during gestation, and suppression of lactation until parturition. The relative importance of each of these functions varies in different species as does the absolute requirement for progesterone. For instance, levels of progesterone are almost undetectable in elephants, and implantation and early development occur in ovariectomized guinea-pigs in the apparent absence of the steroid. In many species much more progesterone is produced than is required and during pregnancy, progesterone withdrawal by whatever mechanism can be somewhat compensated for if it occurs slowly.

Although relatively little is known concerning species differences in substrate utilization, it is likely that most mammals use circulating lipoprotein as the major source of cholesterol. For instance, it is clear that the human placenta utilizes maternal low-density lipoprotein (LDL) and that only little cholesterol is synthesized *de novo*. Uptake of LDL into trophoblast is receptor-mediated. Furthermore, LDL receptors have been identified in human placental tissue at as early as 6 weeks of gestation. Initial studies of LDL receptor mRNA levels suggest that receptor levels remain essentially constant throughout gestation.

It is important to note that, although *de novo* synthesis of cholesterol from 2 carbon units is slow under physiological conditions, this does not preclude *de novo* synthesis from playing a potentially important role under unusual circumstances. The rate-limiting step in *de novo* cholesterol synthesis is that catalysed by 3-hydroxy-3-methylglutaryl CoA (HMGCoA) reductase. The levels and activity of this enzyme are influenced by the intracellular concentrations of cholesterol itself. Therefore, under conditions of reduced LDL uptake or presumably low intracellular cholesterol concentrations, levels of HMGCoA reductase may increase. Thus, in a case of low circulating levels of β-lipoprotein in a mother, it was noted that there was a higher activity of HMGCoA reductase in placenta than might have been expected normally suggesting the development of just such a compensation in the face of low LDL levels.

Cholesterol is oxidized by cytochrome $P450_{scc}$ to produce pregnenolone which is subsequently converted to progesterone by the action of 3β-HSD. Little information presently is available on species differences in the activity of these two key enzymes.

It is accepted that the rate of formation of cholesterol is rate-limiting in C_{21} steroid synthesis, either by way of side-chain cleavage activity or by the rate of transfer of cholesterol across the outer mitochondrial membrane. This appears to be true even for the human placenta, which is exceptional in its daily production of progesterone, because the very high levels of 3β-HSD activity in microsomal fractions would allow 10 times greater amounts of progesterone to be synthesized from pregnenolone if sufficient substrate were available. While these data provide compelling evidence that placental $P450_{scc}$ activity is central to the rate of placental progesterone production under physiological conditions, it is not necessarily the sole determinant of the rate of progesterone synthesis, nor is placental progesterone production itself necessarily indispensable for pregnancy maintenance. In humans, genetic deficiencies of LDL, $P450_{scc}$ and 3β-HSD have all been identified with clinically recognizable consequences that occur at later developmental or postnatal stages. The direct consequences of these deficiencies on placental steroidogenesis have not been well studied, but the success of such pregnancies suggests that the lack of enzyme activity was compensated for in some way. Several possibilities exist to explain these observations. Mutant proteins may retain a small fraction of the activity of the normal enzyme. Alternative sources of substrate such as maternal adrenal pregnenolone sulfate may be utilized. Finally, other organs may provide enough progesterone to support the pregnancy; a surviving corpus luteum, for example, which is still capable of steroid secretion near term.

Oestrogen synthesis

Oestrogens can be considered to be the final synthetic product of the steroidogenic pathways leading to the production of sex steroids. They are formed from C_{19} (androgenic) precursors by aromatization in a reaction catalysed by aromatase cytochrome P450 complex. Androgens, in turn, are synthesized from C_{21} (progestogenic) precursors, either progesterone or pregnenolone, by 17α-hydroxylase/17,20 lyase

cytochrome P450. Therefore, oestrogen synthesis is dependent on the activity of these two enzymes in the placenta or the provision of aromatizable substrate in the absence of 17α-hydroxylase/17-20 lyase cytochrome P450. This aspect of oestrogen synthesis as it occurs during pregnancy will be discussed in relation to species differences.

Oestrogens are thought to be important mediators of implantation, uterine growth, mammary duct development, pelvic and cervical relaxation, the induction of myometrial oxytocin receptors, and maternal behaviour. These functions are species-specific. It is noteworthy, however, that most species synthesize oestrogen in significant amounts at some stage of pregnancy and that maternal systemic levels of oestrogens generally increase as pregnancy advances (Fig. 3.6). In ruminants an early peak of oestrogen is noted in fetal fluids (up to 475 ng/mL in allantoic fluid of bovine fetuses at 133 days of gestation and 14 ng/mL in ovine fetuses at 45 days gestational age). In cattle this peak of oestrogen corresponds to the peak in expression of 17α-hydroxylase in placental tissues at the end of the first trimester. Maternal plasma oestrogen concentrations, however, do not reach significant levels in either cattle or sheep until term when oestrone sulfate concentrations approach 1–2 ng/mL. The pig exhibits a peak of oestrone sulfate secretion that is detectable in maternal systemic plasma at levels as high as 3 ng/mL by day 30 of gestation, which is equivalent to levels seen at term. Free oestrogen concentra-

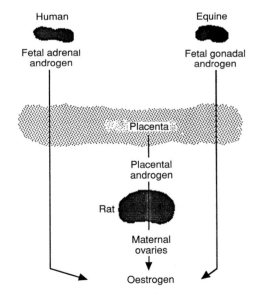

Fig. 3.6 Sources of C_{19} steroid substrate for placental aromatisation and oestrogen production in human, equine species, and rat provide an illustration of the various interorgan communication strategies in steroidogenesis during pregnancy.

tions, however, are also significant and comparable to the sulfoconjugates at term. The precise source of the brief burst of oestrogen secretion seen at day 30 of porcine pregnancy is not well documented. Although it undoubtedly involves aromatization in the placenta, the porcine placenta is relatively deficient in 17α-hydroxylase for most of gestation so that androgen precursors may arise from other fetal tissues such as the developing adrenal gland.

Various strategies appear to have evolved to enable oestrogen synthesis to take place, all involving placental participation, even in those cases where the placenta alone is incapable of the task independently. In cattle, sheep, and pigs, particularly in the final stages of gestation, the placenta can be considered to be a 'complete' steroidogenic organ possessing all the enzymes necessary to synthesize oestrogens successfully. However, certain other species lack particular key enzymes in the placenta. For instance, the human placenta lacks detectable cytochrome $P450_{17\alpha}$ activity and the placenta of the rat lacks cytochrome $P450_{arom}$. In the human, therefore, C_{21} steroids cannot be converted to C_{19} androgens. Instead, dehydroepiandrosterone sulfate (of both maternal and fetal adrenal origin) and 16α-hydroxydehydroepiandrosterone sulfate (of fetal adrenal and liver origin) provide the substrate for desulfurylation and subsequent aromatization. In rats, the placenta synthesizes and secretes androgens during the later stages of pregnancy that are then aromatized in the maternal ovary in the few days prior to parturition. Finally, the synthesis of oestrogens in the horse placenta presents something of a unique case. Studies have not been performed to characterize the presence and relative levels of steroidogenic enzymes in the placenta of this species. The ability of the equine blastocyst to synthesize oestrogens and the demonstrated ability of the equine placenta to produce progesterone suggest, however, that it may also be capable of expressing a full complement of steroidogenic enzymes. However, it is now well documented that the major oestrogens secreted by the placenta during pregnancy in the horse are the B-ring saturated C_{18} steroids, equilin and equilenin. The synthesis of these steroids in large amounts (10 times higher than oestradiol-17β) parallels the dramatic increase in size of the fetal gonads that begins at the end of the first trimester and decreases in the third trimester toward term. Therefore, although it seems likely that the equine placenta is steroidogenically a 'complete' gland, the majority of secreted oestrogens originate as C_{19} steroid precursors in the fetal gonads and are aromatized in the placenta.

In human placenta, oestrogen synthesis is enormous in the later stages of gestation. As with progesterone, there is as yet no convincing evidence for regulation of oestrogen synthesis in the human placenta, although it has been sug-

gested that the level of placental sulfatase activity may be limiting. Certainly, sulfatase deficiency, as in congenital ichthyosis, results in a dramatic decrease in circulating oestrogen levels. Although such pregnancies progress to term, an increased incidence of complications at the time of delivery suggests that the deficiency may have important consequences for the birth process. Whether or not these difficulties can be directly attributed to oestrogen itself is impossible to determine at present. However, a recent report describing a successful pregnancy involving a conceptus with an aromatase deficiency casts further doubt on the absolute necessity for oestrogen synthesis during pregnancy in humans.

Parturition in the ewe has emerged as an important example of the regulation of placental oestrogen secretion in relation to pregnancy. Fetal adrenal participation in the process of parturition was first recognized when it was noticed that the congenital absence of the adrenal glands in the fetal lamb was associated with prolonged gestation. Subsequent studies have since established that adrenal glucocorticoid synthesis and secretion can induce parturition. Furthermore, because increased fetal glucocorticoid levels precede the periparturient rise in maternal oestrogen concentrations and exogenous oestrogen itself induces parturition, it has been proposed that glucocorticoids induce placental oestrogen production leading to parturition. Recent studies have demonstrated that parturition, either natural or induced, is associated with increased levels of cytochrome $P450_{17\alpha}$. However, in vitro studies have not convincingly demonstrated an effect of glucocorticoids equivalent to that seen in vivo. Although some investigators have demonstrated a doubling of enzyme activity of sheep placental explants when exposed to glucocorticoids, similar experiments have failed to show an induction of cytochrome $P450_{17\alpha}$ protein by Western immunoblot analysis. Furthermore, although cytochrome $P450_{17\alpha}$ activity, protein, and mRNA increase in response to glucocorticoids administered to ewes to induce parturition, these changes are seen only hours prior to parturition itself. It is not yet clear whether or not the effect of glucocorticoid exposure on placental enzyme induction is direct.

Fluctuations in the levels of placental cytochrome $P450_{17\alpha}$ are not restricted to the periparturient period. Recent studies indicate that at least two peaks of activity occur in at least some domestic animal species (Fig. 3.7). For instance, levels of cytochrome $P450_{17\alpha}$ are elevated in bovine placental tissue early in gestation. Temporally, it appears that placental expression follows early activation of the fetal adrenal, further supporting the existence of an association between fetal glucocorticoid and placental oestrogen synthetic capacity. Although early activation of the fetal adrenal has been demonstrated in sheep, the con-

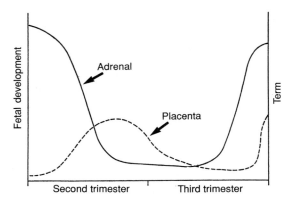

Fig. 3.7 Diagrammatic representation of the relationship between 17α-hydroxylase expression in fetal bovine adrenal glands and placenta during the second and third trimesters of pregnancy. A similar pattern of fetal adrenal 17α-hydroxylase expression is seen in the ovine fetus and in ovine placentomes around the time of parturition. The regulation of 17α-hydroxylase expression is proposed to be a key step in the endocrine events leading to oestrogen secretion during pregnancy and to parturition.

sequences on placental enzyme expression have not yet been investigated. From studies in the cow, it seems that neither placental cytochrome $P450_{scc}$ nor 3β-HSD levels fluctuate dramatically at any stage of pregnancy. Again, the exact consequences of this early enzyme activation are not known.

Steroid metabolism

Both the site and the rate of progesterone metabolism can have important physiological consequences. In humans, progesterone metabolism occurs in the liver and in peripheral tissues, and it has been suggested that the formation of active metabolites in peripheral tissues, 5α-reduction for instance, may mediate endocrine disorders of various poorly defined types in pre- and postmenopausal women. In domestic animal species, the rate of progesterone metabolism in the uterus, even if small, may influence the action of progesterone on both the endometrium and the myometrium. In the past, conclusions about the comparative abilities of placentae to synthesize progesterone have been made by considering the ability of an animal to maintain pregnancy following ovariectomy or removal of the corpus luteum. Results of these types of studies in ewes, cows, and sows have led to the conclusion that the ovine placenta is more capable of progesterone synthesis than are those the cow or the sow. However, these conclusions assume that the placenta and the corpus luteum are the only sources of progesterone during pregnancy, and that the rate

of progesterone metabolism is similar among species. The rate of progesterone metabolism can affect circulating progesterone levels as well as the levels of active steroid in placental tissue. Bovine placental tissues metabolize progesterone at a faster rate *in vitro* than ovine tissues. This may, in part, explain the greater ability of the ovine placenta to maintain pregnancy in the absence of the ovaries. However, when the activity of $P450_{17\alpha}$ is increased in ovine placental tissues, at parturition or following the infusion of dexamethasone, the rate of metabolism increases. Therefore, the induction of placental cytochrome $P450_{17\alpha}$ activity in sheep not only enhances the potential for oestrogen synthesis from pregnenolone via the Δ^5 pathway, but may also be accompanied by an increase in progesterone metabolism thereby contributing further to progesterone withdrawal and parturition. Goats also rapidly metabolize progesterone to 5β-pregnanes and are incapable of maintaining pregnancy at any stage of gestation in the absence of an ovarian source of progesterone. It appears, from the available data, that all mammalian species are probably capable of placental progesterone synthesis, but that considerable species differences exist in placental progesterone production that may be attributable, at least in part, to differing rates of progesterone metabolism *in vivo*.

Finally, it is prudent to note that oxidoreductases, which catalyse reactions at several positions of the C_{21} skeleton, and sulfotransferase or glucuronidases, catalysing conjugation of Δ^5-C_{21} and C_{19}, as well as C_{18} steroids, may also have a major influence on steroid products. Unfortunately, less is known about these enzymes to date. The relative levels and activities of various conjugating enzymes and oxidoreductases could dramatically influence the production of both oestrogen and progesterone by the placenta, or influence their release into the maternal circulation. Further studies are needed to determine their role, if any.

Further reading

Chene, N., Martal, J., and Charrier, J. (1988). Ovine chorionic somatomammotropin and foetal growth. *Reproduction, Nutrition and Development*, **28**, 1707–30.

Conley, A.J. and Mason, J.I. (1990). Placental steroid hormones. *Baillière's Clinical Endocrinology and Metabolism*, **4**, 249–72.

Faria, T.N., Deb, S., Kwok, S.C.M., Talamantes, F., and Soares, M.J. (1990). Ontogeny of placental lactogen-I and placental lactogen-II expression in the developing rat placenta. *Developmental Biology*, **141**, 279–91.

Fenstermaker, R.A., Farmerie, T.A., Clay, M., Hamernik, D.L., and Nilson, J.H. (1990). Different combinations of regulatory elements may account for expression of the glycoprotein hormone α-subunit gene in primate and horse placenta. *Molecular Endocrinology*, **4**, 1480–7.

Flint, A.P.F. (1983). Factors controlling placental endocrine function in domestic animals. *Current Topics in Experimental Endocrinology*, **5**, 75–95.

Jackson-Grusby, L.L., Pravtcheva, D., Ruddle, F.H., and Linzer, D.I.H. (1988). Chromosomal mapping of the prolactin/growth hormone gene family in the mouse. *Endocrinology*, **122**, 2462–6.

Jones, C.T. (1988), Endocrine function of the placenta. *Baillière's Clinical Endocrinology and Metabolism*, **3**, 755–80.

Miller, W.L. (1988). Molecular biology of steroid hormone synthesis. *Endocrine Reviews*, **9**, 295–318.

Mochizuki, M. and Hussa, R. (ed.) (1988). *Placental protein hormones*. Elsevier Science Publishers, Amsterdam.

Morgan, G., Wooding, F.B.P., Beckers, J.F., and Friesen, H.G. (1989). An immunological cryoultrastructural study of a sequential appearance of proteins in placental binucleate cells in early pregnancy in the cow. *Journal of Reproduction and Fertility*, **86**, 745–52.

Ogren, L. and Talamantes, F. (1988). Prolactins of pregnancy and their cellular source. *International Review of Cytology*, **112**, 1–65.

Ohlsson, R. (1989). Growth factors, protooncogenes and human placental development. *Cell Differentiation and Development*, **28**, 1–16.

Porter, D.G., Heap, R.B., and Flint, A.P.F. (1982). Endocrinology of the placenta and the evolution of viviparity. *Journal of Reproduction and Fertility*, **31**, 113–38.

Ringler, G.E. and Strauss III, J.F. (1990). In vitro systems for the study of human placental endocrine function. *Endocrine Reviews*, **11**, 105–23.

Schuler, L.A., Shimomura, K., Kessler, M.A., Zieler, C.G., and Bremel, R.D. (1988). Bovine placental lactogen: molecular cloning and protein structure. *Biochemistry*, **27**, 8443–8.

Segal, S.J. (ed.) (1980). *Chorionic gonadotropin*. Plenum Press, New York.

Simpson, E.R, and MacDonald, P.C. (1981). Endocrine physiology of the placenta. *Annual Reviews of Physiology*, **43**, 163–88.

Stewart, F., Thomson, J.A., Leigh, S.F.A., and Warwick, J.M. (1987). Nucleotide (cDNA) sequence encoding the horse gonadotrophin α-subunit. *Journal of Endocrinology*, **115** 341–6.

Stewart, H.J., Jones, D.S.C., Pascall, J.C., Popkin, R.M., and Flint, A.P.F. (1988). The contribution of recombinant DNA techniques to reproductive biology. *Journal of Reproduction and Fertility*, **83**, 1–57.

Sugino, H., Bousfield, G.R., Moore Jr, W.T., and Ward, D.N. (1987). Structural studies of equine glycoprotein hormones. *Journal of Biological Chemistry*, **262**, 8603–9

Talamantes, F. and Ogren, L. (1988). The placenta as an endocrine organ: polypeptides. In *The physiology of reproduction* (ed. E. Knobil, J. Neill, G.S. Greenwald, and L.L. Ewing), pp 2093–144. Raven Press, New York.

Warren, W.C., Liang, R., Krivi, G.G., Siegel, N.R., and Anthony, R.V. (1990). Purification and structural characterization of ovine placental lactogen. *Journal of Endocrinology*, **126**, 141–9.

4. Metabolic activity of the placenta
William W. Hay, Jr and Randall B.Wilkening

The placenta is a specialized organ of exchange, interposed between the fetal (umbilical) and maternal (uterine) circulations to provide nutrients and oxygen to the fetus and excrete waste products from the fetus. The placenta also synthesizes and transforms a variety of hormones necessary to maintain pregnancy and to prepare for lactation. Because of the essential role of the placenta in exchange and interest in the metabolism of the fetus, there has been little investigation into the metabolism of the placenta. Nevertheless, experiments on placental tissue *in vitro* and the uteroplacenta *in vivo* have demonstrated marked metabolic activity of placental tissue, with oxygen and glucose consumption rates approaching or exceeding those of brain and tumour tissue. An impressive array of metabolic activities has been documented in placental tissue including glycolysis, gluconeogenesis, glycogenesis, carbon substrate oxidation, protein synthesis, amino acid interconversion, triglyceride synthesis, and chain lengthening and shortening of individual fatty acids. Thus, it is important to consider the metabolism of the placenta in its own right and how this metabolism contributes to fetal growth and development.

Placental morphology and development

During the embryonic period the placenta develops from a highly vascularized membrane known as the chorion. This membrane attaches to discrete or diffuse areas of the epithelium of the uterine mucosa, interacting with, or actually invading, this tissue to varying degrees. Discrete implantation usually produces a single trophoblast tissue mass (as in humans) while diffuse implantation usually produces multiple smaller masses of trophoblast tissue such as the placental cotyledons in sheep. With discrete implantation, chorionic interaction begins over much of the uterine epithelium but regresses over gestation to produce a discrete mass covering less than 40 per cent of the uterine epithelial surface. Failure of this regression, as in the condition 'placenta membranacea', produces abnormal placental function, often leading to hydrops fetalis (total body oedema) and preterm delivery. The degree of chorionic interaction and/or invasion appears to be species-specific.

Three major types of interaction between the placenta and the uterine mucosal epithelium have been defined.

1. *Epitheliochorial.* In this type of placental development the placenta (chorionic epithelium) attaches to the uterine mucosal epithelium by the growth of interdigitating microvilli from both placental and uterine epithelial surfaces. This type of placental membrane is found among equine, porcine, ovine, and bovine species (Fig. 4.1 (a)).
2. *Endotheliochorial.* In this type of placenta the uterine mucosal epithelium has been eroded away and the placental or chorionic epithelium interacts directly with the uterine mucosal interstitial space and endothelium of the uterine capillaries. This type of placenta is found primarily among the carnivores (Fig. 4.1 (b)).
3. *Haemochorial.* In rodents and most primates, including humans, the chorionic epithelium invades and destroys both the uterine mucosal epithelium and the uterine capillary endothelium. Vascularized villi of the chorionic epithelium thus are bathed by maternal blood (Fig. 4.1 (c)).

Although earlier investigators suggested that placental function was directly related to the degree of chorionic invasion of the uterine mucosa, current understanding of placental function is quite different. Many other factors contribute to placental function including total placental blood flow, intraplacental blood flow patterns, thickness of the various placental membranes, genetically determined transport properties (e.g. transport proteins and ion channels), the surface area and structure of the membranes, the physical–chemical properties of the membranes, the metabolic activity of the placenta, and various mechanisms of transfer (for example, simple diffusion, carrier-mediated transfer, and active transfer).

Placental growth and its effects on fetal growth

In most species placental growth continues throughout gestation. However, the growth rate of the placenta varies among species. Placental growth is most rapid in early gestation and either continues at a slower rate until term as in man or is completed before the end of gestation as in sheep.

Figure 4.2 shows growth curves for human placentae derived from measurements of placental weights at delivery at a variety of gestational ages. These data demonstrate two important characteristics of placental growth. First, they show that fetal weight is directly related to placental weight. Experimental data have also been obtained in the pregnant sheep in which, prior to pregnancy, uterine caruncules were removed by electrocautery (a caruncle is a specific placental implantation site in the uterine mucosa); this process reduced the number of placental implantation sites and the growth of the placenta. Data in Table 4.1 document the reduced placental size in these experiments and the corresponding reduction in fetal weight, umbilical blood flow, and placental-to-fetal transfer rates of glucose and oxygen. Second, the data provided by the study of human placentae (Fig. 4.2) and corroborated by the experimental carunclectomy studies in sheep (Table 4.1) show that small, average, and large fetuses are

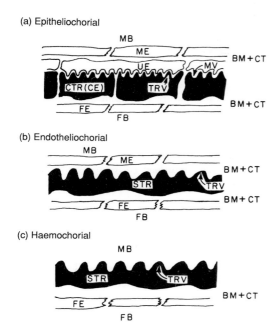

Fig. 4.1 Diagrammatic representation of the three main anatomical types of placenta, classified according to histological structure. (a) Epitheliochorial, which has both fetal and maternal endothelial and epithelial structures, the latter apposing an intervening cytotrophoblast or chorionic epithelium. This type of placenta is found in sheep and swine. (b) Endotheliochorial, which has both fetal and maternal endothelia that appose a syncytiotrophoblast (STR). This type of placenta is common among the carnivores. (c) Haemochorial, which has a fetal endothelium and a syncytiotrophoblast but no uterine epithelium or endothelium. The trophoblast has eroded through the maternal (uterine) epithelium and endothelium, allowing maternal blood to circulate through the intervillous space, bathing directly the maternal-facing villous membrane of the syncytiotrophoblast. This type of placenta is found in humans. MB, maternal blood; FB, fetal blood; ME, maternal endothelium; BM, basement membrane; CT, connective tissue; UE, uterine epithelium; MV, maternal epithelial microvilli; TRV, trophoblast microvilli; CTR, cytotrophoblast; CE, chorionic epithelium; FE, fetal endothelium. (Adapted from: Battaglia, F.C. and Meschia, G. (1986) *An introduction to fetal physiology*. Academic Press, Orlando, Florida; Friess, A.E., Sinowatz, F., Skolek-Winnisch, R., and Trautner, W. (1980). *Anatomic Embryology* 158, 179–91; Steven, D.H. (ed.) (1975). In *Comparative placentation: Essays in structure and function*, p. 25. Academic Press, New York; Freese, V.E. (1972). In *Respiratory gas exchange and blood flow in the placenta*, DHEW Publication (NIH) (ed. L.D. Longo and H. Bartels), pp. 73–361. DHEW, Washington, DC.)

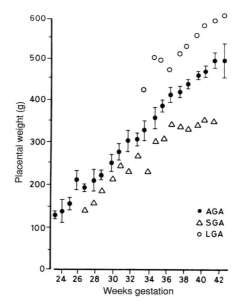

Fig. 4.2 Weights of human placentae are shown to increase with gestational age and to fit three categories, AGA (average or appropriate for gestational age), SGA (small for gestational age), and LGA (large for gestational age), which also correspond to fetal weight/gestational age categories. (Reproduced with permission from Molteni, R.A. Stys, S.J., and Battaglia, F.C. (1978). *Journal of Reproductive Medicine*, **21**, 327–34.)

Table 4.1 Placental and fetal metabolic effects of removal of placentation sites in sheep*

	Controls	Carunclectomy (Growth retarded fetuses)
Placental weight (g)	485 ± 105	197 ± 91
Fetal weight (g)	3720 ± 807	2198 ± 653
Maternal arterial blood glucose concentration (mM)	1.84 ± 0.46	2.41 ± 0.45
Fetal arterial blood glucose concentration (mM)	0.84 ± 0.24	0.57 ± 0.11
Fetal oxygen consumption		
(mg/min)	1.208 ± 0.488	0.748 ± 0.215
(mmol/min/kg fetal weight)	0.325 ± 0.131	0.340 ± 0.100
Fetal glucose consumption		
(mg/min)	18.4 ± 2.7	11.1 ± 4.3
(mg/min/kg fetal weight)	4.9 ± 0.7	5.1 ± 2.0
Estimated placental glucose transfer capacity: fetal glucose consumption (mg/min)/maternal–fetal arterial plasma glucose concentration difference (mg/dL)	0.036	0.018

Data given as mean ± SD.
*Adapted from: Owens, J.A., Falconer, J., and Robinson, J.S. (1987). *Journal of Developmental Physiology*, **9**, 225–38.

directly related to small, average, and large placentae, suggesting that the fetus does not outgrow the placenta. In the ovine carunclectomy group, however, even among the growth-retarded fetuses (those that were smaller than expected for gestational age), the fetal-to-placental weight ratio was increased, indicating the possibility that an increase in functional capacity of the placenta may compensate for the reduction in placental size. Thus, placental weight remains a major determinant of fetal size but is not exclusive of its functional capacity for promoting fetal growth.

The capacity for functional adaptation of the placenta has been documented in most detail in sheep. In this species placental weight is maximal at 50 per cent of gestation and even declines slightly towards term. Fetal growth continues progressively until term, however, demonstrating a marked change in functional capacity of the placenta to meet the nutrient needs of fetal growth. Thus, late gestation is marked by a progressive increase in the functional transport capacity of the placenta that is quite separate from placental weight. In this regard, when

the placenta is of normal size and function, any slowing of fetal growth observed in late gestation in apparently normal fetuses is unlikely to be due to placental limitation of nutrient supply. More probably, normal developmental processes have greater effects on fetal growth at this period of gestation. Furthermore, these functional changes appear independent of the placental morphological type. For example, as shown in Fig. 4.3 in which villus surface area (S_T) is compared with the sum of placental and fetal weights (V_T), the villus surface area is well correlated with fetal weight from species as small as the rat to as large as the elephant, regardless of whether the placental type is compact (discrete) or diffuse. At comparable fetal weights,

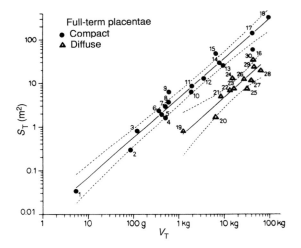

Fig. 4.3 Villous surface area, (S_T in m^2), plotted in logarithmic coordinates against the sum of placental and fetal weight, (V_T, kg), The dashed lines show the 95 per cent confidence limits of the two regression lines for compact (discrete) and diffuse placentae. Compact placentae: 1, rat; 2, guinea-pig (*Cavia*); 3, cat; 4, German sheep dog; 5, sloth (*Choloepus didactylus*); 6, crab-eating monkey (*Macaca fascicularis*); 7, douc langur (*Pygathrix nemea*); 8, Guereza (*Colobus polycomos*); 9, leopard; 10, chimpanzee; 11, gorilla; 12, human; 13, dwarf zebu; 14, seal (*Phoea vitulina*); 15, sea lion (*Zalophus californianus*); 16, giraffe (*Giraffa camelopardalis tippelskirchi*); 17, European domestic cow; and 18, African elephant. Diffuse placentae: 19, European domestic pig; 20, dwarf hippopotamus (*Choeropsis liberiensis*); 21, dolphin; 22, Llama; 23, pony; 24, Sardinian dwarf donkey; 25, zebra; 26, Somalian wild ass (*asinus asinum somalicus*); 27, bactrian camel; 28, Indian rhinoceros (*Rh. unicornis*); 29, horse (thoroughbred); and 30, hippopotamus (*H. amphibicus*). (Reproduced with permission from Battaglia, F.C. and Meschia, G. (1986). *An introduction to fetal physiology*. Academic Press, Orlando, Florida.)

however, diffuse placentae have approximately seven times less villus surface area than the compact (discrete) placentae. Thus, placental size and functional capacity are two important characteristics of placental development during gestation that directly determine the rate and magnitude of fetal growth.

Functional maturation of the placenta has been observed from two perspectives. The first involves morphometric studies that, in a relatively large number of species, have shown that the villus surface area that is visible under light microscopy increases progressively throughout gestation. An example of this information is presented in Fig. 4.4 for the human placenta. In addition to this increase in surface area, similar studies by light microscopy have demonstrated a decrease in the thickness of the placental membranes over gestation. In the human the cytotrophoblast tends to disappear and the syncytiotrophoblast thins considerably. These changes are disproportionate to changes in placental DNA content (which increase in some species, or do not occur in others) indicating that these maturational changes involve alteration in structure in addition to change in placental growth rate.

Functional maturation of the placenta has also been quantified by estimates of transfer capacity. As one example, in pregnant sheep placental permeability (or diffusing capacity) to urea increases dramatically over the second half of gestation, commensurate with the increase in fetal weight but in contrast to the decrease in placental weight characteristic of the sheep placenta in late gestation. More

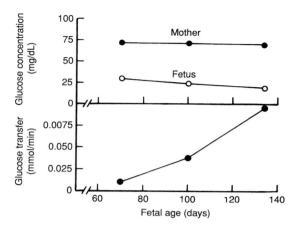

Fig. 4.5 Upper panel: Glucose concentration tends to decrease in fetal sheep over the second half of gestation; maternal concentration was fixed for study purposes at 70 mg/dL plasma. Glucose transfer to the fetus from the mother via the placenta increases about eightfold over this same period; this increase in transport reflects transport capacity, as 60 to 70 per cent is not accounted for by the increase in transplacental glucose concentration gradient. (Adapted from Molina, R.D., Meschia, G., Battaglia, F.C., and Hay, W.W., Jr (1991). Gestational maturation of placental glucose transfer capacity in sheep. *American Journal of Physiology*, **261**, R697–R704.)

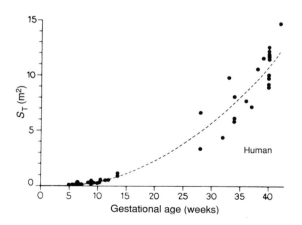

Fig. 4.4 Growth of the surface area, S_T, of the placental membrane in humans. The surface area in m^2 is plotted against gestational age in weeks (S_T/A). (Reproduced in modified form with permission from Battaglia, F.C. and Meschia, G. (1986). *An introduction to fetal physiology.* Academic Press, Orlando, Florida.)

recent studies in sheep have focused on the change in placental glucose transport capacity (PGT) over the second half of gestation which increases approximately 10-fold from mid- to late-gestation. Part of the increase in PGT is due to increased fetal glucose consumption and part is due to the increase in maternal-to-fetal glucose concentration gradient produced by an increase in the rate of placental glucose consumption (absolute rate, as well as on a placental weight-specific basis), even though the proportion of uterine glucose uptake consumed by the placenta, relative to that of the fetus, actually declines. As shown in Fig. 4.5, however, approximately 60 per cent of the increase of PGT is accounted for by the increase in transport capacity *per se*. This could result from an increase in the maximal velocity of transport (V_{max}) or in the Michaelis constant (K_m or $V_{max}/2$) for glucose transport (the latter representing a more rapid rate of transfer at lower glucose concentration gradients). Recent experiments in pregnant sheep have demonstrated that the Michaelis constant does not change over the second half of gestation, whereas the maximal velocity of transport increases at least eightfold. These results suggest that the increase in PGT capacity could be explained primarily by an increase in the number of glucose transporters. *In vitro* evidence from sheep placentae indicates an increase in placental glucose transporter number over the latter part of gestation. It is not clear, however, whether this is due to an increase in concentration per unit of membrane surface area or simply an increase in the surface area of the placenta that occurs concomitantly.

The mechanisms regulating placental growth remain obscure. In part, this regulation may be due to the size and nature of the implantation area as suggested by many clinical observations as well as the experimental carunclectomy studies already described. Placental blood flow increases with, but does not necessarily determine, placental size.

A large variety of growth factors have been associated with various aspects of placental growth. Some growth factors (e.g. fibroblast growth factor, platelet-derived growth factor, insulin-like growth factor I) may enhance cell replication and trophoblast cellular growth. Other growth factors (e.g. epidermal growth factor) may enhance hormonal production. Still other growth factors may enhance trophoblast differentiation (e.g. from cytotrophoblast to syncytiotrophoblast), in opposition to factors (e.g. chorionic gonadotrophin) that suppress differentiation. In addition, protease and adhesion molecules are produced, promoting trophoblast cellular migration into the decidua. Thus, there appears to be a very complex set of growth factor activities in the placenta, with multiple checks and balances, that help to regulate placental trophoblast development, proliferation, and implantation.

Endocrine factors may also be of some importance, particularly during the phase of rapid placental growth. Experiments in both monkeys and sheep have demonstrated compensatory growth when a portion of the placenta is removed or its vascular supply obstructed. In those situations in which placental growth returned to normal size, fetal weight followed accordingly. In rabbits, pregnancy can be prolonged by appropriate endocrine treatment; the fetuses of such prolonged pregnancies are markedly increased in size without a significant increase in placental weight. Also, in the rhesus monkey the fetus can be removed by Caesarean section and the placenta will be retained even beyond term; however, it stops growing in size. The syncytiotrophoblast continues to secrete placental lactogen but fetal capillaries degenerate and the cytotrophoblast disappears suggesting that placental growth and the maintenance of some placental tissues depend upon the continuous supply of trophic factors by the fetus. The haemochorial placentae of mice that are genetically dissimilar from the mother grow approximately 25 per cent larger than the placentae of fetuses whose parents are from the same genetic strain indicating that placental growth may also depend on antigenic differences between mother and fetus. Finally, the environment itself can influence placental growth, as shown by slower placental growth rates and smaller placental sizes at term in pregnant sheep that are relatively hyperthermic during pregnancy.

Methods used to measure placental metabolism

A variety of methods have been used to measure placental metabolism. All have provided important information but not one is wholly satisfactory. *In vivo* measurements using Fick principle and tracer methodology have provided the most reliable data for normal metabolic rates but such rates are not specific to the trophoblast because they also include uterine tissues. *In situ* and *in vitro* perfusion models have suffered from inaccuracies produced by variations in the composition of the perfusion medium, particularly of oxygen content, as well as disruption of the tissues as they are removed from their normal circulations and tissue attachments. *In vitro* tissue incubations have helped to identify specific enzymes and metabolic pathways but rates of metabolism through such pathways and enzymes are far from normal. It is important, therefore, to consider these methods in detail before discussing the results of metabolic rate measurements in placental tissues from a large variety of methods and species.

In vivo *measurements of placental metabolism*

Most *in vivo* measurements of placental metabolism have been made in large animals, primarily the pregnant sheep, comparing the net flux of substrates between maternal blood and the pregnant uterus with the net flux of substrates between the placenta and fetal blood. This comparison provides information about the metabolic activity of the physiological entity known as the 'utero-placenta'. The utero-placenta is a group of tissues that includes the myometrium, endometrium, placental (trophoblast) mass, and the extraplacental chorionic membrane. Studies on the distribution of uterine blood flow and the transplacental diffusion of glucose and oxygen indicate that placental metabolism accounts for the major component (more than 75 per cent) of the metabolic activity of the utero-placenta. Nevertheless, this methodology does not isolate the trophoblast itself or measure its specific metabolic activity.

To quantify utero-placental metabolism, *in vivo* catheters are placed into a maternal artery, a uterine vein, a fetal artery, and the umbilical vein. Following recovery of the animal from surgery and the resumption of adequate feeding, uterine and umbilical blood flows are measured simultaneously. A variety of methods for measuring uterine and umbilical blood flows have been used including the transplacental steady-state diffusion technique, electromagnetic flow transducers, and ultrasonic Doppler. During a steady-state period in which blood flows are measured, blood is sampled simultaneously from each of the four catheterized vessels and concentrations of the measured substances are determined separately. Using the Fick principle, one can then calculate uterine uptake, umbilical uptake, and their difference, utero-placental uptake, for each substance studied (Table 4.2). Figure 4.6 presents typical numerical data across the sheep utero-placenta for oxygen transfer and consumption. This methodology has the advantages of being conducted *in vivo* in conscious healthy animals, including measurement of blood flow, and the potential for regulating and determining the effect of blood flow on placental consumption and transfer of substances. An additional advantage includes the capacity to regulate and quantify the effect of uterine and/or umbilical plasma concentrations of substrates on placental transfer and metabolism. Furthermore, the Fick principle provides measurements of net fluxes and rates of consumption useful for calculating metabolic balances. Several disadvantages of the Fick principle methodology are also apparent. Small animals are extremely difficult to study with this methodology due to technical problems with catheterization. Similar *in vivo* studies of placental metabolism cannot be conducted yet in humans for obvious ethical reasons. Measurement of substrate transfer and metabolism, in general, can be accomplished only for substances that have a high extraction ratio across the uterine and/or umbilical circulations relative to the accuracy of their measurement. For example, for some amino acids the umbilical extraction ratio is approximately 10 per cent, not significantly different from the coefficient of variation of measurement of their plasma concentrations; thus, an accurate measurement of umbilical veno-arterial concentration difference cannot be obtained.

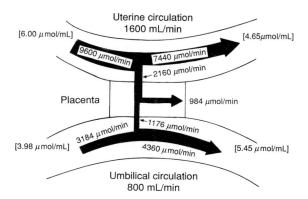

Fig. 4.6 Numerical example of oxygen transport across the placenta of the sheep as measured and calculated using the Fick principle.

In situ *and* in vitro *perfusion models*

The placentae of small animals have been perfused *in situ* by catheterization of the umbilical vessels while the maternal circulation either remains intact or is perfused by catheterization of uterine arteries and veins. The perfused circulations can be connected to a recirculating system in which case the perfusate is moved by a control pump and transport and metabolic characteristics of the placenta are assessed by Fick principle calculations. The perfusion system can also be connected to a non-recirculating system, in which the perfusate leaving the placental tissues can be collected over a timed interval for quantification of transport and metabolic characteristics of placental tissue.

With *in vitro* perfusion systems, usually using placentae of larger species including the human, the placenta is taken at delivery and placed into a specially prepared chamber. In human placentae, the fetal circulation is perfused through the umbilical vessels, while the maternal circulation is perfused via a catheter inserted directly through the basal plate into the intervillous space surrounding the chorionic villi. Uterine circulation drainage is collected by gravity. In epitheliochorial placentae the uterine circulation can be perfused by vascular catheterization. In either case the circulation may be perfused by recirculating or non-recirculating systems.

In general, the perfusate consists of a solution that contains the major serum salts buffered with bicarbonate, albumin, free amino acids at physiological concentrations, 95 per cent oxygen, 5 per cent CO_2, a pH of 7.4, 37°C, and a flow rate per gram of tissue comparable to that *in vivo*. Various perfusion models have been used to study the qual-

Table 4.2 Fick principle calculations of utero-placental metabolism

Uterine uptake of substances S	Uterine blood flow × uterine arterial – venous blood concentration difference of S.
Umbilical (fetal) uptake of S	Umbilical blood flow × umbilical venous – arterial blood concentration difference of S.
Utero-placental uptake of S	Uterine uptake of S – umbilical uptake of S.

Table 4.3 Mean placental O_2 and glucose consumption rates at term

| | Consumption rates (mmol/min/kg) | | |
	Oxygen	Glucose	References*
In vivo (utero-placenta)			
Cow	0.50	0.12	1
Mare	0.45	0.20	2
Sheep	1.98	0.42	3
Perfused (placenta)			
Human	0.43	0.50	4
Human	0.38		5
Human	0.30	0.43	6
Human		0.12	7
Human	0.13	0.09	8
Human	0.22	0.15-0.16	9
Human	0.16	0.15	10
Incubation (placenta)			
Human	0.10–0.19	0.13	11

*References for Table 4.3. Those already included in the 'Further reading' are identified only by author and date.

1. Comline, R.S. and Silver, M. (1986). *Journal of Physiology*, **260**, 571–586.
2. Silver, M. (1985). *Equine Veterinary Journal*, **16**, 227–33.
3. Meschia *et al.* (1980).
4. Schneider et al.(1981).
5. Bloxam (1985).
6. Hauguel, S., Challier, J.C., Cedard, L., and Olive, G. (1983). *Pediatric Research*, **17**, 729–32.
7. Nesbitt, R.E.L., Rice, P.A., and Rourke, J.E. (1973). *Gynecological Investigations*, **4**, 243–53.
8. Brandes, J.M., Tavoloni, N., Potter, B.J., Sarkozi, L., Shepard, M.D., and Berk, P.D. (1983). *American Journal of Obstetrics and Gynecology*, **146**, 800–6.
9. Illsley *et al.* (1984).
10. Miller, R.K., Wier, P.J., Maulik, D., and di Sant'Agnese, P.A. (1985). *Contributions to Gynecology and Obstetrics*, **13**, 77–84.
11. Villee, C.A. (1953).

itative aspects of placental transport and metabolism. Quantitative aspects remain the subject of considerable debate and experimentation based on the concern that the condition and performance of the perfused placenta (either *in situ* or *in vitro*), after isolation from the maternal circulation (either surgically or by delivery) and perfused with or without a medium containing red blood cells, may not be normal. Assessment of the biological function of the preparation is most important because the transport and metabolic characteristics of the placenta, being extremely rapid, might be expected to deteriorate quickly from ischaemic injury both during collection and as a consequence of inadequate or at least variable perfusion. For example, placental tissue ATP concentrations fall quickly during ischaemia (about 50 per cent in 1 min) in the guinea-pig placenta. Similar observations have been made of changes in ATP/ADP ratios and energy charge in the human placenta after delivery. This has not been a universal observation and several recent studies, using improved perfusion techniques, found little deterioration of the energy charge during 30 min after delivery, with stability during perfusion for up to 3 h based on relatively constant energy substrate concentrations and glucose and oxygen consumption rates. In most of these studies there has been no change in ultrastructure during the procedure, although some have found cell swelling and vacuolation of the trophoblast after only 1 hour of open-circuit perfusion. Such tissue changes can be prevented by the addition of autologous fetal blood in the medium, demonstrating the importance of oxygen content in the perfusion medium. Additional studies that have demonstrated changes in energy levels and the balance of metabolic regulators have not, in general, shown adverse changes in oxygen and glucose consumption rates or the rate of lactate production. Similarly, these placentae maintain the normally high concentrations of amino acids in the cytotrophoblast cells or the syncytiotrophoblast and the normal maternal-to-fetal gradient for amino acids. Thus, the validity of the perfusion models for quantifying placental transport and metabolic characteristics remains controversial. Table 4.3 presents representative comparative data of placental oxygen consumption rates from several species studied *in vivo* and *in vitro*.

In vitro *tissue incubations*

In this technique pieces of fresh placenta weighing 200–300 mg are incubated in a bicarbonate buffered salt solution (for example, Krebs or Earl's solutions containing glucose, 5 per cent CO_2, 95 per cent O_2, and the variable presence of protein and amino acids). These incubations are carried out in Warburg-type flasks allowing collection of expired CO_2 and the addition to the incubation medium of test substances such as radioactive tracers. Incubation medium can be varied in its concentration of metabolic substrates and analysed for metabolic products. This methodology has been useful for defining certain metabolic characteristics of the placenta such as the production of glycogen, the uptake and metabolism of fatty acids, the production of lactate and ammonia, and the concentration and exchange of amino acids. This methodology and such studies are not useful for investigating directional transport mechanisms and, in general, have not been appropriate for quantification of normal metabolic rates. According to several experiments using this methodology, individual placentae appear to differ markedly in their capacity to consume substrates such as glucose, and to produce substances such as ammonia and lactate. Such marked variability and metabolic characteristics may be indicative of functional differences among apparently similar but normal placentae or may represent variability induced by the method itself. For example, individual fragments or slices may contain quite different amounts and proportions of vascular, interstitial, cytotrophoblast, and syncytiotrophoblast cell types, each having different metabolic characteristics.

Metabolic activity of the placenta

Oxygen consumption

The placenta has one of the highest rates of oxygen consumption of all tissues in the body. For example, in sheep the utero-placental tissues consume approximately half of the uterine oxygen uptake; the balance is transported to and consumed by the fetus. On a weight-specific basis, however, oxygen consumption by the utero-placental tissues is four- to fivefold greater than that of the fetus, which itself has a metabolic rate approximately twice that of the adult. While these average values for utero-placental oxygen consumption are quite high, they are probably considerable underestimates of oxygen utilization by placental cellular tissue alone. Based on an average placental weight in sheep at term of 360 g, or approximately one-third of the utero-placental mass, and the estimate that 75 per cent of utero-placental metabolism is accounted for by the placental cellular tissue alone, it can be estimated that placental cellular tissue would have a metabolic rate as high as that in brain, liver, kidney, and many tumour cells (oxygen consumption of about 1.5 to 2 μmol/min/g).

Table 4.3 compares the quantities of oxygen that are taken up and consumed by the utero-placental tissues obtained from *in vivo* studies in several species of large animals. Oxygen consumption rates by the utero-placental tissues in more recent *in vivo* studies are considerably greater than those from earlier studies, even within the same species. This suggests that technical aspects of measuring utero-placental oxygen consumption account for the increase. Earlier efforts to measure the rate of oxygen consumption of the utero-placental tissues *in vivo* involved quite unstable physiological preparations including stress-

es such as exteriorizing the uterus or fetus, anaesthesia, umbilical cord occlusion, mechanical ventilation of the fetal lungs via a fetal endotracheal tube, maternal exsanguination (to eliminate oxygen consumption of the placenta from the maternal circulation), and fetal demise by air embolism (quantifying utero-placental oxygen consumption as the difference between uterine oxygen uptake before and after fetal death). Together, these crude attempts at measuring utero-placental oxygen consumption rates provided estimates that were at least half of those produced from stable, unstressed, conscious animals. Such lower estimates indicate that the oxygen consumption of the utero-placenta can be reduced markedly by acute, stressed perturbations such as anaesthesia or by arresting either maternal or fetal perfusion. Other manipulations such as cold stress and physical handling may have been equally important but their effect is less well understood.

Utero-placental oxygen consumption increases over the second half of gestation both in absolute value and on a weight-specific basis. Many synthetic and transport activities of the utero-placenta increase dramatically over the same gestational period. Presumably such activities and the increase in metabolic rate of the utero-placental tissues are directly linked. Thus, although the proportion of uterine oxygen uptake accounted for by the fetus also increases considerably over the second half of gestation, this by no means indicates a senescence of utero-placental metabolic activity.

Table 4.3 also compares oxygen consumption values for placental tissue from studies using perfusion techniques and *in vitro* tissue slice incubations. The oxygen consumption values related to wet weight of tissue for the placental slice incubations are estimated from the dry weight/wet weight ratio. Estimation of metabolic rate relative to dry tissue weight is important to consider because water content varies from placenta to placenta, even within a species, and also varies over gestation. In humans, for example, the dry weight/wet weight ratio nearly doubles between 10 and 40 weeks of pregnancy. On the other hand, the value of placental oxygen consumption in the tissue-slice incubation studies is only 10 per cent of that provided from the *in vivo* and the *in situ* or perfused models. It is doubtful, therefore, that the tissue-slice incubation method provides an accurate estimate of placental cellular metabolic rate and this makes suspect the conclusions about the metabolic characteristics of placental tissues derived from such measurements. For example, based on a decreasing oxygen consumption value per gram of dry weight of human placental tissue studied by tissue-slice incubation techniques, the placenta has been described as tending towards 'senescence' at term. Clearly, such data and conclusions conflict with more current *in vivo* data and data derived from the more physiological perfusion models, and therefore do not provide accurate representations of normal placental metabolic activity.

There also is considerable variation among the values reported for placental oxygen consumption from perfusion models. Part of this variability relates to the composition of the perfusate. For example, the addition of red cells to the perfusate has been observed to increase placental protein synthetic rate approximately twofold. However, the effect on metabolic rate of adding red cells to the perfusate has not been evaluated as rigorously. In one recent study, however, using the human placental perfusion model following Caesarean section delivery (thus eliminating the vagaries of physiological stresses provided by labour and delivery), the lactate-to-pyruvate ratio increased markedly (two- to threefold) over a very short time (10–20 min) after delivery. This change indicated a rapidly reducing cytoplasmic redox state as endogenous oxygen was consumed. This process appeared to be halted and partially reversed during the first few minutes of perfusion. The perfusate consisted of a buffered salt–dextran solution oxygenated with a mixture of 95 per cent O_2–5 per cent CO_2. With continued perfusion the lactate-to-pyruvate ratio remained relatively constant but at about twice the estimated *in vivo* value. At the same time ATP levels were more or less constant but at about half the concentration of that expected from *in vivo* estimates. Furthermore, the ATP/ADP ratio, which recovered slightly during perfusion, decreased after 20 min. These results indicated that the perfused placenta may not deteriorate over a reasonable period of perfusion of one to a few hours but its metabolic characteristics are quite unlike those of the *in vivo* state. In another study, however, the maternal venous perfusion medium contained 18 kilopascals (kPa) of oxygen (compared with an oxygen tension of about 13 kPa in arterial blood *in vivo*) and there was an accumulation of oxygen in the fetal perfusate. Although these values were considered indicative of adequate oxygen supply to the placenta, the measured oxygen consumption rate (approximately 0.4 μmol/min/g wet weight) was considerably less than that found *in vivo* for utero-placental tissues in the sheep (Table 4.3). Thus, it seems reasonable to conclude that the lower energy state of the perfused placenta is accounted for by a relatively low oxygen consumption rate, apparently the result of some aspect(s) of the perfusion model itself. It is also possible that some other factor, such as a different substrate or hormonal milieu, could be responsible for these differences from the results *in vivo*. For example, one study demonstrated an oxygen consumption rate of 0.12 μmol/min/g wet weight in human placental tissue perfused without glucose. Among different experiments, placental oxygen consumption appeared to vary directly with placental glucose supply. In contrast, *in vivo* studies in the sheep have not demonstrated consistently a dependence of placental oxygen consumption on glucose concentration or placental glucose consumption. On balance, therefore, it appears that many of the studies in the perfusion model are providing data of a marginally functional placenta limited by and thus dependent upon the supply of oxygen and metabolic substrates.

Glucose metabolism

In view of the very high oxygen consumption rates of placental tissue, it is necessary that a large source of carbon is readily available and consumed by the placenta to support the rapid rate of oxidation. Among all placentae, glucose most consistently provides the largest amount of carbon. Not all placentae take up lipids, and amino acids are largely transported to the fetus; placental amino acid consumption rates, in general, are too small to provide enough carbon for even a small fraction of placental oxidation. An exception may be glutamate, which is taken up by the placenta from the fetus, at least in the sheep, and oxidized significantly by the placental tissues. Average values for utero-placental glucose consumption *in vivo* and for placental glucose consumption *in vitro* from perfusion or slice techniques at normoglycaemia are shown in Table 4.3. From all of these studies it is clear that utero-placental or placental glucose consumption is dependent upon glucose concentrations in both the maternal and fetal circulations.

The uptake of glucose by the placenta and its transfer to the fetus occur by facilitated diffusion in which perfusion has little role. This process is mediated by specific transporter proteins that have specificity for hexose molecules and differentiation among hexose molecules that favours glucose molecules. Glucose transporters are found both on the maternal facing microvillus brush border of the trophoblast as well as on the fetal-facing basal lateral surface. The number of glucose transporters appears to increase with gestation. This increase is probably due to an increase in placental surface area. However, it has not been determined whether an increase in glucose transporter concentration also occurs. The increase in glucose transporter number over gestation probably accounts for most, if not all, of the increase in placental fetal glucose transport capacity, as well as for the increase in placental glucose consumption. The activity of placental glucose transporters on either the maternal or fetal surface appears dependent on glucose concentration alone; changes in insulin concentration in either the maternal or fetal circulation in all species studied to date, *in vivo* and *in vitro*, appear to have no effect on placental glucose transport or placental glucose consumption. Molecular studies have found only Glut 1 and Glut 3 glucose transporters in placental tissue, neither of which is insulin-sensitive in other tissues.

According to the transporter-mediated, facilitated diffusion model of placental glucose uptake and transfer, the glucose transfer rate approaches a maximum as maternal and fetal glucose concentrations are increased beyond physiological limits. This process was first described according to the following equation for placental-to-fetal glucose transfer (PGT):

$$PGT = V_{max}\left(\frac{G_A}{G_A + K_m} - \frac{G_a}{G_a + K_m}\right)$$

where V_{max} represents the maximal flux of glucose and K_m the Michaelis constant is the concentration of glucose in the maternal arterial plasma (G_A) at which the transport mechanisms are half saturated $V_{max/2}$. G_a is the concentration of glucose in the fetal arterial plasma. This model assumes that the placenta acts simply as a diffusion membrane and does not consume glucose. This model is inaccurate however, as placental glucose consumption in late gestation can account for 50–70 per cent of uterine glucose uptake (Table 4.3). The effect of glucose consumption on placental glucose transfer has been tested *in vivo* showing that the relation between placental glucose transfer and the maternal-to-fetal plasma glucose concentration gradient has a negative intercept. The magnitude of this negative intercept is equal to net fetal-to-placental glucose transfer (and thus net placental glucose consumption) when the maternal and fetal glucose concentrations are equal. Under these conditions glucose from the fetal circulation accounts for approximately 75 per cent of utero-placental glucose uptake and consumption. This observation has been supported by tracer studies that showed that as much as 40 per cent of placental glucose consumption could come from the fetal plasma glucose pool under normoglycaemic conditions. Thus, the placental glucose transport model must include a negative intercept accounting for net placental glucose consumption:

$$PGT = V_{max}\left(\frac{G_A}{G_A + K_m} - \frac{G_a}{G_a + K_m}\right) - \dot{q}_p$$

where $\dot{q}_p$ is the net placental glucose consumption from fetal-to-placental glucose transfer when maternal and fetal arterial plasma glucose concentrations are equal.

The physiological importance of placental glucose consumption and its effect on fetal glucose supply can be demonstrated by an example using data from sheep. At experimentally determined values of $K_m = 70$ mg/dL, i.e. a maternal arterial glucose concentration of 70 mg/dL, $\dot{q}_p = 30$ mg/min, $V_{max} = 209$ mg/min, and umbilical glucose uptake (placental-to-fetal glucose transport) = 20 mg/min, fetal arterial glucose concentration would be equal to about 24.5 mg/dL. If $\dot{q}_p$ were '0', a fetal glucose concentration would equal about 47.5 mg/dL. Thus, net placental glucose consumption contributes significantly to the physiological hypoglycaemia of the fetus and helps to establish the glucose concentration gradient by which both placenta and fetus compete for glucose molecules.

The equation for glucose transport by the placenta indicates that changes in the maternal or fetal concentration of glucose will affect transport and thus affect uterine glucose uptake and placental glucose consumption. The effects of maternal glucose concentration combined with changes in fetal glucose concentration are shown in Fig. 4.7 from recent studies in term pregnant sheep. These data were derived from glucose-clamp experiments in which steady-state fluxes of glucose were quantified by the Fick principle at different but controlled concentrations of glucose in the maternal and fetal circulations. Uptake and net consumption of glucose by the utero-placenta demonstrated saturation kinetics with a V_{max} of about 41 mg/min. K_m (maternal arterial glucose concentration at $V_{max}/2$) and K_s (the maternal arterial glucose concentration at which V_{max} is reached) were 19 and 145 mg/dL, respectively, not significantly different from the same parameters for uterine glucose uptake. Similar data have been obtained from pregnant sheep under fasting hypoglycaemic conditions, fed normoglycaemic conditions, and glucose-infused hyperglycaemic conditions. Furthermore, by using a variety of *in vitro* perfusion experiments, investigators have demonstrated the direct dependence of glucose consumption by the placenta on the perfusate glucose concentration.

In contrast to these studies in which both maternal and fetal glucose concentrations co-varied, other *in vivo* studies have demonstrated that, at a constant maternal glucose concentration, placental glucose consumption is directly related to fetal glucose concentration (Fig. 4.8). In fact, as discussed above, when the transplacental glucose concentration is abolished, approximately 75–80 per cent of the glucose consumed by the utero-placenta is supplied by the fetal circulation. This observation implies that the fetal side of the utero-placenta is markedly more permeable to glucose than the maternal side (approximately eightfold) and indicates that changes in fetal glucose concentration should have a strong influence on placental glucose flux and metabolism. This point is illustrated in Fig. 4.9. An increase in fetal glucose concentration that is independent of the maternal glucose concentration regulates, separately and significantly, utero-placental glucose consumption; in contrast, the total entry of glucose into the uterus, utero-placenta, and fetus is regulated by maternal glucose concentration.

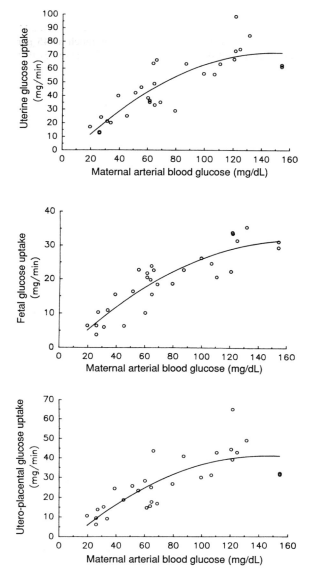

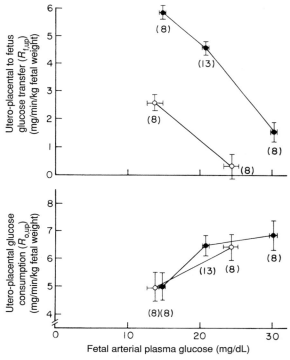

Fig. 4.8 Net rate transfer of glucose to fetus from utero-placenta and net rate of utero-placental glucose consumption in sheep, both expressed per kilogram of fetus, are plotted against fetal arterial plasma glucose. Closed circles: values measured while maternal arterial plasma glucose was clamped at about 70 mg/dL. Open circles: values measured while maternal arterial plasma glucose was clamped at about 50 mg/dL. Error bars represent +/– SE. Numbers of observations are shown in parentheses. These data show that maternal glucose concentration determines the rate of glucose entry into the utero-placenta and fetus but that actual utero-placental glucose consumption is regulated largely by the fetal glucose concentration. (Reprinted with permission from Hay WW, Jr, Molina, R.A., DiGiacomo, J.E., and Meschia, G. (1990). *American Journal of Physiology*, **258**, R569–R577.)

Fig. 4.7 Glucose infused into pregnant sheep after an overnight fast to produce a large variety of maternal arterial blood glucose concentrations. Fick principle measurements were then made of net uterine, fetal, and utero-placental glucose uptake rates. All three relationships show saturation kinetics with an approximate K_m value in the physiological range of maternal glucose concentrations (about 50–60 mg/dL). These results show that maternal glucose concentration determines the rate of entry of glucose into the uterus, utero-placenta, and fetus. (Reproduced with permission from Hay, W.W., Jr and Meznarich, H.K. (1988). *Proceedings of the Society for Experimental Biology and Medicine*, **190**, 63–9.)

A potential clinically important effect of fetal glucose concentration on placental-to-fetal glucose transfer and on net utero-placental consumption has been demonstrated by recent observations in chronically hypoglycaemic pregnant sheep in which fetal glucogenesis developed. The fetal glucogenesis contributed glucose molecules to the fetal glucose pool, sustaining fetal glucose concentration. As a result the placental-to-fetal concentration gradient and the placental glucose transfer rate were relatively reduced.

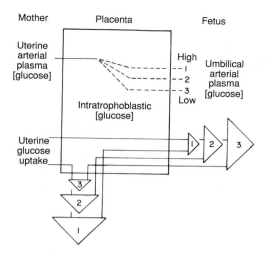

Fig. 4.9 Schema of effects of fetal glucose concentration on placental glucose transfer to the fetus and utero-placental glucose consumption, adapted from data in sheep. A decrease in fetal glucose concentration relative to maternal glucose concentration promotes maternal-to-fetal glucose transfer at the expense of utero-placental glucose consumption. These findings demonstrate a reciprocal relationship between placental glucose transfer and utero-placental glucose consumption that is determined by the fetal glucose concentration. (Reproduced with permission from Hay, W.W., Jr (1991). *Proceedings of the Nutrition Society*, **50**, 321–36.)

Under these circumstances, utero-placental glucose consumption was maintained at near normal rates for the level of maternal glycaemia. Thus, fetal glucose production may be able to compensate for a reduced maternal glucose supply, sustaining not only fetal glucose utilization requirements but also those of the placenta. This phenomenon is a striking example of how fetal metabolic autonomy may be able to protect fetal as well as placental survival without further demand on a limited maternal nutritional supply.

Most *in vivo* studies have shown that placental oxygen consumption does not change significantly over wide ranges of glucose concentration and placental glucose consumption. If placental oxygen consumption is indeed constant over a broad range of placental glucose consumption rates, then it would appear that a reciprocal relationship must exist between placental glucose oxidation and placental oxidation of other substrates. Such an interesting hypothesis remains to be proven and the nature of such substrates (for example, trophoblast glycogen, lactate, amino acids, keto acids, fatty acids, etc.) remains to be determined.

Glucose carbon consumed by the placenta produces CO_2 and H_2O by oxidation and, if all of the consumed glucose is oxidized, it could account for as much as 45 per cent of placental oxygen consumption. This assumes that all of the lactate production by the placenta is derived from net placental glucose metabolism, but this estimate may be in error in that lactate production by the placenta has not been shown to be directly related to placental glucose consumption either *in vivo* or *in vitro*. In the sheep, amino acids are likely alternative sources of placental lactate production. These observations are consistent with *in vitro* studies using the rat epitrochlaris muscle preparation that showed that only 20–40 per cent of the lactate and pyruvate released by the epitrochlaris muscle could be accounted for by glucose utilization, suggesting that amino acid metabolism was a likely source for production of the remainder of the lactate and pyruvate.

Glucose also produces placental glycogen. Rates of production and regulation of this process are not known, although placentae from pregnancies complicated by hyperglycaemia (e.g. diabetes) may be higher in glycogen content. Previous *in vitro* studies demonstrated that placental glycogen is synthesized rapidly in the placenta early in pregnancy and that this synthetic rate decreases towards term.

In vitro *studies of placental glucose metabolism*

Although *in vitro* studies of placental glucose consumption have demonstrated markedly lower rates than those measured *in vivo*, *in vitro* experiments have demonstrated several important features of placental metabolism. Glucose utilization rates by term human placentae studied *in vitro* using the tissue-slice incubation method have demonstrated large interplacental differences in glucose utilization. In vitro placental perfusion studies have also documented as much as a fourfold difference in glucose utilization rates among placentae. Part of this variability can be accounted for by differences in glucose concentration from study to study. Major differences in glucose consumption by placental tissue incubated *in vitro*, however, appear to be due to unique metabolic characteristics of individual placentae, which appear to differ markedly in their ability to utilize glucose in an *in vitro* system. In one study, for example, the intraplacental variability for glucose consumption was only 14 per cent compared with the variability among placentae of 62 per cent suggesting that the differences reflect biological variability in metabolism among placentae rather than simply differences among tissue samples obtained from one placenta. The sources of such biological variability have not been determined but may include developmental characteristics as well as differences in condition of the placentae following delivery.

Lactate metabolism

Lactate is produced by the placenta in all species and in large amounts (averaging approximately 28 μmol/min/kg placental weight in sheep); this lactate then enters the maternal and fetal circulations (at approximately equal rates in sheep). Placental lactate production is a normal process that occurs during aerobic metabolism and is unaffected by small to moderate changes in maternal or fetal blood PO_2, blood oxygen content, or uterine and/or umbilical blood flows. *In vitro* perfusion models have demonstrated marked lactate production that increases rapidly and linearly following delivery unless perfusion with oxygenated medium or red blood cells is provided, at which point lactate production decreases but remains at a value considerably above estimated *in vivo* production rates. Placental glucose consumption *in vivo* in sheep also does not appear to alter placental lactate production indicating the presence of reciprocal pathways of placental glucose metabolism (perhaps glycogen formation) as well as possibly high rates of lactate production from amino acids. In sheep, the amount of lactate taken up by the fetus is about half of net fetal glucose uptake at normal glucose concentrations and about one-third of fetal lactate utilization. Thus, the majority of fetal lactate is produced by the fetus, although it is clear that lactate produced in the placenta is a relatively large net source of carbon for fetal utilization.

In other studies in the pig placenta, experiments using *in vitro* incubation techniques demonstrated placental oxidation and fatty acid production from both glucose and lactate. The contribution of lactate was as important quantitatively as was glucose for utilization through these metabolic pathways. Utilization of lactate was striking, relative to glucose concentration, in that it was present at only one-fifth of the concentration of glucose in the incubation medium. A comparable study in rats demonstrated that lactate was oxidized at only one-third the rate of glucose oxidation in the rat placenta. In the pig placenta, lactate carbon was preferentially partitioned into fatty acids, whereas glucose was used primarily for triglyceride. The same observation has been made in the human placenta. Lactate metabolism thus appears to be responsible for at least as much of the oxidative activity in these placentae as glucose. The ability to metabolize lactate to meet placental energy requirements may thus permit other potential oxidative substrates to be spared for transport to the fetus. Lactate may be one of the substrates essential to the hypothesis that a reciprocal relationship exists between placental glucose oxidation and the oxidation of other substrates.

Fructose metabolism

Placental production of fructose is unique to ruminants and does not occur to any appreciable extent in humans. In the sheep, quantitative aspects of placental fructose production and fetal fructose metabolism have been investigated using Fick principle and tracer methods. In well fed normoglycaemic sheep net placental fructose production rates average about 1.3 ± 0.7 mg/min/kg fetal weight and fructose enters the umbilical vein exclusively. Tracer fructose infused into the fetus enters into both fetal and placental tissues but not into the maternal circulation. Tracer studies have also shown placental oxidation of fructose but at a very slow rate. Fructose does produce a small amount of lactate in the placenta but there is no measurable net conversion to glucose. Contribution of fructose to placental glycogen formation has not been examined. Simultaneous tracer-derived fetal fructose utilization rates of 1.0 ± 0.1 mg/min/kg fetal weight are not different from normal net umbilical fructose uptake rates demonstrating no significant evidence for fetal fructose production.

Amino acid metabolism

At term the placenta contains a large variety of enzymes capable of metabolizing amino acids through a diverse number of metabolic pathways including gluconeogenesis, glycogen synthesis, protein synthesis, amino acid oxidation, and ammoniagenesis. Amino acid flux through these pathways has been demonstrated *in vitro*. Other protein requirements include an undetermined amount for synthesis of secreted protein products. Quantitative aspects of placental amino acid requirements are not well understood because the net uptake of amino acids by the placenta provides an extraction ratio equal to, or less than, the accuracy of their plasma concentration measurements. Based on placental nitrogen content at term in the human placenta, however, it has been estimated that placental growth over gestation would require about 10.6 g of nitrogen or 66 g of protein. This calculation provides a minimum estimate of nitrogen and protein requirements for placental growth. In fact, placental amino acid consumption for other metabolic purposes may be much higher in order to provide carbon for oxidation (by direct entry into the Krebs cycle or indirectly via production of lactate with subsequent conversion to pyruvate), nitrogen for ammonia production, and amino acids for conversion to other amino acids or to proteins that are excreted into the umbilical or uterine circulations. Also, infusions of certain radioactively labelled amino acids (e.g.

lysine, leucine, glycine, aspartate, and α amino isobutyric acid) into the *in situ* perfused guinea-pig placenta have shown that 12–16 per cent of the label can be incorporated into placental proteins. This label incorporation can be inhibited almost completely with cycloheximide, a specific blocker of protein synthesis, demonstrating that the placenta can actively synthesize proteins from amino acids.

Early work on placental amino acid metabolism demonstrated higher amino acid concentrations in fetal tissue and plasma than in maternal plasma which were assumed to result from the even higher active (energy-dependent) concentration of amino acids in the trophoblast cells. Based on this information, it was also assumed that all fetal amino acids were transported by this mechanism from the mother to the fetus. Studies on net amino acid uptake in sheep have shown that this clearly is not the case. In fact, there was no measurable uptake by the umbilical circulation for some amino acids such as serine and aspartate, and glutamate was leaving the fetal circulation and entering the placenta. Thus, directional transport of the amino acids across the placenta and into the fetus is clearly more complex and must involve unique directional amino acid transport characteristics as well as unique placental metabolic pathways for some of the amino acids.

It is now clear that placental amino acid and nitrogen balance involves selective cycling of certain amino acids. Several such cycles have been observed *in vivo* in pregnant sheep. (Fig. 4.10); similar cycles have been found in the *in vitro* perfused human placenta. For example, the placenta actively produces ammonia which is delivered into both uterine and umbilical circulations. This appears to be a normal process in mammalian placental metabolism, occurring in all species studied to date. In fetal sheep, this process occurs over a large part of gestation; for example, the absolute rate of net utero-placental ammonia production at 73–79 (50–65 per cent) days of gestation (about 25 μmol/min) is as high as that estimated from data collected in sheep near term. This process is consistent with *in vitro* evidence of negligible urea cycle activity in sheep placental tissue over the entire length of gestation. Such a relatively high placental ammonia production in midgestation probably contributes to the higher concentration of ammonia (approximately twofold) found in fetal blood in mid-gestation, at which point the fetus is much smaller and its ammonia clearing capacity is markedly reduced relative to term.

Another example of placental-fetal interaction involving amino acid metabolism is demonstrated by placental-fetal hepatic relationships between certain amino acids. For example, there is a net uptake of glutamine and glycine by the fetus from the placenta and by the liver from the umbilical vein, whereas their metabolic products,

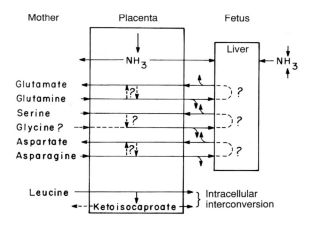

Fig. 4.10 Schema of placental cycles resulting from protein metabolism and amino acid metabolism in the pregnant sheep, as presented by Battaglia and colleagues (see 'Further reading: Battaglia 1992; Bell *et al.* 1989; Carter *et al.* 1991; Holzman *et al*, 1977; Marconi *et al.* 1989). Top: Ammonia (NH₃) is the end–product of placental amino acid deamination; it is taken up by the umbilical vein and can be used for protein synthesis plus urea formation in the fetal liver. Middle: Although serine and glycine may not be taken up appreciably by the placenta from the uterine circulation, placental production of glycine, fetal hepatic uptake of glycine and production of serine, and serine uptake by the placenta indicate (as for asparagine–aspartate and glutamine–glutamate) active fetal–placental metabolic cycles. Bottom: Continued deamination of leucine in the placenta produces α-ketoglutarate which then is taken up by the umbilical circulation. (Reproduced with permission from Hay, W.W., Jr (1991). *Proceedings of the Nutrition Society*, **50**, 321–36.)

glutamate and serine, respectively, are produced in net by the fetal liver and taken up in net by the placenta. A similar but less definite relationship has been found for net hepatic uptake of asparagine and release of aspartate with a reciprocal exchange across the umbilical circulation in the placenta. Quantitative aspects of these cycles are just being developed. Preliminary studies in sheep indicate that, for serine, about 25 per cent of fetal hepatic glycine uptake is used to produce serine and approximately 60 per cent of fetal glycine is derived from fetal serine. These data indicate that the inter-organ cycling of glycine and serine between the placenta and the fetal liver is quantitatively significant, contributing importantly to energy and protein balance in the feto-placental unit. Furthermore, fetal serine production and uptake by the placenta from the umbilical circulation appears to be much larger at mid-gestation than near term, suggesting an active developmental process regulating this inter-organ cycling. This cycle may be important for the supply and conservation of the methyl group in the fetus, and thus may be important for control of fetal growth. Quantitative aspects of glutamine–glutamate cycling are less certain. Placental uptake of glutamate may be important not only to produce glutamine for supply to the fetus where it can promote growth, but also to provide another substrate for placental oxidation (over 60 per cent of placental glutamate uptake from the fetus in sheep is oxidized to CO_2). This process may also help control fetal plasma glutamate concentration and modulate its potential neurotoxicity as an excitatory amino acid.

A further example of placental-fetal amino acid cycling involves the metabolic function of relatively high concentrations and activities of branch-chain amino-acid aminotransferases found in the placenta (at least in the sheep and human). Studies in sheep suggest that net placental uptake and transamination to the corresponding ketoacid can occur for leucine. The importance of this process to placental and fetal amino acid metabolism and fetal growth remains to be determined.

While other placental-fetal metabolic interactions are possible, at least these examples specific to amino acid metabolism demonstrate that the regulation of fetal and placental metabolism by interactive pathways is far more complex and sophisticated than previously conceived. It is also clear that the placental free amino acid pool, derived as it is from protein breakdown in the placenta, from the maternal circulation, from the fetal circulation, and from intraplacental interconversions, cannot be treated as a single homogeneous pool.

In vitro incubation studies of placental tissue have provided less certain data regarding amino acid and protein metabolism. Unless amino acids are added to the incubation medium, a net production of amino acids is observed; this amino acid production is consistent with protein hydrolysis and, in general, is of sufficient magnitude to account for the simultaneous production of ammonia. Ammonia production also can be increased by the addition of glutamine to the incubation medium. Glutamine is the only amino acid that is consistently utilized by placental tissue under these conditions, leading to glutamate production as well as ammonia and perhaps lactate production. Because the cycle of glutamine with glutamate is reversed in these *in vitro* incubations from the *in vivo* and *in vitro* perfusion studies, it should be questioned whether or not the tissue incubation model provides valid data about amino acid metabolism.

Additional *in vitro* studies in the perfused model have focused on protein synthetic rate. In these studies the rate of uptake of a ^{14}C-labelled amino acid (lysine, for example) by the placental tissue was determined and its specific activity in placental protein related to the maternal plasma or trophoblast intracellular free amino acid specific activity. In two separate studies, perfused human placental tissue demonstrated a fractional protein synthetic rate of 40 per cent per day. This value was approximately 60 per cent of the rate observed *in vivo* in the placental cotyledons of sheep and in the placenta of the anaesthetized guinea-pig. Such variability is not surprising, given the major differences among techniques of perfusion and sampling as well as the tracer models used.

Lipid metabolism

Lipid transport and metabolism by the placenta have received less investigation than have those for glucose and amino acids. At least one major reason is that the principal animal model for studying nutrient transport from the maternal to the fetal circulation, the sheep, does not transport lipids to any appreciable extent across the placenta. The amount and type of fatty acid or lipid moiety transported by the placenta varies among species; it is greatest in the haemochorial placenta of the human, guinea-pig, and rabbit, and least in the epitheliochorial placenta of the ruminant and the endotheliochorial placenta of the carnivores. Furthermore, it is interesting that the fat content of the fetus at term varies among species directly with the lipid transport capacity of the placenta.

There are many lipid substances in the plasma that are transported across the placenta (even of sheep) that are essential for placental and fetal development even if they do not contribute to nutritional or energy metabolism. And brown fat is common to all fetuses; it is essential for postnatal thermogenesis, even if the neonate is not 'fat' with white adipose tissue. Also, many lipid substances entering the fetus are qualitatively different from those taken up by the uterus and placenta, indicative of active placental metabolism of individual lipid substances. Placental lipid metabolism, therefore, is qualitatively, as well as in some cases quantitatively, important for fetal development and growth, even if the fetus does not get very fat.

A schema of lipid uptake, metabolism, and transport in the human placenta is shown in Fig. 4.11. After entering the placenta, fatty acids may be used for triglyceride synthesis, cholesterol esterification, membrane biosynthesis, direct transfer to the fetus, or for oxidation. Factors regulating the flux of lipid carbon into these various pathways of transport and metabolism are not known. At least one factor appears to be the maternal plasma concentration of free fatty acids (FFA). For example, placental triglyceride content increases in women who are fasting, who deliver preterm infants, or who have diabetes mellitus, all condi-

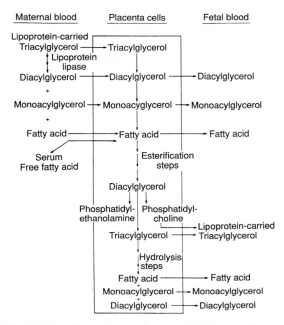

Maternal blood   Placenta cells   Fetal blood

Fig. 4.11 Possible pathways of placental lipid metabolism and transport to the fetus. Lipoprotein-carried triglycerides, phospholipids, and cholesterol esters, and partial glycerides that have been released by placental lipase may be taken up and undergo further hydrolysis within placental cells. Maternal serum free fatty acids and the fatty acids that have been released by placental lipoprotein lipase may enter the fetal circulation. Within placental cells, the free fatty acids taken up directly and those released by intracellular hydrolysis may be transported to the fetal circulation. Alternatively, these fatty acids may be re-esterified to form triglycerides which are transiently stored. The endogenously synthesized triglycerides may then be hydrolysed and the released fatty acids may be transported to the fetus. A less likely possibility is that partial glycerides or endogenously synthesized triglycerides are released as a component of a placental lipoprotein. (Reproduced with permission from Coleman, R.A. (1986). *Federation Proceedings*, **45**, 2519–23.)

tions in which maternal plasma FFA concentrations are increased. Fatty acids may be transferred directly to the fetus according to specific carriers or after modification by processes such as desaturation, elongation, or partial oxidation. Although most fatty acids appear to be transported by direct carrier-mediated mechanisms, some studies suggest that chain-altering metabolism in intratrophoblast peroxisomes may lead to a greater transfer of medium-chain fatty acids into the fetal circulation. This would be advantageous to fetal lipid metabolism, which is limited in its capacity for oxidation of long-chain fats by low carnitine concentrations. Fatty acids also may be esterified to di and triglycerides and to phospholipids, which may be stored transiently or long term as well as hydrolysed by phospholipases and acylglycerol lipases to release fatty acids to the fetal circulation.

Fatty acids also may be synthesized in the placenta in free form or as lipoprotein particles. Recent studies have demonstrated significant fatty acid synthesis in placental tissue and cultured trophoblast cells from human, sheep, and pig placentae. Primarily long-chain fatty acids (oleic, palmitic, and palmitoleic acids) were formed. Evidence for fatty acid synthesis within the placenta also comes from studies that have shown that radiolabelled palmitate can be incorporated rapidly into triglycerides in human placental slices. Similar studies have been conducted in 21-day pregnant rats showing incorporation of [14C]palmitate bound to albumin into placental triglycerides and phospholipids. Still other studies of human trophoblast cells in culture have shown that incubation with albumin-bound [14C]oleate produced labelled cellular triglycerides. After labelling, the fatty acids were released into the culture media but only 80 per cent of the original 14C label was recovered suggesting some oxidation of the oleic acid or its products.

Additional evidence for the hypothesis that transient esterification occurs during transport of fatty acid by the placenta comes from measurements of activities of enzymes required for the isolation of glycerol-3-phosphate. Such enzymes are located in the endoplasmic reticulum and have been identified in placental tissue. Specific activity of enzymes and their total tissue content do not reflect metabolic flux but certainly the presence of these enzymes suggests that fatty acid esterification followed by hydrolysis may be involved in placental lipid metabolism and transport. The regulation of such a process is not known.

There is conflicting evidence regarding the synthesis of lipoproteins by placental tissue and their use for transport of fatty acids to the fetal circulation. By day 21 of gestation in the rat, the placenta appears to synthesize apoprotein B which is normally an integral part of the major plasma lipoprotein carriers of triglycerides. Such observations have not been made in the human placenta, which may secrete lipoproteins only during the early part of gestation, if at all. However, small amounts of low-density lipoprotein (LDL) and very low-density lipoprotein (VLDL) have been identified in human amniotic fluid. Their source has not been determined.

Cultured trophoblast cells incubated with [14C]acetate do not label arachidonic acid. Also delta-5 and delta-6 desaturase activities in human placental microsomes are absent at least early in gestation. This suggests that fetal arachidonic acid is either transferred directly from the mother or synthesized from linoleic acid in the fetal liver. Support for direct transport also comes from studies in the sheep, which does not, in general, transport fatty acids to the fetus but does appear to transport directly linoleic and arachidonic acids.

LDL cholesterol is taken up by endocytosis into trophoblast cells and degraded by lysozymes. It appears to be the major precursor for placental production of progesterone and oestrogen. High-density lipoprotein (HDL) cholesterol also contributes to placental progesterone production in cultured human trophoblast cells. Some of this cholesterol is transferred directly to the fetus although the role of this transport is uncertain, given the large capacity and rate of cholesterol synthesis in the fetal liver.

Conclusions

Clearly the placenta is more than a passive membrane. While it serves the vital role of transporting substances between the uterine and umbilical circulations, it is also an extremely active metabolic organ, accounting for more than half of the uptake of oxygen and glucose by the conceptus (uterus, placenta, and fetus). In fact, a large portion of the gradients that passively and facilitatively drive oxygen and glucose from the uterine to the fetal circulations is accounted for by placental oxygen and glucose consumption. The energy produced by placental oxidative metabolism is also used for the concentration of amino acids in placental cells to very high levels from which the amino acids then diffuse facilitatively into the fetal plasma. Other metabolic pathways in the placenta for oxygen, carbohydrates, amino acids, and lipids are diverse and complex. They serve not only unique metabolic requirements in the placenta but they also modify the delivery to, and concentration in, the umbilical circulation of nearly all essential nutrients. Placental metabolism thus plays a major role in providing for the nutritional requirements of fetal growth and metabolism.

Acknowledgements

This work was supported by grants DK35836, HD20761, and HD28794 from the National Institutes of Health, USA.

Further reading

Battaglia, F.C. (1992). New concepts in fetal and placental amino acid metabolism. *Journal of Animal Sciences*, **70**, 3258–63.

Battaglia, F.C. and Meschia, G. (1986). *An introduction to fetal physiology*. Academic Press, Orlando, Florida.

Battaglia, F.C. and Meschia, G. (1988). Fetal nutrition. *Annual Review of Nutrition*, **8**, 43–61.

Baur, R. (1977). Morphometry of the placental exchange area. *Advances in Anatomy, Embryology, and Cell Biology*, **53**, 3–65.

Beer, A.E., Billingham, R.E., and Scott, J.R. (1975). Immunogenetic aspects of implantation, placentation and fetoplacental growth rates. *Biology of Reproduction*, **12**, 176–89.

Bell, A.W., Kennaugh, J.M., Battaglia, F.C., and Meschia, G. (1989). Uptake of amino acids and ammonia at midgestation by the fetal lamb. *Quarterly Journal of Experimental Physiology*, **74**, 635–43.

Bell, A.W., Wilkening, R.B., and Meschia, G. (1987). Some aspects of placental function in chronically heat-stressed ewes. *Journal of Developmental Physiology*, **9**, 17–29.

Bloxam, D.L. (1985). Human placental energy metabolism: its relevance to in vitro perfusion. *Contributions to Gynecology and Obstetrics*, **13**, 59–69.

Bloxam, D.L. and Babinski, P.M. (1984). Energy metabolism and glycolysis in the human placenta during ischemia and in normal labour. *Placenta*, **5**, 381–94.

Carroll, M.J. and Young, M.J. (1987). Observations on the energy and redox state and protein synthesis rate in animal and human placentas. *Journal of Perinatal Medicine*, **15**, 21–32.

Carter, B.S., Moores, R.R., and Battaglia, F.C. (1991). Placental transport and fetal and placental metabolism of amino acids. *Journal of Nutritional Biochemistry*, **2**, 4–13.

Cetin, I., Fennessey, P.V., Sparks, J.W., Meschia, G., and Battaglia, F.C. (1992). Fetal serine fluxes across fetal liver, hindlimb, and placenta in late gestation. *American Journal of Physiology*, **263**, E786–E793.

Cetin, I., Sparks, J.W., Quick, A.N., Jr, Marconi, A.M., Meschia, G., Battaglia, F.C., and Fennessey, P.V. (1991). Glycine turnover and oxidation and hepatic serine synthesis from glycine in fetal lambs. *American Journal of Physiology*, **260**, E371–E378.

Christie, W.W. and Noble, R.C. (1982). Fatty acid biosynthesis in sheep placenta and maternal and fetal adipose tissue. *Biology of the Neonate*, **42**, 79–86.

Coleman, R.A. (1986). Placental metabolism and transport of lipid. *Federation Proceedings*, **45**, 2519–23.

Coleman, R.A. and Haynes, E.B. (1987). Synthesis and release of fatty acids by human trophoblast cells in culture. *Journal of Lipid Research*, **28**, 1335–41.

Contractor, S.F., Eaton, B.M., Futin, J.A., and Bauman, K.F. (1984). A comparison of the effects of different perfusion regimes on the structure of the isolated human placental lobule. *Cell and Tissue Research*, **237**, 609–17.

DiGiacomo, J.E. and Hay, W.W., Jr (1989). Regulation of placental glucose transfer and consumption by fetal glucose production. *Pediatric Research*, **25**, 429–34.

Hay, W.W., Jr (1988) Placental metabolism of glucose in relation to fetal nutrition. In *Fetal and neonatal development. Research in perinatal medicine*, Vol. 7 (ed. P. W. Nathanielsz) pp. 58–67. Perinatology Press, Ithaca, New York.

Hay, W.W., Jr and Meznarich, H.K. (1988). Effect of maternal glucose concentration on uteroplacental glucose consumption and transfer in pregnant sheep. *Proceedings of the Society for Experimental Biology and Medicine*, **190**, 63–9.

Hay, W.W., Jr, Molina, R.A., DiGiacomo, J.E., and Meschia, G. (1990). Model of placental glucose consumption and glucose transfer. *American Journal of Physiology*, **258**, R569–R577.

Holzman, I.R., Lemons, J.A., Meschia, G., and Battaglia, F.C. (1977). Ammonia production by the pregnant uterus. *Proceedings of the Society for Experimental Biology and Medicine*, **156**, 27–30.

Holzman, I.R., Philipps, A.F., and Battaglia, F.C. (1979). Glucose metabolism, lactate, and ammonia production by the human placenta in vitro. *Pediatric Research*, **13**, 117–20.

Illsey, N.P., Harmonde, J.G., Penfold, P., Bardsley, S.E., Coade, S.B., Stacey, T.E., and Hytten, F.E. (1984). Mechanical and metabolic viability of a placental perfusion system 'in vitro' under oxygenated and anoxic conditions. *Placenta*, **5**, 213–25.

Jones, C.T. and Rolph, T.P. (1985). Metabolism during fetal life: a functional assessment of metabolic development. *Physiological Reviews*, **65**, 357–430.

Lemons, J.A. (1979). Fetal–placental nitrogen metabolism. *Seminars in Perinatology*, **3**, 177–90.

Lemons, J.A., Adcock, E.W., III, Jones, M.D., Jr, Naughton, M.A., Meschia, G., and Battaglia, F.C. (1976). Umbilical uptake of amino acids in the unstressed fetal lamb. *Journal of Clinical Investigation*, **58**, 1428–34.

Loy, G.L., Quick, A.N., Hay, W.W., Jr, Meschia, G., Battaglia, F.C., and Fennessey, P.V. (1990). Feto–placental deamination and decarboxylation of leucine. *American Journal of Physiology*, **259**, E492–E497.

Marconi, A.M., Sparks, J.W., Battaglia, F.C., and Meschia, G. (1989). A comparison of amino acid arteriovenous differences across the liver, hindlimb and placenta in the fetal lamb. *American Journal of Physiology*, **257**, E909–E915.

Meschia, G., Hay, W.W., Jr, Sparks, J.W., and Battaglia, F.C. (1980). Utilization of substrates by the ovine placenta in vivo. *Federation Proceedings*, **39**, 245–9.

Molteni, R.A., Stys. S.J., and Battaglia, F.C. (1978). Relationship of fetal and placental weight in human beings: fetal/placental weight ratios at various gestational ages and birth weight distributions. *Journal of Reproductive Medicine*, **21**, 327–34.

Owens, J.A., Falconer, J., and Robinson, J.S. (1987). Effect of restriction of placental growth on fetal and utero-placental metabolism. *Journal of Developmental Physiology*, **9**, 225–38.

Schneider, H. and Dancis, J. (ed.) (1985). *In vitro perfusion of human placenta tissue*. Karger, Basel.

Schneider, H., Challier, J.C., and Dancis, J. (1981). Transfer and metabolism of glucose and lactate in the human placenta studied by a perfusion system in vitro. *Placenta 2* (suppl.), 129–38.

Simmons, M.A., Battaglia, F.C., and Meschia, G. (1979). Placental transfer of glucose. *Journal of Developmental Physiology*, **1**, 227–43.

Villee, C.A. (1953). The metabolism of human placenta in vitro. *Journal of Biological Chemistry*, **205**, 113–23.

Wilkening, R.B., Battaglia, F.C., and Meschia, G. (1985). The relationship of umbilical glucose uptake to uterine blood flow. *Journal of Developmental Physiology*, **7**, 313–19.

5. Growth factors in fetal growth
Victor K.M. Han and David J. Hill

Fetal growth is not a uniform progression of cell replication, but a series of fundamentally different anabolic processes that are precisely integrated. Development prior to placental implantation involves rapid cell division and pattern formation, which becomes apparent when primitive germ layers separate and presumptive organs can first be identified. Metabolism at this time is anaerobic, and cells communicate by cell–cell contacts and by the release of autocrine and paracrine factors. The implantation of the placenta allows aerobic respiration and the development of energy-expensive enzyme systems necessary for differentiative cell function. The appearance of a feto-placental circulation improves the delivery of nutrients to the tissues and, consequently, the rate of cell multiplication rises. The placenta acts as an endocrine organ influencing both fetus and maternal physiology; it also acts as an immunological barrier between the fetus and the mother to allow the 'foreign' fetus to develop within the maternal immunological environment. During early embryonic development, the pattern of growth is largely dictated by the fetal genome but, as the body size increases, the fetus becomes constrained by maternal and environmental influences such as uterine blood flow, maternal size, and maternal disease.

General Considerations

Regulation of embryonic and fetal development

Growth and development of the mammalian embryo and fetus are determined by many factors, which can be categorized into either genetic or epigenetic factors. The genetic factors include the fetal genotype and sex, maternal genotype, and paternal genotype; the epigenetic or environmental factors include the general maternal environment, immediate maternal environment, maternal age and parity, and other unknown factors. In 1974 Polani estimated that approximately 38 per cent of the variation in birth weight can be explained by genetic factors and the remainder (62 per cent) by epigenetic (environmental) factors. Of the epigenetic factors, maternal factors could explain half of the variation, but the remaining half are due to unknown factors.

Hormonal (endocrine) regulation

Maternal hormones that regulate cellular proliferation and differentiation, including growth hormone, thyroid hormones, insulin, prolactin, and steroid hormones, do not cross the placenta in physiologically important quantities. The placenta is permeable to cortisol, but this is predominantly converted to a biologically inactive form, cortisone, by the placental 11β-hydroxysteroid dehydrogenase. Factors regulating fetal growth are therefore largely contained within the developing fetus. However, growth and development of the fetus are intimately related to the normal development of the maternal reproductive tissues, as well as the placenta, because adequate transfer of substrates (nutrients and oxygen) from the mother to the fetus and the efficient removal of waste material from the fetus are crucial for normal growth and development (Fig. 5.1)

Classical hormones produced by the fetus have little or no role in regulating its growth. Growth hormone (GH) is present in the human pituitary from at least 12 weeks of gestation and, in midtrimester, can reach circulating levels in excess of 100 ng/mL. Despite ample GH and its receptors in fetal tissues, there is little or no evidence that GH plays a role in the regulation of fetal growth. Animal studies showing that fetal hypophysectomy does not lead to growth failure and clinical evidence of anencephalic fetuses having normal size, support the view that fetal growth is independent of GH. The role of thyroid hormones is species-dependent. Thyroidectomy of fetal sheep leads to growth failure, but human fetuses with agenesis of the thyroid gland are normal size when born, except for a slight delay in skeletal maturation. However, small amounts of thyroid hormones can still be produced by ectopic thyroid tissues, and it is not clear whether these may be adequate to support normal fetal growth and development.

The role of insulin as a growth-promoting factor in the fetus is still controversial. As adequate cellular nutrition is essential for normal fetal growth, insulin, by virtue of its function in the transport of glucose and amino acids into cells, plays an important role in cellular growth. Fetal overgrowth in fetuses of poorly controlled diabetic mothers may not necessarily be due to fetal hyperinsulinism causing cellular hyperplasia, but is more likely due to

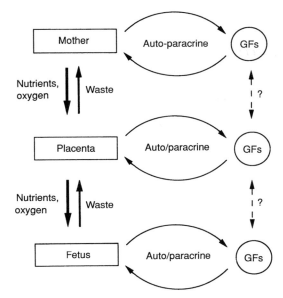

Fig. 5.1 The transfer of substrates and waste between mother and fetus is, for the most part, unidirectional. Adequate transfer of substrates (nutrients and oxygen) from the mother to the fetus via the placenta, and efficient removal of waste from the fetus to the mother are essential for normal fetal growth. Growth factors (GFs) are synthesized in the maternal reproductive tissues, placenta, and fetus and act locally in an autocrine/paracrine manner. Endocrine transfer of growth factors between the mother, placenta and fetus has not been shown to occur.

Local (autocrine/paracrine) regulation

The lack of a central regulation of fetal growth suggests that local regulatory mechanisms must exist to control and co-ordinate the complex processes of embryogenesis and fetal development, which involve not only cellular proliferation (hyperplasia) and differentiation (hypertrophy), but also other cellular events (Fig. 5.2). This integrated process requires precise intercellular communication. The local cell-to-cell and cell-to-matrix interactions occur through several effector molecules including those of the extracellular matrix, intercellular recognition molecules, and peptide growth factors. This chapter will consider the roles of peptide growth factors in embryogenesis and fetal development.

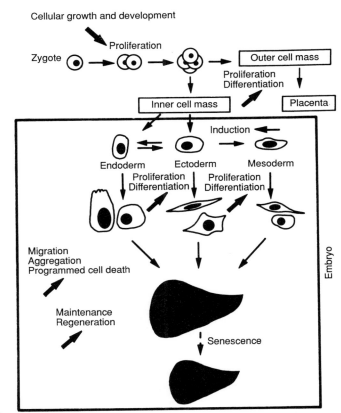

Fig. 5.2 Different cellular processes involved in growth and development of tissues and organs. The processes that may be regulated by growth factors are indicated by arrows. The growth and development of both the embryo and placenta are regulated by endogenously synthesized growth factors, the gene expression of which are dependent on stage of development.

excess adiposity caused by enhanced fetal fat deposition. Attempts to induce somatic overgrowth by experimental fetal hyperinsulinaemia have resulted in, at best, small increases in tissue mass. In contrast, fetal hypoinsulinaemia is consistently associated with intrauterine growth retardation (IUGR), an effect that has been duplicated in some experimental animals by pancreatectomy or administration of pancreatic β-cell toxins such as streptozotocin or alloxan. At the very least, insulin plays an important role in fetal growth by its primary function of regulating cellular nutrition, and perhaps by interacting with other factors. One such interaction may be the negative regulation of insulin on insulin-like growth factor binding protein-I (IGFBP-I), the serum concentration of which has been shown to be inversely correlated with fetal size. Other hormones such as corticosteroids and sex hormones may have functions in the development of specific organs and tissues, but have little or no role in general fetal growth.

Mechanisms of regulation of cellular growth and differentiation

Cellular proliferation and differentiation are just two aspects of a complex process by which the conceptus (organs and tissues) increases in size and functional complexity (Fig. 5.2). In addition, considerable cell migration occurs in early embryogenesis to bring different cell types into intimate contact, which may at a certain phase of development lead to tissue induction (e.g. induction of mesoderm from ectoderm by endoderm). Following blastulation, epithelium–mesenchymal interactions become particularly important in shaping tissue form. With further proliferation, differentiation, migration, aggregation, and programmed cell death, a mature organ or tissue is formed. Even after the formation of a mature functioning organ, cells continue to die and require replacement with new cells (maintenance). Due to certain pathological conditions, parts of the developed organ may be damaged and in need of repair (regeneration). In the majority of organs, the regenerative capacity of an organ is greatest during embryonic and fetal life as compared to any other stage of life.

At least three different control mechanisms govern early tissue interactions and responses: (1) the deposition and subsequent modification of extracellular matrix molecules; (2) the temporal expression of cellular recognition molecules; and (3) the appropriate expression of intercellular messengers such as peptide growth factors. These three basic control mechanisms are by no means independent, but occur in an extremely interactive and interdependent manner. For example, growth factors regulate expression of genes encoding many, if not all, extracellular matrix proteins, and the latter serves as a storage depot for many growth factors such as transforming growth factor-β (TGF-β); the expression of many genes encoding cellular recognition molecules is regulated by growth factors.

Extracellular matrix and intercellular recognition molecules

Extracellular matrix consists of two major classes of molecules: proteins such as collagens, laminin, fibronectin, and tenascin; and mucopolysaccharides such as hyaluronic acid, heparan sulfate, and chondroitin and keratin sulfate. Most epithelia are tightly attached to a basal lamina or basement membrane that is rich in type IV collagen and laminin. Fibronectin and tenascin govern cell shape and proliferation rate *in vitro*, and they may also determine cell migration and the onset of differentiation *in vivo*. Tenascin, in particular, has been shown to appear along migration tracts before the onset of the migration of neural

crest cells from around the neural tube, and is removed immediately afterwards. Hyaluronic acid, a hydrophilic mucopolysaccharide, also forms an important component of the migratory tracts and forms a coating over the migrating cells.

The cellular recognition molecules are utilized by cells to recognize their positions within developing structures and to sort out the major cell types (e.g. epithelial and mesenchymal cells). Two categories of recognition molecules exist: (1) extracellular matrix (ECM)-cell recognition molecules or integrins, which are utilized for interaction between ECM and cells; and (2) the intercellular recognition molecules which are utilized for interaction and recognition between cells. There are two forms of the latter molecules: (1) the cell adhesion molecules (CAMs) that do not require calcium for their biological action, and (2) the cadherins that are calcium-dependent. Several subclasses exist within each group according to their principal or first-observed sites of expression (e.g. epithelial cadherin, neural cadherin, and placental cadherin). Identical recognition molecules are expressed on the membranes of homotypic cells during tissue condensation, or between heterotypic cells during epithelial–mesenchymal or other tissue interactions. Communication of cadherins with actin filament bundles within the cytoskeleton suggests that cells can respond directly to these molecular interactions, or lack of them, with movement. In addition, the cytoskeletal network may be utilized to initiate intracellular biochemical events following cell–cell interaction via cadherins on the cell surfaces.

Peptide growth factors

Peptide growth factors are proteins of less than 30 kDa molecular weight that are widely synthesized and act locally within tissues as paracrine or autocrine factors (Fig. 5.3); they regulate various aspects of cellular growth and development including proliferation, differentiation, induction, migration, aggregation, programmed cell death, maintenance, and regeneration (Fig. 5.2). Growth factors that are expressed in developing tissues and have been shown to have biological actions on developing cells are listed in Table 5.1. Growth factors that have been identified as acting on the different aspects of cellular growth and development are listed in Table 5.2.

Unlike hormones, which are secreted by specific endocrine organs, each growth factor may be synthesized by many cells and tissues. They are generally not stored intracellularly and their release is largely dependent on *de novo* synthesis. The capacity of many different developing cells to synthesize growth factors is a requisite of their autocrine or paracrine mode of action (Fig. 5.3) Many are released into the circulation, but it is doubtful if any of them has an endocrine mode of action. Since they are polypeptides or glycoproteins in nature, peptide growth factors are hydrophilic or lipid-insoluble, and therefore have to interact with specific protein receptors located on the cell membrane of target cells and communicate with second messenger systems by conformational changes. This often involves the autophosphorylation of tyrosine residues located on the intracellular domain of the receptor. The second messenger systems utilized are diverse and include changes to intracellular calcium levels, cyclic AMP, cellular alkalinization, and phosphoinositol metabolites. For most growth factors, however, the system is not well defined. The net biological events are remarkably consistent and include a rapid stimulation of amino acid transport, glucose uptake and utilization, and RNA and protein synthesis. For many, but not all growth factors, this is followed by DNA synthesis and cell replication. Some growth factors, such as TGF-β and tumour necrosis factor, act predominantly as growth inhibitors.

Stimulation of DNA synthesis and mitogenesis are the basic biological functions of many growth factors. The basic mechanism by which this action is mediated is best described in studies on the cell cycle of fibroblasts. It has been shown that some growth factors act in the G_1 phase

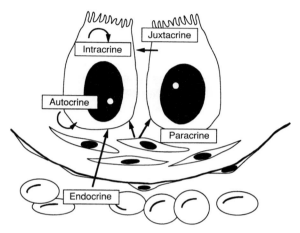

Fig. 5.3 Modes of action of growth factors at the cellular level in developing tissues. The majority of growth factors act in an autocrine and/or paracrine fashion and, to a lesser degree, have an endocrine mode of action. Some growth factors may exert their actions without being secreted (intracrine) and some growth factor precursors may have a transmembrane spanning domain that allows them to act on adjacent cells without being fully secreted into the extracellular milieu (juxtacrine).

Table 5.1 Growth factors that are expressed during development

Insulin-like growth factors (IGF)
Epidermal growth factor (EGF)
Transforming growth factor-α (TGF-α)
Transforming growth factor-β family (TGF-β)
Platelet derived growth factor (PDGF)
Fibroblast growth factors (FGF)
Nerve growth factor family (NGF)
Other growth factors:
    Haemopoietic growth factors
    Interferons
    Bombesin and gastrin-releasing peptide
    Müllerian inhibitory substance
    Inhibin/activin family

Table 5.2 Biological actions of growth factors on various developmental processes

Developmental process	Growth factor
Proliferation	IGF-I and IGF-II
	EGF
	TGF-α
	TGF-β
	FGF
	PDGF
	Haemopoietic growth factors
Differentiation	IGF-I and IGF-II
	EGF
	FGF
	NGF
Induction	FGF
	TGF-β
Migration and aggregation	Cell adhesion molecules and chemotactic factors
	GFs?
Programmed cell death	NGF
	TGF-β
	Inhibins
	Müllerian inhibitory substance
Maintenance	IGF-I and IGF-II
	TGF-α
	TGF-β
Regeneration	IGF-I and IGF-II
	TGF-α
	TGF-β
	NGF
	PDGF

Regulation of cell cycle by growth factors

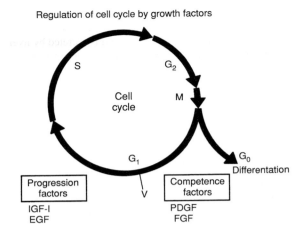

Fig. 5.4 One mechanism by which growth may regulate DNA synthesis during a cell cycle. They may act either in early G_1 phase (competence factors) or in late G_1 phase (progression factors). Many mitogenic growth factors, however, can progress the quiescent cells (G_0 phase) to S phase without requirement of additional growth factors. It is possible that these growth factors may stimulate the synthesis of endogenous competence or progression factors to achieve this effect.

of the cell cycle. This is the phase in which the cell acquires the necessary nutrients, proteins, and enzymes to begin its preparation for DNA synthesis (S phase). Growth factors may act in the first or second half of this phase (Fig. 5.4). The midpoint of the two halves of the phase is known as the 'v' point. Growth factors that take the cells from the quiescent G_0 phase to 'v' point are called competence factors. Fibroblast growth factor (FGF) and platelet-derived growth (PDGF) are competence factors. Growth factors that take the cells from 'v' point to S phase are called progression factors. Epidermal growth factor (EGF) and insulin-like growth factor-I (IGF-I) act as progression factors in this cell system. However, in many other cell systems, the roles of different growth factors are not as well delineated. Various growth factors have been shown to act both as competence and progression factors.

Tissue induction is an important action of growth factors. TGF-β and FGF have been shown to promote the induction of mesoderm from primitive ectoderm in the early embryo. Such induction in the amphibian embryo depends on diffusible morphogens from the vegetable pole ectoderm causing mesodermal development from the animal pole ectoderm. FGF has been shown to be the inducer for ventral mesoderm and TGF-β for the dorsal mesoderm.

Another important function of growth factors is to augment or antagonize the onset of differentiation. A potentiation of differentiation need not exclude a mitogenic role since an immediate stimulation of cell proliferation may be followed by a later promotion of differentiation. Separate growth factors can co-operate or compete during the onset of differentiation, while some show synergism with endocrine hormones. In some growth factors, such as the insulin-like growth factors (IGF), the induction of mitogenesis or differentiation may be a result of different concentrations. For example, physiological levels of IGF-I promote proliferation of myoblasts, whereas higher concentrations promote differentiation into myotubes. Many cells continue to express growth factors even after they have differentiated, suggesting that growth factors continue to be involved in the maintenance of certain differentiative functions of the cells.

As more than one growth factor may be synthesized locally within a developing organ or tissue, it is likely that significant interactions exist amongst different growth factors. Such interactions have been demonstrated in many cell systems, the most prominent of which are muscle (IGFs and TGF-β), cartilage and skeletal system (IGFs, FGFs, and TGF-β), brain (IGFs and EGF/TGF-α), integumentary system (IGFs and TGF-α), and respiratory system (IGFs, FGFs, and PDGF). The final result of such interactions is the programmed development of a mature functioning organ (Fig. 5.5)

Cellular growth balance

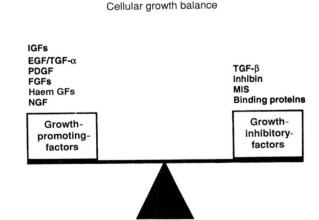

Fig. 5.5 The end result of cellular growth or differentiation is dependent on the interaction of various growth-promoting and growth-inhibitory factors that are endogenously synthesized. Haem GFs: haemopoietic growth factors.

Growth factors, oncogenes, and proto-oncogenes

Oncogenes are components of a normal genome (proto-oncogenes) that, through an inappropriate degree of expression, translocation, or point mutation, are no longer under precise genetic control. Expression of such genes can lead to an inappropriate or unregulated growth such as that observed in cancer or malignancy, or such genes may be expressed during development. Throughout phylogeny, most, if not all, oncogenes have been derived from normal cellular genes (proto-ongenes) via retroviruses, which are responsible for mutations as well as reintroduction into the mammalian genomes. A pivotal finding was that retroviral oncogenes (denoted by v-) were homologous to normal cellular genes (denoted by c-) involved in control of growth and differentiation. The first published example of this was that the v-*sis* oncogene encodes a peptide related to the B chain of platelet-derived growth factor (PDGF). Oncogenes and proto-oncogenes can be categorized based on structure and function of the normal cellular gene product; growth factor (v-*sis* and PDGF), growth factor receptor (v-*erb* and epidermal growth factor (EGF) receptor; c-met and hepatocyte growth factor receptor), intracellular second messenger components (c-*src* and c-*ros* and tyrosine-specific protein kinases; c-*ras* and GTP binding protein), and transcriptional factors (c-*fos* and c-*myc*).

Many growth-factor-related proto-oncogenes are expressed in developing tissues at specific times; for example c-*src* is expressed during development of the eye and brain, c-*ras* in lung and brain. In addition, many growth factors may induce expression of proto-oncogenes during development; for example, FGF can induce a transient expression of c-*myc* and c-*fos* at an early G_1 phase of the cell cycle, and EGF can induce c-*ras* in a later phase of G_1.

Insulin-like growth factors

Peptides

Chemistry

Insulin-like growth factors (IGFs) are peptides that have structural homology with proinsulin, and have a structural configuration of B and A chains, connected by a C peptide. Two types of IGFs, IGF-I and IGF-II, have been purified from the serum and tissue fluids of many species and their complementary DNAs have been cloned. In addition to the B, C, and A domains, the precursor peptides have a hydrophobic signal peptide in the amino terminus and variable lengths of D and E peptides in the carboxyl termi-

nus. IGF-I, previously known as somatomedin-C, is a 70-amino-acid basic peptide (relative molecular weight (M_r), 7648 Da; isoelectric point (pI), 8.8) and is secreted by liver and many tissues in postnatal life in response to growth hormone. In fetal life, its synthesis is low and is independent of growth hormone. Nutritional factors are important regulators of IGF-I secretion in both fetus and growing animals. It is a very potent mitogenic and differentiation-promoting factor in many cells, and is the mediating factor for the action of growth hormone on growth plates in the pubertal growth spurt. IGF-II is a 67-amino-acid neutral peptide (M_r 7469 Da; pI, 6.7) and has 60 per cent homology with IGF-I. It is also secreted by many different tissues but, unlike IGF-I, its secretion is not regulated by growth hormone. In humans, IGF-II concentrations continue to be higher than IGF-I in the serum of an adult. In many other species, however, the concentrations decrease after birth, except in the cerebrospinal fluid (CSF). In the fetal serum of such species, IGF-II concentrations are significantly higher than in adults, suggesting that IGF-II may be more important during fetal life. It is a mitogen but, in most cell types, it is less potent than IGF-I. It also has insulin-like metabolic actions and is more potent than IGF-I in this regard.

Complementary DNAs and precursor molecules

On the basis of complementary DNA (cDNA) cloning and purification studies, variant forms of both IGF-I and IGF-II have been shown to exist. There are two different IGF-I precursors, one derived from exons 1, 2, 3, and 5 (Ea IGF-I) and the other from exons 1, 2, 3, and 4 (Eb IGF-I), which differ in the E domain sequences. (An exon is a segment of an interrupted gene that is represented in the messenger RNA and the E-domain is the portion of IGF precursor peptide closest to the COOH end). IGF-I, which is lacking in the first three N-terminal residues (des(1–3)-IGF-I), has 100-fold less affinity for IGF binding proteins and is present in the human fetal and adult brain, porcine uterus, bovine colostrum, and human platelet lysates. Several variant forms of IGF-II also exist. One form identified by cDNA cloning is a variant in which the serine residue in 29 position is replaced by the sequence Arg–Leu–Pro–Gly, most probably as a consequence of a nine-nucleotide insertion by alternative RNA splicing.

Gene structure

The human IGF-I gene consists of six exons spread over 90 kbp (kilobase pairs) on the long arm of chromosome 12. By alternate processing of messenger RNA (mRNA) this gene encodes two variant forms of IGF-I mRNAs, Ea IGF-I (formerly called IGF-Ia) and Eb IGF-I (formerly called IGF-Ib) mRNAs. The rat IGF-I gene comprises at least six exons, five of which are analogous to those of the human gene, and spans approximately 50 kbp.

The human IGF-II gene consists of nine exons and spans 30 kbp on the short arm of chromosome 11, downstream of the insulin gene. The structure of the human IGF-II gene and the complex transcriptional process is shown in Fig. 5.6. Transcription can be activated by four different promoters (P1 to P4; a region of DNA involved in the binding of RNA polymerase to initiate transcription) which precede the non-coding exons 1 to 6. The rat and mouse genes comprise six exons (homologous to human exons 4 to 9) that span 12 kbp of the rodent genome and are activated by three promoters P1 to P3 located in regions similar to those occupied by human promoters P2 to P4. The use of multiple promoters, alternate mRNA splicing, (removal of introns and joining of exons in RNA), and differential polyadenylation signals (ends of eukaryotic mRNA at different sites) leads to the generation of multi-

ple transcripts of sizes varying from 1.8 to 5.3 kb (kilobase) during fetal development Most of the transcripts, except for the 1.8-kb one, contain the coding regions for the mature IGF-II peptide (exons 7, 8, and 9 in humans, and exons 4, 5, and 6 in rodent). Transcripts derived from P3 are most abundant in human tissues, whereas those derived from human P4 equivalent (rodent P3) predominate in the rat tissues. In the human adult liver, a 5.3-kb transcript, arising from activation by promoter P1 and containing the non-coding exons 1, 2, and 3 and coding exons 7, 8, and 9, is the predominant mRNA. The developmental switch occurs in the pig and sheep, but not in rodents. No promoter analogous to P1 has been identified in rodent; this may explain the rapid decline in IGF-II postnatally in rodents and not in human.

IGF receptors

IGFs bind with high affinity to two distinct types of receptors: the type 1 or IGF-I receptor, and the type 2 or IGF-II/mannose-6-phosphate receptor. Both types of IGF receptors have been identified in membrane preparation and/or cultured cells from virtually every fetal tissues, in the placenta, and in the maternal reproductive tissues, indicating their biological importance during fetal life

The IGF-I receptor has a subunit structure similar to that of the insulin receptor, which consists of a heterodimer of two α and two β subunits linked by disulfide bridges. Affinity cross-linking studies on cells or membranes with high-affinity receptor have yielded a 350 000 M_r protein that, upon reduction, generates a 135 000 M_r protein (α subunit) and a 95 000 M_r protein (β subunit). Receptor binding studies show this receptor to prefer IGF-II and to weakly bind insulin. Immunological studies using monoclonal antibodies have confirmed the similarity between IGF-I and insulin receptor and also indicated the presence of hybrid receptors. cDNA cloning and sequencing of IGF-I receptor revealed extensive homology with insulin receptor. The IGF-I receptor precursor consists of 1337 amino acids, and comprises the α subunit protein in the amino terminal portion of the precursor separated from the carboxyl terminal β subunit by an Arg–Lys–Arg–Arg sequence. The α subunit contains a single cysteine-rich region and 11 potential N-linked glycosylation sites. The β subunit contains a single 24-amino-acid hydrophobic region (transmembrane domain) and the tyrosine kinase domain, which is 84 per cent homologous with that of the insulin receptor. The IGF-I receptor cloned from a human placental cDNA library has been expressed in CHO (Chinese hamster ovary) cells and the resultant protein displays a high-affinity binding to IGF-I and a ligand-stimulated autophosphorylation.

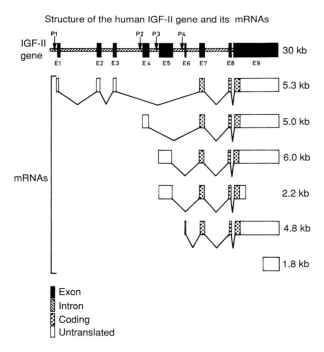

Fig. 5.6 The structure of the human IGF-II gene and the complex mechanism of alternate splicing, differential promoter usage, and polyadenylation by which mRNAs of various sizes are transcribed.

The IGF-II receptor is a single transmembrane protein with an M_r of 260 000–265 000 as demonstrated by affinity cross-linking and sodium dodecyl sulfate–polyacrylamide gel electrophoesis (SDS–PAGE). IGF-II receptor binds IGF-II with a higher affinity than IGF-I, but does not bind insulin. cDNA cloning has predicted the IGF-II receptor protein, which includes a 23-residue hydrophobic region (transmembrane domain), a short carboxyl terminal cytoplasmic domain, and a large extracellular portion (92 per cent of the protein). The extracellular domain consists of 15 conserved repeats of 150 residues, which are 20 per cent identical but share a highly conserved pattern of eight cysteine residues. The predicted amino-acid structure of the IGF-II receptor cDNA revealed 80 per cent identity with that of the bovine cation-dependent mannose-6-phosphate (man-6-P) receptor and 99.4 per cent identity with the human man-6-P receptor, suggesting that IGF-II receptor and man-6-P receptor are the same protein. This has been confirmed by binding, affinity purification, and immunological studies. Man-6-P receptor recognizes mannose-6-phosphate residues on acid hydrolases and targets these enzymes to lysosomes. In addition to the membrane-bound form, a circulating form of IGF-II/man-6-P receptor has been identified in the serum and certain tissue fluids particularly during fetal life. The lack of homology with known protein kinases in the cytoplasmic domain has raised the question of the role of this receptor in signal transduction following ligand binding. Current evidence suggests that most of the well recognized biological actions of IGF-II are mediated by IGF-I or insulin receptor, and that IGF-II/man-6-P receptor may be involved in specific signal transduction using other mechanisms (e.g. via G protein) or in the removal of excess hormone from the circulation and ligand degradation.

IGF binding protein (IGFBPs)

At least six distinct species of IGF binding protein (IGFBPs) exist, and four IGFBPs (IGFBP-1 to IGFBP-4) have been identified in human fetal plasma. IGF binding proteins not only serve as carrier proteins for IGFs but also modulate their biological actions by competing for and interacting with the receptors.

All six IGFBPs have a similar core protein structure with 18 cysteine residues whose alignment is preserved, indicating similar tertiary structure. IGFBP-1 and IGFBP-2 contain an Arg–Gly–Asp (RGD) sequence and IGFBP-4 contains a modified Asp–Gly–Asp (DGD) sequence near the carboxyl terminus. Similar sequences have been shown to be present in extracellular matrix protein (fibronectin, laminin), which mediate the integrin-associated cell surface attachment. RGD sequences in IGFBPs may confer on them the capacity to associate with cell surface integrins and exert thier biological actions. Alignment of the deduced amino acid sequences of the six rat and human IGFBPs showed conserved sequences to be located at the carboxyl terminal third and a portion of the amino terminal third regions, whereas the middle portion of the molecules are the most divergent. *In vitro* mutagenesis studies showed that deletion of the c-terminal 20 amino acid or introduction of frame shifts in this region result in the loss of IGF binding and, for some mutants, in the formation of dimeric IGFBP-1 molecules. A point mutation introduced in the Cys226 to tyrosine completely abolishes IGF binding, suggesting that the IGF binding domain of IGFBP-1 may be located in the vicinity of the intramolecular disulfide bond formed by Cys226.

Chromatographic studies have shown that IGFs migrate in two discrete complexes: a small 50 000 M_r complex and a larger 150 000 M_r complex. The large IGFBP is growth-hormone-dependent and is the major binding protein found in postnatal serum. This is a complex of three subunits: the acid-labile α subunit of 60 000–80 000 M_r, the acid-stable β subunit of 40–60 000 M_r which is the IGFBP-3, and the γ subunit which is the IGF peptide. The smaller peak identified in chromatography has been shown to consist of the other IGFBPs 1, 2, 4, 5, and 6, which are largely unsaturated.

IGFs, IGF receptors, and IGFBPs in fetal development

IGFs

Both IGF-I and IGF-II are present in human fetal plasma from as early as 15 weeks of gestation, and both peptides are present in human fetal tissue extracts from 12 weeks. IGF-II concentrations are higher than those of IGF-I in all tissues examined, varying from 1.6-fold in the muscle to 7.8-fold in the thymus, with an average of 3.3-fold. Although the level of both IGF-I and IGF-II may increase as much as twofold from 32 weeks of gestation to term, the concentrations in human fetal sera and umbilical cord blood remain low compared to maternal plasma. This contrasts with other species in which IGF-I levels steadily rise to adult values through gestation and in which IGF-II levels in late pregnancy are high compared with adult values. A particularly striking example is that of fetal rat serum which, in late gestation, contains 20–100 times more IGF-II than maternal serum. In sheep and guinea-pigs, serum IGF-II levels are higher in the fetus than in adults, but the difference is not as pronounced as in rodents.

Although high serum IGF-II levels in the fetus suggest that the peptide is an important regulator of growth during fetal life, serum IGF-II levels have not been correlated with fetal weight or length. In contrast, serum IGF-I levels have been positively correlated with fetal size, and they are depressed in intrauterine growth retardation in humans and animals. These findings appear to question the importance of IGF-II in fetal growth, or may indicate that serum levels of IGF-II do not correlate with the biological importance of IGF-II in fetal growth at the tissue level. The latter appears to be a more plausible explanation as the elimination of endogenous IGF-II gene expression by homologous recombination in transgenic mice resulted in growth-retarded new-borns; this strongly supports the hypothesis that IGF-II is an important fetal growth factor.

IGF-II mRNA has been identified in the mouse embryo as early as the two-cell stage and IGF-I mRNA has been identified in the post-blastocyst stage. Both IGF-I and IGF-II are expressed in four to six different mRNA transcripts in both species. Both IGF-I and IGF-II mRNAs are detectable in various fetal tissues of rodents, sheep, and humans, but the relative abundance of IGF-II mRNAs is significantly higher than IGF-I mRNAs in every tissue. Ontogenetic studies in fetal and postnatal animals demonstrate that: (1) both IGF-I and IGF-II mRNAs are expressed in many tissues of fetal rodents from as early as 11 days of gestation; (2) the levels of IGF-II are high in the fetus and low or non-detectable in adult tissues, and the reverse is true for IGF-I; (3) the pattern and the levels of IGF-II gene expression are species-specific (that is, in rodents, IGF-II mRNAs are non-detectable in most tissues except in the brain (choroid plexus), whereas in sheep and man, IGF-II mRNAs are detectable in many tissues such as liver, kidney, lung, muscle, and choroid plexus); (4) the structure and promoter organization of the IGF-II gene of rodents is different from that of sheep and humans (i.e. the rodent IGF-II gene does not have the promoter P1 and exon 1 of human and sheep (Fig. 5.6), which transcribe the 5.3-kb mRNA, the predominant transcript in adult tissues); (5) the promoters P2, P3, P4, which are active in the fetus, are present in rodents, sheep, and humans. The difference in genomic structure may explain the absence of IGF-II gene expression in adult rats, except in the brain.

In human and ovine fetal tissues, IGF mRNAs have been localized in connective tissues and cells of mesenchymal origin in every organ and tissue examined and in the epithelium of certain tissues (Fig 5.7). A similar pattern of IGF-II mRNA expression occurs in rodent fetuses. In early organogenesis, IGF-II may be expressed in other non-mesenchymal cells, for example, fetal hepatocytes, bronchial epithelia, and some endodermal cells. However, one of the most intriguing findings is the lack of correla-tion between the cellular sites of IGF mRNA expression and peptide immunoreacitivity in developing fetal tissues. This is consistent with our hypothesis that these peptides are not stored, but are released immediately after synthesis. Our recent studies indicate that the cellular localization of IGF peptides is determined by the presence of an IGFBP in specific cell types

In contrast to the situation after birth, IGF gene expression in the fetus and embryo is not controlled by pituitary GH. Anencephaly in humans and experimental decapitation or hypophysectomy in most experimental animals have no effect on fetal size. In primary human fetal mesenchymal cells and hepatocytes, placental lactogen promotes IGF-I release and stimulates DNA synthesis. However, in humans placental lactogen does not appear to be a major fetal growth factor as infants born to women with no placental lactogen are normally grown. Insulin is another potential regulator of IGF gene expression, but most probably it does this via its regulation of substrate transport. Fetal pancreatectomy in the sheep causes a significant decrease in circulating IGF-I levels and reduction in fetal size, despite the maintenance of fetal euglycaemia. Replacement of insulin in such fetuses enhanced fetal growth and partially corrected serum IGF-I levels.

IGF receptors

IGF receptor mRNAs have been identified in mouse embryos from as early as the six-cell stage, suggesting that the biological action of IGFs is exerted early in embryogenesis. In humans, plasma membrane preparations of fetal organs and placenta from early mid-trimester to late gestation bind specifically to [^{125}I]IGFs, and both IGF receptors are present. IGF receptors have also been identified in cultured cells from many different fetal tissues.

The expression of type 1 IGF receptor mRNA occurs in cells of multiple embryonic origins, and is particularly pronounced in myotomes, vertebral sclerotomes, mesonephros, and in the bowel walls and throughout the developing nervous system, liver, heart, and pituitary. The type 2 IGF/mannose-6-phosphate receptor is very abundant in the fetus, comprising 0.1 to 0.4 per cent of 16-day gestation fetal rat protein and 1.7 per cent of rat placental protein.

IGF binding proteins

As in postnatal animals, IGFs in the circulation and tissue fluids of fetuses do not exist as free peptides, but are bound to binding proteins (IGFBPs). The type and levels of different IGFBPs in fetuses are species-specific. In rodent fetuses IGFBP-2 appears to be the only serum

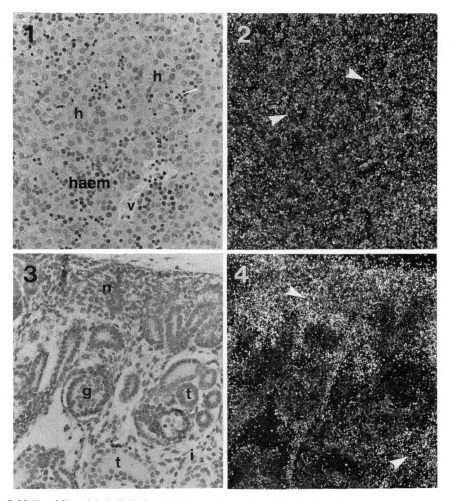

Fig. 5.7 Bright-field (1 and 3) and dark-field (2 and 4) photomicrographs of *in situ* hybridization of liver (1 and 2) and kidney (3 and 4) from a 14-week gestation human fetus identifying the distribution of IGF-II mRNA. Dark-field photomicrographs (2 and 4) of the same field (1 and 3) easily identify the location of IGF-II mRNA hybridized with a ^{35}S-labelled IGF-II cRNA probe (shown by white arrows). IGF-II mRNA is present in the hepatocytes (h) but not in the haemopoietic cells (haem) of the liver, and in the nephrogenic zone cells (n) and renal mesenchymal cells (i), but not in the glomerular (g) and tubular (t) epithelial cells of the kidney. Such information on sites of growth factor gene expression provides important clues to the action and function of these autocrine/paracrine factors.

IGFBP, whereas in human and sheep at least four species of IGFBPs (IGFBP-3, IGFBP-1, and IGFBP-4) have been identified, with IGFBP-3 the predominant IGFBP. IGFBP-1 is abundant in amniotic fluid; the levels are high in early gestation and gradually decrease with fetal maturity. Interestingly, IGFBP-1 is identical with placental protein 12 (PP12), which is synthesized in the maternal decidua. The level of this protein in maternal blood has been shown to be negatively correlated with fetal size. IGFBP-1 mRNA has only been demonstrated in the fetal liver of the rat, sheep, and human suggesting that the liver is the principal source in the fetus. However, immunoreactive IGFBP-1 is present in many fetal tissues, suggesting an endocrine mode of distribution. Malnutrition and prolonged hypoxia induce hepatic IGFBP-1 gene expression in the fetus.

IGFBP-2 is a fetal IGFBP in most species, and the expression of its gene is observed in most fetal tissues, the most abundant expression being seen in the liver. In the fetal rat, IGFBP-2 is the major, if not the only IGFBP. In fetal sheep, IGFBP-2 mRNA is expressed in every tissue before 0.5 gestation (< 75 days; term, 147 days), but becomes limited to the liver, kidney, and choroid plexus in later gestation and after birth. A similar pattern of mRNA expression may be present in humans.

The expressions of other IGFBP genes have not been extensively studied but it appears that all IGFBP genes are widely expressed in every tissue of the human fetus at 14–18 weeks of gestation, with the greatest abundance of each binding protein observed in different tissues.

Biological actions of IGFs and IGFBPs in the embryo and fetus

Studies of cultured embryonic and fetal cells have indicated that IGFs may be involved in several stages of cellular growth and development in the embryo and fetus including proliferation, differentiation, maintenance, and regeneration (Table 5.2).

Both IGF-I and IGF-II are mitogenic for isolated fetal rat or human mesenchymal cells such as fibroblasts, myoblasts, chondrocytes and osteoblasts, fetal hepatocytes, glial cells, and adrenal cells. There are synergistic interactions between IGFs and other peptide growth factors such as fibroblast growth factors (FGFs) or epidermal growth factor. The relative mitogenic response to IGF-I and IGF-II is also cell-type-dependent. In human fetal fibroblasts and myoblasts, IGF-I and IGF-II are equipotent, but in human fetal chondrocytes and rat fetal adrenal cells IGF-II is more potent than IGF-I. Since IGF-II receptor is not thought to mediate the biological actions of IGFs in normal fetal tissues, it is possible that IGF-II actions may be mediated by the capacity of the peptide to bind to a different binding site on the IGF-I receptor or a hybrid IGF receptor.

In addition to their mitogenic actions, IGFs may also influence the rate of cellular differentiation in muscle, cartilage, bone, brain, and adrenal cells from the fetus and new-born. In muscle, IGF-I or IGF-II stimulate myoblasts to differentiate into contractile myotubes. In bone, IGFs enhance glycosaminoglycan synthesis and sulfate incorporation in embryonic chick pelvic cartilages and extracellular matrix synthesis of human fetal chondrocytes and rat calvarial cells. In the nervous system IGF-I stimulates differentiation of dopaminergic cells of the developing mesencephalon, promotes neuritic outgrowth of rat sensory and sympathetic neurone by inducing tubulin gene expression, stimulates the release of acetylcholine from rat cortical slices, and triggers the differentiation of catecholaminergic precursors in dorsal root ganglia. IGF-II mRNA has been co-localized with 3β-hydroxysteroid dehydrogenase immunoreactivity, indicating that IGF-II is expressed in differentiated steroidogenic cells. Such a finding is a strong indication for an autocrine/paracrine role for IGF-II in steroidogenesis or differentiation of these cells.

As the biological actions of IGFs are modulated by IGFBPs in fetal blood and tissue fluids, acute changes in IGFBP levels may be an important mechanism by which IGF actions may be controlled in the developing fetus. IGFBPs enhance the half-life of IGFs in circulation. IGFBP-1 and IGFBP-2 contain RGD motifs close to their carboxyl termini, which may be utilized for association with cell surface integrin receptors and may modulate binding of IGFs to their receptors. Although many studies have shown inhibitory effects of IGFBPs on IGF-induced mitogenesis, several other studies have shown enhancement of IGF action by IGFBPs. Such contradictory results suggest diverse effects of IGFBPs in modulating IGF action; these may be dependent on: (1) cell type; (2) the developmental stage of the cell; (3) the relative molar concentrations of IGFs and IGFBPs in the extracellular milieu; and (4) the type of IGFBP present. Further studies using molecular manipulations of endogenous IGFBP synthesis through the use of gene targeting or antisense transfection may lead to a better understanding of the role of locally synthesized IGFBPs in the regulation of fetal growth.

IGFs and IGF binding proteins in intrauterine growth restriction (IUGR)

Small-for-gestational-age infants often have decreased serum IGF-I levels compared to normally grown infants, and serum IGF-I levels are positively correlated with birth size. In an animal model of IUGR, serum IGF-I levels and hepatic IGF-I concentrations were decreased and there was a marked increase in hepatic IGFBP-1 mRNA. This implicates IGFBP-1 as an inhibitor of IGF action and mediator of the fetal growth restriction. A recent study of human growth-restricted infants demonstrated a similar change in the profile of IGF binding proteins indicating that IGF binding proteins may be involved in the aetiology of fetal growth restriction. In prolonged fetal hypoxia IGFBP-1 levels in the blood and IGFBP-1 gene expression in the fetal liver are increased, with decreased hepatic IGF-II mRNA levels, suggesting that a combination of IGF-II and IGFBP-1 serum and tissue levels may mediate the decreased growth observed under these conditions.

Epidermal Growth Factor (EGF) and Transforming Growth Factor-α (TGF-α)

EGF activity was first discovered when a fraction that did not contain nerve growth factor (NGF) during its purification from mouse submaxillary glands was found to cause specific biological responses when injected into new-born mice. These included precocious opening of eyelids, eruption of incisors, inhibition of hair growth, and 'stunting' of body growth. It was termed epidermal growth factor when it was found to cause growth and thickening of skin or epidermis.

Transforming growth factor-α (TGF-α) was discovered when EGF-like receptor binding and biological activity were noted in conditioned media from retrovirally transformed cells. When subsequently purified, it was shown to have a distinct, but EGF-like, sequence. Later, EGF-like growth factors were described to be encoded from pox virus genes (VGF), and present in media from phorbol ester-treated tumour cells (amphiregulin).

Peptides

Chemistry

The EGF family of peptides share similarities in amino acid sequence and in their ability to mimic EGF-like activities by binding to the EGF receptor. Analysis of the amino acid sequence reveals the presence of 11 residues that are invariant and six cysteine residues that form three disulfide linkages. Overall, the level of sequence identity amongst the family members is about 20 per cent. Although the sequence identity for each subgroup amongst species is ~ 50 per cent for EGF, and 90 per cent for TGF-α, the sequence conservation for TGF-α within any one species is substantially lower — 42 per cent for humans and 36 per cent for rats.

Important insights have been gained from structure–function studies of the EGF family of peptides by employing site-directed mutagenesis (mutation of a DNA sequence at a specific site) on either the highly conserved residues or residues indicated by nuclear magnetic resonance (NMR) to be involved in β-sheet structures, secondary protein structures built up from a combination of several regions of the polypeptide chain. Substitutions at the highly conserved leucine at position 47 drastically reduced growth factor activity. Replacement of a highly conserved tyrosine 37 with another aromatic amino acid, phenylalanine, was without effect on activity, but the reduction of the side chain (alanine or glycine) markedly decreased activity.

Molecular structure

EGF

Analysis of the cDNA sequence and the deduced amino-acid sequence of the mouse and human EGF precursors shows that the mature EGF molecule of 53 amino acids (6 kDa) is encoded by a large precursor molecule of 1200 amino acid residues (130 kDa). The precursors for the two species are similar (66 per cent) and the EGF sequences are 70 per cent homologous, indicating that the entire precursor molecule is conserved as the biologically active growth factor sequence. The sequence for the mature EGF is located near the C-terminus of the precursor (residues 971–1023). Two sections of hydrophobic sequences are present in the EGF precursors. One is at the N-terminus, which is possibly the signal sequence, and the other near the C-terminus following the sequence for mature EGF. The C-terminus hydrophobic section is probably a transmembrane sequence, suggesting that the EGF precursor is a transmembrane protein anchored in the membrane with sites for N-linked glycosylation. The extracellular domain of the precursor has eight EGF-like repeated sections. Many of these occur in tandem, and are organized into large groups separated by approximately 260 amino acids. Although this kind of structure suggests the generation of low-molecular-weight EGF-like molecules, the absence of basic amino acids at each end of these repeated sections suggests that this is unlikely. In addition to the generation of mature EGF, the precursor molecule may be involved in: (1) biological activity — as a regulator of transport of Na^+ and Cl^- at the proximal convoluted tubular epithelium of the kidney; and (2) binding to the EGF receptor — possibly for cell-to-cell recognition and as a cell adhesion molecule. The cytoplasmic domain has no known function or sequence similarity to other proteins, and has the lowest preserved identity amongst species.

TGF-α

The amino acid sequences of the human TGF-α precursors have been derived from cDNAs, and were shown to be similar (92 per cent) to each other. The TGF-α precursors are much smaller than EGF precursors, and contain no EGF-like repeats. Like the EGF precursor, the TGF-α precursor contains two hydrophobic sequences — the N-terminal signal sequence and the C-terminal transmembrane sequence. The precursor molecule is N-linked glycosylated at one site and palmitoylated at the other. These posttranslational modifications may be necessary for the membrane association. The membrane-bound precursor molecules may also be biologically active suggesting a juxtacrine mode of action for the growth factor precursor (Fig. 5.3). High-molecular-weight forms of TGF-α, corresponding to the extracellular domain of the TGF-α precursor, have been shown to be released into the media around transformed cells in culture and may be biologically active.

Gene structure and expression

EGF

The EGF gene is localized in the long arm of chromosome 4 of humans, and in chromosome 3 of mice. It spans approximately 120 kb and includes 24 exons and 23 introns, all of which interrupt the protein coding sequences. The mature EGF is encoded by two exons (20 and 21) which interrupt it at the Asn32 (asparagine at 32 position).

The primary transcript of EGF is predicted to be ~ 110 kb, while the mature EGF mRNA detected in Northern gels is about 4.7–4.9 kb. Factors regulating EGF gene expression are still being elucidated, but it is well known that the EGF gene is expressed in a tissue-specific manner. The human EGF gene contains a CAAT site (a conserved nucleotide sequence located upstream of the start point of eukaryotic transcription units) and TATA box (a conserved A–T rich septamer found about 25 bp before the start point of each eukaryotic RNA polymerase transcription unit) close to the transcription start site, suggesting that transcription of this gene involves mechanisms known in other systems for RNA polymerase II. The EGF mRNA is most abundant in the submaxillary gland and kidney of the male mouse, with 10-fold lower mRNA levels in the female mouse. *In situ* hybridization studies have shown that, in the submaxillary glands, EGF mRNA is localized to the proximal convoluted tubules and is enhanced by androgens and T3.

TGF-α

The human TGF-α gene is located in the short arm of chromosome 2, and contains six exons. The coding sequence for mature TGF-α is contained in two exons (3 and 4) which interrupt the molecule at Val32. This is the exact point at which the coding sequence for mature EGF is interrupted, suggesting that the two genes arise from a single ancestral gene.

The promoter region of the TGF-α gene does not contain any CAAT or TATA sequences, but transcription appears to start at a unique site approximately 62 bases 5′ to the start site. Thus transcription of the TGF-α gene may involve as yet unrecognized sequences for the efficient positioning of the RNA polymerase. Upstream from the transcription start site are several SP-1 binding sites and AP-2 binding sites indicating that transcription may be regulated, at least in part, by known transcription factors. The mature TGF-α mRNA is ~ 4.8 kb. TGF-α mRNA has been detected in a wide variety of transformed cell lines and in carcinomas and in a few non-transformed cells (keratinocytes and activated macrophages).

EGF receptor

Chemistry

The EGF receptor is a glycoprotein of 170 kDa containing a high-affinity binding site for EGF and other EGF-related molecules. The carboxyl terminus possesses a tyrosine kinase domain. The EGF receptor itself is a substrate for kinase activity at specific tyrosine residues. The EGF receptors of humans and other species are about 78 per cent identical.

The extracellular ligand binding domain of the EGF receptor contains a high content of N-linked oligosaccharides, and a high content of cysteines (10 per cent) that could give rise to 25 disulfide bonds. Using affinity cross-linking studies, it has been suggested that [^{125}I]EGF binding is located between the two cysteine-rich regions. Following the cysteine-rich region is the hydrophobic transmembrane domain which anchors the protein to the cell membrane. The C-terminal half of the protein that encodes cytoplasmic tyrosine kinase sequences has a high homology with the v-*erb* B oncogene, which is a viral oncogene with homology to epidermal growth factor receptor gene.

Gene structure and expression

The EGF receptor gene is located on the short arm of chromosome 7 in humans, and on chromosome 11 in the mouse. The gene covers approximately 100 kb of DNA and includes 26 exons.

The region -151 to 120 bases upstream to the transcription start site has been shown to be essential for promotion and regulation of transcription. In addition, the larger region (-481 to -16 bases), as well as untranslated sequences within the first exon, may contribute to gene expression. The promoter region is G (guanine)- and C (cytosine)-rich (88 per cent) with several transcription start sites, but there are no characteristic CAAT or TATA sequences.

The EGF receptor is expressed in nearly all adult tissues with the exception of cells in the haemopoietic system. The primary transcript is approximately 100 kb, and the mature RNA detected in Northern blots is 5.8 and 10.5 kb in length. An additional 2.8-kb mRNA is detectable in the A431 cell line (used in EGF receptor binding studies and assays) and appears to be a truncated EGF receptor gene product that gives rise to a protein corresponding to the external domain of the EGF receptor.

EGF, TGF-α, and EGF receptor in development

EGF and TGF-α

Although the EGF receptor is widely expressed in fetal organs and tissues (see below), very little evidence exists as to the synthesis or the availability of EGF in fetal tissues, suggesting that, if EGF is indeed an important growth regulator during development, it has to be transported from exogenous sources (e.g. maternal decidua). It has not been possible to detect EGF mRNA in pre-implantation mouse embryos; it has not been identified in any tissues in the mouse until after 2 weeks of postnatal age when the mRNA was detected in epithelial cells of the straight and early portions of the distal convoluted tubules of the kidney. Studies that demonstrated EGF-like activity in fetal mouse tissues (between 11.5 and 17.5 days of gestation) by radioreceptor assays, but not by EGF radio-immunoassays, imply that TGF-α may be the peptide detected.

Unlike EGF, TGF-α mRNA has been detected in embryos as early as at the pre-implantation stage. TGF-α mRNA is present in the unfertilized oocytes of mice (maternal transcripts), but disappears after fertilization and appears again at the four-cell stage. This expression increases and peaks at about 9 days. However, subsequent studies failed to demonstrate abundant TGF-α mRNA in fetal tissues except at a low level in localized areas such as branchial arch derivatives, developing mesonephric tubules, and placenta between 9 and 10 days of gestation. The predominant expression of TGF-α mRNA during development occurs in the maternal decidua, and the conceptus is not necessary for induction of the TGF-α gene. Taken together, these findings indicate that: (1) EGF is not synthesized in developing embryos or fetuses at any stage; (2) TGF-α is synthesized early but only at discrete sites; (3) EGF- and TGF-α-like bioactivity detected in developing fetuses has to be transported from extra-embryonic sources. The transplacental passage from the mother remains unresolved.

During fetal life, EGF receptor mRNA is expressed in many tissues including placenta, but detailed studies have not yet been performed. Functional EGF receptors have been localized to the human fetal small intestine and colonic epithelium at as early as 12 weeks of gestation, but in the mouse they are present only postnatally.

Biological actions during development

One of the major cellular events that occurs after binding of EGF to its receptor is the induction of expression of either growth regulatory genes or non-growth regulatory genes. The former control cellular proliferation or are regulated by growth status. The EGF-regulated genes that participate in cellular proliferation are activated rapidly (within 30–60 min) and are independent of protein synthesis. These include the glucose transporter gene, the Knox-24 gene encoding the zinc finger putative transcription factor (a transcriptional factor with an intrinsic Zn atom that interconnects the cysteine residues to form a finger-like structure) and the nuclear antigen cyclin. Amongst a wide variety of genes induced by EGF are c-fos and c-myc genes, which encode nuclear proteins involved in nuclear transcription, and the c-jun gene, which encodes an AP-1-like transcription factor. The non-growth regulatory genes encode for ornithine decarboxylase, an ADP/ATP carrier, and fibronectin. In addition, EGF regulates a wide variety of physiological functions of various tissues unrelated to growth, such as stimulation of prolactin production in the pituitary, increasing prostaglandin E_2 production by amnion, and decreasing lung liquid secretion.

EGF stimulates DNA synthesis in some fetal tissues *in vitro* and stimulates differentiation in a wide variety of developing tissues *in vivo*. The induction of premature eyelid opening and incisor eruption in new-born mice injected with EGF are well established effects of EGF on differentiation. EGF also influences: (1) craniofacial development (synthesis of palatal mesenchymal synthesis of basement membrane proteins; inhibition of the normal breakdown of the medial epithelium which is a pro-grammed cell death); (2) development of the gastrointestinal tract (induction of brush border enzyme activities and mucosal growth); (3) brain development (stimulation of DNA synthesis of astrocytes); and (4) skin development (hypertrophy of sebaceous and sweat glands, thickening of the epidermis).

Placenta and fetal membranes contain a large number of high-affinity EGF receptors suggesting that EGF may regulate fetal growth mainly by its influence on placental development. The presence of EGF receptors in the placenta is dependent on gestational age. Specific EGF binding sites have been found in placenta at as early as 4 weeks and the numbers increase as gestation progresses; EGF receptor mRNA is concurrently increased. In human amniotic fluid, EGF concentrations are also dependent on gestational age (i.e. higher near term than at mid-gestation) but are not correlated to placental weight, pulmonary maturity, or maternal EGF concentrations. The source of this amniotic fluid EGF is unknown.

Fibroblast growth factors (FGFs)

Peptides

FGFs are a family of structurally related heparin-binding growth factors that have the capacity to stimulate fibroblast proliferation. They have a wide range of biological activities including angiogenesis, and were first partially purified from brain and pituitary extracts. There are two distinct molecules with different isoelectric points, one with a high pI mainly from the pituitary (basic FGF) and one with a low pI from the brain (acidic FGF).

Chemistry

Basic FGF (bFGF) is a single-chain polypeptide of 146 amino acids with a molecular weight of 16.5 kDa. Variants with longer amino terminal extensions of 155 amino acid and 157 amino acids have also been described. Acidic FGF (aFGF) is a 140-amino-acid protein with a molecular weight of 15.5 kDa. Mature FGF is a fragment of a larger precursor known as endothelial cell growth factor (ECGF). Basic and acidic FGFs are about 55 per cent homologous. The three-dimensional structure of human bFGF, as determined by X-ray crystallography, resembles that of interleukin-1 and a large group of cytokines.

FGFs are now recognized as belonging to a family of at least five structurally related proteins, consisting, in addition to the two growth factors, of gene products of *hst/ks* oncogene, *int*-2 oncogene, and FGF-5. Two additional FGFs, FGF-6 and KGF/FGF-7, have been described recently.

The biological activities of the FGF family, with the exception of basic and acidic FGFs, are not known, and it is most likely that subtle differences in sequence amongst the family members may result in different actions. One important difference amongst family members is the presence or absence of the signal sequence. Basic and acidic FGFs, and *int*-2 proteins do not have any obvious signal sequence, whereas the others, the *hst/ks* protein and FGF-5 protein, do. The larger-molecular-weight forms of bFGF and *int*-2 are translated from CUG codons upstream from the normal AUG initiation codon, and these forms are preferentially translocated to the nucleus of some cells.

cDNAs and precursor molecules

The cDNA structures of FGF from many species are highly homologous. In humans the genomic structure for FGF family proteins is similar, consisting of three exons and two large introns, suggesting that they evolved from the same ancestral gene. The bFGF gene is located on chromosome 5 and that of aFGF on chromosome 4.

Northern blot analysis revealed basic FGF to be encoded by three transcripts (7.0, 3.7, and 1.4 kb), and acidic FGF by a single 4.8-kb mRNA. However, the signal sequence in the two proteins, the mechanism by which FGF can be detected outside the cell, and whether they have an extracellular receptor or whether they are associated with the extracellular matrix are still unknown. In addition, the mRNA levels as detected by Northern analysis are extremely low in most tissues even though the peptide concentrations are relatively high, suggesting that FGF mRNA is either relatively unstable or that technical problems still exist in the detection of the mRNAs.

Gene expression

Many of the cells that respond to FGFs also synthesize these growth factors. They include, fibroblasts, vascular and capillary endothelial cells, smooth muscle cells, granulosa cells, adrenocortical cells, and astrocytes in culture. The primary precursor products of basic and acidic FGFs are 155-amino-acid proteins, which are then processed into 154-residue mature non-glycosylated polypeptides by the modification of the amino terminus.

Large amounts of basic FGF can be purified from pituitary extracts and in small amounts from many other tissues. Acidic FGF is present in high concentrations in neural tissues (brain and retina) and to a smaller extent in the kidney, myocardium, and bone. The presence of FGF in serum and tissue fluids is still controversial. The detection of FGF-like material in rat serum and vitreous and synovial fluid suggests a possible function in these tissues. It is also possible that they may be inactive forms.

FGF receptor

The actions of FGFs are mediated by specific FGF receptors, which are single-chain polypeptides with molecular weights ranging from 110 to 150 kDa, depending on the cell type. They possess tyrosine kinase activity and bind FGF with high affinity (K_D = 10–80 pM). The number of receptors in each cell ranges from 2 000 to 80 000.

In most cells, FGF receptor binds basic FGF with higher affinity than acidic FGF. Two binding sites with different affinities have been described, but only the high-affinity sites are capable of transmitting the FGF signal to the cell.

The low-affinity site is sensitive to heparinase treatment and its binding activity can be blocked by exogenous heparin. The basic FGF receptor mRNA (*flg*) encodes for a protein with a molecular weight of approximately 100 kDa. Both human and mouse *flg* cDNAs encode receptors recognizing bFGF, aFGF, and K-FGF (Karposi sarcoma FGF) and are now referred to as the FGFR1 group. A second receptor group, called FGFR2, includes the chicken peptides *bck* and *cck3* as well as a variant receptor with high binding affinity for keratinocyte growth factor (KGF). Two additional FGF receptor genes were isolated from human leukaemia cell lines and designated FGFR3 and FGFR4. All of the FGFR classes bind both bFGF and aFGF, except for FGFR4 which does not bind bFGF. It seems certain that other FGFR types remain to be identified that have high specificity for other ligands in this growth factor class.

Following binding of FGF to its receptor, *fos* and *myc* proto-oncogenes are rapidly induced. Several signal transduction mechanisms are involved. These include: (1) the cAMP or cGMP pathway; (2) the phospholipase pathway with production of diacylglycerol and activation of protein kinase-C and influx of calcium; and (3) activation of FGF-receptor-associated tyrosine kinase. In addition, FGF binding to its receptor is associated with increased protein phosphorylation, specifically of a unique 90-kDa protein and the S6 ribosomal protein. Following receptor binding, FGF is internalized and slowly degraded within the cell. In certain cells, FGF has been shown to be translocated into the nucleus and to accumulate within the nucleolus. As FGFs are not secreted like other growth factors, it is possible that the synthesized FGF may also have direct effects inside the cell.

Biological activities

bFGF is 10–1000-fold more potent than aFGF in most cells, reflecting the higher affinity of bFGF for the receptor. Rarely, some cells may respond only to one type of FGF and, in some, aFGF may be equipotent.

FGFs are mitogens for many cells. Initiation of the mitogenic cascade by FGF includes reversible changes of cell morphology, stimulation of cell transport systems, polyribosome formation, RNA stabilization and synthesis, protein synthesis, and inhibition of protein degradation. Enzymes, such as ornithine decarboxylase, that are required for the cell to proceed through the process of division are also induced.

One of the characteristics that distinguishes FGF from many other growth factors is their relatively high affinity for heparin. FGFs are protected from inactivation by enzymes, high temperature, or low pH by the addition of heparin or related highly sulfated glycosaminoglycans (GAGs) and cell-derived heparan sulfate. This suggests that FGFs can be localized, bound to heparan sulfate on the cell surface, and locally available for the target cell. In addition, these interactions can protect the growth factors from the proteolytic degradation that characterizes tissue remodelling and neovascularization, thus allowing the growth factor to operate in a metabolically 'hostile' environment. Interaction of bFGF with a glycosaminoglycan, such as heparin or its cell-surface-associated variant, heparan sulfate, is obligatory before the growth factor will bind to its high-affinity receptors.

FGFs were originally discovered as mitogens for fibroblasts, but their actions are more widespread on many cell types of mesenchymal and neuroectodermal origin. In addition to their mitogenic actions, FGFs possess activities such as chemotaxis, induction or suppression of cell-specific protein synthesis or secretion, differentiation, and modulation of endocrine functions. One of the most important functions of FGFs is stimulation of angiogenesis. This putative role of FGFs may be important in both physiological and pathological conditions. Examples of the former include embryo implantation and embryogenesis, ovulation and capillary network formation, and vascularization during development of various organs such as brain, kidney, and adrenal cortex. Pathological regulation of FGFs in angiogenesis has been most extensively studied in tumour angiogenesis, in neovascularization following ischaemia, and in wound healing and tissue repair.

FGFs and FGF receptors in development

One of the most important functions of FGF during embryogenesis may be its role in the induction of mesoderm. FGF induces the appearance of embryonic muscle actin mRNA in animal cap cells, this effect being potentiated by TGF-β. The significance of this finding is supported by the fact that a cDNA encoding a protein homologous to bFGF could be isolated from *Xenopus* oocyte cDNA. By inhibiting FGF action via expression of a dominant negative mutant form of the FGF receptor it has been demonstrated that explants of *Xenopus* embryos expressing this mutation failed to induce mesoderm in response to FGF; the resulting phenotypic changes include defects in gastrulation and posterior development. These defects could be 'rescued' by overexpressing the wild-type normal receptors. These experiments demonstrate the crucial role of FGF signalling pathways in early embryogenesis. Other direct evidence exists of a role for FGFs in morphogenesis. Antisense oligonucleotides complementary to *int*-2 mRNA or antibodies against the *int*-2 peptide block the development of the chick otic vesicle.

FGF activity has been demonstrated in embryonic carcinoma-cell lines and extracts from whole chick embryos, developing limb buds, and unfertilized eggs. However, the failure of FGF mRNA to be detected in early pre-implantation mouse embryos suggests that synthesis of FGF occurs later in embryogenesis, perhaps in early post-implantation as it has been implicated in mesoderm induction. Demonstration of the ontogeny of expressions of RNAs encoding aFGF, bFGF, K-FGF, and FGF-5 in the mouse embryo supports this assumption. Acidic FGF was expressed in both undifferentiated and differentiated teratocarcinoma cell lines, and continued to be expressed in the embryo between 10.5 and 15.5 days of gestation. In contrast, KFGF mRNA was expressed in abundance in undifferentiated teratocarcinoma cells, and FGF-5 in the differentiated counterparts, but both mRNAs were extremely low in abundance in older embryos. Basic FGF mRNA appears in the embryonic period. FGF-related int-2 proto-oncogene mRNAs have been identified in the developing cerebellum, inner ear, and the mesenchyme of the developing teeth of the mouse embryo. In the 18-day rat fetus, FGF immunoreactivity has been localized to the basement membranes of epithelia, cartilage and bone, and skeletal and smooth muscle cells. This implies that different members of the FGF family have distinct roles at specific times of development, and this idea is supported by studies of FGF receptor expression. Both FGFR1 and -2 mRNAs are barely detected in the primitive ectoderm of the egg-cylinder stage of mouse embryo, but later are widely distributed in mesoderm- and neuroectoderm-derived mouse and human fetal tissues. Generally, FGFR2 transcripts are found in epithelial structures while FGFR1 is expressed in mesenchyme. Within the nervous system, FGFR1 expression is developmentally regulated and alternates with expression of the nerve growth factor receptor. FGFR3 mRNA has been localized to several regions of fetal and adult brain as well as fetal kidney, lung, and intestine. FGFR4 is not expressed in the embryonic central nervous system (CNS) but is found in endodermal and mesenchymal human fetal tissues such as the adrenal gland, lung, kidney, and pancreas. FGFR4 mRNA is present in pre-myogenic areas that will form skeletal muscle. FGF-5 mRNA has also been localized to the developing muscle. FGFR5 may therefore be important for the embryonic differentiations of skeletal muscle and endodermally derived organs.

Basic FGF, in particular, seems closely related to skeletal development in the embryo. Both bFGF mRNA and peptide are present during limb-bud formation in the chick and mouse, and levels are greatest within the limb when the rate of cell proliferation is highest. During craniofacial development of the chick embryo, bFGF potentiates mitogenesis and cartilage formations of the frontonasal mass structure, but not that of the other two facial primordia, the moscilla and mandible.

Although their precise roles have not been identified, FGFs have been implicated in endothelial cell proliferation and migration during angiogenesis. Capillary endothelial cells not only synthesize, but also proliferate in response to FGF. In addition to angiogenesis, both FGFs may be involved in brain development, as they induce neurotrophic factors and promote the survival and differentiation of neurones in culture and stimulate the production of IGF binding proteins by astroglia. Both FGFs also repress skeletal muscle differentiation by preventing the acquisition of a postmitotic phenotype, repressing transcriptional activation of skeletal-muscle-specific genes, and preventing formation of myotubes.

Platelet derived growth factor (PDGF)

PDGF is the major mitogen for connective tissue cells and may be important in wound healing and tissue fibrosis. It is produced by platelets, transformed cells, arterial endothelial and smooth muscle cells, activated macrophages, and by stimulated fibroblasts and trophoblasts.

Peptide

Chemistry

PDGF can be assembled in at least three isoforms — PDGF-AA, PDGF-BB, and PDGF-AB. Each of these binds to a different receptor phenotype which vary in their tissue distribution, suggesting that the isoforms may have tissue-specific functions.

PDGF from platelets has multiple molecular weight forms ranging from 28 to 35 kDa, each containing 16 cysteine residues. Active PDGF is composed of two distinct but homologous polypeptide chains (A and B) linked by disulfide bonds. The major part of PDGF purified from human platelets occurs as a heterodimer of the A and B chains, 70 per cent being PDGF-AB heterodimer and the remainder being PDGF-BB homodimer. Other possible dimeric forms occur in smaller amounts.

cDNA and precursor molecules

cDNA cloning of A and B chains revealed a 50 per cent homology and the conservation of all eight cysteine residues within the mature peptides, indicating a similar tertiary structure and a common ancestral gene. The maintenance of this tertiary structure by intrachain and/or interchain disulfide bonds is necessary for the biological activity of PDGF. The mature B chain of 109 amino acids is also homologous with the oncogene product of simian sarcoma virus p28^{sis}.

Both A and B chains of PDGF are synthesized as precursors. In addition to the hydrophobic signal sequence in the N-terminus, both A and B chains have a stretch of 12 hydrophobic amino acids, 28 and 34 residues, respectively, from the N-terminus of the mature A and B chains. The mature PDGF is glycosylated (N-linked) possibly at the Asn–Thr–Ser sequence at residue 48 of the mature A chain and there is possibly O-linked glycosylation of the B chain.

A significant difference between the A and B chains is that the mature B chain contains no tyrosine and therefore only the A chain can be iodinated. Cysteines, particularly at positions 16, 49, 60, and 97 are required for maintenance of the transforming activity and interaction with the PDGF receptor.

cDNA analysis of PDGF-like molecules reveals a highly conserved structure from *Xenopus* to mammals, with some species differences in the relative amount of the homo/heterodimers. For example, PDGF-BB is the predominant form in pig and in many other mammals except primates. A functional homologue of PDGF and its receptor may be present in more primitive animals.

Gene structure

A-chain. The PDGF-A chain gene consists of at least seven exons spanning 22 to 24 kb of genomic DNA, with the mature peptide being encoded mostly by exons 4 and 5. This gene is located on the long arm of chromosome 7. Northern analysis reveals three PDGF-A chain transcripts of 1.9, 2.3, and 2.8 kb from human tumour cell lines and normal endothelial cells in culture. An unusual feature of the A-chain mRNA is the presence of a long 5′ untranslated region containing three AUG triplets upstream of the authentic initiation codon, and each followed by a stop codon.

The reason for PDGF-A chain mRNA heterogeneity is not completely known. Analysis of the PDGF gene indicates that the existence of multiple transcriptional start sites is an unlikely explanation. Alternate use of polyadenylation sites or alternate splicing of exons (other than exon 6) are the more likely possibilities.

B chain. The B chain gene consists of seven spanning 24 kb of genomic DNA. The first exon contains the long 5′ untranslated region and the signal sequence. Most of the mature peptide coding regions are contained within exons 4 and 5. The B chain is encoded by the c-*sis* proto-oncogene, the cellular homologue of the oncogene transduced by both the simian sarcoma virus and feline sarcoma virus. The gene is located in the human chromosome 22.

PDGF receptors

Using recombinant PDGF peptides, it is now established that there are at least two classes of PDGF receptors — an abundant form that binds only PDGF-BB (B receptor) and a much less abundant receptor form that binds all three isoforms (A/B receptor). The high-affinity PDGF receptor is a dimer recruited from separate pools of 'a' subunits (which bind A and B chains), or 'b' subunits (which bind only B chain). In the absence of PDGF, the subunits exist as separate monomers or as unstable dimers. Two receptor subunits together can form a high-affinity complex with PDGF, presumably with one receptor subunit binding each of the chains of the dimeric ligand. No other growth factor or peptide tested to date is able to bind to the PDGF receptor with high affinity. Only basic proteins, such as histones, poly-lysine, and protamine, and the drug Suramin inhibit PDGF binding.

cDNA

The 'b' subunit of PDGF receptor cDNA from mouse was cloned from placental tissue, and the human cDNA from fibroblasts. Human PDGF mRNA consists of an abundant 5.7-kb form, and a less abundant 4.8-kb form. The gene is located on the long arm of the human chromosome 5. The 'a' subunit cDNA encodes a mRNA of 6.4 kb. The gene is located on the long arm of chromosome 4. Cross-linking studies showed PDGF receptor complex to be about 190 kDa (receptor = 160 kDa).

PDGF and PDGF receptor in development

In pre-implantation mouse embryos, PDGF-A chain mRNA is detectable in unfertilized oocytes (via maternal gene expression), lost following fertilization, and becomes detectable again after the four-cell stage. PDGF peptide is also detectable by immunocytochemistry, suggesting that the mRNAs detected by reverse transcription–polymerase chain reaction (RT-PCR) are translated into proteins.

PDGF mRNA is expressed in multiple cell types of epithelia, connective tissue, muscle, neural, and haemopoietic origin and the expression is tightly regulated. Endothelial cells from bovine aorta or human umbilical veins contain only 10 and 1.3 per cent, respectively, as much B-chain mRNA as do the same cells grown in culture. Once they organize into tube-like structures in culture, similar to the *in vivo* state, the mRNA decreases. This indicates that PDGF gene expression is induced by perturbations associated with tissue culture. The expression of PDGF gene and secretion into the media is constitutive.

The level of gene expression is dependent of the age of the cells when isolated *in vivo*, i.e. younger animals secrete more PDGF than older animals.

PDGF activity, in promoting cellular proliferation and migration, is a suitable growth factor for regulating embryogenesis and development. To date, most studies involve investigating the activity of PDGF in the cytotrophoblasts of placenta, aortic smooth muscle cells from growing rats, embryonic carcinoma cells, and astrocytes. The time course of appearance of PDGF-A chain mRNA in the brain coincides with gliagenesis during brain development.

In situ hybridization studies show that PDGF mRNA is localized in highly proliferative and invasive trophoblasts, and not in the differentiated syncitiotrophoblasts. These cytotrophoblasts also express c-*myc* mRNA. The PDGF-B chain mRNA decreases approximately 10-fold between early first trimester and term. More detailed studies have shown that PDGF-A and -B chains, as well as PDGF receptor 'b' subunit mRNA, are present throughout pregnancy, the abundance being greatest in the second trimester. Placental c-*fos* mRNA, a gene rapidly induced after interaction of PDGF and its receptor, showed a similar temporal increase. At all gestational ages, PDGF-A chain mRNA is more prevalent than B chain mRNA.

Biologically active PDGF has been purified from term human placentae. An ontogenic study of mRNAs encoding PDGF and its receptor has shown that both mRNAs are expressed in a co-ordinated fashion throughout pregnancy, especially in mid-gestation when both PDGF and its receptor are very abundant. PDGF receptors are present in the term placenta as demonstrated by binding studies as well as by the capacity of term cytotrophoblasts to transcribe the PDGF receptor gene. However, the major part of placental PDGF receptor is localized in the endothelial cells, suggesting a paracrine interaction between trophoblasts and the endothelium.

Evidence is now accumulating for a role of PDGF in embryonic morphogenesis. The Ph/+ mutant mouse (a bacteriophage mutant) is viable in the heterozygous condition, but in the homozygous state gives rise to grossly malformed fetuses consistent with a failure of neural crest cell migration. Such embryos can exhibit an open neural tube, clubbed limbs, a lack of thymus, no dermal layer to the skin, lack of connective tissues within the organs, and a failure of craniofacial development. The condition has been associated with a deletion of the PDGF 'a' receptor, suggesting an important role for PDGF in neural crest migration. Both PDGF A peptide and PDGF 'a' receptor are known to be highly expressed in the neural crest areas of the *Xenopus* embryo.

Transforming growth factor-βs (TGF-βs)

Peptides

Chemistry

TGF-βs are a group of related peptides that belong to the TGF-β supergene family, which includes, in addition to TGF-βs, the inhibins and activins, Müllerian inhibitory substance, the *Drosophila* decapentaplegic gene complex, the amphibian gene *Vgl*, and bone morphogenetic proteins. All members of this supergene family are encoded as larger precursors and have seven conserved cysteine residues. Common to all of these peptides is their ability to regulate developmental processes.

TGF-βs are isolated from a wide variety of transformed and non-transformed cells and five members (TGF-β_1 to TGF-β_5) have been identified. Only TGF-β_1 and TGF-β_2 have been isolated and purified from natural sources. The other three (the novel TGF-βs), have been isolated by screening cDNA libraries and by demonstration of mRNAs in tissues. TGF-β_1 was originally isolated from human platelets as a 25 000 M_r peptide and TGF-β_2 from a variety of sources including platelets, bone, glioblastoma cells and placenta. The relative amounts of TGF-β_1 and TGF-β_2 present in tissues or cells are dependent on the cell type as well as the stage of development. TGF-β_3 has been identified in cDNA derived from placenta, ovary, glioblastoma, and chondrocytes. Interestingly, it is the most abundant mRNA expressed by chick embryos. TGF-β_4 and -β_5 cDNAs have been cloned from chicken chondrocyte and frog oocyte libraries.

All members of the TGF-β family share significant sequence homology (60–80 per cent amongst the members) and characteristics which include: (1) translation as larger precursors that share a region of high homology near the N-terminus and show conservation of three cysteine residues in the portion of the precursor that is subsequently processed; (2) sites for N-linked glycosylation; (3) the presence of a cellular recognition site for fibronectin/vitronectin Arg–Gly–Asp (except TGF-β_2); (4) conservation of nine cysteine residues in the processed peptide; (5) conservation of the C-terminal sequence Cys–Lys–Cys–Ser–COOH, with the exception of TGF-β_5 in which asparagine residue replaces the lysine; and (6) similar biological activities. The unique feature of the TGF-β_4 precursor is the lack of signal peptide sequences suggestive of a unique biological activity of functioning within the cell. The existence of many different forms of TGF-βs with similar biological activities may be the result of differential regulation under the control of unique promoters and the capacity to be synthesized by many different cells. In addition, proteolytic processing of the TGF-β precursors may be one additional step by which the synthesis of the multiple forms of TGF-β may be regulated.

Gene structure

The human TGF-β_1 gene is encoded by seven exons, the structure of which is conserved for the most part in the TGF-β_2 and TGF-β_3 genes, suggesting that the various TGF-β genes originate from duplication of the common ancestral gene. Although the mature TGF-β mRNA contains approximately 1200 nucleotides of coding sequence, the mRNA species range from 1.7 to 6.5 kb with extensions at both 5' and 3' ends. TGF-β_1 is encoded by a single mRNA of 2.5 kb, whereas TGF-β_2 is encoded by three transcripts of 4.1, 5.1, and 6.5 kb. In humans the TGF-β_1 gene is located in the long arm of chromosome 19, the TGF-β_2 gene in the long arm of chromosome 1, and the TGF-β_3 gene in the long arm of chromosome 14. In the mouse, the TGF-β_1 to -β_3 genes are located in chromosomes 7, 1, and 12, respectively.

Human TGF-β_1 gene consists of two major transcriptional start sites and two promoter regions. Within the upstream promoter, two distinct negative regulatory regions, an enhancer-like region, and a positive regulatory region have been identified. The positive regulatory region contains several binding sites for known transcriptional factors including nuclear factor 1 (NF-1), SP1, and AP1.

TGF-β receptors

Three classes of high-affinity receptors have been identified. Class I receptor is approximately 65 kDa in all species, whereas class II receptors range from 85 kDa in rodents to 95 kDa in humans. Class I and II receptors are glycoproteins, whereas the class III receptor is a proteoglycan consisting predominantly of heparan sulfate glycosaminoglycan chains with a smaller amount of chondroitin or dermatan sulfate attached to a core protein of approximately 110–140 kDa that contains the binding site. TGF-β_1 binds with higher affinity than TGF-β_2 to both classes of receptors, whereas both forms bind to class III receptor with equal affinity. TGF-β receptors do not contain an integral tyrosine kinase activity.

TGF-β and receptors in development

TGF-β is secreted from cells in a biologically inactive or latent form that requires activation by chemical, physical, or enzymatic means before it can bind to it receptor. As most cells contain TGF-β receptors, this step of activation

may control its biological action. The latent form of the peptide consists of the mature TGF-β dimer, a modulator protein (125–160 kDa), and a latency protein (75-kDa glycoprotein). In serum, an additional latent form consisting of a complex with α_2-macroglobulin may be present.

The biological actions of TGF-β depend on many cellular parameters including the type of the cell, state of differentiation, growth conditions, and the presence or absence of other growth factors. One of the most important biological features of TGF-β is its association with and control of the extracellular matrix (ECM). In vitro studies suggest that the latter action involves a complex interaction between the formation and degradation of the ECM. This includes: (1) activation of gene transcription and increased synthesis of matrix proteins (collagen, fibronectin); (2) a decrease in the synthesis of proteolytic enzymes that degrade matrix proteins (serine, thiol, and metalloproteases) and an increase in the synthesis of inhibitors of these proteases (plasminogen activator inhibitor (PAI), tissue inhibitor of metalloproteinases (TIMP)); and (3) an increase in the synthesis and processing of integrins. The cellular receptor of matrix proteins in the musculoskeletal system is one of the major sites of action of TGF-β. It inhibits myoblast differentiation and may play a role in muscle regeneration by prevention of precocious fusion of embryonic myoblasts.

TGF-β has been extracted from whole mouse fetuses and term placentae, and their genes are expressed in a wide variety of tissues during embryogenesis. TGF-β mRNA has not been detected in mouse oocytes but appears as a product of the embryonic genome. Following fertilization TGF-β is detectable as both the mRNA and peptide in the blastula. Messenger RNA encoding the related gene, Vg-1, is present in Xenopus during blastulation as a maternally-derived transcript.

TGF-β_1 and β_3 are widely expressed in tissues of the mouse and human embryo. In the 11–18 day gestation mouse embryo, TGF-β_1 peptide is associated predominantly with mesenchyme such as connective tissues, cartilage, and bone, and with tissues derived from neural crest mesenchyme such as palate, larynx, nasal sinuses, meninges, and teeth. Staining for TGF-β_1 was most intense during morphogenesis, and at points of critical epithelium–mesenchyme interaction as in hair follicles, teeth, and salivary glands. TGF-β_1 was also associated with tissue remodelling in the heart valves, digit formation, and palatal development. When the distribution of TGF-β mRNA was assessed by in situ hybridization, abundant expression was seen in embryonic bone and the megakaryocytes of liver, and in a number of epithelial cell types at sites of morphogenesis. These epithelial cells included thymus and thyroid, hair follicles, and tooth buds, and the sites of TGF-β_1 mRNA expression were often adjacent to mesenchymal sites of TGF-β_1 peptide localization. This suggests a sequestration of the growth factor, perhaps in basement membranes and on the cell surfaces of mesenchymal cells. TGF-β_2 and β_3 mRNAs are also widely expressed during periods and on sites of maximal expression that are different for each isoform. This is particularly apparent in the skeleton where TGF-β_1 mRNA is abundant in perichondral osteocytes involved with membraneous calcification. However, differentiated cartilage contains little TGF-β_1 mRNA, although this is a site of peptide accumulation. In contrast, the expression of TGF-β_3 mRNA is abundant in the intervertebral discs of the spine, and in perichondrial tissue. During the development of the secondary palate in the mouse embryo, TGF-β_3 mRNA is localized to the epithelial component of the vertical palatal shelf during the initial stages of palatal growth. Later, as the horizontal shelf develops, TGF-β_1 mRNA predominates, and is lost as the epithelium disrupts and fusion of the shelves occurs. TGF-β_2 mRNA is localized to the palatal mesenchyme under the medial edge epithelia of the horizontal palatal shelves. Given the profound effects of TGF-β isoforms on the proliferation of epithelial cells and the production of extracellular matrix, these findings imply that different TGF-β isoforms have distinct roles in secondary palate formation.

The putative role of the TGF-β superfamily in prenatal skeletal development is supported by experimentation. TGF-β_1 and -β_2 are analogous to factors, previously known as cartilage-inducing factors -A and -B, with the ability to induce the appearance of phenotypic chondrocytes from embryonic mesenchymal explants. Other members of the TGF-β family, namely bone morphometric proteins -2B and -3 (osteogenin), have similar actions.

TGF-β may also play an important role in the development and modelling of the placenta as it was first purified from the placenta. The human TGF-β cDNA was isolated from a placental cDNA library and the mRNA was identified in the placenta. One of the more intriguing functions of TGF-β in the placenta is its role in feto-maternal immunity and regulation of the invasion of fetal trophoblasts into maternal tissues. The biological actions of TGF-β that may be important in this function include: (1) suppression of lymphokine-activated killer cell activity; (2) inhibition of interleukin-1-dependent T and β cell proliferation; and (3) decreased cytotoxity associated with macrophage activation.

Conclusions

The expression of genes encoding different growth factors, their binding proteins, and receptors by a variety of developing embryonic and fetal cells at specific times of embryogenesis and fetal development, the diversity of growth factors and their capacity to interact with each other, the multifunctional nature of the biological actions of each growth factor, and the intimate relationship with other molecules of significance in embryogenesis (the extracellular matrix and intercellular recognition molecules) indicate that growth factors play a major role in the paracrine regulation of embryogenesis and fetal development. These observations also highlight the great complexity by which the gene expression of growth factors is regulated and their physiological role is enacted. Unravelling these complex interactions and correlating molecular and physiological events are just two of the exciting fundamental issues facing researchers.

Further reading

Amaya, E., Misci, T.I. and Kirschner, M.W. (1991). Expression of a dominant negative mutant of the FGF receptor disrupts mesoderm formation in Xenopus embryos. *Cell*, **66**, 257–70.

Baird, A. and Bohlen, P. (1991). Fibroblast growth factors in *Peptide growth factors and their receptors I* (ed. M.B. Sporn and A.B. Roberts, pp. 369–418. Springer-Verlag, New York.

Bell, G.I. Fong, N.M., Stempie, N.M., Wormsted, M.A., Caput, D., Ku, L., Urdea, M.S., Rall, L.B., and Sanchez-Pescador, R. (1986). Human epidermal growth factor precursor: cDNA sequence, expression in vitro and gene organization. *Nucleic Acids Research*, **14**, 8427–46.

Cohen, S. (1962). Isolation of a mouse submaxillary gland protein accelerating incisor eruption and eyelid opening in the newborn animal. *Journal of Biological Chemistry*, **237**, 1555–62.

Daughaday, W.H. and Rotwein, P. (1989). Insulin-like growth factors I and II. Peptide, messenger ribonucleic acid and gene structures, serum and tissue concentrations. *Endocrine Reviews*, **10**, 68–91.

DeChiara, T.M., Efstratiadis, A., and Robertson, E.J. (1990). A growth deficiency phenotype in heterozygous mice carrying an insulin-like growth factor II gene disrupted by targeting. *Nature*, **345**, 78–80.

Derynck, R., Rhee, L., Chen. E.Y., and Van Tilburg, A. (1987). Intron–exon structure of human transforming growth factor-α precursor gene. *Nucleic Acids Research*, **15**, 3188.

Derynk, R., Roberts, A.B., Winkler, M.E., Chen, E.Y., and Goeddel, D.V. (1984). Human transforming growth factor-α: precursor structure and expression in E. coli. *Cell*, **38**, 287–97.

Gospodarowicz, D. (1974). Localization of a fibroblast growth factor and its effect alone and with hydrocortisone on 3T3 cell growth. *Nature*, **249**, 123.

Han, V.K.M. and Hill, D.J. (1992). The involvement of insulin-like growth factors in embryonic and fetal development. In *The insulin-like growth factors, structure and biological functions* (ed. P.N. Schofield), pp. 178–220. Oxford Medical Publications, Oxford.

Han, V.K.M., Lu, F., Nassett, N., Yang, K.P., Delhanty, P.J.D., and Challis, J.R.G. (1992). Insulin-like growth factor-II (IGF-II) messenger ribonucleic acid is expressed in steroidogenic cells of the developing ovine adrenal gland: Evidence of an autocrine/paracrine role for IGF-II. *Endocrinology*, **131**, 3100–19.

Han, V.K.M., Lund, P.K., and D'Ercole, A.J. (1987). Cellular localization of somatomedin (insulin-like growth factor) messenger RNA in the human fetus. *Science*, **236**, 193–7.

Han, V.K.M., Lund, P.K., Lee, D.C. and D'Ercole, A.J. (1988). Expression of somatomedin/insulin-like growth factor messenger ribonucleic acids in the human fetus: Identification, characterization and tissue distribution. *Journal of Clinical Endocrinology and Metabolism*, **66**, 422–9.

Hill, D.J., Strain, A.J., and Milner, R.D.G. (1987). Growth factors in embryogenesis. In *Oxford reviews of reproductive biology*, Vol. 9 (ed. J.R. Clarke) pp. 398–455. Clarendon Press, Oxford.

Hynes, R.O. (1922). Integrins: versatility, modulation, and signalling in cell adhesion. *Cell* **69**, 11–25.

Jansen, M., van Schaik, F.M., Ricker, A.T., Bullock, B., Woods, D.E., Gabbay, K.H., Nussbaum, A.L., Sussenbach, J.S., and Van den Brande, J.L. (1983). Sequence of a cDNA encoding human insulin-like growth factor I precursor. *Nature*, **306**, 609–11.

Lassare, C., Hardouin, S., Daffos, F., Forestier, F., Frankene, F., and Binoux, M. (1991). Serum insulin-like growth factors and insulin-like growth factor binding proteins in the human fetus. Relationships with growth in normal subjects and in subjects with intrauterine growth retardation. *Pediatric Research*, **29**, 219–25.

Lee, D.C. and Han, V.K.M. (1990). Expression of growth factors and their receptors in development. In, *Peptide growth factors and their receptors II, Handbook of experimental pharmacology*, Vol. 95 (ed. M.B. Sporn and A.B. Roberts), pp. 611–54. Springer-Verlag, New York.

Lee, D.C., Rose, T.M. Webb, N.R., and Todaro, G.J. (1985). Cloning and sequence analysis of a cDNA for rat transforming growth factor-α. *Nature*, **313**, 489–91.

Leutz, A. and Graf, T. (1990). Relationships between oncogenes and growth control. In *Peptide growth factors and their Receptors II, Handbook of experimental pharmacology*, Vol. 95 (ed. M.B. Sporn and A.B. Roberts), pp. 655–704. Springer-Verlag, New York.

Martin, G.R. and Sank, A.C. (1990). Extracellular matrices, cells and growth factors. In *Peptide growth factors and their receptors II, Handbook of experimental pharmacology*, Vol. 95 (ed. M.B. Sporn and A.B. Roberts), pp. 463–74. Springer-Verlag, New York.

McCusker, R.H. and Clemmons, D.R. (1992). The insulin-like growth factor binding proteins: structure and biological functions. In *The insulin-like growth factors, structure and biological functions* (ed. P.N. Schofield), pp. 110–50. Oxford Medical Publications, Oxford.

Moses, A.C., Nissley, S.P., Short, P.A., Rechler, M.M., White, R.M., Knight, A.B., and Higa, O.Z. (1980). Elevated levels of multiplication-stimulating activity, an insulin-like growth factor, in fetal rate serum. *Proceedings of the National Academy of Sciences of the United States of America*, **77**, 3649–53.

Moxham, C. and Jacobs, S. (1992). Insulin-like growth factor receptors. In *The insulin-like growth factors, structure and biological functions* (ed. P.N. Schofield), pp. 80–109. Oxford Medical Publications, Oxford.

Oshima, A., Nolan, C.M., Kyle, J.W., Grubb, J.H., and Sly, W.S. (1988). The human cation-independent mannose 6-phosphate receptor: cloning and sequence of the full-length cDNA and expression of functional receptor in cos cells. *Journal of Biological Chemistry*, **263**, 2553–62.

Polani, P.E. (1974). Chromosomal and other genetic influences on birth weight variation. In *Size at birth* (ed. K. Elliot and J. Knight), pp. 127–59. Elsevier–Excerpta Medica, North Holland, Amsterdam.

Rappolee, D.A., Brenner, C.A., Schultz, R., Mark, D, and Werb, Z. (1988). Developmental expression of PDGF, TGF-a, and TGF-b genes in preimplantation mouse embryos. *Science*, **241**, 1823–5.

Rappolee, D.A. and Werb, Z. (1991). Endogenous insulin-like growth factor II mediates growth in preimplantation mouse embryos. In *Modern concepts of insulin-like growth factors* (ed. E.M. Spencer), pp. 3–8. Elsevier, New York.

Rechler, M.M. and Nissley, S.P. (1990). Insulin-like growth factors. In *Peptide growth factors and their receptors I* (ed. M.B. Sporn and A.B. Roberts), pp. 263–368. Springer-Verlag, New York.

Rotwein, P., Pollock, D.M., Didier, D.K., and Krivi, G.G. (1986). Organization and sequence of the human insulin-like growth factor I gene: alternative RNA processing produces two insulin-like growth factor I precursor peptides. *Journal of Biological Chemistry*, **261**, 4828–32.

Shimasaki, S. and Ling, N. (1991). Identification and molecular characterization of insulin-like growth factor binding proteins (IGFBP-1, -2, -3, -4, -5, and -6). *Progress in Growth Factor Research*, **3**, 243–66.

Stiles, C.D., Capone, G.T., Scher, C.D., Antoniades, H.N., Van Wyk, J.J., and Pledger, W.J. (1979). Dual control of cell growth by somatomedins and platelet derived growth factor. *Proceedings of the National Academy of Sciences of the United States of America*, **76**, 1279–83.

Todaro, G.J. and De Larco, J.E. (1976). Transformation by murine and feline sarcoma viruses specifically blocks binding of epidermal growth factor to cells. *Nature*, **264**, 26–31.

Ullrich, A., Gray, A., Tam, A.W., Yang-Feng, T., Tsurbokawa, M., Colins, C., Henzel, W., Le Bon, T., Kathuria, S., and Chen, E. (1986). Insulin-like growth factor I receptor primary structure: comparison with insulin receptor suggests structural determinants that define functional specificity. *EMBO Journal*, **5**, 2503–12.

Van den Brande, J.L. (1992). Structure of the human insulin-like growth factors: relationship to function. In *The insulin-like growth factors, structure and biological functions* (ed. P.N. Schofield), pp. 12–44. Oxford University Press, New York.

Wakefield, L.M., Smith, D.M., Masui, T., Harris, C.C. and Sporn, M.B. (1987). Distribution and modulation of the cellular receptor for transforming growth factor-beta. *Journal of Cell Biology*, **105**, 965

6. Fetal metabolism and energy balance
Abigail L. Fowden

The nutrient requirements of the fetus differ from those of the adult or new-born animal. There is a relatively large energy requirement for intrauterine growth while little or no energy is expended by the fetus as compared with adult or new-born animals on processes such as movement, digestion, or temperature regulation. Nutrients taken up by the fetus are therefore either: (1) oxidized to provide energy for those organs and tissues that are metabolically active *in utero*; or (2) accumulated in the fetal carcass in the form of fuel reserves and new structural tissue. Hence, the total nutrient requirement of the fetus depends on its growth rate, body composition, and rate of oxidative metabolism.

If the fetus is to grow and develop normally, sufficient nutrients must be provided to meet the combined requirements of oxidation and tissue accretion. In normal conditions the mother is the main source of substrates for the fetus although some specific nutrients are provided by the placenta and the fetus itself. If these supplies of nutrients fail to match the total requirements, tissue accretion is compromised and the fetus will be growth-retarded as a consequence. There is therefore a fine balance between the supply and demand for nutrients *in utero* that governs many aspects of fetal development.

Experimental techniques

Fick principle

Measurements of the rates of uptake and utilization of oxygen and the different metabolic fuels by the fetus are required if fetal metabolic balance is to be evaluated quantitatively in any given species. In principle, estimation of these consumption rates depends on the law of conservation of mass (Fick principle) which can be expressed for any organ by the following equation.

Rate of substrate = Blood flow × Arteriovenous (AV)
uptake (or output)   concentration
                       difference in the
                       substrate

Precise measurements of oxygen and metabolite utilization by the fetus can therefore be made, provided that: (1) sam-
ples of umbilical venous and arterial blood can be obtained; (2) the blood concentration of substrate can be determined sufficiently accurately to allow calculation of reliable arteriovenous (AV) concentration differences across the umbilical circulation; and (3) umbilical blood flow can be measured.

These provisos mean that quantitative studies of fetal metabolism have been confined mainly to large domestic animals, notably the sheep, in which the fetus can be catheterized easily and studied in the conscious state. Studies on anaesthetized or small laboratory animals are not ideal as it is difficult to obtain undisturbed blood flows or normal substrate concentrations in these conditions. Similar criticisms can be applied to data obtained from restrained primates and from humans at the end of labour or during Caesarean section. Even in the larger, conscious animals care must be taken to ensure that they are unstressed, well nourished, and catheterized correctly, if the observations are to be representative of the normal fetus *in utero*.

Measurement of AV concentration differences across the umbilical circulation is another technical limitation to quantifying fetal metabolism and is the principal source of random error in determining metabolic fluxes by Fick principle. Some substances, e.g. oxygen and carbon dioxide, are highly labile and their concentrations may change during sampling itself. Other substances (e.g. urea, fructose) have such a small percentage change in concentration across the umbilical circulation (coefficient of extraction ≤1 per cent) that their uptake or excretion is difficult to calculate with the analytical techniques currently available. However, umbilical uptake of substances such as oxygen and glucose, which have extraction coefficients of about 40 and 10 per cent, respectively, can be readily quantified.

Umbilical and uterine blood flows can be measured in a variety of different ways. Flowmeters, which give a direct reading of blood flow, have been used on uterine vessels and on fetal vessels such as the femoral artery. They provide continuous recordings and are ideal for monitoring rapid changes in flow. However, they are attached to only one artery and may therefore underestimate total blood flow to the organ. Blood flows can also be measured using radioactively labelled microspheres that are trapped within

the capillary bed of the tissue in proportion to the fractional distribution of blood flow. This is the only technique
that allows calculation of blood flow to several individual
tissues and fetal organs simultaneously but has the disadvantage that flow measurements are limited to a few points
in time that may not coincide with the steady-state period
over which the substrate concentrations were measured.

All other methods of blood flow measurement are indirect and rely on modifications of the Fick principle.
Steady-state conditions are required and the measurements
are more time-consuming. Blood flow is measured with an
inert freely diffusible material (e.g. antipyrine, ^3H_2O,
ethanol) using either the diffusion equilibrium or steady-
state diffusion method. The latter method is used more
routinely as it requires less blood and can be combined
with administration of labelled substrates (see next section). Once equilibrium has been achieved with the steady-
state diffusion method (60–90 min), any number of flow
determinations can be made by measuring the umbilical
and uterine AV concentration differences in the test substance. Since the rate of infusion (Rf) is approximately
equal to the rate of loss across the placenta, blood flow =
Rf/(AV). A number of corrections can be made to the
numerator to allow for accumulation and metabolism of
the test substance but these do not normally alter the
values by more than 5–6 per cent.

Tracer methodology

Infusions of radioactively labelled substrates (e.g.
[^{14}C]glucose) have been used to quantify the flux of substrate carbon between the various pools and metabolic
pathways of the fetus and mother. The main assumption of
these methods is that, at steady state, the labelled and unlabelled substrates, and their products, are treated identically
by the transport mechanisms and biochemical pathways of
the cells. Generally, labelled substances are infused at a
constant rate into the fetus and/or mother until steady state
is obtained. The specific activity of the labelled substrate
in fetal and/or maternal blood is then measured and used to
calculate the turnover rate of the substrate. These measurements can be further refined to provide specific rates of
substrate utilization by using the Fick principle to establish
the amount of infused tracer lost across the placenta or
utero-placental tissues. For example, the rate at which substrate is used specifically by the fetal tissues can be calculated using the following equation

    Rate of fetal substrate utilization
    = [Rate of infusion of tracer into the fetus – Rate of loss
      of tracer from fetus to placenta] – Fetal arterial specific
      activity of the substrate.

Substrate utilization calculated in this way should be equal
to or higher than the umbilical uptake derived by the Fick
principle.

Measurements of fetal utilization rate made by tracer
methodology therefore require determination of: (1) the
umbilical blood flow; and (2) the concentration of labelled
and unlabelled substrate in both the umbilical venous and
arterial blood. One of the main sources of error with tracer
methods is the measurement of labelled substrate concentration. For example, since glucose can be converted to
lactate and fructose in the fetus (see sections on lactate and
fructose), labelled glucose levels will be overestimated if
the various carbohydrates are not separated before the
radioactivity is counted.

Metabolic quotients

The contribution made by the different fuels to oxidative
metabolism can be assessed by comparing the amounts of
oxygen and substrate consumed by the whole fetus or any
individual organ. Substrate: oxygen uptake ratios can be
determined without measuring blood flow provided that
the AV differences in oxygen and substrate are measured
in the same samples and expressed on a molar basis. The
contribution to total oxygen consumption made by aerobic
substrate metabolism in the fetus is therefore calculated by
the equation

$$\text{Substrate: } O_2 \text{ quotient } = \frac{k \times \Delta \text{ substrate}}{\Delta O_2}$$

where Δ substrate and ΔO_2 are the umbilical AV concentration differences measured simultaneously and k is the
number of moles of oxygen required for complete oxidation of 1 mole of the substrate. For example, this number
would be 6 for glucose and 3 for lactate. In assuming that
all the substrate taken up by the umbilical circulation is
oxidized, no allowance is made for the carbon incorporated into new fetal tissue. The sum of all the metabolic
quotients should therefore be greater than 1 in a growing
fetus since there will be both accretion and oxidation of
the substrates.

Fetal metabolites and their consumption

Oxygen consumption

Fetal arterial PO_2 values are lower than those in the
mother in all species studied so far. The magnitude of the
transplacental PO_2 gradient varies between species and
depends on a variety of factors including placental

vascular architecture, the rate of placental oxygen consumption, and the oxygen affinity of the fetal blood. In the sheep the drop in PO_2 across the placenta is 40–50 mmHg which results in a fetal arterial PO_2 of about 20–25 mmHg. However, these relatively low PO_2 values do not imply fetal hypoxia. An adequate oxygen supply to the fetal tissues is ensured by relatively high tissue perfusion and an oxygen affinity that is generally higher than that of the mother. The fetus can also increase the oxygen-carrying capacity of its blood and redistribute its circulation, so that the oxygen supply to essential tissues (brain, heart, placenta) is maintained even during adverse conditions.

Oxygen consumption by the fetus and uterus as a whole can be calculated by the Fick principle and has been shown to remain within a relatively narrow range despite wide fluctuations in blood flow. Oxygen uptake by the uterus is generally higher than that of the fetus because about 50 per cent of the oxygen leaving the uterine circulation is consumed by the utero-placental tissues before it reaches the fetus. Placental oxygen consumption therefore occurs at a higher rate than in the fetus, when values are expressed on a weight basis, and will significantly reduce the amount of oxygen available to the fetus.

Fetal oxygen consumption is about 300 μmol/min/kg and is remarkably uniform in the different species when expressed per kg fetal body weight (Table 6.1). It also varies very little with changes in nutritional state. Small increases (10–15 per cent) in fetal oxygen consumption have been observed when fetal glucose availability is increased by exogenous infusion of glucose or insulin into the sheep fetus. But there is no apparent change in fetal oxygen uptake when fetal glucose availability is restricted by fasting the ewe. Larger increases (20–30 per cent) in fetal oxygen consumption have been reported in response to fetal administration of noradrenaline and tri-iodothyronine but the doses of hormone used were supraphysiological.

Table 6.1 The mean umbilical uptake of oxygen and various substrates in fetuses of different species in fed mothers during late gestation (≥ 85 per cent of gestation)

| Species | Mean umbilical uptake (μmol/min/kg fetal body weight) | | | |
	Oxygen	Glucose	Lactate	Acetate
Sheep	315	25	22	17
Cow	300	30	50	25
Horse	315	45	17	—
Human*	350	45	10	—

*Data obtained from infants at elective Caesarean section.

Table 6.2 The absolute rates of substrate uptake or output by individual tissues of the sheep fetus in fed ewes near term and the proportion (per cent) of umbilical substrate uptake that can be accounted for by the rate of uptake in each tissue

| | Oxygen | | Glucose | | Lactate | |
	Rate*	%	Rate*	%	Rate*	%
Brain	180	10	25.0	16	1.3	5
Heart	400	10	22.0	5	73.0	20
Liver	180	17	0.6	0	37.0	70
Gut	50	6	3.5	4	7.7	13
Carcass† (non-visceral tissue)	22	51	3.0	75	−1.2‡	—

* Absolute rates of substrate uptake or output are given in μmol/min/100 g.
† Calculated from measurements made on the fetal hind limb.
‡ Output of substrate.

Tissue oxygen uptake has been measured for a number of individual organs in the sheep fetus (Table 6.2). Oxygen consumption was highest in the fetal heart and was double the values found in the fetal brain and liver on a weight-specific basis. Together, these organs account for 35–40 per cent of the total oxygen consumed by the sheep fetus (Table 6.2). The largest proportion of fetal oxygen consumption is likely to be used by the fetal carcass (skin, bone, and skeletal muscle). Estimates of oxygen uptake by these tissues, based on measurements of the AV concentration difference in oxygen across the fetal hind limb and the blood flow to the lower carcass, indicate that they could account for about 50 per cent of the total fetal oxygen consumption (Table 6.2). These observations are supported by the finding that paralysis of fetal skeletal muscle can reduce oxygen consumption by as much as 30 per cent in the sheep fetus as a whole.

Glucose

Glucose is a major oxidative substrate *in utero* (Table 6.3). Its concentration in the fetus depends on that in the mother and is always lower than the corresponding maternal value in normal circumstances. The absolute concentrations and the ratio of fetal to maternal glucose levels both vary with species. In the sheep, fetal glucose levels are 25–30 per cent of those in the mother (2.25–3.50 mmol/L) while in other, non-ruminant species, such as the horse, pig, and primate, fetal glucose levels are normally about 60–70 per cent of the maternal value (>3.50 mmol/L). In part, these species differences may be due to structural differences in

Table 6.3 Metabolic (Substrate: O_2) quotients in fetuses of different species

Species	Glucose	Lactate	Amino acids	Acetate
Sheep	0.55	0.25	0.30	0.10
Cow	0.57	0.43	0.10*	0.16
Horse	0.73	0.14	0.12*	—
Human	0.80	0.10	0.15*	—

*Calculated from rate of urea production.

the placenta but they may also reflect differences in the placental transport and consumption of glucose.

Supply

Over the normal range of uterine and umbilical blood flows observed in the sheep, the passage of glucose across the placenta appears to be limited by its diffusion characteristics and not by blood flow. Transport occurs by facilitated diffusion and, hence, the main factors governing the umbilical uptake of glucose under normal conditions are the transplacental glucose concentration gradient and the rate of glucose consumption by the placenta itself. The placenta has a high rate of glucose utilization and uses about 60–75 per cent of the glucose leaving the uterine circulation in a number of different species. Placental glucose consumption therefore significantly reduces the amount of glucose reaching the fetus.

Under normal conditions, the main influence on the transplacental glucose concentration gradient would be changes in the maternal glucose levels caused by variations in maternal diet or nutritional state. However, changes in the fetal glucose level can occur independently of the maternal concentration under adverse or pathological conditions and do alter the umbilical uptake of glucose. The relationship between umbilical glucose uptake and maternal glucose concentrations is linear over the normal range of maternal levels but eventually plateaus at a glucose uptake of about 160–220 μmol/min. This represents the maximum rate at which the ovine placenta can transport glucose and is similar to the values observed in other species when expressed on a weight basis.

The rate of glucose uptake into the umbilical circulation is the rate of exogenous glucose entry into the fetus and is analogous to the dietary supply of glucose in the adult animal. It is not necessarily equal to the rate of fetal glucose consumption as the fetus may produce glucose endogenously either from other substrates or from glucose reserves such as hepatic glycogen. Measurements of glucose utilization made using tracer methodology in the sheep fetus have shown that endogenous glucose production is negligible in unstressed circumstances (Fig 6.1).

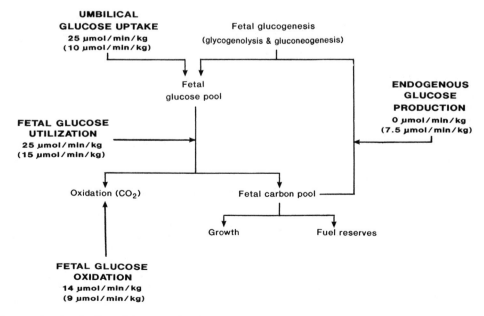

Fig. 6.1 A diagram showing the flux of glucose carbon in the sheep fetus during late gestation ($\geq$ 85 per cent gestation). Fluxes that can be measured are shown in bold type together with their values in fed mothers. Values observed during maternal fasting are shown in parenthesis.

Hence, in normal conditions fetal glucose consumption can be determined by measuring umbilical glucose uptake, labelled glucose disposal, or by determining the uptake of non-metabolizable glucose analogues such as 2-deoxyglucose.

Utilization

Using one of these methods, fetal glucose consumption has been measured in a variety of species and shown to vary between 20 and 70 μmol/min/kg fetal body weight. Fetal ruminants (sheep, cow) tend to have lower rates of glucose uptake than non-ruminant species although the rates are higher than those observed in adults (Table 6.1). In the sheep fetus normal rates of glucose utilization vary between 20 and 40 μmol/kg/min depending on the maternal glucose level but can rise as high as 50–60 μmol/kg/min during experimental conditions that increase fetal glucose availability. Glucose uptake by the sheep fetus is therefore normally limited by the rate of glucose supply and not by the capacity of the fetal tissues to use glucose.

Measurements of the fetal glucose:oxygen quotient also indicated that non-ruminants have a greater dependence on glucose metabolism than the ruminant species (Table 6.3). In the fetal lamb and calf glucose:oxygen quotients of 0.4–0.6 have been measured which means that glucose could account for 40–60 per cent of the total oxygen consumed by the fetus if all the umbilical supply of glucose was oxidized. In non-ruminant species glucose:oxygen quotients are higher although still less than 1 (Table 6.3). Glucose is therefore a major but not the sole oxidative fuel in the fetus even in those species with the highest glucose requirements.

Under physiological conditions, not all the umbilical supply of glucose is oxidized to CO_2. Glucose carbon is also incorporated into new structural tissue and into fuel reserves such as glycogen and fat (Fig 6.1). The partition of glucose carbon between the oxidative and non-oxidative pathways has been quantified in the sheep by measuring the production of labelled CO_2 from labelled glucose infused into the fetus. These studies show that glucose oxidation accounts for 50–60 per cent of the glucose used by the fetus while the non-oxidative pathways consume the remaining 40–50 per cent (Fig 6.1). Glucose oxidation therefore actually accounts for only 25–30 per cent of the total oxygen consumption in the sheep fetus.

Most of the glucose utilized by the fetus is consumed by the carcass. The glucose:oxygen quotient of the fetal hind limb in a normally fed animal is 0.8–1.2 compared with a value of 0.4–0.6 for the fetus as a whole. Estimation of glucose uptake by the fetal carcass indicates that it could account for 75 per cent of fetal glucose utilization in the sheep (Table 6.2). Glucose uptake has been measured in some fetal visceral organs but only accounts for a small proportion of the total umbilical glucose uptake (Table 6.2). Even when the additional requirements of hepatic glycogen accumulation are considered, the known rates of glucose uptake by the individual visceral organs total little more than 30 per cent of the umbilical supply in the sheep fetus (Table 6.2). The fetal brain is an obligatory glucose consumer and has a glucose:oxygen quotient of 0.98 in the sheep fetus. However, in late gestation cerebral glucose utilization still only accounts for 15 per cent of the normal umbilical glucose uptake, although this value may rise to 35 per cent during maternal undernutrition.

Adverse conditions

During fasting and other adverse conditions that compromise the transplacental nutrient supply, the fetus uses less glucose. The fall in fetal glucose utilization may not be as great as the reduction in umbilical supply, as the fetus may produce glucose endogenously under these circumstances (Fig. 6.1). Fetal glucogenesis occurs most readily in late gestation when hepatic glycogen reserves are high and glucogenic enzymes are present in the fetal liver and kidney. In late gestation, the liver of the fetal sheep can produce glucose at a rate of 8–15 μmol/min/kg fetal body weight, which, although insufficient to maintain the normal rate of fetal glucose utilization, will help ameliorate the effects of a reduced umbilical supply of glucose (Fig. 6.1). Fetal hypoglycaemia, hypoxia, and changes in the fetal glucagon:insulin ratio have all been shown to evoke endogenous glucose production in the sheep fetus near term although the relative contributions made by glycogenolysis and gluconeogenesis to hepatic glucose output have not yet been determined. Glycogenolysis appears to be activated first in the fetal sheep, which is consistent with the findings in many species that hepatic glycogen levels are low after adverse conditions *in utero*. Certainly, significant gluconeogenesis is only observed in fetal rabbits when the hepatic glycogen reserves are depleted.

Lactate

After glucose, lactate is quantitatively the next most important carbohydrate fuel in the fetus. Significant levels of lactate are found in the fetal circulation and, while some lactate may be produced anaerobically, its presence does not imply a major anaerobic component to fetal metabolism overall. Early studies on anaesthetized fetuses and infants at delivery showed no significant concentration difference in lactate across the umbilical circulation. However, more recent observations on conscious, unstressed animals (sheep, cow, mare) have shown a significant output of lactate from the placenta into the fetal circulation.

Supply

The lactate produced by the placenta is derived from fetal fructose and from glucose taken up from both the fetal and maternal circulations. Its production rises with increases in umbilical glucose uptake and can account for about one-third of the glucose consumed by the placenta in the sheep and cow. The lactate produced by the placenta is released into the uterine as well as the umbilical circulation with 60–75 per cent of the total production going to the fetus. Since fetal plasma lactate levels are higher than those in the mother, even in unstressed conscious animals, the release of lactate from the utero-placental tissues does not follow simple diffusion kinetics. Several *in vitro* studies have suggested that there is a specific carrier system for lactate but relatively little is know about lactate transport across the placental membranes *in vivo*.

In the sheep lactate is also produced by the fetal tissues at a fairly high rate (45 μmol/min/kg fetal body weight) even in normal unstressed conditions. About 60–70 per cent of this endogenous lactate supply is derived from glucose with the remaining lactate carbon coming from other sources. Measurements of substrate flux across the hind limbs of the sheep fetus have established that the carcass is one of the sites of net lactate production *in utero*. Its lactate output, estimated from the AV lactate concentration difference across the hind limb and the blood flow to the lower carcass, could account for about 40–50 per cent of the lactate produced by the whole sheep fetus in late gestation (Table 6.2).

Utilization

The lactate taken up by the umbilical circulation makes a significant contribution to fetal metabolism in a number of species and, if completely oxidized, would account for 14–45 per cent of the fetal oxygen consumption depending on the species (Table 6.3). In the sheep, the true rate of fetal lactate utilization (65 μmol/min/kg) is three fold higher than the rate of umbilical uptake as lactate is also supplied by the fetal tissues. Of the total amount of lactate consumed by the sheep fetus, 70 per cent is oxidized to CO_2 while the remainder enters the carbon pool and is either recycled or incorporated into new fetal tissue. Lactate oxidation therefore accounts for a greater proportion (40–50 per cent) of fetal oxygen consumption than predicted by the lactate:O_2 quotient in the sheep fetus (Table 6.3).

Several individual tissues have been shown to consume lactate in the sheep fetus (Table 6.2). It is the main carbohydrate taken up by the fetal liver and its consumption there accounts for about 25 per cent of the lactate used by the whole sheep fetus (Table 6.2). It is a major precursor of fetal hepatic glycogen although little circulating glucose is normally produced from lactate by the fetal liver.

Fructose

Fructose is found in the fetal plasma of several species (ruminants, pig, horse) at concentrations 3–4 times higher than those of glucose. *In vitro* studies have shown that it is produced from glucose in the placenta but its rate of production *in vivo* is very low. There is no significant umbilical AV concentration difference in fructose in the sheep, pig, or cow, which implies that fetal fructose consumption is low in these species. Studies with labelled fructose in the sheep fetus have shown that the rate of fructose utilization is about one-fifth that of glucose and can account for only 6 per cent of the total CO_2 produced *in utero*. Labelled fructose is also converted to lactate in the sheep fetus and has been shown to be incorporated into hepatic glycogen in the fetal rat and rabbit although not in the fetal sheep. However, the role of fructose in these processes is quantitatively small compared to the other carbohydrates. Fructose therefore makes very little contribution to fetal metabolism overall, either as a source of energy or of carbon for accretion of new tissue.

Amino acids

In adult and new-born animals, amino acids are used for protein synthesis, oxidation, and for other non-oxidative processes that require carbon or nitrogen. In the fetus amino acids are required for protein accretion and may also provide substrate for oxidation. Certainly, the fetal metabolic quotients suggest that substances other than the carbohydrates must contribute to CO_2 production *in utero* (Table 6.3).

Supply

Essential amino acids, which cannot be synthesized *de novo*, must have a maternal origin in the fetus while the non-essential amino acids could be derived either transplacentally or by synthesis within the feto-placental tissues. In sheep, fetal amino nitrogen levels are higher than those in the mother although the ratio of fetal to maternal concentrations is less in conscious (1.5–2.0) than in anaesthetized animals (2.0–3.0). In other conscious pregnant animals (mare, sow) this ratio is much closer to one.

Transfer of amino acids across the placenta therefore appears to be active, at least in the sheep, although relatively little is known about the mechanisms of placental amino acid transport *in vivo*. Amino acid levels in ovine placental tissue are higher than those in either the fetal or maternal plasma which suggests that it is the uterine and not the umbilical uptake that is the active process.

Umbilical and uterine AV concentration differences in individual amino acids have only been measured in unstressed conditions in the sheep. These studies have shown a net fetal uptake of the neutral and basic amino acids but not of the acidic ones. In fact, there was a significant output of glutamate from the fetus to the placenta despite its accumulation in fetal protein (Fig. 6.2). Similarly, acute and *in vitro* experiments have shown little net transport of glutamate across the placenta in the human, monkey and rat. Glutamate may therefore be used to excrete nitrogen from the fetus or provide substrate for amino acid synthesis within the placenta itself. Certainly, more glutamine is taken up by the sheep fetus than is removed from the uterine circulation, and significant amounts of ammonia are released by the placenta into both

the fetal and maternal circulations in the sheep. The placenta is therefore not only transporting amino acids but is metabolizing them by deamination and transamination reactions.

Utilization

The uptake of amino acids into the umbilical circulation of the sheep exceeds that required for accretion of new tissue by about 50 per cent (Fig. 6.2). Hence, a large proportion of the fetal amino acid uptake is available for catabolism and oxidative metabolism. The first attempts to quantify these processes were made by measuring the fetal production of urea, the main product of amino acid deamination *in vivo*. Fetal urea concentrations are higher than those in the mother and lead to a transplacental urea gradient which varies with species but which is stable over a wide range of maternal concentrations in any individual animal. Urea is therefore produced by the fetus and excreted across the placenta at a steady rate. The umbilical AV concentration difference in urea is too small for the transplacental excretion rate to be calculated directly using the Fick principle. However, it can be estimated from measurements of placental urea clearance and the urea concentration difference across the placenta. Since relatively little urea is excreted by the fetal kidney, the transplacental rate of urea excretion provides a good estimate of the rate of urea production *in utero*.

Fetal urea production, calculated in this way, ranges from 0.2 to 0.4 mg/min/kg fetal body weight in the cow, horse, and human but gives higher values (0.4–0.7 mg/min/kg) in the fetal sheep. These rates of fetal urea production are greater than those in the respective new-born or adult animal and could account for 10–35 per cent of the fetal oxygen consumption depending on the species (Table 6.3). The contribution made by amino acid catabolism to fetal oxidative metabolism therefore varies with species and appears to be less than that of the carbohydrates in well fed animals.

More direct measurements of amino acid oxidation have been made in the sheep fetus using labelled essential and non-essential amino acids. Production of labelled CO_2 has been demonstrated from labelled alanine, glycine, tyrosine, leucine, and lysine but has only been quantified for the latter two amino acids. Lysine has a low oxidation rate and the majority of labelled lysine infused into the sheep fetus is incorporated into body tissue. Leucine, on the other hand, is oxidized at a higher rate, which could account for about 30 per cent of the total rate of leucine utilization in the sheep fetus. These observations are consistent with the findings that the umbilical uptake of lysine only exceeds its accretion rate in protein by 20 per cent while that of leucine has a safety margin of 40–50 per cent (Fig. 6.2).

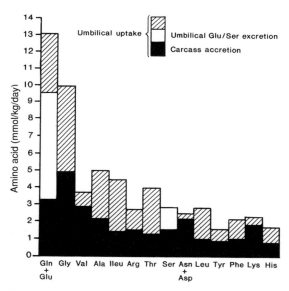

Fig. 6.2 A comparison of the carcass accretion rates (solid bars) and umbilical uptakes (hatched bars) of individual amino acids in fetal sheep in the fed state during late gestation (≥ 80 per cent of gestation). The open bars are the excretion rates of glutamate and serine from the fetus into the placenta. The cross-hatched area therefore represents the delivery of amino acids in excess of their accretion rate in fetal tissue. (Lemons and Schreiner, **244**, 459–66, reproduced with permission from the *American Journal of Physiology*.)

In all animals there is a continuous turnover of body protein but, during growth, the rate of protein synthesis must exceed the rate of degradation if there is to be a net gain of protein mass. The fractional rate of protein synthesis in the sheep fetus decreases with increasing gestational age but is high compared to adult values even in late gestation. The fractional rate of protein synthesis also exceeds the fractional rate of accretion for both labelled lysine and leucine in the sheep fetus. Protein turnover is therefore rapid *in utero* and may provide 40–50 per cent of the leucine used by the sheep fetus in the fed state. In these circumstances the energy cost of fetal protein synthesis will be high and could account for as much as 25 per cent of the oxygen consumed by the sheep fetus in late gestation.

Relatively little is known about nitrogen flux across individual fetal organs. An uptake of α amino nitrogen has been demonstrated across the gut of the sheep fetus, which could account for 6 per cent of the α amino nitrogen taken up by the umbilical circulation. Similarly, significant uptakes of individual amino acids have been shown to occur across the hind limb of the sheep fetus. This uptake is accompanied by a significant output of ammonia, which suggests that fetal skeletal muscle catabolizes amino acids. Calculation of the nitrogen efflux:oxygen quotient indicates that amino acid catabolism could account for about 5 per cent of the oxygen consumed by the hind limbs of the sheep fetus in adequately nourished ewes and a higher proportion during maternal fasting.

Adverse conditions

During maternal undernutrition or other types of fetal nutrient restriction, amino acid catabolism increases in the fetus. Fetal urea production may double in these circumstances and could account for the majority of fetal oxygen consumption during maternal fasting. The specific rates of fetal leucine utilization and oxidation increase by 30–50 per cent during maternal fasting in the absence of any significant change in the umbilical uptake of leucine. Since there is also no apparent change in the umbilical supply of any of the other amino acids, increased fetal amino acid catabolism can only occur during maternal fasting at the expense of fetal protein accretion. In undernourished sheep and rats, reduced fractional rates of fetal protein synthesis and accretion are observed which are accompanied by small or no reductions in the fractional rates of fetal protein degradation. Utilization of body protein stores may therefore occur in the fetus under these conditions which would result in a negative nitrogen balance.

Total amino acid uptake by the hind limbs of the sheep fetus decreases during maternal fasting, although the uptake of specific amino acids such as leucine increases. There are also significant effluxes of the gluconeogenic amino acids, glutamine and alanine, from the fetal hind limbs during maternal fasting, which suggests that fetal skeletal muscle not only oxidizes more amino acids in these conditions but also transaminates them to provide gluconeogenic precursors. Certainly, the mean arterial concentration of the main gluconeogenic amino acids rises in the sheep fetus during adverse conditions in which fetal glucogenesis is known to occur.

Fats

Fat is a heterogeneous group of compounds that includes the triglycerides (TG), phospholipids, free fatty acids (FFA), and volatile, short-chain fatty acids. All of these substances are found in the fetal circulation although the absolute amounts and relative proportions of each vary between species and with nutritional state. In new-born and adult animals fats are used for oxidation and to provide carbon and structural lipids for tissue accretion and repair. In the fetus lipids are essential for growth but appear to have a more equivocal role in fetal oxidative metabolism.

Supply

The FFAs detected in the fetal circulation may come from the mother by direct placental transfer, from *de novo* synthesis in the placenta or other fetal tissues, or from breakdown of TG and phospholipids. Fetal FFA levels are lower than those in the mother in a number of different animals but, despite the gradient, placental permeability to FFA varies widely between species. In some animals (rats and rabbits), labelled FFA injected into the mother appears rapidly in the fetal circulation and fetal FFA levels rise in parallel with maternal concentrations. However, in other species (cow, sheep, horse), only a very small net AV concentration difference in FFA can be detected across the umbilical circulation. Larger concentration differences are observed in human infants delivered by elective Caesarean section although maternal FFA levels may be raised by fasting and stress under these circumstances. Placental transfer of maternal FFA is, therefore, likely to contribute to the fetal FFA pool in the human and must supply at least some of the essential FFAs in other species. However, even in the human infant there are other sources of FFA as maternal FFA can provide only about 20 per cent of the fatty acids stored as TG in the body at term.

Synthesis of FFA has been demonstrated in the placenta, fetal liver, brain, and adipose tissue in *in vitro* experiments on a number of different species. The enzymes required for FFA and TG synthesis are present at high levels in the fetal liver of human, monkey, rat, and guinea-pig near term. However, there are some species (e.g. pig), which have little fat at birth, that have a very limited capacity for hepatic lipogenesis *in utero*.

Very little is known about the breakdown and recycling of TG and phospholipids in the fetus. No significant AV concentration difference in TG was observed in human infants at delivery, which suggests that the TG present in the fetal circulation has been synthesized *in utero*. Small increases in fetal FFA levels, independent of any maternal changes, have been observed in fetal sheep in response to specific stimuli such as asphyxia and adrenaline infusion. Stored TG may therefore be mobilized in the fetus but there is little evidence to suggest that the liberated FFAs can be used to derive energy *in utero*.

Utilization of free fatty acids

The low coefficient of extraction of the FFA across the umbilical circulation and the high molecular weight and carbon content of these substances make quantifying their role in fetal metabolism difficult. A small AV concentration difference across the umbilical circulation may represent a large and physiologically significant uptake of substrate. Fatty acid oxidation, however, does appear to be limited in the fetus. Fatty acids are not taken up in excess of the fetal requirements for growth and the enzymes needed for FA oxidation are low in activity in the fetal liver. *In vitro* experiments have shown that in some species palmitate can be oxidized to CO_2 by fetal tissues but only to a limited extent. It may be that there is insufficient carnitine, an essential co-factor in FA oxidation to permit oxidation at any significant rate in fetal tissues.

Utilization of acetate

In ruminants the volatile fatty acid, acetate, does contribute to fetal metabolism. In sheep and cows there is a significant AV concentration difference in acetate across the umbilical circulation that results in a fetal acetate uptake of 15–35 μmol/min/kg fetal body weight (Table 6.1). This rate of acetate uptake could account for 10 and 16 per cent of the oxygen consumption of the fetal lamb and calf, respectively (Table 6.3). Umbilical acetate uptake increases linearly with the maternal acetate concentration and is less that the uterine uptake in both species. In ruminants significant amounts of acetate are, therefore, used by the utero-placental tissues as well as by the fetus.

There have been no studies of fetal acetate utilization or of the partition of acetate carbon between the oxidative and non-oxidative pathways. Measurements of acetate flux across individual fetal organs have shown that the liver of the fetal sheep produces acetate while the hind limbs consume it, although to a lesser extent than the maternal carcass. Hence, the sheep fetus is acquiring acetate both endogenously and transplacentally, which means the true rate of fetal acetate consumption is likely to be higher than the rate of umbilical supply in this species. Administration of radioactively labelled acetate to the fetus leads to labelling of a wide range of compounds including fatty acids, steroids, and membrane lipid but the flux of acetate into these substances has not been quantified for the fetus as a whole.

Utilization of other lipids

The only other lipid components that could contribute to fetal metabolism are glycerol and ketones. Both are derived from the mother and are found in the fetal circulation at low concentrations under normal conditions. Concentrations rise during maternal undernutrition and it is in these conditions that they are likely to contribute most significantly to fetal metabolic balance. Hydroxybutyrate carbon can be incorporated into amino acids in the fetal brain and has also been shown to be oxidized by a variety of fetal tissues in the rat during late gestation. By contrast, the main metabolic role of glycerol is as a gluconeogenic substrate. The key enzyme required for its conversion to glucose is present in the fetal liver at term and injection of labelled glycerol into rats and rabbits leads to the production of labelled glucose and glycogen in the fetal liver. Glucose derived from glycerol may also contribute to the glucose produced endogenously in the sheep fetus during maternal undernutrition, as glycerol will be readily available in these circumstances.

Control of fetal metabolism

Nutrient uptake by the fetal tissues is regulated partially by the fetal hormonal environment and partially by the circulating nutrient levels *in utero*. By comparison, relatively little is known about the regulation of nutrient production by the fetal tissues or about the control of fetal metabolism at a cellular level although mechanisms must exist to regulate the flux of nutrients through the various biochemical pathways.

Hormone concentrations

In adult animals intermediary metabolism is controlled by the hormonal environment and, in particular, by the pancreatic hormones. The endocrine pancreas is functional *in utero* and responds to metabolic and other stimuli by late gestation in a wide variety of species. Insulin and glucagon can both be detected in the fetal circulation from early in gestation and their concentrations vary with intrauterine stress and fetal nutrient availability.

Glucagon injection into the fetus causes hyperglycaemia and endogenous glucose production in a variety of species but the concentration required to stimulate glucogenesis in the sheep fetus is higher than that needed in the adult. However, high plasma glucagon levels are observed *in utero* during stressful conditions such as labour or acute hypoxia when fetal glucose levels are known to be raised. During fasting there is little change in the fetal glucagon concentration, although the molar ratio of insulin to glucagon levels may alter significantly. Certainly, endogenous glucose production only appears to occur in the sheep fetus when there is a detectable change in the insulin-to-glucagon concentration ratio or an elevation in catecholamine concentrations. Glucagon, therefore, appears to have a relatively minor role in regulating fetal metabolism in normal conditions but becomes more important during fetal stress.

Insulin, on the other hand, is involved in controlling fetal metabolism in both normal and adverse circumstances. Its injection into the sheep fetus causes hypoglycaemia and stimulates the umbilical uptake, utilization, and oxidation of glucose by the whole fetus. It also increases the glucose:oxygen quotient of the fetal hind limbs. Conversely, specific fetal insulin deficiency induced by pancreatectomy or diabetagenic drug treatment leads to fetal hyperglycaemia and a 40–50 per cent reduction in the rates of umbilical uptake, utilization, and oxidation of glucose in the sheep fetus. Similar reductions in glucose metabolism are seen in the whole fetus and its hind limbs when fetal hypoinsulinaemia is induced by fasting the ewe. Fetal glucose production can also be detected in certain other hypoinsulinaemic conditions, although this may not be due solely to low insulin levels. Physiological changes in fetal plasma insulin, therefore, cause parallel changes in the rates of fetal glucose utilization and oxidation but do not alter the distribution of glucose between the oxidative and non-oxidative pathways for metabolism. Insulin therefore stimulates cellular glucose uptake in the fetus but appears to have little effect on the metabolism of glucose once it has entered the cell.

Much less is known about the role of insulin in the uptake and utilization of the other fetal nutrients. There is little or no change in fetal oxygen consumption over the normal range of insulin levels observed *in utero*. Insulin can stimulate the deposition of fat in adipose tissue and enhances the cellular uptake of amino acids into a variety of fetal tissues. It has also been shown to enhance protein synthesis in skeletal muscle of the fetal sheep by reducing the rate of protein degradation. The overall effect of insulin on fetal metabolism is therefore anabolic.

Other hormones such as thyroxine and cortisol can influence fetal metabolism although their actions are more indirect than those of the pancreatic hormones. For instance, cortisol is known to stimulate fetal glycogen deposition and enhance the activity of the glucogenic enzymes in the fetal liver and kidney towards term. It may also induce tissue receptors for hormones such as the catecholamines that have more direct effects on fetal metabolism in late gestation.

Metabolite concentrations

In the adult, metabolite concentrations *per se* can influence the cellular uptake and utilization of particular nutrients. Since changes in fetal metabolite concentrations generally precede any endocrine changes *in utero*, nutrient levels may also have an important role in regulating fetal metabolism. Greater increases in umbilical glucose uptake are observed in response to insulin when its hypoglycaemic effect is prevented by exogenous glucose infusion. Similarly, umbilical glucose uptake is higher in hyperglycaemic pancreatectomized fetuses than in hypoglycaemic fetuses with a similar degree of hypoinsulinaemia caused by maternal fasting. In addition, experimental increases in the plasma glucose level at a constant insulin concentration increase glucose utilization and oxidation by the sheep fetus. Insulin and glucose are therefore both involved in regulating glucose metabolism in the sheep fetus. They may act independently of each other, but in normal conditions their concentrations vary together and they have an additive effect on fetal glucose metabolism. Over the normal range of concentrations observed *in utero* it is glucose that appears to have the more dominant effect, although insulin is still a significant influence on fetal glucose metabolism in these circumstances.

Fetal metabolic balance

Nutrients taken up by fetal tissues are used for two main purposes: first, to fuel oxidative metabolism; and, second, to build new tissue. The relative contributions made by particular nutrients to oxidative metabolism can be assessed either by calculating metabolic quotients or more directly by measuring the production of radioactively labelled CO_2 from labelled substrates supplied to the fetus. The difference between the rates of utilization and oxidation measured simultaneously is assumed to be the rate of substrate flux into the non-oxidative metabolic pathways and, ultimately, into fetal growth (Fig. 6.1).

Oxidation

Measurements of metabolic quotient suggest that carbohydrates (glucose and lactate) are the major oxidative fuels *in utero* and could account for 80–90 per cent of fetal oxygen consumption if all of the umbilical carbohydrate supply were completely oxidized (Table 6.3). However, tracer methods have shown that carbohydrate oxidation is not complete in the sheep fetus and actually accounts for only 50 per cent of the fetal oxygen consumption when the conversion of glucose to lactate is taken into account. There is therefore a significant non-oxidative disposal of carbohydrate carbon in the sheep fetus. The remaining 50 per cent of the umbilical oxygen uptake is probably used to oxidize acetate and amino acids. Significant amounts of oxygen are consumed in producing urea in the sheep fetus and, if all the amino acids supplied in excess of the fetal growth requirements were completely oxidized, amino acid catabolism would account for more than half the oxygen consumed by the sheep fetus. Amino acids, therefore, make a significant contribution to oxidative metabolism in the sheep fetus, although they may be relatively less important in other species.

The amount of energy derived from oxidative metabolism can be calculated from the rate of fetal oxygen consumption and the calorific equivalent of the oxygen consumed. The latter value is determined by the type of substrate oxidized which may vary with species and nutritional state. For the sheep fetus in the normally fed ewe, using a mixture of carbohydrates and amino acids, the

calorific value of oxygen is about 4.9 kcal/L. With an oxygen consumption of 315 μmol/min/kg (Table 6.1), the sheep fetus will derive approximately 50 kcal/kg/day from oxidative metabolism.

Growth

The nutrient requirement for fetal growth is more difficult to estimate as it depends on the growth rate and body composition of the fetus. In late gestation, fetal growth rate varies between species and ranges from 1.5 to 7.0 per cent per day (Table 6.4). Similar species variation occurs in body composition, particularly in body fat content (Table 6.4). Since the calorific value of fat (9.3 kcal/g) is higher than that of non-fat tissues (0.70–0.95 kcal/g), the rate of fat deposition in the fetus towards term is an important factor in determining the energy requirement for growth in late gestation. In species such as the human, which has a high fat content at term, fat accumulation in late gestation accounts for 70–75 per cent of the calorific accretion rate (Table 6.4). This compares to a value of 10–15 per cent in the fetal pig which is only 1 per cent fat at birth (Table 6.4). Because of these differences in fat deposition, the total energy requirement for tissue accretion is similar in the human and pig fetus despite a two-fold difference in their growth rates during late gestation (Table 6.4). The precise body composition and absolute growth rate are therefore both important factors in determining the total nutrient requirement in late gestation.

Total energy balance

The total energy requirement of the fetus can be calculated for any given species from its oxygen consumption,

Table 6.4 The growth rates, body weight, fat content, and energy requirements of fetuses of different species during late gestation ($\geq$ 85 per cent gestation)

Species	Birth weight (kg)	Growth rate (g/kg/day)				% body fat at birth	Energy requirement kcal/kg/day		
		Total	Water	Non-Fat dry wt	Fat		Oxidation*	Growth	Total
Guinea-pig	0.09	68.0	36.1	21.2	10.7	11.7	62	193	255
Pig	1.20	32.8	24.7	6.6	0.5	1.0	46	36	82
Sheep	3.00	36.0	28.7	6.5	0.8	2.0	50	32	87[†]
Human	3.50	15.0	9.2	2.3	3.5	16.0	56	43	99
Horse	52.00	9.2	5.9	2.8	0.4	2.6	50	17	67

*Assuming calorific value of oxygen of 4.9 kcal/L.
[†]See Table 6.5.

Table 6.5 The metabolic balances for energy, carbon, and nitrogen in fetal sheep during late gestation ($\geq$ 85 per cent)

	Fed ewes			Fasted ewes		
	Energy kcal/kg/day	Carbon g/kg/day	Nitrogen g/kg/day	Energy kcal/kg/day	Carbon g/kg/day	Nitrogen g/kg/day
Requirement						
Accumulated in carcass	32.0	3.2	0.65	16.0*	1.6*	0.33*
Excretion						
As CO_2	0	4.40	0	0	4.40	0
As urea	2.0	0.20	0.36	3.6	0.36	0.66
As glutamate	3.2	0.39	0.09	1.4	0.17	0.04
Oxidation	50.0	0	0	50.0	0	0
Total	87.2	8.19	1.10	71.0	6.53	1.03
Supply						
Umbilical uptake						
As glucose	27.4	2.90	0	13.7	1.40	0
As lactate	14.0	1.40	0	14.0	1.40	0
As acetate	7.0	0.56	0	7.0	0.56	0
As amino acids	38.0	3.08	1.34	33.0	2.98	1.03
Total	86.4	7.94	1.34	67.7	6.34	1.03

*Assumes a growth rate half of that in the fed animal.

growth rate, and body composition provided values are expressed in the same units and measured at the same gestational age. For example, the sheep fetus near term accumulates approximately 32 kcal/kg/day as new tissue, requires 50 kcal/kg/day to support oxidative metabolism, and excretes about 5.2 kcal/kg/day into the placenta as urea and glutamate. This makes a total requirement of about 87.2 kcal/kg/day. Similar calculations can be made by expressing values as g carbon or g nitrogen required per day. By comparing these requirements with the umbilical supply of nutrients, the metabolic balance of the fetus can be estimated.

Only in the sheep fetus are there sufficient data to allow the metabolic balance to be calculated with any degree of accuracy. The total requirements for energy, carbon, and nitrogen and the known umbilical uptake of these substances in the sheep fetus are shown in Table 6.5 for animals in the fed and fasted states. Amino acids are the largest source of energy *in utero*, irrespective of nutritional state. They provide 40–45 per cent of the total energy requirement in the fed state and this proportion may rise during fasting (Table 6.5). They also provide all the nitrogen and 35–45 per cent of the carbon required by the sheep fetus each day (Table 6.5). After the amino acids, the next most important energy source is glucose. It provides about 30 per cent of the total energy and 35 per cent of the daily carbon requirement in the fed state although its contribution to the total supply of energy and carbon falls significantly during fasting (Table 6.5). Lactate and acetate also make small contributions to the fetal energy and carbon balances, which do not appear to alter with nutritional state (Table 6.5).

When all the known umbilical nutrient uptakes are summed, the total supply of energy and carbon is slightly less than the total amounts required each day by the sheep fetus in both fed and fasted ewes (Table 6.5). This suggests that there must be other sources of carbon and energy in the fetus that have not been identified yet. Given the difficulty of quantifying fat metabolism, it seems most likely that lipid utilization accounts for the deficits in the fetal carbon and energy balances. However, any contribution made by lipid to the metabolic balance will be relatively small compared to those of the carbohydrates and amino acids.

Further reading

Battaglia, F.C. and Meschia, G. (1978). Principal substrates of fetal metabolism. *Physiological Reviews*, **58**, 499–527.

Battaglia, F.C. and Meschia, G. (1986). *An introduction to fetal physiology*. Academic Press, London.

Beard, R.W. and Nathanielsz, P.W. (ed.) (1984). *Fetal physiology and medicine*. Butterworths, London.

Char, V.C. and Creasy, R.K. (1976). Acetate as a metabolic substrate. The fetal lamb. *American Journal of Physiology*, **230**, 357–61.

Charlton, V.C., Reis, B.L., and Lofgren, D.J. (1979). Consumption of carbohydrates, amino acids and oxygen across the intestinal circulation in the fetal sheep. *Journal of Developmental Physiology*, **1**, 329–36.

Fisher, D.J., Heymann, M.A., and Rudolph, A.B. (1980). Myocardial oxygen and carbohydrate consumption in fetal lambs *in utero* and in adult sheep. *American Journal of Physiology*, **238**, H399–H405.

Fowden, A.L. (1989). The endocrine regulation of fetal metabolism and growth. *Research in Perinatal Medicine*, **VIII**, 229–43.

Gleason, C.A., Rudolph, C.D., Bristow, J., Itskovitz, J., and Rudolph, A.M. (1985). Lactate uptake by the fetal sheep liver. *Journal of Developmental Physiology*, **7**, 177–83.

Hay, W.W., DiGiacomo, J.E., Mezharich, H.K., Hirst, K., and Zerbe, G. (1989). Effects of glucose and insulin on fetal glucose oxidation and oxygen consumption. *American Journal of Physiology*, **256**, E704–13.

Hay, W.W., Myers, S.A., Sparks, J.W., Wilkening, R.B., Meschia, G., and Battaglia, F.C. (1983). Glucose and lactate oxidation rates in the fetal lamb. *Proceedings of the Society for Experimental Biology and Medicine*, **173**, 553–63.

Hay, W.W., Sparks, J.W., Quissell, B.J., Battaglia, F.C., and Meschia, G. (1981). Simultaneous measurements of umbilical uptake, fetal utilization rate and fetal turnover rate of glucose. *American Journal of Physiology*, **240**, E662–8.

Jones, M.D., Burd, L.I., Makowski, E.L., Meschia, G., and Battaglia, F.C. (1975). Cerebral metabolism in sheep: a comparative study of the adult, the lamb and the fetus. *American Journal of Physiology*, **229**, 235–9.

Lemons, J.A. and Schreiner, R.L. (1983). Amino acid metabolism in the ovine fetus. *American Journal of Physiology*, **244**, E459–66.

Meier, P.R., Peterson, R.G., Bonds, D.R., Meschia, G., and Battaglia, F.C. (1981). Rates of protein synthesis and turnover in fetal life. *American Journal of Physiology*, **240**, E320–4.

Meyer, H. and Aliswede, L. (1978). The intrauterine growth and body composition of foals and the nutrient requirements of pregnant mares. *Animal Research and Development*, **8**, 86–112.

Meznarich, H.K., Hay, W.W., Sparks, J.W., Meschia, G., and Battaglia, F.C. (1987). Fructose disposal and oxidation rates in the ovine fetus. *Quarterly Journal of Experimental Physiology*, **72**, 617–25.

Moustagaard, J. (1962). Foetal nutrition in the pig. In *Nutrition of pigs and poultry*. (Ed. J.T. Morgan and D. Lewis), pp. 57–89. Proceedings of the Easter School in Agricultural Science, University of Nottingham, Books Demand UHI.

Nathanielsz, P.W. (1987). *Metabolism. Animal Models in Fetal Medicine VI*. Perinatology Press, Ithaca, New York.

Rattray, P.V., Garrett, W.N., East N.E., and Hinman, N. (1974). Growth, development and composition of the ovine conceptus and mammary gland during pregnancy. *Journal of Animal Science*, **38**, 613–26.

Sparks, J.W., Girard, J.R., Gallikan, S., and Battaglia, F.C. (1985). Growth of fetal guinea pig: physical and chemical characteristics. *American Journal of Physiology*, **248**, E132–9.

Sparks, J.W., Hay, W.W., Bonds, D., Meschia, G., and Battaglia, F.C. (1982). Simultaneous measurements of lactate turnover rate and umbilical lactate uptake in the fetal lamb. *Journal of Clinical Investigation*, **70**, 179–92.

van Veen, L.C.P., Teng, C., Hay, W.W., Meschia, G., and Battaglia, F.C. (1987). Leucine disposal and oxidation rate in the fetal lamb. *Metabolism*, **36**, 48–53.

7. Fetal growth and fetal growth retardation

Jeffrey S. Robinson, Julie A. Owens, and Philip C. Owens

Defining the conditions that support optimal growth has to encompass much of fetal physiology and the maternal adaptations to pregnancy. Deviations from these conditions are more likely to retard fetal growth; however, there are a small number of fetal and maternal conditions that cause acceleration of fetal growth. Fetal growth, strictly speaking, is limited to the period from the end of organogenesis (the embryonic period) to the moment of birth. Centuries ago, Galen reached a similar conclusion remarking that 'growth belongs to that which has already been completed in respect to its form'.

Growth is measured in terms of change in weight and size in relation to age. This definition of growth was used by a subcommittee of the International Federation of Obstetrics and Gynaecology (FIGO) in a report on the measurement and recording of infant growth in the perinatal period. Their report highlighted the need for serial measurements to determine growth of individuals rather than the much more common practice of using single measurements of fetal size or birth weight as surrogates for fetal growth. However, cross-sectional measurements of fetal size have been used to study normal growth of fetuses in a variety of animals and it is only in humans that these measurements are seriously flawed by the problems of the pathophysiological events that cause or lead to premature delivery. Since ultrasound was introduced some 40 years ago by Ian Donald, it has been possible to make accurate serial measurements of individual fetuses, and detailed accounts of the growth of the human fetus are available.

Early literature on fetal growth was reviewed most extensively in 1929 by Scammon and Calkins. However, it was not until after the description of retarded fetal growth in 1947 that accurate percentiles for birth weight were established in 1963. Since that time the normal relationships between birth weight and duration of pregnancy, often called birth-weight curves, have been published for many different societies.

Over the last 25 years a detailed account of the physiological conditions associated with normal and retarded fetal growth has been obtained from studies in small laboratory animals. Combined with studies of chronically catheterized fetal sheep, these results have provided more detailed char-acterization of the requirements for fetal growth. Recent evidence has revealed many similarities between human and experimental intrauterine growth retardation.

Definition of growth and growth retardation

Normal growth is the expression of the genetic potential to grow which is neither abnormally constrained nor promoted by internal or external factors. Although this accurately defines the concept of normal growth, it is difficult to apply to an individual subject in a clinical or experimental setting. Experimentally, normal growth of an individual within a population is readily defined using standard statistical methods such as measurement of weight or size that falls within two standard deviations of the mean for that population. This can also be used clinically but, more commonly, the cut-off point is taken to be the 10th percentile, and babies with measurements below this are growth retarded by definition. However, this classification fails to allow for the fetus that has grown normally but has a lower genetic potential for growth or for the fetus whose growth rate has deteriorated but whose birthweight is within normal limits. Ultrasound measurements of such fetuses crossing to lower centiles as pregnancy proceeds have been reported. Similar problems of definition occur when describing the large-for-dates fetus. Examples of small-for-dates and large-for-dates babies are shown in Fig. 7.1.

Measurement of growth

Experimental animals

The normal growth of fetuses in domestic and experimental animals has been defined by killing large number of animals at different gestational ages. While this approach continues to provide valuable information on the effects of interventions that may alter growth, it does not allow for serial measurement *within* individual animals and therefore has experimental limitations.

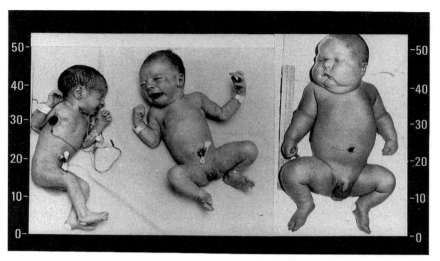

Fig. 7.1 Examples of small-for-dates and large-for-dates babies compared to an appropriately grown baby at term. The small-for-dates weighed 2050 g at birth and was delivered by Caesarean section because of fetal distress (scalp blood sample pH 7.11). The large fetus on the right was also delivered by Caesarean section for cephalo-pelvic disproportion. His birth-weight was 6600 g. A normally grown term infant (3540 g) is shown for comparison.

The direct methods for serial measurement of fetal growth include the crown–rump length measuring device of D. Mellor and implantable ultrasound transducers. Indirect assessment of fetal growth can be obtained from measurements of blood volume, but this may not be as useful as originally suggested. More recently, external ultrasound has been used successfully to describe both fetal and placental growth in the sheep. The latter technique has the advantage that it can be successfully used in the field or in the laboratory.

Human fetus

Until the 1970s growth of the human fetus had to be inferred from measurement of the symphysial-fundal height or from a clinical impression of uterine size obtained by palpation of the uterus. Although the former improved the detection rate of the growth-retarded fetus, errors in the estimation of fetal size were (and are) common. The development of accurate measurements of the linear dimensions of the fetus by ultrasound allowed its growth to be monitored throughout most of pregnancy. Early in pregnancy, measurements of curved crown–rump length can be readily made by ultrasound and have proved useful in the determination of gestational age. By mid-pregnancy, most organs of the fetus can be imaged by this procedure and charts of their normal increase in size with age are available. The ultrasound measurements most commonly used to assess growth of the fetus are the

biparietal diameter, femur length, and abdominal circumference. Two scans performed in pregnancy can detect the majority of fetuses that grow slowly or abnormally. When used to estimate fetal weight, they have an accuracy of about ± 10–15 per cent. Direct comparison of the usefulness of ultrasound and symphysial-fundal height measurements have shown that ultrasound is a much better method for detection of fetal growth retardation.

Birthweight charts

As noted earlier, many charts of centiles for birth weight-for-gestational age have been produced. In the USA the first chart, which was widely used before others became available, suffered from the disadvantage of being derived from a population living at altitude, which can reduce the growth rate of the fetus. Accurate assessment of gestational age in women was a problem but has been addressed statistically with correction for a bimodal distribution of gestational age. Inaccuracies were assumed to be due to errors in calculation of gestational age from the last menstrual period or from other episodes of bleeding. At present, calculation of the duration of pregnancy is doubtful in almost 30 per cent of women. Nowadays, the almost routine practice of ultrasound examination of the fetus at 16–20 weeks obviates the problems of uncertain dates and growth for the majority.

An international reference chart has been made available by FIGO. This is a compilation from birth-weight charts obtained for developing and developed societies and includes babies electively delivered prematurely when the mother or fetus is at risk. Factors adversely affecting fetal growth are more likely to be present when elective preterm delivery is necessary. Thus, such birth-weight charts probably do not provide an accurate estimate of normal fetal growth and are more likely to be erroneous from early in the second half of pregnancy.

Ultrasound has enabled estimation of the intrauterine weight of the fetus and, from these estimates, growth curves have been generated for the normal fetus which is born at term. Unlike the standard birth-weight charts this method describes a decline in growth velocity for the human fetus during late gestation, analogous to that for other species. Another conclusion from such examinations is that about 30 per cent of babies born prematurely have weights that are below 2 standard deviations of the mean for the fetus of the same age that remains *in utero* until term.

Normal growth of the fetus and placenta

The pattern of fetal growth with gestational age is similar in many species. Early in pregnancy the weight of the placenta exceeds that of the fetus. However, fetal growth proceeds more rapidly than that of the placenta so that, by mid-gestation, fetal weight is greater than placental weight. In some species, there is evidence for slowing of fetal growth towards the end of pregnancy. Generally, placental weight increases with gestational age; however, in sheep, placental weight reaches a maximum by 90 days of pregnancy and declines slightly thereafter.

A significant relationship between placental and fetal weight at term is present in many species, but this is not found earlier in pregnancy. The development of this relationship near term suggests that throughout most of pregnancy the placenta has a significant functional reserve. The more rapid growth of the fetus in late gestation reduces the magnitude of this reserve and, in late pregnancy, placental size may limit fetal growth. When restrictions are placed on placental growth and the functional reserve is greatly reduced, this relationship becomes significant at an earlier gestational age.

Alteration of fetal growth in experimental animals

Many factors influencing fetal growth that may act through all or part of gestation have been identified by experimen-

tal intervention (Table 7.1). Many were proposed and tested following identification of conditions associated with altered fetal growth in women or other species. The maternal factors that constrain fetal growth will be considered first; then emphasis will be placed on how placental size, blood flow, and function influence fetal growth. Only a brief reference will be made to a few of the internal fetal constraints to its own growth.

Constraint to growth acting through the mother

Many years ago Walton and Hammond clearly demonstrated that maternal size influenced fetal size at birth more than paternal size. They chose the extreme example of the small Shetland pony mated to the very large Shire horse. Many studies in a variety of species and using modern embryo transfer techniques have confirmed this elegant and classic study on the control of fetal growth.

Maternal undernutrition has been used extensively to reduce fetal growth. Moderate maternal undernutrition for most of pregnancy reduces fetal growth, but sometimes can have an opposite effect on placental growth. Indeed, placental growth may be enhanced by moderate undernutrition in the first two-thirds of pregnancy and this could represent a compensatory response to maintain substrate supply by the placenta to the fetus. Maternal undernutrition in late pregnancy retards fetal growth more severely than similar undernutrition limited to early or mid-pregnancy.

Chronic hypoxia also reduces fetal weight in laboratory animals, but its impact on placental growth appears variable between species, with a reduction in sheep and an increase in women. In studies of several species, a graded effect on fetal growth was observed with increasing severity of hypoxia. In sheep, normobaric hypoxia for up to 28 days had no effect on fetal growth, whereas hypobaric hypoxaemia for 21 days or for most of pregnancy reduced fetal weight. These different results may relate to the degree of hypoxia achieved, since mean PO_2 was reduced to 19 mmHg in the former study and to 12.7 mmHg in the latter. In both studies, the catheterized fetuses responded to hypoxaemia with an increase in the haemoglobin concentration in blood, which stabilized at a new equilibrium after a period of 7–10 days. This together with an increased oxygen extraction would help to maintain oxygen supply to fetal tissues.

Placental control of fetal growth

Placental growth, metabolism, function, and their ontogeny are described elsewhere in this volume. This chapter will focus on placental limitation of fetal growth. Interference with the utero-placental circulation by ligation of a single uterine artery induces fetal growth retardation in the rat and in the guinea-pig. The fetuses growing at the cervical end of the uterus, and closest to the ligated uterine artery, have their growth most severely restricted. Placental infarction is found in this region and death of the fetus nearest the cervix is not uncommon. Hypoxia and reduction in nutrient concentrations in the fetus each contribute to the failure of fetal growth. The uterine arterial cascade is also supplied by the ovarian vessels and this ensures survival and continued growth of the fetuses in the tubal end of the uterine horn. Experimentally, ligation of a single uterine artery has an advantage in that the fetuses in the contralateral horn serve as controls for the growth-retarded ones.

In sheep, injection of microspheres (15 μm diameter) into the uterine arteries to block placental arterioles and hence reduce uterine blood flow also causes late-onset growth retardation associated with hypoxaemia and hypoglycaemia of the fetus. The pattern of fetal growth is altered with

Table 7.1 Pathophysiological factors and their associated clinical conditions that result in growth retardation of the fetus. The experimental methods for induction of growth retardation by similar mechanisms are included for comparison; however, these may not be the only mechanisms retarding fetal growth. For details see text

Pathophysiological factors	Clinical conditions	Experimental methods
Placental failure		
Reduced utero-placental blood flow	Pre-eclampsia; recurrent idiopathic growth retardation	Embolization of uterine/umbilical circulation
Smaller surface area for exchange	Lupus obstetric syndrome; uterine malformation; fibroids; placental infarction	Clamp uterine artery; ligate interplacental vessels or uterine artery
Altered maternal substrate concentrations		
Hypoglycaemia	Malnutrition; undernutrition; anorexia nervosa; bulimia	Undernutrition
Hypoxia	Altitude; anaemia; maternal cyanotic heart disease	Hypobaria or hypoxia
Reduced fetal drive/support for growth		
Chromosomal anomalies	Trisomies 13, 18, and 21 Chromosome 15 with loss of IGF-I receptor	Gene deletion
Gene defects	Phenylketonuria	
Endocrine disorders	Pancreatic agenesis	Fetal pancreatectomy
	Fetal thyrotoxicosis	
	Athyroid fetus	Thyroidectomy
Fetal anomalies	Renal agenesis	Nephrectomy
	Congenital heart disease	Clamp ascending aorta
External or internal inhibition of fetal growth		
Maternal smoking; alcohol or drug abuse		
Therapeutic drugs	Cytotoxic agents	
Irradiation	Radiation therapy	
Maternal disease	Severe infections; malignancy	
Fetal infections	Viral, e.g. rubella, cytomegalovirus, toxoplasmosis	Viral infection
Maternal age	< 16 or > 35 years	'Prepubertal' pregnancy

relative sparing of brain growth, whereas other organs, such as the liver and gut, are disproportionately small. These changes in organ size parallel the redistribution of cardiac output in the growth-retarded fetus. Similarly, a sudden reduction of umbilical blood flow can retard fetal growth. This is readily achieved by ligation of one of the two umbilical arteries in sheep or by ligation of the inter-placental vessels in the rhesus monkey, which in the latter destroys the blood supply to the secondary placental disc of the bipartite placenta. However, if the interplacental vessels are ligated too early in gestation, the remaining placenta is able to compensate because fetal growth continues normally. Ligation in late pregnancy is more likely to result in sudden fetal death from asphyxia.

Embolization of the umbilical circulation in sheep also induces fetal growth retardation. An increase in vascular resistance within the umbilical circulation and altered flow velocity waveforms occur and these are similar to those associated with idiopathic growth retardation in women. Embolization of the umbilical circulation with 8–16 million 15 μm microspheres daily for 9 days reduced total umbilical blood flow by 33 per cent and reduced placental weight (283 ± 76 g in embolized versus 405 ± 137 g in controls) leading to fetal hypoxaemia with a fall in arterial oxygen content of 39 per cent. Unlike many other experiments altering placental size or vasculature, there was no compensatory increase in concentration of haemoglobin in the fetus. Fetal weight was reduced by 20 per cent but over this short period there was no difference in crown–rump length.

Embolization of the maternal or fetal circulations or ligation of placental blood vessels in late pregnancy in sheep results in the sudden onset of fetal growth retardation after a period of normal growth. This is unlikely to be similar to the circumstances commonly leading to intrauterine growth retardation in women, where the adverse influences on growth are usually present from early in pregnancy, even though growth failure may not be recognized until late in pregnancy.

Placental growth can be restricted from the beginning of pregnancy in sheep by limiting the number of implantation sites. The sheep placenta consists of 40–90 placentomes (cotyledons), which form on specialized areas of the non-pregnant uterus known as caruncles. Excision of the majority of caruncles before pregnancy restricts total placental size by limiting the number of placentomes in a subsequent pregnancy. Currently, this is probably the best characterized example of experimental intrauterine growth retardation and it will be described in some detail.

Severe restriction of placental growth can lead to very severe fetal growth retardation and early death before 60 days of pregnancy. Most of the experimental studies have focused on less extreme but still substantial growth failure

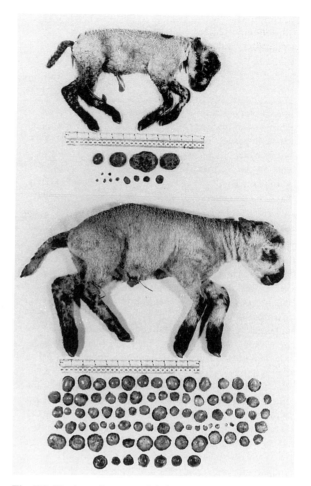

Fig. 7.2 Singleton fetuses and their placentae. The upper fetus weighed 1186 g and was delivered at 127 days of pregnancy from a ewe that had endometrial caruncles removed before pregnancy. Its placenta has 14 placentomes and weighed 71 g. The lower, control fetus was delivered a day earlier in pregnancy and weighed 3840 g, and the placenta, which has 88 placentomes, weighed 446 g. (Harding, 1982, reprinted with permission.)

(Fig. 7.2). Usually fetuses have been categorized as growth-retarded when their weights were below 2 SD of the mean for control fetuses of the same flock. In these circumstances an approximate timing of events can be drawn up indicating when deviations from normal can be recognized (Table 7.2).

The individual placentomes of the sheep placenta reach their maximum diameter early in pregnancy. Recently, an

Table 7.2 Approximate timing of events when placental growth is restricted by prepregnancy excision of endometrial caruncles. For details see text

	Gestational age (days)
Compensatory growth of the placentomes	> 60
Reduced utero-placental blood flow	< 80
Chronic hypoxaemia	<110
Chronic hypoglycaemia	<110
Acute-on-chronic hypoxaemia	>120
Fetal 'wasting'	>120
Fetal death common	>130
Term	145–150

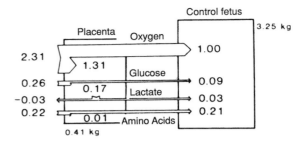

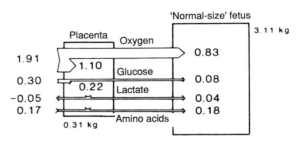

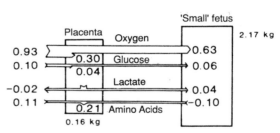

Fig. 7.3 Partition of substrates between the fetus and the utero-placental tissues (placenta) in sheep. Rates of flux of substrates between mother, placenta, and fetus are given in mmol/min. Placental growth was restricted in the 'normal' and 'small' groups by pre-pregnancy excision of endometrial caruncles. The 'normal'-sized fetuses have restricted placental growth and birthweight within 2 SD of the mean for controls. The weights of the growth-retarded fetuses ('small') were more than 2 SD below that mean. (Owens *et al*, 1988, reprinted with permission.)

ultrasound method has been developed that allows the diameters of placentomes to be followed with time. This method has shown that, after excision of caruncles, the placentomes grow at a faster rate but still reach their maximum diameter by 70 days of pregnancy. Successful implantation of chronic fetal catheters has been achieved as early as 75 days of pregnancy and have been used to show that utero-placental blood flow is reduced by 80 days of pregnancy in the growth-retarded animals (term is 145–150 days in the sheep). Oxygen tension is similar in control and growth-retarded animals at this stage of pregnancy, but the growth-retarded fetuses (weight more that 2 SD below the mean of controls) have become hypoxaemic and polycythaemic by 110 days of pregnancy. Less severely growth-retarded fetuses may gradually become hypoxaemic and polycythaemic over the last 20 days of pregnancy. In the growth-retarded sheep fetus, oxygen is redistributed from the placenta to the fetus and placental oxygen consumption per unit mass falls to about one-third of that of the control placenta (Fig. 7.3). The apparent margin of safety of the supply of oxygen to the fetus is also decreased (Fig. 7.4).

Thus, it is hardly surprising that by 120 days of ovine pregnancy additional acute episodes of hypoxia are superimposed on the chronic hypoxaemia (i.e. acute-on-chronic hypoxia). In sheep, these have been recorded by continuous measurement of oxygen saturation or tension. The episodes occur when the uterus contracts (uterine contractures) and temporarily reduces uterine blood flow and hence oxygen delivery. Hypoxic episodes also occur when the fetus becomes active and increases its oxygen consumption. Episodes of either type can exceed the margin of safety for the supply of oxygen. These acute-on-chronic falls in oxygen availability in the growth-retarded fetus are frequently accompanied by heart rate decelerations. These

changes imply a response to the acute hypoxic episode triggered either by a chemoreceptor response or perhaps, even by a direct depressive effect on the myocardium. These episodes may be present intermittently for many days before fetal demise or delivery occurs. In normally grown fetal sheep, experimental acute hypoxia causes a redistribution of cardiac output with increased blood flow to the brain and adrenal glands. The attendant endocrine

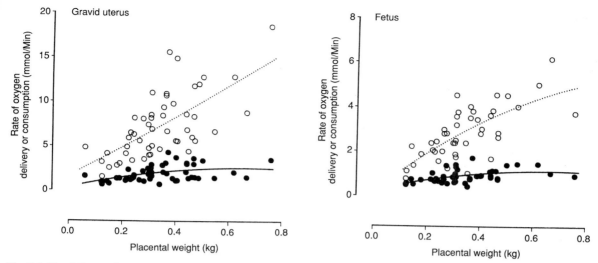

Fig. 7.4 The delivery of oxygen (dotted line) is the product of blood oxygen content and blood flow for the uterine and umbilical circulations. Oxygen consumption (solid line) represents the concentration difference for oxygen in the uterine (arteriovenous) and in the umbilical (veno-arterial) circulations multiplied by utero-placental and umbilical blood flow, respectively. The margin between the delivery of oxygen and its consumption of oxygen decreases rapidly with placental weight making the fetus vulnerable to further acute episodes of hypoxia.

stress response with large increases in blood concentrations of catecholamines, ACTH, and cortisol, also develops rapidly with acute-on-chronic hypoxia in the growth-retarded fetus. In normally grown fetuses, fetal activity or uterine contractions also cause falls of oxygen saturation or tension, albeit from higher baseline values, but are not associated with heart rate decelerations. In addition, these changes in uterine activity may be sufficient to change the fetal behavioural state from low-to-high-voltage electrocortical activity.

The changes in glucose delivery and consumption within the conceptus are similar to those for oxygen when placental and fetal growth is restricted in sheep (Fig. 7.3). Chronic hypoglycaemia is present by 110 days; however, fetal glucose consumption is maintained constant per unit mass of fetus by redistribution of glucose from the placenta to the fetus and by increased endogenous production of glucose. Presumably the growth rate of the fetus is modified to match the supply of glucose to the fetus and this is likely to be mediated by endocrine changes resulting from the altered nutrient concentrations. Lactate concentrations are more variable but usually become elevated in the growth-retarded fetus with advancing gestational age. Shortly before death of the fetus, lactate concentrations become very high and an increasing acidaemia develops.

Amino acid supply is more complex. In late pregnancy, there is a reversal of flux of amino acids so that the branched-chain and some non-essential amino acids are taken up by the placenta from the umbilical and uterine circulations rather than being extracted from umbilical blood by the fetus. This is accompanied by a reduction in the output of branched-chain keto acids and an increase of lactate output from the placenta to the fetus. This loss of amino acids from the fetus could be sustained by wasting of fetal tissues, such as liver and muscle. As evidence of this wasting, loss of amino acids from the fetal hind limb of a normally grown fetus occurs within 3 days of starvation of the mother.

The chronic restriction of substrate and (acute-on) chronic hypoxaemia are the probable causes of the wide variety of endocrine changes that have been documented for the growth-retarded fetus. These endocrine changes, in turn, modify the rate and pattern of fetal growth and of fetal and placental nutrient utilization and exchange as noted above. Hormones that support or promote growth and whose concentrations are reduced in the blood of growth-retarded fetuses include insulin, thyroxine, triiodothyronine, and the insulin-like growth factors-I (IGF-I) and-II. The concentrations of insulin and IGF-I correlate closely with plasma glucose. Insulin-like growth factor-II concentrations are only reduced in the most severely growth-retarded fetuses. The actions of the IGFs are modulated by their binding proteins so that determining the precise role of the IGFs will also require examination of the changes in the concentrations of IGF binding

Table 7.3. Blood gas, metabolic, and endocrine changes associated with intrauterine growth retardation.

Decreased	Unchanged	Increased
PO_2	pH	PCO_2
O_2 content		O_2 carrying capacity (Hb conc.)
Glucose		Lactate
Thyroxine	ACTH	Noradrenaline
Reverse T3	Placental	Adrenaline
T3	lactogen	Cortisol
	Growth	β-endorphin
Insulin	hormone	
IGF-I		
IGF-II		

proteins that accompany growth retardation. In late gestation the plasma concentrations of cortisol, catecholamines, and glucagon are high in the growth-retarded fetus. In addition to very high baseline concentrations of cortisol, noradrenaline, and adrenaline in the hypoxaemic and growth-retarded fetus, further rises of their concentrations occur when the fetus is subjected to further, acute hypoxia. The endocrine changes are summarized in Table 7.3.

Retardation of fetal growth by limitation of placental size is accompanied by an alteration in the pattern of growth, which is similar to the asymmetrical growth retardation of the human fetus. Similar disproportionate growth of fetuses is produced by other methods of reducing placental size, for example, by restriction of uterine or umbilical blood flow in species with short gestation where the insult is present for just a few days or even in species of long gestation in which the effects of the insult evolve slowly throughout pregnancy or where the insult is suddenly imposed in mid- or late-pregnancy. The most severe restriction of fetal growth is found in organs such as the spleen, thymus, liver, and gut. Histomorphometric examination of the gut shows that growth of all layers of the small intestine is reduced and villus height is significantly less than normal, suggesting that absorption of nutrients postnatally may also be altered.

Other organs, such as the heart and the kidney, grow in proportion to body weight. In late pregnancy, acceleration of growth of the fetal adrenal gland occurs earlier than in normally grown fetuses and reflects the premature activation of the adrenal cortex probably by repeated episodes of spontaneous hypoxia.

In the growth-retarded fetal sheep the brain is relatively spared so that it becomes larger relative to body weight than in normally grown animals. However, it must be emphasized that the brain of the growth-retarded fetus weighs less and is also structurally different from normal. Nerve cells remain similar in size relative to controls in the cerebrum and cerebellum but contain more mitochondria. However, there is a significant reduction in the thickness of the motor and visual cortices and an increase in the numerical density of neurones in these areas in the growth-retarded fetuses, indicating a significant reduction in the growth of the neuropil. In the growth-retarded fetus the numerical density of synapses in layer 1 of the visual cortex is greatly reduced. In the optic nerve there is a significant reduction of the myelin sheath relative to the diameter of the axon, while in a peripheral nerve (trochlear nerve) the diameters of both the myelin sheath and the axon are reduced. The significance of these changes on long-term functioning has not been investigated in the sheep. In other species, including the human, the growth-retarded fetus has a greatly increased risk of neurological handicap later in life.

Intrinsic constraint to fetal growth

Fetal growth retardation can result from internal constraints to growth, some of which are reviewed elsewhere in this volume. Ablation of fetal organs such as the kidney, thyroid, or pancreas all retard fetal growth, although not in all species. Timing of the ablation or excision of fetal organs is important; for example, early thyroidectomy or nephrectomy in sheep causes severe growth retardation, whereas excision in late pregnancy does not, even when due allowance is made for the shorter duration of the insult. However, maturation of bone development is delayed.

The role of the fetal pituitary is also discussed elsewhere. Removal of the pituitary in late pregnancy is not associated with restriction of growth of the fetus.

Human growth retardation

Clinical associations

Many studies have examined the associations of low birthweight and/or growth retardation with socio-economic and clinical factors. This section describes these very briefly and the reader is referred to more comprehensive reviews of this subject elsewhere. The percentage of babies who have birth weights of less that 2500 g varies widely between and within different communities. Typically, the incidence of low-birth weight babies is about 5–6 per cent in developed communities. However, the incidence of low-birth weight babies varies with time. For example, recently there has been a significant increase in low-birth weight babies in New York, which has been attributed to the rapid increase in substance abuse in that city. Transient increases in the incidence of low-birth weight babies have also occurred with catastrophic events such as famine. The Dutch famine of the 1940s led to reduced birth-weight of babies exposes *in utero* to famine in late pregnancy. A second-generation effect was observed for children of women who were exposed *in utero* to the famine conditions in the early part of their gestation even though their own birth weight had not been affected. This second-generation effect may be more generalized, since it has been shown that women who were small-for-dates are at increased risk of giving birth to a small-for-dates infant (odds ratio 2.21, 95 per cent confidence interval 1.41–3.48). In developing countries, a high rate of low birth weight (< 2500 g) of 40 per cent or more at term has been associated with low blood glucose concentrations in the mother.

National and hospital-based studies have shown that the major factors associated with reduction in birth weight include maternal size (height or weight-for-height), history of smoking in pregnancy, parity and pre-eclampsia, previous delivery of a low-birth weight infant, and weight gain in pregnancy. However, poor maternal weight gain is not a good predictor of intrauterine growth retardation.

Chronic hypoxaemia probably retards fetal growth in women to a similar degree to that observed in experimental animals. Several studies of infants born to women living at high altitude have shown that birth weight is reduced when compared to infants born at sea level after allowance for socio-economic differences. A similar mechanism may be responsible for the mild fetal growth retardation that commonly occurs in women with cyanotic heart disease.

Placenta and intrauterine growth retardation

The first quantitative description of the changes in the human placenta in pregnancies complicated by growth retardation was published just 25 years ago. The 10 infants were more than two standard deviations below the birth weight-for-gestational ages from 36 weeks to term. The mean volume of the affected placentae was 350 mL compared to 488 mL in normal pregnancies. The proportion of villous to intervillous tissue was similar in both groups but the villous surface area was reduced to 6.4 m^2 compared with 11.0 ± 1.3 (SD) m^2 in controls. A similar reduction

in fetal capillary surface area has been reported. Unfortunately, these descriptions were not always accompanied by comment on the aetiology of growth retardation and, in at least one study, a significant number of the mothers smoked during pregnancy. Recently, emphasis has been placed on the interaction of the endovascular trophoblast and the spiral arterioles. In normal pregnancy approximately 120 spiral arterioles are recruited to form the maternal vascular supply to the placenta. In growth retardation this number is reduced to 80.

The 'physiological' changes in the spiral arterioles that accompany normal pregnancy include invasion of the spiral arteriole with replacement of the intima and muscular coats of the vessels by trophoblasts and later withdrawal of trophoblastic tissue. This leaves a greatly dilated spiral arteriole extending into the myometrium with a 700-fold reduction in vascular resistance to blood flow. In pregnancies accompanied by idiopathic growth retardation, this 'physiological' adaptation of the spiral arterioles either does not occur or is less extensive and confined to the decidual portion of the arteriole. Late in pregnancy a further wave of trophoblast invasion occurs and appears to partially plug these narrow vessels, which then resemble those found in women with severe pre-eclampsia. The pathogenesis of pre-eclampsia is thought to involve an imbalance of the vasodilator and vasoconstrictor prostanoids favouring vasoconstriction. These changes in prostanoids can be mimicked in *in vitro* culture of placental tissue in a hypoxaemic environment. These findings form part of the theoretical basis for treatment with low-dose aspirin of women at risk of a pregnancy complicated by intrauterine growth retardation or pre-eclampsia.

The 'fetal' component of the placenta is also smaller or compromised in association with growth retardation. Infarction is common and other focal lesions that can reduce placental function include placental abruption and haematomas. Placental abnormalities such as placenta praevia, placenta membranacae, and circumvallate placentae are also associated with growth retardation of the fetus. In recent studies, histomorphometric analysis has shown reduced numbers of tertiary stem villi and their associated arterioles, which are the principal resistance vessels within the placenta. These changes closely correlate with alterations in ultrasound measurements of flow velocity waveforms in the umbilical arteries.

Ultrasonography has been utilized to measure placental volume and fetal growth serially from 16 weeks in women. Placental volume is estimated by making parallel transverse sections and by measuring the area on each scan and the distance between each section. Placental and fetal growth followed a sigmoid or nearly linear pattern. Abnormal development with slowing of growth of the placenta always preceded intrauterine growth retardation by at least 3 weeks.

Metabolic and endocrine state of the human growth-retarded fetus

Our knowledge of the physiology of the biochemical, metabolic, and endocrine characteristics of the growth-retarded human fetus has rapidly increased due to the advent of the technique of cordocentesis, which is also called percutaneous umbilical blood sampling (PUBS). This became possible with the development of high-resolution, real-time ultrasound. It enables the investigator to insert a needle directly into the umbilical artery or vein with remarkable safety as evidenced by a fetal loss rate of < 1 per cent in a series of 606 patients undergoing cordocentesis. However, a higher fetal loss rate may occur when the pregnancy is complicated by intrauterine growth retardation. Fetal bradycardia is not uncommon during the procedure, particularly when the umbilical artery is punctured. This sometimes requires immediate delivery by Caesarean section because of fetal compromise. Presumably, this fetal bradycardia is initiated by spasm of the punctured vessel, a process that may alter some of the variables of interest such as blood gas values.

Normal ranges have been obtained from fetuses that were sampled because of a risk of a fetal infection, alloimmunization (isoimmunization), or other potential problems but were subsequently shown to be healthy. Oxygen tension falls progressively with gestational age in the umbilical vein and is accompanied by a rise in haemoglobin concentration so that there is little change in oxygen content. No significant changes with age were found for pH or PCO_2.

Several studies have shown that the growth-retarded human fetus that does not have structural abnormalities is likely to be hypoxaemic and polycythaemic. Reduction in glucose and increases in lactate and PCO_2 in blood correlated with the severity of the hypoxaemia. However, low blood pH was only found with very low PO_2 values when abnormal cardiotocographic findings were also present. In these circumstances, 8 of 10 fetuses sampled had a low pH. The lowest oxygen tensions observed in the umbilical vein of the growth-retarded fetus were less than 10 mmHg. At this level, fetal blood contains little oxygen, leaving little or no margin of safety for the supply of oxygen to the fetus, even if increased oxygen extraction occurs.

The metabolic changes in the growth-retarded fetus include hypoglycaemia and elevated blood levels of triglycerides, which correlate with fetal hypoxia and with the maternal-fetal gradient for glucose. The alteration in metabolism also correlates with abnormal flow velocity waveforms in the umbilical artery. A curvilinear relationship between umbilical lactate concentrations and waveforms defined by the pulsatility index adds weight to this measurement as a test of fetal well-being.

Alpha amino nitrogen concentrations in the blood of the growth-retarded fetus are significantly lower than in controls. Irene Cetin and her colleagues found that most of this difference could be accounted for by a reduction in the concentrations of the branched-chain amino acids, valine, leucine and isoleucine. These changes are in keeping with a substantial increase in utilization of amino acids for energy metabolism as glucose becomes limiting. The source of these amino acids in the circulation of the growth-retarded human fetus may be similar to that noted above for the growth-retarded sheep fetus. There is no veno-arterial concentration difference for amino acids in the umbilical circulation of the growth-retarded human fetus unlike the situation in the normally grown human fetus. Studies using stable isotopes suggest that transplacental transport supplies only about 60 per cent of leucine in the fetal circulation. It has also been reported that there is a reduction in the concentration of essential amino acids in fetal blood and changes in the concentrations of non-essential amino acids. These recent findings, utilizing cordocentesis, are in keeping with earlier reports obtained from cord blood samples at delivery. Significant fetal uptake of amino acids occurs in early gestation, while, at term, no significant uptake of non-essential amino acids could be demonstrated by measurement of umbilical veno-arterial differences. These changes in the growth-retarded fetus are compatible with increased utilization of amino acids for energy metabolism and the mobilization of fetal nitrogen and protein stores leading to fetal wasting, which has been observed by serial ultrasound examinations of the growth-retarded fetus.

Cordocentesis has also been used to determine the endocrine changes associated with growth retardation. While this avoids the effects of labour and/or delivery, it is still possible that endocrine changes may occur due to the sampling procedure. Circulating concentrations of hormones that support fetal growth, such as insulin and thyroid hormones, are reduced. 'Stress' hormones are elevated, including cortisol, ACTH, and the catecholamines.

Supplementation

Supplementation has been used in an attempt to improve fetal growth. The topic has been reviewed in depth recently by Jane Harding and Valerie Charlton and only one aspect will be discussed here. Too often, studies on supplementation have not included adequate control groups to allow assessment of the effects of supplementation and in only one study has supplementation been given to restore to normal an identified reduction in oxygen tension in the fetus.

Hyperoxia has been used to overcome the hypoxaemia in the growth-retarded fetus detected by cordocentesis. When the mothers breathed humidified oxygen (55 per cent), umbilical arterial or venous PO_2 increased into the normal range. This hyperoxia was given for periods of 5 days to 9 weeks. Five fetuses survived with a minimum of morbidity, but there was no objective evidence that fetal growth was improved. Short-term effects of hyperoxia in the human fetus include a redistribution of blood flow within the circulation, since resistance to flow increases in the cerebral circulation and falls in the peripheral vessels. Thus, alteration in the supply of scarce nutrients to essential organs may be one consequence of this therapy. Failure of this adaptation to hyperoxia identified a group of fetuses at very high risk of fetal distress requiring operative delivery.

Genetic influences on fetal growth

The paternal genome is essential for trophoblast development, and normal embryonic development requires the presence of the maternal genome. In the presence of two copies of the paternal genome, trophoblastic tumours occur. Inactivation of single genes can also lead to alteration of fetal and placental growth. This imprinting phenomenon was clearly demonstrated by the gene-targeting experiments that resulted in the loss of an IGF-II gene. For normal fetal and placental growth, the IGF-II gene has to be inherited from the father. In contrast, the IGF-II receptor gene has to be inherited from the mother or else fetal death occurs. Overexpression of the IGF-II gene experimentally in mice, or spontaneously occurring paternal disomy in humans, leads to fetal overgrowth. Maternal disomy (inactivation of the IGF-II gene) leads to proportionate dwarfism in mice and it will be intriguing to see if this occurs in humans. This exciting molecular genetic approach is likely to expand and offers great potential towards the elucidation of the mechanisms regulating fetal and placental growth.

Conclusion

Further definition of the conditions associated with failure of fetal growth in women may lead to novel methods for enhancing the rate of growth of the fetus. However, it is also possible that fetal growth retardation becomes irreversible after a period of constrained growth. This has been described after restriction of fetal growth by maternal undernutrition in sheep for a period as short as 21 days. The determination of mechanisms responsible for this irreversibility is an important research goal.

Most important of all will be the careful transformation of new ideas generated by experimental and clinical physiology into rational and effective care for the pregnant woman and her fetus. For example, David Rush noted that 'attempts at nutritional supplements, while well intentioned, have not always had the desired effect'. Dietary supplementation with added protein in women to improve fetal growth has been associated with a decrease rather than increase in mean birthweight. Even more serious consequences followed the use of a synthetic oestrogen (stilboestrol) which was given to women to enhance fetal growth. Not only was this therapy unsuccessful in its initial aims but, later in adult life, it also resulted in vaginal carcinoma and reproductive failure, a harsh result of a poorly tested and evaluated treatment.

The series of studies on the role of low-dose aspirin in pregnancy is reassuring that well designed studies may provide a valid basis for effective care. Initial epidemiological evidence suggested that aspirin may protect against pre-eclampsia and growth retardation. Later small trials supported this conclusion and noted enhancement of fetal growth. Currently, large-scale trials are in progress to determine the effectiveness and safety of this new use for an old drug. Similar experimental designs will be required to transfer some aspects of the recent explosion of knowledge of intrauterine life to effective and safe clinical practice.

Further reading

Aherne, W. and Dunnill, M.S. (1966). Quantitative aspects of placental structure. *Journal of Pathology and Bacteriology*, **91**, 123–9.

Arduini, D. Rizzo, G., Mancuso, S., and Romanini, C. (1988). Short-term effects of maternal oxygen administration on blood flow velocity waveforms in healthy and growth-retarded fetuses. *American Journal of Obstetrics and Gynaecology*, **159**, 1077–80.

Barcroft, J. (1946). *Researches of pre-natal life*. Blackwell, Oxford.

Boyd. P.A. (1987). Placenta and umbilical cord. In *Fetal and neonatal pathology*. (ed. J.W. Keeling), pp. 45–76 Springer-Verlag, Berlin.

Cetin, I., Marconi, A.M. Bozzetti P., Sereni, L.P., Corbetta, C., Pardi, G., and Battaglia, F.C. (1988). Umbilical amino acid concentrations in appropriate and small for gestational age infants: a biochemical difference *in utero*. *American Journal of Obstetrics and Gynecology*, **158**, 120–6.

Creasy, R. (1989). Intrauterine growth retardation. In *Maternal-fetal medicine: principles and practice* (ed. R.K. Creasy and R. Resnik), pp. 547–64. Saunders, Philadelphia.

Creasy, R.K., Barrett, C. T., de Swiet, M., Kahanpaa, K.V., and Rudolph, A.M. (1972). Experimental intrauterine growth retardation in the sheep. *American Journal of Obstetrics and Gynecology*, **112**, 566–73.

Dawes, G.S. (1968). *Foetal and neonatal physiology*, pp. 42–59. Year Book Medical Publications, Chicago.

DeChiara, T.M. , Robertson, E.J., and Efstratiadis, A. (1991). Parental imprinting of the mouse insulin-like growth factor II gene. *Cell*, **64**, 849–59.

Economides, D.L. and Nicholaides, K.H. (1990). Metabolic findings in small-for-gestational age fetuses. *Contemporary Reviews in Obstetrics and Gynaecology*, **2**, 75–9.

Ferguson-Smith, A.C., Cattanach, B.M. Barton, S.C. Beechey, C.V., and Surani, M.A. (1991). Embryological and molecular investigations of parental imprinting on mouse chromosome 7. *Nature*, **351**, 667–70.

Harding, J.E. (1982). *Placenta and growth of the foetus*. DPhil thesis, University of Oxford.

Harding, J.E. and Charlton, V. (1989). Treatment of the growth-retarded fetus by augmentation of substrate supply. *Seminars in Perinatology*, **13**, 211–23.

Harding, J.E. and Charlton, V. (1990). Experimental nutritional supplementation for intrauterine growth retardation In *The unborn patient* (ed. M.R. Harrison, M.S. Golbus and R.A. Filly), pp. 598–613. Saunders, Philadelphia.

Henry, I. Bonaiti-Pellie, C., Chehensse, V., Beldjord, C., Schwartz, C., Uterman, G., and Junien, C. (1991). Uniparental paternal disomy in a genetic cancer-predisposing syndrome *Nature*, **351**, 665–7.

Manning, F.A. (1989). General principles and application of ultrasound. In *Maternal-fetal medicine: principles and practice* (ed. R.K. Creasy, R. Resnik), pp. 195–243. Saunders, Philadelphia.

Manning, F.A., Menticoglou, S. and Harman, C. (1989). Fetal assessment by biophysical methods: ultrasound. In *Obstetrics* (ed. A. Turnbull and G. Chamberlain), pp. 383–98. Churchill Livingstone, Edinburgh.

Owens, J.A., Falconer, J. and Robinson, J.S. (1989). Glucose metabolism in pregnant sheep when placental growth is restricted. *American Journal of Physiology*, **257**, R350–R357.

Owens, J.A., Owens, P.C., and Robinson, J.S. (1988). Experimental fetal growth retardation: metabolic and endocrine effects. In *Advances in fetal physiology: reviews in honour of G.C. Liggins* (ed. P.D. Gluckman, P.W. Nathanielsz, and B.M. Johnston), pp. 263–86. Perinatology Press Ithaca, New York.

Robertson, W.B., Khong, T.Y., Brosens, I., Wolf, F.D., Sheppard, B.L., and Bonnar, J. (1986). The placental bed biopsy: review from three European centres. *American Journal of Obstetrics and Gynecology*, **155**, 401–12.

Robinson, J.S. (1989). Fetal growth. In *Obstetrics* (ed. A. Turnbull and G. Chamberlain), pp. 141–50. Churchill Livingstone, Edinburgh.

Rush, D. (1989). Effects of changes in protein and calorie intake during pregnancy on the growth of the human fetus. In *Effective care in pregnancy and childbirth* (ed. I. Chalmers, M. Enkin, and M.J.N.C. Keirse), pp. 255–80. Oxford University Press, Oxford.

Scammon, R.E. and Calkins, L.A. (1929). *The development and growth of the external dimensions of the human body in the fetal period*. University of Minnesota, Minneapolis.

Soothill, P.W., Nicholaides, K.H., and Campbell, S. (1987). Prenatal asphyxia, hyperlacticaemia, hypoglycaemia, and erythroblastosis in growth retarded fetuses. *British Medical Journal*, **294**, 1051–6.

8. Development of the cardiovascular system

Kent L. Thornburg and Mark J. Morton

Early development of the heart

Cardiac development is an especially exciting area of research today. Though the descriptive features of cardiac morphogenesis have been known for most of this century, new techniques in experimental embryology and molecular biology have shed new light on the processes that regulate the formation of the heart.

Even with the considerable differences among species, the general principles of heart development have allowed the categorization of cardiac morphogenesis into distinct periods and stages for both avian and mammalian hearts. We use the term 'growth' loosely to indicate tissue alteration. The formation of the heart involves rapidly changing and complex cellular and extracellular interactions. Thus, cell hyperplasia, hypertrophy, migration, and death are all normal features of cardiac tissue modification during heart development and all are included in our use of the word 'growth'.

Precardiogenesis (chick < 30 h incubation; human < 16 postovulation days)

At the cephalic end of the early embryo in the immediate postgastrulation phase, two paired lateral areas in the splanchnic mesoderm contain crescent-shaped areas of cells that represent the precursors (anlage) of the future heart. As these two areas fold inwardly ventrally 'beneath' the embryo proper, their cells become organized into bilateral simple tubular structures. These paired tubes grow more closely together as the tissue folds and forms the cavity that becomes the primitive foregut (Fig. 8.1(a)–(c)).

Fusion (chick 30–33 h; human 21 days)

These heart anlage then fuse to make a single, larger tube with mesodermal remnants 'above' forming the dorsal mesocardium. If fusion of the paired heart primordia is prevented through experimental interference, two functional paired hearts will develop nevertheless. After fusion, the heart becomes a simple tubular structure composed of three concentric layers: an endocardial lining (the

'inner tube'); an outer myocardium; and a filling between the two of a sticky gelatinous material (cardiac jelly). Cardiac jelly is rich in glycosaminoglycans, collagen, and other noncollagenous glycoproteins and is believed to be important: (1) in providing a highly permeable extracellular compartment; and (2) in allowing mechanical alterations during the looping process described next.

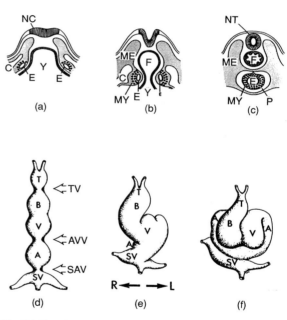

Fig. 8.1 Development of the embryonic heart. (a)–(c) Transverse sections of the early embryo; (d)–(f) looping processes and the arrangement of the chambers just prior to septation. C, coelomic cavity; E, endocardial tubes; F, foregut; ME, mesoderm; MY, myocardium; P, pericardial cavity; NC, neural crest; NT, neural tube; Y, yolk sac; A, atrium; B, bulbus cordis; SV, sinus venosus; T, truncus arteriosus; AVV, atrioventricular valve; SAV, sinoatrial valve; TV, truncal valve. (Reproduced with permission from Reller, M.D., McDonald, R.W., Gerlis, L.M., and Thornburg, K.L. (1991). *Journal of the American Society of Echocardiography*, **4**, 519–32.)

After fusion, the heart becomes surrounded by a primitive pericardial cavity; individual regions, which become specific areas in the fully formed heart, can already be identified. Figure 8.1 (d) shows the following structures diagrammatically, from the cephalic to caudal end: truncus arteriosus; bulbus cordis; ventricle; atrium; and sinus venosus. The atria and truncus are outside the pericardium leaving only the bulboventricular tube within.

Looping (chick 33–40 h; human 25–30 days)

As development proceeds, the primitive heart continues to grow and undergoes a series of shape changes known as looping. The midportion of the tube bulges rightward to form a C-shaped structure which then becomes S-shaped and is further transformed so that the bulboventricular portion becomes a U-shaped loop (Fig. 8.1 (e), (f)). During looping the dorsal mesocardium becomes unattached leaving the bulboventricular tube free to change position. The heart continues to change shape so that the atrium that was once caudal occupies a position cephalic to the ventricle. The atrium becomes a bilobate structure that forms the right and left atrial precursors. The bulbus cordis and atria are juxtaposed when looping is complete with the ventricle located in a medial and inferior position. The apposed walls of the 'U' consisting of bulbus and ventricle are gradually diminished, leaving a single bulboventricular chamber that is later divided into separate chambers by the formation of the interventricular septum. The primitive ventricle gives rise to the muscular portion of the left and right ventricle; the proximal portion of the bulbus cordis becomes the right ventricle. The middle portion of the bulbus will form the outflow tract for the ventricles, and the distal portion will form, along with truncus arteriosus, the aortic and pulmonary roots.

The rapid structural transformation during this period has attracted considerable attention, yet the mechanical forces underlying the process are not understood. Evidence is lacking for the importance of most mechanisms proposed thus far. Mechanisms such as intraluminal or intracardiac jelly hydrostatic pressures, differential rates of cell division and/or differentiation, or specialized extracellular helical fibres could be responsible for looping. One recent model is based on the hypothesis that anisotropy is present in the myocardial wall due to the helical arrangement of myofibrillar components and that these are responsible for the observed rotation of the heart tube. In this model, discontinuities in the structural components would underlie the bending of the heart that occurs simultaneously with rotation. While such models may have shortcomings in their simplicity, they offer a way forward by providing hypotheses to test.

Septation (chick 44+ h; human 30+ days)

In order for the heart to operate as a four-chambered series pump after birth, the heart must be partitioned by the formation of septa (Fig. 8.2). Septation involves several processes: partitioning of the atria and atrioventricular canal; inclusion of the sinus venosus and pulmonary veins into the right and left atria, respectively; separation of bulbotruncal tissue into aortic and pulmonary roots; formation of separate ventricular chambers derived from the single ventricle; and development of the valves.

Atrial septa

The individual atria become distinct entities as two septa dividing the single primitive atrial cavity are formed. The first, the septum primum, grows from the cephalic 'roof' of the atrium toward the atrioventricular canal, obliterating the remaining open interatrial channel (the ostium primum) as it progresses (Fig. 8.2). Fenestrae (holes) are then formed in the septum primum, which coalesce to form the ostium secundum, and interatrial communication is maintained. A second membranous septum, the septum secundum, arises from the wall of the atrium just to the right of the primary septum, and moves toward the atrioventricular canal. Like the primary septum it is an incomplete septum, however, leaving an oval-shaped window (foramen ovale) near the opening of the inferior vena cava. The membranous portion of the septum secundum functions as a flap-valve, preventing left to right blood flow during fetal life. After birth, upon the development of an interatrial pressure gradient, the foramen ovale becomes permanently closed. There are interesting anatomical variations in the interatrial septal valve. The membranous 'foramen ovale' of the sheep has its origin well inside the inferior vena cava and protrudes into the left atrium, giving it a windsock appearance. The left horn of the sinus venosus becomes part of the coronary sinus. The right horn is incorporated into the right atrium.

Interventricular septum

The myocardium continues to mature throughout the looping process. In the bulboventricular region, the endocardium invades the cardiac jelly and the myocardium takes on specialized features as the cardiac jelly gradually disappears in a region-specific manner. The outer layer of myocardium becomes more compact and recognizable as typical myocardial tissue. The inner layer develops into spongy trabeculated tissue characterized by complex folds and interstices lined by endocardium. While most of the trabeculae are ephemeral, some remain in the form of papillary muscle. The origins of the interventricular

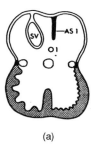
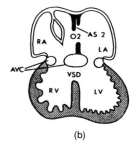
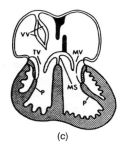

(a)   (b)   (c)

Fig. 8.2 Cardiac septation. (a) Early atrial and ventricular septation. (b) Atrial ostium primum closed and ostium secundum forming. Ventricular septation incomplete. (c) Septation complete and membranous septum formed. Atrioventricular valves formed by cushion tissue and myocardial delamination. AS1, atrial septum primum; AS2, atrial septum secundum; AVC, atrioventricular cushion; LA, left atrium; LV, left ventricle; MS, membranous septum; MV, mitral valve; O1, atrial ostium primum; O2, atrial ostium secundum; P, papillary muscle; RA, right atrium; RV, right ventricle; SV, sinus venosus; TV, tricuspid valve; VSD, ventricle septal defect; VV, venous valves. (Reproduced with permission from Reller, M.D., McDonald, R.W., Gerlis, L.M., and Thornburg, K.L. (1991). *Journal of the American Society of Echocardiography*, **4**, 519–32.)

septum also appear to be related to the rearrangement of tissue during trabeculation. This septum arises from the fusion of trabeculae and eventually extends from the apex and reaches toward the atrioventricular canal forming the muscular portion of the septum but leaves an interventricular channel open (Fig. 8.2). This channel (which shifts in position as the ventricle changes shape) is later closed by the membranous septum, which is derived from surrounding tissue, the bulbar ridges, and nearby cushion tissue.

Atrioventricular canal

In the region of the communication between the atria and ventricles (atrioventricular canal) the cardiac jelly thickens on the dorsal and ventral sides even while it is disappearing in the bulboventricular region. These thickenings, known as endocardial cushions, eventually meet in the central lumen and fuse to form the cushion septum which then obstructs the central portion of the atrioventricular canal (Fig. 8.2). Two lateral channels, the right and left atrioventricular orifices, are thus formed. The atrioventricular valves are formed from the cushion tissue surrounding the atrioventricular regions.

Partitioning of the truncus arteriosus

Two longitudinal, spirally arranged ridges form and fuse within the bulbotruncal region of the heart. After fusion, the partitioning structure becomes the aorticopulmonary septum, which separates anatomically, forming two distinct vessels, the aorta proper and the pulmonary trunk (Fig. 8.3). These vessels intertwine according to the placement of the septum. Simultaneously with the septation, the bulbus cordis is gradually incorporated into the ventricular walls.

Normal septation (partitioning) in the conotruncal region can occur only in the presence of neural crest cell infiltration. Neural crest cells are found laterally along the axis of the closing neural tube in early embryogenesis. Cells from the region of the mid-otic placode to somite 3 (Fig. 8.4) migrate along aortic arches 3, 4, and 6 where they support the structural formation of the great vessels. Some crest

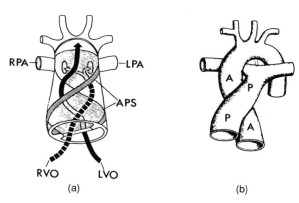

(a)   (b)

Fig. 8.3 Septation of truncus arteriosus. (a) The diagram shows spiral aorticopulmonary septum separating systemic arterial flow (solid dark line) from pulmonary arterial flow (broken line). (b) Post-septation arrangement of mature pulmonary artery and aorta. Note persisting spiral relationship. A, aorta; APS, aorticopulmonary septum; LPA, left pulmonary artery; LVO, left ventricle outflow pathway; P, pulmonary trunk; RPA, right pulmonary artery; RVO, right ventricular outflow pathway. (Taken with permission from Reller, M.D., McDonald, R.W., Gerlis, L.M., and Thornburg, K.L. (1991). *Journal of the American Society of Enchocardiography*, **4**, 519–32.

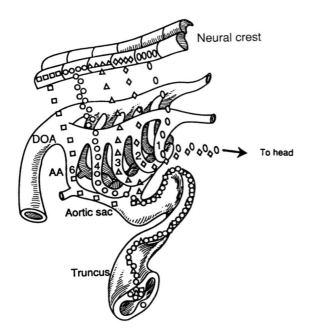

Fig. 8.4 Cranial neural crest migratory pathways. Neural crest cells from pharyngeal arches 3, 4, and 6 migrate from the pharyngeal region to the truncus arteriosus where they are important in the formation of the aorticopulmonary and truncal septa. DOA, dorsal aorta; AA, aortic arch. (Taken with permission from Kirby, M.L. and Waldo, K.L. (1990). *Circulation* **82**, 332–40.)

cells migrate beyond the arches to the heart where they contribute ectomesenchyme tissue to the aorticopulmonary and truncal septa. Outflow tract anomalies inevitably result if neural crest migration is prevented. Parasympathetic postganglionic neurones of the atrial and bulbar regions are also derived from neural crest cells.

Epicardium

Unlike the endocardium, the visceral epicardium is not formed simultaneously with the development of the early heart tube. Instead, epicardial cells originate from the mesothelium of the sinus venosus and migrate over the surface of the heart during looping.

Function of the embryonic heart

Chick

Through a number of intricate techniques, heart rate, determinants of cardiac output, blood pressure, and resistance to flow have all been studied in the early chick embryo. At

the ultrastructural level, the early chick heart has neither T-tubules nor sarcoplasmic reticulum; its myofibrils are arranged in odd configurations. Furthermore, the early embryonic heart has no officially designated valves, though the cushion tissue is believed to behave in valve-like fashion until authentic valves are formed. Yet the number of features of the embryonic heart that are similar in behaviour to the mature heart is surprising.

The heart begins to contract by 1.5 days' incubation and the rate is set by the fastest cells in the sinus venosus. Ventricular cells are also capable of spontaneous depolarization but at a slower pace. Heart rate in the embryo is very sensitive to changes in temperature and increases rapidly with early development and then less rapidly until hatching. This finding is in contrast to the mammalian heart rate that decreases over the last two-thirds of gestation.

Although the embryonic heart is not anatomically mature, its physiological behaviour is similar to that of the adult heart in that its output is regulated by the degree of filling (preload), the pressure against which it ejects (afterload), its strength of contraction, and the rate at which it beats.

Rat

The physiological responses of the embryonic hearts of the rat and chick have been compared at similar days of gestation. Figure 8.5 shows changes in blood pressure in the rat embryo compared to that of the chick. Both pressures increase with gestation as has been shown for the sheep

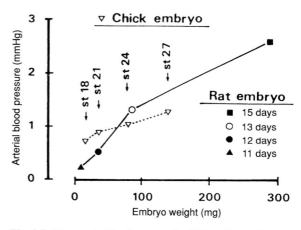

Fig. 8.5 Changes in blood pressure in chick and rat embryos with respect to wet weight of embryo. Stage (st) 18 is equivalent to 3 days; stage 21, 3.5 days; stage 24, 4.5 days; and stage 27, 5.5 days. (Taken with permission from Nakazawa, M., Miyagawa, S., Ohno, T., Miura, S., Takao, A. (1988). *Pediatric Research*, **23**, 200–5.)

fetus. Pressure profiles have striking similarities to adult tracings even though the valves have not fully formed. Although the pressure tracings from the chick and rat are similar at similar anatomical stages, it has been shown that each responds differently to the administration of caffeine, acetylcholine, and isoproterenol, indicating, perhaps, species-dependent differential development of receptor–effector systems in the myocytes themselves.

Early development of the circulatory system

The early embryonic circulation is characterized by a bilateral symmetry both for venous and arterial channels. Profound rearrangements of the primitive vascular structures result in the establishment of fetal and adult circulations.

Venous system

The early venous system is composed of an anterior and posterior cardinal system (Fig. 8.6), an umbilical venous system that drains the chorion, and a vitelline system from the yolk sac. Essentially all adult venous structures cephalad to the heart are derived from the anterior system; note that the superior vena cava is a remnant of the right anterior cardinal vein. Unlike the anterior system, the posterior cardinal system is largely replaced rather early in development (except for the azygous vessels) by subcardinal and supracardinal systems (Fig. 8.6).

Initially, the paired umbilical veins and vitelline veins carry blood directly to the sinus venosus from the placenta and yolk sac. But, as development proceeds, the vitelline vessels form anastomoses throughout the developing liver that are joined by anastomoses from the umbilical veins. This arrangement is replaced so that eventually a single vitelline vein enters the liver and joins liver sinusoids; the right umbilical vein is obliterated and a channel, the ductus venosus, is formed that allows much of left umbilical venous flow to bypass the liver vasculature and to deliver 'placental' blood directly to the inferior vena cava. In fetal sheep two umbilical veins drain separate portions of the placenta but join to form a common umbilical vein just inside the fetal body wall. Humans have but a single umbilical vein while keeping two umbilical arteries. As the pulmonary venous vasculature is becoming established, it drains into the anterior and posterior cardinal

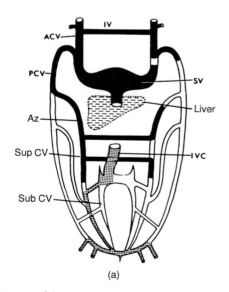
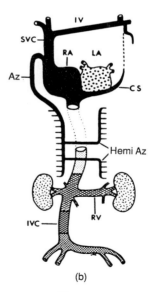

(a)   (b)

Fig. 8.6 Development of the systemic veins. (a) The constituent embryonic components. Those vessels that persist and drain into the superior vena cava are shown in black. Those that form the inferior vena cava are shaded. (b) The mature systemic venous pattern. The constituent portions of the inferior vena cava are indicated. ACV, anterior cardinal vein; AZ, azygos vein; CS, coronary sinus; Hemi Az, hemiazygos vein; IV, innominate vein; IVC, inferior vena cava; LA, left atrium; PCV, posterior cardinal vein; RA, right atrium; RV, renal vein; Sub CV, subcardinal vein; Sup CV, supracardinal vein; SVC, superior vena cava. (Taken with permission from Reller, M.D., McDonald, R.W., Gerlis, L.M., and Thornburg, K.L. (1991). *Journal of the American Society of Echocardiography*, **4**, 519–32.)

veins at several sites. With the development of individual lung buds and their associated airways, these plexiform veins are lost and replaced by two veins from each lung, which join in a separate common pulmonary chamber that drains into the left side of the unseptated atrium. The chamber is incorporated into the left atrium, which then accepts two right and left pulmonary veins.

Arterial system

During the early stages of cardiac looping, the paired dorsal aortae fuse and form a single dorsal aorta that persists throughout life (Fig. 8.7). The dorsal aorta supplies

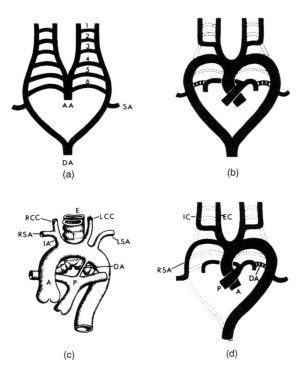

(a)   (b)

(c)   (d)

Fig. 8.7 (a) The basic primitive aortic arch pattern. (b) The hypothetical 'embryonic' double arch pattern. (c) The mature great arteries, normal pattern. (d) Embryological diagram of the constituent elements of the normal arteries. Nos 1–6, primitive embryonic aortic arches; AA, anterior aorta; A, aorta; DA, ductus arteriosus; E, oesophagus; EC, external carotid artery; IA, innominate artery; IC, internal carotid artery; LCC, left common carotid artery; LSA, left subclavian artery; P, pulmonary artery; RCC, right common carotid artery; RSA, right subclavian artery; T, trachea. (Taken with permission from Reller, M.D., McDonald, R.W., Gerlis, L.M., and Thornburg, K.L. (1991). *Journal Of the American Society of Echocardiography*, **4**, 519–32.)

blood to the developing embryo including the head via the internal carotid artery and the yolk sac and placenta via the vitelline and umbilical arteries. The midline arteries are part of the midline splanchnic system and eventually become the coeliac and mesenteric arteries. The umbilical arteries become branches of the iliac arteries even though they were the principal caudal arteries of the early embryo.

Blood flowing from the embryonic heart travels through the truncus arteriosus, through the aortic arch system, and is delivered to the distal tissues via paired dorsal aortae (Fig. 8.7 (a), (b)). The aortic arches are vessels that run through the pharyngeal arches (or branchial arches that supply respiratory gill surfaces in amphibians). The arches are columns of tissue that run in a nearly dorsoventral direction and are delineated by outpouchings of pharyngeal tissue. Each is supplied by a single aortic branch. In the hypothetical primitive arrangement, six aortic arches are symmetrically placed; in some species all of the six are found at some point of development though not necessarily simultaneously. In humans, most of the arch system is lost with remnants of the third (proximal common carotid artery) and fourth (aortic arch) arches being functional throughout life. The pulmonary arteries are branches of the sixth arch and run in the caudal direction toward the lung. A remnant of the sixth arch (ductus arteriosus) connects the pulmonary artery to the aortic arch during fetal life (Fig. 8.7).

Formation of the coronary vessels

The coronary arterial system is formed by the outbudding of primitive vessels off the aorta near left and right Valsalva sinuses; the coronary venous system originates from buds in the dorsal wall of the sinus venosus. Vascular channels form within the developing compact layer of the myocardium; some of these channels form continuities with the sinusoids of the trabecular layer which remain as Thebesian vessels in the adult.

Structural features unique to the fetus

The fetal heart, although four-chambered, does not perfuse two separate circulatory systems (pulmonary and systemic) as is characteristic of the adult circulation. Instead, the fetal circulation is arranged as a parallel system where right ventricular and left ventricular outputs mix (Fig. 8.8). A series of four shunts, each designed to close at birth, makes this arrangement functional during fetal life. These include the ductus venosus, the foramen ovale, the ductus

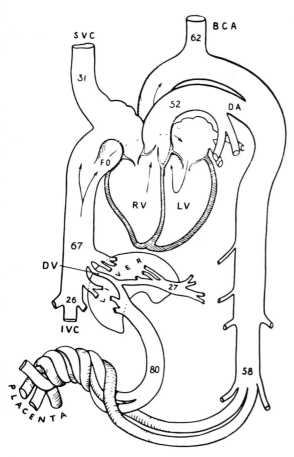

Fig. 8.8 Circulation of the mature fetal lamb. Numbers indicate mean oxygen saturation (per cent). RV, right ventricle; LV, left ventricle; SVC, superior vena cava; BCA, brachiocephalic artery; FO, foramen ovale; DA, ductus arteriosus; DV, ductus venosus. (Taken with permission from Born, G.V.R., Dawes, G.S. and Mott, J.C. (1956). *Journal of Physiology*, **134**, 149–66.)

left-to-right atrial pressure gradient. In most, but not all individuals, the foramen flap-valve fuses with the interatrial septum soon after birth preventing further interatrial shunting. It is becoming increasingly recognized that a persistently patent foramen ovale in adults is a mechanism for venous clots to embolize to the arterial system, producing stroke.

The ductus arteriosus carries the output of the right ventricle, except for the small portion of the flow that perfuses the lung, to the descending aorta. The ductus closes within 48 h of birth under the influence of oxygen-mediated prostaglandin inhibition. The placental circulation is lost as soon as the placenta is delivered and the umbilical cord severed. Tissue remnants of each of these shunts can be found in the adult. The ligamentum venosum and the ligamentum arteriosum are derived from the ductus venosus and ductus arteriosus, respectively. The fossa ovalis is the remnant of the foramen ovale. The umbilical vein and arteries leave behind the ligamentum teres and the lateral umbilical ligaments, respectively.

Common cardiovascular malformations

The variety of congenital defects found in the cardiovascular system is enormous. But there are many rather common abnormalities that occur as a result of faulty embryonic development that can be broadly categorized and understood in light of the above discussion of normal cardiogenic processes. Recent reports indicate that a primarily genetic cause (e.g. chromosomal or gene mutation) can be found in about 8 per cent of congenital heart disease patients. Environmental aetiologies (e.g. maternal rubella or alcohol abuse) are responsible for ~ 1 per cent. Unknown multifactorial inheritance factors underlie the remaining 90 per cent of cases. It has been reported that the incidence of cardiovascular malformations is just under 1 per cent of live births.

Atrial septal defects

Atrial septal defects (ASD) may occur because of improper growth of either the primum or secundum septae or improper incorporation of the sinus venosus. These produce ostium primum defect, ostium secundum defect, or sinus venosus ASD. A small-sized foramen ovale may retard left-sided cardiac growth *in utero*. Postnatally, a large ASD is associated with overfilling of the pulmonary circulation which worsens with the normal postnatal drop in pulmonary vascular resistance.

arteriosus, and the umbilical circulation. The ductus venosus directs the well oxygenated umbilical venous flow into the inferior vena cava where it passes preferentially across the foramen ovale. As mentioned before, the foramen ovale allows blood to pass from the right side of the circulation to the left. Therefore, oxygen-rich umbilical blood can travel from the inferior vena cava to the left ventricle, which supplies the two fetal organs that are critically dependent on a continuous oxygen supply, namely, the heart muscle and the brain. The foramen ovale closes mechanically at birth due to the establishment of a

Ventricular septal defects

Multiple defects occur during ventricular septation and are due to abnormal trabeculation, improper placement or growth of endocardial cushions, or misalignment of the conal (infundibular) septum. A ventricular septal defect (VSD) allows the shunting of blood in either a right-to-left or left-to-right direction depending on the differential pressures during the cardiac cycle. If circulatory beds are otherwise normal and valvular or subvalvular obstruction is not present, the higher pressures on the left side will cause left-to-right shunting and overfilling of the pulmonary circulation.

Endocardial cushion defects

The formation of a complete cushion septum is prerequisite to proper ventricular septation and valve formation. The leaflets of the tricuspid and mitral valves are derived from cushion tissue. Common cushion defects include VSD with anterior mitral leaflet cleft, complete atrioventricular canal with fused tricuspid, and mitral anterior or posterior valve leaflets. Cross-extension of chordae tendineae (these tether valve leaflets to papillary muscles) from one ventricle to the other through a VSD is also found. Cushion defects may also contribute to major heart anomalies such as mitral or tricuspid atresia (absence of atrioventricular valve orifices).

Persistent truncus arteriosus

A 'common' or persistent truncus arteriosus is the result of inadequate formation of the aorticopulmonary septum. While a number of anatomical variations are found, the formation of a single vessel that serves the ascending aorta and the pulmonary arteries is most common. In rare cases, the pulmonary arteries may even be absent. This defect can be caused experimentally by the removal of neural crest tissue before it migrates to the heart.

Transposition of great vessels

If the spiralling of the aorticopulmonary septum is inadequate, the pulmonary artery and aorta may be transposed so that systemic venous return is pumped through the right ventricle via the aorta to the body while pulmonary venous return is pumped back to the lungs by the left ventricle. Under these circumstances the pulmonary and systemic circuits are complete and separate unless a postnatal shunt persists. This condition is compatible with life after birth only to the extent that there are persistent shunts that allow adequate mixing of oxygenated blood.

Tetralogy of Fallot

The tetralogy of Fallot is a lesion associated with at least three embryonic defects: right ventricular outflow obstruction; ventricular septal defect; and an abnormally placed (dextroposed) aorta that sits directly 'above' the ventricular septal defect. This complex tetralogy can be partially explained by a deviation of the conal septum, misalignment of the conotrabecular septum, and underdevelopment of the infundibular (conal) portion of the right ventricle. The defect is usually associated with a thickened right ventricle stimulated by excess pressure within the ventricle.

Persistent ductus arteriosus

The ductus arteriosus is necessary in fetal life as a shunt for the portion of right ventricular output that does not perfuse the high-resistance pulmonary vascular bed. If, however, it remains patent (open) after birth when the pulmonary vascular resistance is low, a significant portion of the left ventricular output may travel through the ductal shunt and overperfuse the lungs. Patent ductus is a common anomaly that may require surgical closure if prostaglandin production inhibitors are ineffective.

Aortic arch abnormalities

Narrowing or coarctation of the aorta is most commonly found near the isthmic aortic junction with the ductus arteriosus. Coarctations may be preductal or postductal. Narrowing is believed to be related to inadequate flows during fetal life as the result of other anomalies. The aortic arch may be incompletely formed with occasional interruptions at predisposed sites. Such defects are commonly associated with ventricular septal defects. It should also be noted that the ascending aorta may be narrow and therefore obstructive and that this anomaly may be associated with aortic stenosis (valve narrowing) or aortic atresia.

Venous anomalies

The persistence of the primitive venous drainage system leads to a series of defects known collectively as total anomalous pulmonary venous connection (TAPVC). The variety of anomalies is endless. Among the abnormal arrangements are cor triatriatum (improper incorporation of the pulmonary venous chamber), persistent portions of anterior cardinal veins, and unusual drainage of the hemiazygos veins (these normally drain into the coronary sinus in ovine species but not in humans).

Cellular development of the myocardium

Ultrastructure of the mature myocardium

The mature myocardium is composed of a mosaic of cell types including myocytes, endothelial cells, fibroblasts, macrophages, nerve cells, blood cells, etc. In fact, the mature myocardium may consist of fewer than 30 per cent myocytes by cell number even though the majority of the volume is 'muscle'. There are differences in structure between ventricular, atrial, and conduction myocytes. The

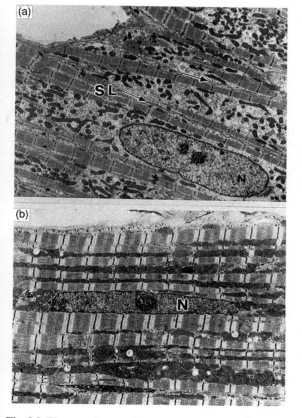

Fig. 8.9 Electronmicrographs of heart muscle from immature cat. (a) In the neonatal cat five cells are seen separated by sarcolemma (SL). (b) In the adult heart one cell is seen. Contractile material is found at the periphery in neonate cells but throughout the cytoplasm in the adult cell. The nucleus (N) in the neonate is centred within the cell and mitochondria are scattered throughout the central cytoplasm. In the adult, mitochondria are sandwiched between the layers of myofibrillar material. (Modified with permission from Maylie, J.G., (1982). *American Journal of Physiology*, **242**, H834–43.)

ventricular myocyte features a compact array of myofibrils with mitochondria sandwiched between fibrillar layers (Fig. 8.9). Elongated nuclei are centrally located and surrounded by tightly arranged organelles. The myofibrils demonstrate Z lines demarcating the sarcomeres with light isotropic and dark anisotropic bands forming striations. The plasma membrane (sarcolemma) invaginates extensively, forming T tubules at the level of Z lines to greatly expand the surface area of the cell and to provide access of the inner regions of the cell to the extracellular ionic environment. Upon excitation, the T tubules carry the wave of sarcolemmal depolarization deep into the cell interior. The sarcolemma, which has a well developed cell coat, gives rise to a number of coated and uncoated vesicles in addition to T tubules. The intercalated disc is a specialized region of the sarcolemma of each myocyte and is composed of several junction types. A particularly noticeable feature of the intercalated disc is the long gap-junctions, which allow electrical continuity between myocytes.

The function of the mature myocyte depends upon its constituent intracellular organelles. The sarcoplasmic reticulum (SR) is a fine network of smooth membrane throughout the cell which stores and releases calcium ion with each contraction cycle. Because the SR is able to release its calcium stores only in response to membrane depolarization (excitation), specialized structures are required to carry the message between the T system and the SR. The structures that couple the T tubules and the SR are known, appropriately, as 'couplings', and the physiological process by which membrane excitation leads to muscle contraction is known as 'excitation–contraction coupling'. Golgi, centrioles, lipid droplets, microtubules, and, in the atria, granules containing atrial natriuretic peptide (ANP) are additional important organelles. The myocyte also contains glycogen particles to varying degrees, depending on species and physiological state. In quantitative terms, the internal volume of the mature myocyte is occupied by myofibrillar material (> 50 per cent), by mitochondria (> 30 per cent), nuclei (a few per cent), sarcoplasmic reticulum (a few per cent) with all other organelles composing the crowded remainder.

Ultrastructure of developing myocardium

Myocyte development

The mature myocyte is a descendant of cells of mesodermal germ layer origin; its developing fine structure has been studied in several species. From the time of differentiation until the late postnatal period, the cardiac myocyte continually refines its ultrastructure until it fully resembles the mature cell. Although there are wide species variations in timing, the organelle maturation sequence is common enough to describe a general plan for all mammals.

The nascent myocyte can be first recognized under the electron microscope by the appearance of myofibrillar material in the form of polymerized actin–myosin 'filaments'. As these become attached to the dense Z material, a primitive sarcomere is formed. The early chick heart is seen to contract even before fibrils are evident by light microscopy. However, the electron microscope reveals the presence of occasional myofibrils within the newly beating myocyte that are apparently random in their orientation and that give the false impression that such disarray could lead only to inconsequential contraction.

The ultrastructural features of myocardial development have been studied in sheep, making use of standard morphometric techniques to quantitate growth patterns of hearts from age 115 days' gestation (term=147 days) to adulthood. The relative cytoplasmic matrix volume (the intracellular volume lacking organelles) decreases during development and myofibrillar and mitochondrial volumes increase, expecially during the prenatal period.

Myocyte size

Before birth, the heart grows primarily by cell division (hyperplasia). However, in the early postnatal period, the myocyte loses its ability to divide so that heart growth from that point on is primarily by cell enlargement (hypertrophy). During the prenatal period, changes in cell size do not necessarily predict changes in myocardial mass because of rapid hyperplasia. In the mature postnatal heart, changes in heart mass are necessarily related to changes in myocyte volume. The haemodynamic environment and the resultant ventricular wall stresses are believed to be important regulators of heart growth throughout life.

What are the physiological constraints on myocyte function as heart cells become larger? The myocyte is programmed to grow during postnatal life to keep heart size proportional to body size. Mature adult myocytes of various species are about the same size regardless of body size. Therefore, the number of myocytes present at birth must anticipate the ultimate size of the adult animal with enough size reserve to allow for physiological adaptations to exercise fitness. Accordingly, the number of myocytes in the neonate, across species, is related to the ultimate size of the animal and is the same for both ventricles. In the human, left ventricular myocyte volume increases 30- to 40-fold during normal growth from neonatal to adolescent life. Superbly trained athletes have left ventricular masses that may be 50 per cent greater than those of the normal population suggesting a small reserve for physiological hypertrophy. With severe hypertrophy in response to chronic pressure or volume loading, heart cells may enlarge up to twofold above normal adult levels and left ventricular masses correspondingly double, whereupon function deteriorates. It appears that a twofold increase in myocyte volume above the normal adult size is about the limit if the heart is to maintain normal function. The cellular mechanisms that restrain myocyte function with hypertrophy are not know, but reduced blood flow reserve has been demonstrated in pathological hypertrophy.

It is noteworthy that the two ventricles appear to be histologically different even before birth — though they develop in nearly identical haemodynamic environments. In fetal sheep, right ventricular myocyte diameter, capillary lumen area, and average intercapillary distance are all larger than their left ventricular counterparts but the left ventricle has a higher capillary density. Whether differential work loads or wall stresses of the two ventricles are determinants of differential cell size and myocardial development has yet to be determined. The fact that, in sheep, rats, and humans, left ventricular myocytes grow more rapidly in cross-sectional area after birth than do right ventricular myocytes is consistent with the notion that the increasing left ventricular wall stress that accompanies postnatal increases in systemic pressure stimulates myocyte enlargement (Fig. 8.10). However, studies in guinea pigs failed to find differences in interventricular myocyte size during the postnatal period.

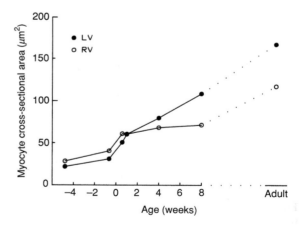

Fig. 8.10 Myocyte cross-sectional area during development of sheep heart. During fetal life, right ventricular (RV) myocytes have larger cross-sectional area than do left ventricular (LV) myocytes. During the early postnatal period, left ventricular myocytes become larger than right ventricular myocytes. (Taken with permission from Smolich, J.J., Walker, A.M., Campbell, G.R., and Adamson, T.M. (1989). *American Journal of Physiology*, **257**, H1–9.)

In sheep the average diameter of the left ventricular myocyte remains nearly constant throughout fetal life. Following birth the diameter increases until it reaches the level of the adult. Fetal myocyte length increases during gestation from about 30 μm at day 50 to more than 50 μm at birth. The length of adult left ventricular myocyte is about 150 μm, which means that the myocyte volume increases many-fold.

Development of sarcoplasmic reticulum and T tubules

There is considerable variation among species as to the relative stage of development when sarcoplasmic reticulum (SR) and transverse tubules (T tubules) appear in myocytes, though their development is undoubtedly coordinated. T tubules do not appear until after birth in cat, rat, mouse, rabbit, dog, and hamster but develop prenatally in monkey, cow, guinea-pig, sheep, and humans. Figure 8.9 shows the immature appearance of the new-born cat heart. SR has been detected at 51 days gestation in sheep but not at 29 days. That the immature cardiac myocyte is able to function without significant SR or T tubules is

worthy of note. It is assumed that the sarcolemma carries out the function of the SR until the latter becomes adequately mature.

Couplings

Unfortunately, the cellular processes involved in the formation of couplings between the sarcolemma (T tubules) and the SR have not been thoroughly investigated in heart but recent findings in skeletal muscle may be relevant. The excitation–contraction coupling phenomenon depends on the presence of voltage-sensitive channels in the membrane that somehow communicate with the calcium release channels of the sarcoplasmic reticulum. In skeletal muscle, the ontogeny of suspected components of the excitation–contraction coupling system has been investigated. The developmental localization of specific markers for T tubules (TS28), voltage-gated Ca^{2+} channels (dihydropyridine receptor — DHPR), and the sarcoplasmic Ca^{2+} release channel has led to the development of a model for the subcellular formation of the excitation–contraction 'organelle' in skeletal muscle (Fig. 8.11). Such a model needs to be studied in the mammalian heart.

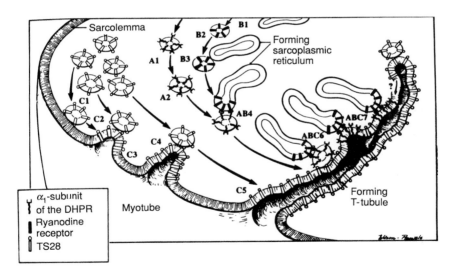

Fig. 8.11 A scheme for the biogenesis of T tubules and triads in developing skeletal muscle. First, α_1-dihydropyridine receptor (A1) and ryanodine receptors (B1) are assumed to be synthesized on membrane-bound polyribosomes, incorporated into unique transfer vesicles, and distributed (A2 and B2, respectively). Next, TS28-containing vesicles (C1) fuse first with the sarcolemma to form a caveolae (C2). Sequential fusions of TS28-containing vesicles to the caveolae result in the formation of a tubular invagination into the myotube (C3–C5). As this occurs, ryanodine receptor-containing vesicles are incorporated into forming SR (B3), followed by complex formation between ryanodine receptors in SR and the α, DHPR-containing transfer vesicles (AB4) in the outer zone of the cytosol. In turn, the α_1-DHPR containing vesicle of this complex fuses with TS28-containing forming T-tubules (ABC6), thus incorporating the α_1-DHPR-containing vesicles into a discrete region of the forming T-tubules (ABC7) and a junctional complex between the T-tubule and the SR. (Taken with permission from Yuan, S., Arnold, W., and Jorgensen, A.O. (1991). *Journal of Cell Biology*, **112**, 289–301; drawn by Linda Wilson-Pauwels.)

Nucleation

In the rat nearly all cells have a single nucleus at the time of birth; binucleate cells begin to appear by about postnatal day 4. By 15 days of gestation > 80 per cent of the cells are binucleate. Mycocytes form the nuclear material for the binuclear myocyte by synthesizing new DNA rather than fusing with a second cell. An even more interesting finding is that binucleation appears to be prerequisite for cell enlargement in preparation for the growth hypertrophy phase of postnatal development in rats. Human and dog myocytes behave somewhat differently. The proportion of binucleated cells increases during the postnatal period to about 40 per cent at maturity in dogs, whereas the percentage of binucleate cells is highest in young children at about 30 per cent but decreases to < 13 per cent in adult humans.

The matter of nucleation is complicated by the fact that the amount of DNA per nucleus is variable among species and is dependent on stage of development. In humans, the cardiac myocyte is primarily diploid at postnatal ages up to the age of 7 years; in adults, 60–80 per cent of myocytes contain tetraploid nuclei.

Capillarity

Several methods have been devised to quantify the degree to which the myocardium is supplied with capillary blood. The average intercapillary distance, the average 'diffusion distance', the capillary-to-myocyte (fibre) ratio, and the capillary volume fraction are examples. The prenatal development of capillary supply (capillarity) has not been studied as extensively as in postnatal development. It appears that the myocyte-to-capillary ratio decreases from as high as 6 in the new-born to 1 in adult man, rabbit, and rat. The intercapillary distance and the average capillary diameter decrease during the same period. In sheep, the capillary density of the ventricles increases during the last few prenatal weeks but decreases sharply during the postnatal period.

Interestingly, in sheep the capillary luminal area was found to be larger for the right ventricle than for the left ventricle but only during fetal life. Capillary density was the opposite, greater for the left ventricle than right ventricle before birth, but less than the right ventricle after birth. It is not know to what degree the haemodyamic environment determines capillary size and density. The largest capillary luminal area and the lowest capillary density were found in the ventricle with the largest wall stress. These patterns reversed themselves with postnatal development as the relative wall stresses were reversed. Evidently, wall tension is not the primary determinant of the capillary/fibre ratio decrease described above. The postnatal decrease in the myocyte-to-capillary ratio found in several species is found also in sheep but, in the case of sheep, the ratio decrease is only in the immediate period preceding birth. This occurs before transitional haemodynamic cues are present.

Fetal coronary blood flow is regulated much as in the adult. Myocardial flow resistance is dependent on oxygen content. However, experiments demonstrating pressure–flow curves at constant myocardial work have not been performed. In sheep, total myocardial blood flow is about 140 ml/min/100 g in both the fetal and new-born periods. This level is nearly twice as great as adult levels. However, there are important differences in flow before and after birth. Fetal free wall flow and oxygen delivery are greater for the right ventricle (RV) than for the left ventricle (LV) in accordance with their relative work loads, but these are reversed in the immediate postnatal period as the work loads are reversed. Reller and colleagues have shown that myocardial blood flow increases in response to acute pressure loading of the right ventricle. Interestingly, left ventricular flow also increased more than twofold (though left ventricular work was not increased) so that the RV:LV flow ratio was maintained. Fetal oxygen delivery remains constant in response to a 50 per cent reduction in fetal arterial oxygen content due to increased coronary blood flow. Thus, fetal myocardial blood flow responds to: (1) increased oxygen demand under conditions of increased work or (2) hypoxaemia.

Development of contractile proteins

The contractile proteins include myosin, actin, tropomyosin, and three types of troponin. Molecular myosin is composed of two heavy chains and four light chains, two of which can be phosphorylated and two which cannot. The ATPase activity is carried on a heavy-chain site and exists in two isoforms (alpha, beta). In mammals, three heavy-chain isoenzymes have been identified (V1, V2, V3) according to their mobility under electrophoresis. V3 has the least mobility. The V1 isoenzyme is composed of an α/α homodimer and the V3 of a β/β homodimer. V2 is the α/β heterodimer. The ATPase activity of V1 is some five fold higher than V3. In small rodents V1 is the predominant isomyosin except under conditions of hypertrophy where V3 can predominate. In large mammals and humans ventricular muscle contains primarily the V3 form and is little affected by hypertrophic stimuli.

While there is much to learn regarding developmental patterns for the contractile proteins, it is known that fetal myocardium, both atrial and ventricular, contains a myosin light chain (nonphosphorylatable) that is unique to the fetus. But the level of expression depends on developmental age. In fetal life for all mammals the slow V3 isoform is most abundant. In large animals, V1 is expressed after birth but, as already noted, V3 is the most abundant form in these animals.

In atria and ventricles, actin can be found in two forms, the α-skeletal and the α-cardiac isoforms. In small rodents, the cardiac form is found in the ventricle but both cardiac and skeletal forms of actin are found in the fetal heart. It is also interesting to note that both cardiac and skeletal forms are found in the human ventricle and the human atrium.

A thorough understanding of developmental gene expression may be more important than once thought. Several laboratories have shown that hypertrophic or failing myocardium may express genes that have not been expressed since embryonic life. This includes the skeletal α actin and β myosin heavy-chain genes, atrial natriuretic peptide genes, and others. Therefore, it is becoming increasingly clear that understanding developmental gene expression patterns may elucidate pathophysiological mechanisms in the adult heart.

Fetal cardiac function

The purpose of the circulation is to transport oxygen from the environment to tissues, and wastes to appropriate exit sites. The oxygen consumption and wastes are roughly linked and it is useful to consider cardiac function with respect to oxygen consumption. In the adult animal oxygen consumption at rest is fairly constant over the duration of life. However, activity is necessary for feeding, breeding, defence, and play. Cardiac output may also be necessary for thermal regulation. Oxygen consumption in well trained humans can increase 20–25-fold for short periods of time. This increase is effected by a 3–4-fold increase in heart rate, a doubling of stroke-volume, and a tripling of arteriovenous O_2 difference. The increase in stroke-volume is effected by a 20 per cent increase in end diastolic volume (Starling mechanism) when exercise is performed in the upright posture and a 30 per cent decrease in systolic volume due to increased contractility and reduced impedance to ejection. The increased contractility is trivially related to heart rate and occurs almost entirely by adrenergic stimulation. Decreased impedance is due to the four- to five-fold decrease in vascular resistance. Arterial compliance, the other component of impedance, has not been studied during exercise but may respond to adrenergic stimulation.

The adult human at rest is throttled back with resting vagal tone reducing heart rate; resting cardiac adrenergic tone is negligle and, in the upright posture, preload is sub-

maximal. The heart is ideally suited at this idling state to respond to the body's needs for increased flow. The fetus, in contrast, does not need to feed, breed, defend itself or its temperature, or, arguably, to play. While movement of the limbs, trunk, and respiratory muscles are frequent, the oxygen cost of this activity is quite small compared to the activity of the adult animal. Paralysis of skeletal muscle results in a 17 per cent reduction in oxygen consumption of fetal sheep. In contrast, pregnant ewes exercising maximally on a treadmill increase their oxygen consumption almost sixfold. However, what the fetal heart must respond to is the growth of the body, not physical exercise. Oxygen consumption of the body increases roughly linearly with body weight throughout gestation. Other things being equal, this requires that cardiac output increases linearly with weight and, with minor extrapolation, this means that heart volume must increase linearly with fetal weight. In fetal sheep then, the cardiac output and cardiac volume increase fivefold between days 100 and 140 of gestation when fetal weight increases from 1 to 5 kg. Thus, heart rate, preload, afterload, and contractility, the important *short-term* regulators of cardiac output in fetal and adult animals, are trivial regulators of output in the fetus compared to cardiac growth. Fetal cardiac growth is the most important determinant of cardiac output. In the near-term ovine fetus, cardiac output increases approximately 4 per cent per day with no measurable change in preload, afterload, contractility, or heart rate. This increase is accomplished entirely by cardiac growth.

Having recognized the importance of cardiac growth to function in the fetus, we must acknowledge our ignorance regarding growth's regulation, the cues for growth, and the pathophysiology of what must be a wonderful control mechanism. The short-term regulators of function have been much more completely investigated. Frequently, fetal short-term regulators are compared to neonatal or adult parameters. While this is important to the understanding of the fetal response to stress *in utero* and during the transition from fetal to neonatal life, this practice may not increase our understanding of the control of fetal cardiac output over any significant period of time.

Glossary of terms

Afterload	Force per unit cross-sectional area of myocardium during contraction
Compliance	Change in volume per unit change in transmural pressure
Contractility	Intrinsic strength of contraction in myocardium, independent of pre-load or afterload

End diastolic volume	Volume of the ventricle just preceding contraction, primary determinant of preload
Stroke volume	Difference between end diastolic and end systolic volume
Ejection fraction	Stroke volume divided by end diastolic volume
Filling pressure	End diastolic ventricular pressure, frequently approximated by mean atrial pressure; an important determinant of preload
Fractional shortening	Change in ventricular dimension divided by end diastolic dimension; analogue to ejection fraction
Impedance	Resistance of the vascular tree offered to ventricular output with pulsatile flow; includes vascular resistance and compliance
Preload	Force per unit cross-sectional area at end diastole; used interchangeably with end diastolic length or volume
Wall stress	Equilibrium force per unit cross-sectional area of myocardium that counteracts transmural pressure

Contractility

The force–velocity and length–tension relations

The classic studies of myocardial muscle mechanics show that strips of isolated muscle can be studied *in vitro* in order to quantify the performance of the muscle. Figure 8.12 shows the relationship of length to tension during isometric (constant length) contraction and passive stretching. At short muscle lengths, both active and passive tensions are low. As the muscle is stretched, analogous to diastolic filling of the heart, passive tension rises slowly while active tension, the force generated during systole, increases rapidly. The diastolic muscle length that produces the maximum systolic tension is called L_{max}. Near L_{max}, passive tension increases rapidly reducing net tension development. The relationship between diastolic length and systolic tension is the basis of the Frank–Starling mechanism that relates end diastolic volume of the ventricle to stroke volume. The molecular basis of the length–tension relationship may rest on the sliding filament hypothesis of the interaction of actin and myosin, although the biochemical mechanisms regulating the length–tension relationship are controversial. Nevertheless, the thin actin filaments are connected at each end

of the sarcomere to Z bands. In the middle, the actin filaments overlap thick myosin filaments. The interaction of actin, myosin, and calcium produces tension generation and movement of myosin along the actin filaments. Because there are repeating myosin heads along the filament, an optimal length exists for the Z bands that produces the maximum overlap of actin and myosin and, therefore, the maximal force generation. Maximal force generation occurs at a Z band length of approximately 2.2 microns in adult heart muscle, corresponding to L_{max} in Figure 8.12. L_{max} in fetal and adult sheep is not different.

Cardiac muscle strips can also be studied during isotonic (constant force) contraction and the velocity of contraction related to initial muscle length and afterload (Fig. 8.13). The isotonic load during systole is called the afterload. The muscle length during diastole is determined by the passive tension or preload on the muscle. Thus, cardiac muscle can either shorten quickly at low tensions or generate very high tensions but shorten slowly. Increasing preload raises the velocity of shortening at any afterload, except at the lowest afterloads where the velocities converge to a theoretical maximum (V_{max}). V_{max} is thought to represent an intrinsic property of myocardium (contractility) that represents the strength of the muscle contraction independent from all other factors, for example, preload or afterload.

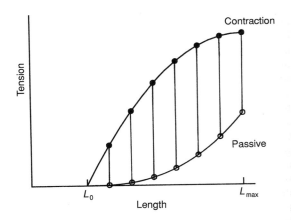

Fig. 8.12 The relationship of length to tension for passive (open circles) and contracting (closed circles) cardiac muscle strips is shown schematically. The vertical lines connect the diastolic length–tension point to the systolic length–tension point for an isometric contraction. L_0 is the unstressed muscle length and L_{max} is the muscle length that produces the maximum tension during isometric contraction. The rapid increase in active tension at muscle lengths above L_0 is the basis for the Frank–Starling mechanism.

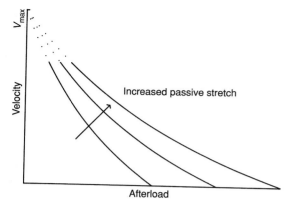

Fig. 8.13 The relationship of velocity of contraction to systolic tension or afterload is shown schematically for isotonically contracting cardiac muscle strips. Velocity and extent of contraction are inversely related to afterload. Increased preload or passive stretch and increased contractility increase velocity of shortening at any given afterload. V_{max} is the extrapolated intercept velocity at zero afterload and is relatively constant at different preloads allowing it to be used as an index of contractility. The reciprocal relationship between extent of shortening and afterload is the basis for the reciprocal relationship between arterial pressure and stroke volume.

Geometric considerations

Ideally, to truly understand contractility we would like to know the contraction behaviour of a single unit of actin and myosin. Since this is technically impossible at this time, we must deal with finite muscle strips and, indeed, with whole ventricles and hearts; therefore, we must account for the effects of geometry. The first consideration for the fetus is the percentage of myocardium that is actually fibril. As noted above, the percentage of fibril in ventricular myocardium increases progressively through gestation to comprise approximately 50–60 per cent of adult myocardium. The second consideration is the cross-sectional area of the muscle strip under investigation. The third consideration is the length of the muscle strip. Finally, the orientation of the fibrils is important since force is a vector. If we assume that fibrils are arranged along the axis of the muscle strip in parallel, then force generation for the muscle strip will be proportional to the cross-sectional area of the strip and the fraction of the strip that is fibril. At the same time, actual shortening of the series-linked fibrils will be proportional to the number of sarcomeres in series. Therefore, we can account for different muscle lengths by normalizing for unstressed muscle length or muscle length at L_{max}.

Diastole must be considered separately from systole. It has been assumed that the fibril is nearly freely extensible. Thus, the parallel elasticity responsible for the curvilinear passive length–tension relationship in Figure 8.12 may not be constant throughout gestation. Increasing fibril content through gestation might not only increase force generation, but may also increase the extensibility of muscle strips.

What are the parameters in a three-dimensional structure such as a ventricle that can be measured and that relate to the classic properties of isolated muscle strips, which allow contractility to be assessed? A thin-walled sphere is a useful physical model to understand the relationships between geometry and wall stress. As noted above, wall stress is an important variable since systolic force generation must be expressed as force per unit area (stress), and diastolic distending forces in the ventricle are likewise expressed as stress. For a thin-wall sphere (Fig. 8.14) circumferential wall stress σ is

$$\sigma = Pr/2h \qquad (1)$$

where r is the radius of the sphere, h is the wall thickness, and P is the transmural pressure. By imagining muscle strips arranged circumferentially around the sphere in Figure 8.14, it is possible to construct a pressure–volume

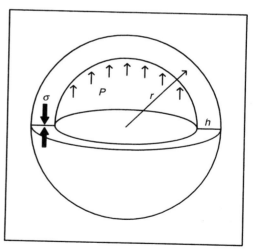

Fig. 8.14 Analysis of the equilibrium state for a thin-walled sphere with radius r, wall thickness h, transmural pressure P, and circumferential wall stress σ provides a framework for understanding ventricular function based on the known physiology of muscle strips. Circumferential wall stress opposes the transmural pressure where $\sigma = Pr/2h$. Wall stress is therefore proportional to transmural pressure and the radius-to-wall-thickness ratio. Circumferential wall stress is analogous to preload during diastole and afterload during systole.

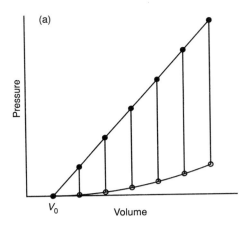 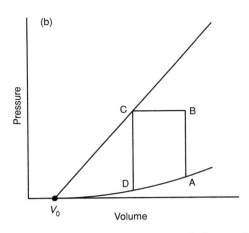

Fig. 8.15 Pressure–volume relationships for fetal cardiac muscle. (a) By analogy to Fig. 8.12, systolic and diastolic lines can be constructed that represent the pressure at end diastole (open circles) and the peak pressure during an isovolumic beat (closed circles). These lines originate from near the unstressed volume of the ventricle, (V_O) and form the boundaries within which the ventricle must operate as shown in (b). (b) A normal cardiac contraction is neither isometric nor isotonic but contains elements of both. Point A represents the pressure and volume at end diastole. With the onset of systole, pressure increases rapidly with volume constant until the semi-lunar valve opens (point B). The ventricle ejects blood, losing volume until the valve closes at end systole (point C). Point C is on or very near the line created by isovolumic contractions noted in (a). The ventricle then relaxes isovolumetrically until the atrioventricular valve opens (point D) and ventricular filling begins. From D to A, the ventricle fills, for the most part along the passive pressure–volume line noted in (a).

relationship (Fig. 8.15 (a)) of an idealized ventricle from the tension–length relationships in Fig. 8.12. The open circles (Fig. 8.15(a)) represent the pressure and volume of the ventricle during diastole. The curvilinear relationship is a consequence of the non-linear relationship between stress and strain in all biological materials and is evidenced in the muscle strip in Fig. 8.12. If, at each diastolic volume, we cause the ventricle to contract isovolumetrically, we will achieve a peak pressure shown by the corresponding closed circle. The points are analogous to the points on the active tension curve in Fig. 8.12 or the X-intercept points in Fig. 8.13. Because hearts don't normally contract isovolumetrically, the curves in Fig. 8.15(a) become the boundaries within which the ventricle must operate during systole and diastole. Figure 8.15(b) shows the typical pressure–volume relationship of a ventricle contracting, ejecting, relaxing, and filling within the boundaries described in Fig. 8.15(a). The upper left-hand corner of the pressure–volume loop (C) abuts the line of maximum pressure generation. Thus, the end systolic pressure–volume relationship can be generated by varying filling pressures or arterial pressures, while a series of pressure–volume loops are recorded. Because the end systolic pressure–volume line represents the projection of the length tension relationship to our three-dimensional ventricle, it can also be used to assess intrinsic strength of contraction

or contractility. Agents that increase contractility increase the slope of the end systolic pressure–volume relationship. Agents that decrease contractility reduce the slope of this relationship.

Fetal cardiac contractility

Almost 20 years ago it was found that heart muscle from fetal sheep produces less tension and lower rates and extent of shortening at constant tension than adult heart muscle. However, velocity of shortening at very low levels of tension was thought to be not different at different ages. Thus, it was concluded that contractility was not importantly different between the fetus and the adult but that the difference in function was accounted for by the difference in myofibrillar mass. When the wall stress produced by fetal and adult myocardium was normalized for the fraction of myofibrillar mass (30 per cent versus 60 per cent adult), the performance of fetal and adult myocardium was not significantly different.

Subsequent studies of the maturation of myocardial function have focused on important biochemical discoveries. These include changes in the contractile proteins, proteins regulating contraction, and the segregation of cytosolic calcium. The biochemical determinants of contractility may be simply viewed as: enzymatic speed and power of myosin; the sensitivity of the contractile proteins to activation by calcium; and the availability of cytosolic calcium. The contractility of myofibrillar protein has at least in part been linked to its ATPase activity. Alpha (VI) myosin and beta (V3) myosin are isoforms with high and low ATPase activity, respectively. ATPase activity is positively correlated with unloaded velocity of muscle contraction or contractility. In the rat, for example, ventricular myocardium changes from V3 to V1 from the neonatal to adult period and then returns to V3 in old age. However, this pattern is not consistent across species; for example, the V3 isoform is the predominant protein in adult humans and other large mammals. The regulatory proteins tropomyosin and troponin also vary during development and, because of this, alter the relationship between calcium concentration and actin inhibition and thus contraction. Finally, the change from sarcolemmal calcium transport in early gestation to sarcoplasmic reticulum release and sequestration in the adult as mechanisms for contractile protein activation suggests important differences in contractile function during development. It is not too surprising then, when examined carefully, that investigators have been able to show maturational increases in contractility in sheep, rabbit, and dog that were not evident in earlier experiments on fetal lambs.

In fetal sheep post-extrasystolic potentiation has been measured *in vivo* and *in vitro* to detect maturational increases in contractility. As a result of such studies, it has been proposed that contractility may be measured as the ratio of a change in force or pressure per unit time (dF/dt or dP/dt) in a maximally potentiated beat versus a control beat in isolated or *in vivo* myocardium. Clearly, this ratio is an expression of the functional contractile reserve of the muscle that can be activated by increasing cytosolic calcium. From 83 days of gestation to 62 days of postnatal life, these ratios increase, indicating a maturational

increase in myocardial reserve or contractility. Interestingly, the ratio drops for a few days after birth at a time when baseline dP/dt and fractional shortening are increased. It has been argued that this represents utilization of contractile reserve by the increased catecholamines present at birth. On the other hand, despite the statistically significant increases in contractile reserve documented in these studies, there is no significant change in fractional shortening from about 27 days before birth to 122 postnatal days except for the previously mentioned increase at birth. Thus, the maturational change in contractile reserve does not appear to have a major impact on cardiac output during gestation and postnatally. However, it has been proposed that the maturational increase in contractility provides a reserve that the mature fetus can draw upon for the stresses of birth. In addition, increased contractility may be necessary for the wide range of cardiac output necessary during postnatal life.

Sheep are known to be relatively mature at birth and, thus, it may be harder to show maturational changes in contractility in this species and particularly the significance of these changes. Rabbits are less mature at birth and are similar to mice and rats. The relationship between calcium concentration, developed tension, and myofibrillar ATPase activity has been studied in rabbit myocardium from 18 days gestation (0.58) through to adult life. The maximal tension developed at the optimal calcium concentration is shown in Table 8.1. The results are shown according to age and are normalized for the weight of myocardium and for percentage of myocardial weight taken up by myofibrils.

In the rabbit, the heart weight–body weight ratio was 0.34 g/100 g at 18 days gestation and 0.4 g/100 g in 5-day-old neonates. During this period, body weight increased from 1.25 to 85 g. Thus, heart weight increased almost as much as the 68-fold increase in body weight.

Table 8.1 Cardiac contraction in the rabbit*

Gestation (days)	Maximal developed tension (g/g cardiac muscle)	% of cardiac muscle that is myofibril	Maximal developed tension (g/g cardiac muscle)/% of cardiac muscle that is myofibril
18	18	13	1.4
21	22	15.3	1.4
28	40	24.3	1.6
New-born	68	39	1.7
Adult	78	50	1.6

*Adapted from Nakanishi, T., Seguchi, M., and Takao, M. (1988). *Experientia*, **44**, 936–44.

Table 8.2 Rabbit contractile protein ATPase activity*

	Fetus	Newborn	Adult
% VI myosin	38	52	5
Myofibrillar ATPase (n mol/mg protein/min)	100	130	160
Myosin ATPase + actin + tropomyosin + troponin (n mol/mg protein/min)	170	220	300

*Adapted from Nakanishi, T., Seguchi, M., and Takao, A. (1988). *Experientia*, **44**, 936–44.

Table 8.2 shows the percentage V1 myosin and the ATPase activity of intact myofibrils compared with purified myosin and myosin in the presence of actin and actin plus tropomyosin and troponin.

Myofibrillar ATPase activity is greatest in the adult and least in the fetus consistent with the force-generating capability shown in Table 8.1. However, the fraction of the V1 isoform, which has the highest ATPase activity, is greatest in the new-born and least in the adult. The explanation for this apparent discrepancy may exist in the development of the regulatory proteins. In the presence of troponin and tropomyosin, enzyme activity closely mimics the developmental activity increase seen for the intact fibril. These findings have led to the conclusion that developmental changes in myocardial force generation were, at least in part, due to changes in total myofibrillar ATPase activity and not in VI isoform fraction. The changes in myofibrillar activity during development appear to be due to age-related differences in the interaction of myosin with troponin and tropomyosin rather than alterations in the myosin isoform itself.

Table 8.3 Effects of lanthanum and ryanodine on maximal rate of tension development $(dT/dt)_{max}$ in rabbit myocardium*

Gestation (days)	Per cent $(dT/dt)_{max}$	
	Ryanodine	*Lanthanum*
18	90	4
21	86	6
28	70	21
New-born	57	30
Adult	43	38

*Adapted from Nakanishi, T., Seguchi, M., and Takao, A. (1988). *Experientia*, **44**, 936–44.

Finally the processes which control calcium concentration during excitation–contraction coupling undergo a revolution during development with both structural and functional manifestations. Table 8.3 shows the effects of lanthanum or ryanodine on contraction of rabbit myocardium during development.

Ryanodine selectively inhibits calcium uptake and release from sarcoplasmic reticulum. In contrast, lanthanum blocks sarcolemmal calcium transport but does not enter the cell. Thus, in concert with the development of the sarcoplasmic reticulum and T tubules, there is a shift in the source of activator calcium from the external milieu in the immature heart to the sarcoplasmic reticulum in the mature heart. This is expressed by the lack of effect of ryanodine on contraction in immature hearts that do not use sarcoplasmic reticulum calcium to activate myosin; in contrast, immature hearts are almost totally blocked by lanthanum indicating that activator calcium must cross the sarcolemma in these hearts. Adult myocardium is most affected by ryanodine and least affected by lanthanum indicating that the sarcoplasmic reticulum is an important source of activator calcium in the adult. Functionally, the transition from sarcolemma to sarcoplasmic reticulum as the source of calcium for activation has two important features. First, paired electrical stimulation, which causes increased contractility in adult hearts has progressively less effect on contractility of immature hearts. It has been proposed that the inotropy of paired electrical stimulation comes from intracellular calcium release from sarcoplasmic reticulum and the T tubules. Second, the slow calcium channel blockers, such as verapamil, that affect sarcolemma calcium movement have profound effects on contractility in immature as opposed to mature hearts.

In conclusion, many of the factors which affect contractility are changing during development. Consistent evidence supports a mild increase in contractility from early embryonic through to adult life. The mechanisms behind this increase in contractility are unclear but may reside in the proteins that regulate excitation–contraction coupling or in the regulation of cytosolic calcium level rather than in the myosin itself. Differences in cardiac function during development can largely be explained by the amount and orientation of myofibrils within the myocyte with important modulation of contractility through varied biochemical mechanisms.

Preload

Preload determines end diastolic sarcomere length. Since we cannot measure sarcomere length in intact systems, diastolic muscle length or ventricular volume would suffice for acute studies but would be inadequate for chronic studies because of possible slippage or growth. Frequently, end diastolic pressure or mean atrial pressure are used as estimates of preload and are justified by the reproducibility of the diastolic pressure–volume relationship and the relationship between mean atrial pressure and end diastolic pressure in the ventricle. However, the utility of mean atrial pressure as an indicator of preload and sarcomere length has serious limitations. For example, changes in myocardial stiffness alter the mean atrial pressure–end diastolic pressure relationship. Ventricular pressure–volume relationships are altered by changes in myocardial stiffness, pericardial and thoracic restraint, or contralateral ventricular or atrial chamber volumes. Therefore, the use of mean atrial pressure as a preload indicator should be made with caution and only for a given ventricle over an appropriate time period. From the standpoint of the circulation, the only way it can regulate preload and therefore affect stroke volume is by changing mean atrial pressure. Thus, the relationship between mean atrial pressure and stroke volume (the cardiac function curve) has achieved importance for understanding heart function. From the standpoint of the heart, the relationship between end diastolic volume and stroke volume is fairly linear and is affected only by important changes in afterload or contractility. This is Starling's law, i.e. the strength of the heart's contraction varies with end diastolic filling.

The questions regarding preload, and Starling's law in the fetus are several. Is Starling's law applicable to the fetus; is preload similar in adult and fetal hearts; is the relationship between pressure and volume different in adult and fetal hearts; do fetal hearts have preload reserve different from adults? The answer to the first question is a resounding yes; Starling's law does apply to the fetal heart. Furthermore, Starling's law holds for cases ranging from stage-18 chick embryos with a heart consisting of a muscle-wrapped bent tube to the fully developed near-term fetal sheep heart (Fig. 8.16).

Sarcomere lengths at resting end diastolic volumes may not be the same in fetal and new-born hearts. While right and left ventricular end diastolic sarcomere lengths are not known throughout development, we can speculate that the acute changes in cardiac pressure at birth alter preload. In the fetus, right and left atrial pressures are equal over a wide range of filling pressures and the equality is independent of arterial pressure and cardiac output. With *in utero* ventilation (Fig. 8.17) or with birth, a left-to-right interatrial and interventricular pressure gradient develops. Whether this means the left ventricle has a higher preload than the right ventricle after birth or whether they are equal after birth and the left ventricle was relatively unloaded prior to birth is open to speculation. Considering the marked increase in left ventricular stroke volume with birth, without large changes in con-

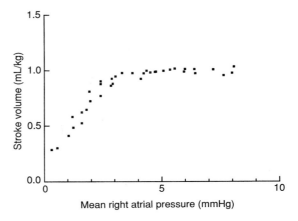

Fig. 8.16 The stroke volume for the near-term fetal sheep right ventricle measured by electromagnetic flowmeter is shown as a function of right atrial minus pericardial pressure. Control right atrial pressure is approximately 3 mmHg. During rapid haemorrhage the stroke volume falls rapidly as right atrial pressure drops. During reinfusion of blood and saline, stroke volume increases little at mean right atrial pressures above control.

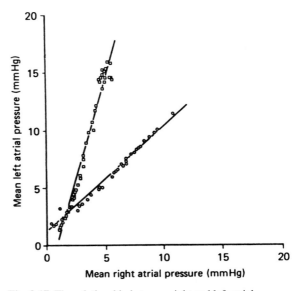

Fig. 8.17 The relationship between right and left atrial pressures is shown before and after *in utero* ventilation. Before ventilation, the pressures are not different over a wide range. During ventilation, pulmonary blood flow is increased and left atrial pressure exceeds right atrial pressure establishing a left–right atrial and ventricular diastolic pressure gradient. (Taken with permission from Morton, M.J., Pinson, C.W., and Thornburg, K.L. (1987) *Journal of Physiology*, **383**, 413–24.)

tractility despite an increase in afterload, the latter explanation is more likely.

Fetal hearts are generally thought to be stiffer (i.e. less
compliant) than adult hearts. The mechanism for this may
be the smaller relative amounts of contractile element,
which is thought to be freely extensible, in the cells. Thus,
one study found developmental reduction in stiffness of the
right and left ventricle from fetal to adult life. The implications of these findings are that increases in filling pressure
in the fetal right and left ventricle might produce less sarcomeric stretch than in the adult. Another major difference
between the adult and the fetus is the common diastolic
pressure in the fetus due to the interatrial shunt. The importance of ventricular interaction on fetal diastolic pressure–
volume relationship has been investigated and found to be
significant. Furthermore, in sheep the impact of changes of
right ventricular volume on left ventricular pressure–
volume relationships is most marked in the fetus compared
with the new-born and adult. From the curves in Fig. 8.18,
we see that it would be very difficult for any physiological
filling pressure to achieve end diastolic volumes in the fetal
left ventricle, with equal right and left ventricular pressures, similar to those that could be achieved in the neonate
with lower right ventricular pressure. The establishment of
a left–right diastolic pressure gradient at birth may increase
the left ventricular preload at any given left atrial pressure.
Increased left ventricular preload may underlie the large
increase in left ventricular stroke volume at birth.

As noted above, external cardiac restraint affects
preload. External cardiac restraint may be different in the
fetus than adult and can affect diastolic volume pressure
relationships. In the fetal lamb, the transatrial pressure at
rest has been measured at 2.5 mmHg with atrial pressure
referenced to an amniotic fluid pressure of 3.2 mmHg.
Thus, the pericardium and thorax contribute a 0.7 mmHg
restraint. In the adult, thoracic pressure is negative due to
the negative intrapleural pressure. Thus the cardiac distending pressure may be greater than the intracavitary
pressure referenced to air.

In the fetus when the blood volume is rapidly expanded
by infusion, transatrial pressure rises more slowly than the
intraatrial pressure referenced to thoracic pressure, with a
regression coefficient of 0.5. The restraint was at the level of
the pericardium. These studies of fetal pericardial
physiology were conducted in chronically instrumented
lambs with intact pericardium and fluid-filled catheters.
However, no difference in pressures was measured between
fluid-filled and balloon catheters in chronically instrumented
fetal sheep when the normal amount of pericardial fluid was
present. Other authors disagree with this finding and propose that no increase in transmural filling pressure can be
achieved in the fetus above control values due to pericardial

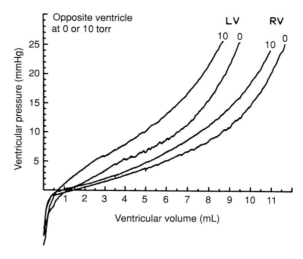

Fig. 8.18 Right and left ventricular pressure–volume
relationships are shown for a potassium-arrested, near-term
fetal sheep heart. Each ventricle is totally emptied and then
filled with a syringe pump with the contralateral ventricle at 10
or 0 mmHg and the pericardium in place. Right ventricular
(RV) volumes are always greater than left ventricular (LV)
volumes at common filling pressures. Both ventricular
volumes are importantly affected by contralateral ventricular
pressure. Establishment of a left–right diastolic ventricular
pressure gradient would markedly increase left ventricular
volume, for example, at birth (see Fig. 8.17). (Taken with
permission from Thornburg, K.L., Morton, M.J., Pinson, C.W.,
Reller, M.D., and Reid, D.L. (1987). In *Perinatal development
of the heart and lung*, Proceedings of the 1st International
Christie Conference, (ed. J. Lipshitz, J. Maloney, C. Nimrod,
and G. Carson). Perinatology Press, Ithaca, New York.)

and thoracic restraint. Further studies will be necessary to
resolve these issues. The question is important, however, in
that atrial and ventricular pressures are used to generate
function curves to quantify preload effects. Obviously, if no
increase in transcardiac pressure can be achieved in the
fetus, then no further stretching of the sarcomere can occur
above control levels. This does not mean that the heart
muscle itself cannot respond to increased preload, merely
that the intact heart cannot respond to elevated filling pressures. Thus the final question of whether fetal hearts have
preload reserve different from adult hearts is topical.

The issue of fetal and neonatal preload reserve was raised
to help explain the fragility of the neonatal circulation in the
presence of pathology. Studies of neonatal and mature
lambs were thought to show important maturational differences in preload reserve. Preload reserve means the ability
of the heart to increase its stroke volume with increases in
filling pressure. In addition, cardiac function curves in the
fetus showed limited ability to increase stroke volume at

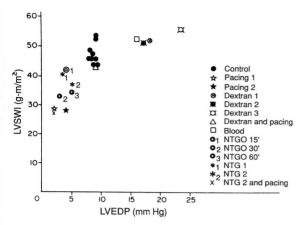

Fig. 8.19 Parker and Case plotted stroke work (gram-meters per square meter) as a function of left ventricular end diastolic pressure (LVEDP) to generate function curves in supine normal human subjects. Left ventricular stroke work index (LVSWI) was measured under conditions which altered LVEDP, as listed at right (NTGO, nitroglycerine ointment, NTG, sublingual nitroglycerine). With minimal changes in arterial pressure, stroke work index and stroke volume index are comparable (stroke work = stroke volume × arterial pressure). The left ventricular function curve for the supine human also has a steep ascending limb and a plateau. (Taken with permission from Parker, J.O. and Case, R.B. (1979). *Circulation*, **60**, 4–11.)

right atrial pressures above control levels. However, the data documenting abundant preload reserve in adult animals were largely collected in acute experiments with anaesthetic agents present. Preload reserve in chronically instrumented adult animals is limited. In addition, studies in supine humans show a limited preload reserve (about 20 per cent) during volume infusion (Fig. 8.19). Most importantly, very little preload reserve is utilized during maximal supine exercise. End diastolic volume increases by approximately 26 per cent at peak exercise and this would be expected to produce a similar increase in stroke volume if afterload and contractility were constant. Upright posture places the human down on the function curve and exercise is associated with a significant utilization of preload reserve; end diastolic volume increases 36 per cent from control to peak exercise. Thus, in the absence of changes in contractility or afterload, preload in the adult has only a modest effect on increasing stroke volume and therefore cardiac output, in the order of 20–25 per cent, within the physiological range of filling pressures.

Does the fetus respond differently? One study has shown a 17 per cent increase in combined stroke volume above control with a volume infusion associated with an increase in right atrial pressure of 3.5 mmHg. Another

study, in which each fetal ventricle was tested separately, found that neither increased their stroke volume much in response to increases in right or left atrial pressure. The regression coefficient for atrial pressures above control values is 0.02 ml kg^{-1} mmHg^{-1} for both the right and left ventricles. Thus, a 5 mmHg increase in mean atrial pressure would produce a 7 per cent increase in right or left ventricular stroke volume. Thus, the fetal heart appears to be operating near the top of its function curve when atrial pressure is used as reference. However, there may be no meaningful difference between the appearance of the adult and fetal function curves (Fig. 8.20). Each is characterized

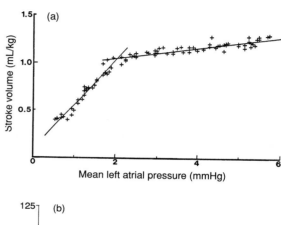

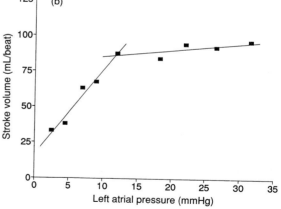

Fig. 8.20 The relationship between stroke volume and mean left atrial pressure is shown for (a) a near-term fetal sheep and (b) an adult sheep. Both studies were peformed in chronically instrumented animals with electromagnetic flow probes on the ascending aorta. Both function curves show a steep ascending limb and plateau, although the breakpoint of the adult animal is at a higher filling pressure. (a) Taken with permission from Thornburg, K.L. and Morton, M.J. (1986). *American Journal of Physiology*, **251**, H961–8. (b) Unpublished data, courtesy of Dr George Giraud, Oregon Health Sciences University.)

by a steep relationship between filling pressure and stroke volume at filling pressures below control levels and a relatively flat relationship above control. Whether the plateau in the relationship is related to the curvilinear pressure–volume relationship of the ventricle, or to the extent of external restraint, or to the length–tension relationship of the sarcomere is unclear at this time. However, studies relating end diastolic chamber size to shortening show a strong linearity in fetal and adult hearts over a wide range of filling pressures. These data suggest that the plateau lies not with the sarcomere but rather with the non-linear relationship between intracavitary pressure and ventricular volume. The precise contribution to ventricular pressure–volume relationships of external restraint remains to be quantified. From the viewpoint of cardiovascular control, however, it is clear that increased cardiac output due to increased filling pressure is not a powerful mechanism in the fetus.

Afterload

Afterload in the intact circulation is a difficult quantity to measure and study. Most investigations of the effects of afterload on the fetal heart have studied altered arterial pressure in the fetal sheep. A recent study of the effects of methoxamine on the cardiac function curve showed that increased arterial pressure produced by methoxamine depressed the function curve. Subsequently, we showed in fetal sheep that the majority of the reduced stroke volume associated with elevated arterial pressure during phenylephrine infusion was due to depression of right ventricular stroke volume (Fig. 8.21). Interestingly, right ventricular stroke volume increased significantly at lower arterial pressures produced by nitroprusside. Because pulmonary arterial pressure drops at birth, the latter finding suggests a mechanism for right ventricular stroke volume increase at birth despite the establishment of the left-to-right pressure gradient that may reduce right ventricular preload.

The finding of differential sensitivities to arterial pressure between the fetal right and left ventricle was surprising since the two fetal ventricles share common loading conditions *in utero*. Pinson and colleagues offered an anatomical explanation. They showed that, although fetal right and left ventricular weights were similar, the fetal right ventricular volume was greater than left ventricular volume and the right ventricular radius-to-wall thickness ratio was greater than left ventricular radius-to-wall thickness ratio. It follows that right ventricular wall stress would be greater than in the left ventricle at equal arterial pressures. Indirect evidence in support of this hypothesis comes from morphometric studies showing that fetal right

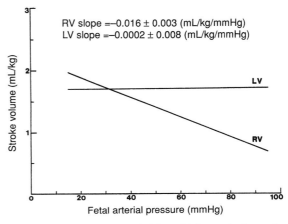

Fig. 8.21 The relationship of right and left ventricular stroke volumes to mean arterial pressure was studied in chronically instrumented fetal sheep by increasing (phenylephrine) and decreasing (nitroprusside) systemic vascular resistance at elevated filling pressures. The left ventricle was relatively insensitive to pressure changes within the range expected in the fetus and neonate. Right ventricular stroke volume was inversely related to mean arterial pressure, increasing at neonatal levels of pulmonary artery pressure and decreasing at neonatal levels of systemic arterial pressure. (Taken with permission from Thornburg, K.L. and Morton, M.J. (1986). *American Journal of Physiology*, **251**: H961–8.)

ventricular myocytes were on average large than left ventricular myocytes. This relationship reverses after birth with the increase in left ventricular pressure and decrease in right ventricular pressure. Further evidence to support an anatomical explanation is that mild pressure loading decreases right ventricular radius-to-wall thickness ratio and improves right ventricular arterial pressure sensitivity. In an entirely different experiment it has been shown that right ventricular sensitivity to arterial pressure improves with chronic fetal hypoxia. While this response might seem unexpected, chronic hypoxia was associated with a 20 per cent increase in mean arterial pressure. Thus, chronic hypoxia was another method of training the fetal right ventricle with mild hypertension. Why the fetal right ventricle is programmed to accept higher wall stress and increased arterial pressure sensitivity is unclear. It appears that, despite similar haemodynamic cues, the fetal right ventricle is anticipating its postnatal function as a low-pressure pump.

The effect of afterload on the shape of the fetal atrial pressure–stroke volume relationship has received much attention and is controversial. It has been recognized that changing arterial pressure was an unavoidable consequence of haemorrhage and that reinfusion could be a method of generating function curves. Rudolph and colleagues concluded that the plateau of the function curve relating left atrial pressure to left ventricular stroke volume in fetal and new-born lambs was due to increases in arterial pressure and afterload. However, they did not show the absence of a plateau at normal arterial pressure, but a shift to the right of the ascending limb of the function curve at high arterial pressures. These experiments too, lack the ability to keep arterial pressure constant at *normal pressure* while varying filling pressure. Thus, the true shape of the function curve at a constant *normal arterial pressure*, like the true transmural distending pressure,

remains to be definitely described. Lacking definitive data, we offer the following three-dimensional view of fetal ventricular stroke volume and of atrial and arterial pressure relationships (Fig. 8.22).

The implications of fetal cardiac sensitivity to changes in arterial pressure are important in the adaptation to birth. As described above, reduced pulmonary arterial pressure after birth may aid in increasing right ventricular stroke volume. Conversely, the relative insensitivity of the left ventricle to increased arterial pressure (minus 0.5 per cent stroke volume/mmHg) may be important in allowing the left ventricle to increase volume at higher pressures after birth.

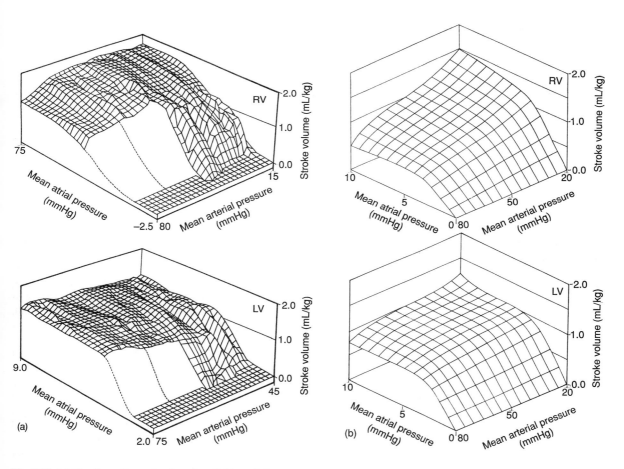

Fig. 8.22 (a) The three-dimensional surface linking all data regarding atrial and arterial pressures with stroke volume for a chronically instrumented fetal sheep with a pulmonary artery (RV) or aortic (LV) electromagnetic flow probe. Atrial pressure was varied by rapid haemorrhage and reinfusion and arterial pressure by nitroprusside or phenylephrine infusion. (b) The idealized right and left ventricular function surfaces. These surfaces were created from the data obtained from chronically instrumented fetal sheep (Courtesy of Dr Andrew Zivin *et al.*, Oregon Health Sciences University) with simultaneous right and left ventricular flow measurements and a descending aortic occluder to increase arterial pressure.

Cardiac output regulation: heart rate versus stroke volume

The pacemaker cells of the heart set the intrinsic rate of the heart but this rate can be increased by β-adrenergic stimulation or slowed by cholinergic muscarinic tone via vagal outflow. The factors that regulate the intrinsic rate during life are not known. The *intrinsic* rate is rather slow when myocytes first begin to beat but increases to its basic 'prenatal rate' early in embryogenesis and remains constant for the duration of fetal life. However, there is a typical pattern of heart rate change over the life of the sheep (Fig. 8.23). Fetal heart rate decreases during late gestation partly under the influence of autonomic tone. The rate increases dramatically at birth and then decreases over the first months of life until adulthood when it averages 80 beats/min. It appears that the progressive decrease in rate during neonatal and adult life is related to both increases in vagal tone and to a slowing of the intrinsic rate.

In the human, fetal heart rate also decreases with gestational age (Fig. 8.24). The maternal administration of atropine at any age returns heart rate to the 'intrinsic' 14-week rate of about 160 beats/min, again suggesting that parasympathetic tone is increasing as gestational age proceeds.

The interesting feature of heart rate development is that the inherent rate of the embryo and young fetus appears to be rather similar among animal species regardless of size. It has been suggested that this is because embryonic tissue has a similar metabolic rate regardless of embryo size and that the ratio of heart size to body size is relatively similar among species regardless of size. Therefore, at a given

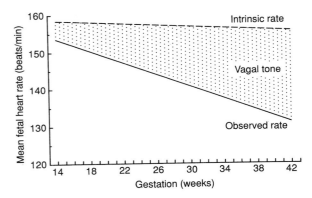

Fig. 8.24 Mean heart rate in the human fetus as a function of gestational age. The shaded portion of the figure is the difference between the observed heart rate and the 'inherent' heart rate, the heart rate measured following the maternal administration of atropine. It appears that the intrinsic heart rate is rather constant throughout gestation but that heart rate decreases as a function of increasing vagal tone. (Data from Schifferli, P.U. and Caldeyro-Barcia, R. (1973). In *Fetal pharmacology* (ed. L.O. Boréus), pp. 259–79. Raven Press, New York.)

inherent heart rate, oxygen delivery would be expected to be similar per unit weight in all species. During the postnatal period the specific metabolic rate is reduced as the animal matures, with larger animals eventually having a relatively lower metabolic rate.

Thus cardiac output should fall with specific metabolic rate. Because the heart size/body size ratio and ejection fraction are relatively constant across species, the stroke volume/body size ratio is also constant. Thus, heart rate must decline with metabolic rate. This suggests that both heart size and heart rate are finely regulated to keep stroke volume near the breakpoint of the function curve even across species that vary greatly in size.

Since ventricular output is merely the product of heart rate and stroke volume, it should not be difficult to determine which variable or to what extent each variable is used by the fetus to regulate cardiac output. However, the major intellectual hurdle is to recognize that the circulation, not the heart, regulates cardiac output. Thus, attempts to independently alter heart rate or stroke volume may not be very instructive regarding the mechanisms for cardiac output regulation in the fetus. The problem is, for example, that increased heart rate produced by pacing over a wide range of physiological rates does not affect cardiac output in either the fetus or the adult animal. There is simply a reciprocal drop in stroke volume as rate is increased. The problem is different for increases or decreases in stroke volume. Clearly, the function curves shown in Fig. 8.25

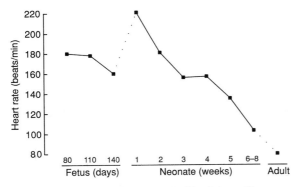

Fig. 8.23 Heart rate changes over the life of sheep. Heart rate decreases generally during fetal life until birth at which time it increases and then decreases throughout life until it reaches about 80 beats/min. (Modified with permission from Assali, N.S., Brinkman, C.R. III, Wood S, R.Jr., Dandavino, A., and Nuwayhid, B. (1978). In *Fetal and newborn cardiovascular physiology* vol. 1 (ed. L.D. Longo and D.D. Reneau), pp. 47–91. Garland STPM Press, New York.)

for the fetal sheep ventricles show a powerful effect of filling pressure or venous return on stroke volume at filling pressures less than control. However, at filling pressures greater than or equal to control values, not much stroke volume increase and therefore cardiac output increase can be achieved by increasing venous return alone. Thus, it would appear that neither increased venous return nor increased heart rate *alone* will increase cardiac output. Decreased venous return and decreased heart rate will both decrease cardiac output, and the fetus frequently does both in response to stress.

To understand how the fetus increases its cardiac output we must look at spontaneous changes in cardiac output in chronically instrumented animals. When this is done in the sheep fetus, we see that stroke volume remains fairly constant while cardiac output reflects changes in heart rate (Fig. 8.26). The fetal stroke volume–filling pressure points cluster around the breakpoint of the function curve (Fig. 8.26(a)). Thus, when venous return is increased and heart rate is elevated, ventricular output is increased. Anderson and colleagues assessed each ventricle separately in the fetal lamb to determine the effects of rate and contractility

Fig. 8.25 Average function curves are shown for the chronically instrumented fetal sheep right and left ventricle. The breakpoints are not significantly different from control values of right atrial pressure and right ventricular stroke volume and left atrial pressure and left ventricular stroke volume, respectively. Thus, at filling pressures less than control, mean atrial pressure has a powerful effect on stroke volume, but a lesser effect at filling pressures above control. (Taken with permission from Reller, M.D., Morton, M.J., Reid, D.L., and Thornburg, K.L. (1987). *Pediatric Research*, **22**, 621–6.)

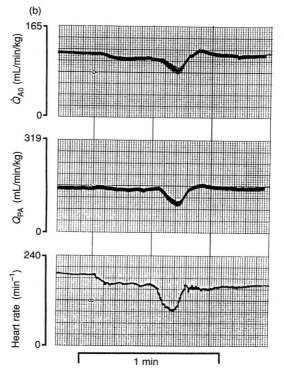

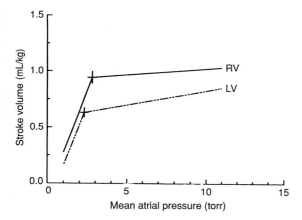

Fig. 8.26 (right) (a) Twelve 5-s averages of stroke volume and mean transmural right atrial pressure during control conditions (closed squares) superimposed on the right ventricular function curve (open circles). The fetal circulation is almost always found working at or near the breakpoint of the function curve. Thus, spontaneous increases in venous return have little effect on cardiac output without a corresponding increase in heart rate or contractility. (b) A spontaneous deceleration of heart rate is shown for the same fetus with simultaneous measurements of right and left ventricular outputs by electromagnetic flow probes. Right and left ventricular outputs mirror heart rate except during the recovery when output returns to baseline faster than heart rate. This may indicate an adrenergically mediated increase in contractility induced by hypotension during the bradycardia. $\dot{Q}_{PA}$, pulmonary artery blood flow. $\dot{Q}_{A0}$, aortic blood flow.

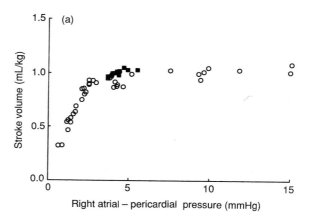

on cardiac output. They found that heart-rate variability within the physiological range was associated with a small but statistically significant effect on contractility. Thus, if venous return increases with spontaneous increases in heart rate, cardiac output will be augmented both by a heart rate increase and a mild increase in contractility.

Cardiac metabolism

During ovine fetal life oxygen consumption is about 8 mL/min/kg body weight. This level of oxygen consumption is nearly one and one-half times higher than found in the adult sheep and one-half the oxygen consumption found in the immediate new-born period. The relatively high oxygen demand found in the fetus requires a high fetal cardiac output particularly because fetal blood oxygen content is lower than for adults. Fetal myocardial oxygen needs are met through a 60 per cent greater resting myocardial blood flow than the adult and greater oxygen extraction. Fetal right ventricular myocardial blood flow is larger than left. It appears that the two fetal ventricles differ in their oxygen needs even though filling pressures and arterial pressures are the same. This may be related to the fact that the fetal right ventricle has a larger chamber volume, a larger stroke volume and, because of its larger radius to wall thickness ratio, a larger wall stress. Unfortunately, oxygen consumption for the right ventricle has not been measured because right ventricular myocardial venous outflow cannot be collected for the measurement of oxygen extraction. If O_2 extraction rates are similar for the two ventricles, their relative blood flows will indicate their relative differences in oxygen consumption.

The preferred fuel of the adult heart is long-chain fatty acids with glucose and lactate being minor fuels under aerobic conditions. However, the myocardium of the fetus does not appear to be capable of metabolizing free fatty acids. Several lines of evidence point to a deficiency of the mitochondrial enzyme, carnitine palmitoyl Co-A transferase (CPT), which prevents the fetal myocardium from metabolizing free fatty acid. However other enzyme deficiencies in the fetal myocyte may also occur. In general, the fetal heart prefers lactate over all other fuels and consumes lactate to account for two-thirds of its energy consumption with glucose providing the remainder of the energy required. It is interesting that the heart becomes able to utilize free fatty acids as a fuel shortly after birth. Free fatty acids are then the preferred fuel throughout life except under conditions of tissue hypoxia.

Fisher and colleagues investigated myocardial oxygen consumption and carbohydrate metabolism under conditions of hypoxaemia in the fetal sheep. The fetus was made hypoxic by reducing the fraction of inspired oxygen

to the mother. Oxygen content of blood supplying the coronary vessels was reduced by about 50 per cent and lactate and pyruvate concentrations were marginally increased. Fetal myocardial oxygen consumption was not changed in these experiments. Coronary blood flow to the left ventricular free wall increased from about 180 to 340 mL/min/100 g. The oxygen content of the venous effluent, obtained by sampling the coronary sinus blood, was lower in the hypoxaemic animals and the arterial venous oxygen difference was also lower in the hypoxaemic fetuses. Because of the increased coronary flow, myocardial oxygen consumption was not altered in these experiments. This indicates that the fetal myocardium is

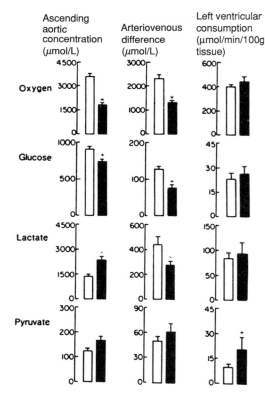

Fig. 8.27 Oxygen and metabolite concentrations in the ascending aorta, arteriovenous difference across left ventricular myocardium, and myocardial consumption of oxygen, glucose, lactate, and pyruvate of 13 normoxic (open squares) and 10 spontaneously hypoxic (black squares) fetal sheep *in utero*. Values given means ± SE.; +, p = 0.05; *p < 0.01. (Taken with permission from Fisher, D.J., Heymann, M.A., and Rudolph, A.M. (1982). *American Journal of Physiology*, **243**, H959–63.)

'protected' from moderate hypoxaemia by autoregulation of the coronary vascular bed. The autoregulatory limits of the coronary tree have not been investigated. Even under hypoxic conditions the primary fuels of the fetal heart were lactate and glucose indicating that these carbohydrates are still the preferred fuel of the heart under hypoxic conditions (see Fig. 8.27).

Why it is advantageous for the fetal myocardium to metabolize carbohydrate rather than free fatty acid is not clear. Some have argued that there is an oxygen consumption advantage. However, even if true, this effect would be very small. Perhaps a more reasonable argument is that lactate is a substrate in generous supply as it is produced continuously by the placenta and released into the fetal circulation, whereas fetal levels of free fatty acids are low. Furthermore, it can be argued that free fatty acids are necessary for the formation of membrane and other structural components and should not be used as a fuel. Therefore, the myocardium with its high substrate use takes advantage of the readily available carbohydrate, rather than free fatty acids that are needed for other purposes.

Fetal circulation

Normal vascular pressures

The fetal circulation is characterized by arterial systems (pulmonary artery and aorta) that operate in parallel and that have similar pressures (Table 8.4), with the pulmonary artery pressure being slightly but consistently higher. The two atria have similar pressures of about 2–3 mmHg (referenced to pericardial pressure) which change simultaneously when fetal blood volume is changed. (Fig. 8.17).

Table 8.4 Normal fetal haemodynamic values (sheep)*

Heart rate (beats/min)	162
Pressure (mmHg)	
Brachiocephalic artery	45
Pulmonary artery	48
Abdominal aorta	43
Right atrium	2.4
Left atrium	2.9
Output (mL/kg/min)	
Right ventricle	277
Left ventricle	185

*Source: Anderson D.F., Bissonnette, J.M. Faber, J.J., and Thornburg, K.L. (1981). *American Journal of Physiology*, **241**, H60–H66.

Arterial pressure

The physiological mechanisms underlying the maintenance of arterial blood pressure in the fetus are not known. It is noteworthy that fetal arterial blood pressure is lower than at any other time in the organism's life. The fetal circulation is characterized by a very low resistance shunt in the form of the placenta, which is lost at birth. The role of the kidney in maintaining arterial blood pressure is not known. However, it is known that a reduction in blood flow to the kidney during descending aortic occlusion causes hypertension in the upper body and a hypotension in the lower. Second, it is known that removal of the fetal kidneys will cause a substantial reduction in fetal arterial blood pressure over many weeks of time.

Haemodynamics revisited

Vascular resistance

The fetal heart provides the energy for the movement of blood through the circulation. The anatomical arrangement in the fetus is sufficiently different from the adult that one cannot assume that the fetal circulation is merely a miniature adult version. However, the well described physical principles of blood flow through adult vessels apply to fetuses as well. The mean pressure differential between the great arteries and the atria is the driving force for blood through the fetal circulation. The resistance to flow, R, of any organ is defined as the ratio of the driving pressure, $P^{art} - P^{atr} = \Delta P$, and flow, Q, through that organ:

$$R = (\Delta P)/Q \qquad (2)$$

One may reasonably assume that changes in resistance are due to one of the several parameters in Poiseuille's law. The law actually applies only to continuous laminar flow of a Newtonian fluid through long straight tubes — conditions rarely met in the circulation. Nevertheless, the law is useful for approximating resistance changes. According to Poiseuille's law,

$$R = 8\eta l / \pi r^4 \qquad (3)$$

where l is vessel length, r is vessel radius, and η is blood viscosity. From this we learn that resistance is directly proportional to vessel length but inversely proportional to the fourth power of the radius. This drives home the point that even small decreases in the radius of a resistance vessel will be associated with large (fourth power) increases in resistance to flow. Through the regulation of the relative resistances in individual organs, the fetal body is able to distribute its cardiac output between organs as needed for oxygen delivery (autoregulation).

Because arterial pressure remains relatively constant during late fetal life, organ resistances must drop in proportion to the increase in flow. Thus, organ vascularity increases in concert with organ mass. Two organ beds in the fetus are particularly interesting. The lung is vasoconstricted and flow is proportional to oxygen content and is therefore not autoregulated. The placenta is a passive bed that does not autoregulate or respond to adrenergic stimulation.

Vascular impedance

It has long been known that the pulsatile nature of blood flow through compliant blood vessels complicates the analysis of fluid energetics in the circulation. The concept of resistance being the ratio of mean pressures and flow, though useful as an approximation, is inadequate to explain the true complexity of circulatory function. A more precise description of the resistive nature of a vessel system with pulsatile pressures and flows is found in the measurement of vascular impedance, Z. The concept of impedance may require a brief review. Because the heart ejects blood into the vascular tree in a pulsatile manner, one stroke volume at a time, the physical principles that describe flow through the arteries are considerably more complex than if its output were steady. As each 'beat' is suddenly ejected into the arterial tree, the wall of the elastic vessel is stretched and pressure/flow waves are set into motion down the elastic tree. The amplitude and shape of the waves change as they progress down the arterial tree; the pressure wave gains in amplitude and narrows. From a single beat, where high-fidelity measurements are made, pressure and flow profiles can be displayed and the profiles dissected mathematically. Fourier analysis can be used to analyse the profiles at multiples (nth harmonic) of the fundamental harmonic until a close approximation of the wave form is described mathematically from the sums of the amplitudes at each frequency.

The total volume flow, $\dot{Q}_t$, is expressed as

$$\dot{Q}_t = \dot{Q}_m + \Sigma \dot{Q}_n \sin(n\omega t + \theta_n) \qquad (4)$$

and, for pressure,

$$P_t = P_m + \Sigma P_n \sin(n\omega t + \beta_n) \qquad (5)$$

where $\dot{Q}_m$ is the mean flow, $\dot{Q}_n$ the amplitude for the nth harmonic, ω is the fundamental angular frequency, t is the length of the sequence, θ_n is the phase angle of the nth harmonic, P_m is the mean pressure, P_n is the amplitude of the nth harmonic, and β_n is the phase angle of the nth harmonic.

Using these equations, pulsatile pressure and flow can be analysed and a graphical representation of the pressure and flow profiles produced. However, the most useful application of the equations is the graphical display of the impedance moduli and the phase angle as functions of frequency. The impedance modulus, Z_m, is the quotient of mean pressure and flow, $P_m/\dot{Q}_m$, and is equivalent to the vascular resistance (eq(2)). As shown in eq(4) and (5), additive pressure, P_n, and flow, $\dot{Q}_n$, amplitude can be determined for each harmonic. The ratio of these amplitudes, $P_n/\dot{Q}_n$ is the impedance for that harmonic. The sum of the individual harmonic impedances and the mean impedance gives the total impedance, commonly known as the input impedance, Z_i. The phase angle for pressure and flow can be determined for each harmonic and subtracted ($\beta_n - \theta_n$) to give the net phase angle. In this way an impedance profile of the arterial tree can be formed.

Several types of impedance have been defined. The input impedance, Z_i, is the ratio of P_t and $\dot{Q}_t$ as defined above and is the sum of the mean impedance modulus plus all the impedance moduli at all higher frequencies. Z_i is the impedance that blood energy must overcome to travel down the vascular tree and includes the significant effect of pressure waves reflected from vessel branches downstream. It is known that reflected waves are dependent upon vessel lengths and arrangement and that they contribute importantly to the input impedance of any vascular tree. Reflected waves resist ventricular ejection if they return to the ventricle during systole since pressure reflected pressure summates, with ejection pressure increasing the ventricular pressure and therefore wall stress. The characteristic impedance, Z_0, is the theoretical impedance measured in a branchless tube of infinite length and without wave reflections. It is estimated by averaging the higher frequencies that are believed to be resistant to influence by reflected waves. The characteristic impedance is importantly affected by vascular compliance. In summary, input impedance is determined by vascular resistance, vascular architecture, and vascular compliance.

Fig. 8.28 shows a graph of impedance moduli versus frequency in normal (control) dogs and dogs with pulmonary hypertension. The control curve shows an impedance profile typical of the arterial tree of many animals. At 0 frequency, the modulus is the mean impedance, Z_m, as defined above and is vascular resistance. At each frequency above 0, the impedance value is the $P_n/\dot{Q}_n$ ratio at that frequency. A higher impedance value represents a greater hindrance to blood flow at that frequency. Note that, in these typical impedance profiles, the impedance moduli decline to a minimum, and reach a plateau at higher frequencies. The whole curve shifts upward in the pulmonary circuit of dogs with chronic pulmonary hypertension, indicating an increase in pulmonary vascular resistance, Z_m, and characteristic impedance, Z_0. Z_0 is

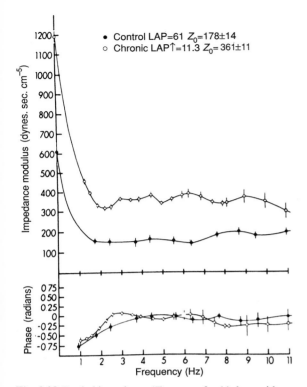

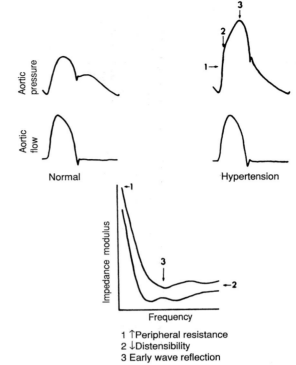

Fig. 8.28 Pooled impedence (Z) spectra for 11 dogs with chronic pulmonary venous hypertension (open circles) and 12 control normal dogs (filled circles). Note dramatically elevated impedence moduli at every frequency in the chronic pulmonary venous hypertension animals. LAP, left atrial pressure. (Taken with permission from Hopkins, R.A., Hammond, J.W., Jr, McHale, P.A., Smith, P.K., and Anderson, R.W. (1980). *Circulation Research*, **47**, 902–10.)

Fig. 8.29 The top panel shows the effects of hypertension on the pressure wave (upper tracing) and flow (lower tracing) in the ascending aorta. The bottom shows effects of hypertension on the relationship between impedence modulus (vertical axis) and frequency (horizontal axis) in the ascending aorta. (Taken with permission from O'Rourke, M.F. (1990). In *McDonald's blood flow in arteries* 3rd edn, (ed. W.W. Nichols and M.F. O'Rourke), pp. 389–420. Lee and Febiger, Philadelphia.)

importantly affected by vascular compliance. Therefore, in this model of pulmonary hypertension, the analysis of impedance allows us to conclude that both the resistance vessels and the capacitance vessels are altered. Fig. 8.29 provides the rationale for such changes.

In the published literature, one can only speculate on those features that are unique to the fetus and that affect impedance. The most obvious relate to Z_i for the pulmonary artery. Before birth, pulmonary vascular resistance is very high and the right ventricle ejects predominantly into the ductus arteriosus and the descending aorta. Because right ventricular output increases and pulmonary arterial pressure decreases after birth, Z_m must fall. The impact of ductal closure and pulmonary arterial dilatation on Z_o remains to be determined as does the profound

change in vascular architecture on Z_i. Left ventricular output nearly doubles at birth but mean arterial pressure increases by only about 50 per cent; therefore Z_m for the aorta decreases but not as markedly as for the pulmonary artery. The changes in vascular compliance and architecture after birth are more subtle for the aorta, but are partly due to elimination of the placental circulation.

There are important relationships between body size and cardiac efficiency in adult animals because pulse wave velocity does not vary greatly among animals for a given artery. This means that reflected waves return to the heart sooner after ejection in small animals than in large ones where branches are further away. Therefore, one might expect that the heart rate of different-sized adult animals would be 'adjusted' to operate at a frequency where impedance due to reflected waves would be minimized during ejection. The inverse relation between body size and heart rate across the adults of different species is well known. Heart rate is appropriately slower in large animals and appears, in teleological terms, to take advantage of the impedance spectra which change with size. How does the heart rate versus size phenomenon work for the fetus? As mentioned above, there is no such relationship among young fetuses of different species, which have similar heart rates regardless of size. This puts the fetus in a unique position of not necessarily operating at an optimal input impedance. How this affects heart function while the fetus is growing in size by orders of magnitude has yet to be explored.

Venous return

While it is commonly known that the heart imparts energy to the blood through its muscle action, the role of the circulation in regulating the function of the blood pump is less often considered. The role of the circulation is, in its simplest form, intuitively obvious. The heart must have ample blood returning from the body to maintain its output. We know from haemodynamic principles that the driving pressure through the systemic circulation is the mean arterial–right atrial pressure gradient,

$$\dot{Q} = (P_{art} - P_{ra}) / \text{resistance}. \quad (6)$$

And, under steady-state conditions, the blood flow leaving the heart (cardiac output) must be equal to the blood flow returning to the heart (venous return) and therefore $\dot{Q}$ may be used to represent either.

If mean right atrial pressure (the downstream component of the driving pressure through the circulation in eq(6)) is raised while arterial pressure and resistance are kept constant, the driving pressure for flow through the body is reduced and flow returning to the heart (so-called venous return) is reduced accordingly. We know from Fig. 8.16 that the mean filling pressure of the heart is a key determinant of ventricular preload and therefore, stroke volume. This means that the downstream pressure of the circulation, right atrial pressure, is the same pressure that determines the filling of the ventricle and affects its output through the ventricular function curve.

If the heart is stopped in the adult dog without spinal reflexes, pressures in the circulation equilibrate everywhere so that a 'mean systemic pressure' of about 7 mmHg is found throughout the circulation. If one then pumps blood out of the venous circulation into the arterial tree, as any good heart does, the right atrial pressure will drop as the arterial pressure increases, the degree of change of each depending upon the compliance and capa-

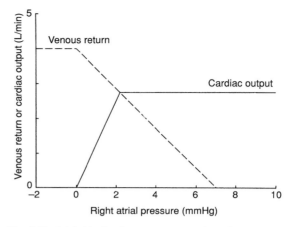

Fig. 8.30 Adult idealized venous return and cardiac output curves on the same axis. Notice that the venous return curve and the cardiac output curve intersect at the breakpoint of the cardiac output curve and that cardiac output is then determined by the mean right atrial pressure, which is the downstream pressure of the venous return curve and the mean filling pressure for the right ventricle.

citance of the two beds. Blood will then flow back through the capillary beds to the venous system. Experiments show that the rate of blood return to the right atrium is proportional to the mean systemic pressure–right atrial pressure difference. When pumping begins, right atrial and mean systemic pressures are equal, and no flow returns. As blood is pumped out of the venous system, right atrial pressure drops and venous return increases as shown in Fig. 8.30.

Several features of the curve should be noted:

1. The driving pressure for blood flow $\dot{Q}_t$, returning to the heart is the difference between mean systemic pressure, P_{ms}, and right atrial pressure, P_{ra}, so that venous return, $\dot{Q}$, is determined by the quotient of driving pressure and resistance to venous return, R_v, such that

$$\dot{Q}_t = (P_{ms} - P_{ra}) / R_v \quad (7)$$

2. The returning flow, $\dot{Q}_t$, is minimum (0) when the right atrial pressure is at mean systemic pressure, about 7 mmHg.

3. Venous return is maximal when the right atrial pressure is zero. Because the large veins collapse at negative pressures, the driving pressure for flow back to the heart (eq(7)) cannot be increased by reducing the right atrial pressure below zero.

4. The position of the venous return curve is altered by changing the blood volume in the circulatory system.

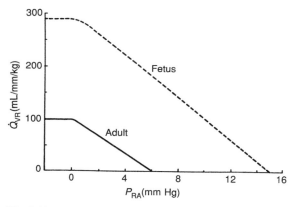

Fig. 8.31 Idealized venous return curves for the adult and the fetus. Note that the fetal venous return curve intersects the abcissa at a mean systemic filling pressure of about 15 mmHg, whereas the adult curve intersects at about 6 mmHg. Also note that for any mean right atrial pressure, flow back to the heart ($\dot{Q}_{VR}$) is considerably higher in the fetus than it is in the adult. The slopes of the two curves are about equal. (Taken with permission from Gilbert, R.D. (1978). In *Fetal and newborn cardiovascular physiology* (ed. L.D. Longo and Reneau D.D.). pp. 299–316. Garland STPM Press, New York.)

This occurs because blood volume and P_{ms} are linearly related. Thus, greater filling of the circulation is accompanied by a rightward shift in the venous return curve and an increase in mean systemic pressure, the *x*-intercept.

5. The slope of the venous return curve is altered by changes in peripheral vascular resistance.

Experiments performed in fetal sheep that were similar to those performed in adult dogs showed that the fetal circulation is different from the adult circulation. Figure 8.31 shows the venous return curve of the fetus. These data demonstrate that the mean systemic venous pressure of the fetus is considerably higher than for the adult (16 versus 7 mmHg, Fig. 8.31), and that resistance to venous return does not change measurably with increased sympathetic activity. This arrangement results in a high blood flow returning to the heart and a consequent high fetal cardiac output.

Distribution of cardiac output

In their classic study, Rudolph and Heymann described the distribution of cardiac output in chronically prepared fetal sheep. While earlier studies showed the general pattern of the fetal circulation with the exteriorized fetus under anaesthesia, Rudolph and Heymann were able to study the normal

physiological state by allowing postoperative recovery before performing experiments. They determined the following approximate distributions of combined ventricular output for the near-term fetus using the radiolabelled microsphere technique together with electromagnetic flow sensors: placenta, 40 per cent; carcass, 35 per cent; brain, gut, and heart, 5 per cent each; lungs, 4 per cent; kidney, liver, and spleen, 2 per cent each.

In the parallel circulatory arrangement of the fetus (Fig. 8.8) the right ventricle contributes 60–70 per cent of the biventricular cardiac output with the remaining 30–40 per cent coming from the left ventricle (ignoring coronary flow). The biventricular output is distributed so that about 55 per cent goes through the ductus arteriosus, 34 per cent through the foramen ovale, and 12 per cent through the aortic isthmus.

Blood returning from the placenta has a high oxygen content with greater than 80 per cent saturation and a PO_2 of greater than 30 mmHg. This blood enters the fetal circulation via the ductus venosus where it joins lower body venous blood and it is preferentially channeled through the foramen ovale to the upper body including heart and brain. About 45 per cent of the umbilical venous blood is distributed to the liver itself with the remainder continuing to the heart where 22 and 24 per cent, respectively, are distributed to the placenta and carcass. The left liver lobe receives most of its blood supply from the umbilical circulation whereas the right lobe receives blood largely from the portal vein.

Control of heart and circulation

Autonomic nervous system

In adults, autonomic innervation of the heart is used to modify heart rate and contractility to meet the daily demands of exercise and stress. Altricial animals (rat, rabbit) are born in a relatively immature state, whereas precocial animals (guinea-pig) are well developed at birth. At birth, the guinea-pig has concentrations of noradrenaline in the heart (an indicator of the development of nerve endings) that rival those of the adult but rat pups require 5 postnatal weeks before adult levels are found. Nerve endings are present in the heart of the sheep fetus by mid-gestation but reach adult levels only several days after birth.

Baro- and chemoreflexes

By mid-gestation the fetal sheep has a functional central and peripheral nervous system and arterial baroreflexes and chemoreflexes are present. Afferents run from the carotid body, a major chemo-sensing tissue, to the medulla via the sinus nerve, a branch of the glossopharyngeal nerve (IX). Chemosensitive areas, found in tissue surrounding the aorta, have afferent input via the vagus nerve (X). 'Stretch-sensitive' baroreceptor areas are found in the carotid vasculature and along the aorta. Chemo- and baroreflexes arise from the heart and lungs as well as from the great vessels and so-called low-pressure sensors are present in the venous system and within the fetal heart. The roles played by these systems in fetal cardiovascular homeostasis are not clear. The baroreceptor reflex shows interesting maturational changes, though the relative power of the reflex is still controversial. Based on carotid sinus nerve recordings, the stimulus–response curve relating mean arterial pressure to baroreceptor discharge shifts as blood pressure increases with age, keeping the basal baroreceptor discharge rate in the normal blood pressure range. It appears that both dynamic and steady-state pressure sensitivities are reduced as the fetus matures. The autonomic nervous system does not appear to be mature in the sheep fetus even by the time of birth. In contrast to the adult, ganglionic blockade, the administration of adrenergic and cholinergic blocking drugs, and denervation of the sinus and aortic nerves do not affect heart rate or blood pressure very much. Yet the administration of the β-adrenergic agonist, isoproterenol, increases heart rate dramatically and α-adrenergic stimulants will profoundly increase blood pressure and elicit a reflex slowing of the fetal heart. In fact the fetal sheep heart appears to be more sensitive to catecholamine stimulation than does the adult heart but the opposite is true for immature dog heart. The adrenal medulla of the mature sheep fetus appears to be directly stimulated by hypoxia so that blood catecholamine levels are increased many-fold under hypoxaemic conditions by increased sympathetic stimulation and by direct action on the adrenal gland.

The α and β-adrenoceptors appear to be independently regulated as do their subtypes. β-adrenoceptors increase with gestational age well into the postnatal period. However, the α-adrenoceptors become greater in number just before birth and decline in number soon after birth. α-adrenoceptors may be important in regulating myocyte size during the perinatal period. α_2-adrenoceptors may also play a role in augmenting atrioventricular conduction and myocardial function during periods of hypoxia. Following the neonatal period, the α-adrenoceptors decline and few can be found in adult myocardium.

Fetal responses to stress

As mentioned above, about 40 per cent of the fetal cardiac output is directed to the placenta. Umbilical venous blood

returning to the fetus has a PO_2 of about 30 mmHg, a PCO_2 of 38 mmHg, and a pH of about 7.35. Because most of the stresses that a fetus encounters *in utero* lead to fetal hypoxaemia, fetal hypoxaemic stress (usually for 1–2 h) has been studied most often. In sheep, four different models have been particularly helpful: (1) low maternal inspired oxygen; (2) fetal haemorrhage; (3) umbilical cord compression; and (4) uterine blood flow reduction.

Hypoxaemia

Fetal hypoxaemia can be induced by lowering the fraction of inspired oxygen to the ewe. When fetal arterial PO_2 is acutely reduced to about 12 mmHg (without changing CO_2 levels) fetal arterial pressure increases by 15 per cent and heart rate decreases by about 25 per cent. As shown in Fig. 8.32, blood flow and estimated oxygen delivery

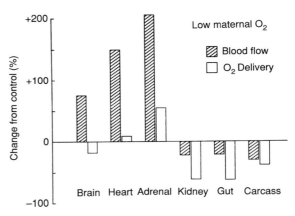

Fig. 8.32 Changes in blood flow and oxygen deliveries to selected sheep fetal organs during an acute fetal hypoxaemic episode induced by allowing the ewe to breathe air containing a reduced fraction of oxygen (Cohn, H.E., Sacks, E.J., Heymann, M.A., and Rudolph, A.M. (1974). *American Journal of Obstetrics and Gynecology*, **120**, 817–24.). Oxygen deliveries were estimated from published relationships between organ flows and blood content for various organs and the estimated contents from similar experiments in other laboratories (Longo, L.D., Wyatt, J.F., Hewitt, C.W., and Gilbert, R.D. (1978). In *Fetal and newborn cardiovascular physiology* (ed. L.D. Longo and D.D. Reneau), p. 259–87. Garland STPM Press, New York; Peeters, L.L.H., Sheldon, R.E., Jones, M.D., Jr., Makowski, E.L., and Meschia, G. (1979). *American Journal of Obstetrics and Gynecology*, **135**, 637). (Taken with permission from Thornburg, K. L. (1991). In *The childhood environment and adult disease*, CIBA Foundation Symposium, no. 156 (ed. G. R. Bock and J. Whelan). John Wiley & Sons, New York.)

change considerably from control conditions. Oxygen delivery to the heart and brain are maintained at near control levels by large increases in blood flow and oxygen delivery to the adrenal gland is increased. Blood flow to kidneys, gut, carcass, lungs, and spleen are all reduced under these hypoxaemic conditions. These studies show that maintenance of oxygen delivery to the brain and heart is made at the expense of other organs.

Cord compression

When umbilical blood flow is reduced by 50 per cent during umbilical cord occlusion, Rudolph's group found that upstream umbilical venous pressure rose to 32 mmHg (up from about 12 mmHg) and cardiac output was reduced by 20 per cent; heart rate decreased by 15 per cent and mean arterial pressure by 10 per cent. Descending aortic PO_2 dropped to nearly 15 mmHg. The distribution of blood flow under these conditions is shown in Fig. 8.33. Note that oxygen delivery to the brain and heart are maintained during cord compression as with the other types of hypoxaemia. Oxygen deliveries to kidney, gut, and carcass are decreased although blood flows to these organs were not decreased.

Haemorrhage

The alterations in organ blood flow and oxygen delivery following fetal haemorrhage are shown in Fig. 8.34. In a series of experiments from Rudolph's laboratory, fetal haemorrhage was induced by removing 22 per cent of the fetal blood volume. This reduced cardiac output by 30 per cent with no change in heart rate or arterial blood pressure. Aortic blood gases did not change. Oxygen delivery to the brain and heart were slightly reduced. The reduction in oxygen delivery to the heart was commensurate with an apparent workload reduction. Oxygen delivery to the heart and brain was maintained at the expense of the fetal kidney, gut, and carcass.

Uterine flow reduction

Fetal hypoxaemia has been induced by constricting the uterine artery to reduce uterine blood flow. Fig. 8.35 shows fetal blood flows and oxygen deliveries after a 40 per cent decrease in fetal arterial oxygen saturation during uterine flow restriction (measurements were made about 48 h after the constriction began). Arterial CO_2 concentration rose and pH dropped. The fetus responded by increasing heart rate and blood pressure after a transient

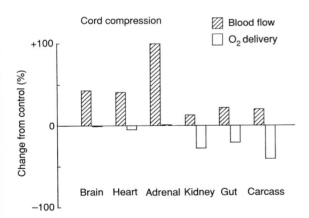

Fig. 8.33 Changes in the blood flow and oxygen deliveries to selected fetal organs during a 50 per cent reduction in umbilical cord occlusion with an inflatable occluder. Note large increases in blood flow to heart, brain, and adrenal gland preventing loss of oxygen flow to those organs at the expense of other organs (Itzkovitz, J., LaGamma, E.F., and Rudolph A.M) (1987). *American Journal of Physiology*, **252**, H100–9). (Taken with permission from Thornburg, K.L. (1991). In *The childhood environment and adult disease*, CIBA Foundation Symposium, no. 156. (ed. G.R. Bock and J. Whelan). John Wiley & Sons, New York.)

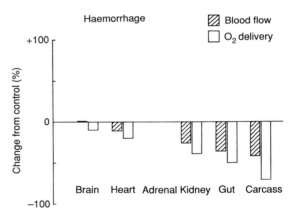

Fig. 8.34 Changes in blood flow and oxygen deliveries to selected fetal organs after loss of fetal blood volume by fetal haemorrhage (approximately 20 per cent blood loss). Note that, unlike Fig. 8.32, brain and heart show small decreases in oxygen delivery following haemorrhage (see Itzkovitz, J., Goetzman, B.W., and Rudolph, A.M. (1982). *American Journal of Physiology*, 242, H543). (Taken with permission from Thornburg, K.L., (1991). In *The childhood environment and adult disease*, CIBA Foundation Symposium, no. 156, (ed. G. R. Bock and J. Whelan). John Wiley & Sons, New York.)

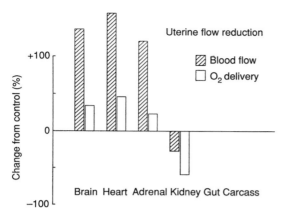

Fig. 8.35 Changes in blood flow and estimated oxygen deliveries to several fetal organs during reduction in uterine blood flow produced by a mechanical occluder on uterine artery. (From Bocking, A.D., Gagnon, R.,White, S.E., and Homan, J. (1988). *American Journal of Obstetrics and Gynecology* **159**, 1418–24). Note substantial increase in blood flow to brain, heart and adrenal gland which preserved estimated oxygen flow to those organs. Oxygen deliveries were estimated as in Fig. 8.32. Blood flows to gut and carcass were not measured in this study, however nuchal muscle blood flow was measured and did not change. (Taken with permission from Thornburg, K.L. 1991). In *The childhood environment and adult disease*, CIBA Foundation Symposium, no.156, (ed. G.R Bock and J. Whelan). John Wiley & Sons, New York.)

bradycardia. Under these conditions brain, heart, and adrenal blood flows were increased dramatically and oxygen delivery to these organs was greater than under control conditions. These responses are made at the expense of flow and delivery to the fetal kidney and presumably to the gut and carcass also (not reported). This model may mimic human fetal stress better than the other models mentioned because many cases of human fetal distress are believed to be the result of 'utero-placental insufficiency' or uterine ischaemia.

There are several physiological mechanisms that come into play during fetal hypoxic stress.

1. Autoregulation is the ability of an organ to locally alter its flow needs according to the oxygen content of the blood perfusing it. It appears that the heart and brain are powerful autoregulators, able to vasodilatate rapidly under hypoxaemic conditions to ensure adequate oxygen delivery at all times.

2. Autonomic nervous control is an important regulator of the cardiovascular system during periods of stress. Adrenergic vasoconstriction is responsible for a great portion of the reduction in blood flows and oxygen

deliveries to kidney gut and carcass during stress. However, the nature and power of the response is fetal-age-dependent. In mid-gestation, when autonomic reflexes are just beginning to appear, hypoxaemic stress causes an increase in heart rate rather than a decrease as for the mature fetus.

Some indication of the organ-specific roles of autoregulation and autonomic control is evident from the responses of animals that have been chemically denervated before hypoxaemia is induced. In these cases, blood flow to the heart, brain, and adrenals will increase as expected while flows to the gut, kidney, and carcass are increased.

3. The third factor is the role of circulating vasoactive compounds. It is known that plasma levels of catecholamines, vasopressin, angiotensin II, atrial natriuretic peptide, and cortisol all increase in response to acute hypoxaemia. However the relative roles of these various important substances in regulating the distribution of blood flow during stress needs further study. These hormones undoubtedly play a role in the regulation of the fetal circulation under normoxic conditions. However, their roles are so subtle and difficult to study that we know little about them at this time.

Birth transition

The transition from fetal to neonatal life poses formidable tasks for a fetal circulation that most investigators agree is operating near its maximum *in utero*. Mechanical, neurohumoral, and metabolic changes occur; they have been investigated and are fairly well understood. The precise cardiac adaptation to extrauterine life remains to be fully elucidated.

The mechanical events that occur include pulmonary ventilation, severing of the umbilical cord thereby eliminating the placental shunt and the ductus venous shunt, and the establishment of a left–right interatrial pressure gradient with elimination of foramenal shunting. The ductus arteriosus shunt, which flows left to right after birth, is usually eliminated within 48 h due to oxygen-induced local vasoconstriction. Thus, all of the fetal shunts are eliminated within 48 h of birth. Pulmonary ventilation reduces pulmonary vascular resistance both through mechanical effects and an oxygen inhibition of pulmonary arteriolar vasoconstriction (Fig. 8.36). The drop in pulmonary vascular resistance is probably the cause of the establishment of the left-to-right interatrial pressure gradient due to increased venous return to the left atrium from the lungs. This thesis is supported by the fact that *in utero* ventilation without birth will establish a left-to-right interatrial pressure gradient without disturbing the placental circulation, although

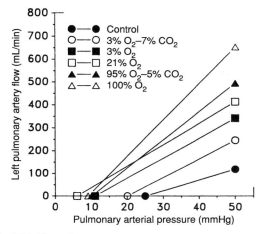

Fig. 8.36 The pulmonary artery pressure–flow relationships are shown for chronically instrumented fetal sheep. Fetuses were studies under control conditions and with ventilation with various gas mixtures as noted. At a mean pulmonary artery pressure of 40 mmHg pulmonary blood flow can be increased sevenfold by ventilation with 100 per cent oxygen and no carbon dioxide. Approximately half of this increase can be achieved by ventilation with a gas mixture that does not change fetal arterial blood gases alone. (Taken with permission from Reid, D.L., and Thornburg, K.L. (1990). *Journal of Applied Physiology*, **69**, 1630–6.)

Tabel 8.5 Regional blood flow in near-term fetal sheep*

	Regional blood flow (mL/min/100g)	
Tissue	Controls	After oxygen ventilation
Lung	159	1040
Left Ventricle	265	102
Brain	151	58
Carcass	15	8
Adrenal	397	167
Brown fat	66	13
Kidney	245	151
Placenta	138	89

*Source: Teitel, D.F. (1988). *Seminars in Perinatology*, **12**, 96–103.

Table 8.6 Ventricular output (in sheep)*

	Ventricular output (mL/min/100g)	
	Prenatal	Postnatal
RV	340	425
LV	170	425

RV, Right ventricle; LV, left ventricle.
*Source: Teitel (1988). *Seminars in Perinatology*, **12**, 96.

oxygen ventilation will increase placental vascular resistance. Neurohumoral changes include increased adrenergic tone which is responsible for increased heart rate and contractility. In addition, circulating thyroid hormone concentrations are increased, which may contribute, albeit on a longer time scale, to circulatory adaptation. Metabolic changes include the shift from glucose and lactate to fatty acid as the primary fuel for the heart. The trigger for these metabolic changes and their significance is unclear.

A striking feature of the transition is the increase in cardiac output and alteration of output distribution. Regional blood flows determined by the microsphere technique before and after pulmonary ventilation with oxygen are shown in Table 8.5. The increase in pulmonary blood flow is striking and is due to the combined mechanical and biochemical effects of ventilation with oxygen. The drop in flow to other tissues probably represents autoregulation since arterial oxygen contents are approximately doubled during these manipulations. With birth, the neonate's oxygen consumption increases about twofold due to muscle use and thermoregulation. This increase in oxygen consumption is most certainly the driving force for the increase of right and left ventricular output noted at birth and shown in Table 8.6.

Prior to birth, left and right ventricular stroke volumes cannot be much raised by volume infusion. However, after birth, left ventricular stroke volume nearly doubles and right ventricular stroke volume increases by approximately 50 per cent. These changes are not blocked by β-adrenergic blockade. The increase in left ventricular stroke volume at birth can be duplicated by *in utero* ventilation, but not the birth-related increase in right ventricular stroke volume (Fig. 8.37). It may be that sufficient time is necessary for pulmonary vascular resistance to fall in order that the afterload sensitivity of the right ventricle can be manifest as an increased stroke volume after birth. To date, *in utero* ventilation experiments have not been carried out for long enough to adequately test this hypothesis. The mechanism for left ventricular stroke volume increase with either *in utero* ventilation or birth is unclear. Both arterial pressure and systemic vascular resistance increases, indicating that left ventricular afterload is elevated. Sonomicrometer studies indicate that end diastolic volume increases at birth. Thus, the neonate may have a left ventricular preload reserve unavailable to the fetus. One theory that explains how preload may become available after birth is based on ventricular interaction (i.e. the effect of right ventricular pressure on left ventricular pressure–volume relationships).

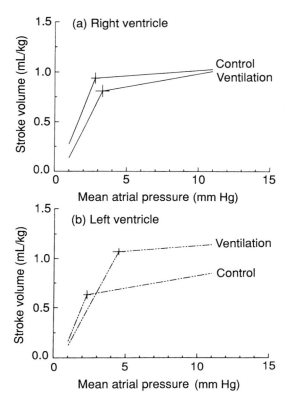

Fig. 8.37 For fetal sheep, *in utero* ventilation with oxygen produces a rapid shift upward of (b) the left, but not (a) the right ventricular function curve. The establishment of a left–right interatrial pressure gradient may be responsible for these changes. (Taken with permission from Reller, M.D., Morton, M.J., Reid, D.L., and Thornburg, K.L. (1987). *Pediatric Research*, **22**, 621–6.)

According to this theory, the left ventricle cannot fill before birth to the same extent as after birth because filling pressures in the two ventricles are equal before birth. However, left ventricular filling pressure becomes much larger than right ventricular filling pressure as the lungs become inflated at birth. Whether ventricular interaction explains increases in left ventricular function at birth remains to be tested.

Further reading

Anderson, P.A.W. (1990). Myocardial development. In *Fetal and neonatal cardiology* (ed. W.A. Long), pp. 17–38. Saunders, Philadelphia.

Battaglia, F.C. and Meschia, G. (1986). *An introduction to fetal physiology*. Academic Press, New York.

Clark, E.B. and Takao, A. (ed.) (1990). *Developmental cardiology: morphogenesis and function*. Futura Publishing, Mt Kisco, New York.

Cohn, H.E., Sachs, E.J., Heymann, M.A., Rudolph, A.M. (1974). Cardiovascular response to hypoxemia and acidemia in fetal lambs. *American Journal of Obstetrics and Gynecology*, **120**, 817–24.

Faber, J.J., Anderson, D.F., Morton, M.J., Parks, C.M., Pinson, C.W., Thornburg, K.L., and Willis, D.M. (1985). Birth, its physiology and the problems it creates. In *Physiological development of the fetus and newborn* (ed.C.T. Jones), pp. 371–80. Academic Press, London.

Fisher, D.J. (1984). Oxygenation and metabolism in the developing heart. *Seminars in Perinatology, 8*, 217–25.

Friedman, W.F. (1972). The intrinsic physiologic properties of the developing heart. *Progress in Cardiovascular Disease*, **15**, 87–111.

Gilbert, R.D. (1982). Effects of afterload and baroreceptors on cardiac function in fetal sheep. *Journal of Developmental Physiology,4*, 299–309.

Icardo, J.M. (1984). The growing heart: an anatomical perspective. In *Growth of the heart in health and disease* (ed. R. Zak), pp. 41–79. Raven Press, New York.

Iwamoto, H.S., Teitel, D., and Rudolph, A.M. (1987). Effects of birth-related events on blood flow distribution. *Pediatric Research*, **22**, 634–40.

Klitzner, T.S. (1991). Maturational changes in excitation–contraction coupling in mammalian myocardium. *Journal of the American College of Cardiology*, **17**, 218–25.

Nakanishi, T. Seguchi, M., and Takao, A. (1988). Development of the myocardial contractile system. *Experientia*, **44**, 936–44.

Pinson, C.W., Morton, M.J., and Thornburg, K.L. (1987). An anatomic basis for fetal right ventricular dominance and arterial pressure sensitivity. *Journal of Developmental Physiology*, **9**, 253–69.

Reller, M.D., Morton, M.J., Reid, D.L., and Thornburg, K.L. (1987). Fetal lamb ventricles respond differently to filling and arterial pressures and to *in utero* ventilation. *Pediatric Research, 22*, 621–26.

Reller, M.D., McDonald, R.W., Gerlis, L.M., and Thornburg, K.L. (1991). Cardiac embryology: basic review and clinical correlations. *Journal of the American Society of Echocardiography*, **4**, 519–32.

Rudolph, A.M. and Heymann, M.A. (1973). Control of the foetal circulation. In *Foetal and neonatal physiology*. Cambridge University Press.

Smolich, J.J. (1987). The morphology of the developing myocardium. In *Research in perinatal medicine, (V). Perinatal development of the heart and lung* (ed. J. Lipshitz, J. Maloney, C. Nimrod, and G. Carson), pp. 1–22. Perinatology Press, New York.

Teitel, D.F. (1988). Circulatory adjustments of postnatal life. *Seminars in Perinatology*, **12**, 96–103.

Thornburg, K.L. and Morton, M.J. (1986). Filling and arterial pressures as determinants of left ventricular stroke volume in unanaesthetized fetal lambs. *American Journal of Physiology*, **251**, H961–8.

9. Fetal oxygenation, carbon dioxide homeostasis, and acid–base balance

Daniel W. Rurak

Oxygen is a critical factor in prenatal development, and an inadequate supply of oxygen can lead to fetal death or to long-term physical and/or mental disabilities after birth. Consequently, there has been a great deal of animal research on fetal oxygenation, in both normal and pathophysiological conditions. Until recently, ethical and technical constraints limited the study of the human fetus, except at delivery. However, blood samples can now be obtained from the human fetus *in utero*, and this, along with non-invasive techniques, has added greatly to our knowledge of fetal oxygen homeostasis in the human. The purpose of this chapter is to outline current knowledge of fetal oxygenation under normal conditions and when fetal oxygen delivery is reduced. Also, since the latter condition can be associated with, and exacerbated by, respiratory and metabolic (lactic) acidaemia, feto-maternal carbon dioxide exchange and fetal acid–base balance will also be discussed.

Maternal-fetal oxygen transfer and fetal vascular PO_2

Fetal oxygen supply comprises three main elements: the delivery of oxygen from maternal lungs to the placenta via the maternal circulation; diffusion across the placenta down a physicochemical gradient; and delivery of the oxygen to fetal tissues by the fetal cardiovascular system. Transplacental oxygen transfer is affected by the mean difference in dissolved oxygen concentration (or PO_2) between maternal and fetal placental blood, the permeability of the placenta to oxygen, the area of the exchange surface, and the relative orientations of maternal and fetal placental blood flows. Data from several species indicate that maternal-fetal O_2 transfer is normally not limited by placental permeability. Oxygen is thought to equilibrate completely or nearly completely between maternal and fetal bloods during transit of blood through the placenta. However, a striking feature of fetal oxygenation is the low PO_2 of fetal blood. For example, arterial PO_2 in the fetal lamb is only ~ 20 per cent of the value in the mother (Table 9.1), and there is a ~ 20 mmHg difference between PO_2 values in uterine and umbilical venous blood. This large difference is not a consequence of a placental barrier to oxygen; as noted above, placental permeability to oxygen is high. Rather the maternal-fetal PO_2 difference appears to result from at least two other factors. First, the placenta itself consumes oxygen at a high rate, which would lower PO_2 on the fetal side. Also, there appear to be

Table 9.1 PO_2 values in blood sampled from maternal artery (MA), uterine vein (UtV), and fetal umbilical artery (UA) and vein (UV) in several species. Also given are the P_{50} (the PO_2 at which haemoglobin is 50% saturated with oxygen) values for maternal (P_{50m}) and fetal (P_{50f}) bloods, and the difference between the two ($P_{50m} - P_{50f}$)

Species	PO_2 (mmHg)				P_{50} (mmHg)		
	MA	UtV	UV	UA	P_{50m}	P_{50f}	$P_{50m} - P_{50f}$
Sheep	95	57	35	20	34	17	17
Cow	98	59	38	26	31	22	9
Pig	84	52	31	21	33	22	11
Horse	95	50	49	33	31	27	4
Human	95	50*	41	29	26	22	4

Data from Comline and Silver (1974), Faber and Thornburg (1983), and Soothill *et al.* (1986) (full references are given in the 'Further reading').

* Blood sampled from the intervillous space.

local inequalities in the distribution of maternal and fetal blood flows within the placenta. This is equivalent to ventilation/perfusion inequalities in the lung, and may account for 50 per cent of the maternal–fetal PO_2 difference in sheep.

Table 9.1 gives maternal and fetal PO_2 values in late pregnancy for several species. The values in the cow and pig are similar to those in the sheep, whereas in the horse fetal arterial PO_2 is 7–11 mmHg higher, and there is only a ~ 2 mmHg difference between uterine and umbilical venous PO_2 values. The higher fetal PO_2 in the horse is associated with evidence for a countercurrent arrangement of maternal and fetal placental blood flows, which would increase the efficiency of placental O_2 transfer, in terms of the maternal-fetal PO_2 gradient. In the sheep and cow, there is no evidence for a counter–current flow arrangement. The human values are of particular interest, since the previous estimates of human fetal vascular PO_2 were obtained from umbilical cord blood at delivery, and were similar to those observed in fetal sheep *in utero*. The more recent data in Table 9.1 were obtained using transabdominal blood sampling of the human fetus *in utero* and indicate substantially higher arterial and umbilical venous PO_2s, although not as high as in the horse. The reason for the higher PO_2 in the human is unclear, since it is unlikely that a countercurrent maternal-fetal vascular arrangement is possible within the intervillous space of the human placenta.

Fetal oxygen delivery and consumption

Although fetal vascular PO_2 is undoubtedly the most commonly measured variable related to fetal oxygenation, it indicates little about the actual rate of fetal oxygen supply. For this, estimates of fetal O_2 delivery and consumption are required. Figure 9.1 illustrates the method most commonly employed in animal species. Fetal oxygen delivery is the rate at which oxygen is delivered from the placenta to the fetus, i.e. the product of umbilical blood flow and the oxygen content of umbilical venous blood. Only a portion of this oxygen is consumed by the fetus, and, by the Fick principle, this is equal to the product of umbilical blood flow and the difference in O_2 content between the umbilical vein and artery. The ratio between oxygen uptake and oxygen delivery is termed fractional oxygen extraction; it can also be calculated as the umbilical veno-arterial oxygen difference divided by umbilical venous O_2 content. Similar calculations can be performed for total uterine O_2 delivery and consumption. The difference between total uterine and fetal O_2 uptakes in the rate of O_2 consumption of the uterine tissues and the placenta.

Table 9.2 gives values for some of the variables important in determining O_2 delivery and O_2 consumption in the fetal lamb: umbilical venous O_2 variables and

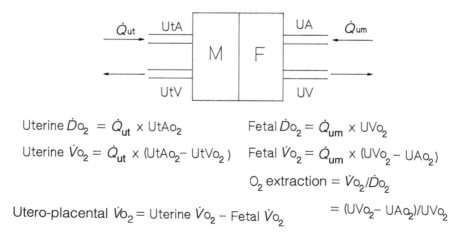

Uterine $\dot{D}o_2 = \dot{Q}_{ut} \times UtAo_2$

Uterine $\dot{V}o_2 = \dot{Q}_{ut} \times (UtAo_2 - UtVo_2)$

Fetal $\dot{D}o_2 = \dot{Q}_{um} \times UVo_2$

Fetal $\dot{V}o_2 = \dot{Q}_{um} \times (UVo_2 - UAo_2)$

O_2 extraction $= \dot{V}o_2/\dot{D}o_2$

Utero-placental $\dot{V}o_2 =$ Uterine $\dot{V}o_2 -$ Fetal $\dot{V}o_2$

$= (UVo_2 - UAo_2)/UVo_2$

Fig. 9.1 Diagramatic representation of maternal (M)-fetal (F) oxygen exchange, with equations for fetal and uterine oxygen delivery ($\dot{D}_{O_2}$), oxygen consumption ($\dot{V}_{O_2}$), fetal oxygen extraction, and utero-placental oxygen consumption. $\dot{Q}$ut, uterine blood flow; $\dot{Q}$um, umbilical blood flow. UtA_{O_2} and UtV_{O_2} are, respectively, the uterine artery and uterine vein blood oxygen content. UA_{O_2} and UV_{O_2} are, respectively, the umbilical artery and vein blood oxygen content.

Table 9.2 Variables important in determining oxygen delivery and consumption in fetal and adult sheep. Blood O_2 variables are from the umbilical vein in the fetus, and the aorta in the adult. Flow is umbilical blood flow and cardiac output in the fetus and adult, respectively. Oxygen consumption ($\dot{V}_{O_2}$) by utero-placental tissues is also given

	Fetus	Adult
PO_2 (mmHg)	35	95
O_2 saturation (%)	85	95
[Haemoglobin] (g/100 mL)	12	9
O_2 content (mM)	5.5	6.8
Flow (ml/min/kg)	200	96
O_2 delivery (μmol/min/kg)	1100	650
Arteriovenous O_2 difference (mM)	1.7	2.1
O_2 consumption (μmol/min/kg)	340	195
O_2 extraction (%)	31	43
Uteroplacental $\dot{V}_{O_2}$ (μmol/min/kg)	1750	

Fetal and adult data are from van der Weyde *et al.* (1992) and Nesarajah *et al.* (1983), respectively (see 'Further reading' for full references); utero-placental $\dot{V}_{O_2}$ data comes from M. van der Weyde and D. Rurak (unpublished data).

haemoglobin concentration, umbilical blood flow, and the umbilical veno-arterial difference in O_2 content. Values for the equivalent parameters in adult sheep are given for purposes of comparison. These include systemic arterial O_2 and haemoglobin concentrations, cardiac output, and the arteriovenous difference in O_2 content across the systemic circulation.

Haemoglobin oxygen affinity in the fetus

Vascular PO_2 in the fetus is low in comparison to that in the adult, and this would be expected to result in lowered values for blood oxygen saturation and content. However, oxygen saturation and content in umbilical venous blood are only marginally lower in the fetus than in adult arterial blood (Table 9.2). Two factors are responsible for this. One is the higher haemoglobin concentration in the fetus, which increases the O_2 carrying capacity of fetal blood. But more important is the higher affinity for oxygen of fetal blood; the O_2 dissociation curve for fetal blood is shifted to the left in relation to the adult (Fig. 9.2). Thus, even with the low PO_2 values, fetal blood has a relatively high oxygen saturation and total O_2 content. The high O_2 affinity could limit the unloading of oxygen from blood to fetal tissues. But this is minimized by the steepness of that portion of the dissociation curve over which the fetus normally operates. Even with the relatively small drop in PO_2 that occurs between arterial and systemic venous bloods, there is a large fall in O_2 saturation (Fig. 9.2).

There are species variations in the magnitude of the fetal-adult difference in haemoglobin O_2 affinity, which can be expressed by the adult-fetal difference in P_{50}, with P_{50} being the PO_2 at which haemoglobin is 50 per cent saturated with oxygen. In some species, such the cat, there is no difference at all. This leads to the question of the actual

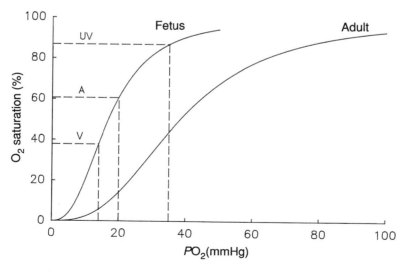

Fig. 9.2 O_2 dissociation curves for fetal and maternal blood in sheep. The lines intersecting the fetal curve denote PO_2/O_2 saturation relationships for umbilical venous (UV), umbilical arterial (A), and fetal systemic venous (V) bloods.

survival value of the increased O_2 affinity of fetal blood. It may be that this mechanism is most important in species such as the sheep, cow, and pig, which have a large maternal/fetal PO_2 gradient, a relatively low fetal vascular PO_2, and a maternal-fetal difference in P_{50} ranging from 9 to 17 mmHg. In the sheep acute replacement of fetal blood with maternal blood results in the expected decreases in blood oxygen saturation and content, but this is associated with reduced fetal oxygen delivery and consumption and metabolic acidaemia, suggesting that the fetal sheep cannot survive *in utero* with adult blood. In contrast, with intrauterine blood transfusion to treat erythroblastosis in the human fetus, almost complete replacement of fetal red cells by adult cells is apparently well tolerated. However, human fetal vascular PO_2 is higher than in the sheep and there is a smaller maternal-fetal difference in P_{50} (Table 9.1). The same P_{50} difference is found in the horse. The higher PO_2 values in the horse and human may minimize the requirement for having a high haemoglobin–oxygen affinity.

Oxygen delivery and consumption in the fetus and adult

The relatively high umbilical venous O_2 content in the fetal sheep is associated with a much higher weight-specific rate of placental blood flow compared to pulmonary flow in the adult. Consequently, O_2 delivery/kg of body weight in the fetus is nearly double that in the adult, and this is also the case for oxygen consumption. This contrasts with the lower value for fetal O_2 extraction. Also given in Table 9.2 is the rate of oxygen consumption for the utero-placental unit; the value is very much higher than for the fetus itself. Even in late gestation, when fetal weight (3–4 kg) is much greater than placental weight (0.3 kg), the utero-placental tissues account for nearly 50 per cent of total utero-placental O_2 consumption (see Chapter 4).

Fetal O_2 consumption been measured in the species listed in Table 9.1, as well as in several other species. The values are all similar, ranging from 260 to 360 μmol/min/kg fetal weight (6–8 mL/min/kg). Thus, there is not the inverse relationship between body size and O_2 consumption (i.e. metabolic rate) that is present after birth. However, the higher metabolic rate of smaller mammals after birth reflects their greater relative rate of heat loss to the environment and hence the need for greater rates of energy production to maintain body temperature. This is not a requirement for the fetus, since the temperature of the fetus and its intrauterine environment are essentially the same. Oxygen in the fetus is largely used for tissue maintenance functions and growth, rather than for heat generation.

Oxygen delivery to fetal tissues

In the fetus, because of the mixing of oxygenated umbilical venous blood with deoxygenated blood returning from tissues, the PO_2, O_2 saturation, and O_2 content of systemic arterial blood are substantially lower than in the umbilical vein (Table 9.1). Moreover, because of the preferential distribution of umbilical venous blood to the ascending aorta, PO_2 and O_2 saturation and content are 10–20 per cent higher than in the descending aorta. Nonetheless, O_2 contents in both the ascending and descending aortas are substantially lower than in the adult (Table 9.3). However, for most tissues, there is a high blood flow, so that oxygen delivery is similar or higher than in the adult. For those organs for which estimates of O_2 consumption are available, the rate is similar in the fetus and adult. Hence, O_2 extraction across organs in the fetus is similar to or lower than that in the adult. An exception is the kidney; both O_2 delivery and consumption are higher after birth, which reflects the greater glomerular filtration rate and sodium reabsorption postnatally (see Chapter 13). Overall though, the low PO_2 and O_2 concentration in fetal arterial blood is accompanied by high organ blood flows, as a consequence of a high cardiac output (see Chapter 8). This appears to be the most important compensatory mechanism for the low vascular PO_2 and allows O_2 delivery to meet total and regional demands. Moreover, it also permits O_2 extraction across most fetal organs to be lower than in the adult. This may serve to minimize t!✸ drop in PO_2 that occurs in fetal

Table 9.3 Representative values for blood flow and oxygen delivery in various tissues from fetal and adult sheep. Arterial blood O_2 contents used to calculate O_2 delivery are 2.9, 2.6, and 6.2 mm in fetal ascending aorta, fetal descending aorta, and adult, respectively

	Blood flow (mL/min/100g)		O_2 delivery (μmol/min/100g)	
	Fetal	Adult	Fetal	Adult
Cerebral cortex	167	62	491	396
Myocardium	320	134	942	855
Adrenal gland	385	194	963	1237
Kidney	172	689	454	4396
Intestine	123	72	340	459
Spleen	324	190	877	1212
Skeletal muscle	12	4	33	25

Fetal and adult data from Rurak *et al.* (1990) and Nesarajah *et al.* (1983), respectively (full references are given in the 'Further reading').

blood during transit through tissues, thereby maintaining an adequate gradient for oxygen diffusion.

Spontaneous fluctuations in fetal PO_2 and O_2 consumption

It was once thought that fetal oxygen consumption and vascular PO_2 were constant, a view that was based upon intermittent and infrequent measurements of both variables. However, when measured with indwelling oxygen sensors, fetal vascular PO_2 in pregnant sheep was found to fluctuate continuously, with frequent, transient drops of 3–4 mmHg. These fluctuations are due to non-labour uterine contractions (contractures) and periods of increased fetal skeletal muscle activity in the form of breathing and body movements. With the former events, PO_2 falls because of a transient decrease in uterine blood flow, while the fall in PO_2 with increased skeletal muscle activity is associated with a 15–30 per cent increase in fetal O_2 consumption, at least during episodes of fetal breathing. Conversely, when fetal activity is abolished by a neuromuscular blocking agent or general anaesthesia, O_2 consumption falls by 8–23 per cent and this is associated with a rise in vascular PO_2. Thus, in the fetus there is an inverse relationship between O_2 consumption and arterial blood oxygen levels, a situation not normally present in the adult. After birth increases in tissue O_2 demands are matched by elevations in cardiac output, ventilation, and pulmonary O_2 uptake. The equivalent responses in the fetus would require that uterine blood flow fluctuate to match transient changes in fetal oxygen uptake and this does not occur. Hence, when fetal O_2 consumption rises, more oxygen is removed from maternal blood as it flows through the placenta. The PO_2 in maternal blood leaving the exchange area is thus lower and, as there is equilibration between maternal and fetal placental blood, umbilical venous PO_2 also falls. The opposite changes occur when fetal O_2 consumption is lowered. The result is continuous fluctuation in fetal arterial PO_2, a situation far different from that in the adult.

Fetal carbon dioxide homeostasis

CO_2 is produced by fetal tissues at about the same rate as oxygen is consumed, and, to maintain CO_2 balance, the rate of fetal CO_2 production must equal the rate of CO_2 loss to the mother. Carbon dioxide in blood is hydrated in a reaction catalysed by the enzyme carbonic anhydrase to form carbonic acid which spontaneously dissociates to bicarbonate and a proton:

$$CO_2 + H_2O \leftrightarrow H_2CO_3 \leftrightarrow H^+ + HCO_3^- \qquad (1)$$

Carbonic acid levels in the blood are very low, so CO_2 and HCO_3^- are the dominant species, with the latter ion accounting for most of the carbon dioxide in blood; there is also CO_2 reversibly bound to haemoglobin.

Table 9.4 gives maternal and fetal blood PCO_2 and pH values from pregnant sheep in late gestation. Although transplacental bicarbonate exchange may occur via an anion exchanger, data from sheep suggest that the bulk of fetal-maternal carbon dioxide transfer occurs via diffusion of dissolved CO_2. As with oxygen, the permeability of the placenta to CO_2 is high, and equilibration between maternal and fetal placental venous bloods would be expected. However, as indicated in Table 9.4, umbilical venous PCO_2 is about 4 mmHg higher than in uterine venous blood. This differences is probably due to the same factors that are responsible for the maternal-fetal PO_2 differences discussed earlier, namely local inequalities in maternal and fetal placental blood flows and placental CO_2 production. The umbilical-uterine PCO_2 gradient is lower than the maternal-fetal PO_2 gradient in pregnant sheep (Table 9.1); and this is also the case in other species. Hence, in the horse, which has a fairly small transplacental PO_2 difference (Table 9.1), there is virtually no difference between uterine and umbilical venous PCO_2 values. In the fetal sheep, arterial PCO_2 is ~ 15 mmHg above maternal arterial values (Table 9.4), and this increased CO_2 concentration is associated with a slightly lower pH. Fetal PCO_2 is affected by maternal arterial PCO_2 and thus is altered by maternal hypo- and hyperventilation, which increase and decrease

Table 9.4. Mean (± standard error) values for PCO_2 and pH from maternal artery, uterine vein, and umbilical vein and artery in nine pregnant sheep at 125–135 days gestation

	Maternal artery	Uterine vein	Umbilical vein	Umbilical artery
PCO_2 (mmHg)	33.8 ± 1.1	38.3 ± 1.0	42.2 ± 0.8	47.3 ± 1.0
pH	7.47 ± 0.02	7.43 ± 0.01	7.40 ± 0.01	7.37 ± 0.1

Data from van der Weyde *et al.* (1992) (full reference in 'Further reading').

fetal PCO_2, respectively. Fetal PCO_2 can also be increased acutely by decreases in uterine and umbilical blood flows. The effects of elevated fetal CO_2 levels (i.e. hypercapnia) have been most commonly studied by increasing maternal inspired CO_2 for ~ 1 h. The resulting fetal hypercapnia elicits effects similar to those in adults: respiratory stimulation and an increase in cerebral blood flow. There is also a rise in fetal vascular PO_2, apparently due to Bohr shifts in the oxygen dissociation curves of maternal and fetal blood. Overall, the effects of raised fetal CO_2 levels are relatively benign.

Fetal acid–base balance

The basic principles of fetal acid–base regulation are the same as those in the adult. Hydrogen ions are continuously produced in the fetus via metabolic processes, and buffering systems are necessary to minimize the resulting intracellular and extracellular pH changes. In blood, the carbonic acid–bicarbonate system and haemoglobin are the most important buffering elements, with the former system being of particular importance to the regulation of acid–base balance. This is because, in the adult, bicarbonate and CO_2 (and thus PCO_2) are regulated by the kidney and lungs, respectively. The reactions that relate CO_2 and bicarbonate are given in eq (1). The working of the system can be illustrated with the example of metabolic acidosis due to lactic acid accumulation. This is especially relevant to the fetus, since lactic acidaemia is one of the consequences of reduced fetal oxygen delivery. When lactic acid levels increase, there is a corresponding increase in protons in the blood, resulting in a fall in pH. The reactions in eq (1) run to the left with a fall in bicarbonate and a rise in dissolved CO_2 and PCO_2. The magnitude of these changes can be calculated using the Henderson–Hasselbach equation for the bicarbonate system:

$$pH = pK' + \frac{\log [HCO_3^-]}{[CO_2]} \qquad (2)$$

where pK′ is the negative logarithm of the equilibrium constant for the CO_2/bicarbonate reaction.

In the adult, the initial response to the acidaemia is hyperventilation which lowers PCO_2 (and hence raises pH). However, in the fetus, the analogous response would require an increase in uterine blood flow and this does not occur. Thus, when metabolic acidosis is induced in the fetal sheep by infusion of acid, PCO_2 rises, and this limits the ability of the fetus to restore the acid–base balance. A slower response to metabolic acidosis involves increased renal reabsorption of bicarbonate and increased excretion

of acid. The kidney of the fetal sheep can reabsorb bicarbonate and excrete protons, and can therefore increase its acid excretion in response to an acid load. But these abilities are less than in the adult and they may be decreased further with reduced fetal oxygen delivery, which could impair renal function. Hence, the ability of the fetus to regulate acid–base balance via the CO_2/bicarbonate system is limited when compared to that of the adult.

Fetal hypoxaemia and hypoxia

Fetal oxygen supply can be reduced via a number of maternal, fetal, or placental factors. Before summarizing these, three terms commonly and often interchangeably used to describe reduced oxygenation will be defined: hypoxaemia; hypoxia; and asphyxia. Hypoxaemia is a decrease in blood O_2 concentration, whereas hypoxia is a fall in tissue oxygen consumption resulting from a reduced rate of O_2 delivery. Asphyxia is a fall in blood O_2 concentration associated with CO_2 accumulation and acidaemia. It is important to note that the fetus can be hypoxaemic without being hypoxic. By postnatal standards, the fetus is always hypoxaemic, and transient reductions in vascular PO_2 are a normal feature of prenatal life. Fetal hypoxia results when umbilical oxygen delivery falls below that required to meet normal metabolic requirements. This can occur via a number of mechanisms, as summarized in Table 9.5. With the obvious exception of reduced umbilical blood flow, these factors decrease fetal O_2 delivery by lowering umbilical venous O_2 content. The common causes of reduced O_2 delivery in the human fetus appear to

Table 9.5 Factors that can reduce fetal oxygen delivery

Maternal

Reduced arterial PO_2
Anaemia
Decreased haemoglobin O_2 affinity
Reduced placental blood flow

Placental

Reduced placental size or exchange area
Premature separation of maternal and fetal placental tissues (placental abruption)

Fetal

Anaemia
Decreased haemoglobin O_2 affinity
Reduced umbilical blood flow

be decreased uterine and/or umbilical blood flows, which can occur during labour and can lead to fetal asphyxia. It appears that long-term reductions in uterine blood flows and oxygen delivery can occur in pregnancies complicated by maternal hypertensive disorders and associated with impaired fetal growth.

The effects of the factors listed in Table 9.5 have been examined in numerous animal studies, particularly in the fetal lamb. A number of compensatory responses have been identified. Many of these appear to be reflex in nature, and involve neural, cardiovascular, and endocrine mechanisms. But probably the most important mechanism results from the relationship between oxygen delivery and consumption,

$$O_2 \text{ consumption} = O_2 \text{ delivery} \times O_2 \text{ extraction.} \quad (3)$$

A rise in the rate of O_2 extraction can allow the fetus to maintain O_2 consumption in the face of a fall in O_2 delivery, and this applies whether we consider the fetus as a whole or in terms of its individual organs. However, there is a limit to the extent to which this can compensate for reduced oxygen delivery, and to understand the reasons for this it is useful to refer to the equation for fetal O_2 consumption given in Fig. 9.1,

$$\dot{V}_{O_2} = \dot{Q}_{um} \times (UV_{O_2} - UA_{O_2})$$

where $\dot{V}_{O_2}$ is oxygen consumption, $\dot{Q}_{um}$ is umbilical blood flow, and UV_{O_2} and UA_{O_2} are, respectively, umbilical vein and umbilical artery blood oxygen content. Irrespective of whether a fall in O_2 delivery occurs via a decrease in blood flow or UV_{O_2}, arterial O_2 content must also fall if O_2 consumption is to be maintained. It is the drop in arterial O_2 content, which is accompanied by a fall in PO_2, that results in the increase in O_2 extraction. However, this arterial blood also perfuses fetal tissues and there must be a lower limit below which arterial PO_2 must fall before there is an impairment of O_2 diffusion into tissue. This probably limits the extent to which O_2 extraction can increase to compensate for reductions in O_2 delivery. In pregnant sheep, reductions of up to 50 per cent can be accommodated via an increase in extraction, with no fall in $\dot{V}_{O_2}$. If the decrease is more than this, oxygen consumption falls and lactic acidaemia develops, indicating hypoxia in at least some fetal tissues. The mechanism and consequences of the lactic acidaemia are discussed in the next section.

Fetal responses to acute hypoxaemia

The fetal cardiovascular, metabolic, and behavioural responses to reductions in O_2 delivery have largely been elucidated in animal studies involving short-term (eg. 1–2 h) reductions in O_2 delivery. But, more recently, the use of *in utero* blood sampling and Doppler ultrasound blood flow monitoring techniques have confirmed that some of the responses also occur in the human fetus. Fig. 9.3 illustrates some of the acute responses that have been observed. There is complete or near complete suppression of fetal breathing and body movements; this inhibition appears to involve activation of neural elements in the midbrain, and an inhibition of spinal reflexes. A redistribu-

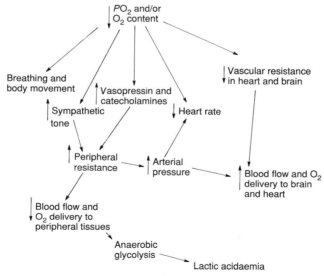

Fig. 9.3 Fetal responses to acute reductions in fetal O_2 delivery ($\uparrow$ increase, $\downarrow$ decrease).

tion of fetal cardiac output occurs to favour the heart, brain, and adrenal gland at the expense of less vital organs. This redistribution is achieved via vasodilatation in the heart, brain, and adrenal gland as a result of local mechanisms. Pulmonary vasoconstriction also occurs, and, if the reduction in O_2 delivery is severe enough, there is also reduced perfusion in other vascular beds, such as the carcass (skin, muscle, bone), kidney, and gastrointestinal tract. These changes result from activation of peripheral chemoreceptors and are mediated, in part at least, by increased sympathetic tone and elevated blood concentrations of catecholamines and vasopressin. The vasoconstriction increases peripheral resistance and hence fetal arterial pressure, and this is associated with transient bradycardia. This is followed by tachycardia if the duration of hypoxaemia is more than ~1 h, and then there may also be a modest increase in cardiac output. However, the fetus has a very high resting cardiac output and it appears to have a limited ability to increase this further, so that this mechanism is probably of limited importance in compensating for a reduction in O_2 delivery. Although not indicated in Fig. 9.3, increased plasma levels of adrenocorticotrophic hormone and cortisol also occur and plasma insulin concentrations are reduced. As a consequence of this latter change and the elevated catecholamine levels, the plasma glucose concentration rises due to breakdown of glycogen stores.

The responses to acute hypoxaemia just noted serve at lease two purposes. The suppression of skeletal muscle activity should result in a small decrease in O_2 demand. Second, the cardiovascular adjustments result in the preferential distribution of the available oxygen to those organs whose normal functioning is vital for short-term survival — the heart and brain — or important in the endocrine responses to hypoxaemia — the adrenal gland. However, although O_2 delivery to the heart and brain may be maintained at levels sufficient to sustain oxidative metabolism, this is not the case for the other regions of the body, particularly when perfusion to some of those organs is reduced. This will exacerbate the initial fall in O_2 delivery to these tissues and it can fall below that required to maintain mitochondrial electron transport and oxidative phosphorylation. A shift to anaerobic pathways occurs during acute hypoxaemia and this appears to be the factor that leads to the accumulation of lactic acid in the fetus. As noted previously, the fetus has a reduced ability to compensate for acid–base disturbances. The acidaemia will cause a rightward (Bohr) shift in the haemoglobin O_2 dissociation curve which further lowers blood O_2 content and which facilitates O_2 delivery to tissues. Thus, with > 50 per cent reductions in fetal O_2 delivery, there is a continuous cycle of decreased O_2 delivery to peripheral tissues, leading to acidaemia and further decreases in O_2 supply, until ultimately myocardial function is impaired, leading to circulatory collapse and death.

Fetal responses to chronic reductions in O_2 delivery

The process of progressive deterioration resulting from severe reductions in fetal O_2 delivery occurs over several hours. What remain to be discussed are the fetal compensatory responses to less severe (< 50 per cent) decreases in oxygen delivery, associated with hypoxaemia, which are sustained for days or weeks. As noted earlier, this appears to occur in human pregnancies complicated with fetal growth retardation. In this situation, cerebral, myocardial, adrenal, and umbilical blood flows may be increased, although this has not been consistently observed. However, there is not a decreased perfusion to other organs, nor is there the complete suppression of breathing movements that is observed with acute hypoxaemia. Heart rate and arterial pressure are unchanged. Fetal erythropoietin levels are increased, which leads to elevations in haemoglobin concentration and haematocrit, thereby increasing blood O_2 content. Oxygen extraction is increased to compensate for the fall in O_2 delivery, but there is also a ~ 20 per cent fall in total body O_2 consumption, at least when oxygen delivery is chronically reduced in fetal sheep by lowering umbilical blood flow. However, this reduction in O_2 consumption appears to be an adaptive response as it is not associated with metabolic acidosis, and may result from the reduced growth that has been observed in many animal studies of long-term fetal hypoxaemia. However, intrauterine growth retardation is also associated with reduced plasma levels of glucose and other anabolic substrates, which probably contribute to the impairment of growth. Thus, with a sustained reduction in O_2 delivery, two factors would appear to be important for the maintenance of oxygen homeostasis: an increase in O_2 extraction and a reduction in O_2 demands resulting from a slowing of growth. However, the increased extraction reduces the ability to compensate for further reductions in O_2 delivery, and this is most likely to occur during labour.

Long-term consequences of reduced fetal O_2 delivery

The most serious consequence of reduced oxygen delivery to the fetus is, of course, intrauterine death resulting from a significant decrease in O_2 delivery (> 50 per cent), tissue hypoxia, and progressive metabolic acidosis that is of sufficient duration to lead to myocardial failure. With less severe hypoxia, there is the possibility of damage to other organs such as the lung, liver, gastrointestinal tract, and kidney. However, the most serious long-term consequence of fetal hypoxia is brain damage, including cerebral palsy. In many cases, this probably involves a chronic, severe reduction in O_2 delivery and/or repeated, transient reductions, such as would occur with intermittent cord compression. Finally, even in the absence of specific organ damage, there are other consequences of reduced fetal O_2 delivery and hypoxaemia that can put the new-born at risk. As noted earlier, acute fetal hypoxaemia elicits glycogen breakdown to increase plasma glucose levels. The glycogen stores that are depleted, along with lipid stores, serve the important function of sustaining catabolic, energy-yielding processes in the immediate new-born period before milk intake becomes sufficient to meet metabolic needs. Thus, sustained or intermittent decreases in O_2 delivery prior to birth can result in insufficient energy reserves in the new-born and an inability to maintain body temperature. The resulting hypothermia can impair neurological functions and contribute significantly to neonatal death, particularly in animals.

Further reading

Anderson, D.F., Parks, C.M., and Faber, J.J. (1986). Fetal O_2 consumption in sheep during controlled long-term reductions in umbilical blood flow. *American Journal of Physiology*, **250**, H1037–H1042.

Battaglia, F.C. and Meschia, G. (1986). *An introduction to fetal physiology*. Academic Press, Orlando, Florida.

Bocking, A.D., Gagnon, R., Milne, K.M., and White, S.E. (1988). Behavioral activity during prolonged hypoxemia in fetal sheep. *Journal of Applied Physiology*, **65**, 2420–6.

Bocking, A.D., Gagnon, R., White, S.E., Homan, J., Milne, K.M., and Richardson, B.S. (1988). Circulatory responses to prolonged hypoxemia in fetal sheep. *American Journal of Obstetrics and Gynecology*, **159**, 1418–24.

Carter, A.M. (1989). Factors affecting gas transfer across the placenta and the oxygen supply to the fetus. *Journal of Developmental Physiology*, **12**, 305–22.

Clapp III, J.F., Peress, N.S., Wesley, M., and Mann, L.I. (1988). Brain damage after intermittent partial cord occlusion in the chronically instrumented fetal lamb. *American Journal of Obstetrics and Gynecology*, **159**, 504–9.

Comline, R.S. and Silver, M. (1974). Placental transfer of blood gases. *British Medical Bulletin*, **31**, 25–31.

Daniel, S.S., Baratz, R.A., Bowe, E.T., Hyman, A.I., Morishima, H.O., Sarcia, S.R. and James, L.S. (1972). Elimination of hydrogen ion by the lamb fetus and newborn. *Pediatric Research*, **6**, 584–92.

Edelstone, D.I. (1984). Fetal compensatory responses to reduced oxygen delivery. *Seminars in Perinatology*, **8**, 184–91.

Faber, J.J. and Thornburg, K.L. (1983). *Placental physiology*. Raven Press, New York.

Harding, R., Sigger, J.N., and Wickham, P.J.D. (1983). Fetal and maternal influences on arterial oxygen levels in the sheep fetus. *Journal of Development Physiology*, **5**, 267–76.

Jacobs, R., Robinson, J.S., Owens, J.A., Falconer, J., and Webster, M.E.D. (1988). The effect of prolonged hypobaric hypoxia on growth of fetal sheep. *Journal of Development Physiology*, **10**, 97–112.

Kitanaka, T., Alonso, J.G., Gilbert, R.D., Siu, B.L., Clemons, G.K., and Longo, L.D. (1989) Fetal responses to long-term hypoxemia in sheep. *American Journal of Physiology*, **256**, R1348–R1354.

Longo, L.D., Delivoria-Papadopoulos, M., and Forster, R.E. II. (1974). Placental CO_2 transfer after fetal carbonic anhydrase inhibition. *American Journal of Physiology*, **226**, 703–10.

Low, J.A., Robertson, D.M., and Simpson, L.L. (1989). Temporal relationships of neuropathologic conditions caused by perinatal asphyxia. *American Journal of Obstetrics and Gynecology*, **160**, 608–14.

Nesarajah, M., Matalon, S., Krasney, J., and Farhi, L. (1983). Cardiac output and regional oxygen transport in the acutely hypoxic conscious sheep. *Respiratory Physiology*, **53**, 161–72.

Nicolaides, K.H., Economides, D.L., and Soothill, P.W. (1989). Blood gases, pH, and lactate in appropriate- and small-for-gestational-age fetuses. *American Journal of Obstetrics and Gynecology*, **161**, 996–1001.

Perlman, J.M. (1989). Systemic abnormalities in term infants following perinatal asphyxia: relevance to long-term neurologic outcome. *Clinics in Perinatology*, **16**, 475–84.

Richardson, B.S. (1989). Fetal adaptive responses to asphyxia. *Clinics in Perinatology*, **16**, 595–611.

Rurak, D., Selke, P., Fisher, M., Taylor, S., and Wittman, B. (1987). Fetal oxygen extraction: comparison of the human and sheep. *American Journal of Obstetrics and Gynecology*, **156**, 360–6.

Rurak, D.W., Cooper, C.C., and Taylor, S.M. (1986). Fetal oxygen consumption and PO2 during hypercapnia in pregnant sheep. *Journal of Developmental Physiology*, **8**, 447–59.

Rurak, D.W., Richardson, B.S., Patrick, J.E., Carmichael, L., and Homan, J. (1990). Blood flow and oxygen delivery to fetal organs and tissues during sustained hypoxemia. *American Journal of Physiology*, **258**, R1116–R1122.

Soothill, P.W., Nicolaides, K.H., Rodeck, C.H., and Campbell, S. (1986). Effect of gestational age on fetal and intervillous blood gas and acid–base values in human pregnancy. *Fetal Therapy*, **1**, 168–75.

Towell, M.E. (1988). The rationale for biochemical monitoring of the fetus. *Journal of Perinatal Medicine*, **16**, 55–70.

Walker, D.W. (1984). Brain mechanisms, hypoxia and fetal breathing. *Journal of Developmental Physiology*, **6**, 225–36.

Wilkening, R.B., Boyle, D.W., and Meschia, G. (1989). Fetal neuromuscular blockade: effect on oxygen demand and placental transport. *American Journal of Physiology*, **257**, H734–H738.

van der Weyde, M.P., Wright, M.R., Taylor, S.M., Axelson, J.E., and Rurak, D.W. (1992). Metabolic effects of ritodrine in the fetal lamb. *Journal of Pharmacology and Experimental Therapeutics*, **262**, 48–59.

10. Development of the respiratory system
Richard Harding

This and the next chapter describe the prenatal preparation of the respiratory system for its postnatal function. Unlike many other organ systems, the respiratory system is not vital for survival during fetal life, but from the moment of birth it must have reached a sufficient degree of development and maturity to take over from the placenta the all-important role of gas exchange. Inadequate development of the respiratory system in the fetus can result in suboptimal respiratory function in the new-born, resulting in such problems as hypoxia, cyanosis, and recurrent apnoea. It may also contribute to respiratory disorders later in life. This chapter focuses on the prenatal development of structural and functional aspects of the respiratory system, while the next chapter deals with the final maturation of the lung.

The respiratory system is made up of several components of diverse structure. In this chapter, the major divisions that will be considered are the lungs and airways, the respiratory muscles, and the neural centres and pathways that initiate and control breathing movements. It is now apparent that during prenatal life these components develop the ability to function as an integrated system, capable of producing rhythmical, orderly breathing movements leading to normal lung development, and reaching a state of functional competence at term. That this is achieved is remarkable given that before birth the lungs are liquid-filled, breathing movements are intermittent, and the lungs, which have only a small blood flow, play no role in gas exchange.

Lung growth and development

This section briefly summarizes the major stages in the structural development of the lungs of the mammalian fetus, with an emphasis on the human. The process of fetal lung growth is, by convention, divided into four or five phases: an embryonic phase, a pseudoglandular (or glandular) phase, a canalicular phase, and a terminal sac phase. In man and some other species, a fifth, alveolar, stage can be recognized. The timing and major developmental features of these overlapping periods are summarized in Fig. 10.1 and Table 10.1. It should be recognized that the relationships between these stages of lung development and stages of gestation vary between species. Lung development is a continuous process that proceeds well into postnatal life, involving not only an increase in cell numbers but also complex processes of cellular differentiation and tissue remodelling, the intricacies and mechanisms of which are slowly becoming understood.

Embryonic phase

This sees the first appearance of the lung as an entity. The primitive lung first appears as an outgrowth of the foregut 22–26 days after fertilization in the human. The endodermal tissue of this out-pouching soon divides to form left and right lungs and eventually develops into the epithelial

Table 10.1 Stages in fetal lung development

Stage	Human (weeks)	Rat (days)	Rabbit (days)	Sheep (days)	Major features
Embryonic	0–6	0–13	0–18	0–40	Lung buds formed; epithelial–mesenchymal interactions; formation of major airways
Pseudoglandular	6–16	13–18	18–24	40–80	Formation of bronchial tree, acini, and cartilage and smooth muscle
Canalicular	16–26	18–20	24–27	80–120	Formation of the lung periphery; epithelial differentiation into type I and II cells; formation of air–blood interface
Saccular	26–term	20–term	27–term	120–term	Thinning of epithelial cells; formation of saccules; expansion of air space

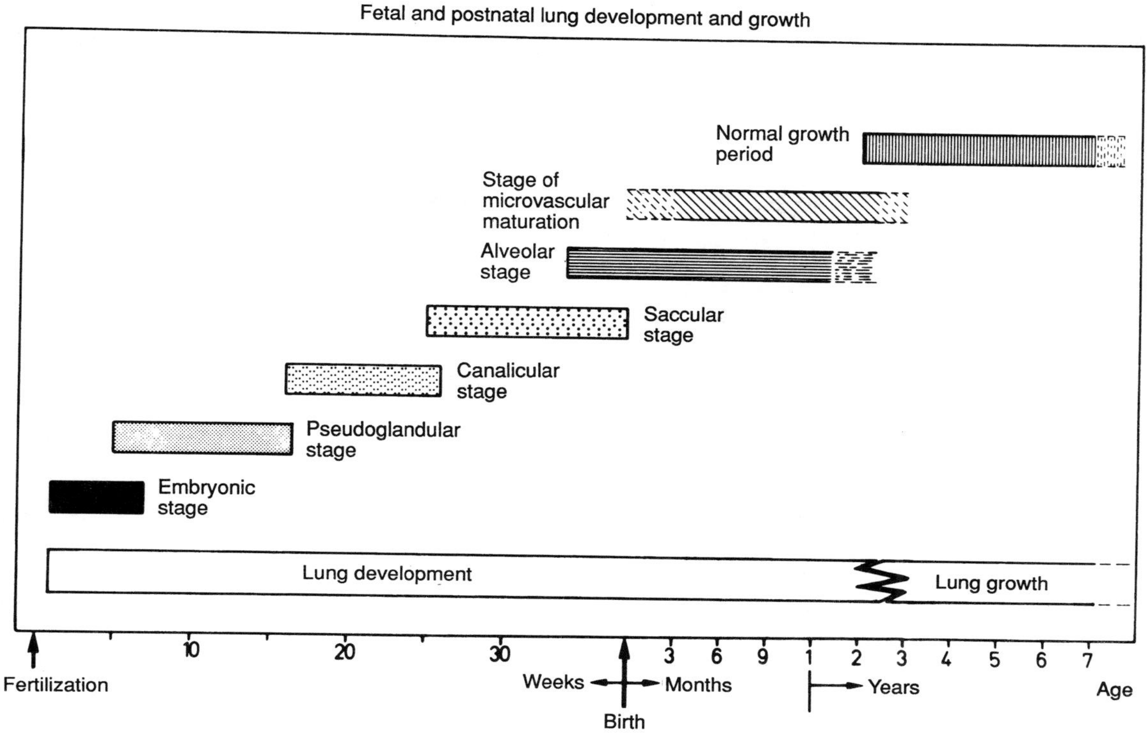

Fetal and postnatal lung development and growth

Fig. 10.1 Phases of lung development in the human fetus and neonate. (Taken with permission from Zeltner, T.B. and Burri, P.H. (1987). *Respiration Physiology*, **67**, 269–82).

lining of the pulmonary 'airways' (Fig. 10.2). Repeated divisions of the original paired 'bronchi' invade the surrounding splanchnic mesoderm and become invested with mesenchymal tissue. By 34 days the major bronchopulmonary segments have been formed. The primitive 'airways' are lined with tall columnar cells, rich in glycogen granules. The splanchnic mesenchymal tissue surrounding the 'airways' eventually gives rise to non-epithelial structures such as pulmonary blood vessels and lymphatics, cartilage, smooth muscle, and connective tissue. At maturity more than 40 different cell types will have evolved from the primitive endodermal and mesenchymal cells.

Pseudoglandular phase

From 5–17 weeks, the human fetal lung resembles a typical secretory gland in histological sections. During this period, the major airways (bronchi) form and the respiratory units of the lung (acini) become recognizable. Both of these processes depend on the invasion of mesenchymal tissue by repeated branching of the distal extremity of the

tubes of epithelial (endodermal) cells; at this stage these tubes are blind ending (Fig. 10.3). Branching of the distal 'airway' is stimulated by the interaction between the epithelial cells and surrounding mesenchymal tissue and probably involves the release of locally acting mitogenic agents. Throughout the pseudoglandular stage the liquid-filled 'airways' are embedded in abundant mesenchyme that has only a modest blood supply. The absence of a dense capillary network and of terminal air sacs precludes the possibility of the lung functioning as an organ of gas exchange at this stage.

The epithelial cells lining the airways begin to differentiate, starting centrally at the hilum of the lung and progressing towards the periphery (Fig. 10.3). The primitive, simple, epithelial cells (columnar in proximal regions and cuboidal in peripheral regions) differentiate into goblet cells and mucous glands, and flatter epithelial cells with cilia appear by 11–13 weeks. The appearance of definitive acini is paralleled by growth of the pulmonary circulation, each of the major respiratory tubules (airways) being accompanied by a major artery.

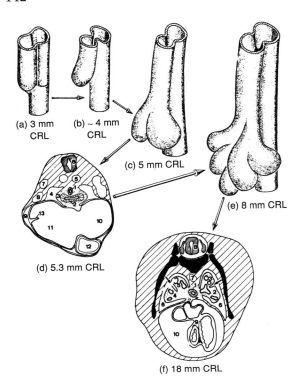

Fig. 10.2 Early stages of lung development in the human embryo. (a) and (b) Formation of primitive lung as a diverticulum from foregut at crown–rump length (CRL) of 3–4 mm (24–26 days post-conception). (c) Separation of left and right divisions at 5 mm CRL (26–28 days). (d) Cross-section of thorax at 5.3 mm CRL (28 days) showing location of primitive airway and other organs. 1, Oesophagus; 2, trachea at point of bifurcation; 3, pulmonary mesenchyme; 4, pericardio-peritoneal canal; 5, dorsal aorta; 6, neural tube; 7, postcardinal vein; 8, common cardinal vein; 9, pericardial cavity; 10, left atrium; 11, right atrium; 12, conus cordis; 13, right venous valve. (e) Further divisions of the epithelial tube form major lobes of the lungs at 8 mm CRL (about 4.5 weeks). The grooves between foregut (oesophagus) and trachea deepen and eventually completely separate the two tubes. (f) Cross-section of thorax at CRL of 18 mm, about 46 days post-conception, showing location of major lobes of lungs. 1, Oesophagus; 2, left upper lobe of lung; 3 and 4, lower lobes; 5, middle lobe; 6, pleural cavity; 7, aorta; 8, neural tube; 9, heart; 10, pericardial cavity. Small arrows indicate oblique fissures of the lungs. (Taken with permission from Burri, P.H. (1984). *Annual Review of Physiology*, **46**, 617–28.)

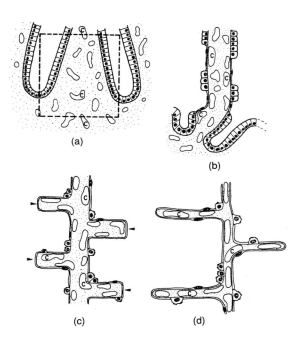

Fig. 10.3 Stages of development of lung parenchyma in fetus and neonate. (a) Pseudoglandular stage, showing epithelial tubes invading the mesenchyme which contains a loose network of capillaries (labelled in all parts of diagram as 'c'). (b)–(d) Further development of structures shown in the frame outlined in (a). (b) Canalicular stage, showing thinning of mesenchyme and arrangement of capillaries around widening and lengthening epithelial tubes. Except in peripheral parts of the lung, epithelial cells differentiate into type I and type II cells. Potential air–blood interface is formed where capillaries come into close contact with the epithelial basement membrane adjacent to flattened (type I) epithelial cells. (c) Terminal sac stage, showing development of secondary septa (arrowheads) from primary septa. Septa are primitive and contain a double capillary network. Capillaries are embedded in a thick layer of connective tissue with few elastic fibres. (d) Mature lung (postnatal) showing single capillary network within thin interalveolar septa. (Taken with permission from Burri, P.H. (1984). *Annual Review of Physiology*, **46**, 617–28.)

Canalicular phase

This phase is characterized by the expansion of the airways as a result of their lengthening and widening, and the thinning of mesenchymal tissue. This occurs as a result of repeated branching of airways and the widening of their terminations. These changes lead to a large increase in the proportion of potential air space (Fig. 10.4), and hence in the volume of liquid within the future air spaces. The process of canalization proceeds in a centrifugal manner, and overlaps in time with the preceding phase of development. The canalicular phase is also characterized by the establishment of recognizable respiratory units (lobules), each composed of a cluster of terminal bronchioles, several 'respiratory' bronchioles and a peripheral cluster of closely branched buds that subsequently form terminal saccules. Along with growth and widening of the airways, capillaries of the mesenchyme invest the tubules, coming into close contact with the epithelial layer (Fig. 10.3).

During the canalicular phase, the distance between the luminal epithelial surface and blood cells in pulmonary

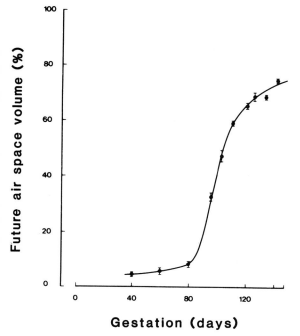

Gestation (days)

Fig. 10.4 Estimated air space, based on morphometric analysis, in fetal sheep lung during the last two-thirds of gestation. Potential air space increases greatly during the canalicular stage (80–120 days). (From Alcorn, D.G., Adamson, T.M., Maloney, J.E. and Robinson, P.M. (1981). *Anatomical Record*, **201**, 655–67. Copyright © 1981, John Wiley and Sons, Inc. Reprinted by permission of Wiley-Liss, a division of John Wiley and Sons, Inc.).

capillaries becomes smaller, as a result of branching of blood vessels, enlargement of the airways, and thinning of the epithelial layer. Cells destined to become type I epithelial cells from progenitor cells become flatter as their cytoplasm becomes attenuated. These processes mark the beginning of the development of a functional air-blood interface, and respiratory gas exchange becomes possible during the latter part of the canalicular phase. In the human, surfactant-containing lamellar bodies first appear in type II epithelial cells at this stage.

Terminal sac (saccular) phase

This period sees a progressive enlargement of the potential air spaces in the periphery, giving rise to the characteristic spongy appearance of the lung. Late in gestation, this process leads to a pronounced increase in the volume of liquid within the future air spaces. The development of transient alveolar ducts or sacs from terminal saccules occurs during this last stage of fetal lung development (24–40 weeks). At term, the human fetal lung is still undergoing growth and maturation and the formation of definitive gas exchange units (alveoli) occurs predominantly after birth. Elastic fibres are formed by interstitial cells in the septa between ducts and saccules, and it has been proposed that production of this elastic tissue is important in the formation of alveoli.

Major changes in the epithelial lining of the terminal airways take place during this period, as discussed in detail in the next chapter. Briefly, the formation of pulmonary surfactant in the lamellar bodies of type II epithelial cells is increased, thereby facilitating the chances of postnatal survival.

The capillary network (circulatory and lymphatic) around the developing terminal saccules proliferates during this period. The separation between the epithelial lining and blood is reduced further as a result of the further formation of 'air spaces' and the thinning of mesenchymal tissue, enhancing the ability of the lung to function as an organ of gas exchange. Postnatally, the effectiveness of gas exchange improves as a result of further attenuation of the tissue separating adjacent air spaces and a further increase in the proportion of air spaces in the lungs.

Formation of alveoli

The alveolus is the definitive gas-exchanging unit of the lung, characterized by its size and shape, its very thin epithelium richly endowed with blood capillaries, and its separation from adjoining alveoli by narrow septal walls. Alveoli are formed from terminal saccules by remodelling of terminal saccules or transitory ducts; the formation of septal divisions is an important part of this process

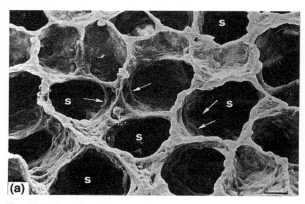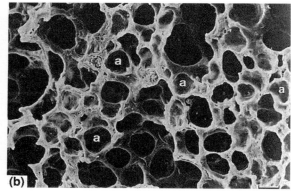

Fig. 10.5 Scanning electron micrographs of postnatal rat lungs at (a) 6 days and (b) 17 days. The scale bar represents 20 μm. (a) Large, smooth-walled saccules become progressively partitioned into smaller sacs by the formation of low ridges, which are the secondary septa (shown by arrows). (b) Secondary septa increase in height and subdivide the original saccules into alveoli (a). (Taken with permission from Burri, P.H. (1991). In *The lung: scientific foundations* (ed. R.G. Crystal and J.B. West), pp. 677–87. Raven Press, New York.)

(Fig. 10.5). It is widely accepted that, in altricial species such as the rat, only terminal sacs are present at birth and that definitive alveoli develop only after birth. In the human, however, alveoli do develop before birth, although there is some controversy as to the extent to which this occurs. It seems apparent that, although several million (up to 20–50 million) alveoli may develop before birth, the majority are produced postnatally to reach the adult number of about 300 million. Thus, an 'alveolar' stage of lung development can be recognized in the human fetus between 36 weeks of gestation and 1–2 years after birth.

Regulation of pulmonary tissue growth

During the period of embryonic and fetal lung development, the lung changes from a simple structure containing a small number of primitive cells types to a highly complex organ containing many different types of specialized cells. Like other organs derived from more than one type of embryonic tissue, the structural development of the lung is determined by interactions between cells derived from these tissues (i.e. endodermal and mesodermal). For example, as in the development of the salivary glands, interactions between these two tissues facilitate the branching of the conducting tubules (i.e. the future airways). Cells of mesenchymal origin (fibroblasts) have the important function of laying down structural fibrous material such as collagen and elastin. Mesenchymal cells are also important sources of mitogenic peptides such as the insulin-like growth factors (IGF-I and -II) which are thought to act in a paracrine fashion to stimulate the replication of adjacent cells such as those of the epithelium.

In the lung, particular interest has been directed at the interactions between cells of mesenchymal origin (i.e. interstitial fibroblasts) and epithelial cells. It is the structural and metabolic maturation of epithelial cells that largely determines the functional success of the lung as a gas-exchanging organ, as these cells not only form one side of the gas-exchange surface (type I cells) but also secrete surfactant (type II cells), which is necessary for alveolar stability. Specialized intercellular contacts between mesenchymal and epithelial cells may also be involved in co-ordinating the growth and differentiation of different types of cells. With increasing gestation, foot processes develop on the basal surface of cuboidal epithelial cells and come into close apposition with the fibroblast cell membrane. These cells, which are destined to become type II epithelial cells, show increasing evidence of surfactant synthesis as the number of cell-to-cell contacts increases. In contrast, newly formed type I epithelial cells, also derived from cuboidal epithelial cells, have few contacts with fibroblasts and do not produce surfactant. It has been proposed that direct cell-to-cell communication with fibroblasts may be involved in the specialization of epithelial cells to produce surfactant. It has also been proposed that fibroblasts secrete a polypeptide (fibroblast pneumocyte factor) that facilitates lipid synthesis by type II epithelial cells.

The developing lung contains large numbers of richly innervated neuroendocrine cells that produce potentially trophic factors. Gastrin-releasing peptide (GRP) is one such factor; the amount of mRNA encoding this peptide is greatly increased during the canalicular phase of lung development in human fetuses. Receptors for epidermal growth factor (EGF), another potent mitogen, are also found in the fetal lung, and this peptide is thought to play a role in the growth and maturation of epithelial cells and the synthesis of pulmonary surfactant.

Development of the pulmonary circulation

Structural development

As each new generation of airways is formed, it is accompanied by a branch of the pulmonary artery. By this process, pulmonary arteries follow the bronchial tree to the capillary bed around the alveoli. In contrast, veins do not follow the airways but pass between adjacent lobules. In the peripheral parts of the lungs, many short arteries branch off the airway-associated arteries to supply the network of alveolar capillaries.

In the human fetus, the adult pattern of branching of the major pulmonary arteries is complete by 19 weeks. This follows the adult pattern of airway branching to the level of respiratory bronchioles, which is complete by 16 weeks. Following these times, there is no further increase in the numbers of generations of airways and arteries; further lung growth is accompanied by the formation of more terminal respiratory units with their vascular beds.

The formation of an effective air–blood interface involves the creation of a dense capillary bed in close apposition to the alveolar epithelium. In mature lungs, the septal divisions between adjacent alveoli contain more capillary than intercapillary space, allowing a virtual sheet of blood to surround each alveolus (Fig. 10.3 (d)). Early in development, pulmonary capillaries form a loose network near the developing airways. By the canalicular phase, the capillaries have greatly increased in number and have made contact with the airway epithelial cells. With subsequent development into the terminal sac stage, the capillaries surround the saccules in a dense network. The septa between adjacent sacs become thinner, leading to an increase in the volume of potential 'air space' in the lungs and to a reduction in separation between the airways lumen and blood (Figs 10.3 and 10.5).

The lung has a second system of blood vessels, collectively termed the bronchial circulation. Bronchial arteries arise from the upper descending aorta and the upper intercostal arteries, and run within the bronchial walls to the terminal bronchioles. They do not, however, supply the alveolar capillaries; thus they do not supply the gas exchanging parts of the lung. Bronchial arteries terminate in capillary networks supplying the muscular, glandular, and other tissues of the airways. Bronchial veins drain, via other major veins, into both left and right atria.

Regulation of pulmonary blood flow

In the fetus, only about 10 per cent of the total cardiac output passes through the lungs, the remainder passing into the systemic circulation via the aorta and ductus arteriosus (see Chapter 8). This arrangement allows adequate perfusion of the placenta, the fetal organ of gas exchange. In the fetus, only a small pulmonary blood flow is needed to provide the metabolic and growth-related requirements of the lungs.

In fetal sheep, vascular resistance in the lungs is high compared to that after birth. Blood flow through the fetal lungs is correspondingly small, and in fetal sheep only about 30 mL/min/100 g of tissue flows through lung tissue at 100–120 days of gestation. This flow gradually increases during the last third of ovine gestation, reaching about 125 mL/min/100 g tissue. This increase is due to a substantial fall in pulmonary vascular resistance over the last third of gestation that may be due to formation of new blood vessels. However, the resistance is still much greater (approximately eight-fold) than that after birth. The mechanisms maintaining pulmonary vascular resistance at a high level during fetal life and causing it to fall after birth are not fully understood. Pulmonary vascular resistance is affected by blood oxygen tension, being increased during fetal hypoxia and reduced in the presence of elevated oxygen tension (hyperoxia). It can also be lowered experimentally by acetylcholine, bradykinin, and prostaglandins (PGE_2 and PGI_2), each of which is a potent vasodilator in the fetus.

There is now evidence that endogenous vasoconstrictor agents, leukotrienes (LT) C_4 and D_4, may be involved in the maintenance of high pulmonary vascular resistance in the fetus. They are thought to play a role in the hypoxic increase in this resistance, and may also have a tonic action on the pulmonary circulation. In fetal sheep, LT receptor blockade or inhibition of LT synthesis substantially increases pulmonary blood flow.

The onset of pulmonary ventilation at birth signals a progressive and large reduction in pulmonary vascular resistance and an increase in pulmonary blood flow. This may be partly triggered by a rise in blood oxygen tension, but gaseous expansion of the lungs seems to be the principal factor. Although the precise mechanisms are not clear, lung expansion with air probably leads to a reduction in vascular resistance by a combination of chemical and physical processes.

The creation of a gas–liquid interface causes an increase in alveolar surface tension (from zero in the liquid-filled lung), which may lead to pulmonary vasodilatation. The nature of the gas or its effects on blood gas content are not important, and it is thought that the creation of a gas–liquid interface alters the geometry of the pulmonary vessels such that pulmonary capillaries are dilatated and recruited. Expansion of the lungs with liquid, without oxygenation, causes an increase in pulmonary vascular resistance and reduces blood flow.

Lung expansion with gas could also release cyclooxygenase products of arachidonic acid, such as prostacyclin (or prostaglandin I_2), into the pulmonary circulation; this process is not stimulated by oxygenation alone. The blockade of prostaglandin (PG) production by indomethacin prior to the onset of rhythmical lung expansion reduces the fall in pulmonary vascular resistance that normally follows it. Prostaglandin D_2, which is released from pulmonary mast cells and which has a vasodilator action in the newly born, may also be involved. There is also evidence that the vasodilator bradykinin is released by ventilation of the lungs. It is likely that the pulmonary circulation of the fetus and neonate is not regulated by one factor alone, but by a range of chemical and physical factors. It is apparent that pulmonary vasoconstriction is actively supported in the fetus, while active vasodilatation occurs after the onset of air-breathing. A summary of factors possibly involved in the establishment of pulmonary vasodilatation at birth is given in Fig. 10.6.

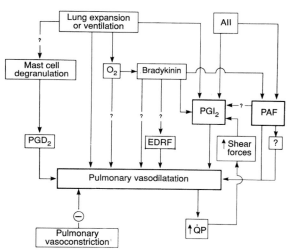

Fig. 10.6 Hypothetical scheme showing factors and mechanisms in postnatal dilation of the pulmonary blood vessels. AII, angiotensin II; PGD_2, prostaglandin D_2; PGI_2, prostaglandin I_2; EDRF, endothelium-derived relaxing factor; $\dot{Q}$P, pulmonary blood flow. (Taken with permission from Heymann, M.A. (1989). In *Advances in fetal physiology* (ed. P.D. Gluckman, B.M. Johnston, and P.W. Nathanielz), pp. 55–68. Perinatology Press, Ithaca, New York.)

Fetal lung liquid

Lung liquid secretion

The fetal lungs do not develop in a collapsed state, but are partially expanded by a luminal liquid. It was once thought that this liquid (fetal lung liquid) was inhaled amniotic fluid. Although inhalation of small amounts of amniotic fluid may occasionally occur in healthy fetuses, the composition of lung liquid is normally quite distinct from that of amniotic fluid, as shown in Table 10.2. In healthy fetal sheep, lung liquid is a clear fluid, isotonic with plasma, with a low viscosity, similar to that of water. The major differences between lung liquid and plasma are the high chloride content of lung liquid, its low protein and bicarbonate contents, and its low pH (Table 10.2).

Lung liquid is secreted into the pulmonary lumen in fetal sheep from at least mid-gestation, although the stage of lung development at which its secretion starts is not known. Around mid-gestation in the ovine fetus (term c. 145 days) lung liquid is secreted at a rate of 1–2 mL/h/kg body weight. By the last third of gestation, the rate has increased to 3–4 mL/h/kg body weight (Fig. 10.7). The approximate doubling in the rate of lung liquid secretion, adjusted for body weight, may be due, in part, to the large increase in the epithelial surface area during the canalicular phase of lung development. Secreted lung liquid flows up the trachea and escapes via the larynx into the pharynx from where it may be swallowed or pass into the amniotic sac. Normally, the rate of production is constant, but it may be altered by a variety of factors including hypoxia and fetal hormones (see below).

The production of fetal lung liquid is an active process, which is a property of the epithelial cells of the entire respiratory tree, and as such can be blocked by metabolic poisons such as cyanide. Pulmonary epithelial cells of the fetus are connected by 'tight' junctions that exclude the

Table 10.2 Composition of lung liquid, amniotic fluid, and plasma in fetal sheep

	Lung liquid	Amniotic fluid	Plasma
Osmolality (mOsm/L)	290–300	265	290
Sodium (mM/L)	140–150	115	150
Potassium (mM/L)	5–7	7–8	4–5
Chloride (mM/L)	140–150	90–100	105–110
Bicarbonate (mM/L)	3	—	25
Total protein (g/100mL)	0.05	0.1	4–5
pH	6.2–6.8	—	7.35

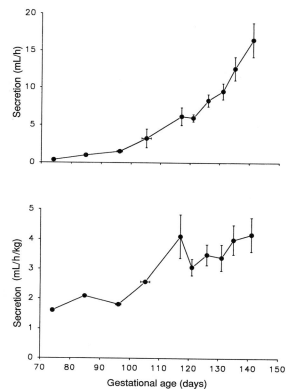

Fig. 10.7 Changes in the rate of secretion of lung liquid in fetal sheep measured by indicator dilution during the latter half of gestation. The lower panel shows the secretion rate corrected for fetal body weight. The approximate doubling of secretion rate, relative to body weight, between 100 and 120 days of gestation is probably due to a large increase in the surface area of the pulmonary epithelium at this time. (Data on fetuses less than 110 days old are from Olver, R.E., Schneeburger, E.E. and Walters, D.V. (1981). *Journal of Physiology*, **1981**, 395–412.)

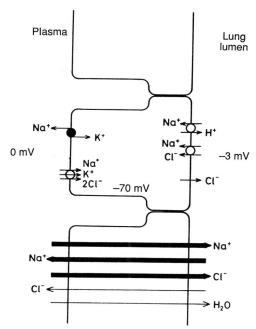

Fig. 10.8 Possible mechanism whereby fetal lung liquid is secreted by pulmonary epithelial cells, based on a model for active Cl^- in other epithelia. The activity of Na^+K^+ ATPase on the basolateral surface of epithelial cells generates a gradient for Na^+ which enters the cell via the basolateral membrane via a $Na^+–K^+–Cl^-$ co-transporter. Chloride ions then move down their electrochemical gradient out of the cell, via selective Cl^- channels in its apical membrane, into the airway lumen. This results in the net movement of Cl^- from plasma to lumen, as shown in the lower part of the diagram. Water moves into the lung lumen owing to the osmotic gradient thus created. Movement of H^+ into lung lumen, due to $Na^+–H^+$ antiport mechanism, results in low pH of lung liquid.

passage of molecules larger than mannitol, thereby effectively preventing the movement of protein and other solutes from plasma or extracellular fluid into the lung lumen.

Secretion of lung liquid into the lumen generates a potential difference of 1–10 mV between lung liquid and plasma, i.e. lung liquid is negatively charged with respect to plasma. This is due to the extrusion of Cl^- into the lumen by a mechanism similar to that proposed for a number of secretory epithelia. The driving force for this process is apparently derived from Na^+ K^+ ATPase activity on the basolateral membrane of the epithelial cells (Fig. 10.8). It is proposed that Na^+ diffuses into the cell along an electrochemical gradient, and that Cl^- is coupled to it (Cl^- moving against its electrochemical gradient). Having entered the cell, Cl^- diffuses into the lung lumen down the electro-chemical gradient. This results in a net movement of Cl^- from plasma to lung lumen. Water moves from plasma, via the epithelial cells, into the lumen along an osmotic gradient generated by the movement of Cl^-.

During labour, lung liquid secretion is initially reduced and then becomes reversed, resulting in fluid reabsorption by the pulmonary epithelium. The reabsorption of lung liquid is apparently determined by the activation of Na^+ channels on the apical (luminal) surface of the epithelial cells. It is proposed that these channels are only present, or only able to be activated, during late gestation (i.e. after 130–135 days in the sheep fetus). They are activated by β-adrenergic agonists (e.g. adrenaline) and vasopressin, which are released during labour, leading to a net movement of Na^+ from the luminal liquid into the epithelial cells. Sodium ions are then extruded from the cell via the ATPase on the basolateral surface into the intracellular fluid and plasma. The number of Na^+ channels that are activated determines the degree of reduction of secretion and the extent of liquid reabsorption. An overview of fluid, ionic, and molecular movements involved in lung liquid secretion and reabsorption is shown in Fig. 10.9.

Hormonal control of lung liquid secretion and reabsorption

Several hormones affect fetal lung liquid secretion and may be involved in the modulation of secretory and absorptive processes under conditions of fetal stress or

during labour. The influence of these factors on lung liquid secretion and reabsorption increases markedly during late gestation.

Catecholamines

β-adrenergic agonists (e.g. adrenaline, isoprenaline, isoxsuprine, terbutaline) are known to inhibit the secretion of lung liquid and to reduce the volume of liquid in the 'air spaces' of fetal lungs. Pulmonary responsiveness to β-agonists, which develops late in gestation (after 130 days in fetal sheep), can be blocked by β-antagonists such as propranolol; the receptors involved are of the $β_2$ subgroup. Administration of adrenaline to fetal sheep during late gestation causes an increased electrical negativity of the

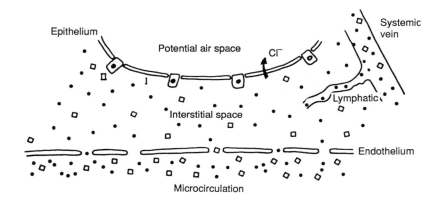

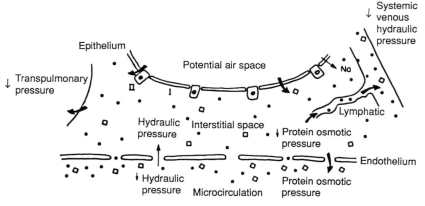

Fig. 10.9 Schematic diagrams showing fluid compartments of fetal lung when secreting liquid (upper diagram) and reabsorbing liquid (lower diagram). Upper: Note the tight epithelial barrier to protein movement from the interstitial space. The vascular endothelium is more permeable, allowing movement of albumin (closed circles) more readily than globulins (open squares); both are returned to the blood circulation via lymph vessels. Lower: Fluid reabsorption at birth involves net movement of water into interstitial space, and then into lymph and blood vessels. (Taken with permission from Bland, R.D. (1987). *Advances in Pediatrics*, **34**, 175–222).

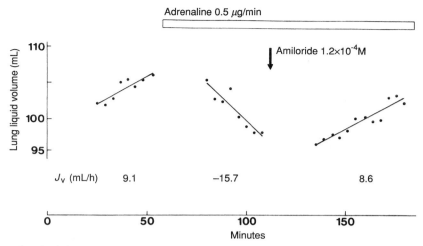

Fig. 10.10 The effects, in a fetal sheep, of an intravenous adrenaline infusion on lung liquid secretion, and of amiloride on the reabsorption of fluid induced by adrenaline. In the absence of both agents, lung liquid secretion rate (J_V) was 9.1 mL/h (shown as an increase in lung liquid volume). Adrenaline infusion caused reabsorption of lung liquid. Amiloride, when added to lung liquid, blocked the absorption of lung liquid caused by adrenaline. (Taken with permission from Olver, R.E., Ramsden, C.A., Strang, L.B. and Walters, D.V. (1986). *Journal of Physiology*, **376**, 321–40.)

pulmonary luminal liquid, with respect to plasma, indicating that removal of a cation (Na^+) may be involved, rather than a reduction of Cl^- transport into the lumen. Use of the Na^+ channel blocker, amiloride, has shown that the reduced secretion and initiation of reabsorption of lung liquid involves the activation of Na^+ channels in the epithelial cell membrane (Fig. 10.10). Amiloride, when added to lung liquid, has no effect on pulmonary secretion under basal conditions, indicating that Na^+ channels on the luminal surface of the epithelial cells are not involved in the normal secretory process. The stimulatory effect of adrenaline on Na^+ channels is probably mediated by increased intracellular production of cyclic AMP (cAMP). Administration to lung liquid of a synthetic analogue of cAMP (which readily enters the cell) has an inhibitory effect on secretion late in gestation that is similar to that of adrenaline.

It seems likely that the maturation of secretory and reabsorptive sensitivity of the fetal lung to β-agonists is under the influence of adrenocortical and thyroid hormones. In this respect, the maturation of secretory-reabsorptive processes resembles maturation of the surfactant system. Thyroidectomy of fetal sheep not only impairs their secretory-reabsorptive responsiveness to adrenaline but the effect of cAMP is also diminished. Thus, thyroid hormones must influence a component of the intracellular signalling process controlled by β-adrenergic stimulation which occurs after the involvement of cAMP. It is possible that the formation of Na^+ channels may be dependent on thyroid hormones. The effects of adrenalectomy are

unknown. However, the administration of a corticosteroid potentiated the effects of a thyroid hormone (triiodothyronine, T_3) in causing maturation of the reabsorptive effects of adrenaline. This supports the view that, in addition to thyroid hormones, corticosteroids, which are increasingly released into fetal plasma during late gestation, also play a role in maturation of the reabsorptive process.

Vasopressin

In fetal sheep, the administration of arginine vasopressin (AVP) causes a slowing of lung liquid secretion. This effect is weak before 130–135 days, and increases as term approaches such that, close to term, AVP may cause cessation of secretion or reabsorption of lung liquid. The increase in pulmonary sensitivity to AVP is paralleled by the increase in plasma concentrations of cortisol (Fig. 10.11). It is likely, therefore, that cortisol is involved in the formation of AVP receptors on the lung epithelium or in mediating the intracellular effects of receptor activation.

During fetal stresses, such as hypoxia and labour, circulating concentrations of catecholamines and vasopressin, among other hormones, are increased. It is likely that AVP may potentiate the inhibitory and reabsorptive action of adrenaline, in particular, during fetal hypoxia or labour. The intracellular mechanisms underlying the effect of AVP on secretion/reabsorption of lung liquid are likely to share a common component with that mediating the effect of adrenaline. The blockade of Na^+-channels by intralum-

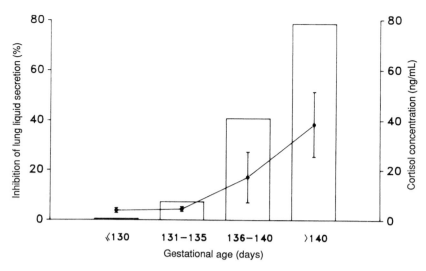

Fig. 10.11 The effect of gestational age on the inhibitory action of arginine vasopressin (AVP) on lung liquid secretion in fetal sheep. The degree of inhibition is shown by open bars. Mean (± SEM) plasma cortisol concentration, prior to AVP infusion, is shown by closed circles. Plasma cortisol concentrations increase in parallel with the increased inhibitory effect of AVP on lung liquid secretion. (Taken with permission from Wallace, M.J., Hooper, S.B., and Harding, R. (1990). *Journal of Applied Physiology*, **258**, R104–R111.)

inal amiloride administration, which inhibits the effects of adrenaline on lung liquid secretion, also blocks the inhibitory effect of AVP on lung liquid secretion. Hypoxia and asphyxia can inhibit lung liquid secretion independently of adrenaline or AVP. It is possible that this effect, which can occur in the immature fetus prior to pulmonary sensitivity to hormones, is due to a reduction in oxygen delivery to the lung, exacerbated by hypoxia-induced pulmonary vasoconstriction.

Disorders of lung liquid reabsorption

Failure of the clearance of liquid from the airways soon after birth can impair gas exchange and alter the respiratory rhythm. This condition has been referred to as 'wet lung disease' or 'transient tachypnoea of the new-born' and usually persists for several days after birth. It is characterized by elevated respiratory rates (up to 120/min) in the presence of normal blood gases. Chest radiographs show evidence of mild pulmonary oedema and enlargement of pulmonary lymphatics, suggestive of retarded lymphatic drainage of the lungs. Disorders of lung liquid clearance are more common following Caesarean birth, particularly if labour has not been experienced, than after vaginal delivery. The lack of both physical compression and the endocrine changes associated with labour (e.g. release of catecholamines and vasopressin) would be expected to delay the reabsorption of lung liquid.

Lung liquid volume and lung growth

As stated above, the fetal lungs are maintained in a partially expanded state by their luminal liquid. The change in lung liquid volume in the ovine fetus with increasing gestation is shown in Fig. 10.12. Prior to the canalicular phase of lung development, the volume of pulmonary luminal liquid is very small. In fetal sheep, this volume is 4.3 mL/kg (about 1 mL) at mid-gestation (74 days). With canalization and increasing mesenchymal thinning the volume increases to 11 mL/kg (about 10 mL) at 96 days and 27 mL/kg (about 35 ml) at 105 days. During the last third of ovine gestation the rate of growth of the pulmonary airways is approximately proportional to the increase in body weight and the lung liquid volume is 35–45 mL/kg body weight. Near term, the lungs of a 4–5 kg ovine fetus may contain up to 200 mL of lung liquid.

Close to term, before the onset of labour, the luminal volume of the lung increases (Fig. 10.12). This may be due to the action of cortisol, which is present in increasing concentrations in the fetal circulation as term approaches. Cortisol is known to cause the thinning of alveolar walls and an increase in the potential air space of the fetal lungs. Some studies indicate that there is a reduction in lung expansion close to term, prior to the onset of labour. Such observations may be due to alterations in the endocrine state of the fetus near term, as the lung epithelium becomes increasingly sensitive to hormones such as

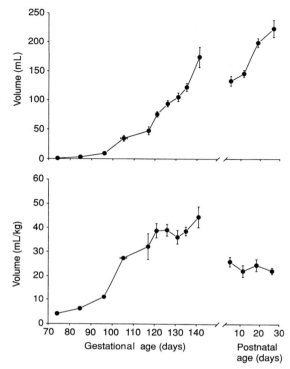

Fig. 10.12 Changes in lung liquid volume, measured by indicator dilution, in fetal sheep during the latter half of gestation and in functional residual capacity (FRC), measured by helium dilution, in new-born lambs. Upper panel shows absolute volumes (mL) and lower panel shows data corrected for body weight (mL/kg). A large increase in fetal lung liquid volume occurs during the canalicular period FRC in the neonate is lower than in the fetus owing to the effect of surface forces acting within the air-filled lung. (Data on fetal sheep less than 110 days old are taken from Olver, R.E., Schneeburger, E.E., and Walters, D.V. (1981). *Journal of Physiology*, **1981**, 395–412.)

The degree of lung expansion with liquid is an important determinant of fetal lung development. If the volume of fetal lung liquid is reduced for long periods, the growth and structural maturation of the lungs are retarded. Pulmonary hypoplasia, a reduced number of cells in the lungs, can occur if liquid is continuously drained from the trachea or if the thoracic (pleural) cavity contains liquid or abdominal contents, thereby reducing lung liquid volume and preventing the lung from expanding. This is a local effect, and is restricted to the affected lung or part of the lung. The prolonged drainage of lung liquid by gravity in fetal sheep leads to a greatly reduced rate of DNA synthesis and a virtual cessation of lung growth. It has recently been shown that the expression of the gene controlling the synthesis of insulin-like growth factor II (IGF-II) is reduced under these conditions. These experiments indicate that the unexpanded fetal lung has a low intrinsic potential for growth, and requires expansion by lung liquid to grow normally.

Conversely, a prolonged period of overexpansion of the fetal lung causes an increase in the rate of DNA synthesis, tissue growth, and IGF-II gene expression in the lungs. This may occur if the outlow of lung liquid is obstructed experimentally or by a congenital malformation of the trachea or larynx. Under these conditions, the lungs are not only larger, but the thinning of septal walls between alveoli is accelerated. Increased tissue growth can occur very quickly in response to lung expansion; for example, when the trachea is totally obstructed in late-gestation fetal sheep, causing the accumulation of luminal liquid, the DNA content (or number of lung cells) can double within 7 days. Experiments such as these have given rise to the concept that fetal lung liquid, the volume of which is actively maintained, acts as an internal splint or mould around which pulmonary tissue is modelled.

Development of breathing movements before birth

The major postnatal function of the respiratory system is gas exchange, and to achieve this the respiratory muscles must be rhythmically activated. This process, which has its origins early in fetal life, depends on phasic activation of brainstem neurones resulting in the orderly activation of respiratory muscles. Fetal breathing movements (FBM) are typically discontinuous, play no part in fetal gas exchange, and represent a cost to the fetus in terms of oxygen and substrate utilization. It is now recognized that FBM are important in the preparation of the respiratory system for its postnatal function of gas exchange, which must operate effectively from the moment of birth.

adrenaline and vasopressin that can cause the reabsorption of lung liquid and a reduction in its volume. These hormones are increasingly released into the fetal circulation during labour, as well as during periods of fetal hypoxia or asphyxia. Changes in fetal posture and uterine muscle activity are also likely to cause a reduction in lung expansion as parturition approaches. Postural changes have been shown to markedly influence lung volume in the fetus; with increased spinal flexion, which can be caused by uterine contractions or reduced amounts of amniotic fluid, the lungs would be expected to contain less luminal liquid.

Ontogeny of fetal breathing

As a result of extensive studies of the fetal sheep *in utero* and observations of the healthy human fetus, it is now recognized that episodes of rhythmical breathing movements are a normal feature of fetal life. These movements are caused by contractions of the diaphragm, each resulting in a brief reduction in intrathoracic pressure. In animals, they can be identified as fluctuations in intrathoracic, or intra-airway, pressure or from electromyograms of respiratory muscles. In humans they can be detected with ultrasonic scanning devices that display movement of the diaphragm, thoracic, or abdominal walls; with each inspiratory effort, the diaphragm descends, the anterior chest wall moves in, and the abdominal wall moves out (Fig. 10.13).

In fetal humans and animals at least three types of thoracic movement can be recognized. The most common are due to rhythmic activation of the diaphragm and occur in episodes throughout the latter part of gestation; these will be referred to as FBM because they are considered to be analogous to postnatal breathing movements. A second type of inspiratory effort, which occurs in episodes throughout much of fetal life, is the hiccup. Gasping, a third type of inspiratory effort, is induced by severe asphyxia or hypoxia and is not detected in healthy fetuses. Gasping movements involve intense activation of inspiratory muscles, including the dilator muscle of the upper respiratory tract. As is the case with hiccups, gasping is

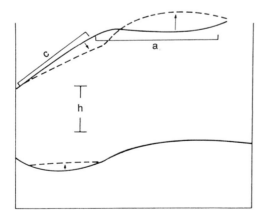

Fig. 10.13 Diagram showing chest (c) and abdominal (a) wall movements associated with human fetal breathing movements. The position of the heart is shown by 'h'. With each inspiratory effort, the anterior chest wall moves inwards and the abdominal wall moves outwards, as shown by dashed line. (Taken with permission from Patrick, J., Fetherston, W., Vick, H., and Voegelin, R. (1978). *American Journal of Obstetrics and Gynecology*, **130**, 693–9.)

probably co-ordinated centrally by neurones separate from those responsible for generating the rhythm of breathing movements.

FBM are first detectable at around 10 weeks of gestation in humans, a little later than the onset of detectable body movements, and at around 50 days in the sheep (i.e. *c.* 1/3 of term). Their occurrence must depend upon the ability of the brainstem respiratory neurones to generate a co-ordinated output, on effective conduction and transmission along descending nerve pathways, and on the ability of these nerves to cause contraction of the skeletal muscles of the thorax and upper respiratory tract. Prior to the development of neural pathways from the brainstem, it is likely that the respiratory muscles are activated by spinal motor neurones. However, such movements cannot be regarded as FBM as they are not co-ordinated with other muscles involved in breathing.

Even after the development of pathways from the brainstem respiratory neurones to phrenic and intercostal motor neurones not all contractions of the diaphragm, or reductions in intrathoracic pressure, are due to descending drive from the respiratory neurones of the brainstem. The phrenic motor neurones innervating the diaphragm receive inputs other than those from the brainstem respiratory neurones and are activated during certain types of gross body movements such as postural adjustments. Early in ovine gestation it is difficult to dissociate between postural and respiratory movements of the diaphragm; this becomes easier with development, as respiratory movements become larger and more rhythmical. The observation of muscle activities associated with respiratory movements, such as those of the upper respiratory tract (see below), assists in the recognition of FBM.

In humans the incidence of FBM increases with gestational age from about 2 per cent (of time) at 10 weeks, to 6 per cent at 19 weeks, 11–13 per cent (depending on time of day) at 22–24 weeks, 12–14 per cent at 24–29 weeks, and to 31 per cent at 30 weeks. Between 30 and 40 weeks, the mean incidence of FBM remains at about 30 per cent, with individual values ranging from 17 to 65 per cent. The hourly incidence of FBM varies throughout the day in both humans and sheep. A major part of this variation can be attributed to fluctuations in fetal plasma glucose concentrations which are related to maternal feeding (see below). Some of the variation in fetal breathing in sheep is apparently related to diurnal changes in circulating melatonin concentrations in the ewe which are due, in turn, to the light–dark cycle (see Chapter 25).

The frequency of FBM within an episode is typically highly variable. At 24–28 weeks in the human, mean frequencies of 42–44 inspiratory efforts/min have been reported, while slightly higher mean frequencies

(45–57/min) occur at 34–38 weeks of gestation. In the sheep fetus, electromyographic (EMG) recordings from the diaphragm muscle at one-third of term (~ 50 days) show frequent periods of sustained activity, much of which is coincident with activity in other muscles (e.g. intercostal, abdominal, extraocular). This early type of diaphragmatic activity may be due to spinal segmental influences, as much of it persists after section of the upper cervical spinal cord. During the last third of ovine gestation, the mean frequency of FBM (within a bout) decreases from 32/min at 99–106 days to 22/min at 120–128 days. Over this period, the mean amplitude of tracheal pressure fluctuations is 3.3–4.1 mmHg.

Relationship to behavioural states

From the time that ovine FBM can first be recorded, they are episodic, but the pattern changes with age. With increasing maturation of the central nervous system periods of respiratory inactivity (apnoea) become longer and FBM periods become less fragmented. After 120–125 days of ovine gestation, when periods of high-voltage, slow waves first appear in the fetal sheep electrocorticogram (ECoG), these periods coincide with an absence of FBM (i.e. fetal 'apnoea'). FBM occur only during periods of low-voltage electrocortical activity during which rapid eye movements are present (Fig. 10.14). Breathing movements also occur during a state of heightened motor activity resembling fetal arousal or wakefulness, which, in fetal sheep near term, occurs with an incidence of 2–5 per cent. In fetal sheep, these periods are characterized by a low-voltage ECoG, the presence of eye movements, and variable amounts of postural muscle activity, particularly in the dorsal neck (nuchal) muscles. Episodes of swallowing (sucking) and micturition also occur during these short periods of 'arousal' as they do in postnatal lambs.

There is evidence that, during periods of high-voltage ECoG, FBM are inhibited by descending nerve traffic from higher centres of the brain. Transverse sections of the brainstem in fetal sheep, at the level of the upper pons, can lead to continuous breathing movements; that is, FBM are no longer inhibited during periods of high-voltage ECoG. The site of origin of the inhibitory neural traffic is not known.

Through the use of non-invasive monitoring techniques, fetal behavioural states have been identified in the human (see also Chapter 24). Cycles of rest and activity can be recognized as early as 28 weeks of gestation from recordings of eye movements, body movements, and heart rate variability. Late in gestation, FBM occur mainly during 'active' periods, when rapid eye movements and gross body movements are more frequent and when the variability in fetal

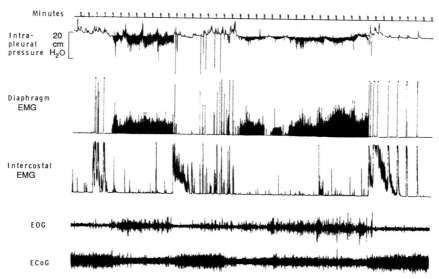

Fig. 10.14 Polygraph recording from a fetal sheep showing fetal breathing episodes and associated sleep states. Fetal breathing, as indicated by fluctuations in intrapleural pressure and bursts of electromyographic (EMG) activity of the diaphragm (integrated), coincide with inactivity of the intercostal muscles, episodes of rapid eye movements (electro-oculogram, EOG), and low-voltage electrocortical (ECoG) activity. Periods of fetal apnoea are associated with sustained periods of EMG activity in the intercostal muscles, few eye movements, and high-voltage ECoG activity. (Taken with permission from Harding, R. (1980). *Sleep*, **3**, 307–22.)

heart rate is increased (Fig. 10.15). Of considerable interest is the mechanism whereby breathing movements, which are intermittent and linked to behavioural states in the fetus, become continuous, regardless of behavioural state, after

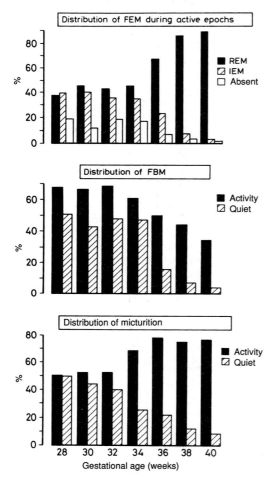

Fig. 10.15 Distribution of fetal eye movements (FEM), fetal breathing movements (FBM), and micturition in human fetuses during the last trimester. Top panel: Distribution of fetal eye movements (REM, rapid eye movements; IEM, intermittent eye movements) during fetal active periods. Active periods were identified by heart rate variability, and by the presence of heart rate accelerations and fetal body movements. Middle and bottom panels: Distribution of FBM and micturition episodes during periods of fetal activity and during quiet (non-active) periods. (Modified from: Arduini, D., Rizzo, G., Giorlandino, C., Valensise, H., Dell'aqua, S., and Romanini, C. (1986). The development of fetal behavioural states: a longitudinal study. *Prenatal Diagnosis*, **6**, 117–24. Copyright © 1986, John Wiley and Sons, Ltd. Reprinted by permission of John Wiley and Sons, Ltd.)

birth. As discussed below, the onset of continuous breathing may be due to a variety of factors including the withdrawal at birth of one or more inhibitory factors of placental origin, increased sensory stimulation, increased metabolic rate (and hence CO_2 production), and altered sensitivity of respiratory neurones and chemoreceptors.

Muscles involved in fetal breathing

Information on the participation of respiratory muscles in FBM has been derived almost entirely from studies with fetal sheep. Pairs of fine wire electrodes can be sewn into individual muscles to yield recordings of electromyographic (EMG) activity that can be compared with fluctuations in intrathoracic pressure. Such recordings have shown that FBM, like postnatal breathing movements, are associated with the co-ordinated activity of thoracic and upper respiratory tract muscles.

The principal thoracic muscle activated during FBM is the diaphragm (Fig 10.16). In human fetuses, ultrasonography has revealed that the dome of the diaphragm can be seen to descend towards the abdomen during each inspiratory effort. Because these movements are essentially isovolumetric (i.e. there is no substantial change in intrathoracic volume) and because the chest wall is highly compliant in the fetus, the anterior thoracic wall is drawn towards the spine, and the anterior abdominal wall moves away from it. Bilateral section of the phrenic nerves in fetal sheep abolishes or greatly attenuates intrathoracic pressure fluctuations, indicating the predominant role of the diaphragm. Early in ovine gestation the intercostal muscles contract in phase with the diaphragm, whereas after the establishment of well-defined electrocortical states, intercostal involvement during inspiration is virtually absent. This is probably a consequence of the descending inhibition of postural motor neurones that occurs during the low-voltage electrocortical (rapid eye movement (REM) sleep) behavioural state; postnatally, motor neurones supplying intercostal muscles, and most postural muscles, are tonically inhibited during REM sleep.

Dilator muscles of the upper respiratory tract are phasically activated during FBM. The principal dilator muscles of the larynx (posterior cricoarytenoid, PCA), which are innervated by the vagal and recurrent laryngeal nerves, are increasingly active during each inspiratory effort in the ovine fetus (Fig. 10.16). Their activation results in rhythmical widening of the glottis. In human fetuses, the glottis has been observed to rhythmically widen during each inspiratory effort, as it does in association with postnatal breathing. After birth and throughout postnatal life, activation of the glottic dilator muscles is essential to permit the unobstructed flow of air into the trachea during inspiration,

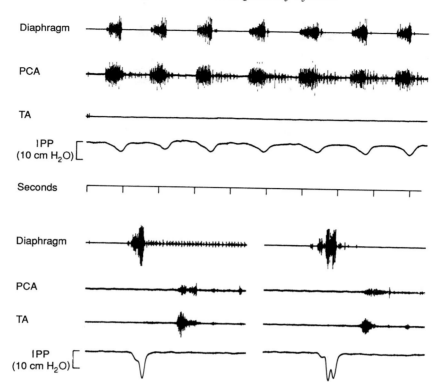

Fig. 10.16 Two types of inspiratory effort in the sheep fetus, and the involvement of laryngeal muscles. Upper panel: Rapid, irregular breathing during the REM, low-voltage electrocortical state. Recordings show electromyographic activity of the diaphragm, posterior cricoarytenoid muscle (PCA), and thyroarytenoid muscle (TA), and intrapleural pressure (IPP). Each diaphragmatic contraction is associated with increased activity in the laryngeal dilator muscle (PCA). Lower panel: Deep inspiratory efforts during the non-REM, high-voltage electrocortical state. Laryngeal dilator muscle activity does not occur during these diaphragmatic contractions. (Taken with permission from Harding, R. (1980). *Sleep*, **3**, 307–22.)

by preventing collapse of the larynx when luminal pressure falls below atmospheric pressure.

The dilator muscles of the nostrils (alae nasi) can also become active during the inspiratory phase of FBM in sheep particularly during augmented breathing efforts. The genioglossus (tongue) muscle, which is active during inspiration in many species after birth, is not rhythmically active in phase with FBM in the sheep fetus. In sheep, this muscle does not normally show inspiratory activity in the new-born or adult, unless the upper airway is obstructed and inspiratory efforts are augmented.

Mechanics of fetal breathing

The principal reason that the respiratory mechanics of FBM differ from those of postnatal breathing is that the lungs are filled with liquid (which has a very high viscosity relative to air) and that the upper airway effectively

presents a high resistance to the inspiratory flow of liquid; therefore, individual FBM cause little change in thoracic volume. Although the basal degree of lung expansion in the late-gestation fetus is greater than that after birth (Fig. 10.7), the ingress of liquid during FBM in the late-gestation ovine fetus is normally less than 0.5–1 ml, less than 1 per cent of functional residual capacity (FRC). In contrast, the tidal volume of the new-born is approximately 20 per cent of FRC.

In the absence of inspiratory muscle activity, the pressure within the fetal 'airways' is equal to, or slightly greater (by 1–3 mmHg) than amniotic sac pressure, while pressure in the pleural space is equal to, or slightly below (0.5–1 mmHg) this pressure (Fig. 10.17). Both the small positive pressure, relative to amniotic sac pressure, in the airways and the small standing negative intrapleural pressure can be attributed to elastic recoil of the fetal lungs, as these pressures are not affected by skeletal muscle paraly-

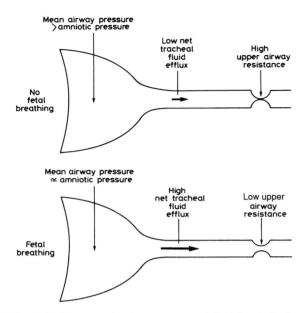

Fig. 10.17 Diagrams showing pressures and fluid flow in fetal respiratory tract. Upper diagram: In the absence of fetal breathing, airway pressure is greater than amniotic sac pressure, due to high resistance of upper airway (larynx) and elastic recoil of the lungs. The high resistance of the upper airway results in a low rate of fluid efflux from the lungs. Lower diagram: During episodes of fetal breathing, basal (expiratory) airway pressure is close to amniotic sac pressure due to lowered resistance of upper airway (larynx). Laryngeal dilatation allows increased efflux of fluid from lungs, driven by pulmonary elastic recoil.

sis. Lung recoil in the fetus is due to the inherent elasticity of pulmonary tissue, and is not as powerful as it is postnatally because the substantial contribution of alveolar surface forces (at the air–liquid interface) is absent.

Fetal breathing and tracheal fluid movement

Inspiratory muscle activity during late ovine gestation typically reduces intra-airway pressure by up to 5 mmHg below amniotic pressure; however, the amplitudes of individual FBM are very variable, ranging from barely detectable (~ 1 mmHg) to 10–20 mmHg. Intuitively, one would expect that such pressure fluctuations would cause a substantial movement of liquid into the lungs. However, owing to the relatively high viscosity of water (compared to air) and to the limited amount of fluid available for inhalation from the fetal pharynx, this does not occur. Furthermore, the resistance to fluid influx via the nostrils and mouth is high in the fetus, thereby retarding the flow of liquid between the amniotic sac and lungs. If the resist-

ance of the upper respiratory tract is abolished, by the creation of a tracheo-amniotic shunt, episodes of FBM cause the influx of substantial amounts of amniotic fluid into the lungs (Fig. 10.18). Hence it may be argued that the upper respiratory tract, by restricting the ingress of amniotic fluid, protects the lower airways and pulmonary epithelium from potentially harmful contact with amniotic fluid. Several studies have shown that amniotic fluid, especially if contaminated with meconium, can interfere with surfactant synthesis in alveolar epithelial cells if it enters the lungs. By virtue of its resistive and valvelike actions, the upper airway also helps to maintain lung liquid volume within narrow limits, particularly in the absence of FBM (Fig. 10.18). Bypass of the upper airway, by the creation of a tracheo-amniotic shunt, causes a considerable efflux of lung liquid during periods of apnoea (Fig. 10.18) and a substantial reduction in lung expansion.

The absence of a net 'inspiratory' flow of amniotic fluid into the lungs during FBM in the intact fetus is verified by the constancy of the composition of lung liquid (which is quite distinct from that of aminotic fluid), and by continuous recordings of fluid flow in the fetal trachea. Such

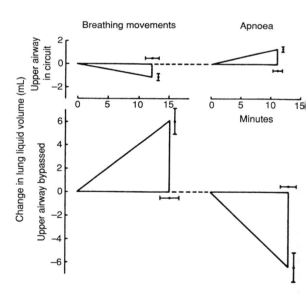

Fig. 10.18 Mean changes in lung liquid volume in the ovine fetus when the airway circuit is intact (upper diagram) and when the upper airway is bypassed by the creation of a tracheo-amniotic shunt (lower diagram). Diagrams show mean changes in volumes (and SEM) caused by episodes of fetal breathing movements (left-hand side) and periods of apnoea (right-hand side). (Taken with permission from Harding R. and Bocking, A.D. In *Handbook of Human Growth and Development Biology*, vol. III Part B. (ed. E. Meisami and P.S. Timiras) pp. 131–48. CRC Press, Boca Raton.)

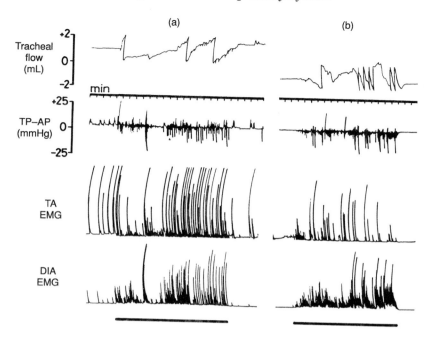

Fig. 10.19 Tracheal fluid flow in a fetal sheep caused by two episodes of fetal breathing movements (a) and (b) which are indicated by recordings of tracheal pressure minus amniotic fluid pressure (TP – AP), by integrated electromyographic activity of the diaphragm (DIA emg) and by bars beneath these tracings. Tracheal flow away from the lungs is indicated by upward deflection of integrated flow signal; the integrator resets to zero after 2 mL of liquid has flowed away from (positive) or towards (negative) the lungs. Recording of integrated electromyogram of thyroarytenoid muscle (TA emg) shows low-level tonic activity when breathing movements are absent. In (a) fetal breathing is associated with a net efflux of fluid from the lungs; this is the most common pattern. In (b) fetal breathing causes a net influx of liquid. Little liquid flow occurs in the absence of FBM.

recordings, using a flowmeter sensitive to low flow rates, show that episodes of FBM are normally associated with a net efflux of liquid from the lungs (Fig. 10.19(a)), although periods of influx may occur during some periods of vigorous FBM and changes in fetal posture (Fig. 10.19(b)). Thus, episodes of FBM normally cause a small, but measurable, reduction in lung volume (Fig. 10.18). The reason for this is probably that the reduction in upper airway resistance that occurs during FBM episodes, due to active, rhythmical laryngeal dilatation, allows fluid to escape from the lungs under the influence of pulmonary elastic recoil pressure. During periods of fetal apnoea lung volume is reestablished (Fig. 10.18). This occurs because the rate of lung liquid production exceeds its rate of efflux via the trachea, as shown in Fig. 10.19(a). Owing to the sustained activity of laryngeal constrictor muscles (e.g. thyroarytenoid, TA) in association with an absence of laryngeal dilator muscle activity, the resistance of the upper respiratory tract to the escape of fluid from the lungs is increased and the efflux of liquid is thereby decreased (Fig. 10.17).

Other types of inspiratory effort

Hiccupping occurs episodically in the healthy non-ruminant fetus and has been reported in the human to be particularly common during early gestation. The incidence of hiccupping in the human fetus declines with age, from 8–10 per cent of recording time during early gestation (14–22 weeks) to 1–2 per cent near term.

Episodes of hiccupping have also been detected in fetal pigs (incidence of 5 per cent near term), guinea-pigs, rabbits, and baboons. An analogous, but slower and less frequent, type of inspiratory effort is seen in fetal sheep (Fig. 10.16, lower). As is the case postnatally, fetal hiccups are due to a rapid contraction of the diaphragm (usually < 200 ms), probably in the absence of active glottic widening. Postnatally, the rapid fall in tracheal pressure causes an abrupt closure of the glottis, resulting in the characteristic sound. A similar but less intense phenomenon is laryngeal stridor (noisy, partially obstructed inspiratory efforts), resulting from ineffective glottic widening during inspiration. Neither the function of fetal hiccups, nor events leading to their initiation, are understood. Postnatally, however, bouts of hiccupping commonly follow feeding, and nerve traffic from the oesophagus or stomach may be involved in their initiation. It is of interest that in the human fetus, hiccupping bouts commence only during the 'active' state, when bouts of ingestion are common. It is possible that hiccupping, which appears to be initiated in response to gastric or oesophageal distension, may be an evolutionary remnant of a primitive regurgitative reflex.

Gasping movements in the fetus can be initiated by severe hypoxia or asphyxia. They can be identified as large reductions in intrathoracic pressure, with intense activation of the diaphragm and other inspiratory muscles such as the external intercostal and laryngeal dilator muscles. Trains of gasping movements are often seen prior to hypoxic or asphyxial death in the fetus, resulting for example, from cord occlusion, reduced uterine blood flow, or placental separation. Gasping appears to be a primitive reflex, co-ordinated in the brainstem, aimed at achieving pulmonary ventilation in the presence of life-threatening hypoxia or asphyxia. It is thought that recurring gasping *in utero* may lead to inhalation of amniotic fluid into the lower airways. The inhalation of amniotic fluid, which may be contaminated with meconium (fetal faeces) released in response to asphyxia, is potentially detrimental to the pulmonary epithelium. Recent evidence indicates that individual gasping events in the severely asphyxiated fetus do not move significant amounts of amniotic fluid into the lungs. In contrast, a substantial influx of amniotic fluid occurs in association with the vigorous FBM and inhibition of lung liquid secretion that are caused by prolonged hypoxia.

Control of fetal breathing movements

In recent years, many studies in humans and animals (principally sheep) have been directed at understanding how fetal breathing is regulated. The clinician is interested in this topic because human fetal breathing can be readily monitored by ultrasonic devices and is used in the assessment of fetal health and in the prognosis of postnatal respiratory function. The physiologist is interested because information on the control of breathing movements before birth should lead to a greater understanding of the ontogeny of basic respiratory regulatory mechanisms, and may lead to an understanding of problems of respiratory control in the new-born (such as recurrent apnoea and the sudden infant death syndrome). Many factors, both endogenous and exogenous, influence fetal breathing, including fetal behavioural states, as mentioned above. Only those factors that have been well documented and are of physiological or clinical importance will be mentioned here.

Carbon dioxide

As is the case after birth, the partial pressure of CO_2 in the fetus has a profound effect on the rhythmic activation of respiratory neurones. A direct relationship between PCO_2 and the incidence of FBM has been observed in human and sheep fetuses (Fig. 10.20). In response to an elevation in fetal PCO_2 when the mother inhales a gas mixture containing CO_2 (e.g. 5 per cent) the incidence of FBM is increased and trains of FBM become more regular in amplitude and frequency. In fetal sheep, the incidence of the low-voltage ECoG state is also increased. Conversely, reductions in fetal (and maternal) PCO_2 produced by maternal hyperventilation are associated with a reduced incidence of FBM. Thus, as in the adult, CO_2 appears to be a major determinant of the rhymicity of central respiratory neurones, and these observations provide evidence of the fundamental role of central chemoreceptors early in development. The central effect of CO_2 is probably mediated by hydrogen ions as acidification of the fetal blood or cerebrospinal fluid (CSF) increases the incidence of FBM, whereas increasing the pH of CSF with bicarbonate has the opposite effect.

It is of interest that the normal arterial pH and PCO_2 in the fetus, if present in the adult, would strongly stimulate breathing. Some resetting of the central chemoreceptors, or the removal of an inhibitory influence, must take place after birth.

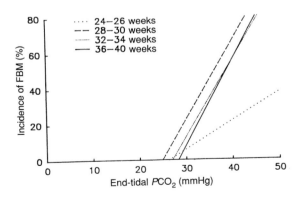

Fig. 10.20 Relationship between the incidence of fetal breathing movements in human fetuses at different gestational ages and maternal end-tidal PCO_2. The slope of the response in the 24–26 week old group is significantly different from those for older fetuses. (Taken with permission from Connors, G., Hunse, C., Carmicheal, L., Natale, R., and Richardson, B. (1989). *American Journal of Obstetrics and Gynecology*, **160**, 932–8.)

Hypoxia and asphyxia

A reduction in oxygen delivery to the fetus causes hypoxaemia; asphyxia occurs when hypoxaemia is combined with an increased blood PCO_2 and a decreased pH. In the short term, fetal hypoxia rarely occurs naturally, but is valuable as an experimental tool. In the adult, acute hypoxia stimulates breathing, whereas in the fetus it has the opposite effect. Hypoxia can be induced experimentally in the fetus by the mother breathing a low-O_2 gas mixture (e.g. 10 per cent O_2 in N_2). Asphyxia can be induced by restriction of uterine blood flow or by restriction of flow in the umbilical vessels. In sheep each of these procedures rapidly leads to a reduced incidence or abolition of FBM, a reduced incidence of the low-voltage ECoG (REM) behavioural state, and the inhibition of other fetal body movements, resulting, at least in the short term, in a reduction in fetal oxygen consumption. This effect may last for up to 8–12 h if the hypoxia or asphyxia is maintained, but, if the treatments are maintained for longer than this, a progressive adaptation takes place and by 16–18 h the incidence of FBM may have returned to normal (Fig. 10.21). It is of interest that the incidence of FBM can be apparently normal in fetuses that are chronically hypoxic (and possibly growth-retarded) due to placental insufficiency.

The effects of asphyxia on FBM are essentially the same as those of isocapnic hypoxia. This indicates that the inhibitory effect of a low O_2 tension on FBM overrides any potential stimulatory effect of hypercapnia or acidaemia.

Although the fetus may be considered to be in a chronically hypoxic condition relative to the adult (i.e. tissue PO_2 is lower), oxygen availability is probably not a limiting factor for many fetal activities, including FBM. In healthy fetuses, the elevation of arterial PO_2 (e.g. by maternal inhalation of 50 per cent O_2) has little effect on the incidence of FBM. Transection of the brainstem above the upper pons, or destruction of lateral pontine nuclei, abolishes the inhibitory effect of hypoxia, leading to the belief that inhibitory neurones in, or above, these regions are activated (or disinhibited) by hypoxia. Recently, recordings have been made from neurones in the upper pons of postnatal lambs that are stimulated by hypoxia. The inhibitory effect of hypoxia on FBM can be replicated by administration of inhibitors of mitochondrial ATP production such as adenosine. Thus, it is likely that oxidative processes are necessary for the normal activation of respiratory neurones in the fetus.

Peripheral chemoreceptors are unlikely to play a major role in the hypoxic inhibition of FBM because this effect persists even after the denervation or ablation of aortic and carotid bodies. There is still considerable uncertainty as to the role of carotid bodies in the fetus, even though afferent nerve recordings have shown them to be active and sensitive to hypoxia. There is evidence that their activation may increase the rapidity of the inhibitory effects of hypoxia. Denervation of the carotid chemore-

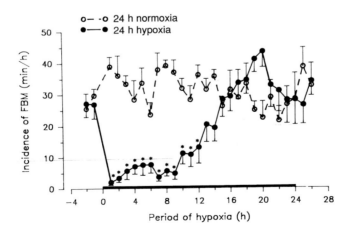

Fig. 10.21 The effect of prolonged hypoxia, induced by reduced uterine blood flow, on the incidence of fetal breathing movements in fetal sheep. Hypoxia causes a profound reduction in the incidence of FBM, which gradually returns to control levels after 8–12 h. Asterisks show values that are significantly different to mean values in normoxic fetuses. (Taken with permission from Hooper, S.B. and Harding, R. (1990). *Journal of Applied Physiology*, **69**, 127–35.)

ceptors leads to a reduction in the incidence of FBM in sheep.

Glucose availability

Glucose is a major substrate for metabolism in the fetus (see Chapter 6). Maternal fasting, sufficient to reduce plasma glucose levels in the fetus, can greatly reduce the incidence of FBM. Fetal insulin infusions have a similar effect. Conversely, glucose administration, to the mother or fetus, increases the incidence of fetal breathing. This effect can be seen in the human fetus at mid-gestation, although the effect is more pronounced nearer term. Fetal body movements are not affected by glucose levels, suggesting that the effect may be restricted to FBM. The inhibitory effect of hypoglycaemia on FBM may be mediated by circulating prostaglandins because fetal hypoglycaemia is associated with elevated placental production of PGE_2 which is known to inhibit FBM.

Prostaglandins

Attention was first drawn to the effects of prostaglandins, particularly PGE_2, on respiratory control by its ability to induce episodes of apnoea in infants with congenital heart disease to whom PGE_2 was administered in an attempt to maintain patency of the ductus arteriosus. In fetal life, PGE_2 is produced by the placenta and by several sites in the fetal central nervous system (CNS) and other tissues, and is present in relatively high concentrations in the fetal circulation. Administration of PGE_2 into the circulation inhibits FBM in fetal sheep and inhibits breathing postnatally in piglets and lambs, as well as in human infants. This effect is probably mediated by an action on the brainstem respiratory neurones. It is likely that PGs may function by limiting the release of catecholamines from adrenergic neurones. In fetal life, centrally released noradrenaline has an excitatory effect on FBM.

Administration of a prostaglandin synthase inhibitor (indomethacin or meclofenamate) to fetal sheep causes a profound increase in the incidence of FBM, suggesting that FBM are normally tonically suppressed by prostaglandins. The site of action of PGE_2 on FBM seems to be within the brainstem, because the stimulation of FBM by inhibition of PG synthesis occurs after transection of the upper brainstem, and after the peripheral chemoreceptors have been denervated. It is of interest that the central action of PGs appears to be restricted to neurones involved in respiratory control. Suppression of PG synthesis greatly stimulates FBM but it does not consistently alter electrocortical states or the activities of non-

respiratory muscles. This selective effect differs from that of other neuromodulators that augment FBM.

Catecholamines

Circulating concentrations of catecholamines in the fetus are increased during hypoxia (or asphyxia) and during the period leading up to parturition. This release of catecholamines, principally adrenaline and noradrenaline, is essential for the adaptation to, and survival of, fetal stress. In fetal rhesus monkeys, adrenaline infusions had no effect on FBM, whereas noradrenaline led to their inhibition. Studies in fetal sheep have shown that systemic catecholamine infusions cause a slight increase in FBM, which may be attributable to an increase in the rate of O_2 consumption or CO_2 production.

Arginine vasopressin

An increased release of AVP from the fetal pituitary occurs in response to hypoxia (or asphyxia) and during parturition. When infused into fetal sheep at sufficient doses to inhibit the secretion of lung liquid, AVP causes a transient reduction (< 1 h) in the incidence of FBM. AVP may act by altering fetal sleep states, but this has not been demonstrated in fetal sheep.

Temperature

Cooling the skin of fetal sheep, *in utero* or *ex utero*, induces slow, deep breathing movements that persist for the duration of cooling. This effect is not, apparently, dependent upon a reduction of fetal core temperature and is probably due to the stimulation of cutaneous thermoreceptors. The fetal behavioural state during cutaneous cooling has not been precisely defined. Cold stimuli applied to the skin would be expected to elicit CNS arousal (as it would postnatally), but one study suggested that fetuses moved into a state resembling quiet sleep (i.e. high-voltage electrocorticogram, non-REM). Cooling the skin of the newly born may provide a stimulus for the onset of breathing after birth.

Increasing the body temperature of fetal sheep in a saline bath can induce FBM resembling panting for 5–10 min. The induction of fetal hyperthermia secondary to maternal hyperthermia can also affect FBM, but the effect is most striking during the period following the cessation of heating during which FBM occur throughout each behavioural state, including the high-voltage electrocortical state. Raising the core temperature of fetal sheep by 0.8–2.0°C, without causing maternal hyperthermia, led to a short-term (2–3 h) stimulation of the incidence and amplitude of FBM, followed by a depression of both. As with maternal hyperthermia, fetal hyperthermia alone caused fetuses to breathe through the high-voltage electro-cortical state. In human pregnancy, fetal hyperthermia may occur during severe maternal exercise, exposure to high ambient temperatures, or fever, and each of these may be expected to affect FBM.

Uterine motility and labour

Throughout the latter half of gestation, low-level, non-labour contractions of the myometrium occur and have been found to affect FBM and fetal behavioural states. In sheep these contractions, which last for 5–8 min, occur 1–4 times per hour and raise intrauterine pressure by 2–5 mmHg. They begin at about one-third of term (50–60 days) and continue until the onset of labour. In humans, these contractile events, known as Braxton–Hicks contractions, are often undetected by the mother. In both sheep and humans, non-labour uterine contractions are often associated with a reduced incidence of FBM and fetal body movements. Contractions of a similar amplitude and duration induced by administration of oxytocin also inhibit FBM. In fetal sheep, the abolition of FBM by uterine

contractions is usually accompanied by a change of behavioural state, from REM low-voltage ECoG to non-REM, high-voltage ECoG (Fig. 10.22).

The mechanisms underlying the effect of uterine contractile events on FBM, fetal body movements, and behavioural states probably involve transient reductions in brain oxygenation. It is known that uterine contractions can cause mild hypoxia in the fetus, but quite severe hypoxia, such as that caused by occlusion of a uterine artery, is required to abolish FBM and to cause a change in fetal behavioural state from REM, low-voltage ECoG to non-REM, high-voltage ECoG. Uterine contractions also affect the posture of the fetus, and can cause distortion of the thorax, increased flexion of the spine, and increases in thoracic, venous, and intracranial pressures of up to 5 mmHg.

Simulating some of the effects of uterine contractions by increasing fetal intracranial pressure to a degree similar to that caused by uterine contractions inhibited FBM and led to a change in fetal state from REM, low-voltage ECoG to non-REM, high-voltage ECoG. This observation suggests that the rise in intracranial pressure during uterine contractions may lead to localized brain hypoxia due to alterations in brain perfusion. This may be the cause of the cessation of FBM and body movements and the alteration

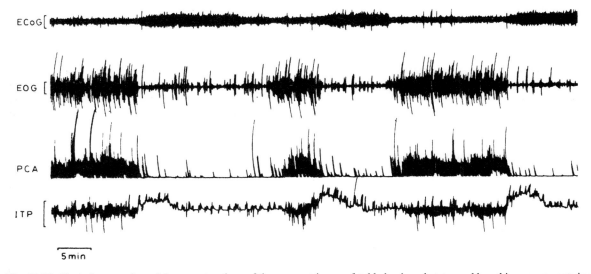

Fig. 10.22 The influence of non-labour contractions of the myometrium on fetal behavioural states and breathing movements in a sheep. Three uterine contractions are indicated on recordings of fetal intratracheal pressure (ITP). Recordings of the fetal electro-corticogram (ECoG) and electrooculogram (EOG) show alternating periods of low-voltage ECoG with rapid eye movements (REM) and high-voltage ECoG with few eye movements. Fetal breathing activity is shown by the integrated electromyogram of the laryngeal dilator muscle (posterior cricoarytenoid, PCA) and tracheal pressure fluctuations. Non-labour myometrial contractions often terminate periods of the REM state and fetal breathing, leading to the onset of the non-REM (quiet) sleep state. (Taken with permission from Nathanielsz, P.W., Bailey, A., Poore, E.R., Thorburn, G.D., and Harding, R. (1980). *American Journal of Obstetrics and Gynecology*, **138**, 653–9.)

in fetal behavioural state that frequently accompany these contractions.

The incidence of FBM in sheep and humans begins to decline 2–3 days before the onset of labour, and they become absent during active labour. The incidence of fetal body movements also declines, approximately in parallel with that of FBM. Ovine FBM are also inhibited during labour induced by intrafetal infusions of ACTH or cortisol.

Many endocrine and metabolic changes occur in the fetus during labour, and it has not yet been determined which is responsible for the inhibition of fetal breathing and body movements. Administration of a hyperoxic gas mixture to ewes in labour increased the incidence of FBM, suggesting that the fetal hypoxia that accompanies active labour may be involved. The increased circulating levels of prostaglandins in the fetus during labour could also play a role, although a recent study has shown that the inhibition of PG synthesis during ovine labour did not prevent the inhibition of FBM. It seems likely that repeated hypoxic episodes, together with increments in intracranial pressure caused by frequent uterine contractions during labour, have the effect of increasing the incidence of the high-voltage ECoG state in the fetus and of reducing the incidence of FBM and body movements.

Opioids

The effects of endogenous and exogenous opioids on FBM have been studied using a range of agonists and antagonists. Blocking the actions of endogenous opioids with naloxone and other, more specific opioid antagonists, has no effect on the amount of FBM in fetal sheep *in utero*, although the episodes of FBM may become fragmented. Other studies showed that a large dose of naloxone given to fetal sheep *in utero* can affect fetal behavioural states, leading to a state of apparent arousal.

Administration of opioids such as morphine and methadone has been shown to affect FBM and fetal behavioural states in sheep, but the effects are dependent on the dose and route and type (i.e. bolus or infusion) of administration. Given intravenously, or into the lateral cerebral ventricles, large doses of morphine induce a short period of apparent REM with large-amplitude FBM, resembling postnatal breathing. These effects on FBM, and the associated changes in fetal behavioural states, are blocked by naloxone indicating that they are mediated by specific opioid receptors. Methadone administration to fetal sheep also stimulates FBM, but the associated behavioural state appears to be one of arousal, characterized by increased motor activity and oxygen consumption.

Although it is evident that exogenous opioids can have profound effects on fetal behavioural states and breathing movements, it is not apparent that endogenous opioids are involved in the regulation of these functions before birth. The mechanisms whereby opioids influence FBM and fetal states are not clear, but recent evidence indicates that they may involve central muscarinic pathways.

Cigarette smoking and nicotine

FBM are inhibited following cigarette smoking by pregnant women, the effect of a single cigarette lasting up to 1 h. It is likely that the supression of FBM is due to fetal hypoxia, as components of cigarette smoke are known to reduce the oxygen content of fetal blood. This effect is partly due to the formation of carboxyhaemoglobin in the fetus and partly to a reduction in utero-placental blood flow due to the vasoconstrictive action of nicotine.

The administration of nicotine to pregnant sheep rapidly produces a prolonged hypoxaemia in the fetus and an inhibition of FBM. In human pregnancy, the use of Doppler ultrasound has shown that cigarette smoking leads to reduced blood flow in the umbilical circulation, which would be expected to cause fetal hypoxia and supression of FBM. It is also likely that cigarette smoking causes circulatory changes in the fetus indicative of stimulation of the sympathoadrenal system; this may occur as a consequence of nicotine and/or the accompanying fetal hypoxia.

Alcohol

Alcohol (ethanol) rapidly passes from the maternal circulation into the fetal circulation, and can readily cross the blood–brain barrier of the fetus. It is well established, both in humans and sheep, that alcohol supresses FBM without affecting other fetal body movements. In women, the duration of FBM suppression is directly related to the quantity of alcohol consumed; it has been estimated that FBM are inhibited for 180 min after the ingestion of 0.25 g alcohol/kg maternal body weight. The human maternal blood alcohol level that causes complete supression of FBM is 0.33 mg/mL (0.033 g/dL).

Alcohol administration to sheep (1 g/kg over 1 h) suppresses FBM for 9 h, and also suppresses both low- and high-voltage electrocortical states and rapid eye movements in the fetus for 3 h. Transection of the fetal brainstem above the pons does not alter these effects, indicating a direct inhibitory action of alcohol on the lower brainstem. Alcohol reduces blood flow, oxygen delivery, and oxygen consumption of the fetal CNS, although it does not affect fetal blood gas tensions or pH. Thus, the inhibition of FBM cannot be attributed to a systemic hypoxia, but it could be due to hypoxia of the CNS.

The mechanisms underlying the action of alcohol on the fetal CNS have been the subject of recent experimentation. It has been proposed that alcohol causes the release of prostaglandins (principally PGE_2) which are known to have an inhibitory effect on FBM. In support of this hypothesis, it is known that, in adult animals, pre-treatment with indomethacin blocks the CNS depressant effects of ethanol. Furthermore, administration of indomethacin to fetal sheep prevented the inhibition of FBM by alcohol, but did not alter the alcohol-induced suppression of electrocortical activity or rapid eye movements. It has been shown that the concentration of PGE_2 in fetal cerebrospinal fluid increases following administration of alcohol to the mother. This effect may be due to an increase in PG synthesis or to an inhibition of its breakdown. In spite of this circumstantial evidence, it remains possible that PGs are not involved in the alcohol-induced inhibition of FBM.

Narcotics and analgesics

FBM are inhibited by many commonly used drugs with narcotic, sedative, or analgesic actions. In sheep, barbiturates, even at low doses that cause only mild sedation of the ewe, profoundly depress or abolish FBM for long periods. This effect is associated with, and may be due to, an increased incidence of the high-voltage electrocortical state. This is supported by the observation that transection of the fetal brainstem, which interupts neural traffic to the brainstem from higher brain centres, abolishes the inhibitory effect of barbiturates.

Diazepam (*Valium*), a widely used tranquillizer, has been found to affect FBM in humans and sheep. In sheep, a maternal dose of about 0.2 mg/kg reduced the incidence of FBM without affecting fetal blood gas tensions or pH. However, prolonged infusions of diazepam increased the incidence of FBM. Meperidine administration to pregnant women led to the inhibition of FBM; it has no inhibitory effect in sheep, although the stimulatory effect of hypercapnia on FBM was abolished.

Caffeine and theophylline

Caffeine and theophylline, both substances that block adenosine receptors, have long been known to stimulate postnatal breathing by an action within the CNS. Both drugs have been used to treat infants with recurring episodes of apnoea. Infusions of caffeine to pregnant ewes increase both the incidence and amplitude of FBM. Theophylline has a similar effect and increases the sensitivity of FBM to CO_2. Adenosine, when infused into fetal sheep, causes an inhibition of FBM, and it has been suggested that it mediates the inhibitory effect of hypoxia on FBM.

The transition from FBM to continuous postnatal breathing

During fetal life, breathing movements are episodic even though the blood gas status of the fetus, if present postnatally, would be expected to strongly stimulate breathing. This suggests either that, during fetal life, breathing activity (and possibly other body movements) are tonically inhibited or that, after birth, stimuli other than hypoxia and hypercapnia provide an important drive for breathing. At present it is not known why breathing movements in the fetus are discontinuous and related to behavioural states, nor is it known how breathing becomes continuous following birth.

There are several lines of evidence suggesting that brainstem respiratory activity, and indeed other forms of motor activity, may be tonically suppressed during late fetal life. (Teleologically, it may be appropriate for inhibitory mechanisms to dominate during fetal life, thereby reducing the magnitude of fetal movements and limiting oxygen consumption.) Transection of the upper brainstem of the ovine fetus can lead to continuous FBM that are not suppressed during the high-voltage electrocortical state. A similar effect can be obtained by increasing the drive to respiration by acidification of the CSF, by large doses of serotonin, and by the inhibition of PG synthesis. These observations indicate that the neural, muscular, and metabolic substrates for prolonged episodes of vigorous breathing are present in the late-gestation fetus, and that the brainstem respiratory rhythm is normally inhibited, such that FBM appear only when the inhibition is least strong.

At birth, a range of possible mechanisms exist to both remove putative inhibitory influences and to provide excitatory inputs to the brainstem respiratory neuronal pool. Important clues as to the nature of such mechanisms have been obtained from studies on fetal sheep *in utero*. Continuous fetal breathing can be initiated, along with signs of arousal, by the administration of 100 per cent O_2 via an endotracheal tube. This effect occurs only in the mature fetus (>135 days) and seems to depend on raising fetal PO_2 and on inducing fetal arousal. These observations suggest that an increase in oxygen delivery to the brain may induce changes leading to cortical arousal and continuous breathing. Increased brain PO_2 may cause a decrease in cerebral blood flow resulting in an increase in brain tissue PCO_2 which may stimulate breathing.

In experiments in which ovine fetuses have been ventilated *in utero* to maintain normal fetal blood gas tensions, occlusion of the umbilical cord can induce continuous FBM. It has been postulated that occluding the cord leads to the removal from the fetal circulation of inhibitory substance (e.g. PGE_2, adenosine), thereby allowing FBM to become continuous. Blood levels of PGE_2 decrease in the fetus during cord occlusion, as they do soon after birth. Receptors for PGE_2 have been located in the vicinity of the brainstem areas considered to be important for respiratory control (tractus solitarius, pontine parabrachial nucleus). It is also possible that PGs originating within the CNS are involved in the control of breathing, as the enzymes for PG synthesis are located within the brainstem. Increased brain oxygenation after birth (or during *in utero* ventilation) may decrease PG production in the brain.

Another neuromodulator that may be involved in the tonic inhibition of FBM is adenosine. In both the fetus and new-born, blockade of adenosine receptors by theophylline leads to a facilitation of respiratory activity and a reduced incidence of apnoea. Theophylline also blocks the hypoxic inhibition of breathing in the fetus and new-born, indicating that adenosine may also be involved in responses to acute, further reductions in fetal oxygenation.

In fetal sheep, the circulating concentration of adenosine ($c.$ 2 μM) is about four times that in the mother, possibly due to an increased rate of adenosine synthesis in relatively hypoxic fetal and placental tissues. Occlusion of the umbilical cord of near-term fetal sheep *in utero*, ventilated via an endotracheal tube, results in an approximate halving of fetal plasma adenosine concentrations, suggesting that a large proportion of fetal adenosine may be of placental origin, the remainder being derived from the fetal liver and other fetal tissues. These findings suggest that adenosine may play a role in the tonic inhibition of fetal CNS activity. Increased production of adenosine in fetal tissues in response to a further lowering of fetal blood PO_2 may be involved in the further inhibition of FBM and fetal motor activity.

Other factors that may be involved in the stimulation and maintenance of respiratory activity after birth, and indeed the increased overall excitability of the CNS after birth, are the very profound changes that occur in sensory stimulation. These include increased intensity of light and sound, cooling of the skin, and inputs from proprioceptors. Inputs from peripheral chemoreceptors may not be important, as fetal sheep previously subjected to chemodenervation (section of the carotid sinus nerves) readily establish continuous breathing after birth. The increased metabolic requirements of the new-born may also be involved in the maintenance of the respiratory rhythm. After birth, the metabolic rate increases, partly to achieve thermoregulation, thereby increasing the rate of CO_2 production.

Another factor that must be considered is the increase in plasma catecholamine concentrations during parturition which is likely to contribute to CNS arousal and stimulation of breathing. Changes in neural inputs from pulmonary stretch receptors and/or juxtaglomerular receptors may also be involved in the maintenance of the respiratory rhythm of the new-born.

There is evidence that both central and peripheral chemoreceptors adapt to blood gas tensions that prevail after birth. That is, these receptors, which must have adapted to fetal conditions, alter their sensitivities after birth, such that they can respond to deviations from postnatal set points.

Influence of fetal breathing on development of the respiratory system

Evidence from observations of human fetuses and from animal experimentation has indicated that FBM are important for the normal development of the fetal lungs. In general, the congenital absence or experimental abolition of FBM has been shown to be associated with impaired lung growth or maturation. However, the mechanisms underlying the effects of FBM are not well understood. Experimentally, it has been established that the abolition of phasic diaphragmatic contractions, caused by an interruption in the phrenic nerves, is followed by reductions in lung liquid volume and in lung growth (i.e. pulmonary hypoplasia). In these studies, the lung hypoplasia (lung dry weight was approximately halved) was probably due to the prolonged underdistension of the lungs with liquid. However, following phrenic nerve section, the denervated diaphragm muscle atrophies and possibly lengthens, as a result of which the lungs partially collapse owing to their elastic recoil, allowing the diaphragm to move increasingly into the thorax.

The experimental abolition of FBM has been refined by sectioning the spinal cord above the level of the phrenic motor neurones, thereby maintaining the trophic effect of the phrenic nerves on the skeletal muscle of the diaphragm and avoiding its atrophy. Studies in fetal rabbits and sheep have shown that, following section of the upper cervical spinal cord and elimination of phasic respiratory movements of the diaphragm, the DNA content of the lungs was reduced. In fetal sheep, pulmonary distensibility (i.e. lung compliance) with air was reduced. Section of the spinal cord below the phrenic motor neurones had less, or no effect on lung DNA content. These observations have further substantiated the role of FBM in the control of fetal lung growth, and have led to the belief that phasic distortion of the lungs during FBM stimulates pulmonary cell division. However, it has been shown recently in fetal sheep that transection of the upper cervical spinal cord, or pharmacological blockade of conduction in the phrenic nerves, leads to a reduction in lung expansion of 25–35 per cent. It is likely that this reduction in lung expansion was due to the unopposed escape of liquid from the lungs during episodes of respiratory muscle activity. After spinal cord transection or phrenic nerve blockade, episodes of central breathing activity continue, but only affect upper airway muscles as their innervation is unaffected. A reduction in lung expansion probably occurs after abolition of the thoracic component of FBM because the tendency for the lungs to collapse during central episodes of FBM, when the resistance of the upper airway is lowered, is now unopposed by contractions of the diaphragm. Thus, it seems likely that some, if not all, of the effects of abolishing FBM by high spinal cord transection, and possibly by phrenic nerve section, are mediated by a sustained reduction in lung expansion. A reduction in fetal lung expansion following high-cervical spinal cord transection causes a reduced content of mRNA for IGF-II in the lungs of fetal sheep (Fig. 10.23).

Another experimental approach has shown that diminishing the pressure fluctuations caused by FBM, by increasing the compliance of the thoracic wall, leads to lung hypoplasia. However, this procedure would be expected to result in reduced lung expansion as there would now be less opposition to the collapsing tendency of the lungs. It is noteworthy that the surfactant content of the lungs (per g of lung tissue) from animals subjected to phrenic nerve section, spinal cord section, or increased

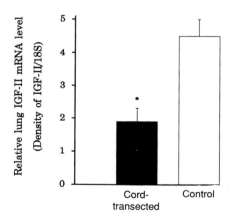

Fig. 10.23 The effect of the abolition of fetal breathing movements by transection of the high cervical spinal cord on the expression of mRNA for insulin-like growth factor II (IGF-II) in the lungs of fetal sheep. (Taken with permission from Harding, R., Hooper, S.B., and Han, V.K.M. (1993). *Pediatric Research*, **34**, 148–53.)

thoracic wall compliance did not differ from that of normal lungs. This reinforces the conclusion that maturation of the surfactant system and lung growth are under independent control.

It is possible that all of the changes in fetal lung growth caused by phrenic nerve section, spinal cord section, and increased chest wall compliance are attributable to long-term reductions in lung expansion. As stated above, prolonged reductions in lung expansion cause corresponding alterations in the rate of lung tissue growth. The same mechanism probably accounts for fetal lung hypoplasia, which follows herniation of the diaphragm, allowing the entry of abdominal contents into the thorax, and leakage of fluid into the pleural cavity, which reduces the intrathoracic space available for the lungs.

In summary, the diaphragmatic component of fetal breathing movements apparently serves to maintain lung liquid volume by retarding the escape of lung liquid that occurs in association with episodes of FBM (Figs 10.17–10.19). Normally, lung liquid leaves the lungs at an increased rate during episodes of FBM, due to the intrinsic elastic recoil of the lungs and the low resistance of the upper respiratory tract. If this process is unopposed by rhythmical contractions of the diaphragm, large reductions in fetal lung expansion occur.

Disorders of fetal lung growth

Abnormalities of fetal lung development are a common cause of respiratory insufficiency in the new-born. This section deals briefly with the fetal origins of some of the more frequent causes of neonatal respiratory insufficiency that can be attributed to disorders of tissue growth. The respiratory distress syndrome, which is due to delayed maturation of the pulmonary epithelium, is dealt with in the following chapter.

Reduced growth of the fetal lungs (pulmonary hypoplasia) is a significant cause of, or contributor to, respiratory distress in the new-born. Recent neonatal mortality data indicate that small lungs, relative to body weight, are present in 15–20 per cent of infants dying soon after birth; pulmonary hypoplasia is the most common single abnormality in these early neonatal deaths. Furthermore, pulmonary hypoplasia, which can be detected radiographically, is likely to be present in a substantial proportion of infants who suffer respiratory distress in the neonatal period but do not die from it.

Fetal lung hypoplasia is associated with a wide range of abnormalities, the majority of which apparently result in a reduction in intrathoracic volume (Table 10.3). It has become apparent that the growth of the lung, unlike the

Table 10.3 Pathologies associated with lung hypoplasia. (Adapted from: Wigglesworth, J.S. (1988). *British Medical Bulletin*, **44**, 894–908)

Oligohydramnios

Prolonged rupture of membranes
Reduced/absent fetal urine production due to renal or urinary
  tract abnormality

Diaphragmatic hernia

Entry of abdominal contents into thorax

Pleural effusions

Immunological hydrops

Thoracic skeletal deformities

Osteogenesis imperfecta
Achondroplasia
Dwarfism — certain types

Diaphragmatic neuromuscular abnormalities

Congenital muscular dystrophy
Anencephaly
Phrenic nerve lesions

Increased volume of abdominal contents

growth of other organs, is particularly sensitive to the physical environment of the developing organ. Perhaps the most important aspect of this environment is the extent to which the lungs are able to be distended, especially during the latter half of gestation.

Oligohydramnios

One of the most common causes of lung hypoplasia is a prolonged reduction in amniotic fluid volume (oligohydramnios). This can be caused by a reduced or absent flow of fetal urine into the amniotic sac (Potter's syndrome) or by the prolonged rupture of membranes surrounding the fetus, allowing leakage of amniotic fluid via the vagina. Fetal urine flow into the amniotic sac can be reduced by the congenital absence of kidneys (renal agenesis) or obstruction of the ureters or urethra. The duration and severity of oligohydramnios are important determinants of the severity of lung hypoplasia, and the effects are greater if oligohydramnios is present during the canalicular phase of lung development when the airways are rapidly expanding.

Oligohydramnios, and other causes of compression disorders of lung development, result in characteristic changes in lung structure. Most notably, there is a reduced

amount of pulmonary DNA, indicative of a preceding reduction in cell proliferation. There is a marked reduction in the amount of potential 'air space' in the lungs, due partly to a reduction in the number of terminal saccules and alveoli. The terminal airways may be relatively unexpanded and their epithelial cells immature, suggesting a lack of distension by lung liquid. There may also be a reduction in the development of elastic tissue in the lungs.

Some studies have indicated that there is a lack of pulmonary surfactant production following oligohydramnios. However, it is likely that the effects on the lungs will differ according to the cause, duration, and extent of the reduction in amniotic fluid volume. The net effect is a reduced ability for the lung to exchange gases postnatally, leading to hypoxia and carbon dioxide retention. This effect may be compounded by reduced lung compliance as a consequence of a surfactant deficiency, as many babies exposed to oligohydramnios following rupture of the membranes are born before term.

The mechanisms whereby oligohydramnios causes lung compression and lung hypoplasia are now becoming clear. Studies in humans and animals have shown that the lack of amniotic fluid causes exaggerated spinal flexion and a reduction in the size and speed of body movements. As is the case postnatally, an increase in spinal flexion will reduce the capacity of the thorax by upward displacement of the diaphragm as a result of compression of the abdominal contents; if prolonged in the fetus, this would be expected to cause lung hypoplasia. It is also possible that oligohydramnios restricts the diaphragmatic excursions associated with fetal breathing movements owing to the close apposition of the uterine wall to the fetal trunk. This may favour the loss of liquid from the lungs, or prevent normal enlargement of the lungs with fetal growth. Recent evidence indicates that, in neonatal lambs previously exposed to oligohydramnios, respiratory insufficiency may be principally due to a decreased compliance of the rib cage rather than to lung hypoplasia alone.

It is of interest that the severity of the postnatal effects of oligohydramnios appears to be related to the observed incidence of breathing movements prior to delivery. In cases of premature rupture of membranes, fetal breathing could be inhibited by the effects of intrauterine infection, which commonly occurs after membrane rupture. Regardless of the cause of reduced fetal breathing in association with oligohydramnios, it is apparent that diminished FBM (incidence or amplitude) can adversely affect the postnatal outcome. It is important to recognize that animal studies have shown that, in an otherwise healthy fetus, the lack of amniotic fluid *per se* does not significantly interfere with the central generation of the respiratory rhythm nor its modulation by fetal behavioural states. However, the physical effects of FBM on the lungs may be diminished by oligohydramnios, thereby contributing to decreased lung growth.

Diaphragmatic hernia

Fenestration of the diaphragm, either unilaterally or bilaterally, results in the entry of abdominal contents into the thorax and a failure of lung growth. Congenital diaphragmatic hernia occurs in about 1 in 5000 live births, but, because many affected fetuses die *in utero*, it has been estimated that the incidence may be as high as 1 in 2000 conceptions. The most common form is a left-sided hernia, and other congenital abnormalities are often present. This syndrome has a high neonatal mortality (up to 50 per cent) and treatments by surgery, high-frequency ventilation, or cardiopulmonary bypass have a high failure rate. Persistent fetal pulmonary circulation (i.e. a high pulmonary vascular resistance) is a major complication of this disorder. In recent years, there have been attempts, some successful, to repair the diaphragmatic defect *in utero*.

Lung hypoplasia following diaphragmatic hernia is usually more severe than that following oligohydramnios. Common findings are a reduced branching of the airways and a reduction in the proportion of 'air space' due to a greatly reduced number of alveoli. There are also marked changes in the pulmonary vascular bed as a result of diaphragmatic hernia, leading to elevated pulmonary vascular resistance and right-to left ductal shunting after birth.

It is believed that this is largely due to a decrease in the total size of the pulmonary vascular bed, and an increase in the muscular content of the resistance vessels.

Further reading

Bland, R.D. and Nielson, D.W. (1992). Developmental changes in lung epithelial ion transport and liquid movement. *Annual Review of Physiology*, **54**, 373–94.

Harding R. (ed. R.G. Crystal and J.B. West), (1991). Fetal breathing movements. In *The lung: Scientific foundations*, pp. 1655–63. Raven Press, New York.

Harding R., Hooper, S.B., and Dickson, K.A. (1990). A mechanism leading to reduced lung expansion and lung hypoplasia in fetal sheep during oligohydramnios. *American Journal of Obstetrics and Gynecology*, **163**, 1904–13.

Hodson, W.A. (1992). Normal and abnormal structural development of the lung. In *Fetal and neonatal physiology* (ed. R.A. Polin and W.W. Fox), pp. 771–82. Saunders, Philadelphia.

Johnston, B.M. and Gluckman, P.D. (ed.) (1986). *Respiratory control and lung development in the fetus and newborn*. Perinatology Press, Ithaca, New York.

Scarpelli, E.M. (ed.) (1990). *Pulmonary physiology: fetus, newborn, child and adolescent* (2nd end). Lea & Febiger, Philadelphia.

Strang, L.B. (1991). Fetal lung liquid: secretion and reabsorption. *Physiological Reviews*, **71**, 991–1016.

Walters, D.V., Strang, L.B., and Geubelle, F. (ed.) (1987). *Physiology of the fetal and neonatal lung*. MTP Press, Lancaster.

11. Fetal lung maturation
Deborah K. Froh and Philip L. Ballard

For the lung to function effectively at birth, there must be appropriate development and function in five major areas: lung growth; airway branching and patency; responsiveness of the pulmonary vasculature; clearance of lung fluid; and differentiation of acinar cells. Immaturity or dysfunction in any of these areas may cause pulmonary disease. The preceding chapter has discussed lung morphogenesis and growth, fluid production, and the pulmonary vasculature. This chapter describes how the cells of the lung parenchyma acquire their specialized functions during development.

Developmental immaturity of the acinar structure reduces the area for gas exchange and increases the oxygen diffusion distance, and deficiency of surfactant production by type II epithelial cells predisposes to atelectasis. In many premature infants this combination of structural and functional immaturity leads to the respiratory distress syndrome (RDS or hyaline membrane disease) and associated complications. Whereas growth of the lung (cell proliferation) is independent of circulating hormones, the rate of acquisition of mature functions (cell differentiation) is regulated by the interactions of several hormones. Thus, this chapter focuses on the influence, mechanisms of action, and therapeutic role of hormones in lung development and disease.

Structural development and cytodifferentiation

Normal development

During the first 16 weeks of human development, the entire branching pattern of airways is established and tissue mass increases many-fold. This growth is reflected in high rates of DNA synthesis, particularly during the glandular period (6–16 weeks). However, the rate of lung growth decreases thereafter, and is superseded by cellular differentiation (e.g. surfactant synthesis by type II epithelial cells). Late in gestation, in animals, when both type I and II epithelial cells are differentiating, DNA synthesis and lung growth virtually cease. This process (differentiation at the expense of growth) also occurs in response to hormone treatment. For example, glucocorticoids decrease lung growth (even to a greater degree than they decrease

overall somatic growth) and accelerate differentiation, whereas androgens seem to promote growth (cell proliferation) and delay differentiation (i.e. surfactant synthesis).

During the canalicular stage of lung development, lung parenchymal cells begin to assume distinct ultrastructural appearances indicative of their eventual functions (Fig. 11.1). At approximately 20–24 weeks of human gestation, epithelial type II cells first become identifiable by the presence of lamellar inclusion bodies, which contain concentric whorls of surfactant lipid (Fig. 11.2). In the following weeks the content of stored glycogen in these cells decreases, presumably due to it being used as substrate for surfactant synthesis, and microvilli develop on the apical surface. Meanwhile, type I epithelial cells appear, with their cytoplasm spread thinly, stretching between adjacent type II epithelial cells and providing a suitably narrow barrier for gas diffusion with the onset of air breathing. The interstitium decreases in volume (mesencyhmal condensa-

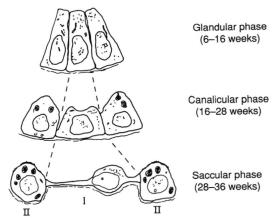

Glandular phase
(6–16 weeks)

Canalicular phase
(16–28 weeks)

Saccular phase
(28–36 weeks)

Fig. 11.1 Cytodifferentiation of terminal respiratory epithelium in human lung. During the glandular phase the epithelial cells lining the peripheral airways are uniformly columnar. These cells differentiate into type II epithelial cells, which are cuboidal and contain lamellar bodies (surfactant secretion), and subsequently into type I cells with their thin cytoplasmic sheets overlying capillaries. (Adapted from Burri, P.H. and Weibel, E.R. (1977). Ultrastructure and morphometry of the developing lung. In *Development of the lung* (ed. W.A. Hodson), p. 215. Marcel Dekker Inc., New York.)

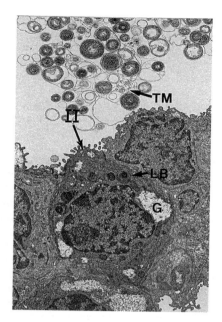

Fig. 11.2 Electron micrograph of fetal rat lung at 21 days of gestation (4000 ×). A type II cell is visible (II) that contains glycogen stores (G) and lamellar bodies (LB). The alveolar lumen contains secreted lamellar bodies and also tubular myelin (TM), the lattice-like form of surfactant that may be an intermediate form in the generation of a surface active film. (Photomicrograph, courtesy of Mary C. Williams.)

tion) as the fluid-filled air spaces expand. Even at birth, however, there is not complete alveolarization of terminal respiratory units; this is accomplished during the postnatal period and throughout childhood. This fact is both fortunate and unfortunate. For infants who must endure the stresses of mechanical ventilation, oxygen therapy or pulmonary infection, there is the risk of disrupting the normal process of alveolarization; however, since the process continues for years, there is opportunity to form new, healthy alveoli undisturbed by these adverse factors.

The developing structures within the lung must be supported by connective tissue elements to afford strength and distensibility. If extracellular matrix is not made and remodelled properly, alveoli may not mature. In fetal mouse lung *in vitro*, for example, inhibition of proteoglycan synthesis decreases both acinar formation and surfactant accumulation. Copper deficiency (either inherited or induced, in animals), in which there is inhibition of cross-linking of collagen and elastin, results in enlarged alveoli and altered lung mechanical properties (decreased elastic recoil and increased compliance, as in emphysema).

In the acinar region, there is a juxtaposition of epithelial (type I and II cells), endothelial (capillary), and mesenchy-

mal cells (fibroblasts). There are important interactions between these cells and their underlying connective tissue matrix that are important for both organogenesis and differentiation of the terminal airways. The mesenchyme is essential for airway branching in early lung development; at sites of active branching there is a high turnover of basement membrane glycosaminoglycans, which are made by epithelium and remodelled by mesenchymal cells. Late in gestation the basement membrane underneath the alveolar epithelium develops gaps, through which 'foot processes' of type II cells may extend, thus providing direct contact with the mesenchyme. This contact may allow for signalling between epithelial and mesenchymal cells (paracrine regulation), providing co-ordinated development of both cell types. Maintaining a differentiated state may also require cellular interaction with matrix; isolated type II cells grown on floating collagen gels or on thick gel matrix retain their cuboidal shape and continue production of surfactant components, whereas those grown on plastic flatten and stop producing surfactant.

Other cell–cell interactions also may play a role in determining lung structure. Neuroepithelial bodies are clusters of neuroendocrine cells along the airways (especially near branch points) having granules containing, in particular, gastrin-releasing peptide (GRP), the mammalian homologue of amphibian bombesin. There are very high levels of GRP and its messenger RNA (mRNA) in human fetal lung, particularly at 16–30 weeks, during the canalicular phase. In studies of fetal mouse lung *in vivo* and *in vitro*, this neuropeptide promoted both DNA synthesis and tissue maturation (reflected by an increased percentage of type II cells and increased synthesis of saturated phosphatidylcholine, the major surfactant lipid). Thus, GRP and other neuropeptides may have a paracrine role in growth and/or differentiation.

The type I–type II cell relationship remains incompletely defined. Classically, the type II cell has been considered to be the progenitor cell for the type I lineage, as is true in lung regeneration following pneumonectomy or injury (type II cells proliferate and give rise to new type I cells). This idea was based on apparent intermediate cell types and on the timing of DNA synthesis and mitosis. Recently, however, this theory has come into question because, in fetal rat lung, cell surface antigens specific for type I cells appear earlier than do type II cells.

Hormonal influences

G.C. Liggins first described an *in vivo* effect of hormones on lung maturation in 1969. During studies on glucocorticoid induction of premature labour, he noticed that treated fetal sheep had better lung expansion than expected at that

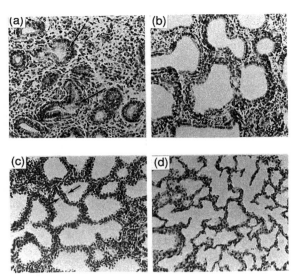

Fig. 11.3 Cortisol effect on lung morphology (160 ×). Fetal rabbits at different gestational ages were treated with cortisol for 2 days before delivery. Panels (a) (control) and (b) (treated), from fetuses delivered at 21 days, show that cortisol induces a decrease in mesenchymal volume and a transition from glandular-like ducts (arrows) to more dilatated ducts lined with low columnar or cuboidal cells. At day 26, (c) control fetuses still have moderately thick septae (arrows), which are much thinner in (d) treated fetuses, whose potential air spaces appear mature. (Taken with permission from Kikkawa, Y., Kaibara, M., Motoyama, E.K., Orzalesi, M.M., and Cook, C.D. (1971). *American Journal of Pathology*, **64**, 423–42.)

Hormone treatments

In 1971, Kikkawa and co-workers published a key paper that detailed histologically the accelerating effect of cortisol on lung structural development. Kikkawa carefully analysed lung morphology in rabbit fetus littermates injected directly with cortisol or saline. Single injections were done at 19 to 25 days of gestation (full term is 30 days) and fetuses delivered 2 to 3 days later. Cortisol precociously induced all the normal morphological and ultrastructural developmental changes, accelerating maturation gestational age. He proposed that dexamethasone treatment had induced early appearance of surfactant and thus stable alveolar expansion. This observation has been repeated in numerous experimental systems and it is now established that glucocorticoids and certain other hormones accelerate development of lung structure and cellular differentiation.

by ~ 1.5 days overall at each gestational age (Fig. 11.3). Subsequent studies in fetal sheep and monkeys found a significant increase in the area of potential air spaces and a decrease in the volume of the interstitium after betamethasone treatment. A recent study of rabbit fetuses treated with maternally administered betamethasone demonstrated an increased proportion of potential air space, increased proportion of type II cells within alveolar epithelium, and increased number of epithelial-mesenchymal communications via 'foot processes'.

Glucocorticoid effects *in vitro* resemble those seen *in vivo*. With the addition of dexamethasone to the medium, cultured fetal lung can be induced to show acinar maturation and type II cell differentiation, both morphological (for instance, lamellar body formation) and biochemical (synthesis of surfactant components).

Although the mechanisms for induction of structural changes are not defined, glucocorticoids have been shown experimentally to increase the content and rates of synthesis of collagen and elastin, at least during the canalicular phase of lung development. In fetal sheep studies, the mechanical properties of the lung (distensibility on inflation and stability on deflation) correlated well with concentrations of elastin and collagen in lung parenchyma. Hormonally induced improvements in fetal lung compliance cannot be entirely explained by effects on surfactant; when pressure–volume studies were performed in saline-filled lungs (thus negating surfactant effects), greater distensibility was found in glucocorticoid-treated than control fetuses. Thus, glucocorticoids affect both connective tissue elements and surfactant to improve lung compliance. Studies in fetal mice found apparent dose and time course differences between dexamethasone stimulation of air space volume and stimulation of lamellar body formation, probably reflecting the complex processes, involving several cell types, associated with structural changes.

The effects of exogenous thyroid hormone (thyroxine, T_4) were tested in fetal rats (via injection into the amniotic sac) by Hitchcock in 1979. She demonstrated acceleration of structural and ultrastructural changes similar to that found with glucocorticoid therapy. Endogenous glucocorticoids enhanced the effects of T_4. This was tested by administering T_4 after surgical adrenalectomy or metopirone treatment of the pregnant mother (metopirone will decrease adrenal steroid production in both the mother and fetus), with or without subsequent replacement of hydrocortisone. This observation provided the first evidence for an interaction between glucocorticoids and thyroid hormones in lung maturation.

Combination hormonal therapy is more effective than single agent therapy, at least in the immature fetal sheep. Animals treated with both thyrotrophin-releasing hormone

(TRH), which increases thyroid hormone and prolactin levels, and cortisol demonstrated greater lung distensibility, alveolar septal thinning, potential air space volume, and alveolar surface area than those treated with single agents. These observations illustrate the complex multihormonal regulation of lung development.

Other hormones also affect structural development of the lung. For example, cyclic AMP (cAMP) analogues and agents that increase endogenous cAMP (e.g. β-adrenergic agonists) cause septal thinning and increase the 'air space' volume in explant cultures of fetal lung, possibly by increasing fluid transport. Catecholamines and other hormones are discussed further in subsequent sections of this chapter.

Endogenous hormones

Experimental observations support a physiological role for endogenous hormones in lung development. In most species there is a substantial increase in fetal cortisol and triiodothyronine (T_3) during late gestation. In fetal sheep, structural maturation (lung distensibility, airspace size, alveolar septal thickness), like concentrations of surfactant lipids, correlates better with the plasma cortisol level than with the gestational age. Additional evidence for the importance of endogenous glucocorticoids comes from ablation studies in which the adrenal production of corticoids is prevented either surgically (by hypophysectomy to eliminate pituitary release of adrenocorticotrophic hormone, ACTH) or chemically (by metopirone to block adrenal steroid production). Sheep fetuses hypophysectomized at 99–122 days gestation and delivered at term (147–150 days) had significantly impaired structural lung development (lung distensibility) and surfactant synthesis and secretion (saturated phophatidylcholine in lung tissue and lavage). Treatment with either ACTH or cortisol counteracted the adverse effect of hypophysectomy on structural morphology.

Endogenous cortisol and catecholamines are responsible at least in part, for the maturational effects of labour and of fetal stress during gestation. In both situations, hormone levels rise, although it is difficult to assess the relative contributions of fetal and maternal adrenal glands. Studies in fetal rabbits indicate that surgical procedures increase fetal plasma corticoids and accelerate lung morphological development, deflation stability, and rates of choline incorporation into phosphatidylcholine (as an indication of surfactant synthesis). In fetal rats, the maturative effect of sham surgery can be blunted by metopirone and, to a lesser extent, by maternal adrenalectomy, suggesting that fetal adrenal corticoids are an important mediator of this effect.

An effect of endogenous corticoids on lung maturation could explain some clinical correlations in humans. First, higher free and total cortisol levels are present during spontaneous labour and delivery than during Caesarean delivery, and there is a lower incidence of the respiratory distress syndrome following vaginal delivery. Second, chronic stress *in utero*, for example due to severe maternal diabetes, pre-eclampsia, or Rhesus (Rh) isoimmunization, is associated with accelerated lung maturity, at least in terms of surfactant secretion. The frequently observed improvement in established RDS 2–3 days after birth may reflect the action of endogenous cortisol, which is elevated during the illness. Finally, data from some studies suggest lower blood cortisol levels in preterm infants, born at 31–35 weeks, who develop RDS as compared to those who do not.

Surfactant biology and function

Of the developmental changes in fetal lung, perhaps the best understood (and most important physiologically) is the ability to produce surfactant. Surfactant is made by type II epithelial cells and is a complex mixture (Fig. 11.4) containing approximately 90 per cent lipid and 10 per cent protein. In the functioning air space, surfactant interfaces with air as a film of tightly packed lipid molecules overlying the liquid phase that covers the alveolar cells. If surfactant were not present, surface tension would cause the alveolus to collapse during exhalation, which would then result in diffuse atelectasis (i.e. collapse of alveoli) and impaired gas exchange.

In 1959 Avery and Mead provided the first data linking surfactant deficiency with RDS in the new-born. They tested the surface-tension lowering ability of lung extracts from infants of varying gestational ages and birth weights who had died of RDS or of other causes. There was a striking finding: samples from all nine infants weighing < 1200 g could not maximally lower surface tension *in vitro*. This was also true of samples from those who had died of respiratory distress (even at a higher birthweight), and from one stillborn infant of a diabetic mother. In contrast, samples from adults, children, and infants weighing > 1200 g without RDS showed good lowering of surface tension (to 5–7 dyn/cm at maximal compression versus 20–30 dyn/cm in the abnormal group). These and other studies linking prematurity, surfactant deficiency, and RDS established the clinical problem as one of developmental immaturity. The incidence of RDS is nearly 100 per cent in infants born at 26 weeks gestation, and drops to about 50 per cent by 30–31 weeks. The syndrome also occurs in other situations where normal lung maturation is delayed, such as in infants of diabetic mothers.

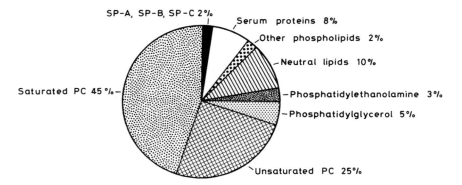

Fig. 11.4 Composition of surfactant as isolated by broncho-alveolar lavage. Saturated PC is the most surface-active and the most abundant component of surfactant. PC, phosphatidylcholine; SP, surfactant protein. (Taken with permission from Ballard, P.L. (1989). *Endocrine Reviews*, **10**, 165–81. © The Endocrine Society.)

Components of surfactant

The component of surfactant with the greatest effect on surface tension is its major lipid, dipalmitoyl phosphatidylcholine (DPPC). This molecule has a polar, water-soluble end containing choline, and a non-polar end with two long fatty acid molecules (palmitate) containing completely saturated carbon–carbon bonds. These fatty acids project out into the air above the alveolar liquid film, and the DPPC molecules can compress tightly against one another because their fatty acid chains are straight. Besides phosphatidylcholine, other phospholipids may be important in surfactant dispersion and

function, particularly the acidic phospholipids phosphatidylglycerol and phosphatidylinositol.

There are at least four known surfactant-associated proteins (Table 11.1). The most abundant is surfactant protein A (abbreviated SP-A), a 28–36 kDa glycoprotein that is hydrophilic. It binds to calcium, sugars, and lipids, and increases the calcium-induced aggregation of surfactant lipids. It has an additive role with two other surfactant proteins (B and C) in speeding the formation of a surface film *in vitro*. Although DPPC is the surface-active component, it does not readily reach and spread over the surface unless surfactant proteins are present. SP-A may also participate in regulating the alveolar concentration of surfact

Table 11.1 Surfactant protein characteristics

Protein	Molecular weight (kDa)	Properties	Proposed functions
SP-A	28–36 (reduced) 650 (native)	Hydrophilic glycoprotein Collagen-like domain Binds sugars, calcium, and lipids Posttranslational modifications	Promotes surfactant film formation Inhibits surfactant lipid secretion Increases surfactant lipid uptake Maintains tubular myelin structure Local immune defence Blocks serum protein inhibition of surfactant activity Decreases serum protein leakage into the alveolus
SP-B	8	Lipophilic	Promotes surfactant film formation and squeeze-out of non-DPPC lipids
SP-C	5	Lipophilic	Promotes surfactant film formation
SP-D	43	Hydrophilic glycoprotein	?Local immune defence

ant lipid, as it inhibits phosphatidylcholine (PC) secretion from isolated type II cells and increases its uptake. In addition, SP-A may have local immune functions. *In vitro* it enhances chemotaxis of macrophages and, at least in some studies, enhances their phagocytosis and killing of bacteria.

Surfactant proteins B and C (SP-B and SP-C) are much smaller (5 and 8 kDa, respectively) and are lipophilic. They play an essential role in the formation of the surface film, and *in vitro* studies have demonstrated that they (at least in combination) greatly enhance the rate at which surfactant lipid enters and spreads within the surface monolayer. One or both of these proteins is present in replacement surfactant prepared from lavage of animal lungs or from human amniotic fluid.

Another protein physically associated with surfactant has been recently identified and named SP-D; however, it may not truly be a 'surfactant protein' in that a role for it in surface film formation or lowering of surface tension has not been documented. SP-D is a 43-kDa glycoprotein made by type II cells and non-ciliated bronchiolar (Clara) cells. It has some structural similarities to SP-A and to serum proteins known to be involved in immune defence (conglutinin and mannose binding protein), and thus SP-D has been postulated to play such a role. Studies *in vitro* have demonstrated that it binds to Gram-negative bacteria

and also enhances the production of oxygen radicals (for microbial killing) in alveolar macrophages from rats.

Surfactant appearance in the fetus

Synthesis of surfactant lipid begins in the human fetus at about 20 weeks of gestation, even before the saccular period. At this time, the lung tissue concentration of saturated PC (a reflection of surfactant levels) begins to rise; at the same time, lamellar bodies begin to appear in type II cells. Surfactant is not secreted *in utero* until about 30 weeks, when it can be recovered from amniotic fluid. This allows for testing of amniotic fluid (obtained by amniocentesis) to predict postnatal alveolar stability or instability based on surfactant content.

The most widely used amniotic fluid tests are the lecithin to sphingomyelin ratio (L/S ratio) and the presence of phosphatidylglycerol (Table 11.2). Their patterns during gestation are depicted in Fig. 11.5. Lecithin is a synonym for PC, and the acetone-precipitable fraction of lecithin represents saturated PC, thus reflecting secreted surfactant. This value is normalized relative to content of sphingomyelin as a non-specific membrane lipid. An L/S ratio exceeding 2.0 (a value normally attained at approximately 35 weeks of gestation) predicts a low risk of RDS, whereas

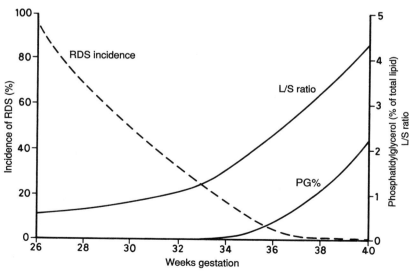

Fig. 11.5 Diagram showing incidence of respiratory distress syndrome (RDS) and predictors of lung maturity. Both L/S ratio and phosphatidylglycerol levels in amniotic fluid rise in late gestation and are used as predictors of lung maturity (see Table 11.2). The approximate incidence of RDS (without prenatal steroid treatment) decreases markedly in late gestation and correlates inversely with L/S ratio and phosphatidylglycerol concentrations.

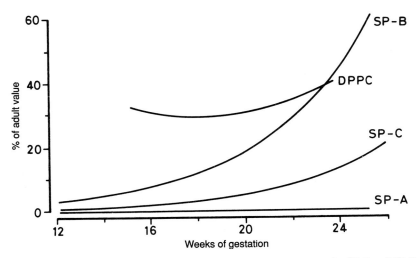

Fig. 11.6 Ontogeny of surfactant proteins (mRNAs) and lipid in human fetal lung. mRNAs for SP-B and SP-C appear as early as 12 weeks of gestation and their content increases during the second trimester, whereas SP-A mRNA is not detected and DPPC (dipalmitoyl phosphatidylcholine) content does not increase until late in the second trimester. (Taken with permission from Ballard, P.L. (1989). *Endocrine Reviews*, **10**, 165–81. © The Endocrine Society.)

Table 11.2 Amniotic fluid tests for lung maturity. The most commonly used tests are the L/S ratio and phosphatidylglycerol contents, but other methods (e.g. biophysical and optical) are useful for rapid screening

Tests	'Mature' value
Biochemical	
L/S ratio	≥ 2.0
PC content	> 0.1 mg/dL
Phospatidylglycerol	Present
SP-A	≥ 2 µg/mL
Biophysical	
'Shake test'	Stable ring of foam
Optical	
Fluorescence polarization (due to surfactant aggregates)	Approx. < 0.3
OD 650 (lamellar bodies)	> 0.15

a ratio less than 1.5 predicts a high risk. This ratio is least helpful when intermediate values are obtained and in diabetic pregnancies, where false-mature L/S values may occur. In these situations in particular, concurrent testing of other indices may be helpful. The appearance of phosphatidylglycerol in surfactant reflects the shift from phosphatidylinositol to phosphatidylglycerol that occurs during gestation (it is felt that these two phospholipids have equivalent surfactant-related functions). At present, the use of phosphatidylglycerol and L/S determination together is as accurate as any other method in confirming lung maturity, but other methods are being sought that will be less cumbersome and perhaps better predictors of RDS when the phosphatidylglycerol content and L/S ratio are low.

The ontogeny of mRNA for surfactant proteins A, B, and C is shown in Fig. 11.6. The appearance of SP-A parallels that of DPPC, appearing in amniotic fluid at about 30 weeks of gestation and then steadily increasing during the third trimester. Less is known about SP-B and -C; however, based on their messenger RNA levels, they may be synthesized in human fetal lung at as early as 13 weeks of gestation, well before the appearance of both surfactant lipid and SP-A. SP-B appearance in amniotic fluid is apparently delayed until after 30 weeks of gestation, however. In the fetal rat *in vivo*, both SP-A and DPPC increase markedly between day 19 of gestation and postnatal day 1, whereas SP-B protein shows a more gradual increase beginning on day 20 of gestation and peaking on postnatal day 4.

Hormonal effects on surfactants

The accepted model for lung maturation *in utero* is a complex one, involving many hormones primarily of fetal origin, which act upon a variety of pulmonary cell types to modulate the rate of development. The synthesis, content, and secretion of surfactant is often used as an index of lung maturation, and these processes have been extensively studied with regard to hormonal effects both *in vivo* and *in vitro*. The following sections describe the effects and mechanisms of action of several hormones.

Glucocorticoids

The proposal, first made by G.C. Liggins, that glucocorticoids accelerate the appearance of surfactant in the fetus has been confirmed by a number of investigators who examined pulmonary mechanical properties, surfactant content, and morphology. The species used have been sheep, rabbits and rats. These investigations were extended to humans in 1972 when Liggins and Howie carried out a trial of antenatal steroid administration to prevent respiratory distress syndrome in infants. Glucocorticoids have now become standard therapy in situations where premature delivery can be anticipated. Established benefits include decreased mortality, decreased incidence and severity of respiratory distress syndrome, and, as recently reported, decreased subsequent development of chronic lung disease. Non-pulmonary benefits include a decreased incidence of patent ductus arteriosus, necrotizing enterocolitis, and intraventricular haemorrhage.

The role of endogenous steroids in modulating surfactant synthesis has been examined in animal studies. Temporal associations have been documented for rising plasma cortisol levels and the onset of surfactant production in many species. Studies in late-gestation fetal sheep demonstrated that several markers of lung surfactant correlated much better with fetal cortisol levels than with gestational age. These markers included: concentration of saturated PC (which is mostly DPPC) as an indicator of surfactant content in lung tissue (intracellular) and in lavage fluid (secreted), lung distensibility (volume after air inflation to 40 cm H_2O), and lung stability at low volumes (volume at deflation to 10 cm H_2O).

To study hormonal effects, *in vitro* culture systems have been developed for fetal lung. These include lung slices, minces, and organotypic cultures. Several investigators have used the explant culture system, in which small pieces of fetal lung tissue are maintained in culture. This system has the advantages of preserving cell–cell interactions and allowing manipulation of the hormonal milieu, as the tissue can be maintained in medium free of serum that could contain hormones or growth factors. Hormonal effects on the surfactant system have been assessed by examining morphology (especially presence of lamellar bodies), content of total PC and saturated PC in lung tissue, choline (and other precursor) incorporation into PC

and saturated PC, activity of enzymes involved in surfactant lipid synthesis, and content of surfactant proteins and their mRNAs.

Phospholipid synthesis and content

Rates of incorporation of choline into PC, an approximation of PC synthetic rate, increase dramatically (two- to fivefold) toward the end of gestation in all species studied thus far. This is reflected by a modest increase in lung tissue saturated PC content and a great increase (10-fold in the rabbit) in the amount of saturated PC in lung lavage fluid. Measurements in whole lung tissue are less sensitive indicators because the tissue contains much non-surfactant lipid.

In vivo, the administration of glucocorticoids can cause the natural rise in choline incorporation to occur earlier than otherwise expected. Glucocorticoid treatment does not enhance incorporation of precursors into phosphatidylethanolamine, a phospholipid that is prominent in membranes, suggesting that the choline incorporation effect does not simply represent a generalized increase in lipid synthesis. In minced rabbit lung, increased choline incorporation is first evident within 6–12 h and is maximal by 48 h after dexamethasone treatment. Maximal stimulation of alveolar surfactant levels occurs somewhat later, as more time is needed for packaging and secretion in addition to synthesis of surfactant. Glucocorticoids do not directly stimulate secretion of surfactant, but they may do so indirectly by increasing the lung content of PC and by inducing β-adrenoceptors, thus increasing the sensitivity of alveolar cells to endogenous catecholamines that stimulate surfactant secretion. The timing of glucocorticoid stimulation of surfactant synthesis is reflected in the 24-h lag to achieve a benefit from prenatal steroid administration in treatment protocols for anticipated premature deliveries.

The glucocorticoid-induced stimulation of choline incorporation into PC is mediated by glucocorticoid receptors and is blocked by inhibitors of either protein or RNA synthesis. Using binding studies and autoradiography, glucocorticoid receptors have been demonstrated in the lungs of many species (fetal and adult rat and rabbit, adult hamster and mouse, fetal sheep, fetal monkey, and fetal human) at concentrations several-fold higher than in most other tissues. Furthermore, the number of cytosolic binding sites for glucocorticoids increases in late gestation, at least in rat, rabbit, and lamb, possibly increasing the sensitivity of cells to these hormones. Glucocorticoid receptor sites have been localized to most lung cells including epithelial cells (such as type II cells), fibroblasts, smooth muscle, and endothelial cells. Localization to type II cells was important in establishing that glucocorticoids could have direct effects on surfactant production. The presence of receptors on lung fibroblasts is consistent with experimental evidence that some effects of glucocorticoids are mediated via the fibroblast, which is thought to secrete a low-molecular-weight protein (fibroblast pneumonocyte factor) that then acts upon the adjacent type II cells to induce phospholipid synthesis.

Enzymes of phospholipid synthesis

Individual enzymes of surfactant phospholipid synthesis have also been examined in the search for targets of hormone actions. The time course for the hormone-induced increase in PC synthesis is consistent with a mechanism involving enzyme induction. So far, four enzymes have been found that show developmental increases in activity, are rate-limiting in their biosynthetic pathway, and are stimulated by glucocorticoids (Fig. 11.7).

Fatty acid synthetase is the major enzyme for synthesis of fatty acids from acetyl coenzyme A (CoA) and malonyl CoA; in later steps, these fatty acids are incorporated into PC. In fetal rat lung, fatty acid synthetase activity increases during the last few days before birth (at the same time that choline incorporation into PC is increasing), and is stimulated by in vivo glucocorticoid treatment. This stimulation is thought to occur via increasing expression of the synthetase gene (increased synthesis of the enzyme), based on immunoprecipitation studies. In cultured human fetal lung (16–22 weeks gestation), the kinetics and dose–response properties of induction are similar for synthetase activity and choline incorporation. In addition, both processes are stimulated to a similar extent by cyclic AMP (cAMP). These parallel reponses both in vivo and in vitro suggest that induction of fatty

acid synthetase may be a key event in the stimulation of surfactant lipid synthesis.

Cholinephosphate cytidylyltransferase catalyses the rate-limiting step in the pathway to incorporate the choline moiety into PC (see Fig. 11.7). The activity of this enzyme increases in rat lung during gestation and after birth, and dexamethasone stimulates its activity both in vivo (rabbit, rat, and mouse) and in vitro (fetal rat and human). However, the amount of the enzyme is not increased by dexamethasone, indicating that the mechanism of stimulation is indirect. If acidic phospholipid (a cofactor for the enzyme) is present in the assay system, there is decreased stimulation of enzyme activity by glucocorticoids. Also, if de novo fatty acid synthesis is inhibited, dexamethasone no longer stimulates cytidylyltransferase activity. Thus it is likely that cytidylyltransferase is not a primary site of glucocorticoid action but rather amplifies PC synthesis secondary to increased availability of phospholipid and/or fatty acids as cofactors.

Phosphatidic acid phosphatase catalyses the conversion of phosphatidic acid to diacylglycerol, which can then form PC when condensed with CDP choline (see Fig. 11.7). This enzyme is associated with lamellar bodies, the surfactant storage organelles, and is present in cytosol and in microsomal fractions (which include endoplasmic reticulum). Enzyme activity undergoes a developmental increase (in rat and rabbit lung and in human amniotic fluid) in parallel with increasing PC content. In vivo studies in rabbits demonstrate a stimulation of activity following corticosteroid treatment, but organ culture studies using fetal rabbit and fetal human lung have not yielded similar results.

An enzyme involved in a final step of surfactant lipid synthesis is lysoPC:acyl CoA acyltransferase, which carries out 'remodelling' to the disaturated form of PC. This remodelling must occur for some PC because an unsaturated fatty acid is often initially added on to carbon 2 of the PC backbone, and it must then be removed (by phospholipase A_2) and replaced with a saturated fatty acid (by the acyltransferase). This enzyme is developmentally regulated and induced by corticosteroid treatment in vivo and by both dexamethasone and cAMP in cultured tissue.

Surfactant proteins

Much attention has been focused on regulation of surfactant proteins A, B, and C in recent years, with availability of cDNA (i.e. complementary DNA) probes for all three and specific antibodies for SP-A and SP-B. Limited information is available on SP-D, although probes and antibodies have been developed.

The SP-A gene contains at least one sequence with high homology to the 'glucocorticoid response element' of

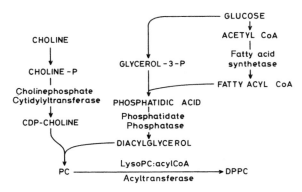

Fig. 11.7 Major pathways of PC (phosphatidylcholine) synthesis in the lung. Key enzymes are indicated (see text for details.) (Taken with permission from Ballard, P.L. (1989). *Endocrine Reviews*, **10**, 165–81. © The Endorine Society.)

other known glucocorticoid-responsive genes. There are two possible sites for cAMP regulation, and possibly one for interferon. The SP-B gene has putative response elements for glucocorticoids and cAMP. The SP-C gene contains two possible cAMP response sites. Experiments are under way to elucidate the molecular mechanisms and sites of hormone action within surfactant protein genes.

Glucocorticoids increase the expression of all three surfactant protein genes. One report has demonstrated that lung SP-D content also increases after *in vivo* treatment with dexamethasone. In particular, the response of SP-A to glucocorticoids has proved quite interesting. In cultured explants of human fetal lung, which at the midtrimester of gestation normally have no detectable SP-A protein or SP-A mRNA, there is spontaneous appearance of SP-A during culture, even without hormones added (in part due to increasing content of endogenous cAMP). With dexamethasone (5–10 nM) present in the culture medium there is an initial stimulatory effect for the first few days, followed by an inhibition compared to control. This dual response to glucocorticoid is unusual and is not observed for SP-B or SP-C mRNA. The mechanism in the case of SP-A and dexamethasone remains unclear, but may involve effects on both gene activation and mRNA stability, possibly mediated by other inducible proteins. *In vivo*, the response of SP-A (but not SP-B or SP-C mRNA) to dexamethasone depends on the gestational age of the tissue and duration of treatment, suggesting differential regulation of these three surfactant proteins. Glucocorticoid effects on the surfactant proteins are receptor-mediated, but may involve different regulatory mechanisms since dose–response relationships and time courses for induction in cultured tissue are different. Thus, although glucocorticoids increase all surfactant components, the response is not regulated in a co-ordinated manner, probably reflecting differences in mechanisms of gene regulation.

Other proteins

In cultured human fetal lung, dexamethasone induces at least 16 and decreases 4 out of approximately 1000 proteins resolved by two-dimensional gel electrophoresis. However, the identity of most of these proteins remains unknown. Overall protein synthetic rate does not change with dexamethasone treatment, indicating that the effect is a selective one, on only a limited number of proteins.

Thyroid hormone

Fetal thyroid hormones have been shown to affect surfactant as well as lung morphology. For example, thyroidectomized fetal sheep delivered prematurely demon-

strate delayed thinning of the alveolar–capillary barrier, decreased numbers of lamellar bodies in type II cells, and decreased amounts of surfactant in lavage fluid compared with euthyroid controls. This effect is not mediated by glucocorticoids, as serum cortisol levels were similar in treated and control groups.

In cultured fetal lung, both T_3 and T_4 stimulate choline incorporation into PC and increase tissue content of PC, although to a lesser degree than does dexamethasone. However, T_3 does not increase the per cent saturation of PC or the synthesis of phosphatidylglycerol, suggesting that thyroid hormones may increase surfactant synthesis indirectly as a consequence of generalized stimulation of lipid synthesis. T_3 does not induce fatty acid synthetase in human fetal lung explants and increases cytidylyltransferase activity only by increasing the amount of its lipid cofactor. Other lipid synthetic enzymes that are stimulated by dexamethasone, namely, lysoPC:acyl CoA acyltransferase and phosphatidic acid phosphatase, are also not affected by thyroid hormone. Similarly, synthesis of surfactant protein SP-A in human fetal lung is not stimulated by T_3, and two-dimensional gel electrophoresis of T_3-treated tissue showed no difference among the resolvable proteins when compared to untreated cultured lung. Thus, the exact site of thyroid hormone action remains unknown. When both dexamethasone and T_3 are added to explant cultures, the stimulation of PC synthesis is supra-additive (Fig. 11.8). This fact, and the different responses elicited (e.g. on enzyme activities), suggest that these hormones act at different biochemical sites to stimulate PC synthesis.

Of the two major thyroid hormones, T_3 is present at lower levels *in vivo* but is more biologically active in stimulating phospholipid synthesis because of a higher affinity for lung nuclear receptors. Fetal lung tissue has only a limited ability to convert T_4 to T_3 by outer ring deiodination. Reverse T_3 (the product of inner ring deiodination of T_4) is present at high levels in fetal blood, but it has low binding affinity and low bioactivity. In the human fetus, T_4 and thyroid-stimulating hormone (TSH) are measurable in plasma beginning at 10–11 weeks gestation, with a significant rise between 11 and 24 weeks; plasma levels of T_3 increase during the last 10 weeks of gestation. *In vivo*, T_3 and T_4 do not cross the placenta efficiently because of binding to thyroid hormone binding globulin in maternal plasma. The developmental pattern of thyroid hormones differs between species; it is interesting to note that in rats, in which the plasma concentration of thyroid hormone increases during the second postnatal week, alveolarization occurs after birth, whereas in humans, rabbits, and sheep, in which thyroid hormones appear much earlier, alveolarization begins *in utero* (although it is not completed until later in humans).

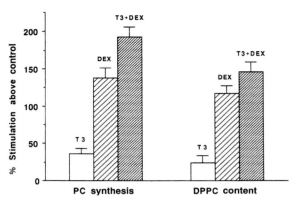

Fig. 11.8 Stimulation of PC (phosphatidylcholine) synthesis and DPPC content by T_3 and dexamethasone. Human fetal lung at 16–25 weeks gestation was cultured for 7 days with T_3 (2 nM) and/or dexamethasone (DEX, 10 nM). PC synthesis was determined by incorporation of [^{3}H]choline into PC, and DPPC (dipalmitoyl PC) content was determined by phosphorus assay after thin-layer chromatography separation. All values are significantly increased over control, and there is an additive response to T3 + DEX. (Data from Gonzales, L.W. and Ballard, P.L. (1989). In *Lung biology in health and disease*, Vol. 41 (ed. D. Massaro), pp. 539–90. Marcel Dekker, New York.)

The effect of thyroid hormone on lung cells is mediated by the interaction of thyroid hormone with nuclear receptors, presumably increasing specific mRNA and protein synthesis. Correlating with the rising levels of hormone during gestation, at least in fetal rabbit and human lung, there is a developmental increase in the number of nuclear T_3 receptor sites as well as percentage occupancy by endogenous hormone. For the rabbit, the net result is a calculated threefold rise in occupied receptor sites.

Thyroid hormones have been linked to lung development in clinical situations. Some studies have correlated lower cord blood levels of T_3 and T_4 (even though still within the normal range) with an increased incidence of respiratory distress syndrome in new-borns, both premature and full term; true hypothyroid infants have an apparently increased incidence of RDS. One study has found that intra-amniotic administration of T_4 was associated with enhanced lung maturity and a lower than expected incidence of RDS; however, this information has not been clinically implemented, primarily because of the impracticalities of using this route of administration. Another therapeutic strategy that is currently being tested is the administration of thyrotrophin-releasing hormone (TRH) antenatally to improve outcome for premature infants. TRH, being a tripeptide, can easily cross the placenta to stimulate release of TSH (and prolactin) by the fetal pitu-

itary gland, thereby increasing fetal thyroid hormones. In clinical trials, TRH has been given along with antenatal glucocorticoid to determine whether additive and earlier effects on surfactant synthesis occur, as seen *in vitro*. An early potentiation of the glucocorticoid effect would be important clinically when, for instance, premature delivery could only be delayed for 12–24 h. Initial studies have found a trend towards a decreased incidence and severity of RDS and a significantly lower incidence of bronchopulmonary dysplasia (chronic lung disease associated with altered lung mechanics and increased airway reactivity). The beneficial effect with regard to chronic lung disease was unexpected and the mechanism is not yet known.

Prolactin

Plasma levels of fetal prolactin increase during mid- to late-gestation, preceding the rise in plasma corticoids and the appearance of surfactant lipid in tracheal fluid, but the role of prolactin in lung maturation has not been definitively established. Lower cord blood levels correlate with a higher incidence of RDS in infants born at approximately 32–37 weeks of gestation. In fetal sheep, however, prolactin alone has no effect on passive lung mechanics or alveolar saturated PC. Nevertheless, it appears to be required for optimal responsiveness to cortisol plus T_3 treatment in fetal sheep < 130 days of gestation (term being 147–150 days) as shown in Table 11.3. Data from most studies using fetal rabbits also show no effect of prolactin administered *in vivo*, and there have been conflicting findings regarding possible permissive effects of prolactin on surfactant synthesis in cultured lung.

Although the role of endogenous prolactin is not clear, one hypothesis is that response of the fetal lung to glucocorticoids and thyroid hormones requires prolactin, levels of which would only be adequate in the latter half of the third trimester. Earlier in gestation, prolactin or TRH, which stimulates both TSH and prolactin, would have to be administered exogenously to assure a good response to the other hormones.

Sex hormones

During pregnancy, maternal plasma concentrations of oestrogens rise progressively, particularly after 32 weeks of gestation. Low fetal levels of oestrogens have been correlated with a higher incidence of RDS after birth. Studies in rats and rabbits indicate that oestradiol stimulates choline incorporation into PC in lung tissue and accelerates morphological development. However, one study in fetal sheep failed to detect any stimulatory effect of oestradiol infusion on lung maturation, as shown by lung and

Table 11.3 Synergistic effects of hormones on lung mechanics and alveolar DPPC (dipalmitoyl phosphatidylcholine) content of fetal sheep at < 130 days gestation. T_3, cortisol, and prolactin (Prl) were infused singly and in combinations. Cortisol was required for benefit versus control, and the presence of Prl produced optimal improvement in lung mechanics as well as DPPC content. (Data from Schellenberg *et al.* (1988): for details see 'Further reading'.)

	Distensibility (V40) (mL air/g lung)	Deflation Stability (V5) (mL air/g lung)	Alveolar DPPC (μg/g lung)
Control	0.48 ± 0.10	0.22 ± 0.08	10 ± 4
Cortisol	0.91 ± 0.12	0.43 ± 0.09	12 ± 3
T_3 + cortisol	1.25 ± 0.18	0.81 ± 0.18*	161 ± 52[†]
T_3 + cortisol + Prl	1.71 ± 0.12[†]	1.16 ± 0.09[†]	156 ± 53[†]

*$p < 0.05$ versus cortisol.
[†]$p < 0.01$ versus cortisol.

tracheal fluid lipids and light and electron microscopic appearance.

Although oestradiol binding sites have been found in fetal lung and in adult airway cells, the available evidence does not support a direct receptor-mediated action of oestradiol in lung. Thus, any effects of oestradiol may be mediated through effects on other endogenous substances or may involve actions of oestradiol metabolites. One recent study using autoradiography of treated explants of fetal rat lung suggested that oestrogens might be taken up by fibroblasts rather than by type II epithelial cells. The hypothesis would be that these fibroblasts then are induced to produce a maturational factor (fibroblast pneumonocyte factor or another substance) that affects adjacent type II cells.

The male disadvantage in the occurrence and severity of RDS is well known. In addition, males show a poorer response to antenatal glucocorticoid therapy. In attempting to understand this sex difference in lung maturation, investigators have examined the effects of androgens and Müllerian inhibitory substance, a glycoprotein produced by the fetal testis.

Plasma levels of testosterone are higher in male than female fetuses in several species and reach a peak during the second trimester, prior to cytodifferentiation of type II cells. In fetal rabbits, L/S ratios of lavage fluid are lower for males than females, and females with neighbouring male littermates have ratios in the intermediate range (Fig. 11.9). This 'neighbour effect' is presumably due to androgen spillover from male to female fetal circulation. When pregnant rabbits are injected with dihydrotestosterone (DHT), the sex difference is obliterated, with L/S ratios in females decreasing to male values. In human fetal lung explants, DHT does not affect surfactant protein

mRNA levels, and there are conflicting reports regarding effects on synthesis of saturated PC. Thus, androgens appear to influence the developmental pattern for surfactant lipid synthesis, but there is little insight yet into the mechanisms involved.

Müllerian inhibiting substance, a product of the Sertoli cell of the testis, also may play a role in the sexual dimorphism of lung development. Recent studies suggest that, at least *in vitro*, Müllerian inhibiting substance may depress accumulation of saturated PC.

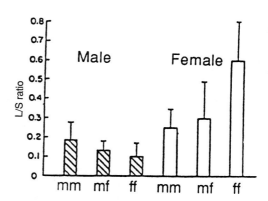

Fig. 11.9 Neighbour effect of fetal gender on fetal rabbit surfactant production. Lavage fluid L/S ratios were determined in 26-day-old fetal rabbits and related to the gender of neighbouring littermates. Male (m) fetuses have overall lower L/S ratios, and female (f) fetuses show a decrease in L/S ratio corresponding to the number of neighbouring male fetuses. (Adapted from Torday, J.S. and Nielsen, H.S. (1987). The sex difference in fetal lung surfactant production. *Experimental Lung Research*, **12**, 1–19.)

Cyclic AMP

Cyclic AMP (cAMP) stimulates both the synthesis and secretion of surfactant. In lung explants, cAMP analogue (e.g. 8-bromo cAMP, dibutyryl cAMP) increase the synthesis of disaturated PC, phosphatidylglycerol, and SP-A. The increase observed in SP-A content is associated with a several-fold increase in transcription of the SP-A gene, which contains putative regulatory sites for cAMP. cAMP also increases levels of mRNA for SP-B and activities of fatty acid synthetase, cholinephosphate cytidlylyltransferase, and lysoPC acyltransferase. As shown in Fig. 11.10 cAMP and dexamethasone have an additive effect on several of these indices. The only surfactant component that does not increase in response to cAMP in cultured explants is SP-C. cAMP also affects lung morphology, with treated explants demonstrating less interstitial connective tissue, larger ductal lumens, more lamellar bodies, and more secreted surfactant products (lamellar bodies and tubular myelin) than untreated explants.

Fetal lung also responds to agents that increase the content of endogenous cAMP by either stimulating adenyl cyclase (prostaglandins, β-agonists, forskolin) or inhibiting cAMP degradation (methylxanthines). Relatively modest increases in tissue cAMP levels (twofold) are sufficient to increase gene expression (SP-A and SP-B) and enzyme activities. During explant culture of human fetal lung in the absence of hormones, the endogenous cAMP content increases spontaneously and is partly responsible for the culture-induced increase in SP-A,

SP-B, and surfactant lipids. This observation suggests that the rate of lung maturation *in vivo* can be influenced by developmental changes in cAMP production and degradation as well as by increasing levels of stimulatory hormones.

β-adrenergic agonists exert their effects via activation of adenylate cyclase to increase formation of cAMP from ATP. Binding of the β-agonist to its receptor on the cell membrane initiates a sequence of events involving GTP/GDP (i.e. guanosine triphosphate/guanosine diphosphate) binding to an intracellular regulatory G protein, which then interacts with adenylate cyclase. β-adrenergic receptors, predominantly of the β_2 subclass, have been identified in membrane preparations from fetal lung in humans and other species, and their concentration increases during gestation and after birth. These binding sites are probably present on both type I and type II alveolar cells, based on autoradiographic binding studies. Glucocorticoids increase the concentration of β-adrenergic binding sites by 60–120 per cent *in vivo* and in explants of fetal rabbit and rat lung; however, in cultured human lung, dexamethasone partially inhibits the spontaneous increase in β-receptors normally seen during culture and decreases both the cAMP content and the isoproterenol-stimulated generation of cAMP. Thus, there may be species differences in the regulation of β-adrenoceptors by glucocorticoids.

β-adrenergic agonists affect three crucial processes that have clinical relevance for the new-born infant: surfactant synthesis; surfactant secretion; and lung fluid reabsorption. Endogenous catecholamines, adrenaline, and noradrena-

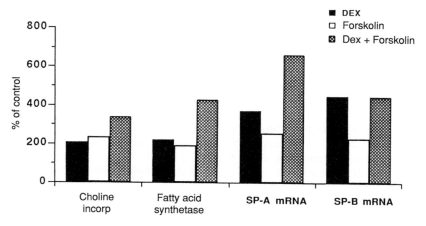

Fig. 11.10 Interaction of glucocorticoid and cyclic AMP (cAMP) in stimulating surfactant synthesis. Dexamethasone (DEX, 10 nM) and/or forskolin (10 μm), which increases cAMP, were added to cultured human fetal lung. Each agent stimulates synthesis of surfactant phospholipid (choline incorporation rate (Choline incorp), fatty acid synthetase activity) and of surfactant proteins A and B, and together they have an additive effect except on SP-B. (Data from Ballard, P.L. (1989). *Endocrine Reviews*, **10**, 165–81.)

line, are released from the fetal adrenal medulla and pulmonary sympathetic ganglia late in gestation, with a surge in plasma levels occurring during labour. Plasma levels are even higher if there is fetal stress during labour and delivery. The rate of PC secretion also increases with gestation, especially during labour, and it can be inhibited in fetal animals by adrenalectomy or by administration of propranolol (a β-adrenergic blocker) or a β-adrenergic antagonist. In human studies, amniotic fluid concentrations of catecholamine metabolites correlate with indices of lung maturation (L/S ratio, presence of phosphatidylglycerol). These observations therefore suggest a role for endogenous catecholamines, along with other hormones, in the maturation of pulmonary functions. In fact, infants born by Caesarean section, especially without having undergone any labour, have an increased incidence of RDS and transient tachypnoea of the new-born.

Exogenously administered β-adrenergic agonists may also affect the fetal lung. These agents, which include terbutaline and ritodrine, are commonly used to treat preterm labour. Several groups have reported a significantly lower incidence of RDS in premature infants born after β-agonist treatment, although these were not rigorously controlled trials. In addition, data from rabbit studies indicate that catecholamines are beneficial for surfactant production after short-term but not long-term treatment. This could occur if β-agonist treatment caused a rapid release of stored surfactant with subsequent transient deficiency and/or a downregulation of β-adrenoceptors and thus diminished responsiveness.

Prostaglandins also act via the cAMP system to stimulate synthesis and secretion of surfactant. In cultured lung, for example, prostaglandin E_2 (PGE_2) and prostaglandin E_1 (PGE_1) increase levels of cAMP, SP-A mRNA, SP-B mRNA, and fatty acid synthetase activity and cause expansion of the ductal lumen by promoting fluid secretion. PGE_2 also stimulates PC release from type II cells $in\ vitro$. Treatment of fetal sheep with inhibitors of prostaglandin synthesis (e.g. indomethacin) decreases DPPC content of tracheal fluid, suggesting a possible physiological role for endogenous prostaglandin. Glucocorticoids inhibit both prostaglandin and cAMP production by lung tissue $in\ vitro$, perhaps by inducing phospholipid-binding proteins (lipocortins) that prevent phospholipase A_2 from initiating the arachidonic acid cascade toward prostaglandin production. However, the importance of this glucocorticoid effect $in\ vivo$ is uncertain since the dominant response to steroid is stimulation of lung maturation.

Methylxanthines, the most notable of which is aminophylline, exert their effects by inhibiting cAMP phosphodiesterase, the enzyme that normally degrades cAMP. In cultured lung, for example, treatment with isobutyl methylxanthine causes a doubling of endogenous cAMP and induces cAMP-responsive proteins. In a study in rabbits, aminophylline given $in\ vivo$ resulted in a modest increase in surfactant phospholipids and in lipid synthesis in general, but this may have been secondary to increased endogenous cortisol. Like other agents that increase cAMP, methylxanthines also promote secretion of surfactant.

Insulin and glucose; diabetes

Infants born to diabetic mothers have an increased risk of RDS, with estimates of as much as a fivefold increase; if the mother's glucose levels are tightly controlled, however, the risk can decrease to normal. Maternal hyperglycaemia leads to fetal hyperglycaemia and, in response, fetal hyperinsulinaemia. It has been proposed that the adverse effects on the fetus, including delayed lung maturation, are due to chronically elevated fetal insulin and/or glucose levels. Indirect effects are also possible, perhaps involving other hormones; decreased levels of prolactin and T_3 have been reported in blood of infants born to diabetic mothers.

In examining amniotic fluid from diabetic pregnancies, early studies found a delay in the increase in L/S ratio (for class A–C diabetes), but some more recent studies have found a normal developmental pattern. Furthermore, 'false mature' L/S ratios are a problem in diabetes, with some infants developing RDS at birth despite a ratio ≥ 2.0. Therefore, qualitative differences in surfactant have been examined as an explanation for the RDS risk. The appearance of phosphatidylglycerol is typically delayed in diabetic pregnancies, and lower levels of amniotic fluid SP-A have been demonstrated in some but not all studies.

Studies in animals have examined genetic models of diabetes and chemically induced diabetes. In fetal mice born to genetically diabetic mothers, there was no inhibition of choline incorporation into PC (in fact, it was greater than in controls), but there was decreased incorporation of glycerol into phosphatidylglycerol. In addition, morphological maturation of acini was delayed, suggesting that the diabetic state may also affect non-surfactant aspects of lung development. This should not be surprising considering the fact that infants of diabetic mothers have altered body growth in general (macrosomia, organomegaly) and a heightened incidence of many different congenital anomalies. Animal studies of induced diabetes have provided conflicting results, however, on pulmonary content of PC, choline incorporation into PC, and activities of various enzymes involved in phospholipid synthesis and glycogen breakdown. Interpretation of results has been clouded because of the severity of the diabetes induced by chemical methods (alloxan or streptozo-tocin) and high rates of fetal death. Other animal studies have used chronic fetal infusions of glucose or insulin to simulate the state of the infant of the diabetic mother. Glucose infusions into fetal sheep delayed the appearance of both PC and phosphatidylglycerol and prevented the simulatory effects of exogenous cortisol. In addition, the concentration of β-adrenoceptors was decreased, possibly explaining decreased surfactant flux.

There is evidence, based on binding studies of adult and fetal rat lung membranes and the presence of receptors, that the lung is a target tissue for insulin at least on adult type II cells. Unlike some adult tissues, the receptors in fetal lung may not be downregulated when insulin levels are high; thus, in a diabetic pregnancy the fetus may experience greater effects than would a hyperinsulin-aemic adult.

The mechanisms for the effects of diabetes on fetal lung development remain unclear. With indications that the quantity of surfactant PC may be normal, current research may focus increasingly on qualitative differences in sur-factant; these could be due to altered metabolism affecting availability of substrates (fatty acids, glucose, glycerol, myoinositol) or altered glycosylation of proteins such as SP-A or lung cell surface proteins. Indirect effects of insulin on other endogenous hormones such as cortisol, prolactin, and T_3 will also undoubtedly be explored more fully.

Epidermal growth factor

Epidermal growth factor (EGF) and members of its family (such as transforming growth factor alpha) are primarily known for promoting cellular proliferation and are among the growth factors that are likely to be important in fetal development. EGF is present in most cell types and body fluids of several mammalian species including humans. In fetal lung, EGF receptors have been demonstrated from early gestation (15 weeks through to 2 weeks postnatal in the human) and have been immunolocalized mainly to epithelium of the distal conducting airways and bronchial mucous glands. Similar studies demonstrated EGF protein in tracheal lining cells from 12–24 weeks and in tracheal submucosal glands and bronchial epithelium from 18–20 weeks onwards. These findings have suggested that EGF may participate in early airway development.

Treatment with EGF accelerates some aspects of lung maturation in fetal rats, rabbits, and sheep, but also causes hyperplasia of airway epithelial cells. Some *in vivo* EGF effects may be secondary, in that EGF alters levels of other circulating hormones, increasing cortisol and decreasing T_4. In cultured fetal rat lung, EGF increases precursor incorporation into all phospholipids, and in human explants EGF increases synthesis of SP-A.

It has been proposed that specific hormones might influence local EGF effects in the lung. In many body tissues, thyroid hormones stimulate synthesis of both EGF protein and its receptor. Dexamethasone has been found to either decrease or increase the synthesis of pulmonary EGF receptors in different experimental systems. Retinoic acid, a vitamin A derivative thought to be important in regulating embryonic differentiation of many tissues, increases synthesis of the EGF receptor in rat lung cells. EGF itself may affect the secretion of other hormones. For example, in fetal sheep, infusion of EGF stimulates release of ACTH; in other animal and cell culture models, EGF stimulates secretion of prolactin, cortisol, growth hor-mone, luteinizing hormone, placental hormones, and sex hormones, and inhibits secretion of T_4. However, as circu-lating levels of EGF are low and the effects of EGF are mostly paracrine and autocrine, the physiological significance of these latter effects is not clear.

Other hormones and factors

A number of other endogenous substances have possible roles in pulmonary epithelial differentiation, but their importance in this regard has yet to be established. These include, for example, transforming growth factor beta (from fibroblasts), retinoic acid, and fibroblast growth fac-tors. The complex interactions between epithelial and mes-enchymal cells, and between cells and matrix, remain a topic of great interest in understanding both early and late events in lung development.

Antioxidant defences

While an essential function of the lung is to facilitate uptake of oxygen from the air to the bloodstream, oxygen also has the potential to cause local cellular damage because of the production of highly reactive metabolites of oxygen, including superoxide radical (O_2^-), hydroxyl rad-ical (OH•), hydrogen peroxide (H_2O_2), singlet oxygen ($^1O_2$), and peroxide radical (ROO•). These metabolites are produced at low levels during normal metabolism, with about 1 per cent of all cellular O_2 contributing toward their production. Production increases markedly, however, under conditions of hyperoxia, and oxygen radicals are released locally in the lung by activated inflammatory cells that use them to kill phagocytized bacteria. Reactive oxygen metabolites may cause cell injury by damaging membranes by lipid peroxidation, inactivating enzymes by

Table 11.4 Proposed cellular defences against reactive oxygen metabolites

Metabolite	Scavenger	Reaction
Superoxide (O_2^-)	Superoxide dismutase	$O_2^- + O_2^- + 2H^+ \rightarrow O_2 + H_2O_2$
	β-carotene Ascorbate Glutathione	} Non-enzymatic free radical acceptors
Hydrogen peroxide (H_2O_2)	Catalase Glutathione peroxidase	} $2H_2O_2 \rightarrow O_2 + 2H_2O$
Peroxide radical (ROO•)	Glutathione peroxidase	$2ROO• + 2H^+ \rightarrow 2ROH + O_2$
	Vitamin E	Non-enzymatic free radical acceptor
Hydroxyl radical (OH•)	Vitamin E Glutathione	} Non-enzymatic free radical acceptors

oxidation of sulfhydryl groups, depolymerizing carbohydrates, and damaging DNA.

Defences against oxidant stress include three major enzymes that detoxify oxygen metabolites: glutathione peroxidase, catalase, and superoxide dismutase (Table 11.4). These enzymes are present in lung and also in other organs such as liver, kidney and heart. Non-enzymatic defences (intracellular and extracellular) include glutathione, vitamin A, α-tocopherol, ascorbic acid, cysteine, caeruloplasmin, uric acid, and haemoglobin. Protection from oxygen-induced lung injury has been consistently correlated with higher levels of antioxidant enzymes in the lung. For example, animals exposed to lethal levels of oxygen (95–100 per cent) die within 2–3 days, whereas those pre-treated with supplemental lung antioxidant enzymes can survive 2–3 days longer in the same hyperoxic environment. In addition, tolerance to 100 per cent oxygen in rats can be induced by prior exposure to 85 per cent oxygen, which causes an adaptive increase in levels of superoxide dismutase and catalase.

At birth, the infant undergoes an abrupt change from the hypoxaemic fetal environment (arterial PO_2 approximately 20 mmHg, and 'alveolar' PO_2 probably less) to the relatively hyperoxic external environment. Antioxidant defences would be especially important in the case of a premature infant who requires oxygen therapy for RDS. It has been suggested that the genesis of bronchopulmonary dysplasia (BPD), a chronic lung disease that occurs in about 20 per cent of infants with RDS, relates in part to toxic effects of oxygen metabolites on cells (pulmonary and endothelial) with subsequent damage secondary to the inflammatory response (for instance, due to prostaglandins or neutrophil elastase).

Development of the antioxidant enzyme system has been studied in the rat, rabbit, hamster, guinea-pig, and sheep. In general, the activity of all three enzymes increases by an average of 1.5- to 2-fold during the last 10–15 per cent of gestation, with some variation between species. Human fetal lung has measurable activities of all three enzymes during the second trimester; however, according to one recent study, only catalase shows higher levels of activity in late gestation. The lack of increases in the other two enzymes obviously raises questions about the significance of the antioxidant system in the transition to air breathing, but further studies are needed to verify and elucidate this issue in human lung.

Dexamethasone treatment of pregnant rats results in a precocious increase in antioxidant enzyme activities, as well as an increase in surfactant. Treatment with metopirone, an inhibitor of adrenal 11β-hydroxylase, to block endogenous glucocorticoid synthesis delays maturation of both the antioxidant and surfactant systems. Preliminary data from human fetal lung explant cultures, however, showed no effect of dexamethasone on antioxidant enzyme levels.

Thyroid hormone suppresses or does not affect antioxidant enzymes, in contrast to its stimulatory effect on lung PC synthesis. Rat fetuses exposed to high doses of T_3 had significantly decreased levels of all three enzymes, but elevated amounts of saturated PC and total phospholipid in pulmonary tissue. In experiments in which endogenous thyroid hormone production was blocked by administering either methimazole or propylthiouracil, the activities of antioxidant enzymes were elevated. In fetal lambs whose mothers were treated with TRH, there was no effect on antioxidant enzyme activity, whether TRH was administered alone or together with cortisol (which did induce catalase and superoxide dismutase).

In animals, the levels of antioxidant enzymes in the lung may be increased by direct administration of the enzymes encapsulated in liposomes and delivered either intravenously (although local levels in the lung have not been measured) or intratracheally. Pulmonary cells have been shown to take up the liposomes and increase enzyme levels both *in vivo* and *in vitro*, and pre-treatment of rats with superoxide dismutase- and catalase-containing liposomes enhances survival in hyperoxia. In one preliminary study of intravenous administration of superoxide dismutase to premature infants requiring oxygen therapy and mechanical ventilation, the severity of subsequent BPD was less than in controls. More clinical trials are likely, possibly including transfection of DNA encoding these enzymes administered via liposomes into the trachea, with

Table 11.5 Efficacy of glucocorticoids in preventing RDS in infants of ≤ 34 weeks gestation. Data were compiled by Crowley* from 12 studies involving a total of 3266 patients. Glucocorticoids significantly lower the incidence of RDS when given for 24 h 7 days before delivery

Treatment	Number of studies	Incidence of RDS in new-borns (%)	
		Treated	Control
24 h 7 days before delivery	12	8.9[†]	21.6
< 24 h	10	21.6	28.5
> 7 days	6	7.4	9.7

*Crowley, P. (1989). In *Effective care in pregnancy and childbirth* (ed. I. Chalmers, M. Enkin, and M.J.N.C. Keirse). pp. 746–64. Oxford University Press, Oxford.
[†] Odds ratio significantly better for treated than control.

the hope of inducing transient gene expression in airway cells. Other studies are currently assessing the usefulness of vitamin E and of diets rich in polyunsaturated fatty acids that could theoretically act as free-radical 'sinks'.

Therapeutic implications

The developing fetal lung is exposed to many different hormones, the levels of which change during pregnancy and in response to stressful situations. Important hormones for lung maturation include glucocorticoids, thyroid hormones, catecholamines, and possibly prolactin and sex hormones. These hormones modulate the timing of lung cellular differentiation, and, consequently, the ability of the lung of the prematurely born infant to support extrauterine life. Basic research has advanced our understanding of the process of lung development and has aided in devising therapeutic strategies and understanding their limitations.

Unfortunately, prematurity is a relatively common problem worldwide, occurring in approximately 4 per cent of pregnancies in the USA. The most significant complication of premature birth is RDS and its sequelae, and this disease occurs in approximately 50 per cent of babies born at 30 weeks of gestation.

Antenatal glucocorticoid therapy was introduced nearly 20 years ago and is now established as beneficial for premature infants (Table 11.5) This treatment is considered to mimic the effect of endogenous corticosteroids are thereby accelerate lung maturation. Effectiveness of therapy requires 24–48 h from administration to delivery (as it requires induction of protein synthesis), but it is often impossible to forestall delivery for this long. Furthermore, the benefit of a course of treatment does not seem to

extend beyond 7 days, presumably because of enzyme 'de-induction', necessitating a repeat course if delivery is further delayed but is still preterm. Responsiveness to glucocorticoids also depends on gestational age: treatment is most helpful at 26–34 weeks of gestation. In male babies and in twins, treatment may be less effective, although this remains controversial.

Even with optimal glucocorticoid therapy, RDS still occurs in some infants. This has prompted research on combined hormonal therapies that might act more rapidly or more effectively. In recent trials, treatment with glucocorticoid plus TRH reduced the severity of RDS and the incidence of subsequent chronic lung disease in premature infants. Ambroxol is a metabolite of bromhexine, an expectorant, and has been used primarily in Europe to induce lung maturation. It increases phospholipid synthesis in the lungs of experimental animals, especially choline incorporation into PC and sphingomyelin. However, studies of cultured rabbit and human fetal lung have not documented any direct stimulatory effect on surfactant production. Some clinical studies have found that ambroxol treatment reduces the incidence of RDS and improves amniotic fluid L/S ratio and the lowering of surface tension. However, treatment for 5–7 days is necessary, limiting the usefulness of the drug to preterm deliveries that can be successfully delayed.

Replacement lung surfactant is now routinely used in combination with glucocorticoids to prevent and treat RDS in new-borns. This combination makes sense because antenatal hormones, in addition to promoting surfactant production and secretion, enhance structural development of the lung. Structural immaturity of the lung will be a major limiting factor in extending premature survival to lower gestational ages, since, even if adequate surfactant is hormonally induced or exogenously supplied, the lungs must be adequately mature in terms of airway, alveolar, and vascular development to allow gas exchange to occur effectively. Additional therapies to treat or prevent respiratory disease in preterm infants may focus on prevention of coexisting problems such as asphyxia, maintenance of proper blood flow and fluid balance in the lung, augmenting immunological defences, and increasing local antioxidant defences.

Further reading

Avery, M.E. and Mead, J. (1959). Surface properties in relation to atelectasis and hyaline membrane disease. *American Journal of Diseases of Children*, **97**, 517–23.

Ballard, P.L. (1986). *Hormones and lung maturation*, Monographs on Endocrinology, Vol. 28. Springer-Verlag, Berlin.

Ballard, P.L. (1989). Hormonal regulation of pulmonary surfactant. *Endocrine Reviews*, **10**, 165–81.

Benson, B., Hawgood, S., Schilling, T., Clements, J., Damm, D., Cordell, B., and White, R.T. (1985). Structure of canine pulmonary surfactant apoprotein: cDNA and complete amino acid sequence. *Proceedings of the National Academy of Sciences of the United States of America*, **82**, 6379–83.

Clements, J.A. (1957). Surface tension of lung extracts. *Proceedings of the Society for Experimental Biology and Medicine*, **95**, 170–2.

Dobbs, L.G., Mason, R.J., Williams, M.C., Benson, B.J., and Sueishi, K. (1982). Secretion of surfactant by primary cultures of alveolar type II cells isolated from rats. *Biochimica et Biophysica Acta*, **713**, 118–27.

Frank, L. and Sosenko, I.R.S. (1987). Prenatal development of lung antioxidant enzymes in four species. *Journal of Pediatrics*, **110**, 106–10.

Gluck, L. and Kulovich, M.V. (1973). Lecithin/sphingomyelin ratios in amniotic fluid in normal and abnormal pregnancy. *American Journal of Obstetrics and Gynecology*, **115**, 539–46.

Gonzales, L.W. and Ballard, P.L. (1989). Hormones and their receptors. In *Lung biology in health and disease*. Vol. 41. *Lung cell biology* (ed. D. Massaro), pp. 539–90. Marcel Dekker, New York.

Gross, I. and Wilson, C.M. (1982). Fetal lung in organ culture. IV. Supra-additive hormone interactions. *Journal of Applied Physiology*, **52**, 1420–5.

Hitchcock, K.R. (1979). Hormones and the lung. I. Thyroid hormones and glucocorticoids in lung development. *Anatomical Record*, **194**, 15–40.

Kikkawa, Y., Kaibara, M., Motoyama, E.K., Orzalesi, M.M., and Cook, C.D. (1971). Morphologic development of fetal rabbit lung and its acceleration with cortisol. *American Journal of Pathology*, **64**, 423–42.

King, R.J., Ruch, J., Gikas, E.G., Platzker, A.C.G., and Creasy, R.K. (1975). Appearance of apoproteins of pulmonary surfactant in human amniotic fluid. *Journal of Applied Physiology*, **39**, 735–41.

Kitterman, J.A., Liggins, G.C., Campos, G.A., Clements, J.A., Forster, C.S., Lee, C.H., and Creasy, R.K. (1981). Prepartum maturation of the lung in fetal sheep: relation to cortisol. *Journal of Applied Physiology*, **51**, 384–90.

Liggins, G.C. (1969). Premature delivery of foetal lambs infused with glucocorticoids. *Journal of Endocrinology*, **45**, 515–23.

Liggins, G.C. and Howie, R.N. (1972). A controlled trial of antepartum glucocorticoid treatment for prevention of the respiratory distress syndrome in premature infants. *Pediatrics*, **50**, 515–25.

McElroy, M.C., Postle, A.D., and Kelly, F.J. (1992). Catalase, superoxide dismutase and glutathione peroxidase activities of lung and liver during human development. *Biochimica et Biophysica Acta*, **1117**, 153–8.

Mendelson, C.R., Johnston, J.M., MacDonald, P.C., and Snyder, J.M. (1981). Multihormonal regulation of surfactant synthesis by human fetal lungs *in vitro*. *Journal of Clinical Endocrinology and Metabolism*, **53**, 307–17.

Meyrick, B. and Reid, L. (1977). Ultrastructure of alveolar lining and its development. In *Development of the lung* (ed. W.A. Hodson), pp. 135-214. Marcel Dekker, New York.

Rooney, S.A., Gobran, L., Gross, I., Wai-Lee, T.S., Nardone, L.L., and Motoyama, E.K. (1976). Studies on pulmonary surfactant. Effects of cortisol administration to fetal rabbits on lung phospholipid content, composition and biosynthesis. *Biochimica et Biophysica Acta*, **450**, 121–30.

Schellenberg, J.C., Liggins, G.C., Manzai, M., Kitterman, J.A., and Lee, C.C. (1988). Synergistic hormonal effects on lung maturation in fetal sheep. *Journal of Applied Physiology*, **65**, 94–100.

Snyder, J.M., Rodgers, H.F., O'Brien, J.A., Mahli, N., Magliato, S.A., and Durham, P.L. (1992). Glucocorticoid effects on rabbit fetal lung maturation *in vivo*: an unusual morphometric study. *Anatomical Record*, **232**, 133–40.

12. Assessment of fetal health

Alan D. Bocking and Robert Gagnon

With the development of ultrasound technology, our understanding of the normal development and physiology of the human fetus has dramatically improved over the past 15 years. The recognition of fetal rest–activity patterns with simultaneous quantification of gross fetal body movements (GBM), fetal breathing movements (FBM), fetal heart rate (FHR) variability, and FHR accelerations have provided the physiological basis of currently used antenatal fetal assessment. This chapter will deal exclusively with the antenatal assessment of fetal health and readers are referred to other textbooks of obstetrics and gynaecology or maternal–fetal medicine for reviews of intrapartum fetal monitoring.

Fetal breathing movements

As discussed in Chapter 10, FBM occur episodically and are present approximately 30 per cent of the time during the last 10 weeks of human pregnancy. When utilizing FBM as an indicator of fetal health it is important to remember that, as a result of normal behavioural patterns, healthy human fetuses may not demonstrate FBM for as long as 120 min at a time. The incidence of FBM in the 30–40 week fetus is strongly affected by maternal meals and particularly glucose ingestion, whereas there does not appear to be the same strong relationship between maternal meals and FBM in the younger human fetus. The time of day, meals, gestational age, alcohol, drugs, carbon dioxide levels, and labour all have an effect on FBM in the normal fetus.

FBM occur during times of low-voltage electrocortical activity in the late-gestation fetal sheep, and are generally associated with rapid eye movements (Fig. 12.1). The inhibitory effect of hypoxia on FBM in the chronically catheterized sheep fetus has led to the incorporation of FBM into the biophysical profile score, which will be discussed subsequently. It is of interest that acute hypoxaemia secondary to maternal hypoxaemia in fetal sheep of less than 114 days of gestation gives rise to only a 50 per cent reduction in the incidence of FBM and, when uterine blood flow is restricted in pregnant sheep between 110 and 120 days of gestation (term ~ 145 days) to produce fetal hypoxaemia for 2 h, the inhibition of FBM is present for only the first hour. These observations suggest that the younger sheep fetus is less responsive behaviourally to

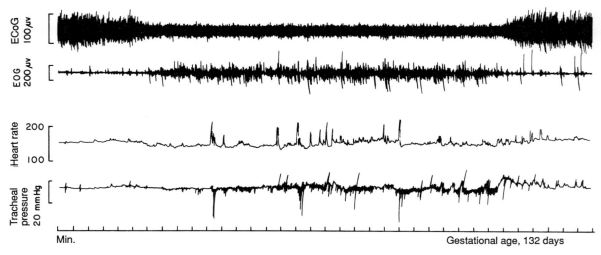

Fig. 12.1 Electrocortical activity (ECoG), electroocular activity (EOG), heart rate, and fetal breathing movements, as indicated by tracheal pressure fluctuations, in a sheep fetus at 132 days of gestation.

acute hypoxaemia when compared to the more mature fetus. Recently, it has been shown by a number of investigators that, when fetal hypoxaemia is maintained in pregnant sheep for more than a few hours in the absence of progressive acidaemia, either by restricting uterine blood flow or inducing maternal hypoxaemia, the incidence of FBM returns to normal after 12–16 h. The importance of these observations as regards the use of FBM in assessing fetal health in the human remains to be determined, and it must be recognized that prolonged fetal hypoxaemia in association with progressive metabolic acidaemia results in the sustained inhibition of FBM.

Fetal gross body movements

GBM in the human fetus can be visualized using real-time ultrasound and are found to occur approximately 10 per cent of the time during the last 10 weeks of pregnancy. GBM represent an important physiological activity that is used in determining behavioural states in the human fetus as discussed in Chapter 24. In contrast to FBM, the incidence of GBM in the human fetus is not altered by maternal meals or glucose concentrations.

In both the sheep and human fetus, accelerations in FHR and fetal body movements are closely related. In the term human fetus, 85 per cent of all accelerations in FHR are associated with GBM and 92 per cent of all GBM are associated with FHR accelerations. It is this close coupling of GBM and FHR accelerations that provides the physiological basis for the non-stress test in clinical practice to assess fetal health, and its incorporation into the biophysical profile score.

Animal studies have demonstrated that acute fetal hypoxia secondary to either maternal hypoxia or restriction of uterine blood flow leads to a reduction in skeletal muscle activity as measured by using either electromyographic electrodes or ultrasound crystals to record forelimb movements. The effect of more prolonged hypoxaemia on fetal movements in experimental animals remains unknown.

Fetal heart rate

As previously mentioned, the close relationship between fetal body movements and accelerations in FHR is well established. During the antenatal period, potential fetal compromise may be identified by a prolonged bradycardia, absence of accelerations, and/or the presence of variable decelerations in the FHR. A decrease in mean FHR may occur in a variety of clinical situations such as intermittent umbilical cord compression, acute hypoxaemia secondary to uterine contractions with placental insufficiency, and/or reductions in uterine blood flow. These acute decreases in FHR are generally reflex responses to hypertension and/or hypoxaemia with activation of baroreceptors and/or chemoreceptors. In the younger sheep fetus, acute hypoxaemia results in either no change or an increase in mean FHR. More prolonged fetal hypoxaemia in pregnant sheep beyond 120 days of gestation does result in an initial bradycardia but this is followed by a tachycardia. There is no significant change in either the number or pattern of FHR accelerations in fetal sheep exposed to prolonged hypoxaemia caused by a restriction of uterine blood flow. Acute hypoxaemia in fetal sheep gives rise to an increase in short-term variability with a subsequent decrease in heart rate variability only when the fetal arterial pH approaches approximately 6.9. Clearly, a decrease in heart rate variability is a sign of advanced compromise in fetal sheep.

Fetal heart rate variability

Patterns of FHR accelerations in healthy fetuses

Until recently, recording of fetal heart rate tracings obtained with Doppler ultrasound was associated with an extremely high signal loss, approaching 60 per cent at 26 to 28 weeks of gestational age. With the advent of advanced FHR monitors employing an autocorrelation technique, signal loss has been markedly reduced in early gestation. Using a microprocessor, it is also now possible to fit, on-line, a baseline to the heart rate record and to recognize automatically FHR accelerations of a predefined amplitude and duration. This technique allows a more objective and reproducible assessment of FHR tracings.

There are a number of important maturational changes in FHR patterns that occur during a critical and narrow period between 28 and 30 weeks of gestation. These may be characterized by: (1) a decrease in baseline FHR of 5 beats/min; (2) an increase in the amplitude of FHR accelerations of 4 beats/min; and (3) an increase in long-term FHR variability. If one measures the interval necessary to detect ≥ 2 FHR accelerations of ≥ 15 beats/min for ≥ 15 s, which is the definition commonly used for a reactive non-stress test, 65 per cent of tracings are non-reactive at 60 min at 26 to 28 weeks of gestation. Most of the FHR accelerations in the young gestational age group (26–28 weeks) are < 15 beats/min and cannot be detected with an increase in the observation time. Using a minimal amplitude of 10 beats/min for ≥ 15 s above the baseline to define an acceleration at 26 to 28 weeks, however, the length of time from the beginning of recordings until two FHR accelerations occur is similar to that at 30 to 40 weeks. Therefore, interpretation of non-stress testing (NST) should take into account gestational age and maturation of the autonomic control of the fetal heart in healthy human fetuses.

Patterns of FHR accelerations in the growth-restricted fetus

Using computerized analysis of human FHR tracings, it has been shown that long-term FHR variability is decreased in growth-restricted fetuses where there is clinical evidence of associated maternal pathology, such as hypertension and proteinuria. It is important that, although the FHR variation is decreased with intrauterine growth restriction, it is usually within the normal range. Only extremely low FHR variability is associated with metabolic acidaemia.

The distribution of FHR accelerations has recently been analysed in 24 growth-restricted fetuses from 30 weeks to term to determine if current criteria used for a reactive non-stress test (FHR accelerations of ≥ 15 beats/min for ≥ 15 s) were applicable. Each fetus with intrauterine growth restriction (IUGR) was matched for gestational age with an appropriately grown (AGA) fetus. Umbilical artery blood gases and pH obtained at birth demonstrated that IUGR fetuses were mildly hypoxaemic (average PO_2 was 3 mmHg less than AGA fetuses) but not acidaemic. The results indicated that the distribution of FHR acceleration amplitudes in IUGR fetuses from 30 to 40 weeks was characterized by: (1) a larger proportion of small amplitude (< 10 beats/min) FHR accelerations; (2) a smaller proportion of large amplitude (> 20 beats/min) FHR accelerations than in AGA fetuses; and (3) a 50 per cent decrease in the number of FHR accelerations per hour. Moreover, a non-stress test was only feasible using 10 beats/min to define the minimum amplitude of an acceleration in IUGR fetuses. Using ≥ 15 beats/min for ≥ 15 s as a definition for an acceleration, there was an unacceptably high rate of falsely non-reactive tests in IUGR fetuses (24 per cent after 60 min of observation). There were no differences in the baseline FHR and the mean number of FHR decelerations between IUGR and AGA fetuses.

Clinical significance

A reduction in FHR accelerations and FHR variability is often associated with fetal hypoxaemia and evidence of nutritional deprivation. Since the introduction of antepartum FHR testing, a decline in the overall stillbirth rate from 10.5 per 1000 in 1971–75 to 4.4 per 1000 in 1981–85

has been reported. The corrected fetal death rate following a normal test is approximately 1.9 per 1000.

Many different protocols of antepartum FHR testing have been described. In our institution, a reactive NST consists of the occurrence of ≥ 2 FHR accelerations of ≥ 15 beats/min for ≥ 15 s in a 20-min period. The test is continued up to 120 min if necessary until criteria of FHR reactivity are met. If the NST is non-reactive for 2 h, an ultrasound examination is carried out to look for fetal structural abnormalities and for amniotic fluid volume assessment. In the absence of maternal drug ingestion, which may affect fetal heart rate control, and major structural malformations, immediate delivery would then be carried out as long as fetal viability is reasonably assured on the basis of gestational age.

Using computerized FHR analysis, a decrease in the mean minute range (a computer-derived index of long-term FHR variability) of less than 20 ms is associated with a high risk of fetal hypoxaemia and/or acidosis and has been suggested as a threshold to proceed to delivery for fetal compromise especially in the presence of intrauterine growth restriction.

Biophysical profile score

Observations in fetal sheep along with observations in the human fetus have led to the development of the biophysical profile score (BPS) to assess fetal health. The biophysical profile is a survey of five discrete biophysical variables including fetal body tone, fetal breathing movements, gross fetal body movements, amniotic fluid volume, and non-stress test, with values assigned as 0 or 2 if the variable studied is abnormal or normal respectively. The progressive decrease in amniotic fluid volume in high-risk pregnancies has been consistently associated with poor fetal outcome. However, it is not clear yet by what mechanism(s) chronic fetal hypoxia may lead to a decrease in amniotic fluid volume.

Technique and interpretation

Using real-time ultrasound, the largest vertical diameter of pockets of amniotic fluid is determined. The presence of FBM (> 30 s duration) and GBM (> 3 body or limb movements) is then coded. Fetal limb tone is defined as one episode of active extension with return to flexion of a fetal limb. Opening and closing of the hand is also considered normal tone. Each variable is then assigned a score of 2 (normal) or 0 (abnormal). A composite score of 8 is considered normal. If the BPS is less than 8, a non-stress test is then performed and, if reactive (≥ 2 FHR accelerations of > 15 beats/min and of at least 15-s duration in 30 min), a score of 2 for non-stress test is assigned. If oligohydramnios is documented (largest pocket < 2 cm) and renal agenesis and premature rupture of membranes have been ruled out, delivery is indicated.

Clinical significance and limitations

Originally, the same power was given for each variable within the BPS to predict outcome. However, more recent analysis indicates that only the non-stress test, fetal breathing movements, and amniotic fluid volume are essential to predict abnormal outcome. Also, no pattern of sequential loss of the different biophysical variables is seen prior to fetal compromise. It is important to note that the interval between an abnormal test and fetal death can be as short as 30 min or as long as 11 days. For these reasons, it is not yet clear what is the best frequency at which to perform biophysical profiles.

Using the rationale that two episodes of gross fetal body movements would normally have been predicted during a 2-h interval, the concept of prolongation of observation time for non-stress testing up to 120 min to define an abnormal test has been developed. Using this protocol, the false positive tests were considerably reduced with a positive predictive value for perinatal outcome of 86 per cent and a negative predictive value of approximately 99 per cent. At 30 min of observation (which is the observation time for BPS), 12 per cent of healthy fetuses had a non-reactive FHR pattern due to episodes of fetal quiescence which normally alternate with periods of fetal activity. Therefore, a significant number of fetuses with equivocal biophysical profiles (BPS = 6 or 4) also need prolongation of observation time or to be followed by non-stress testing.

At present, antepartum fetal assessment tests are based on the observation of normal fetal behaviour. The methods are time-consuming, need expensive equipment, are labour-intensive, and are subject to error due to periodicities in fetal activity. For these reasons, there has been a renewed interest in the use of external stimulation such as sound and vibration to increase fetal body movements and FHR reactivity.

Vibroacoustic stimulation

Fetal auditory development

It is difficult to estimate what the human fetus exactly hears *in utero*, because both the external and middle ear are filled with amniotic fluid. In an extrauterine organism, such a status would result in a hearing threshold increase of 30 to 40 dB. In the fetal sheep, estimates of fetal sound isolation have been made through recordings of the cochlear microphonic, which is a small electrical potential generated at the level of the hair cell epithelium in the inner ear. The difference between the sound pressure necessary to produce equal electrical outputs from the cochlea of the fetus and the new-born was used as an estimate of fetal sound isolation. The data indicate that fetal sound isolation averages 40–45 dB at frequencies of 500 to 2000 Hz, and 15–20 dB at lower frequencies.

The auditory evoked brainstem potential (AEBP) is a well-established index for obtaining an assessment of the auditory pathways of the brainstem and midbrain. The AEBP in fetal sheep consists of five well-defined wave deflections occurring within the first 7 ms following a click stimulus. A fetal AEBP is present at day 115–117 of gestation. With advancing gestational age there is a progressive decrease in each latency value. The higher the stimulus intensity, the more rapid is the neurological transmission of the auditory signal. Using extrauterine sound as a stimulus, an intensity above 100 dB is necessary to obtain fetal AEBPs.

Glucose utilization in the central auditory structures of fetal sheep is higher than in other cerebral structures under normal laboratory sound conditions. Bilateral cochlear ablation results in a global decrease in local cerebral glucose utilization of both grey and white matter of all auditory and non–auditory pathways. These observations suggest that the activity of auditory pathways may play a role in the normal function and development of the fetal brain.

Fetal heart rate

Many different sources of sound and vibration have been used externally to increase FHR reactivity. Vibroacoustic stimulation (VAS) using an electronic artificial larynx (EAL) is now widely used in the assessment of fetal health. This device produces a broad-band noise at a fundamental frequency of 87 Hz, with multiple harmonics up to 15000 Hz. The surface of the instrument also vibrates at all frequencies between 87 and 15000 Hz, and maximum vibration occurs at 450 Hz.

The FHR response to vibroacoustic stimulation of 5-s duration is dependent on gestational age, in that after 30 weeks of gestation there is a prolonged and significant increase in the basal FHR. Near term, the increase in FHR is prolonged for up to 1 h. Between 26 and 30 weeks, the FHR response to stimulation consists of a single prolonged FHR acceleration. Conversely, in fetuses of more than 33 weeks, there is an increase in the number of FHR accelerations between 10 and 20 min following the stimulus.

Fetal body movements

In fetuses between 33 and 40 weeks of gestational age there is an increase in gross fetal body movements that begins 10 min following a 5-s vibroacoustic stimulus and persists for up to 1 h. However, in fetuses between 26 and 32 weeks, gross fetal body movements are not altered by this stimulus.

Fetal behavioural state

Behavioural states were determined 30 min before and after a 3-s stimulus with EAL in fetuses near term. Twenty-six of 28 fetuses exhibited FHR accelerations immediately following the stimulus. The four fetuses in state 1F (quiet sleep state) before the stimulus switched to state 4F (wakefulness), and 11 of 16 in state 2F (active sleep state)

switched to state 4F (wakefulness). These observations demonstrated definitive alterations in human fetal behaviour following stimulation with the EAL (Fig. 12.2).

Clinical significance

With the observations that external VAS produces remarkable changes in FHR and fetal movement patterns, FHR responses to VAS have become a widely investigated indicator of fetal health. Currently available data using VAS for antenatal fetal surveillance suggest that vibroacoustic stimulation tests, like non-stress testing and the biophysical profile, indicate fetal health at the time of the test only and cannot predict events associated with fetal distress during labour nor can they predict a low Apgar score. In addition, the risk of intrauterine fetal death within 7 days of a normal test is 1.9 per 1000 which is no different than

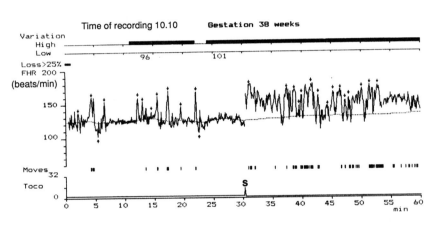

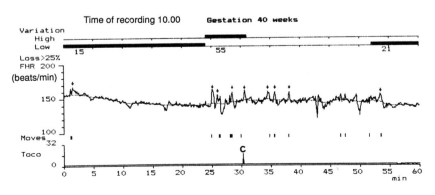

Fig. 12.2 A typical example of the FHR and body movement responses to vibroacoustic stimulation (S; top) in a healthy full-term fetus, indicating a transition from fetal behavioural state 2F (active sleep state) to 4F (wakefulness). An example of the spontaneous transition from state 1F (quiet sleep state) to state 2F occurring at 25 minutes of observation (bottom). 'Toco' (tocodynamometer) refers to the presence or absence of uterine activity (not recorded). (C indicates a control 'stimulus').

that following a spontaneously reactive non-stress test (1.6 per 1000).

Doppler ultrasound blood velocity waveforms

Technique

The flow velocity waveform (FVW) represents the Doppler-shifted signal obtained from the vessel being insonated. A strong correlation has previously been established between abnormal umbilical artery flow velocity waveforms and intrauterine growth restriction. The importance and reproducibility of these measurements have been previously confirmed. Different indices of downstream vascular resistance obtained from the FVWs, which are angle-independent, have been developed including the S/D ratio (peak-systolic flow velocity/end-diastolic flow velocity), the resistance index ([peak-systolic flow velocity minus end-diastolic flow velocity]/peak-systolic flow velocity) and the pulsatility index ([peak-systolic flow velocity minus end-diastolic flow velocity]/mean velocity). Different fetal blood vessels can be assessed including the umbilical artery, the uterine artery, and the internal carotid artery. Figure 12.3 shows a typical normal umbilical artery FVW obtained from a healthy term fetus and from the maternal uterine artery. In the normal state,

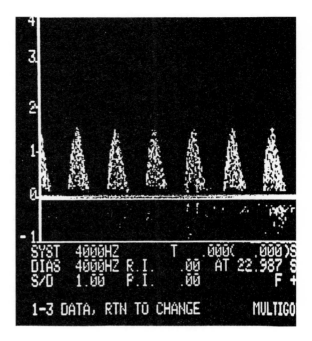

Fig. 12.4 Typical example of flow velocity waveforms obtained in the umbilical artery of a growth-restricted fetus. There was absent end-diastolic flow velocity indicating high placental vascular resistance.

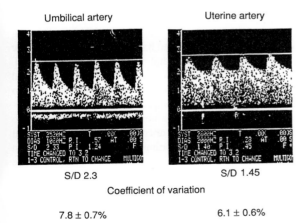

Umbilical artery	Uterine artery
S/D 2.3	S/D 1.45

Coefficient of variation

7.8 ± 0.7%   6.1 ± 0.6%

Fig. 12.3 Typical example of flow velocity waveforms obtained in the umbilical artery (left) and the maternal uterine artery (right). The peak-systolic/end-diastolic ratio is lower (lower vascular resistance) in the uterine artery than the umbilical artery in normal pregnancy. Coefficients of variation (± SEM) are shown.

both vessels have high flow velocity throughout diastole indicating low placental vascular resistance seen in normal pregnancies. In contrast, Fig. 12.4 shows typical umbilical artery FVWs obtained in a severely growth-restricted fetus. There is no end-diastolic flow velocity indicating high placental vascular resistance. The abnormal umbilical artery FVW pattern seen in pregnancies complicated by intrauterine growth restriction is usually associated with a reduction in the number of arterioles in the tertiary stem villi. Studies using cordocentesis to directly measure fetal arterial gases have shown that the growth-restricted fetus with absent end-diastolic velocity is more likely to be hypoxic (85 per cent) or acidotic (50 per cent) than those with normal waveforms.

The fetal internal carotid artery generally has a higher downstream vascular resistance and lower diastolic flow velocity than the fetal placental circulation. There is, however, a lower vascular resistance in this artery in hypoxaemic growth-restricted fetuses due to fetal cerebral vasodilatation which is a well-established fetal adaptive response to hypoxia. Therefore, a complete mapping of the fetal circulation using Doppler technology could be useful to identify a group of fetuses already hypoxaemic and at risk of intrauterine death.

Clinical significance

In prospective trials of high-risk patients, randomized either into a control group without the use of Doppler or into a group in which the umbilical artery Doppler analysis was available to the clinician, there were fewer Caesarean sections performed for fetal distress and fewer cases of fetal distress in labour when the Doppler analysis was made available. There was also a trend towards improved neonatal outcome in the study group (Doppler available), but no individual measures of outcome have achieved statistical significance. Other studies have suggested a decrease in both the incidence of perinatal death and the mean length of antenatal hospitalization when the Doppler results were revealed as opposed to being concealed to clinicians managing the patients' care. Despite these studies, the precise role of Doppler velocimetry in clinical practice is not clearly established yet. However, it could be used as an indicator for increased fetal surveillance, but not as an indication for delivery. Any other intervention or management practice should be made only when conventional test results, such as non-stress testing and biophysical profile, become abnormal.

Case presentations

Numerous studies have demonstrated considerable variability in the visual assessment of antepartum FHR tracings, despite the fact that it is the most widely used method of assessing fetal health. As a result, many investigators have tried to develop new objective methods to analyse antepartum FHR tracings. Recently, it has been shown that expert observers fail to detect approximately 90 per cent of FHR decelerations ($\geq$ 60 s) detected by computerized FHR analysis. In addition, there is a high level of inaccuracy in the visual detection of FHR accelerations. As a result, there is considerable disagreement amongst the same observer and different observers in answering simple, judgement-related questions such as, 'Are you concerned?', or 'Would you advise to continue or stop recording?' As a result we have recently introduced computerized antepartum FHR analysis. This system allows on-line analysis of signal loss, detection of FHR accelerations and decelerations, and the detection of episodes of low FHR variability and high FHR variability that are usually not detectable by visual assessment. The following examples illustrate abnormal FHR tracings and their underlying pathophysiology.

Case 1 (Fig. 12.5)

An 18-year-old woman was referred at 26 weeks of gestation because of intrauterine growth restriction. Initial assessment indicated symmetrical IUGR and absent end-diastolic flow velocity in the umbilical artery, indicative of placental insufficiency. Initial non-stress testing indicated fluctuations between low and high FHR variability and low to normal overall FHR variation. Amniotic fluid volume and biophysical profile remained normal. Fetal movement count maintained by the mother remained normal. At 29 weeks a repeat ultrasound scan indicated no fetal growth over the previous 2.5 weeks, a chest circumference that was 3 standard deviations below that expected for gestational age, compatible with hypoplastic lungs, and an estimated fetal weight of 710 g (50th percentile for 24.5 weeks). A repeat of the non-stress test indicated sudden deterioration of the fetal condition, with the appearance of a flat decelerative FHR tracing (Fig. 12.5). Because of the hypoplastic lungs in the fetus a Caesarean section was not performed and the following day intrauterine fetal death occurred; the autopsy confirmed severely hypoplastic lungs (combined lung weight: 10 g).

Case 2 (Fig. 12.6)

A 29-year-old woman was seen at 34 weeks of gestation with gestational diabetes. During a non-stress test there were no fluctuations between episodes of low and high FHR variability. Umbilical artery flow velocity waveforms were normal. The overall FHR variation was 19.4 ms which was abnormal. The fetal echocardiogram suggested hyperdynamic myocardial contractility. Non-stress testing was continued and remained flat for another 60 min. There was no evidence of fetal anomaly on ultrasound. Biophysical profile indicated normal amniotic fluid volume and the presence of fetal body movements, but no fetal breathing activity. The patient was delivered by elective Caesarean section of a normally grown neonate who was hypotonic, but whose arterial cord gases were normal at birth. Further investigation indicated a haemoglobin concentration of 79 g/L, confirmation of the total absence of respiratory drive, and evidence of brain death on the encephalogram. Karyotype was normal. Placental examination and viral serology were compatible with intrauterine viral infection with parvovirus B-19. Ventilator support was discontinued at 8 days of life due to severe central nervous system dysfunction.

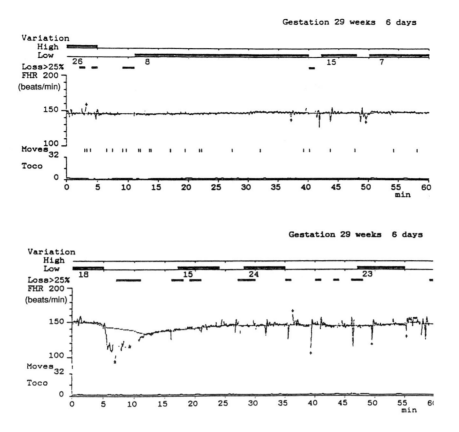

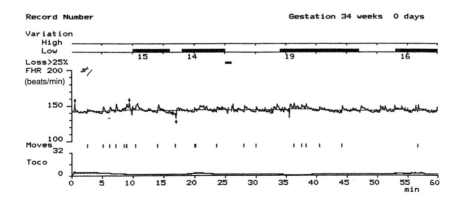

Fig. 12.5 A typical example of a pre-terminal fetal heart rate tracing (two consecutive hours) secondary to placental insufficiency. See text for clinical description. 'Toco' indicates uterine activity; 'moves' indicates body movements.

Fig. 12.6 A 1 h tracing of a flat fetal heart rate secondary to severe central nervous system dysfunction. See text for a clinical description. 'Toco' indicates uterine activity; 'moves' indicates body movements.

Percutaneous umbilical blood sampling

Percutaneous umbilical blood sampling (PUBS) or cordocentesis has recently been introduced to clinical obstetrics as a method of performing a biochemical, metabolic, or haematological evaluation of the human fetus. PUBS is an invasive technique requiring the aseptic placement of a fine needle (usually 22-gauge) under direct ultrasound visualization into the umbilical vein, generally near its placental insertion. The primary indications for this procedure are the need for rapid karyotyping in fetuses with severe intrauterine growth restriction and/or major structural malformations known to be associated with chromosomal abnormalities. PUBS has also been a very useful addition to the management of immune haemolytic disease whereby direct assessment of the fetal haematocrit and blood group and type can be made. This information is then utilized in developing a plan of management that may involve direct intravascular transfusion of severely affected fetuses.

It has also been suggested that PUBS may be of use in human pregnancy, to assess fetal oxygenation and acid–base status. There is only preliminary evidence in the literature regarding the relationship between fetal blood gas measurements and biophysical assessments. The major concern with these studies is that PUBS generally involves the aspiration of blood from the umbilical vein, whereas the clinician is primarily interested in the status of the umbilical artery unless severe acidaemia is present. It is unlikely that the metabolic information gained from PUBS in pregnancies complicated by IUGR would justify the additional risk posed to the fetus, particularly in view of the readily available non-invasive methods of assessing fetal health discussed previously. In selected cases, however, additional karyotypic, biochemical, and haematological information may be extremely useful in determining the aetiology of severe IUGR and may assist in subsequent management decisions.

Hormonal and placental function measurements

Following the observation that the urine of pregnant women contains large amounts of oestrogens originating in the placenta, measurements of the urinary excretion of metabolites of these hormones have been performed in an attempt to provide an index of 'placental function'. The placenta has the enzymatic capacity to convert the precursors of oestriol which are produced by the fetal adrenal glands. This process primarily involves the conversion of 16α-hydroxydehydroisoandrosterone sulfate into oestriol. Unfortunately, the range of variation in the amount of urinary oestriol in human pregnancy is too great to be clinically useful and has been abandoned as a method for the assessment of fetal health.

Other hormonal measurements performed on maternal serum, which have been performed in order to assess placental function, include placental lactogen (HPL) and pregnancy specific β-glycoprotein (SP$_1$). However, with the greater potential for antepartum biophysical testing, biochemical tests are no longer used, mainly due to the cost, the time required for obtaining the results (12–24 h), the lack of predictive value for fetal outcome, and the large variability in the normal population.

Summary and future directions

Currently, non-invasive methods, such as antepartum fetal heart rate testing (with or without acoustic stimulation), ultrasonographic assessment of fetal behavioural activity and amniotic fluid volume, and ultrasound Doppler measurements, are being used as primary methods to determine the need for obstetrical intervention and premature delivery of the fetus in high-risk pregnancies. These methods are labour-intensive and time-consuming, but so far have provided the best predictive value for outcome. Umbilical artery flow velocity waveforms are being used to assess the degree of placental insufficiency and, if abnormal, can identify a fetus at higher risk and the requirement for intensive surveillance. The decision to deliver a premature fetus solely because of absent or reversed end-diastolic flow velocity without other biophysical signs of imminent fetal death is still controversial and not yet universally recommended. Further research is required to determine the relationship between abnormal waveforms and direct measurements of umbilical blood flow, and the degree of fetal oxygenation under physiological conditions using animal models. In addition, a better understanding of the mechanisms leading to placental insufficiency would be required to determine the predictive value of the various methods of assessment of fetal health and the timing of delivery. Studies are ongoing regarding the value of fetal behavioural state assessment and specific movement patterns in determining fetal health. Ideally, in many situations, one wishes to perform a 'neurological' examination of the fetus. This is becoming increasingly possible with improvements in the resolution of ultrasound equipment and our knowledge of behavioural state and fetal activity patterns, of both the healthy and the 'ill' fetus.

Further reading

Abrams, R.M., Hutchison, A.A., McTiernan, M.J., and Merwin, G.E. (1987). Effects of cochlear ablation on local cerebral glucose utilization in fetal sheep. *American Journal of Obstetrics and Gynecology*, **157**, 1438–42.

Abrams, R.M., Ito, M., Frisinger, J.E., Patlac, C.S., Pettigrew, K.S., and Kennedy, C. (1984). Local cerebral glucose utilization in fetal and neonatal sheep. *American Journal of Physiology*, **246**, R608–R618.

Bocking, A.D. (1992). Fetal behavioural states: pathological alteration with hypoxia. *Seminars in Perinatology*, **16**, 252–7.

Bocking, A.D. and Gagnon, R. (1991). Behavioural assessment of fetal health. *Journal of Developmental Physiology*, **15**, 113–20.

Bocking, A.D., Gagnon, R., Milne, K.M., and White, S.E. (1988). Behavioral activity during prolonged hypoxaemia in fetal sheep. *Journal of Applied Physiology*, **65**, 2420–6.

Brown, R. and Patrick, J. (1981). The nonstress test: how long is enough? *American Journal of Obstetrics and Gynecology*, **141**, 646–51.

Dawes, G.S., Redman, C.W.G., and Smith, J.H. (1985). Improvements in the registration and analysis of fetal heart rate records at the bedside. *British Journal of Obstetrics and Gynaecology*, **92**, 317–25.

Devoe, L.D., Murray, C., Faircloth, D., and Ramos, E. (1990). Vibroacoustic stimulation and fetal behavioral state in normal term human pregnancy. *American Journal of Obstetrics and Gynecology*, **163**, 1156–61.

Gagnon, R. (1989). Stimulation of human fetuses with sound and vibration. *Seminars in Perinatology*, **13**, 393–402.

Gagnon, R. and Patrick, J. (1990). Vibroacoustic stimulation as a test for fetal health. *Fetal Medicine Review*, **2**, 159–70.

Gagnon, R., Hunse, C., Carmichael, L, Fellows, F., and Patrick, J. (1986). Effects of vibratory acoustic stimulation on human fetal breathing and gross fetal body movements near term. *American Journal of Obstetrics and Gynecology*, **155**, 1227–30.

Gagnon, R., Hunse, C., Carmichael, L., and Patrick, J. (1989). Vibratory acoustic stimulation in the 26- to 32-week, small-for-gestational-age fetus. *American Journal of Obstetrics and Gynecology*, **160**, 160–5.

Gagnon, R., Hunse, C., Fellows, F., Carmicheal, L., and Patrick, J. (1988). Fetal heart rate and activity patterns in growth-retarded fetuses: changes after vibratory acoustic stimulation. *American Journal of Obstetrics and Gynecology*, **158**, 265–71.

Gagnon, R., Hunse, C., and Foreman, J. (1989). Human fetal behavioral states after vibratory stimulation. *American Journal of Obstetrics and Gynecology*, **161**, 1470–6.

Gagnon, R., Hunse, C., and Vijan, S. (1990). The effect of maternal hyperoxia on behavioral activity in growth-retarded human fetuses. *American Journal of Obstetrics and Gynecology*, **163**, 1894–9.

Manning, F.A. (ed.) (1983). Fetal monitoring. *Clinics in Perinatology*, **16**, 583–783.

Morrison, J.C. (ed.) (1990). Antepartum fetal surveillance. *Obstetrics and Gynecology Clinics of North America*, **17**, 1–273.

Nijhuis, J.G. (ed.) (1992). *Fetal behaviour: developmental and perinatal aspects.* Oxford University Press.

Nijhuis, J.G., Prechtl, H.F.R., Martin, Jr. C.B., and Bots, R.S.G.M. (1982). Are there behavioural states in the human fetus? *Early Human Development*, 6, 177–95.

Patrick, J. and Gagnon, R. (1989). Measurement of fetal activity. In *Maternal–fetal medicine: principles and practice* (ed. R.K. Creasy and R. Resnik), pp. 268–87. Saunders, Philadelphia.

Spencer, J.A.D. (ed.) (1989). *Fetal monitoring: physiology and techniques of antenatal and intrapartum assessment.* F.A. Davis, Philadelphia.

13. Renal function in the fetus

Jean E. Robillard, Francine G. Smith, Jeffrey L. Segar, Edward W. Guillery, and Pedro A. Jose

Knowledge of fetal renal physiology, once mainly of theoretical value, is now essential for the clinical care of premature infants and the management of fetal renal abnormalities. In the human, fetal urine production begins near the eighth week of gestation and increases progressively from 25 to 39 weeks. It is also well accepted that, at least in late gestation, amniotic fluid volume is regulated largely through the production of urine by the fetal kidneys; in addition to fetal urine, fetal lung fluid also contributes to amniotic fluid formation. Studies have suggested that the fetal kidneys influence fetal growth and that, in some instances, absence of fetal kidneys may lead to marked intrauterine growth retardation and anomalies of bone formation. Although the placenta is the major regulatory organ of the fetus, the fetal kidney plays an important role in the regulation of fetal arterial pressure, fluid and electrolyte homeostasis, acid–base balance, and hormonal synthesis.

This chapter reviews our current understanding of the physiology of the fetal kidney.

Renal blood flow

The kidneys of the fetus receive between 2 and 4 per cent of the combined ventricular output during the last trimester of gestation, whereas the kidneys of the new-born receive 15–18 per cent of the cardiac output. In fetal sheep, renal blood flow is of the order of 1.5–2.0 mL/min/g of kidney weight and increases rapidly during the first week of life. This relatively low rate of renal blood flow during fetal life is associated with a high renal vascular resistance and a low filtration fraction when compared to that of new-born animals (Fig. 13.1).

In lambs, there is no immediate increase in renal blood flow at birth. However, there is a redistribution of blood flow to the superficial renal cortex so that the ratio of outer to inner cortical flow increases after birth. In the weeks following birth, renal blood flow increases. This increase in renal blood flow is probably secondary to a decrease in renal vascular resistance concomitant with a rise in arterial pressure. However, since the rise in arterial pressure is of

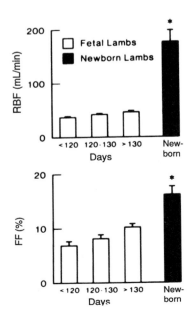

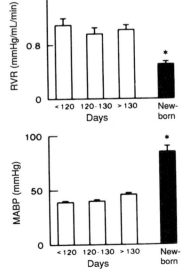

Fig. 13.1 Renal haemodynamics during ovine development. RBF, renal blood flow; RVR, renal vascular resistance; FF, filtration fraction; MABP, mean arterial blood pressure. Values are mean ± SEM. * $p < 0.05$ when compared to fetal values.

lesser magnitude than the rise in renal blood flow, factors other than pressure alone are involved in the postnatal changes in renal haemodynamics.

Factors regulating renal haemodynamics during fetal life

Neuroadrenergic system

Studies in fetal sheep during the last trimester of gestation have shown that neither surgical nor pharmacological renal denervation is associated with significant changes in renal blood flow, suggesting that renal innervation is not an important modulator of renal haemodynamics during resting conditions prior to birth. On the other hand, renal nerves play an important role in modulating fetal renal blood flow during stressful conditions such as hypoxaemia. Recent studies have also shown that the overall decline in renal blood flow and the overall rise in renal vascular resistance secondary to an increase in renal nerve activity are less pronounced in fetal than in older animals (Fig. 13.2). There is evidence, however, that the fetal kidney is sensitive to very small changes in renal sympathetic nerve activity (Fig. 13.2).

In addition to its renal vasoconstrictor function, renal nerve stimulation can also produce a renal vasodilatation during α-adrenoceptor blockade in fetal and new-born animals. The ability of renal nerve stimulation to induce a renal vasodilatation seems to be particular to the developmental period since, in adults, it has not been possible to isolate a neural renal vasodilator mechanism. Interestingly, this renal vasodilatation, which is of greater magnitude in fetal than in new-born sheep, is independent of activation of cholinergic or dopaminergic receptors but is completely blocked by selective β_2-adrenoceptor antagonists. These results are of interest in several respects. First, they demonstrate that neuronally released noradrenaline can induce renal vasodilatation early during development. Second, they provide evidence of an age-dependent neural renal vasodilator mechanism in mammals. Third, the observation that the renal vasodilatation observed during renal nerve stimulation decreases postnatally suggests that maturation of the adrenergic system may produce a downregulation of β-adrenoceptors in renal vessels. Such a mechanism may play an important role in maintaining renal blood flow at the time of birth, a period characterized by large increases in circulating catecholamines.

The role of dopamine (DA) in the regulation of renal haemodynamics during fetal life has also been investigated. Studies in sheep have demonstrated that intrarenal infusion of DA produces renal vasoconstriction during fetal and postnatal life. The absence of renal vasodilatation

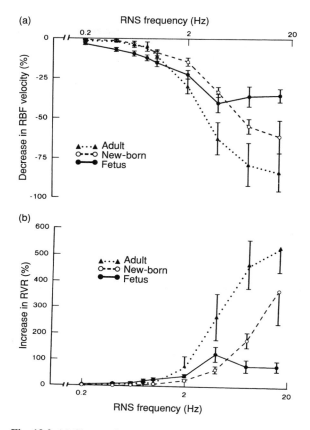

Fig. 13.2 (a) Changes in renal blood flow (RBF) velocity during renal nerve stimulation (RNS) in fetal, neonatal, and adult sheep. (b) Changes in renal vascular resistance (RVR) during renal nerve stimulation (RNS) in fetal, neonatal, and adult sheep.

with low DA doses in fetal and new-born sheep confirms results from other species demonstrating that 'α-adrenoceptor' effects of DA are predominant in immature animals. On the other hand, during blockade of renal α- and β-adrenoceptors, intrarenal DA infusions produce similar renal vasodilatatory responses in fetal, new-born, and adult sheep. The presence of DA-induced renal vasodilatation during α- and β-adrenoceptor blockade in fetal and new-born sheep is somewhat different from the situation in new-born swine and dogs in which no renal vasodilator effect of DA is observed.

DA acts on two specific DA receptors (DA_1 and DA_2) in the adult kidney. DA_1 receptors are localized to the renal vasculature and are present almost exclusively in the renal cortex, more specifically in the proximal convoluted and straight tubules and in cortical collecting ducts. In contrast, DA_2 receptors are present in both renal cortex and medulla and are found in glomeruli and in renal arteries. Characterization of DA receptors in near-term fetal, new-born, and adult sheep has shown that renal DA_1 receptor density and affinity are similar in all age groups, whereas the density of DA_2 receptors decreases with age.

Renin–angiotensin system

The vasopressor response and the renal vasoconstrictor response to infusion of angiotensin II (AII) are smaller in fetal than in adult sheep. However, administration of AII to fetal sheep increases arterial pressure, decreases umbilical flow, and decreases renal blood flow. Glomerular filtration rate remains unchanged, suggesting that AII acts primarily by increasing the tone of the efferent arteriole, as shown previously in adults.

Controversy exists with respect to the role of the renin–angiotensin system in modulating fetal renal haemodynamics during stressful conditions, including hypoxaemia and haemorrhage. Inhibition of AII synthesis using captopril, an angiotensin-converting enzyme inhibitor, or saralasin, an AII antagonist, does not protect against the decrease in renal blood flow and the rise in renal vascular resistance associated with fetal hypoxaemia. On the other hand, the decrease in fetal renal blood flow associated with removal of 20 per cent of feto-placental blood volume is abolished by the administration of captopril, but not by infusion of saralasin. These differences may be due to the fact that captopril, but not saralasin, inhibits the degradation of bradykinin. The subsequent rise in circulating bradykinin may blunt the effect of haemorrhage on renal blood flow.

It is important to note that maternally administered captopril is a fetotoxic agent in sheep and rabbits due to its prolonged depressor effects and low fetal tissue perfusion resulting from a decreased uterine blood flow. It is also associated with decreased prostaglandin production, disruption of the normal preparturient increase in cortisol, and inhibition of increased oxytocin levels induced by AII. These observations suggest that the fetal renin–angiotensin system may play an important role in fetal survival and in the onset of events associated with parturition.

Arginine vasopressin

Arginine vasopressin (AVP) increases blood pressure and decreases heart rate when infused intravenously to fetal sheep during the last third of gestation. A rise in fetal plasma AVP concentration also produces a redistribution of blood flow from the gastrointestinal tract and peripheral organs to the umbilical-placental unit, myocardium, and central nervous system. Renal blood flow and renal vascular resistance are not affected. Furthermore, the specific vascular AVP inhibitor $(d(CH_2)_5Tyr(Me)AVP)$ does not alter the fetal renal haemodynamic response to hypoxaemia. Thus, AVP does not appear to affect fetal renal haemodynamics either at rest or during stressful conditions.

Atrial natriuretic factor

The renal haemodynamic response to systemic infusion of atrial natriuretic factor (ANF) is different when fetal, new-born, and adult sheep are compared. Systemic infusion of ANF to chronically instrumented fetal and new-born sheep produces a decrease in renal blood flow and a rise in renal vascular resistance, whereas no significant changes are observed in non-pregnant adult ewes. It has been suggested that the renal vasoconstriction observed during systemic infusion of ANF in fetal and new-born animals is not secondary to a direct renal vasoconstrictor action of ANF but may depend on a decrease in cardiac output and/or an increase in the activity of the neuroadrenergic system.

Prostaglandins

The presence of high concentrations of the prostaglandins (PG) PGE_2, $PGF_{2\alpha}$, and metabolites of prostacyclin and thromboxane in urine of fetal sheep demonstrates that the fetal kidney has the ability to synthesize PGs. It has also been suggested that PGs produced by the fetal kidney may be involved in the regulation of renal haemodynamics and function since, at least in the pig, fetal renal vessels have a higher capacity to produce and to release PGE_2 and PGI_2 than those of the adult. Furthermore, a transient reduction in fetal renal blood flow has been observed following administration of the PG inhibitors meclofenamate or indomethacin to chronically catheterized fetal sheep.

Kallikrein–kinin

Urinary kallikrein excretion rate, expressed in absolute values, or corrected for kidney weight increases significantly during fetal life and after birth. This increase in urinary kallikrein excretion rate correlates closely with the increase in renal blood flow. Inhibition of kininase II is also associated with a decrease in renal vascular resistance in near-term fetal sheep, whereas administration of $[sar^1],[gly^8]$-AII, a competitive antagonist of AII, does not alter renal vascular resistance.

Autoregulation of renal blood flow during fetal life

The mature kidney is capable of maintaining renal blood flow relatively constant when major changes in perfusion pressure occur, a phenomenon known as autoregulation. Autoregulation has been observed in new-born dogs and in new-born lambs. More recent studies in piglets have suggested that renal blood flow autoregulation is negligible at birth, but increases with postnatal development.

Limited data are available on renal autoregulation in the fetus. However, during infusion of vasopressin or AII to chronically instrumented near-term fetal sheep, no changes in renal blood flow were observed despite a moderate increase in arterial blood pressure. These data suggest that the fetus may autoregulate renal blood flow during modest increases in renal perfusion pressure.

Glomerular filtration rate

Glomerular filtration rate (GFR) is low during fetal life and increases in relation to gestational age. In fetal sheep, GFR increases 2.5-fold during the last trimester of gestation and correlates closely with both fetal age and fetal kidney weight (Fig. 13.3). When expressed as a function of fetal body weight or fetal kidney weight, however, there is no significant change in GFR during the latter period of gestation.

During the first 24 h after birth, the GFR of preterm infants reflects the stage of intrauterine development reached by the infant. A good correlation has been found between GFR and gestational age in new-born infants delivered at between 27 and 43 weeks of gestation. Interestingly, GFR per kg body weight of premature infants (1.07 ± 0.12 mL/min/kg) measured during the first 24 h after birth is similar to that of the near-term fetal sheep (1.14 ± 0.08 mL/min/kg). An active nephrogenesis, changes in renal vascular resistance, increasing function of the superficial nephrons, and modification of forces involved in the process of ultrafiltration have all been postulated to contribute to the maturation of GFR during fetal life.

This rise in GFR after birth starts within the first few hours of postnatal life (Fig. 13.4). Such a rapid increase in GFR indicates the occurrence of functional rather than morphological changes and is consistent with increased glomerular perfusion resulting from recruitment of superficial cortical nephrons.

Tubular function

In fetal sheep, urine (600–1200 mL/day) passes into the amniotic cavity through the urethra, and into the allantoic cavity through the urachus; maintenance of these volumes is thought to be essential for normal fetal development.

The following section describes the handling of water and solutes along the renal tubules during development

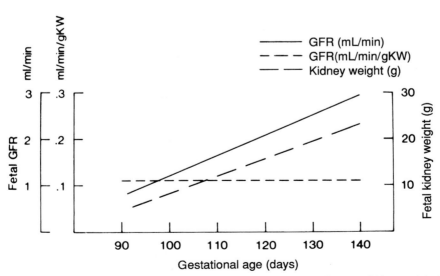

Fig. 13.3 Glomerular filtration rate (GFR) during ovine fetal development. gKW, Gram per kidney weight in grams.

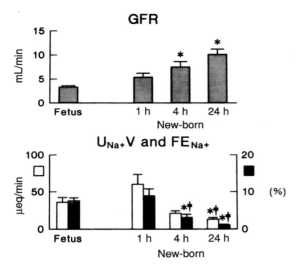

Fig. 13.4 Glomerular filtration rate (GFR), urinary sodium excretion ($U_{Na^+}V$) and fractional excretion of sodium (FE_{Na^+}) during the transition from fetal to new-born life in sheep. Values are means ± SEM. When fetal values are compared to new-born values (†) $p < 0.05$. When newborn values at 1 h are compared to values at 4 and 24 h, (*) $p < 0.05$.

and those factors influencing the volume and composition of fetal urine.

Sodium

In the adult, more than 99 per cent of the filtered sodium load is reabsorbed by the renal tubules. During fetal life, sodium reabsorption is comparatively low so that a greater amount of sodium is excreted than later in life. In the ovine fetus, sodium reabsorption is between 85 and 95 per cent, and increases with gestational age. A similar pattern has been observed when premature infants are compared with term neonates.

Renal tubular immaturity, a large extracellular fluid volume, presence of circulating natriuretic factors, and relative tubular insensitivity to circulating aldosterone have all been postulated as possible factors contributing to the high rate of sodium excretion by the immature kidney. Moreover, studies in fetal guinea-pigs have suggested that a functional glomerulotubular imbalance may be partly responsible for the high fractional excretion of sodium seen prior to birth. On the other hand, recent studies have shown that there is a direct correlation between GFR and proximal sodium reabsorption in fetal sheep arguing against the existence of glomerulotubular imbalance during fetal life.

The effect of birth on sodium excretion has also been investigated in sheep during the transition from fetal to new-born life. It has been shown that, after the first hour of life, there is a rapid decline in urinary sodium excretion (Fig. 13.4). Interestingly, this decrease in urinary sodium excretion happens in the face of a rise in GFR and filtered sodium load. Factors modulating this rapid decrease in urinary sodium excretion after birth have not been studied in detail. However, one may speculate that it may be related to a reduction of extracellular fluid volume. Furthermore, it has been demonstrated that renal nerves may also play a role in the ability of the new-born kidney to increase sodium reabsorption during the first day of life. Finally, other factors such as a rise in circulating catecholamines, changes in different hormonal systems, increased renal O_2 delivery, and changes in $Na^+K^+ATPase$ activity in renal tubular cells may also contribute to the increase in sodium reabsorption observed during the transition from fetal to new-born life.

Factors influencing tubular handling of sodium

The adrenal gland of the fetus has the ability to synthesize and secrete aldosterone in vitro following stimulation by angiotensin II, adrenocorticotrophin (ACTH), or high serum potassium concentration. Elevation of plasma angiotensin II stimulates aldosterone secretion in fetal sheep during the last third of gestation, although to a lesser degree than in the adult ewe. In vivo attempts to stimulate aldosterone secretion by infusing ACTH or potassium have remained unsuccessful. The high rate of sodium excretion during fetal life may be due to either a relative tubular insensitivity to circulating aldosterone or to the distal tubular reabsorption of sodium being already maximally stimulated. A limited delivery of oxygen to the fetal kidney and, hence, a lower renal oxygen consumption, may also contribute to the elevated sodium excretion during fetal life. In support of this hypothesis is the observation that sodium excretion increases during fetal hypoxaemia but not during fetal haemorrhage.

Other factors contributing to the elevated sodium excretion during fetal life include the natriuretic influence of cortisol and atrial natriuretic factor (ANF). Systemic infusion of atrial natriuretic factor to fetal sheep increases the excretion of potassium, chloride, and calcium, as well as free water clearance, and enhances sodium excretion by depressing proximal sodium reabsorption. Cortisol, when infused to near-term fetal sheep, also causes proximal sodium reabsorption to be depressed and distal reabsorption to be enhanced so that, overall, there is no change in total fractional sodium reabsorption. On the other hand, it has been shown that a natriuresis occurs following infusion of cortisol to younger fetal sheep (aged 111 to 120 days). Interestingly, glucocorticoid hormones appear to be an endogenous driving force for the maturation of the sodium pump near term.

The role of renal sympathetic innervation in modulating sodium excretion during fetal life has also been studied. The presence of renal sympathetic nerve activity has been observed in near-term fetal sheep. Furthermore, it has been demonstrated that renal denervation tends to blunt the natriuretic response to volume expansion during fetal life but less so than after birth.

Glucose

It has been found that the *tubular maxima* (Tm) for glucose when corrected for GFR (Tm/GFR) is greater in fetal than in adult sheep. The renal plasma threshold for glucose is also greater during fetal life and increases with gestational age.

The characteristics of sugar transport by the developing kidney have been investigated using α-methyl-D-glucopyranoside (α-MG), a substance that shares the D-glucose carrier present in the proximal tubule brush border membrane but does not interact with the D-glucose transport system at the basolateral border. It has been demonstrated that the transport of α-MG by kidney slices from near-term rat fetuses showed adult characteristics, i.e. the transport was inhibited by phlorhizin, and the uptake of α-MG was sodium-dependent. The development of the Na^+-dependent glucose transport system has also been studied in the fetal rabbit proximal tubule, late in gestation. This transport system has been shown to be stereospecific, electrogenic, cation-specific, and pH-sensitive. Thus, the fetal transport system for glucose is qualitatively similar to that of the adult.

Potassium

Potassium excretion depends on both tubular reabsorption and secretion. Early in gestation potassium secretion is low and increases toward term. This rise in potassium secretion as the fetus approaches term might result from a larger tubular surface area available for potassium secretion relative to body size, an increase in Na^+-K^+-ATPase activity, or an increase in the sensitivity of the fetal nephron to aldosterone. Fractional potassium reabsorption measured in near-term fetal sheep (67.1 ± 9.7 per cent, n=13) is similar to values measured in new-born lambs 24 h or more after delivery by Caesarean section (58.3 ± 4.8 per cent, n=12) and in adult ewes (52 ± 7.6 per cent, n=5). Thus, the secretory pathway for potassium appears to be functional during late fetal life.

Calcium

In humans, sheep, and guinea-pigs the concentrations of total, ultrafiltrable, and ionized calcium are higher in fetal than in maternal plasma. These levels appear to be generated and maintained by the placenta, which transports calcium actively from the mother to the fetus. Calcium-binding protein has been isolated from the placenta of the rat and a calcium-stimulated ATPase has been found in the placenta of the guinea-pig. In fetal sheep, there is an increase in plasma parathyroid hormone levels after intravenous infusion of EDTA, indicating that the fetal parathyroid glands can respond to a hypocalcaemic stimulus. The fetal kidney contributes to the transplacental transfer of calcium by the synthesis of 1,25-dihydroxycholecalciferol. Fetal nephrectomy abolishes the transplacental calcium gradient, which can be restored by intravenous injection of parathyroid hormone. However, the excretion of cyclic-AMP in response to parathyroid hormone and calcitonin is attenuated in fetal rats and new-born infants, suggesting that the decreased sensitivity of the kidney to parathyroid hormone may contribute to the hypocalcaemia of the neonate.

Phosphate

The concentration of inorganic phosphate in fetal plasma is greater than in maternal plasma, and is transported across the placenta from mother to fetus against a concentration gradient. In fetal sheep, fractional phosphate reabsorption varies between 60 and 100 per cent, the concentration of inorganic phosphate in fetal urine being extremely low.

The fetal kidney responds to parathyroid extract, and to an endogenous rise in plasma parathyroid hormone (PTH) by increasing urinary excretion of phosphate and cyclic AMP (cAMP). Thus, relative parathyroid insufficiency may contribute to the low fetal renal clearance of phosphate and subsequent hyperphosphataemia. Acute expansion of the extracellular volume in fetal sheep produces a significant increase in phosphate excretion, independent of plasma calcium concentration and PTH, but related to sodium excretion. Phosphate excretion also increases following severe metabolic acidosis and following the administration of cortisol to the fetus.

Acid–base homeostasis

The fetal kidney is involved in the regulation of acid–base balance. Fetal sheep reabsorb between 80 and 100 per cent of the filtered bicarbonate load, the renal threshold for bicarbonate being lower in fetal than in adult sheep (17.7 ± 1.4 mmol/L in fetuses; 28.7 ± 1.7 mmol/L in non-pregnant adult ewes). This may result from the relatively large extracellular fluid volume normally found in the fetus, since bicarbonate reabsorption increases during fetal volume depletion.

The renal response to metabolic acidosis is limited in the fetus compared with the adult, due primarily to a low excretion rate of phosphate, but also to a limited capacity to synthesize ammonia. Prolonged and severe metabolic acidosis results in an increase in hydrogen ion excretion, indicating that the distal segments of the fetal renal tubule are able to generate a pH gradient.

Concentrating capacity of the fetal kidney

Fetal urine is usually hypoosmotic with respect to fetal plasma, and, in chronically catheterized fetal sheep, urinary osmolality varies between 100 and 250 mOsm/kg H_2O.

The ability of the kidney to concentrate the urine to adult levels is not reached until after birth. The structural immaturity of the medulla, characterized by short loops of Henle, and the relatively high blood flow through the vasa recta may limit the buildup of an osmotic gradient by the fetal kidney. On the other hand, the decreased ability of the fetal kidney to concentrate the urine is not secondary to the inability of the fetus to synthesize and/or secrete vasopressin, which is present in the posterior pituitary of fetal animals and humans. Furthermore, both volume and osmoreceptor controls of AVP secretion are fully functional in the last trimester of gestation in fetal sheep. Increased levels of AVP in the fetal circulation occur following an osmotic stimulus, haemorrhage, hypoxaemia, and diuretic administration.

The fetal nephron is, however, less sensitive to AVP than that of the adult, since the rise in urinary osmolality following the infusion of AVP is three times lower in fetal than in adult animals for the same plasma AVP concentration. Reduced AVP receptor numbers and/or poor coupling between AVP receptor binding and cAMP generation may also contribute to the hyporesponsiveness of the immature kidney to AVP. Studies in sheep have demonstrated that renal AVP receptors in the new-born have lower binding affinity and higher binding capacity than in the adult. These studies, however, do not differentiate between AVP-V1 receptors that mediate the vascular response to AVP and AVP-V2 receptors that are involved in the antidiuretic effect of AVP.

Vasotocin, a neuropeptide that differs from AVP by a single amino acid substitution, and which is present in

fetal plasma, urine, and amniotic fluid, can also increase fetal urinary osmolality. However, this effect seems to be more a reflection of a decrease in solute reabsorption than an increase in water reabsorption.

Renal hormones

In addition to its excretory function, the fetal kidney is able to synthesize various hormones. Components of the renin–angiotensin, prostaglandin, and kallikrein–kinin systems, as well as the enzymatic machinery necessary for vitamin D metabolism, have been identified in the fetal kidney of mammals.

The renin–angiotensin system

Plasma renin activity (PRA) is high during fetal life, the levels being substantially above those present in the maternal circulation. Removal or absence of fetal kidneys is associated with low PRA, indicating that the major source of fetal renin is the fetal kidney.

Fetal PRA and plasma angiotensin II levels increase after stimulation by furosemide, blood volume reduction, fetal hypotension, and hypoxaemia. Noradrenaline release from renal nerves and stimulation of β-adrenoceptors also promote renin release in near-term fetal sheep. Conversely, expansion of fetal blood volume, inhibition of prostaglandin synthesis by indomethacin, vasopressin infusion, and hypertension, are associated with significant decreases in fetal PRA.

It has been suggested that the higher activity of the renin–angiotensin system during fetal and early postnatal life is the result of increased renin gene expression along renal arterial vessels. It has been shown, using in situ hybridization, that renin mRNA gene expression is localized in arcuate and interlobular arteries early during fetal life in rats, with the localization shifting to the classical juxtaglomerular localization during postnatal maturation.

Prostaglandins

Synthesis and catabolism of prostaglandins (PG) occur early during gestation in fetal animals and humans. In the ovine fetus PGF is present in renal homogenates by 40 days of gestation; PGE_2 is first formed by 77 days of gestation and is predominant by 116 days of gestation. PG catabolism via the PG 15-hydroxydehydrogenase pathway is present at 40 days of gestation, while catabolism through the PG 13–reductase pathway occurs by about 110 days of gestation, persisting until term.

The kallikrein–kinin system

Kallikreins are serine proteases that generate kinins from kininogen substrates by limited proteolysis. It has been suggested that, during the latter part of gestation in the ovine fetus, urinary kallikrein excretion is dependent on both age and plasma aldosterone concentration. There is also an inverse relationship between urinary kallikrein excretion and sodium excretion. Furthermore, a rapid rise in fetal arterial PO_2, produced by hyperbaric oxygenation of the ewe, activates the bradykinin-generating system and produces both a fall in plasma kininogen concentration and a rise in the fetal plasma concentration of bradykinin.

Vitamin D metabolism

The fetus also participates in the synthesis of (1,25 dihydroxy vitamin D), fetal plasma levels of $1,25(OH)_2D$ being greater than in adult sheep. In human fetuses, $1,25(OH)_2D$ levels are greater in arterial than in umbilical venous blood. In fetal sheep the renal conversion of $^3H–25(OH)D$ to $^3H–1,25\ (OH)D$ was reported to be about 20 per cent, whereas the conversion rate to $^3H–24,25(OH)_2D$ was only 4 per cent. Kidney homogenates from fetal rats, guineapigs, chicks, and rabbits are also capable of hydroxylating $25(OH)D$ to $1,25(OH)_2D$.

Acknowledgement

Portions of the present work were supported by National Institutes of Health grants HD20576, HL14388, DK-44756, DK38302, and HL35600. F. G. Smith was a C.J. Martin Fellow supported by the National Health and Medical Research Council of Australia. J.L. Segar and E. N. Guillery are recipients of National Research Service Awards from the National Institutes of Health, HL08170 and HL08366, respectively.

Further reading

Beck, J.C., Lipkowitz, M.S., and Abramson, R.G. (1988). Characterisation of the fetal glucose transporter in rabbit kidney. Comparison with the adult brush border electrogenic Na⁺-glucose symporter. *Journal of Clinical Investigation*, **82**, 379–87.

Broughton-Pipkin, F., Symonds, E.M., and Turner, S.R. (1982). The effect of captopril (SQ14, 225) upon mother and fetus in the chronically cannulated ewe and in the pregnant rabbit. *Journal of Physiology*, **323**, 415–22.

Feltes, T.F., Hansen, T.N., Martin, C.G., Leblanc, A.L., Smith, S., and Giesler, M.E. (1987). The effects of dopamine infusion on regional blood flow in newborn lambs. *Pediatric Research*, **21**, 131–6.

Gomez, R.A. and Robillard, J.E. (1984). Developmental aspects of the renal response to hemorrhage during converting-enzyme inhibition in fetal lambs. *Circulation Research*, **54**, 301–12.

Gomez, R.A., Lynch, K.R., Sturgill, B.C., Elwood, J.P., Chevalier, R.L., Carey, R.M., and Peach, M.J. (1989). Distribution of renin mRNA and its protein in the developing kidney. *American Journal of Physiology*, **257**, F850–F858.

Hill, K.J., Lumbers, E.R., and Elbourne, I. (1988). The actions of cortisol on fetal renal function. *Journal of Developmental Physiology*, **10**, 85–96.

Kesby, G.J. and Lumbers, E.R. (1988). The effects of metabolic acidosis on renal function of fetal sheep. *Journal of Physiology*, **396**, 65–74.

Lumbers, E.R. (1983). A brief review of fetal renal function. *Journal of Developmental Physiology*, **6**, 1–10.

Lumbers, E.R., Hill, K.J., and Bennett, V.J. (1988). Proximal and distal tubular activity in chronically catheterized fetal sheep compared with the adult. *Canadian Journal of Physiology and Pharmacology*, **66**, 697–702.

Nakamura, K.T., Felder, R.A., Jose, P.A., and Robillard, J.E. (1987). Effect of dopamine in the renal vascular bed of fetal, newborn, and adult sheep. *American Journal of Physiology*, **252**, R490–R497.

Nakamura, K.T., Matherne, G.P., McWeeny, O.J., Smith, B.A., and Robillard, J.E. (1987). Renal hemodynamics and functional changes during the transition from fetal to newborn life in sheep. *Pediatric Research*, **21**, 229–34.

Page, W.V., Perlman, S., Smith, F.G., Segar, J.L., and Robillard, J.E. (1992). Renal nerves modulate kidney renin gene expression during the transition from fetal to newborn life. *American Journal of Physiology*, **262**, R459–R463.

Robillard, J.E. and Nakamura, K.T. (1988). Neurohormonal regulation of renal function during development. *American Journal of Physiology*, **254**, F771–F779.

Robillard, J.E., Nakamura, K.T., Varille, V.A., Andresen, A.A., Matherne, G.P., and Van Orden, D.E. (1988). Ontogeny of the renal response to natriuretic peptide in sheep. *American Journal of Physiology*, **254**, F634–F641.

Robillard, J.E., Nakamura, K.T., Wilkin, M.K., McWeeny, O.J., and DiBona, G.F. (1987). Ontogeny of renal hemodynamic response to renal nerve stimulation in sheep. *American Journal of Physiology*, **252**, F605–F612.

Robillard, J.E., Smith, F.G., Segar, J.L., Merrill, D.C., and Jose, P.A. (1992). Functional role of renal sympathetic innervation during fetal and postnatal development. *News in Physiological Sciences*, **7**, 130–3.

Robillard, J.E., Smith, F.G., and Smith, Jr, F.G. (1992). Developmental aspects of renal function during fetal life. In *Pediatric kidney disease*, Vol. 1. (2nd edn), (ed. C.M. Edelmann, J. Bernstein, S.R. Meadow, A. Spitzer, and L.B. Travis), pp. 3-18. Little, Brown and Co, Boston.

Robillard, J.E., Weismann, D.N., Gomez, R.A., Ayres, N.A., Lawton, W.J., and VanOrden, D.E. (1983). Renal and adrenal responses to converting-enzyme inhibition in fetal and newborn life. *American Journal of Physiology*, **254**, R249–R256.

Smith, F.G., Nakamura, K.T., Segar, J.L., and Robillard, J.E. (1991). Renal function in utero. In *Neonatal and fetal medicine: physiology and pathophysiology* (ed. R.A. Polin and W.W. Fox), pp. 1185–95. Saunders, Philadelphia.

Wintour, E.M., Brown, E.H., Denton, D.A., Hardy, K.J., McDougall, J.G., Robinson, P.M., Rowe, E.J., and Whipp, G.T. (1977). *In vitro* and *in vivo* adrenal cortical steroid pro- duction by fetal sheep: effect of angiotensin II, sodium deficiency and ACTH. In *Research on steroids*, Vol. VII, Transaction of the seventh meeting of the International Study Group for Steroid Hormones (ed. A. Vermeulen, P. Jungblut, A. Klopper, L. Lerner, and F. Sciarra), p. 475. Elsevier North- Holland Inc, New York.

14. Fetal fluid balance
Robert A. Brace

The dynamics and regulation of fetal fluids are dramatically different from those of the adult. This uniqueness occurs not only because the fetus is surrounded by fluid but also because the fetal body contains a much greater proportion of water than does the adult. In addition, the rates of several fluid movements within the fetal body average 5–10 times more than those in the adult, and there are pathways for fluid movement within the fetus that do not exist in the adult. These facts combined with the intimate interactions between the fetus and its mother make fluid dynamics within the fetus a truly remarkable aspect of life's continuum of changes in body fluid balances. This chapter provides an overview of our current understanding of fluid balance in the mammalian fetus. Collectively, this includes four components: the fluid composition of the fetus; the distribution of this fluid among the various compartments within the fetus; the routes and flows of fluid among the compartments; and the regulatory mechanisms that affect each of these flows.

Water content of the fetal body

Gestational changes in total body water content

At 8 weeks of gestation, the human fetus is composed of 95 per cent water, which is much greater than the normal adult value of 57 per cent. As seen in Fig. 14.1, fetal body water content gradually decreases throughout gestation. The extent of this decrease varies with species, with greater decreases occurring in the guinea-pig than in humans or sheep. In humans a more rapid decrease in fetal water content occurs at roughly the middle of the third trimester, and human fetal water content averages 70 per cent at the time of birth. This accelerated decrease in water content is due to the accretion of significant amounts of body fat, which occurs during late gestation in humans but not in rabbits, sheep, or guinea-pigs. Relative to lean body mass, there are no differences in the gestational changes in body water content between human and ovine fetuses. The rabbit fetus has a higher body water content than the other species throughout gestation and this may be due to the relative immaturity of the rabbit at birth. The difference between the body water content of the ovine and guinea-pig fetuses at term does not appear attributable to differences in maturity at birth

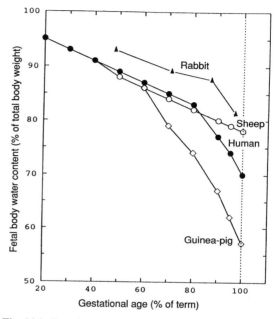

Fig. 14.1 Gestational changes in fetal body water content in rabbits, humans, sheep, and guinea-pigs. Data derived from carcass dehydration or dilutional space of radiolabelled water. (Data from Battaglia, F.C. and Meschia, G. (1986). *An introduction to fetal physiology*. Academic Press, Orlando.)

because both the sheep and guinea-pig are mature and capable of co-ordinated movement such as walking shortly after birth. To date, these species differences are unexplained.

Even though fetal water content decreases throughout gestation, the term fetus is composed of a much larger percentage of water than the adult. Thus, the normal state of fluid balance in the fetus may be considered to be somewhat analogous to whole-body oedema in the adult.

Compartments, distributions, and volumes of fetal fluids

Fetal fluid compartments

The intracellular, extracellular, blood, plasma, and interstitial fluid spaces in the fetus are analogous to those of the adult. There are two or perhaps three additional fluid compartments unique to the fetal period. First, the fetal lungs are unique in that the future airways are filled with a liquid rather than gas. The second unique fetal compartment is the amniotic fluid that surrounds the fetus and, in turn, is surrounded by fetal tissue (i.e. the amnion and the chorion). Thus, amniotic fluid is a fluid compartment of the fetus. Third, in species, such as the sheep and ungulates in general, an allantoic fluid sac exists adjacent to the amniotic sac. The allantoic fluid is surrounded by the allantoic membrane or allantois. Along their common border, the amnion and allantois make direct contact with each other. At their interface with the uterine wall, the chorion surrounds the amniotic and allantoic compartments (Fig. 14.2). The allantoic sac is connected to the fetal urinary bladder via the urachus, a small membranous vessel that passes through the umbilical cord. There is only one anatomical inlet to the allantoic space and no anatomical outlet. This shows that fluid leaving the allantoic space must do so by crossing the allantoic membrane.

Fluid distributions and volumes

Fluid volumes: a methodological note

An understanding of the regulation of fluid distributions within the body depends upon the accuracy of the methods used for volume measurements. Unfortunately, there have been relatively few volume measurements in the fetus and many of these suffer methodological problems. This occurs because, with the indicator dilution techniques used for volume determinations, the indicator labels, which were thought to be restricted to a certain compartment, often spill over into another compartment. For example, extracellular markers may enter the cellular space, or plasma labels may enter the interstitial space, leading to overestimation of plasma and extracellular volumes and thus to underestimation of interstitial and cellular volumes. This is particularly a problem in the fetus because capillary and cell membrane permeabilities are substantially greater than in the adult. Thus, as the distribution and regulation of fluids in the fetus are described, it is important to interpret cautiously data regarding fetal fluids.

Intracellular and extracellular volumes

Concurrent with the decline in total body water content during fetal development (Fig. 14.1), there are major changes in the distribution of the fluid within the fetus as seen in Fig. 14.3. That is, the percentage of the body that is extracellular fluid sharply decreases, while there is a gradual rise in the percentage of body fluid residing within the cells. These changes appear to be attributable to an increase in the density of cells in developing fetal tissue combined with a deposition of ground substance within the extracellular matrix, as well as to the deposition of body fat late in the fetal period.

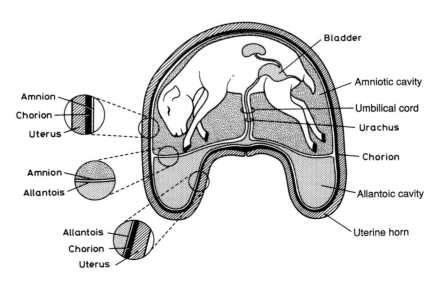

Fig. 14.2 Schematic showing relationships among fetal membranes and extrafetal fluid compartments in sheep. (From Gilbert W.M. (1993). Intramembranous absorption of water from the ovine allantoic cavity. *Journal of Maternal-Fetal Medicine*, **2**, 55–61 Copyright © 1993, John-Wiley and Sons, Inc. Reprinted by permission of Wiley-Liss, a division of John-Wiley and Sons, Inc.)

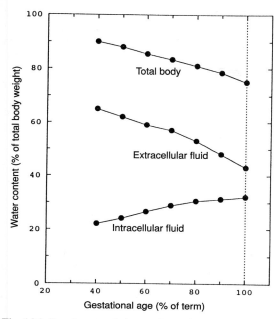

Fig. 14.3 Developmental changes in total, cellular, and extracellular fluids in the human fetus. Data determined from volumes of distribution of radiolabelled tracers.

Plasma and interstitial fluid volumes

The extracellular fluid compartment described above is functionally divided into two major spaces: (1) plasma; and (2) interstitial fluid. At present, the regulation of the distribution of fluid between the plasma and interstitial spaces in the normal human fetus is not well understood because measurements in live fetuses have been made only in experimental animals. In late-gestation fetal sheep, plasma volume averages 76 mL/kg. Considerably higher volumes have been reported, but these values are suspect because the labels used to measure plasma volume rapidly crossed the fetal capillary wall and thus part of the interstitial fluid was included in the estimated plasma volume. The value of 76 mL/kg includes plasma within the fetal body, umbilical cord, and fetal side of the placenta. Only 70 per cent of this, or 53 mL/kg, is within the fetal body, a number only 10–20 per cent higher than plasma volume of the lean adult. The extent to which plasma volume varies with gestational age is unknown although the plasma volume in fetal sheep of 76 mL/kg is independent of fetal weight over the last trimester. Older studies have suggested that fetal plasma volume early in gestation is greater than that of late gestation. However, plasma volume was overestimated in those studies because the plasma label rapidly leaked from the fetal circulation.

Fetal interstitial volume averages three times plasma volume or 235–240 mL/kg of body weight for 1–4 kg ovine fetuses. The ratio of interstitial fluid volume to plasma volume in the sheep fetus is 3:1, as in the adult of a number of species. This ratio can be misleading because, as noted above, roughly 30 per cent of fetal plasma is located outside the fetal body, i.e. in the umbilical cord and placenta. Correcting for this, the ratio of interstitial to plasma volume becomes 4.4:1. This ratio clearly shows that the interstitial space of the fetus is volume-expanded compared to that of the adult. In addition, the fact that fetal plasma volume, normalized for body weight, does not vary with fetal weight over the range of 1–4 kg indicates that the large decrease in fetal extracellular fluid volume with advancing gestational age (Fig. 14.3) appears to be due to decreases in interstitial fluid volume rather than in plasma volume.

Blood volume in the fetus

The volume of blood circulating in the human fetus has been estimated using different techniques under three conditions: (1) in early to mid-gestation, at the time of pregnancy termination; (2) at the time of red cell transfusions for the treatment of fetal anaemia; and (3) immediately postpartum as the sum of blood volume within the new-born plus that retained in the placenta. A value of 168 mL/kg was obtained at pregnancy termination but this is much too high because of loss of the plasma label. Recent studies in fetuses undergoing *in utero* transfusions have determined fetal blood volume to be 90–105 mL/kg. It is known that this is an underestimation of the true volume but the magnitude of the error is unknown. As determined immediately after delivery, human fetal blood volume averages 105 mL/kg. However, in both human and sheep fetuses, blood volume decreases by 10–20 per cent during the process of labour and delivery as estimated from haematocrit changes. Thus, a best estimate would be that blood volume in the human fetus is 110–115 mL/kg but it is not clear whether this varies with fetal development.

In animals, fetal blood volume has been most frequently determined in the chronically catheterized sheep fetus. With careful consideration of potential methodological error, the circulating blood volume in fetal sheep averages 110–115 mL/kg of fetal body weight under normal unstressed and non-labour conditions. It may be fortuitous that circulating blood volumes in the human and sheep fetuses are the same.

Although the average fetal blood volume is 110–115 mL/kg, in individual fetuses, blood volume varies from 95 to 130 mL/kg. In order to understand what contributes to this variability, recall that blood volume is the sum of red cell volume and plasma volume, and that each of these is

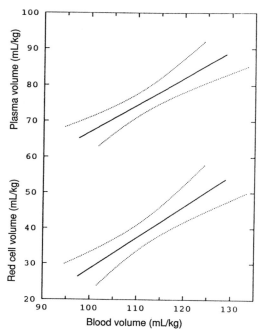

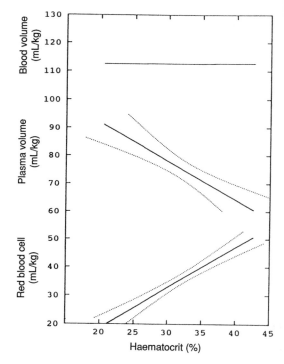

Fig. 14.4 The relationships among red cell volume, plasma volume, and blood volume in the ovine fetus under resting conditions. Means and 95 per cent confidence intervals about the regression lines are indicated. Data are from simultaneous measurements of plasma volume and red cell volume in 25 animals. (Data from Brace, R.A. (1991). Fluid distribution in the fetus and neonate. In *Fetal and neonatal physiology* (ed. R. Polin and W. Fox), pp. 1288–98. Saunders, Philadelphia.)

Fig. 14.5 The dependency on haematocrit of blood volume, plasma volume, and red cell volume in the ovine fetus. (Data from Brace, R.A. (1991). Fluid distribution in the fetus and neonate. In *Fetal and neonatal physiology* (ed. R. Polin and W. Fox), pp. 1288–98. Saunders, Philadelphia.)

regulated separately. The interrelationships between red cell, plasma, and blood volumes in the ovine fetus are shown in Fig. 14.4. In fetuses with a high blood volume relative to body weight, both red cell volume and plasma volume are greater than average. Conversely, fetuses with a low blood volume have a low plasma volume as well as a low red cell volume. Fetal blood volume, however, does not vary significantly with fetal haematocrit (Fig. 14.5). Instead, relative to haematocrit, plasma volume decreases as red cell volume increases (Fig. 14.5). These observations emphasize that fetal blood volume, red cell volume, and plasma volume are each regulated separately.

With fetal growth rates in various species of 1–10 per cent/day, fetal blood volume increases proportionately. The mechanisms that maintain blood volume in the face of rapid growth are twofold: first, new red blood cells and plasma proteins are constantly being formed and released into the fetal circulation, thereby helping to expand intravascular volume; second, the blood vessels them-

selves grow as the fetus grows, thereby increasing vascular compliance. Although an increase in compliance would tend to cause a reduction in blood pressure, the synergistic regulatory mechanisms noted below act to expand blood volume and thus nullify any fall in pressure.

Amniotic fluid volume

In humans, amniotic fluid volume averages 12 mL at 8 weeks gestation when the fetus weighs only about 1 g. As seen in Fig. 14.6, amniotic volume increases rapidly and averages 95 mL at the end of the first trimester, and 340 mL at midpregnancy. From 24 to 39 weeks' gestation, amniotic fluid volume changes little and averages 780 ml. In pregnancies that last more than 40 weeks, amniotic volume decreases, averaging 400 mL at 43 weeks of gestation.

Amniotic fluid volumes have been determined in a number of species besides humans including sheep, baboons, monkeys, and pigs. In sheep, amniotic fluid

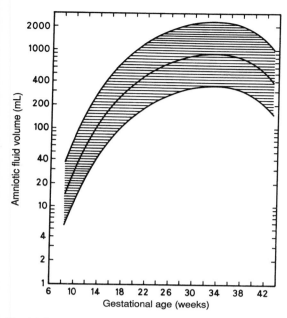

Fig. 14.6 Amniotic fluid volume in normal human pregnancy. The shaded area is the 95 per cent prediction interval about the mean. Data were derived from 705 pregnancies using direct collection or indicator dilution techniques. (Data from Brace, R.A. and Wolf, E.J. (1989). Normal amniotic fluid volume changes throughout pregnancy. *American Journal of Obstetrics and Gynecology*, **161**, 382–8.)

volume changes during gestation are similar to those in humans after correction is made for a gestation lasting 21 weeks rather than 40 weeks. In addition, amniotic fluid volume decreases in sheep near the end of gestation and averages 300–500 mL, half of its maximal volume, at term.

Allantoic fluid volume

Allantoic fluid exists in ungulates and carnivores, but not in primates including humans. Allantoic fluid volumes have been examined most often in sheep and pigs. In both species, allantoic volume increases to an initial peak at 30 per cent of gestation and then decreases. Beginning at 50 per cent of gestation, allantoic volume increases again. In the pig, the allantoic volume peaks at 80 per cent of gestation and then decreases to near zero before term. In sheep, allantoic volume increases steadily throughout the last half of gestation and averages 500–1000 mL at term. Thus, the sum of allantoic plus amniotic volumes in sheep is greater than the amniotic fluid volume in humans.

Routes and rates of fluid movement within the fetus

Routes of fluid movements

In the adult, routes for fluid transfer into or from the body's various compartments include urinary output, swallowing, capillary filtration, lymph flow, and transcellular flow. The interrelationships among these flows and the various fluid compartments are summarized in Fig. 14.7(a). These same pathways and compartments occur in the fetus. Four additional routes for fluid movement also contribute significantly to the overall fluid balance in the fetus and these are summarized in Fig. 14.7(b). Of course, the placenta is perhaps the principal of these four unique fetal pathways, with an overall net fluid movement into the fetus as the fetus grows. The second fluid pathway unique to the fetus involves the lungs. During the fetal period, the lungs are filled with a liquid that originates from an active chloride secretory process. During the last trimester, large volumes of fluid flow each day from the lungs and either enter the amniotic compartment or are swallowed as the fluid exits the trachea. The third pathway is the 'transmembranous pathway' whereby exchange of water and solutes occurs across the amniochorion between the amniotic fluid and the maternal blood that perfuses the wall of the uterus. The fourth unique pathway is the combination of all routes whereby water and solutes directly exchange between fetal blood and the amniotic (and allantoic) fluid. This latter pathway has been termed the 'intramembranous pathway' and includes several components — the exchange that occurs between the amniotic fluid and: (1) the richly vascularized fetal surface of the placenta; (2) the fetal blood that perfuses the fetal membranes (in ungulates, fetal blood vessels are located on the outer surface of the amnion and within the chorion, Fig. 14.2); (3) the fetal skin; and (4) the surface of the umbilical cord.

Rates of fluid movements

As described in detail below, many of the fluid movements in the fetus are regulated in an attempt to maintain blood volume. Hence, the following description of the rates of fetal fluid movements may be viewed from the perspective of blood volume regulation.

Transcapillary flow

The movement of water and solutes across the capillary wall within the fetal body depends upon the available capillary surface area as well as on the capillary's permeability and filtration characteristics. These relationships are usually

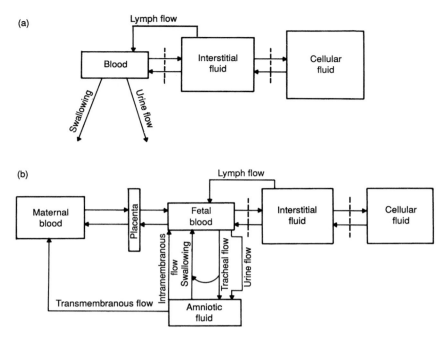

Fig. 14.7 Fluid compartments and routes of fluid movement in (a) adults or (b) in the fetus. Dashed lines represent membranes.

expressed in the form of two equations, the Starling equation (eqn (1)) and the solute flux equation (eqn (2)),

$$J_v = \text{CFC}\,(\Delta P - \sigma\Delta\pi), \tag{1}$$

$$J_s = J_v\,(1 - \sigma)\,(C_p - C_i e^x)/(1 - e^x) \tag{2}$$

where $x = J_v\,(1 - \sigma)/PS$.

For the Starling equation, J_v is the volume *rate* at which fluid is moving across the capillary wall (units of volume/time), CFC is the capillary filtration coefficient (i.e. the filtration capacity per unit mmHg driving force), ΔP is the net hydrostatic pressure gradient between the capillary and interstitial fluid, $\Delta\pi$ is the protein (i.e. colloid) osmotic pressure gradient across the capillary wall, and σ is the reflection coefficient. The reflection coefficient expresses the ability of the capillary wall to restrict the transcapillary movement of dissolved solutes in the presence of a volume flow (with $\sigma = 1$, no conductive solute movement occurs; with $\sigma = 0$, the solute moves as freely as water). Although each solute has a different value for σ, an average value is normally used for the plasma proteins (eqn (1)). The solute flux equation (2) is more complex than the volume flux (Starling) equation because solute movement occurs by both diffusion and convection and because there is an interaction between the diffusive and convective movements. The result is that the solute flux equation is non-linear in that there is an exponential

term in both the numerator and denominator. For this equation, J_s is the net rate at which solute is crossing the capillary wall, C_p and C_i are the plasma and interstitial concentrations of the solute, and PS is the product of capillary permeability (P, the diffusional permeability per unit surface area) and capillary surface area (S).

In the adult, only a fraction of the capillaries are perfused at any given instant so a considerable capillary exchange area is held in reserve. In the fetus, it is unknown whether all fetal capillaries are continuously perfused and thus it is unknown if fetal capillary surface area can be increased. Nonetheless, we do know that the capillary permeability to plasma proteins multiplied by the capillary surface area in the late-gestation ovine fetus is 15 times that of the adult (per unit body weight). Until more is known, it may be speculated that this is due to an increased surface area in combination with an increased permeability. We also know that the filtration coefficient of the capillaries in the fetal body is five times that of the adult and the reflection coefficient for plasma proteins is lower in the fetus. The significance of these observations is that both fluids and proteins move across the capillary wall much more rapidly in the fetus than in the adult, and it is likely that the capillaries are more permeable in younger than older fetuses.

On a short-term basis, transcapillary flow in the fetus is unquestionably the primary determinant of fetal blood volume changes. This occurs because the filtration coefficient of the fetal capillaries is approximately 100 times that of the placenta in the sheep. When vascular volume is increased by infusing isotonic fluids, for example, the ovine fetus retains only 6 per cent of the infused fluid within its circulation 30 min after the infusion, whereas the adult retains 30 per cent under comparable conditions. Under conditions of blood loss, fetal blood volume rapidly returns to normal. For example, after a 30 per cent haemorrhage over 2 h, ovine fetal blood volume is back to normal levels within 3–4 h. In the adult 24–48 h are required for full volume restoration after haemorrhage. Thus, under conditions of both blood volume expansion and reduction, the fetus is much more efficient at maintaining its blood volume near normal. The rapidity of fetal blood volume regulation occurs not only because the fetal capillaries have five times the fluid conductance of the adult but also because the fetal interstitial space is 10 times as compliant as that of the adult (relative to body weight). In other words, the fetal capillaries together with the interstitium allow much more rapid and much more extensive fluid movements to occur.

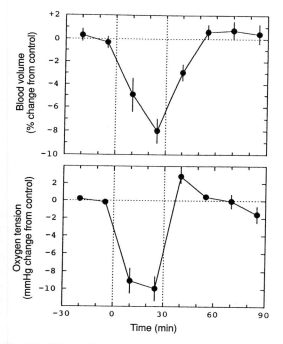

Fig. 14.8 Blood volume responses to acute hypoxia in ovine fetuses. Hypoxia was produced by lowering maternal inspired oxygen content and blood volume changes were calculated from haematocrit. Data are mean ± SE. (Data from Brace, R.A. and Cheung, C.Y. (1987). Role of catecholamines in mediating fetal blood volume decrease during acute hypoxia. *American Journal of Physiology*, **253**, H927–H932.)

The high conductance of the fetal capillaries is also largely responsible for the vascular volume changes that occur during hypoxia (Fig. 14.8). During acute hypoxia in ovine fetuses, blood volume decreases by about 1 per cent for each 1 mmHg decrease in arterial oxygen tension. This hypoxia-induced hypovolaemia occurs secondarily to an increased capillary hydrostatic pressure within the fetal body. Capillary pressure, in turn, rises as a consequence of the hypoxia-induced elevation of arterial pressure and venous pressure (with more severe hypoxia), in combination with a rise in capillary pressure that occurs in several vascular beds as vessels dilate in an attempt to maintain oxygen delivery.

In general, any condition that acutely alters fetal capillary pressure, either directly or secondarily due to altered arterial or venous pressures, will produce rapid alterations in fetal blood volume. For example, intravascular infusions of vasoactive hormones such as arginine vasopressin, angiotensin II, noradrenaline, or atrial natriuretic factor, all produce reductions in fetal blood volume of 10–15 per cent at doses producing high physiological to pharmacological levels of the hormones. In addition, in late-gestation ovine fetuses, the non-labour uterine contractions that occur at roughly 45-min intervals produce a transient reduction in fetal blood volume of 2–4 per cent in association with increases in fetal arterial pressure of 3–4 mmHg and venous pressure of 1–2 mmHg.

Lymph flow

Lymph flow rates in the late-gestation sheep fetus are greater than those in the new-born or adult (relative to body weight). This is consistent with the observation that interstitial volume is greater during the fetal period than during later life, and suggests that fetal lymphatics are quite capable of transporting fluid. In fetal sheep, pulmonary lymph flow in the right thoracic duct is higher than that in new-born lambs or adult sheep. Left thoracic duct lymph flow rate in fetal sheep averages 0.15–0.25 mL/min/kg, whereas in adult sheep, thoracic duct lymph flow rate averages 0.04–0.05 mL/min/kg; thus, fetal thoracic duct lymph flow rates are 4–5 times those of the adult.

Even though fetal lymph flow rates are high, it is becoming clear that there are major developmental differences in the ability of the lymphatic system to pump fluid from the interstitial spaces back to the circulation. These differences are illustrated by the lymph flow function curves in Fig. 14.9. In order to interpret these curves, recall that normal outflow pressure for the lymphatic system is venous pressure. In Fig. 14.9 it can be seen that there is a much greater lymph flow rate at low venous

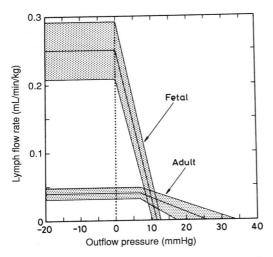

Fig. 14.9 Left thoracic duct lymph-flow function curves in ovine fetus and adult. Data are mean ± 2 × SE. Normal outflow pressure is venous pressure which averages 3–4 mmHg. Measurements were made in animals with chronically implanted lymphatic and vascular catheters. (Taken with permission from Brace, R.A. (1991). Fluid distribution in the fetus and neonate. In *Fetal and neonatal physiology* (ed. R. Polin and W. Fox). Saunders, Philadelphia.)

pressures in fetal than in adult sheep. In addition, thoracic duct lymph flow rate decreases as venous pressure is elevated above zero in the fetus, whereas lymph flow in the adult decreases only as venous pressure is elevated to values considerably greater than normal. In the near-term sheep fetus, venous pressure is the same as in the adult and averages 3–4 mmHg. Thus, normal venous pressure in the fetus appears to be an impediment to basal lymph flow and any increase in venous pressure causes a further reduction in lymph flow rate. Furthermore, lymph flow stops in the fetus when venous pressure is elevated to only 11 mmHg, whereas venous pressure must be increased to 25 mmHg before thoracic duct lymph flow stops in the adult sheep.

Even though basal lymph flow rates are high in the fetus, its lymphatic system is quite capable of responding to changes in interstitial volume. Fetal thoracic duct lymph flow increased following rapid vascular volume loading with 20 mL/kg of saline. The increase in lymphatic volume flow over 30 min is equivalent to 5 per cent of the infused volume, the same percentage as occurs in the adult under similar volume loading conditions. Thus, the fetal lymphatic system appears to be as responsive as the adult's following volume loading. Conversely, following acute haemorrhage, fetal left thoracic duct lymph flow rate decreases as interstitial fluid is reabsorbed across the capillary wall into the circulation. Collectively, these observations show that the

fetal lymphatics are in intimate association with the sites of fluid exchange across the capillary wall.

Placental fluid movements

Most of the fluid within the fetal compartment enters the fetus via the placenta. Because several disease states, including fetal oedema (hydrops fetalis), polyhydramnios (excess amniotic fluid), and oligohydramnios (abnormally small volume of amniotic fluid), are characterized by an abnormal fetal fluid status, scientists and clinicians have long been fascinated by the question, 'How does the fetus acquire fluid from its mother?' Placental solute transfer must occur because of either diffusion, bulk flow (i.e. convection or ultrafiltration), or specialized transport mechanisms. Transplacental water movements, however, occur only via the first two of these, because water is not actively transported. Diffusion requires a concentration gradient across the placenta and ultrafiltration occurs as the result of hydrostatic or osmotic pressure gradients.

Early studies in animal fetuses post-mortem found fetal osmolality to be higher than maternal, suggesting that fluids moved osmotically across the placenta into the fetus. However, for many years it has been known that this elevated osmolality occurs as a consequence of fetal death. In humans and several other species, maternal and fetal osmolalities are equal as are the plasma concentrations of the major electrolytes such as sodium and chloride. Thus, the driving force that leads to the net accumulation of fluid within the fetus is not readily identified. In animals with low placental permeabilities, such as sheep and goats, maternal osmolality (as well as plasma sodium and chloride ion concentrations) is slightly but significantly higher than that of the fetus. This has been termed the 'osmotic paradox' because it appears that the mother should be continually removing water from the fetus by osmosis, a condition clearly incompatible with fetal life. However, it has been surmised that the sheep and goat placentae function as semipermeable membranes with reflection coefficients for sodium and chloride of approximately 0.8. The net result is that the osmolalities that are effective at the placental membrane are essentially equal. Thus, it remains unclear in these species as to what is responsible for the long-term transplacental acquisition of fluid by the fetus.

On a short-term basis, it is clear that changes in maternal osmolality have direct effects on fetal fluid balance. In humans, intravenous fluids are frequently administered to pregnant women in labour. Alterations in fetal osmolality (as determined from cord blood at delivery) occur in parallel with those in maternal osmolality as a consequence of the fluids administered to the mother. In sheep, many studies have shown that hypertonic fluids administered to the mother are effective at elevating the osmolality of fetal blood. However, there is a delay in the fetal responses and 1–2 h are required for fetal osmolality to equilibrate with that of the mother as seen in Fig. 14.10. The rise in fetal osmolality is due at least in part to an osmotic dehydration of the fetus because fetal blood volume decreased by 11 per cent at 15 min after bolus injection of hypertonic NaCl into the ewe (Fig. 14.10). However, it is unclear whether the fetus remains dehydrated for long because blood volume returned to normal within 1 h. In addition, in ewes that were given only small amounts of water for several days, fetal blood volume increased normally even though fetal and maternal osmolalities were elevated. These observations suggest that the long-term rise in fetal osmolality during maternal hyperosmolality may occur largely as a consequence of solute transfer to the fetus rather than of loss of water to the mother. In addition to osmolality changes, there appear to be other fluid interactions between the fetus and its mother. Thus, amniotic fluid volume is low in healthy women who have low plasma volumes and is high in women with high plasma volumes.

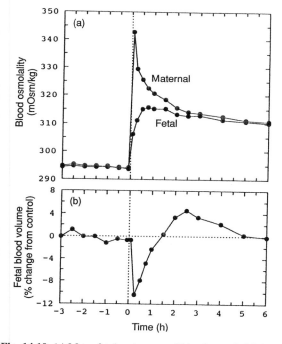

Fig. 14.10 (a) Mean fetal and maternal blood osmolalities following intravenous injection of hypertonic NaCl into pregnant sheep and (b) the induced changes in fetal blood volume. (Data from Woods, L.L. and Brace, R.A. (1986). Fetal blood volume, vascular pressure, and heart responses to fetal and maternal hyperosmolality. *American Journal of Physiology*, **251**, H716–H721.)

Because of difficulties in measuring the small fluxes across the placenta under basal conditions, a number of studies have explored the effects of large excesses of fluid or salts on fetal fluid balance. During an intravascular infusion of 4 L of physiological saline over 4 h into ovine fetuses (i.e. over 100 per cent of body weight over 4 h), fetuses developed either no or mild oedema. Instead, fetal venous and arterial pressures increased by 3 and 7 mmHg, respectively, and 500 mL/h of isotonic fluid crossed the placenta. This requires the permeability of the placenta to be 100-fold greater than previously estimated. In contrast, when sufficient fluids were infused intravenously into the ewe to raise maternal venous and arterial pressures by 7 and 30 mmHg, respectively, no detectable transplacental fluid movements into the fetus occurred. It is presently unclear why the ovine placenta appears to function as a one-way valve, allowing fluid transfer only away from the fetus in the presence of transplacental hydrostatic pressure gradients.

Infusions of concentrated salt solutions also provide unique insights into the role of the placenta in regulating fetal fluid balance. Even though it has been speculated that an excess of NaCl in the fetal compartment would cause fetal oedema and/or polyhydramnios in sheep, an intravenous infusion into the fetus of 5 M NaCl (240 mEq/day each of Na^+ and Cl^-) for 3 days resulted in no excess fluid or salt accumulation in the fetal compartment. Instead, the infused Na^+ and Cl^- were transferred to the maternal compartment either against, or in the absence of, transplacental concentration gradients. In contrast, infusion of equivalent amounts of Na^+-lactate produced severe polyhydramnios. Thus, solutes that have a higher molecular weight than Na^+ or Cl^- may play a significant role in osmotically attracting fluid into the fetus from its mother. In fact, in human fetuses that are severely anaemic, hydrops fetalis (i.e. severe fetal oedema) develops only in those fetuses that have elevated lactate concentrations. Thus, in humans as well as in animal species, the role played by individual solutes in determining fetal fluid balance appears to depend on molecular size.

Urinary output

The topic of fetal urine flow is described in detail in Chapter 13. Fetal urinary output will be discussed here only as it relates to net fluid changes in the fetus.

In the ovine fetus during the last third of gestation, urinary output averages 0.2 mL/min/kg of fetal body weight. Over each 24-h period, this amounts to 25–30 per cent of fetal body weight. Not only is this very much greater than in the adult, but it also emphasizes that only modest changes in fetal urinary output potentially can have large effects on fetal and amniotic fluid balance when consid-

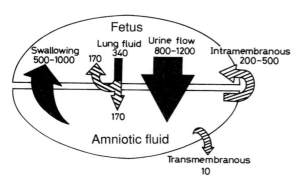

Fig. 14.11 Summary of net flows (mL/day) between the near-term fetus and the amniotic fluid as estimated from literature values. The size of the arrow is proportional to the magnitude of flow. Solid arrows represent measured flows and hatched arrows represent estimated flows. (Taken with permission from Gilbert, W.M., Moore, T.R., and Brace, R.A. (1991). Amniotic fluid volume dynamics. *Fetal Medicine Review*, **3**, 89–104. Cambridge University Press.)

ered over a period of days. The daily contribution of fetal urine to amniotic fluid is compared to the net fluxes across the other pathways in Fig. 14.11. The figure shows that fetal urine is the major source of fluid entering the amniotic space in late gestation.

In the human fetus, urine flow was estimated by ultrasound measurements of changes in bladder dimensions and found to average 15 per cent of fetal body weight per day. However, recent investigations have reported that previous fetal urine flow measurements that utilized ultrasound suffered from a methodological error due to frequent fetal micturition and gave a result that was only about half the true flow. Thus, urinary output during the latter half of gestation in both human and ovine fetuses may average 25–30 per cent of body weight per day. Further, in pregnant sheep that have continuous access to food and water, fetal urinary output undergoes a strong diurnal rhythm, with minimal flow a few hours before noon and maximal flow shortly before midnight. The latter observation shows that estimating net daily urine flow from measurements made during daylight hours would lead to a significant underestimation of total urine flow.

Changes in fetal urine flow in response to both fetal haemorrhage and vascular volume loading show that urine flow changes contribute to the regulation of blood volume. In response to rapid haemorrhage, urine flow decreases as would be expected if the fetus were attempting to restore its blood volume to normal. It is unknown whether fetal urine flow changes in response to a slow haemorrhage, although the fetus rapidly restores its blood volume.

In response to vascular volume loading of the ovine fetus, urine flow may or may not increase, depending on the rate and extent of the volume loading. With slow infusions of isotonic saline solutions over 30–60 min, urine flow changes little. In contrast, rapid infusions over 5 min elicit a transient diuresis. This difference is due largely to the combination of the high fetal capillary filtration coefficient plus the high interstitial compliance described above. With slow infusions over many minutes, very little of the infused fluid remains within the vasculature so there is little stimulus for a diuresis. With rapid infusions, fetal blood volume and pressure increase transiently and the plasma concentrations of arginine vasopressin, renin, and atrial natriuretic factor all change in the direction needed for promoting a diuresis and thus urine flow increases. However, the diuresis is short-lived and urine flow returns to normal levels within 20 min because fetal blood volume returns to normal, thereby removing the stimuli for the diuresis.

During long-term vascular volume loading, the fetal kidneys provide an extremely powerful mechanism that prevents the fetus from becoming volume overloaded. This concept is demonstrated by the urine flow responses of fetal sheep during late gestation to volume loading over 72 h (Fig. 14.12). During intravenous infusions of 1, 2,

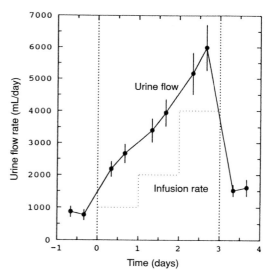

Fig. 14.12 Ovine fetal urine flow responses to infusion over 24 h of 1, 2, and 4 L of physiological saline into the fetal circulation. Data are mean ± SE. (Data from Brace, R.A. (1989). Fetal blood volume, urine flow, swallowing and amniotic fluid volume responses to long-term intravascular saline infusions. *American Journal of Obstetrics and Gynecology*, **161**, 1049–54.)

and 4 L/24-h period, fetal blood volume and extracellular fluid volume increased only slightly while fetal urine flow increased by an amount which exceeded the infusion rate even though the rate of 4 L/24-h was in excess of total fetal body weight/day. This enormous urine flow response to long-term volume loading may explain why polyhydramnios sometimes develops in human fetuses even though the fetal body appears normally hydrated. On the other hand, hydrops fetalis often occurs simultaneously with polyhydramnios in human fetuses. This suggests that the immature human fetus may not be able to undergo the extensive diuresis seen in the late-gestation ovine fetus, most probably because of the immaturity of its kidneys.

Swallowing

An analysis of the composition of the fluid swallowed by the ovine fetus found that pure amniotic fluid is sometimes swallowed. At other times, the fetus may swallow tracheal fluid as it exits the lungs, a combination of tracheal plus amniotic fluids, or a combination of tracheal, amniotic, and oronasal fluids. Two methods have been used for measuring the rate of fetal swallowing. Collection of fluid swallowed by the fetus showed that most of the fluid is swallowed during two to seven bouts of rapid swallowing activity per day with up to 200 mL swallowed during a single bout. The rate of fetal swallowing has also been estimated from the clearance rate of non-permeant markers following their injection into the amniotic sac. However, this technique records only the volume of amniotic fluid that is swallowed and does not include the volume of lung or oronasal fluid that is swallowed if these fluids do not first enter the amniotic space. Collectively, published data from the clearance technique indicate that human and ovine fetuses swallow a volume of amniotic fluid nearly equal to the volume of urine produced each day (Fig. 14.11). In sheep fetuses, the volume of amniotic fluid swallowed each day increases from 15 per cent of body weight at 100 days' gestation (i.e. 150 mL/kg/day) to 50 per cent of body weight at term as seen in Fig. 14.13. This contradicts the observation that fetal urinary output averages 25–30 per cent of body weight over the same range of gestational ages. Herein lies a major problem. That is, rarely are fetal urine flow and swallowing measured simultaneously. When they were, swallowing of amniotic fluid late in gestation averaged 50–200 mL/day less than urine flow.

Tracheal flow and lung fluid volume

Prior to birth, the developing lungs are filled with fluid rather than with gas as after birth. This fluid is formed by an active secretion of chloride ions into the alveolar space, resulting in a progressive accumulation of fluid in the lungs as gestation progresses. Fluid secretion by the fetal lungs and tracheal movements of lung fluid are reviewed in detail in Chapter 10. The present discussion focuses on the contributions of fetal lung fluid to amniotic fluid.

In fetal lambs, tracheal flow rate averages 4.4 mL/kg/h during the last half of gestation or roughly 10–15 per cent of body weight per day (Fig. 14.11). Near term in fetal lambs, lung liquid volume averages 40 mL/kg of body weight or slightly more than one-half of plasma volume. This large volume suggests that lung liquid may be a potential source of fluid to the fetus but the extent to which lung fluid is transferred to the circulation has not been explored.

In human fetuses, much of tracheal outflow enters the amniotic space as evidenced by the fact that surfactants from the fetal lungs are found in the amniotic fluid. In sheep fetuses, several older studies have suggested that most of the lung effluent is swallowed as it exits the trachea. This is supported by the observation that little sur-

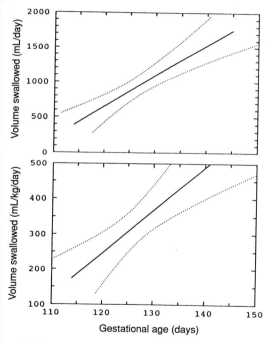

Fig. 14.13 Ovine fetal swallowing of amniotic fluid as a function of gestational age. Dotted lines are 95 per cent confidence intervals around the regression line. Data were obtained from the kinetics of I[125]-labelled disappearance following injection into the amniotic fluid. (Data from Tomoda, S., Brace, R.A. and Longo, L.D. (1985). Amniotic fluid volume and fetal swallowing rate in sheep. *American Journal of Physiology*, **249**, R133–R138.)

factant appears in the amniotic fluid even though it is pre-sent in tracheal fluid. Because recent studies have shown that some of the fluid secreted by the ovine fetal lungs enters the amniotic fluid, further studies are needed to clar-ify the proportions of tracheal outflow that enter the oesophagus and amniotic sac.

Transmembranous flow

The interface of the fetal membranes and the wall of the uterus provides a large surface that may potentially allow the exchange of water and solute between the amniotic fluid and maternal blood. Following death of the ovine fetus, amniotic sodium concentration decreases from 100 to 10 mM/L over 24–48 h and amniotic fluid volume decreases. This observation shows that the transmembranous route clearly functions under select conditions, but the significance of this route under normal conditions is unknown. The observation that ovine amniotic fluid osmo-lality increased when the pregnant ewe was made hyper-tonic was once interpreted to indicate the exchange of fluid by the transmembranous route. More recent studies have shown that maternal hyperosmolality produces fetal hyper-osmolality because of osmotically induced fluid movement across the placenta. The fetus responds by creating hyper-tonic urine and this, in turn, is responsible for the increased amniotic fluid osmolality. This, together with other observa-tions, suggests that, under normal conditions, little net fluid exchange occurs across the fetal membranes. However, the amnion and chorion are permeable to water and a variety of solutes; thus, some transmembranous fluxes must occur. Recent studies have estimated that perhaps 10 mL/day may enter the maternal compartment via the transmembranous pathway in sheep (Fig. 14.11). This, in combination with observations that the transmembranous permeabilities to urea and carbon dioxide are undetectably low, suggests that the transmembranous pathway may play only a minor role in regulating fetal fluid balance under normal conditions.

Intramembranous flow

The intramembranous pathway allows the direct exchange of water and solutes between amniotic fluid and fetal blood at several potential sites: the fetal surface of the pla-centa; the membranes in those species in which the mem-branes are perfused with blood; the surface of the umbilical cord; the fetal skin; and possibly the oronasal cavities. Because of the multicomponent nature of the intramembranous route, its contribution to fetal fluid bal-ance has been difficult to quantify. There are nonetheless many observations that show that the intramembranous pathway is a physiologically significant route. Studies in

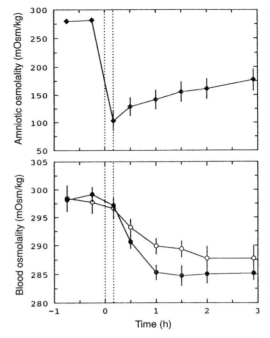

Fig. 14.14 Fetal blood and amniotic fluid osmolalities following infusion of 1.5 L of warmed distilled water into amniotic cavity from 0–10 minutes in ovine fetuses with an intact (open circles) or ligated (filled circles) oesophagus. (Data from Gilbert, W.M. and Brace, R.A. (1989). The missing link in amniotic fluid volume regulation: intramembranous absorption. *Obstetrics and Gynecology*, **74**, 748–54.)

humans, monkeys, and sheep have found that substances, such as amino acids, iodine, arginine vasopressin, and water, that are injected into the amniotic fluid appear in fetal blood much too rapidly to be explained by fetal swal-lowing. In fact, in studies in the monkey and sheep, liga-tion of the fetal oesophagus did not inhibit the vascular uptake of substances injected into the amniotic fluid, and nephrectomy did not inhibit transfer from the vascular to the amniotic compartments. As an example, Fig. 14.14 compares the changes in fetal blood osmolality following infusion of 1.5 L of water at body temperature into the amniotic compartment of sheep. In this case, osmolality decreased more rapidly in the fetuses with a ligated oesophagus. This was interpreted to show that oesophageal ligation increased the fluid conductance of the intramem-branous pathway. Similarly, arginine vasopressin (molecu-lar weight, 1084 Da) appears rapidly in the fetal circulation following intraamniotic injection and, in fetuses with a ligated oesophagus, plasma levels rose to twice that in fetuses with an intact oesophagus. This observation of an increased permeability of the intramembranous pathway following oesophageal ligation is important because it provides an explanation for the observation that oesophageal ligation does not produce polyhydramnios in sheep even though fetal urinary output remains normal.

A second approach to evaluating the significance of the intramembranous pathway has been to add up simultaneously measured flows into and out of the amniotic space. On average, late in gestation, fetal urinary output is 50–200 mL/day greater than fetal swallowing of amniotic fluid. The net absorption by the intramembranous pathway equals this volume plus the volume of any lung fluid or fluid secreted from the fetal head which enters the amniotic space. This large flow through the intramembranous pathway is easily possible due to two factors: (1) the hydraulic conductance (i.e. filtration coefficient) of the intramembranous pathway has been estimated to be 5 per cent of that of the placenta; and (2) a large osmotic force is present for the absorption of amniotic fluid because amniotic osmolality normally is 20–30 mOsm/kg lower than fetal blood osmolality. In addition, the reflection coefficients of the intramembranous pathway for sodium and chloride ions have been estimated to be 0.4. Thus, low to medium molecular weight substances, including arginine vasopressin, are reabsorbed from the amniotic fluid into the fetal circulation by this pathway.

The contribution of the intramembranous pathway to fetal and amniotic fluid balance is only now becoming clear. The reason for this is that the exchange capacity of the fetal blood just beneath the fetal surface of the placenta which contacts the amniotic fluid has not been recognized. In addition, in animals such as sheep, although the macroscopic vessels that run between the amnion and chorion or allantois and chorion have long been recognized, there are extensive microscopic networks of fetal vessels which cover the entire amnion and chorion. When viewed under the microscope, 50 per cent of the surface area of the ovine chorion is covered by microvessels and this does not vary with gestational age. By assuming that these vessels are circular in cross-section, the total surface area of the microvessels in the chorion averages 150 per cent of the surface area of the chorion (i.e. 50 per cent × 3.14159). The microvessels that cover the outer surface of the ovine amnion differ in that the amniotic vessels are longer, less dense, and have fewer interconnections. The amniotic microvessels cover 30 per cent of the surface area of the amnion at midterm and this decreases to 17 per cent at term. This corresponds to a total microvessel surface area of 100 per cent of that of the amniotic membrane at midgestation, decreasing to 50 per cent as term approaches. These vessels are important because there is a large potential for exchange between fetal blood and the amniotic

fluid through these vessels. This can be seen from the perspective that, relative to fetal weight, the surface area of the microvessels in the amnion averages 6 cm^2/g of fetus at midterm and this decreases 10-fold as term approaches. In comparison, the surface area of the microvessels within the chorion is 15 cm^2/g of fetus at midterm, decreasing to 1.5 cm^2/gm at term. A further understanding of the role of these intramembranous vessels in the regulation of fetal and amniotic fluid balance requires that their permeability and transport characteristics be established.

Water as a byproduct of metabolism

It is important to realize that water is a byproduct of metabolism. With a high fetal metabolic rate, water from metabolism may provide 20 per cent of the water needed for growth in rapidly growing species such as sheep, and 50 per cent of that needed in slower growing species such as humans, particularly late in gestation when large quantities of fat are being deposited in the fetal body.

Overall regulation of fetal fluid balance

From the perspective of overall regulation, it is important to recognize that each fluid movement within the fetus is either actively and/or passively regulated, and that many of the regulatory processes interact synergistically for one common purpose. That purpose is to maintain fetal blood volume so that cardiac output is sufficient to promote adequate perfusion of fetal tissues. Thus, under a number of circumstances, the simultaneous changes in fetal swallowing, urinary output, transcapillary flow, lymph flow, tracheal flow, and intramembranous flow occur, at least in part, in order to keep blood volume within its normal limits. The mediators of this commonality of response include the autonomic nervous and endocrine systems as well as the passive responses to alterations in vascular pressures or osmolalities.

An integration of the simultaneous flows and their interactive regulatory mechanisms is required to produce an overall understanding of the regulation of fetal fluid balance. This is not yet possible for the fetus because of the paucity of detailed studies of each and every pathway for fluid movement. The placental, lymphatic, transmembranous, and intramembranous pathways are particularly difficult to study *in vivo* due to their inaccessibility and the difficulty in separating their respective fluxes from others occurring simultaneously. Nonetheless, progress has been steady and it is clear that the fetus is quite efficient at regulating its fluid balance.

It is also clear that, on a short-term basis, the primary movement of fluid movement in the fetus is across its body capillaries because that pathway has a much greater hydraulic conductance than other pathways. In addition, on a long-term basis, the fetus can utilize the amniotic (and allantoic) fluid(s) not only as a source of fluid but also as an overflow reservoir, although the latter may vary with gestational age. The placenta, of course, is the major long-term determinant of fluid status of the fetus. However, the placental and/or fetal mechanism(s) that provide the fine adjustments required for maintaining fetal fluids within their normal ranges are yet to be understood.

Further reading

Battaglia, F.C. and Meschia, G. (1986). *An introduction to fetal physiology*. Academic Press, Orlando, Florida.

Bell, R.J. and Wintour, E.M. (1985). The effect of maternal water deprivation on ovine fetal blood volume. *Quarterly Journal of Experimental Physiology*, **70**, 95–9.

Brace, R.A. (1984). Blood volume in the fetus and methods for its measurement. In *Animal models in fetal medicine* (ed. P.W. Nathanielsz), pp. 19–36. Perinatology Press, Ithaca, New York.

Brace, R.A. (1986). Amniotic fluid volume and its relationship to fetal fluid balance: review of experimental data. *Seminars in Perinatology*, **10**, 103–12.

Brace, R.A. (1991). Fluid distribution in the fetus and neonate. In *Fetal and neonatal physiology* (ed. R. Polin and W. Fox), pp. 1288–98. Saunders, Philadelphia.

Brace, R.A., Ross, M.G., and Robillard, J.E. (1989). *Fetal and neonatal body fluids: the scientific basis for clinical practice*. Perinatology Press, Ithaca, New York.

Faber, J.J. and Anderson, D.F. (1990). Model study of placental water transfer and causes of fetal water disease in sheep. *American Journal of Physiology*, **258**, R1257–R1270.

Faber, J.J. and Thornburg, K.L. (1983). *Placental physiology*. Raven Press, New York.

Friis-Hansen, B. (1961). Body water compartments in children: changes during growth and related changes in body composition. *Pediatrics*, **28**, 169–81.

Gilbert, W.M. and Brace, R.A. (1989). The missing link in amniotic fluid volume regulation: intramembranous absorption *Obstetrics and Gynecology*, **74**, 748–54.

Gilbert, W.M., Moore, T.R., and Brace, R.A. (1991). Amniotic fluid dynamics. *Fetal Medicine Review*, **3**, 89–104.

Harding, R., Sigger, J.N., Poore, E.R., and Johnson, P. (1984). Ingestion in fetal sheep and its relation to sleep states and breathing movements. *Quarterly Journal of Experimental Physiology*, **69**, 477–86.

Lingwood, B., Hardy, K.J., Horacek, I., McPhee, M.L., Scoggins, B.A., and Wintour, E.M. (1978). The effects of antidiuretic hormone on urine flow and composition in the chronically-cannulated ovine fetus. *Quarterly Journal of Experimental Physiology*, **63**, 315–30.

Lingwood, B.E., Hardy, K.J., Long, J.G., McPhee, M., and Wintour, E.M. (1980). Amniotic fluid volume and composition following experimental manipulations in sheep. *Obstetrics and Gynecology*, **56**, 451–8.

Ross, M.G., Ervin, M.G., Leake, R.D., and Fisher, D.A. (1985). Amniotic fluid ionic concentration in response to chronic fetal vasopressin infusion. *American Journal of Physiology*, **249**, E287–E291.

Towstoless, M.K., Congui, M., Coghlan, J.P., and Wintour, E.M. (1987). Placental and renal control of plasma osmolality in chronically cannulated ovine fetus. *American Journal Physiology*, **253**, R389–R395.

Wilbur, W.J., Power, G.G., and Longo, L.D. (1978). Water exchange in the placenta a mathematical model. *American Journal of Physiology*, **235**, R181–R199.

Wintour, E.M., Barnes, A., Brown, E.H., Hardy, K.J., Horacek, I., McDougall, J.G., and Scoggins, B.A. (1978). Regulation of amniotic fluid volume and composition in the ovine fetus. *Obstetrics and Gynecology*, **52**, 689–93.

Wintour, E.M., Laurence, B.M., and Lingwood, B.E. (1986). Anatomy, physiology, and pathology of the amniotic and allantoic compartments in the sheep and cow. *Australian Veterinary Journal*, **63**, 216–21.

15. Development of the gastrointestinal tract

Jeffrey F. Trahair and Richard Harding

Survival of the newly born mammal depends on its ability to extract nutrients from suckled milk soon after birth. Thus, by the time of birth, the gastrointestinal (GI) tract must have become sufficiently developed to allow it to conduct digesta through it, to modify nutrients into a digestible form, and to extract nutrients into the circulation. The stage of maturation reached by the GI tract at birth depends on the duration of intrauterine development. In altricial species, such as rodents, the GI tract at birth is less well developed than in precocial species such as guinea-pigs and sheep. In marsupials, which have very short gestation periods, the GI tract at birth is quite rudimentary, but still capable of fulfilling its basic functions.

Most of our knowledge of the early development of the GI tract is derived from studies on rodents (altricial) and sheep (precocial) and from observations in humans. As this chapter is concerned with the prenatal development of the GI tract, much of it relies on recent information obtained from fetal sheep, which can be studied *in utero*. The reader should be aware that there exists elsewhere a large body of information relating to GI development in the suckling young, particularly of altricial species in which much of the functional development occurs after birth.

The GI tract is one of the most adaptable organs in the body. It can alter its digestive capacity in response to both the diet and the substrate demands of the organism. This adaptability relies on both the genetic potential of the GI system and the action of factors that regulate this potential. These regulatory factors fall into three groups: systemic factors, i.e. factors that circulate throughout the body via the blood stream; luminal factors i.e. factors in the GI lumen; and local factors, i.e. factors or processes arising as a result of interaction between different tissues and cells (including nerve cells) of the GI tract.

In the fetal environment, regulatory factors may be derived not only from the fetus, but also from the mother. Furthermore, the interactions of tissue-derived factors in the fetus are not necessarily analogous to those found in the adult; many are unique attributes of the fetal environment or even specialized functional features. These interactions are the result, therefore, of a unique set of circumstances specific to growing tissues and organ systems. In the case of the fetal GI tract there are special conditions due to the reliance on parenteral (placenta-derived) nutrition, the absence of atmospheric respiration, and the presence of an amniotic-fluid-filled environment.

Morphological development of the gastrointestinal tract

Embryology of the human GI tract

At around 22 days postconception, as a result of head and tail flexion and folding of the embryo, the yolk sac constricts and an endodermal tube is formed from its dorsal portion (Fig. 15.1). At around 25 days, the foregut and hindgut lengthen and the liver develops from extraembryonic mesoderm and yolk stalk mesenchyme within the septum transversum. The foregut gives rise to pharynx, thyroid, thymus, parathyroid, respiratory tract, oesophagus, stomach, upper duodenum, liver, and pancreas. The midgut gives rise to lower duodenum and small and large intestine (to distal third of the transverse colon), with the remainder of the bowel being derived from the hindgut.

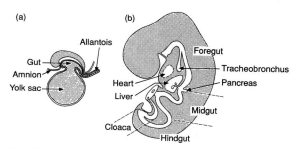

Fig. 15.1 (a) By 22 days the primitive gut develops from the dorsal part of the yolk sac. Successive elongation, folding, and flexion of the embryo results in the formation of the gut endodermal tube (primitive gut). (b) By 25 days the gut tube can be divided into foregut, midgut, and hindgut (including the cloaca). The tracheobronchus has formed. The liver and pancreas appear in the mesoderm.

From 25 to 45 days the foregut and mouth structures develop via a series of transformations of the brachial arch and cleft structures. Late in the fourth week, the lung develops as a diverticulum of the foregut. It is thought that failure of separation of these two structures results in a range of tracheo-oesophageal fistulae, with or without oesophageal atresia.

The stomach develops as a swelling and dilatation of the caudal part of the foregut by about 5 weeks. Gastric pits appear by 7 weeks, first in the fundus, and then later in the pylorus and cardia. The duodenum grows rapidly from the terminal part of the foregut and the cranial portion of the midgut forming a loop at first. The midgut increases in length resulting in a U-shaped loop, then into a more complex arrangement forcing the intestine out of the abdominal cavity, and herniating into the vitelline stalk. Between 10 and 12 weeks the intestine is retracted back into the abdominal cavity, which has grown large enough to accommodate it. During this process and shortly thereafter, a series of rotations fixes the GI tract into the final position in the abdominal cavity.

In the hindgut a ventral diverticulum develops into the allantois. Caudally, the cloaca develops. At around 6 weeks, division of the cloaca into rectum (dorsal) and urogenital sinus (ventral) takes place. At around 8 weeks the cloacal membrane closing the rectum ruptures.

Congenital anomalies of the GI tract are established in the early weeks of development. These include failure of regression of the vitello-intestinal duct, failure in retraction of herniated intestine and/or malrotations, the group of symptoms comprising the VATER association (acronym for concurrence of anomalies that include vertebral and anal malformation, tracheo-oesophageal fistula with oesophageal atresia, and radial or renal dysplasia).

The pancreas and biliary tree arise from endodermal outgrowths of the primitive foregut at around 4 weeks. The pancreas forms from two anlagen that fuse at around 7 weeks. At the same time the duct systems also fuse to form the common pancreatic duct.

Development of the gut wall components

The primitive endodermal gut tube ultimately gives rise to the epithelial elements of the GI tract, while the mesenchyme gives rise to the muscle and fibrous coats and their components. Nerve cells migrate into the gut wall mostly in a caudal direction from proximal portions of the tract, but also cranially from caudal parts. By 5 weeks the vagal trunk reaches the oesophagus. The most caudal regions also receive innervation from pelvic and pre-aortic plexuses. At 6 weeks progenitor nerve cells appear in the stomach wall, by 7 weeks in the midgut, by 8 weeks in

proximal colon, and by 12 weeks in the rectum. Nerves are arranged in ganglia comprising three plexuses: Auerbach's plexus (myenteric) between inner and outer muscle layers of the external muscle coat; Meissner's plexus, close to the muscularis mucosa; and a third plexus (Henle's plexus) on the inner aspect of the inner circular layer.

Despite the common origin of mesenchyme and endoderm, and the generalized similarity of the layered structure of muscularis, submucosa, and mucosa, the architectural development of the GI tract, especially of the mucosa, is highly region-specific. In general, the endodermal tube is comprised of only a single cell thickness of undifferentiated epithelial cells. With successive cell divisions this layer stratifies, and is then transformed into the appropriate specialized arrangements found in each region.

Oesophagus

In the oesophagus the lumen is established by 10 weeks, lined by stratified columnar ciliated epithelium, which is replaced by a stratified squamous epithelium by 5 months.

Stomach

The glandular pits of the stomach appear at 6 weeks with rapid cytodifferentiation of the various cell types occurring over the next few weeks. By 12 weeks parietal, chief, enteroendocrine, and mucous cells have appeared.

Small intestine and the apical endocytic complex

The development of the small intestine has been studied more than any other region of the tract. For species of long gestation (e.g. human), the fetal sheep provides a valuable experimental model for studies on GI development and its control. At 20–30 days, the fetal sheep's small intestinal tube is lined by a single layer of poorly differentiated columnar endodermal cells (Figs 15.2 and 15.3). The gut is herniated outside the body wall. A dense mesenchymal sleeve surrounds the endodermal tube. At the periphery myoblast differentiation is observed. Over the next 10 days the intestine elongates. The endodermal cells rapidly proliferate and form a stratified epithelium that occludes the lumen (Figs 15.4 and 15.5). The muscularis externa develops at this stage, the inner circular layer appearing first, followed by the outer longitudinal layer, together with prominent nerve cells of the developing Auerbach's plexus. Muscle development takes place in a craniocaudal direction. By 45 days the abdominal cavity has enlarged sufficiently for the gut to retract. Villus formation begins first in more proximal regions, proceeding soon

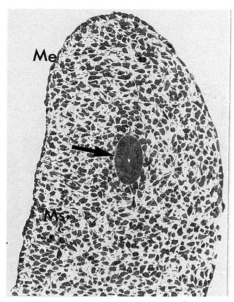

Fig. 15.2 Light micrograph of fetal sheep small intestine at 27 days of gestation. The intestine is still outside the abdomen. The central endodermal tube (arrow) is surrounded by a sheath of mesenchyme (Ms), covered by a mesothelium (Me). × 230. (Reproduced with permission from Trahair, J.F. and Robinson, P.M. (1986). *Anatomical Record*, **240**, 294–303.)

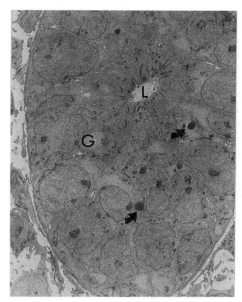

Fig. 15.3 Electron micrographic detail of endodermal tube (Fig. 15.2). A narrow lumen (L) is present. Epithelial cells are undifferentiated with moderate amounts of glycogen (G) and lipid droplets (arrows). × 4500. (Reproduced with permission from Trahair, J.F. and Robinson, P.M. (1986). *Anatomical Record*, **240**, 294–303.)

afterwards in a proximodistal direction (Fig. 15.6). By 60 days goblet cells, absorptive cells, and enteroendocrine cells are present.

Between 60 and 70 days the intestinal lumen becomes patent, presumably as ingestion begins (see below). At the same time a significant transformation within the entero-cytes lining the villi takes place. A complex endocytic apparatus develops, first in the villus tip cells, which are the most mature cells present on the villus, then extending to the whole villus enterocyte population. This complex, termed the apical endocytic complex (AEC, sometimes also called the apical canalicular/tubule system, ACS), is found in enterocytes of every species at some stage during development, and is not normally seen in adult enterocytes (with the notable exception of some species of fish).

As its name implies, the AEC is involved with the uptake (endocytosis) of material from the lumen, and its transport across the epithelium into the blood stream. In some species, its presence after birth is essential for survival as it is involved with the uptake and transfer of colostral antibodies and the establishment of passive immunity, which protects the neonate from disease. Its function *in utero* is not known but, given that swallowed

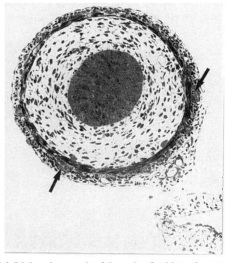

Fig. 15.4 Light micrograph of the ovine fetal intestine at 35 days of gestation. The intestine is still outside the body cavity. External muscle coats have developed and neural infiltration has advanced (areas indicated by arrows). The central endodermal tube has thickened, and the epithelial cells stratified. × 150. (Reproduced with permission from Trahair, J.F. and Robinson, P.M. (1986). *Anatomical Record*, **240**, 294–303.)

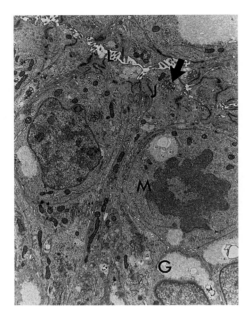

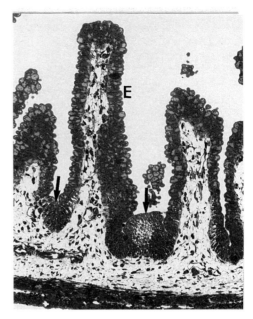

Fig. 15.5 Electron micrograph of enterocytes from Fig. 15.4. The lumen (L) is obliterated by the crowding of actively dividing epithelial cells. Well developed junctional complexes (J) are present at the luminal aspect. A cell undergoing mitosis is demonstrated (M). Epithelial cells still possess considerable cytoplasmic glycogen (G). × 4500. (Reproduced with permission from Trahair, J.F. and Robinson, P.M. (1986). *Anatomical Record*, **240**, 294–303.)

Fig. 15.6 Light micrograph of proximal small intestine at 75 days in a fetal sheep. Villi have developed and the lumen is patent and fluid filled. Epithelial cells (E) are polyhedral, bulging at their apical poles. Areas between villi are still stratified (arrows). × 160. (Reproduced with permission from Trahair, J.F. and Robinson, P.M. (1986). *Anatomical Record*, **240**, 294–303.)

Fig. 15.7 Light micrograph of proximal small intestine in a fetal sheep at 90 days. The villi are covered with columnar epithelial cells. Vacuoles (arrows) have begun to appear in the apical regions of villus tip cells. × 180. (Reproduced with permission from Trahair, J.F. and Robinson, P.M. (1986). *Anatomical Record*, **240**, 294–303.)

fluid contains a wide variety of hormones and factors derived from ingested amniotic fluid, lung liquid, or oronasal secretions, it is likely that these substances find their way into the fetal circulation via this route. The possible significance of this pathway will be discussed in the sections entitled 'The role of fluid ingestion in development of the GI tract' and 'Intestinal transport of macromolecules'.

At 90 days in the ovine fetus the features of the AEC are readily demonstrated (Figs 15.7–15.12). In the supranuclear region, within a prominent glycogen deposit, vacuoles appear. The vacuoles increase in size by fusion with adjacent smaller vacuoles and contain a range of densities of flocculent material, as well as membranous fragments. Between adjacent microvilli, deep pits and channels form the AEC. Numerous vesicular, sacular, and tubular components comprising this network can be visualized. The inner membrane leaflet displays a regular particulate array, often giving a ladder-like appearance in longitudinal section (Figs 15.10 and 15.11). The particles have been identified as being the site of hexosaminidase activity, an enzyme involved in the digestion of milk sugars. Within the same region as the AEC, numerous coated vesicles are also present (Fig. 15.12); these may be important in selective pathways of uptake and transfer or, alternatively, may be part of a membrane recycling process.

Another marker for the membranes of the AEC has also been described in many species. This membrane glycoprotein of molecular weight 55–61 kDa was first isolated from the suckling rat, but also appears to be present (as detected by monoclonal and polyclonal antibodies) in fetal sheep and human intestine. Vacuolated cells develop most fully in more distal regions of the small intestine (and proximal colon, while villi are still present). By 120 days in the ovine fetus the vacuoles occupy most of the cytoplasm of the enterocytes and vacuolated enterocytes cover at least the upper 60 per cent of the villus (Fig. 15.13), while in more proximal regions vacuoles have disappeared and enterocytes are relatively mature (Fig. 15.14).

In sheep, enterocytes with a well developed AEC are still present in the ileum for a few days after birth, corresponding to the time when the gut is still permeable to maternal antibodies. Although the neonatal human intestine is, on the whole, more permeable to whole proteins than the adult, in absolute terms very little macromolecular transport takes place after birth. This does not totally exclude the possible entry of antigens, both dietary and environmental, especially in premature babies where the

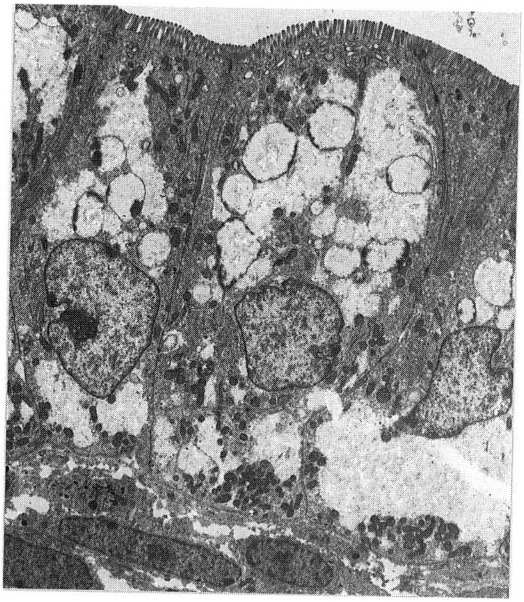

Fig. 15.8 Electron micrograph of proximal small intestine at 90 days in a fetal sheep. Vacuoles have appeared within the extensive intracellular pool of glycogen. × 3200. (Reproduced with permission from Trahair, J.F. and Robinson, P.M (1986). *Anatomical Record*, **240**, 294–303.)

Fig. 15.9 Electron micrograph detail of vacuole and contents at 90 days in ovine fetal proximal small intestine. Vacuoles increase in size by fusion of smaller vacuoles. Vacuoles contain flocculent material and membrane fragments/vesicles. × 24 000. (Reproduced with permission from Trahair, J.F. and Robinson, P.M. (1986). *Anatomical Record*, **240**, 294–303.)

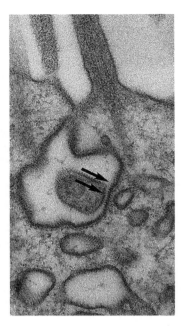

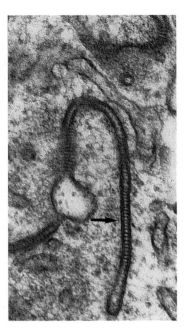

Fig. 15.10 Electron micrograph of apical endocytic complex (AEC). Deep channels and invaginations of the apical plasma membrane form continuities with vesicles and tubules of the AEC. The inner membrane leaflet is lined with a regular particulate array (arrows). × 65 000. (Reproduced with permission from Trahair, J.F. and Robinson, P.M. (1986). *Anatomical Record*, **240**, 294–303.)

Fig. 15.11 Electron micrograph of tubules and vesicles of the AEC in a fetal sheep. The regular array is particularly well demonstrated in a longitudinally sectioned tubule (arrow). × 110 000. (Reproduced with permission from Trahair, J.F. and Robinson, P.M. (1986). *Anatomical Record*, **240**, 294–303.)

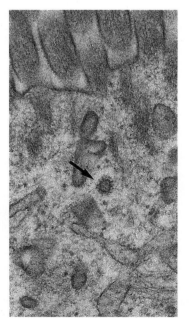

Fig. 15.12 Electron micrograph of the apical cytoplasm demonstrates the presence of numerous vesicles and tubules of the AEC. Many of the vesicles are coated on the outer membrane leaflet (arrow). × 70 000. (Reproduced with permission from Trahair, J.F. and Robinson, P.M. (1986). *Anatomical Record*, **240**, 294–303.)

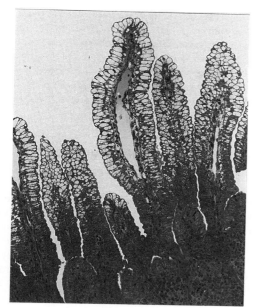

Fig. 15.13 Light micrograph of mid small intestine in a fetal sheep at 124 days. There is a gradient of vacuolation of epithelial cells lining the villi. Cells in crypts and lower villus are non-vacuolated while cells towards the tip are extensively vacuolated. × 110. (Reproduced with permission from Trahair, J.F. and Robinson, P.M. (1986). *Anatomical Record*, **240**, 294–303.)

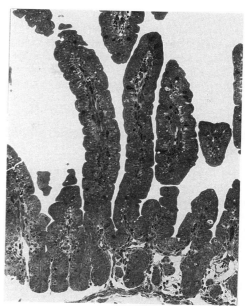

Fig. 15.14 Light micrograph of proximal small intestine in a fetal sheep at 132 days. Well developed crypts are present and the epithelium is morphologically mature. × 130. (Reproduced with permission from Trahair, J.F. and Robinson, P.M. (1986). *Anatomical Record*, **240**, 294–303.)

transfer of whole proteins is greatly increased. It is likely that this increased intestinal permeability is associated with enterocyte immaturity, although the morphological basis has not yet been documented.

Despite these functional differences, the ovine small intestine is morphologically mature well before birth (especially in its proximal regions). A question arises about its function during the weeks preceding the functional load placed on it by the demands of postnatal enteral feeding. As discussed below, it is quite possible that throughout this period the GI tract may be significantly contributing to the substrate input for growth via fetal ingestion and intestinal absorption, especially in rapidly growing fetuses, or under conditions in which placental function is compromised. Thus, enteral feeding *in utero* might represent a significant aspect of digestive function that is yet to be fully explored.

Large intestine

The large intestine develops along a similar pattern to that of the small intestine, including, somewhat surprisingly, the development of villi. These villi are more prominent in the proximal colon and, in the human fetus, reach their maximum development during the fourth or fifth month. The enterocytes covering the villi are morphologically and biochemically very similar to small intestinal enterocytes, including the presence of a well developed AEC. This suggests a similar capacity for endocytosis and hence uptake of intact proteins from the lumen. These villi disappear by a mechanism yet to be defined after the fifth month. Thereafter the epithelium is arranged as crypts, or glands, with goblet cells being the predominant cell type.

Cell kinetics of the gastrointestinal mucosa

In the adult GI tract, the epithelial cell population is constantly being replaced by new cells which are the product of a population of constantly replicating stem cells. It is generally accepted that all epithelial cell types arise from the same pool of stem cells located in the pits (stomach), crypts (intestine), or glands (colon). The constant production of cells results in a diverse range of cell ages being present at any one time, and is thought to be important in maintaining optimal functional capacity and adaptability throughout the GI tract in the face of varying conditions (e.g. acidic, proteolytic, and dietary variability). It has recently been shown that the fetal GI tract also renews itself whilst *in utero*. As the conditions that are thought to favour the need for renewal in the adult do not exist *in utero*, the purpose for this apparent waste of cells is not known.

Patterns of cell renewal in the developing GI tract

Early in development the small intestinal enterocytes appear to be capable of division; however, an important distinction should be made between these early mitotically active cells (often called progenitor cells) and the stem cells of the adult. The daughter cells of progenitor cells are of a different phenotype to the daughter cells of adult stem cells. It is more useful to think of the progenitor–daughter axis itself as undergoing maturation, especially as it is not yet known at which level of control of gene expression various maturative events are effected. By introducing radioactively labelled nucleotides (e.g. tritiated thymidine) into the fetal circulation, it is possible to label the thymine in newly synthesized DNA. By qualitative and quantitative methods it is thereby possible to study the localization of cell proliferation and to make estimates of the number of cells entering the S phase of the cell cycle. Studies of the kinetics of human fetal cells have been limited to short-term, *in vitro* studies of only early gestation tissue. The fetal sheep is the only long gestation species in which such measurements have been made *in vivo* over a wide range of fetal and neonatal ages.

In fetal sheep the highest rate of cell proliferation occurs early in gestation (around 60 days), and, after declining over the next 60 days, begins to rise again late in gestation. The late gestational rise is most likely linked to rising plasma cortisol concentrations which precede the onset of labour. From the earliest appearance of villi, mitotic activity is confined to the intervillus regions (prior to crypt formation) and then to the crypts themselves (after they appear at 90–100 days). This partitioning effectively separates immature cells from mature cells, and establishes an axis of cellular differentiation along the developing villus profiles. It is important to know the age of enterocytes and their maturative profiles in order to be able to understand pathways of both normal and abnormal development. After cells arise from the progenitive zone they migrate along the villus, but the speed of migration is much less than in the adult. The time taken for complete renewal of the fetal villi epithelial lining, depending on the region of the intestine and fetal age, ranges from 8 to 20 days, compared with 2–4 days in the mature intestine.

One important reason for the slower renewal time in the fetus is that new cells are also required for growth. Increasing mucosal dimensions, therefore, account for a significant portion of the output of the progenitor cells' activity. Another reason for increased renewal times might also, by inference from the adult situation, be the absence of the presumed stimulatory effects that the luminal environment imposes.

Factors that influence growth of the gastrointestinal tract

There is a vast literature concerned with the control of growth and cellular differentiation in the GI tract, although most of it describes development in the neonatal period. Little is known about the regulation of GI development *in utero*, no doubt reflecting the difficulty of studying suitable experimental models. With regard to human studies, except for a small number of clinical and post-mortem descriptions, all studies undertaken have, of necessity, used *in vitro* systems. These studies have a number of important limitations. First, the development of human fetal GI explants in culture does not follow the same pathway as described for tissue *in utero*; in general, maturational events are accelerated. Furthermore, tissue for these studies is generally derived from aborted fetuses and so is usually limited to 20 weeks of gestation. Once in culture, tissues can be maintained for a few weeks at most.

Glucocorticoids are powerful agents for induction of terminal differentiation of many cell types. In fetal human small-intestinal explants, hydrocortisone added to the medium increases cell turnover, and lactase and alkaline phosphatase levels, but decreases sucrase levels. In the fetal sheep's small intestine the rate of cell turnover is sensitive to cortisol. The late gestational rise in cell proliferation and migration can be delayed by fetal adrenalectomy or can be prematurely induced by cortisol infusion. Addition of epidermal growth factor (EGF) to explants of human tissue depressed cell turnover and decreased sucrase, trehalase, and glucoamylase levels but increased lactase. In fetal sheep, reduced gastrin levels after antrectomy resulted in altered GI growth. The wide range of factors known to influence GI growth in neonates has yet to be investigated in fetal preparations. Furthermore, owing to the long period of time spent *in utero*, it is likely that hormones and factors (which often have a longer effective half-life in the fetus than in the adult) could have greater biological significance than otherwise might have been predicted, even if present in low concentrations in tissue or fetal fluids.

Absorptive functions of the fetal gastrointestinal tract

With the exception of some data on fetal sheep, most data on uptake from the GI tract have been derived from *in vitro* studies, using segments of intestine from aborted human fetuses.

Water and solutes

Water absorption from the fetal GI tract is slower than in the adult, though a reduced rate of gastric emptying might contribute to this effect. In the adult it has been suggested that the paracellular pathway might be the most important route for solute movement. The amount of material transported by this pathway depends on the luminal glucose concentration, which influences the degree of contraction of the terminal web of the enterocytes thereby opening up the junctional complexes, allowing paracellular movement to occur passively. The extent to which this phenomenon occurs in the fetus is not known, but it should be kept in mind when assessing the relative importance of the various pathways to be described.

Solutes are handled differently in different regions of the GI tract. In the jejunum an electrochemical gradient across the epithelium is generated by $Na^+K^+ATPase$ localized in the basolateral membranes. Sodium flows along this gradient and is pumped out of the cell. Solutes (including organic solutes) accumulate to a high level within the cell and leave the cell via facilitated diffusion. Absorption is generally confined to villus enterocytes, whereas crypt enterocytes are generally secretory.

The ileum and colon are regions of water and electrolyte conservation. However, available data do not present a consistent view as to the most important mechanisms. Three models have been described: uncoupled electrogenic Na^+ absorption; neutral coupled NaCl co-transport; and double anion–cation exchange.

A major component of the establishment of transporting mechanisms in the GI tract is the formation of an epithelial sheet, and the establishment of cellular polarity. In the developing enterocytes, basolaterally partitioned $Na^+K^+ATPase$ and adenylate cyclase appear at as early as 10–11 weeks of gestation in the human, establishing the possibility that transport mechanisms may be functional from very early in development. From the tenth week, specific active transport mechanisms for sugars and amino acids have developed in both the jejunum and ileum, which have all the features of Na^+ co-transporting systems of the adult.

Carbohydrates

At around 10 weeks, the human fetal intestinal wall is capable of transporting glucose. By 18 weeks the proximodistal gradient of glucose transport (as in the mature gut) has been established, that is, glucose transport becomes greater in the proximal parts of the intestine than in distal parts. Studies on brush border membrane vesicle preparations suggest the presence of at least two carrier systems: one with high affinity and low capacity; and another with low affinity and high capacity. These mechanisms are quite different to those in the adult as they are capable of functioning under anaerobic conditions; in the adult, glucose transport declines as oxygen supply is reduced. This may be an important adaptation for the fetus with its oxygen-reduced environment in which episodes of hypoxia, which markedly reduce intestinal blood flow, are common.

The production of monosaccharides from ingested carbohydrates depends on the development of active digestive processes, which include the appearance of the appropriate enzyme (usually found in the brush border), pH optimum, and the presence of the substrate (disaccharide, polysaccharide) in the luminal contents. These processes are not necessarily fetal analogues of those in the adult, as important differences exist in the biochemical and functional capacities of intestinal enterocytes.

Lipids

In the adult, lipases are found in the salivary glands, stomach, and pancreas. Little data about fetal lipase development exist. However, there is a report that fetal human intestine at 10 weeks can absorb corn oil, and another which documents intestinal fatty acid absorption at 24 weeks. The only sequential data on fat absorption are from studies on suckling rats. In neonatal rats neutral fats are absorbed faster than in adults, although the overall lipolytic activity is very low, despite the very high fat content of the diet. Paradoxically, as the diet of the suckling rat shifts to one of high carbohydrate/low fat content, lipase levels rapidly increase.

Biochemical maturation of the gastrointestinal tract

Background

In species of long gestation, biochemical maturation of the gut is achieved early in development, and follows a developmental profile unique to the particular species. However, this profile may differ from that of the adult. For instance, while small intestinal sucrase specific activity is low and lactase specific activity is relatively high in the suckling human, the reverse is true in the adult. The most striking difference between GI function in the fetus and adult is the reliance on parenteral, as opposed to enteral nutrition. However, to the extent that enteral nourishment might be important, it is of significance that, in the developing intestine, intracellular digestive pathways are much more developed than in the adult where luminal digestion is more important. Furthermore, digestive enzymes in fetal or neonatal life may not be in the same form or distribution as the adult. While a significant component of this difference lies in the intrinsic genetic programme of the GI tissues, as in the adult, this programme is constantly modified by the environment and substrate demands of the individual.

Technical considerations

There are many potential difficulties in interpretation of the assay systems used to assess enzyme levels. Measured tissue levels result from the aggregate of many cells, even when specialized isolation procedures have been used. The diversity of cell ages in a growing system can make the interpretation of measured levels and trends very difficult. Cytochemical and immunological techniques might not reflect functional differentiation because substrate specificity can vary in developing systems and epitopes for antibodies might not be specific for functional forms of the enzyme under investigation. Thus extreme must be taken in interpreting quantitative data.

Development of digestive enzymes

Disaccharidases

Sucrase, maltase, isomaltase, and trehalase activity can be found in intestinal tissues and amniotic fluid by 10 weeks in the fetal human and by 50 days in the fetal sheep. By 14 weeks (human), the activity of these α-glucosidases is 70–100 per cent of adult values, with the normal proximodistal orientation being established by 15 weeks.

Lactase is also present by 10 weeks (50 days in the sheep) and its specific activity remains relatively low throughout development. However, in the month before birth lactase levels rise dramatically to levels higher than those found in infants up to 1-year-old.

Alkaline phosphatase

Alkaline phosphatase is detectable from the earliest stages of intestinal brush border differentiation. In the human it is

present by 10 weeks and its specific activity rises rapidly until 23 weeks of gestation (though still well below adult levels). Alkaline phosphatase of intestinal origin can also be found in fetal serum and amniotic fluid and may be of use to diagnose some congenital abnormalities.

Enterokinase

Enterokinase is an important digestive enzyme due to its role in the activation of pancreatic zymogens. It is present after 26 weeks of gestation in the human fetus, but reaches only relatively low levels (about 15 per cent of those in children).

Peptidases

The appearance of a variety of peptidases has been noted by 10 weeks of gestation in the human; they are most likely present earlier, coincident with brush border differentiation, and are maximal in proximal regions. By 12 weeks the tissue-specific activities of many of these peptidases have reached adult levels, but some (leucine aminopeptidase) increase steadily with gestational age after 12 weeks.

Development of gastric proteolysis

Parietal cells and acid secretion

Morphological examination of tissue from the developing stomach suggests that parietal cells first appear at around 11 weeks of gestation in the human, and are present by day 55 in the sheep fetus. However, it is possible that parietal cell differentiation has already occurred by this age as succinic dehydrogenase (a major mitochondrial enzyme found in the mitochondria-rich parietal cells) is strongly localized in the developing pits before recognizable parietal cells are present. Hydrochloric acid is rarely found in fetal human stomach before the 32nd week of gestation, and at term the gastric contents are usually around pH 6, thereafter declining rapidly. As the fetus is hypergastrinaemic, the absence of a stimulatory drive to produce acid is curious, because the sheep, at least, does respond to chronic administration of pentagastrin. (The human neonate does not exhibit this response to a range of secretagogues.) The fetal stomach, therefore, presents an interesting model in which to study mechanisms of acid release relevant to the understanding of GI disease in which overproduction of acid is thought to be involved. When assessed on a body weight basis, the acid output of the neonate is low and, given the high buffering capacity of milk, this suggests that protein digestion is almost entirely dependent on enzymes capable of functioning at near neutral pH, or on entirely different routes for protein uptake and transfer.

Chief cells and pepsinogens

Chief cells produce a range of pepsinogens and are involved with the digestion of proteins. In humans, chief cells containing both a pepsinogen-like substance and adult forms of pepsinogen have been identified at as early as 8 or 9 weeks of gestation. Mucous cells (in the stomach, in Brunner's glands, and in goblet cells) also produce pepsinogens. Thus it is likely that pepsinogens are present from the earliest stages of mucous cell differentiation. Pepsinogens require an acid environment for activation, and hence await the development of gastric acid production to become functional.

Fluid ingestion in the fetus

Mammalian fetuses ingest fluid from an early stage of development, much of it in discrete episodes of repetitive swallowing. Although individual swallowing movements can occur early in gestation (by 11 weeks in the human, 60 days in the sheep), episodes of swallowing are not seen until later. These episodes probably represent the prenatal manifestation of feeding behaviour.

Measurements of fetal swallowing by several different methods have been made in fetal sheep *in utero*. Recordings from flow probes implanted around the oesophagus and collections of swallowed fluid via an oesophageal cannula indicate that ovine fetuses during late gestation swallow 100–1000 mL/day in episodes that occur, on average, every 2h. The amount of fluid ingested and the frequency of swallowing episodes are highly variable, even for an individual fetus.

Electromyographic recordings from laryngeal muscles in fetal sheep during late gestation have shown that swallowing at a buccopharyngeal level is highly organized, as in the postnatal animal. Fetal swallowing involves contraction of the lingual and pharyngeal muscles (causing a brief rise in intrapharyngeal pressure), adduction and elevation of larynx (causing glottic closure), and sequential contraction of the oesophageal muscles. (In sheep, the oesophageal muscle is skeletal and is controlled by vagal postganglionic motor fibres.)

In fetal sheep swallowing movements usually occur in a burst-pause pattern (Fig. 15.15). During 'bursts' of swallowing at frequencies of 2–4/s, ingested fluid accumulates in the oesophagus, and is cleared by peristaltic contractions of the mid and distal oesophagus during the 'pauses'

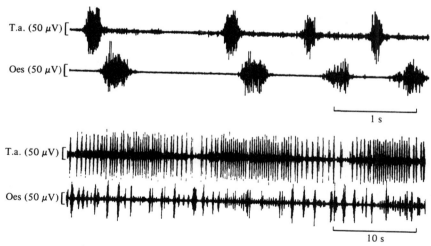

Fig. 15.15 Relationship between swallowing movements, as indicated by laryngeal adduction, and electrical activity of the cervical oesophagus (Oes) in a fetal sheep. Each swallow is indicated by a brief, intense burst of electromyographic activity in the principal adductor muscle of the larynx (thyroarytenoid, T.a.). In the upper panel, when swallowing rates are low (*c*. 1/s), each swallow is followed by contraction of the cervical oesophagus. The lower panel shows a section of recording made during an episode of repetitive fetal swallowing, in which swallowing occurs at a rate of 2–3/s. Contraction of the oesophagus occurs at a lower frequency, during pauses between periods of rapid swallowing. (Reproduced with permission from Harding, R., Sigger, J.N., Poore, E.R., and Johnson, P. (1984). *Quarterly Journal of Experimental Physiology*, **69**, 477–86.)

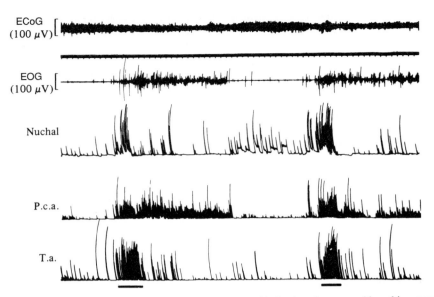

ECoG
(100 μV)

EOG
(100 μV)

Nuchal

P.c.a.

T.a.

Fig. 15.16 The relationship between swallowing episodes (underlined) and behavioural states and breathing movements in a fetal sheep during late gestation. Bouts of rapid swallowing, detected as repeated bursts of electromyographic (EMG) activity in the thyroarytenoid muscle (T.a.), occur during periods of low-voltage electrocortical activity (ECoG) and rapid eye movements, as shown in the electrooculogram (EOG). Swallowing episodes are usually accompanied by periods of emg activity in the muscles of the dorsal neck (nuchal muscles). Breathing activity is shown by emg activity in the laryngeal dilator muscle (posterior cricoarytenoid, P.c.a.). Electromyograms are shown as time averages. (Reproduced with permission from Harding, R., Sigger, J.N., Poore, E.R., and Johnson, P. (1984). *Quarterly Journal of Experimental Physiology*, **69**, 477–86.)

in the train of swallowing. This is similar to events during the suckling of milk after birth. When the fetus swallows at lower rates, each swallow is followed by a wave of peristalsis along the entire oesophagus (Fig. 15.15).

In the late-gestation sheep fetus swallowing movements occur during a state of behaviour resembling arousal or wakefulness. Low-voltage electrocortical activity is present and eye and breathing movements are usually present. The presence of phasic activation of the dorsal neck muscles suggests that fetuses are not in a state of typical rapid eye movement (REM) sleep during ingestive episodes (Fig. 15.16). Individual swallowing movements are highly co-ordinated with breathing movements at least during the last third of gestation. Contraction of the diaphragm is briefly interrupted (for 100–200 ms) during each swallow; after birth this co-ordination is necessary to prevent aspiration of milk during suckle feeding.

Composition of ingested fluid

The fluid swallowed by the fetus is a variable mixture of several components. Amniotic fluid, which is principally formed from fetal urine, and lung liquid are the major con-

stituents, making approximately equal contributions in late-gestation fetal sheep. Secretions from the nasal and oral cavities are also swallowed and probably account for the very high viscosity of ingested fluid due to their high mucus content. Cellular debris, particles of meconium, and hairs are also swallowed.

Amniotic fluid contains a wide range of putative and known growth-modulating substances, some of which may be derived from the fetal lung or salivary secretions. Although these may be present at very low concentrations, the relatively high rates of fluid ingestion and the absence of proteolysis in the fetal gut suggest that swallowed growth factors could play a regulatory role in GI tract development. Epidermal growth factor (EGF) and insulin-like growth factors (IGF), which are present in fetal lung cells and in fetal lung liquid, are likely to be ingested.

Regulation of fetal swallowing

While it has been established that, in fetal sheep, the low-voltage electrocortical state is permissive for ingestive episodes, the factors regulating these episodes and the volume of fluid ingested are poorly understood. In particular

lar, it is yet to be determined whether 'thirst' or 'hunger' signals play any regulatory role. Perhaps fetal ingestion is a prenatal manifestation of appetite.

In order to address some of these questions a range of experiments have been carried out in fetal sheep. Preventing gastric filling, by oesophageal fistulation, does not appear to influence the incidence or duration of swallowing episodes; nor does bilateral vagotomy, which prevents afferent neural information on gastric volume from reaching regulatory neurones in the brainstem. The intravenous infusion of hyperosmotic saline into fetal sheep causes only a transient increase in swallowing; however, it is unlikely that this response to acute hyperosmolality normally plays a role in regulatory fetal swallowing. Other evidence indicates that the renin–angiotensin system, which is involved in the regulation of fluid balance postnatally, does not play a role in the control of fetal ingestion. Furthermore, plasma volume expansion does not diminish, but rather enhances, fetal swallowing. These findings suggest that 'thirst' stimuli are not normally involved in the regulation of fetal swallowing.

Fate of ingested fluid

The fetus passes little (if any) faeces *in utero*; the contents of the distal large intestine accumulate as meconium. Thus, most of the water ingested by the fetus must be absorbed by the gut into the circulation. By term, the gut of the human or ovine fetus is evidently capable of absorbing up to 1 L of fluid per day.

Meconium is present in the distal gut from as early as 16 weeks in the human. Its composition changes during gestation. Early in gestation it is composed of 70 per cent carbohydrate, 13 per cent protein, and 8 per cent lipid and bilirubin (up to 1 mg/g). By late gestation, protein is no longer present, presumably due to increased protease activity of the gut. The passage of substantial amounts of meconium *in utero* is abnormal, and is thought to indicate fetal asphyxia, or possibly infection. The passage of some meconium into the amniotic sac may occur in the healthy fetus. Observations using fetoscopy suggest that fetal defecation may occur as early as 18 weeks. The presence in amniotic fluid of brush border enzymes and other gut-specific markers show that at least some material from the GI tract normally enters the amniotic fluid.

Gastric motility

The transport of ingested material from the oesophagus to the stomach and the accumulation of meconium in the large intestine indicate the existence of effective movement of luminal contents along the GI tract of the fetus.

Evidence for gastric motility in the human fetus is based on the presence of ingested fluid in the small intestine and on radiographic observations of the passage of contrast material injected into the amniotic fluid. The latter technique indicates that gastric motility is present at 20–25 weeks of gestation and that gastric emptying can occur at 31–32 weeks. In fetal sheep and cats, peristaltic movements in the stomach occur at around mid-gestation, corresponding approximately with the time that ingested fluid first appears in the small intestine and with the establishment of the enteric nervous system. Electromyographic recordings from the stomachs of late-gestation fetal sheep show slow waves of contraction passing over the gastric antrum (3–4/min) that are co-ordinated with the proximal duodenum.

Intestinal motility

Observations of small intestinal motility in the fetus have been made in humans, cats, guinea-pigs, and sheep. Such observations indicate that effective motility is present by at least mid-gestation; it has been observed as early as 11 weeks in human fetuses. By 34 weeks radiographic contrast material introduced into the amniotic sac could be detected within the small intestine in less than 1 h and in the colon after 8 h.

Electromyographic recordings from the small intestine of fetal dogs and sheep show three types of activity: (1) disorganized; (2) a 'fetal' type of regular spike activity; and (3) a migrating motor complex pattern (Fig. 15.17). Ages at which similar types of activity patterns were found in preterm human infants were: (1) 27–30 weeks; (2) 30–34 weeks; and (3) 34–35 weeks. A striking correlation with the more mature MMC pattern and the ability to suck has been observed.

Colonic peristalsis has been observed in the 12-week-old human fetus. By this time, Meissner's and Auerbach's plexuses are present. Defecation *in utero* at 18 weeks is also an indication of the early establishment of colonic motility.

The role of fluid ingestion in development of the GI tract

By preventing the fetus from ingesting fluid it has been possible to assess the impact of fluid throughput on the development of the gut. This has been accomplished in fetal sheep and rabbits following oesophageal ligation; in humans, observations have been made of fetuses with oesophageal atresias.

Prevention of fluid ingestion in fetal rabbits and humans is associated with restricted somatic growth, but this does

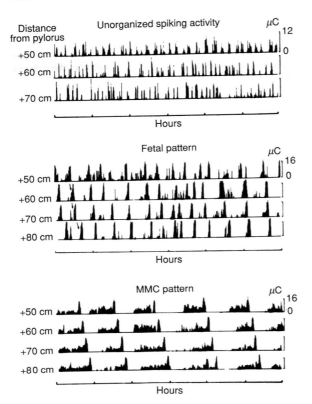

Fig. 15.17 Integrated electromyographic recordings from the small intestine (jejunum) of fetal sheep *in utero*. Top recordings (unorganized) were from a fetus at 0.7 of term (*c*. 100 days), middle recording (fetal pattern) were taken at 0.9 of term (*c*. 130 days), and lower recordings (MMC pattern) were made 2 days before birth at term. The level of spiking activity increased progressively towards term, and aboral propagation of some phases of regular spiking activity became prominent at 0.9 of term. Near term, these phases recurred at approximately hourly intervals and were preceded by a phase of irregular spiking activity, thus resembling the MMC pattern of the adult. (Reproduced with permission from Bueno, L. and Ruckebusch, Y. (1979). *American Journal of Physiology*, **237**, E61–E67.)

not occur in the ovine fetus. These observations indicate that the rabbit and human fetus derive a nutritional benefit from fluid ingestion. On the basis of amniotic fluid composition and swallowing rates, it has been estimated that 10–15 per cent of the requirements for total body protein deposition of the human fetus could be derived from ingestion of amniotic fluid. In the sheep fetus, measurements of arteriovenous concentration differences in nutrients across the small intestine indicate that nutrients are lost, rather than gained, from the circulation of this tissue.

Studies of the GI tract from fetal sheep subjected to oesophageal ligation show that development of the mucosal structures is particularly impaired. In the stomach, the thickness of the external muscle layers was reduced by the prevention of fluid input. Similar changes were observed in the duodenum and proximal small intestine. The length of intestinal villi were reduced and the rate of migration of epithelial cells was also reduced. The observed changes suggest that the small intestinal surface area is reduced when ingested fluid is absent. Recent studies have examined the ultrastructure of intestinal tissues from these fetuses. Enterocyte morphology was drastically affected. There were areas of microvillus effacement, abnormal cell lysis and shedding, and general immaturity (Figs 15.18 and 15.19). Many of these features resembled the appearance noted in malnutrition, suggesting that a significant level of local nutrition for growth might be a direct benefit from swallowed fluid. Alternatively, it is possible that growth-promoting substances (e.g. EGF) in amniotic fluid, lung liquid, or secretions of the upper digestive tract (e.g. saliva) may stimulate mucosal development and maturation. It is also possible that ingestion of fluid may cause the release of potentially trophic agents (e.g. gastrin) from the GI tract and associated organs.

In addition to its effects on fetal nutrition and GI tract development, fetal ingestion is involved in the regulation of fetal fluid balance. It represents a major route by which

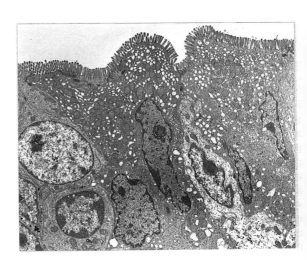

Fig. 15.18 Electron micrograph of groups of enterocytes from fetal sheep small intestine after oesophageal ligation. Enterocytes are extensively vesiculated, presumably undergoing cell death and possible extrusion. × 4100. (Reproduced with permission from Trahair, J.F. and Harding, R. (1992). *Virchow's Archiv A*, **420**, 305–12.)

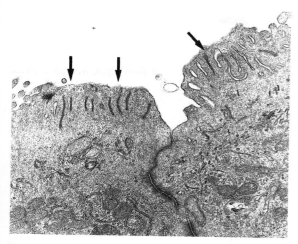

Fig. 15.19 As for Fig. 15.18. In some areas the brush border is absent. Elements of the AEC are arranged perpendicularly beneath the apical cell plasma membrane (arrows). × 18 000. (Reproduced with permission from Trahair, J.F. and Harding, R. (1992). *Virchow's Archiv A*, **420**, 305–12.)

fluid excreted as fetal urine and lung liquid is returned to the fetus. Prevention of swallowing may therefore affect fluid balance. In humans, oesophageal atresia is often (but not always) associated with an excessive volume of amniotic fluid (see Chapter 14).

Immune system

Humoral and mucosal immunity

The lymphoid tissue of the adult GI tract is the largest lymphoid organ in the body. Whilst *in utero*, the fetus is protected from external antigens and the effects of the maternal immune system. Nevertheless, the fetus must prepare itself for independent life, in particular for the postnatal exposure to antigens in food and in the air. Cells of the gastrointestinal-associated and bronchial-associated lymphoid systems (GALT and BALT, respectively) appear very early in development. Lymphocytes present in the mesenchyme of the lamina propria appear at 8 weeks in the human, and intraepithelial lymphocytes are present from 11 weeks.

A further specialization of the GALT and BALT systems is the mucosal immune system, which is especially sensitive to antigens presented via the airways or the GI tract. Because these organs are major sites of antigen entry, it is important that the immune system is functional soon after birth. After antigens cross the epithelium (see below), they interact with the underlying lymphoid cells, evoking an immune response. The mucosal immune response involves production of specific immunoglobulins (IgAs), that pass back into the lumen to protect the mucosa from further antigen exposure. IgA crosses the epithelium as an IgA-secretory component complex via a specialized receptor-mediated process. Thus the mucosal immune response provides a first line of defence against ingested and inhaled antigens. Antigen penetration and transport is increased in specialized parts of the GI tract, in areas where lymphoid aggregations are found.

M cells and Peyer's patches

Intestinal lymphoid aggregates, either nodules, or clusters of many nodules (Peyer's patches), are covered with a specialized epithelium (microfold, or M cells) whose enterocytes possess the ability to transport or provide sites of attachment for whole proteins, microorganisms, viruses, and even whole particles. M cells are found in the fetal human GI tract, as early as 17 weeks, with definitive Peyer's patches being present in the ileum by 20 weeks. In fetal sheep, Peyer's patches appear both in the proximal jejunum and in the ileum. At puberty the more proximal Peyer's patches regress, leaving only the permanent ileal patches remaining. It is not yet clear whether the recently described lymphocyte-filled villi, whose role is yet to be defined, possess an M cell covering.

Passive immunity and closure

Passively acquired immunity (i.e. immunity derived from the mother and independent of direct antigenic exposure) is vital for the well-being of neonates. However, the timing and pathways for acquisition of passive immunity vary considerably. In the fetal human IgG (but not IgA) is transferred across the placenta, with the fetus attaining a normal infant titre of IgG by mid-gestation. IgA does not appear until after IgA-producing plasma cells appear after birth. In the fetal sheep (and other ungulates) there is no prenatal transfer of antibodies; hence the newborn lamb is agammaglobulinaemic. Antibodies are transported after birth from the first (IgG-rich, IgA-low) colostrum feeds. Within hours of feeding, the new-born lamb has attained a normal antibody titre and, in fact, begins to lose IGg from the serum back into the GI tract.

'Gut closure' is the cessation of antibody transfer across the GI tract. While this term has relevance to long-gestation species like sheep, which have a defined neonatal period of antibody transfer (closure at 2–4 days), in humans the term closure is not applicable since there is no period of significant intestinal Ig transfer. This is not to suggest, however, that transfer of other proteins might not be important and that the timing of such transport might not affect the developmental outcome.

Intestinal transport of macromolecules

The developing intestine possesses the capacity to transport vast amounts of whole protein. All species exhibit this feature at some stage during their development. Its importance for the survival of the neonate is not known but, as suggested above, it is vital for some species as the transfer of antibodies and the development of passive immunity occurs solely via the GI tract. Of increasing interest is the possibility that this system also facilitates the transfer of growth factors that are present in swallowed fluid (either amniotic fluid *in utero*, or milk during postnatal suckling).

Two systems for the transfer of intact macromolecules have been described. The first has been documented most fully in the proximal small intestine of the suckling rat. Receptors for the crystallizable (Fc) portion of IgG are located on the luminal aspect of a series of specific cellular structures (coated vesicles and tubules) found between the bases of adjacent microvilli. These structures are readily visualized using morphological techniques. The receptors are pH-sensitive and selective for rat IgG and are responsible for the transfer of maternal IgG from colostrum and milk to the neonatal circulation. There they confer passive immunity on the neonate. Fc receptors isolated from human placenta are likely to be the pathway by which the human fetus acquires passive immunity from maternal serum.

Recent experiments have demonstrated the presence of a fetal human intestinal Fc receptor, raising the possibility that IgG transfer can take place *in utero*, although amniotic fluid contains only very small traces of immunoglobulin. This capacity may be of particular relevance, however, in the management of premature infants in whom maternal milk is but one of the many currently employed nutritional options.

In the sheep, or other species in which Ig transfer occurs primarily after birth, transfer is apparently non-selective and may occur across the vacuolated epithelial cells. The permeability of the epithelium is readily demonstrated by introducing Igs into the lumen of fetal small intestine. Within minutes the introduced Igs appear in the AEC (Fig. 15.20). It is of note, however, that the morphological features of both transferring systems (i.e. the AEC and coated vesicles/tubules) are present within the same cell (see Figs

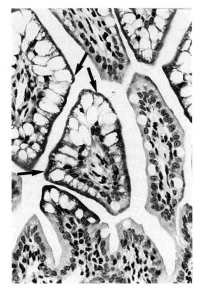

Fig. 15.20 Light micrograph of distal small intestine in a fetal sheep at 136 days, showing immunoperoxidase staining for ovine IgG. IgG was introduced into the lumen of ligated loops for 20 min. IgG was immunolocalized to the AEC region of vacuolated enterocytes, indicating uptake of IgG from the lumen. × 160.

15.10–15.12), raising the possibility that more than one type of transcellular pathway exists within the same cell type.

Vacuolated enterocytes are also capable of transferring hormones/growth factors from the lumen into the circulation via the AEC. These factors could be derived from milk or amniotic fluid, which contains many growth factors. Recent studies on the effects of preventing ingestion *in utero* show that, not only is the GI tract growth retarded, but other organs (liver and pancreas) are also growth-retarded. This strongly suggests that the delivery of intact growth factors via the GI tract might be an important means for controlling not only GI growth, but growth of associated organ systems.

Effects of placental nutrient supply on gastrointestinal tract development

It is now clear that intrauterine growth restriction (IUGR) induced by a reduced supply of nutrients to the fetus is associated with altered development of a number of organs including the central nervous system and respiratory tract. In rats, restriction of maternal dietary protein or unilateral uterine artery ligation produce growth-retarded fetuses. These fetuses, and those of growth-retarded fetal sheep, have small intestines that are disproportionately reduced in weight. Structurally, the proximal regions of the small intestine were affected more than the distal regions. Villi were shorter and less dense than in normally grown fetuses, leading to a reduction in intestinal surface area. The epithelial cells were immature in appearance and contained abnormally large amounts of glycogen. In fetal rats, IUGR led to a decreased presence of some brush border enzymes. In fetal sheep with IUGR it was apparent that the AEC was also retarded in its development.

The mechanisms whereby placental insufficiency or restricted maternal diet impair fetal GI tract development are not clear. They may involve restricted blood supply to the gut, or tissue wasting may occur. Chronic hypoxaemia, hypoglacaemia, and hypoinsulinaemia may be contributory factors.

The changes that have been described in animal studies of induced IUGR suggest that human fetuses with IUGR may suffer from impaired GI tract function after birth. It is, however, possible that the deficiencies seen in the fetal gut may become resolved with adequate postnatal nutrition.

Further reading

Baintner, K. (1984). *Intestinal absorption of macromolecules and immune transmission from mother to young.* CRC Press, Boca Raton, Florida.

Elliot, K. and Whelan, J. (ed.) (1979). *Development of mammalian absorptive processes*, CIBA Foundation Symposia, No. 70 (new series). Excerpta Medica, Amsterdam.

Kretchmer, N. and Minkowski, A. (ed.) (1983). *Nutritional adaptation of the gastrointestinal tract of the newborn.* Raven Press, New York.

Lebenthal, E. (ed.) (1989). *Textbook of gastroenterology and nutrition in infancy*, 2 volumes (2nd edn). Raven Press, New York.

Lebenthal, E. (ed.) (1989). *Human gastrointestinal development.* Raven Press, New York.

Morisset, J.A. and Solomon, T.E. (ed.) (1990). *Growth of the gastrointestinal tract: gastrointestinal hormones and growth factors.* CRC Press, Boca Raton, Florida.

Tanner, M.S. and Stocks, R.J. (ed.) (1984). *Neonatal gastroenterology — contemporary issues.* Intercept Ltd, Newcastle-upon-Tyne.

Trahair, J.F. and Harding, R. (1987). Development of structure and function of the alimentary tract in fetal sheep. In *Animal models in fetal medicine* (ed. P.W. Nathanielsz), pp. 1–36. Perinatology Press, Ithaca, New York.

Walker, W.A., Durie, P.R., Hamilton, J.R., Walker-Smith, J.A., and Watkins, J.B. (ed.) (1991). *Pediatric gastrointestinal disease.* B.C. Decker Inc, Philadelphia.

16. Ontogeny of gastrointestinal regulatory peptides and their function

Arthur Shulkes

The gastrointestinal (GI) tract is the largest endocrine organ in the body. In fact the discipline of endocrinology began in the GI tract when Starling suggested in 1902 that a messenger substance from the duodenum was released into the circulation to alter pancreatic secretion. The substance was called secretin and the general term coined for a blood-borne substance acting on a distance site was hormone from the Greek, 'I excite or arouse'.

Functions of hormones from the GI tract include effects on intake, digestion, and absorption of food and changes in gut secretions, motility, and growth. Since GI hormones are localized in both endocrine cells and nerves, and are present in many organ systems outside the gut, 'regulatory peptide' is probably a more appropriate term than the traditional 'gastrointestinal hormone'. Table 16.1 summarizes the location, stimulus for release, and actions of some of the major gut regulatory peptides. For a more detailed compendium Dockray (1987), Thompson *et al.* (1987), and Walsh (1987) in the 'Further reading' list are recommended.

One of the most dramatic adaptive changes of the newborn is the change from continuous parenteral feeding via the umbilical cord to intermittent oral feeding. There is now considerable evidence that the regulatory peptides have major roles in this adaptation. However, given the rapidity of this adaptation, it is also obvious that much of the preparation of the gastrointestinal tract must occur before birth.

Before considering specific roles of the regulatory peptides in the fetus and neonate, some of the principles of gastrointestinal endocrinology need to be summarized.

Distribution of the regulatory peptides

The regulatory peptides are found in both endocrine cells and nerves. Rather than being concentrated groups of cells as in other endocrine organs such as the thyroid, the endocrine cells are scattered throughout the GI tract. There are two distinct types of endocrine cells in the mucosa of the GI tract. The so-called 'open cells' have microvilli that project into the lumen and can respond to changes in the luminal contents. In contrast, 'closed cells' do not project into the lumen and rely on neural, blood-borne, or local stimuli. The same peptide may be present in both cell types. For example, somatostatin is present in open cells in the gastric antrum but predominantly in closed cells in the fundus. In general, the endocrine cells have been named either on the basis of their morphological appearance or because of their peptide product.

Each peptide has a characteristic distribution in defined areas of the gastrointestinal tract. For instance, gastrin is predominantly located in the antrum while secretin and cholecystokinin are found in the duodenum. However, the distribution in the fetus is often quite different and can include transitory populations of cells.

The regulatory peptides are present both in intrinsic and extrinsic nerves. The extrinsic nerves are parasympathetic and derived from the vagus and pelvic nerve, while the sympathetic nerves arise from the sympathetic ganglia. However the majority of the peptidergic nerves are intrinsic with cell bodies in the myenteric (Auerbach) and submucous (Meissner) plexuses. The one neurone can contain several neuropeptides and coexistence between neuropeptides and the classical transmitters, acetylcholine and noradrenaline, has also been observed. This chemical coding (combination of histochemical markers) has enabled maps of the projections and connections of the different neurones to be constructed. There appears to be a characteristic distribution for each of the chemically distinct classes of neurones. The task remains to correlate the particular projections with function. These functions include transit of luminal contents, secretion and absorption of electrolytes, and regulation of blood flow.

Mechanism of action

The presence of the regulatory peptides in both nerves and endocrine cells and the scattered distribution of the endocrine cells has meant that there are a number of ways in which the regulatory peptides exert their effects. An endocrine mechanism involves release into the circulation

Table 16.1 Summary of location, stimulus for release, and gastrointestinal actions of some of the gut peptides

Peptide*	Location	Stimulus for release	Action
Gastrin	Antrum, duodenum	Food (protein), vagus	Stimulates gastric acid secretion; trophic action on gastric mucosa
CCK	Upper small intestine	Food (protein, fat)	Stimulates pancreatic enzyme secretion; augments action of secretin on pancreatic bicarbonate secretion; causes gall bladder contraction; inhibits gastric emptying
Secretin	Upper small intestine	Acid	Stimulates pancreatic and biliary bicarbonate secretion; augments CCK stimulated pancreatic enzyme secretion; inhibits gastric acid
GIP	Upper small intestine	Food (carbohydrate)	Inhibits gastric acid; potentiates glucose-dependent insulin release
Motilin	Upper small intestine	Alkali, acid	Increases GI tract mobility
Enteroglucagon	Intestine	Food	Increases intestinal growth; inhibits gastric acid secretion
VIP	Stomach, intestine,		Neurotransmitter; stimulates pancreatic water and bicarbonate secretion; inhibits gastric acid secretion
Pancreatic polypeptide	Pancreas	Food (mixed meal)	Inhibits pancreatic bicarbonate and enzyme secretion
Neurotensin	Intestine	Food (fat)	Inhibits gastric acid
Somatostatin	Stomach, intestine, pancreas	Acid	Neurotransmitter/paracrine agent; inhibits peptide release and gastrointestinal functions
Substance P	Intestine		Neurotransmitter; stimulates intestinal smooth muscle contraction

*CCK, cholecystokinin; GIP, gastric inhibitory peptide; VIP, vasoactive intestinal peptide

and so has been the easiest to study. An example is gastrin, which is released from the gastric antrum into the portal circulation and passes via the liver into the systemic circulation. The gastrin reaches the main target organ, the fundic parietal cell, via the arterial system. Since all GI peptides released into the circulation must pass through the liver, the extent of hepatic metabolism is an important consideration. For instance vasoactive intestinal peptide (VIP) is nearly completely destroyed by a single pass through the liver and is only found in the circulation in patients with liver failure or VIP-producing tumours.

A second mechanism is paracrine control, which is the local release of a peptide into the surrounding interstitial fluid to act on adjacent cells. Given the dispersed nature of the endocrine cells, the paracrine mechanism is ideal for integration and co-ordination of responses. The absence of any measurable release into the circulation creates problems in determining whether a peptide has a paracrine function. This mode of action is thought to be the most ancient and probably arose in organisms that lack a well developed circulatory system. Similarly, this paracrine action may be important for various hormone-like factors in embryogenesis, i.e. before organogenesis or the establishment of the circulation.

In the mature animal, a specialized morphology has developed in some cases to allow a more precise paracrine delivery of an active substance. The ubiquitous inhibitory peptide somatostatin is the classic example where the somatostatin-containing D cells of the antrum and fundus have cytoplasmic processes that impinge on to the gastrin cells of the antrum and the acid-secreting parietal cells of the fundus. In this way precise control of the overall output of acid is possible.

The final mechanism is neurocrine which involves the release of the peptide from a neurone. Depending on the stimulus and the neurone type, various combinations of peptides and non-peptide transmitters such as acetylcholine and noradrenaline may be released. Peptide neurotransmitters have more diffuse and longer lasting actions than the classical transmitters. The predominance of the peptidergic nerves is therefore consistent with the nature of gut function, which tends to involve rather slow coordinated responses. Some neuropeptides such as somatostatin and cholecystokinin are also classical hormones. In fact, depending on the stimulus and location, the one peptide (e.g. somatostatin) may function as an endocrine, paracrine, or neurocrine agent. It should be remembered that most of the neuropeptides are also found in the brain and constitute the gut–brain axis.

An appreciation of the phenomenon of potentiation is important for determining particular roles for gastrointestinal hormones. The best example of potentiation is in the control of pancreatic exocrine secretions. Cholecystokinin potentiates the effects of secretin on bicarbonate secretion while secretin potentiates the enzyme stimulating effect of cholecystokinin. Potentiation occurs when secretagogues use different intracellular mediators. In this example, cholecystokinin acts via mobilization of calcium while secretin activates adenylate cyclase.

Synthesis and processing

Most of the regulatory peptides are synthesized as large precursors or prohormones. Posttranslational processing involves a series of proteolytic cleavages to release the bioactive peptide. One result of this processing is that there can be multiple molecular forms of a hormone that are biologically active because they share a common sequence. However the biological activities of these different forms are not necessarily equivalent. For instance, somatostatin-14 is more potent than the N-terminal extended form, somatostatin-28. Other modifications include sulfation of tyrosine residues (gastrin and cholecystokinin), cyclization of N-terminal glutamic acid to increase stability (gastrin), and amidation of the C-terminal amino acid. More than half of the gut peptides are amidated, and for these peptides amidation is essential for biological activity.

Of particular interest is that the extent of this processing can be quite different in the fetus. For instance, the proportion of some hormone precursors is higher in the fetus than at later stages of development. Whether this is the lack of stimuli or the availability of the processing enzymes is not clear. The best studied system is gastrin where there appears to be an accumulation of biologically inactive intermediates at certain stages of development (see below).

Receptors

No matter what the mechanism of action (paracrine, endocrine, or neurocrine), the regulatory peptides need to interact with specific receptors to initiate a biological response. Thus, the ontogeny of receptors must be considered in parallel with the ontogeny of the regulatory peptides. The initial step to activate a response involves binding of the peptide to the receptor. The number and affinity of the receptors for the peptide are important determinants of the cellular response. Most of the receptors are coupled to one of two major pathways: the adenylate cyclase pathway; and the calcium/phosphoinositide second messenger system. Although this separation of pathways is conceptually useful, evidence is accumulating that the two second messenger systems interact within the cell offering a further level of control.

As a result of gene duplication from ancient precursors, a number of the regulatory peptides share similarities in their sequences. Specificity of action is determined in part by the development of distinct receptor populations. The best studied is the gastrin/cholecystokinin family. Gastrin and cholecystokinin share the same carboxyl terminus pentapeptide sequence. The specificity of response is determined by the position and sulfation status of the tyrosine residue. For cholecystokinin the tyrosine (seven amino acids from the C terminus) must be sulfated for full cholecystokinin activity while, for gastrin, sulfation of the tyrosine (six amino acids from the C terminus) has no effect on gastrin biological activity. Other peptide families include the secretin/glucagon/VIP family, pancreatic polypeptide/peptide YY/neuropeptide Y family, and the tachykinin/gastrin-releasing peptide family.

Ontogeny of specific regulatory peptides

Because of the difficulty of studying human specimens, most information originates from animal studies, especially the rat. However, the pattern of development in the rat can be quite different and may not reflect the processes in man. For this reason, data from human studies, as well as information from species other than rats, are included when available.

Gastrin

The major roles of gastrin in the mature animal are stimulation of gastric acid secretion and renewal and growth of the gastric mucosa. The latter function appears to be the most important in the fetus.

The ready availability of assays for amidated gastrin (the biologically active form) and, more recently, the processing intermediates and the ability to measure end-organ response such as acid secretion or growth have meant that the ontogeny of gastrin has been examined in more detail than other gut peptides.

Gastrin is typical of the regulatory peptides in that it is processed from a high-molecular-weight precursor to inactive intermediates, active forms, and metabolites. The extent of the processing and the proportions of biologically active gastrin appear to vary markedly during development so a consideration of the mechanisms involved is appropriate.

A major advance was the isolation of the gastrin cDNA and the subsequent development of antisera against specific parts of progastrin and various intermediates. Our current understanding of the processing of gastrin is illustrated in Fig. 16.1. Conversion of preprogastrin to the mature biologically active species involves removal of the N-terminal signal sequence to form progastrin. Progastrin is then converted to glycine extended intermediates by trypsin-like enzymes cleaving at pairs of basic residues, and a carboxypeptidase B-like enzyme that removes C-terminal basic residues. The amidating enzyme then transfers the amino group of the C-terminal glycine residue to the preceding phenylalanine to form phenylalanine amide. Only gastrin peptides with an amidated C terminus have biological activity.

In adult man the highest concentration of gastrin is in the gastric antrum, predominantly as gastrin 17-amide and the inactive precursor gastrin 17-gly. The duodenum is a significant source of gastrin in man with equal amounts of gastrin-17 and gastrin-34. In the human about one-third of circulating gastrin-17 and gastrin-34 is reported to originate from the duodenum. However species such as rat, dog, pig, and sheep have only a small amount of gastrin in the mature duodenum.

In the human fetus, gastrin is first detected in the duodenum at 10–11 weeks and only appears in the fetal antrum several weeks later. Gastrin-34 amide is the predominant amidated form until 25 weeks when gastrin-17 predominates. The proportion of glycine extended forms in the human fetus is unknown.

Far more extensive studies have been performed in the rat, although the pattern of development is different with much of the maturation occurring between birth and weaning (21 days). At birth, the pancreas and duodenum contain most of the gastrin with only 5 per cent in the antrum. Subsequently, there is a steady increase in antral gastrin, which accelerates in the 3 days before weaning. Duodenal gastrin increases, but at a slower rate than antral gastrin, while pancreatic gastrin declines and is not detectable by weaning. Thus, at weaning more than 95 per cent of amidated gastrin is present in the antrum. The increased gastrin synthesis at and before weaning is accompanied by an increase in the proportion of the glycine extended intermediate suggesting that the increased translation exceeds the amidation capacity. This is despite a stimulatory effect on amidation by the corticosterone surge associated with weaning.

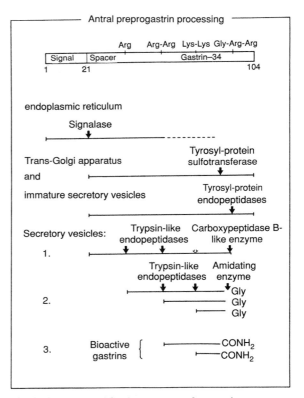

Fig. 16.1 A proposal for the sequence of processing events during maturation of preprogastrin. (Taken with permission from Hilsted, L. and Rehfeld, J.F. (1987). α-Carboxyamidation of antral progastrin. *Journal of Biological Chemistry*, **262**, 16953–7.)

As in other aspects of fetal development, corticosteroids appear to be important in the ontogeny of gastrin. The increase of antral gastrin levels in the rat during the third postnatal week correlates with the rise in plasma corticosterone concentrations. Furthermore, corticosterone administered in the first postnatal week produces a precocious rise in antral gastrin. Plasma gastrin concentrations are high in the new-born human, dog, rat, and sheep. In the sheep circulating gastrin has been detected from 50 days before birth and is present in higher concentrations than in the mother for the last 20 days of gestation and for up to 5 days postpartum. Antrectomy of the fetal sheep blocks the increase in gastrin. In the human, fetoscopy and fetal blood sampling prior to termination of pregnancy at 10–12 weeks gestation has shown that fetal plasma gastrin is about one-third of maternal levels. The only known tissue source of gastrin at this time is the duodenum.

It is likely that different mechanisms are involved in the hypergastrinaemia before and after birth. The postpartum hypergastrinaemia, although not initiated by oral food intake, may be maintained by this, since plasma gastrin is diminished in infants not enterically fed. Food restriction in the first 2 weeks of life causes a 40 per cent decrease in rat antral gastrin. There is little gastric acid secretion in the neonatal period and the resultant absence of feedback inhibition could result in an elevated gastrin production.

The high gastrin levels and low gastric acidity at birth in the human, dog, rat, and sheep and the relative unresponsiveness of the parietal cell to exogenous gastrin at this time is consistent with the absence of gastrin receptors. The gastrin receptor in the developing rat, the only species studied to date, does not appear until day 20 of life and coincides with a positive acid response to pentagastrin administration. This is also the first time that pentagastrin is able to exert a trophic effect on the gastric mucosa. These studies imply that gastrin has no role in the development and function of the fundus between birth until after weaning. However, receptors in other organs such as the pancreas and duodenum have not been looked for.

Investigations of acid secretion before birth are more limited. There is a transient production of acid and responsiveness to pentagastrin between day 19 and 20 of fetal life in the rat that disappears after birth. It is not known whether gastrin receptors are present in the fetus. Thus there appears to be a biphasic pattern of acid secretion in the rat. The mechanism for the postpartum absence of acid secretion has not been defined but may involve the rapid development of gastric somatostatin cells that occurs at this time. The new-born human has hypergastrinaemia and hyposecretion of acid. However, the initial lack of acid is partly attributable to the presence of the alkaline amniotic fluid. Sensitivity of the parietal cell to exogenous gastrin develops 48 h after birth.

The trophic effect of gastrin, as measured by increases in RNA, DNA, and protein synthesis in stomach, intestine and pancreas, is well documented for adult animals. However, there are little data that gastrin has a trophic effect during development. Indeed pentagastrin is ineffective in the rat from birth until around 3 weeks. Fetal ovine antrectomy in the last trimester, which reduced plasma gastrin to one-third of normal, did not produce gross atrophy of the gut mucosa, although it reduced the density of crypts in the small intestine and led to shorter gastric glands. The relatively small inhibitory effect of antrectomy on growth of the gastrointestinal tract confirms the view that gastrin is only one of a series of gut trophic factors.

Cholecystokinin (CCK)

The major sites of synthesis of CCK are the duodenum, jejunum, and brain. CCK is present in multiple molecular forms including CCK-58, CCK-39, CCK-33, and CCK-8. CCK and gastrin share a common C-terminal pentapeptide sequence, the biologically active part of the molecule. Specificity of function for CCK is determined by tyrosine sulfate in the seventh position from the C terminus while gastrin has a tyrosine residue in position 6, which may be sulfated or non-sulfated. The effects of CCK in the adult include stimulation of enzyme secretion, gall bladder contraction, and relaxation of the sphincter of Oddi.

CCK-containing cells are present in the human fetal intestine from week 10. The concentration of CCK in the intestine of the new-born rat is higher than in the adult. The highest concentration is at day 14 of life with a steady decrease thereafter, reaching adult levels by day 30.

The exocrine pancreas of the new-born rat is not responsive to CCK probably because of a relative absence of pancreatic CCK receptors and the immaturity of signal transduction mechanisms. Although exogenous CCK had a trophic effect on pancreatic growth in young (4-week-old) animals, administration of a CCK antagonist did not alter the normal growth of the pancreas suggesting that CCK is not an essential mediator of pancreatic growth. However the role of endogenous CCK in younger animals or in utero has not been tested. The recent availability of specific antagonists means that many of the assumptions based on exogenous administration can now be tested directly.

Secretin

Probably the most important function of secretin in the adult is to maintain a neutral environment in the duodenum. Thus, acid in the duodenum releases secretin, which stimulates pancreatic bicarbonate secretion and also inhibits the further secretion of gastric acid. These functions are probably not relevant to the fetus since the duodenal contents *in utero* are neutral because of the swallowed alkaline amniotic fluid.

In both mature and developing animals the duodenum contains the highest secretin concentration with decreasing concentrations in the jejunum and ileum. Secretin was first detected in the human fetal duodenum at 8 weeks, increased steadily until term, and reached adult levels by 31 weeks. In the rat and guinea-pig, the levels of duodenal secretin are similar to adult levels at birth, decrease for the next week, and then gradually increase to adult levels at the time of weaning.

The high tissue levels at birth in the human are matched by the circulating concentrations that are higher at birth than in adults. Although plausible, there is no evidence that the elevated secretin at birth is responsible for the hypochlorhydria at this time. Similarly, there is no secretin response to a meal until after day 24, and this has been attributed to the comparative lack of acid in the neonatal period.

The presence of substantial amounts of secretin in the circulation and intestine of the new-born and the fact that secretin stimulates the growth of stomach, small intestine, and pancreas of adult animals suggest a possible role for secretin in the postnatal growth of these organs. Exogenous administration of secretin to neonatal rats accelerates the growth of stomach, small intestine, and pancreas indicating that secretin may be one factor regulating growth and development of these organs.

Somatostatin

Although originally isolated from the hypothalamus, the largest source of somatostatin is the GI tract. Somatostatin has a wide spectrum of inhibitory actions in the GI tract including inhibition of release of other regulatory peptides and reduction of gastric and intestinal secretions.

The highest concentration of somatostatin in human fetuses at 15–21 weeks gestation is in the pancreas followed by the duodenum. Levels in the stomach are one-tenth of those found in the pancreas. This contrasts with the adult where the stomach has the highest concentration of somatostatin. However, at term, somatostatin levels in stomach, intestine and pancreas are about 90 per cent of adult levels. In the fetal rat, somatostatin is first detected in the pancreas and intestine 15–18 days postcoitum; gastric somatostatin appears later on postpartum day 4. Somatostatin continues to rise in these organs in the next few weeks with the most rapid increase occurring in the

stomach. It is quite feasible, although not proven, that the increase in somatostatin is responsible, by either a paracrine or endocrine mechanism, for the low neonatal gastric acid secretion. In relation to an endocrine mechanism it is relevant that the human neonate has high circulating levels of somatostatin and that these remain high for the first few months of life. The developmental pattern of somatostatin and gastrin are similar, with an initial intestinal preponderance followed by a predominance in the stomach. However, other than these types of correlations, no definitive role for somatostatin in development has been demonstrated.

Motilin, gastric inhibitory peptide (GIP), neurotensin, and enteroglucagon

These four peptides are located predominantly in intestinal endocrine cells. The major gastrointestinal actions are listed in Table 16.1. Whereas concentrations of motilin and GIP are highest in the duodenum, with lesser concentrations in the jejunum and ileum, neurotensin and enteroglucagon concentrations are highest in the ileum. This distribution pattern is also evident in the fetus.

Motilin, GIP, and enteroglucagon are measurable from 8–11 weeks of gestation in the human fetus but neurotensin was only detectable after 12 weeks. The concentrations of these peptides increase rapidly, reaching a developmental plateau at 20 weeks. A slow increase then occurs until term when concentrations are similar to those found in the adult. As gestation proceeds, the molecular forms change from high-molecular-weight species to the smaller forms present in the adult. These changes probably reflect the maturational state of the processing mechanisms.

Fetal plasma enteroglucagon and GIP are higher than maternal levels at 16–20 weeks of gestation. Concentrations at different time points are not known. Little is known of the functions of these peptides before birth. Measurement of circulating concentrations from birth in normal and ill infants has enabled some tentative conclusions to be made. At birth, the basal plasma concentrations of enteroglucagon, GIP, motilin, and neurotensin are similar to adult values. During the next few weeks the basal levels of these hormones increase and exceed adult concentrations. This increase does not occur in infants deprived of enteral feeding, suggesting that feeding is responsible for the initiation of these postnatal increases. It is not known whether these increases are responsible for the rapid growth of the gut. The rise in GIP may be responsible for the increased insulin release and glucose tolerance described in the infant.

Full-term infants born following fetal distress have increased circulating levels of motilin, GIP, neurotensin, and enteroglucagon. The elevated motilin may increase GI motility and account for the passage of meconium seen in this condition.

Vasoactive intestinal peptide (VIP) and substance P

These two peptides are located in nerves of the gut and consequently are thought to be neurotransmitters. VIP causes relaxation of smooth muscle and may be the mediator of oesophageal and anal sphincter relaxation and gastric receptive relaxation. Other effects include stimulation of pancreatic and intestinal secretion and vasodilatation. Substance P is also a vasodilator but, in contrast to VIP, it stimulates smooth muscle and contracts the oesophageal sphincter.

VIP is present in human fetal intestine and pancreas from 10 weeks of gestation. The concentration increases rapidly in the next month and again near term achieving adult levels by birth. VIP is found exclusively in nerves and, although initially confined to the myenteric plexus, by 25–30 weeks of gestation substantial amounts are present in the submucous plexus. This redistribution may have functional significance since the VIP-containing submucous fibres surround secretory glands of the mucosa involved in water and electrolyte transport. This pattern was also seen in the rat in which VIP immunoreactive nerve fibres first appear in the myenteric plexus 3 days before birth and in the mucosal layer after birth. VIP is at adult levels in the human term fetus. However, in the rat the major increases occur after birth especially at weaning.

In the rat, substance P-containing fibres develop earlier than VIP fibres (i.e. 5 days before birth), increase rapidly until birth, and remain at this level thereafter. The temporal difference in the appearance of substance P and VIP is reflected in function: non-adrenergic, non-cholinergic excitatory responses, presumably mediated by substance P, precede the development of the VIP-mediated inhibitory responses.

Plasma VIP concentrations at birth in the human are higher than fasting adult values although there is no change in response to a meal. It is not known whether the elevated levels are the result of impaired clearance or increased production. The very high plasma VIP levels reported in infants born following fetal distress might be involved in the redistribution of visceral blood flow in distressed fetuses.

Pancreatic peptides: insulin, glucagon, pancreatic polypeptide, and somatostatin

The mammalian pancreas stores insulin, glucagon, pancreatic polypeptide, and somatostatin in four distinct cell types. Neuropeptides such as VIP and substance P are present in nerve fibres innervating the pancreas. As noted earlier, gastrin is present transiently in the neonatal pancreas. The ontogeny of pancreatic somatostatin has been discussed in the section on somatostatin.

Glucagon is detected in the 6-week human fetus while insulin is first present from 10 weeks of gestation. Reflecting the immaturity of the rat at birth, the time of appearance of insulin and glucagon in the rat is later with much of the increase occurring in the perinatal period. In the human fetus, the glucagon concentration remains higher than that of insulin until birth when the concentrations are about equal. Despite the early appearance and predominance of glucagon, there are few receptors for glucagon in fetal tissue and its intrauterine role is not known. This contrasts with insulin receptors which are abundant in fetal tissues. It is now accepted that insulin is a physiological regulator of glucose metabolism and utilization in the fetus. Insulin also affects growth since surgical pancreatectomy in fetal sheep causes severe growth retardation and infants born with diabetes are small at birth.

At birth, plasma glucagon concentrations increase 3- to 5-fold while insulin levels fall, remaining in the basal range. These changes may be mediated by the catecholamine surge occurring at birth. Similarly, feeding initiates glucagon secretion with little change in insulin. These changes at birth are consistent with the switch from *in utero* anabolic processes to activation of catabolism and utilization of endogenous fuel stores.

The function of pancreatic polypeptide in the adult is still an enigma so understandably there is no established role for pancreatic polypeptide in the fetus. Pancreatic polypeptide is localized predominantly in islets but pancreatic-polypeptide-containing cells are also scattered throughout the exocrine parenchyma. It is released by a meal and cholinergic stimuli, and its best established action is the inhibition of pancreatic secretion. There are no data on the effects of pancreatic polypeptide during gestation. Pancreatic polypeptide has been identified in the 8–10 week old human fetal pancreas and develops in parallel with the other pancreatic hormones, increasing progressively through gestation. The pancreatic polypeptide cells are predominantly in the head of the pancreas which originates from the ventral primordium.

Infants of diabetic mothers have an increased number of insulin-containing cells but have a decreased density of glucagon in pancreatic-polypeptide and somatostatin-containing cells. This suggests that hyperglycaemia *in utero* may be affecting the differentiation of the pancreatic endocrine cells.

Overview and clinical significance

Gastrointestinal endocrinology has, until recently, been a descriptive discipline, but the advent of specific peptide antagonists is now enabling functions to be assigned to particular peptides. In the fetus and new-born, which are less amenable to experimental manipulation, gastrointestinal endocrinology is still largely an area of phenomenonology.

However, the dramatic change from continuous parenteral to intermittent enteric feeding provides a unique opportunity to correlate changes in function with the endocrine milieu and to assemble a coherent physiological framework.

The GI tract is highly active in the new-born: the calorie intake of a 1-week old infant is four times as high per kg body weight as that of adults. As described, increased concentrations of many of the GI tract regulatory peptides are present at birth and therefore precede the high food intake. This suggests that the GI tract has been primed *in utero*, perhaps as a result of the fetal swallowing of an amniotic fluid rich in growth factors. It is important to note that endocrine cells appear from a very early stage of development in long-gestation species such as sheep, pig, and man and are present for a large fraction of intrauterine life.

The natural rate of development can be accelerated: prematurely born infants can adapt to enteral feeding up to 10 weeks before normal delivery time. Enteral feeding appears to be the key since, in the first postnatal days in both premature and full-term infants, it stimulates the secretion of postprandial and subsequent basal concentrations of gut regulatory peptides. These increased concentrations in turn promote developmental changes of gut growth, gut motility, gastric and intestinal secretions, and the enteroinsular axis. This proposal is primarily based on the observation that prematurity, the type of food (formula or breast milk), and the route of administration (intravenous or enteric) are major influences on the hormonal profile and maturation of the GI tract. An important outcome of these studies is that enterically administered subnutritional volumes of milk can induce the hormonal surges, thus supporting the concept of enteric feeding, even if only minimal, as a treatment of severely ill infants.

A number of pathological conditions associated with changes in the regulatory peptide profile have been reported. Infants with nesidioblastosis have reduced pancreatic somatostatin content. Infants with acute gastroenteritis have massive elevations of plasma concentrations of enteroglucagon and motilin. Adults with acute infectious diarrhoea also have increases in enteroglucagon and motilin but these are quite modest. Whether the massive increase in enteroglucagon in the infant is an adaptive response to mucosal damage is unclear. However, measurement of enteroglucagon concentrations may give quantitative information of the extent of damage and the response to therapy.

The quality of the intrauterine environment can also be of importance as fetal distress results in increased concentrations of GIP, VIP, glucagon, and pancreatic polypeptide at birth.

An example of dysfunction of the peptidergic nervous system is Hirschsprung's disease in which there is no intrinsic innervation of a segment of the distal colon, leading to a constricted segment. The lack of the tonically active, VIP-containing inhibitory neurones is thought to be responsible for this constriction. Hirschsprung's disease may be one extreme example of a derangement of the neural systems controlling gut function. More subtle developmental disorders may be responsible for a spectrum of motility and absorptive problems in the infant.

Finally, regulatory peptides and their analogues are increasingly being used as therapeutic agents in the adult. The most relevant to the infant is the long-acting somatostatin analogue, octreotide. Possible uses are the control of neonatal hyperinsulinaemia, diarrhoea, and hormone overproduction such as that of VIP in ganglioneuroma and neuroblastoma. Infusions of motilin have been used to restore intestinal motility in infants with pseudo-obstructive ileus. The advent of potent, long-lasting, orally administered motilin agonists should be of benefit for the treatment of gut motility disorders in the infant as well as in the adult.

The outlook is for studies on the ontogeny of GI regulatory peptides to move from phenomenonology to a detailed assessment of the control, mechanisms of action, physiological significance, and therapeutic potential.

Further reading

Aynsley-Green, A. (1989). The endocrinology of feeding in the newborn. *Bailliére's Clinical Endocrinology and Metabollism*, **3**, 837–68.

Bassett, J. (1985). Integration of pancreatic and gastrointestinal endocrine control of metabolic homeostasis during the perinatal period. In *The physiological development of the fetus and newborn* (ed. C.T. Jones and P. Nathanielsz), pp. 124–33, Academic Press, London.

Bryant, M.G., Buchan. A.M.J., Gregor, M., Ghatei, M.A., Polak, J.M., and Bloom, S.R. (1982). Development of intestinal regulatory peptides in the human fetus. *Gasteroenterology*, **83**, 45–54.

Dockray, G.J. (1987). Physiology of enteric neuropeptides. In *Physiology of the gastrointestinal tract* (2nd edn) (ed. L.R. Johnson), pp. 41–66. Raven Press, New York.

Johnson, L.R. (1985). Functional development of the stomach. *Annual Review of Physiology*, **47**, 199–215.

Larsson, L.I. (1988). Endocrine development of the gut in relation to function. In *Fetal and neonatal development* (ed. C.T. Jones), pp. 452–7. Perinatology Press, Ithaca, New York.

Thompson, J.C., Greeley, G.H., Rayford, P.L., and Townsend, Jr, C.M. (ed.) (1987). *Gastrointestinal endocrinology*. McGraw-Hill, New York.

Walsh, J.H. (1987). Gastrointestinal hormones. In *Physiology of the gastrointestinal tract* (2nd edn) (ed. L.R. Johnson), pp. 181–257. Raven Press, New York.

17. The development of the immune system in the fetus

Wayne G. Kimpton, Elizabeth A. Washington, and Ross N.P. Cahill

The lymphoid system is composed of the primary lymphoid organs, the thymus and bone marrow in mammals, and peripheral lymphoid tissues such as the white pulp of the spleen, lymph nodes, and gut-associated lymphoid tissues. The central cell of this system is the lymphocyte and, although lymphocytes are concentrated in lymphoid organs, they are nonetheless widely dispersed throughout all the tissues of the body except the central nervous system. Lymphocytes are not sedentary cells but, on the contrary, most mature lymphocytes continuously recirculate between the various lymphoid organs and other tissues of the body, travelling from the blood stream into lymphoid organs, then to the collecting efferent lymphatics, and back into the blood stream to repeat the cycle. Lymphocytes do not recirculate randomly, but rather a series of closely regulated tissue-specific migration streams exist that direct lymphocytes to particular organs such as peripheral lymph nodes or skin or gut. Such migration streams ensure that a microenvironment that optimizes the interactions between co-operating immune cells and antigen exists in the peripheral lymphoid tissues so that appropriate immune responses are initiated and disseminated throughout the body.

Fetal studies have shown that the ability of lymphocytes to recirculate among blood, tissues, and lymph is an inherent physiological property of lymphocytes and in this chapter we consider the development of the major pathways of recirculation of lymphocytes as an integral part of the development of the fetal lymphoid system. The focus of this chapter is the ovine fetus since a number of factors, including the length of gestation, availability of reagents, and surgical accessibility, have made the ovine fetus an excellent model for studying the development of the mammalian lymphoid system. Immunological studies have shown essentially the same pattern of development for the thymus, spleen, lymph nodes, and Peyer's patches in the human and ovine fetus, which is not surprising since the human infant and lamb have reached a similar stage of immunological development at birth, unlike rats and mice where the peripheral lymphoid organs are poorly developed at birth. The development of the ovine lymphoid system considered in this chapter can be regarded as typical for most mammals, particularly humans, and, wher-ever clear species differences are known to exist, these are referred to in the text. For obvious reasons there are no studies available on the circulation of cells in human fetal lymph but, from the limited studies that have been performed on lymph in adult humans and from immunohistological studies on human fetal tissues, it is believed that the essential features of lymphocyte migration described in the ovine fetus are directly applicable to the human fetus.

The cells of the immune system, which include lymphocytes and cells of the monocyte–macrophage series, develop during embryogenesis and continue to do so throughout life from pluripotential stem cells. They originate from relatively fixed numbers of stem cells, progenitors, and precursors and they accumulate at specific locations with precise, and apparently tightly controlled, chronology during fetal development. Thus, the appearance of lymphocytes is related to the ontogeny of haemopoiesis, and to the acquisition of particular immunological capabilities by the fetus, suggesting that lineage development, the number of cell divisions in early cell embryogenesis, and the steps of differentiation along the way to the fully developed fetal immune system are all closely regulated. The different organs and tissues that make up this system are formed during ontogeny under the influence of environmental stimuli that direct the movement and migration of cells to specific sites in the developing embryo. The exact nature of these environmental stimuli is unknown, although it is known that foreign antigen has nothing to do with these pathways of differentiation; studies on the developing immune system in the sheep fetus have demonstrated that the entire lymphoid and immune apparatus of the sheep fetus develops in the total absence of any foreign antigen.

The ovine placenta is syndesmochorial and protects the developing fetus from exposure to foreign antigen, including maternal immunoglobulins, yet the sheep fetus develops an extensive array of immune capabilities from quite early in gestation (gestation is normally 150 days) and is able, for example, to reject allogeneic skin grafts from around day 70 of gestation. Indeed the sheep fetus provides an unusual opportunity for studying the growth and development of the lymphoid system in a mileu where the fetus remains immunologically virgin. The development in this species of various surgical techniques, such as chronic catheterization of blood vessels and lymphatics *in utero*, has made it an excellent animal model for studying the physiological processes underlying the development of the fetal immune system and for examining the changes that occur in the immune system at birth. The advent of hybridoma technology in recent years has seen the generation of a large number of monoclonal antibodies against cell surface proteins on sheep lymphocytes that define different cell subsets, some of which are shown in Fig. 17.1. These reagents have enabled detailed studies to be made on the appearance and early development of various functional classes of immune cells during fetal life.

Development of the haemopoietic system

The cells in the immune system that are concerned with the discrimination of self and non-self are derived from the haemopoietic system. Haemopoiesis starts within the yolk sac of the sheep fetus 16 days after fertilization and blood cells from the yolk sac pass to the embryo on day 18–19 when the embryonic and vitelline blood vessels join. Haemopoiesis commences in the fetal liver at around 21 days of gestation and the liver remains the principal site of haemopoiesis throughout gestation. Haemopoiesis in the

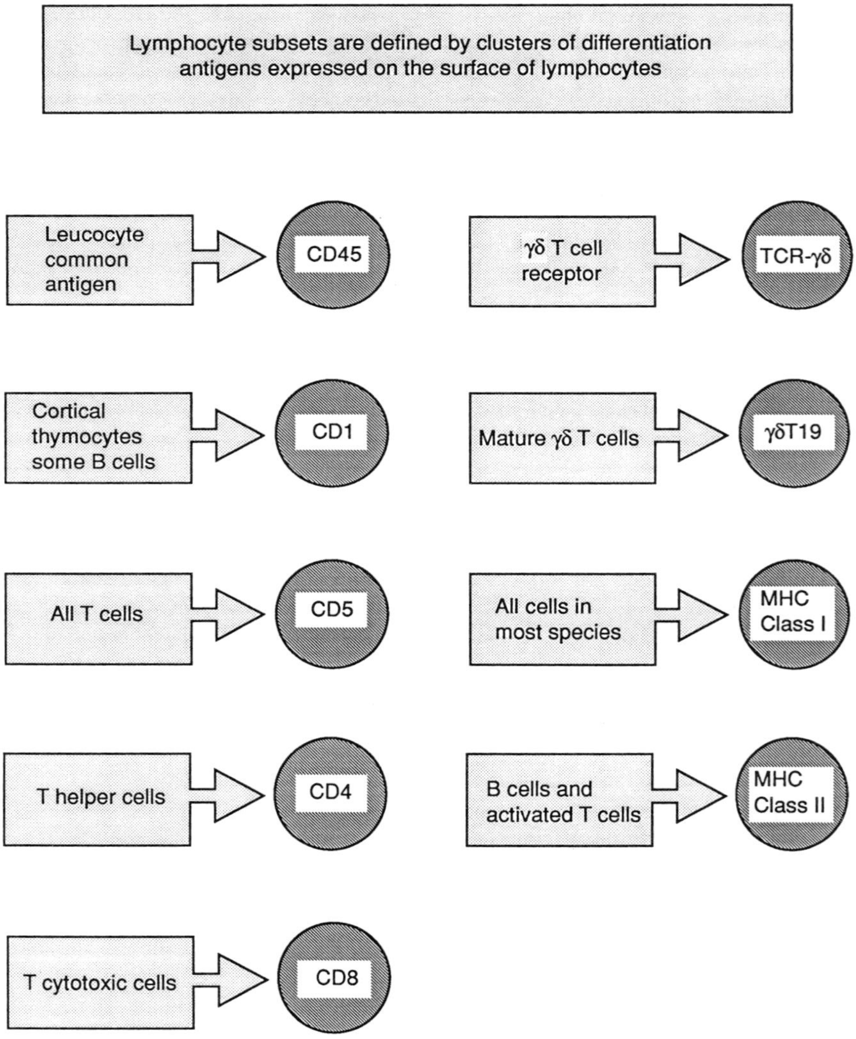

Lymphocyte subsets are defined by clusters of differentiation antigens expressed on the surface of lymphocytes

Leucocyte common antigen	→	CD45
γδ T cell receptor	→	TCR-γδ
Cortical thymocytes some B cells	→	CD1
Mature γδ T cells	→	γδT19
All T cells	→	CD5
All cells in most species	→	MHC Class I
T helper cells	→	CD4
B cells and activated T cells	→	MHC Class II
T cytotoxic cells	→	CD8

Fig. 17.1 Cell surface molecules that define major lymphocyte subpopulations.

liver is largely erythroid with much lower levels of megakaryopoiesis and granulopoiesis. In contrast to the liver, haemopoiesis in the bone marrow does not commence until about 70 days of gestation but, as is the case with fetal liver, lymphopoiesis is present at very low levels. At term lymphocytes make up approximately 5 per cent of nucleated marrow cells, but the majority of these cells are almost certainly in transit and are not derived from marrow lymphopoiesis.

Lymphopoiesis in the fetus

The lymphocyte is the central cell of the immune system and, during embryogenesis and early fetal development, there is a sequential involvement of various tissues in the production of lymphocyte populations of different functions and phenotypic specificities (Fig. 17.2). The order of appearance of lymphocytes in the developing lymphoid system is thymus, blood, spleen, lymph nodes, lymph,

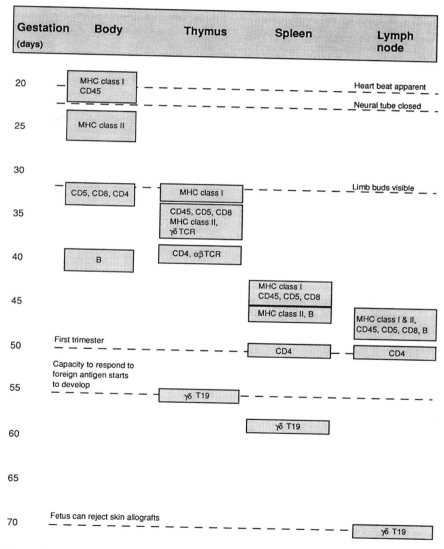

Fig. 17.2 Overall view of the sequential appearance of lymphocyte populations of different functions and phenotypic specificities in various tissues throughout ovine gestation.

bone marrow, and Peyer's patches. Around the same time as that at which lymphocytes first appear and start to accumulate in developing lymphoid tissues another series of developmental events is occurring in the sheep fetus (at around 60–65 days of gestation) which gives rise to a functional lymphatic system that becomes integrated with the lymphoid tissues such as lymph nodes. It is only at this stage, some 30–35 days after the first lymphocytes are detectable in the thymus and some 25–30 days after the first lymphocytes appear in the blood, that specific antibody responses can be elicited.

Development of T cells in fetal thymus

The essential requirement of the immune system is the generation of a large number of lymphocytes with unique antigen-specific receptors that enable the immune system to distinguish between ligands that are self and those that are non-self or foreign. Recognition of non-self ligands activates the immune system to eliminate foreign molecules, and failure to recognize self molecules prevents the immune system from destroying its host. The immune system acquires this knowledge during its development and the sheep fetus acquires a comprehensive array of immune capabilities in the complete absence of any foreign antigen. This is accomplished by generating a unique antigen-specific receptor on each lymphocyte through a novel genetic recombination mechanism, and then selectively deleting or inactivating those lymphocytes bearing anti-self receptors so that the immune system is self-tolerant and does not attack its host.

It is the function of the thymus to produce large numbers of specific antigen-reactive cells and to delete clonally any T cells bearing self-reactive receptors by a process called negative selection. Precursors of T lymphocytes migrate to the thymus where they are induced to divide rapidly and to begin rearranging T cell receptor (TcR) genes. Two sets of genes, $\alpha\beta$ and $\gamma\delta$, encode two different TcRs that are expressed on different populations of mature T cells. The $\alpha\beta$ lineage represents the major developmental pathway and the majority of T cells in man, mouse, and sheep use the $\alpha\beta$ TcR and express either CD4 or CD8 cell surface proteins. The receptors on CD4+ cells recognize peptides derived from endocytosed antigens bound to MHC (major histocompatibility complex) class II molecules and, when activated, help B cells, make antibodies, or activate macrophages in delayed hypersensitivity reactions. The receptors on CD8+ cells generally recognize peptides derived from cytoplasmic antigens (e.g. viral proteins) bound to major histocompatibility complex (MHC) class I proteins and, when activated, become cytotoxic for target cells expressing the foreign antigen.

The thymus is first identifiable at around 33–35 days of ovine gestation and increases in size from around 10 mg at 40 days of gestation to around 30 g at birth. Lymphopoiesis increases in intensity throughout gestation and the increase in thymic size is due largely to the accumulation of lymphocytes. The appearance of cells expressing T-cell lineage and other cell-surface antigens follows a quite specific sequence in the early ovine embryo and thymus which is summarized in Fig. 17.2. CD45 and MHC class I antigens are seen on a small number of cells in the body of the embryo at day 19 of gestation, while MHC class II antigens are first observed at day 25. CD5+ cells are first seen in the mesenchyme of the neck and in the dorsal mediastinum at day 33, although they are not present in the thymus at this time. The first identifiable lymphocytes to appear in the thymus are CD5+ cells which are first seen at day 35 and are soon followed by CD8+ cells and then by CD4+ cells. It should be noted that, whereas colonization of the thymus occurs late in the final third of gestation in the mouse, in humans colonization of the thymus occurs in the first third of gestation, the same as in the sheep. Phenotypic analysis of sheep thymocytes using two-colour immunofluorescence has demonstrated that at 40 days of gestation virtually all the thymocytes are CD45+CD5+, while 6.5 per cent are CD4+CD8+ and 17.5 per cent are CD4–CD8–. By 50 days of gestation, fetal thymocytes are nearly all CD4+CD8+ (i.e. predominantly cortical in type), whereas from 60 days of gestation the staining patterns resemble those found in post natal sheep where four subpopulations of thymocytes with particular thymic localizations can be distinguished: (1) CD4–CD8–, outer cortex (12 per cent); (2) CD4+CD8+, inner cortex (74 per cent); (3) CD4+CD8–, medulla (10 per cent); (4) CD4–CD8+, medulla (4 per cent). This is similar to the situation in humans and mice except for a lower percentage of CD4–CD8– cells in the mouse.

A second developmental pathway is represented by cells of the $\gamma\delta$ lineage. This second, numerically smaller subset of mature T cells, is thought to be involved in a surveillance system for infected or transformed cells at epithelial surfaces. Cells expressing the $\gamma\delta$ TcR never express CD4, very rarely express CD8, and, in sheep, can be divided into two subsets according to whether or not they express the lymphocyte surface antigen T19. T19 appears to distinguish two populations of $\gamma\delta$ cells within the sheep thymus. The immature population is localized to the outer thymic cortex, is T19–, and is MHC class I low or negative, and the mature population is localized predominantly in the inner cortex and medulla, is T19+, and expresses high levels of MHC class I molecules.

As in mice, chickens, and humans, the $\gamma\delta$ T-cell subset develops early in sheep fetal thymic ontogeny, first appearing at 40 days of gestation. At 55 days of gestation, when the thymic medulla is just emerging, the first $\gamma\delta$ T19+ cells appear in the thymic cortex, probably being derived from the $\gamma\delta$ T19– subset. By 60 days of gestation $\gamma\delta$ T19+ cells are predominantly localized in the medulla, and by 80 days the fetal thymus resembles that of postnatal animals, at which stage about 3 per cent of thymocytes express $\gamma\delta$ TcRs, of which roughly 50 per cent are T19+. A remarkable feature of $\gamma\delta$ T-cell ontogeny is the close association they exhibit with Hassall's corpuscles, around which they form large foci, suggesting these structures may play some role in $\gamma\delta$ T-cell development within the thymus.

Development of lymphocytes in fetal spleen

The ovine spleen arises as a localized condensation of mesodermal cells in the dorsal mesogastrium at around day 25 of gestation and lymphocytes are first identifiable at day 43 of gestation. The primordial splenic rudiment at 43–44 days of gestation contains scattered CD45+ cells, reticular cells, and rather few blood vessels. It is surrounded by a thin capsule and contains low numbers of CD1+, CD5+, and CD8+ cells. B cells are first detected at 45–50 days, CD4+ cells at 50–55 days, and $\gamma\delta$ T19+ cells at around 57 days of gestation (Fig. 17.2). B-cell organization precedes that of T lymphocytes with perivascular cuffs of B lymphocytes being first seen in fetuses of 50 days of gestation. T lymphocytes start accumulating within the periarteriolar region at around 55 days of gestation, from which time they start to displace the B cells peripherally. B-cell follicles are not seen before 70 days of gestation and germinal centres are not seen until the second week after birth. At mid-gestation the fetal spleen contains more CD8+ cells than CD4+ cells, but this situation is reversed in the late-term fetus and remains so after birth. The development of the spleen from 70 days onwards is associated with the formation of the white pulp. This forms in association with the splenic arterioles and progressively enlarges during fetal development so that the fetal spleen at term is approximately 50 per cent lymphoid cells. The appearance of germinal centres in the follicles occurs around 2 weeks after birth in both sheep and humans.

Development of lymphocytes in fetal lymph nodes

The most precocious and rapidly growing lymph nodes in the sheep fetus are the prescapular and mediastinal with hepatic, mesenteric, and prefemoral lymph nodes developing several days later. Early fetal lymph nodes exhibit a reticular framework with scattered mesenchymal cells and small aggregations of large lymphoid cells together with other haemopoietic cells. At around 60–65 days of gestation the lymphatic system becomes functional and connected to the developing lymph nodes and blood vessels, and, by 70 days of gestation, there is histological evidence that lymphocytes are already recirculating from blood through fetal lymph nodes into lymph in both sheep and humans. The sequence of appearance of lymphocyte subpopulations in fetal lymph nodes is the same as that for spleen and thymus (Fig. 17.2). CD5+ and CD8+ cells and B cells are present at 47 days of gestation, the first CD4+ cells are found at day 50, and $\gamma\delta$ T19+ cells first appear at day 69 of gestation.

By around 75 days' gestation the lymphocyte population of ovine fetal lymph nodes has enlarged sufficiently to allow a cortex and a medulla to be distinguished. Although the cortex of fetal lymph nodes is very much thinner in relation to the medulla compared with lymph nodes after birth, the ratio of cortex to medulla and the size of the deep cortical units increases steadily throughout fetal life. Small B-cell follicles are found in the peripheral cortex immediately below the subcapsular sinus and CD4+ cells considerably outnumber CD8+ cells in the cortex, while the reverse is true in the medulla. $\gamma\delta$ T19+ cells show a characteristic distribution, lying predominantly under the subcapsular sinus and scattered diffusely through the medulla. In the first 2 weeks after birth, lymph nodes increase in size enormously with germinal centres appearing in the B-cell follicles (which have also increased in size considerably). B-cell development in sheep fetal lymph nodes closely resembles that reported for human fetal nodes.

Lymphopoiesis in fetal Peyer's patches

The growth and development of and lymphopoiesis in fetal and postnatal Peyer's patches of sheep have been studied extensively, and it has been established that the ileal Peyer's patch is a major site of lymphopoiesis and is the principal source of B cells in the sheep. Although Peyer's patches are well developed in the human fetus, there is a species difference in that the bone marrow is the principal source of B cells in the human. Primordial Peyer's patches in sheep are first seen at 60 days of gestation, lymphoid follicles first appear at 75 days, and by 120 days Peyer's patch follicles are histologically mature. By birth they occupy the wall of the terminal 2 m of the small intestine. Peyer's patches weigh about 20 g at birth and are sites of intense lymphopoiesis, although the majority of B cells produced there are not exported since the proportion of these cells in fetal blood and lymph remains constant at around 3–4 per cent. Removal of the terminal ileal Peyer's patch before birth prevents the 10-fold increase in the percentage of B cells in blood and lymph that normally occurs in the first few weeks after birth.

Development of lymphocyte recirculation

Lymphocytes in blood and lymph form part of a large pool of lymphocytes that are continuously recirculating between blood, tissues, and lymph (Fig. 17.3). This process is controlled principally by specialized vascular endothelial cells that line post capillary venules called high endothelial venules in traffic sites such as lymph nodes and gut. These high endothelial venules express adhesion molecules on their surface that bind to complementary ligands called homing receptors expressed on the surface of

circulating lymphocytes. This binding of lymphocytes to endothelial cells is the crucial first step in enabling lymphocytes to leave the blood and enter the tissues. High endothelial venules are present in lymph nodes and gut and they arise in a variety of chronic inflammatory conditions, including rheumatoid arthritis and chronic inflammatory bowel disease. Since tissue-specific migration streams to gut, peripheral lymph nodes, and skin were first described in postnatal sheep, numerous studies in many other species, including humans, have confirmed the existence of tissue-specific migration of lymphocytes. Recently, a number of homing receptors and endothelial adhesion molecules have been defined that direct this tissue-specific traffic. In animals after birth, the recirculation of lymphocytes between various tissues and lymphoid organs is essential for immunosurveillance and these cells play a critical role in the dissemination of immune responses and the transport of immunological memory to previously encountered antigens.

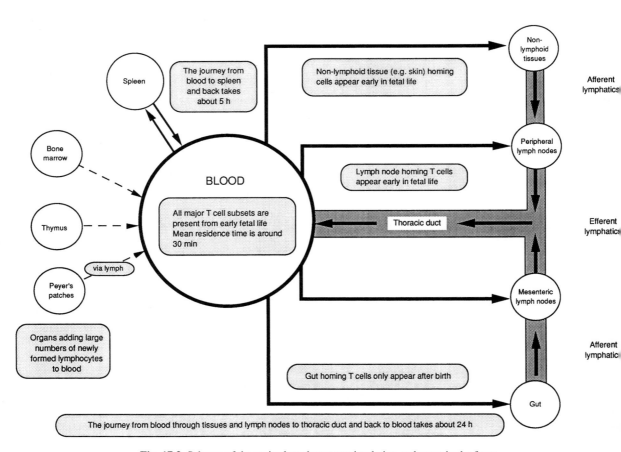

Fig. 17.3 Scheme of the major lymphocyte recirculation pathways in the fetus.

Given the critical role of recirculating cells in immuno-surveillance and memory, for many years it was believed that the presence of antigen was necessary to trigger the development of lymphocyte recirculation. This is not the case, however, since more recent studies have demonstrated that lymphocytes are recirculating between blood and lymph in the sheep fetus from quite early in gestation, indicating that lymphocyte recirculation develops over a protracted period in the fetus and continues to expand and develop throughout the life of the animal.

Lymphocytes start to recirculate from blood to lymph in the sheep fetus at least as early as 75 days of gestation and the size of the recirculating pool of lymphocytes increases exponentially throughout *in utero* life as the fetus grows and new cells are added to the blood, lymph, and lymphoid tissues. Recent immunohistological studies have indicated that lymphocytes also commence recirculating in the human fetus at as early as 12 weeks of gestation. The fact that a large-scale recirculation of lymphocytes is just as much a feature of the fetal immune system as it is in animals after birth only serves to emphasize that pathways of recirculation and the capacity of lymphocytes to recirculate arise as a physiological process independent of antigenic stimulation, since the sheep fetus is not exposed to foreign antigen and its lymphoid system is composed of lympho-cytes which, in an immunological sense, are disputably naïve. In adult animals there is no doubt that recirculating lymphocytes are immunologically competent cells and that they are responsible for the transport of immunological memory and for the dissemination and amplification of the immune response. The presence of a large pool of recircu-lating, immunologically naïve lymphocytes in the fetus suggests that viewing lymphocyte recirculation solely in these terms may be too restrictive. What is required is a more general understanding of the physiological basis of a process that is established so early in fetal development and that is, in some of its aspects, similar to the migration of cells that occurs generally in organogenesis.

Tissue-specific lymphocyte migration streams in the fetus

The size and composition of the pool of recirculating cells has been examined by collecting cells from the fetus in experiments involving chronic cathetherization of blood vessels and lymphatics in fetal lambs *in utero*. The output of lymphocytes from the thoracic duct of fetal sheep increases exponentially from around 1.5×10^6/h at 70 days of gestation to 2.0×10^8/h at term. The size of the fetal pool of recirculating lymphocytes increases from 5.5×10^8 cells at 90–95 days of gestation to 1.2×10^{10} cells at 140–150 days of gestation. The predominant cell recircu-lating in the fetus is the T cell and all the major T-cell sub-sets present in the adult are recirculating through fetal lymph nodes (see Figs 17.3 and 17.4). CD4+ lymphocytes are the major T-cell subset in fetal lymph, and are enriched in fetal lymph compared with blood, whereas CD8+, $\gamma\delta$+, and B cells are not, suggesting that T helper cells are extracted by lymph nodes in the fetus at a faster rate than other T cells and B cells.

The gut-homing T cells that migrate between blood and intestinal lymph and that are such a prominent feature of lymphocyte recirculation in postnatal life are not present in the fetus. Tissue-specific homing of T cells to gut com-mences in the first few weeks after birth and the traffic of T cells through the gut and gut-associated lymphoid tissues increases enormously during the first 6 months of postnatal life. Although gut-homing T cells have not been identified in the fetus, lymph draining the ileal lymph node and ileum of the fetus is unusual because it contains a higher percent-age of CD8+ cells than is found in lymph circulating through any other tissues or organs of the fetus thus far examined. It is not known whether these CD8+ cells are migrating preferentially through fetal ileal lymph nodes and/or ileum, or whether they are being locally produced.

Homing of lymphocytes to skin and lymph nodes in the fetus

Lymphocytes can enter peripheral lymph nodes in one of only two ways. They can enter either directly from the blood via high endothelial venules in the lymph node itself or they can enter via afferent lymph draining skin and peripheral tissues. Regardless of their route of entry in most species, once T cells have entered lymph nodes, they can then only leave via efferent lymph. The input of cells into lymph nodes via these two routes has been measured directly in fetal and postnatal sheep and the results are summarized in Fig. 17.5. The cell input to a lymph node equals the cell output from the node because, in unstimu-lated lymph nodes in both the fetus and adult, less than 2 per cent of cells in efferent lymph is locally produced within the node.

The major difference in the entry of cells into lymph nodes before and after birth is that much larger numbers of skin-homing T cells enter lymph nodes via afferent lymph in the fetus than is the case in postnatal animals. The capacity of cells to home to skin and peripheral tissues is thus established during fetal life so that, rather than being directed by any immunological stimulus, recirculation through skin and afferent lymph is a physiological prop-erty of immunologically naïve T cells. This must be taken into account before attributing any immunological significance to the traffic of T cells through skin.

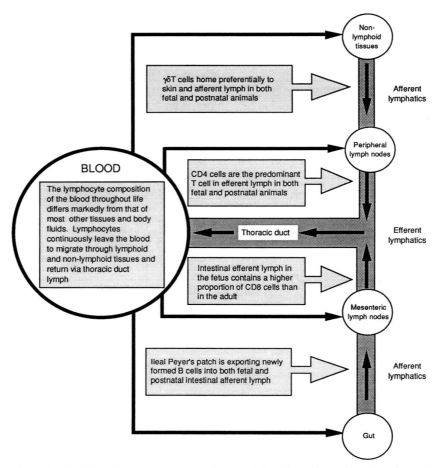

Fig. 17.4 Scheme of the blood and lymph compartments associated with the skin, peripheral lymph nodes, and gut. Different proportions of lymphocyte subsets are found in blood and lymph draining different tissues.

It has been suggested that only memory cells traffic through skin in adult animals, but this proposition needs to be reappraised in view of the large number of skin-homing T cells in the fetus. Recirculating T lymphocytes in the fetus almost certainly use the same basic lymphocyte–endothelium recognition mechanisms to extravasate from the blood into lymph nodes and skin as do T cells recirculating in the adult. For example, fetal gut vascular endothelial cells express tissue-specific adhesion molecules that enable postnatal T cells to migrate preferentially to fetal gut in the same way as they do in postnatal animals. Antigen-stimulated fetal lymph nodes also show the same increased traffic of T cells from blood into efferent lymph as do adult lymph nodes and fetal T cells display the same adhesion molecules and lymph-node homing receptor as do adult T cells. The selective entry of

$\gamma\delta$ T cells into skin and afferent lymph is also established in fetal life and has nothing to do with whether or not they are memory cells or have been recently activated by antigen. At present the selective homing of $\gamma\delta$ or $\alpha\beta$ T cells cannot be related to the presence or absence in lymph of memory T cells, nor can it be related to the expression of any known adhesion molecules. The critical molecules directing the homing of $\alpha\beta$ and $\gamma\delta$ T cells into skin and peripheral tissues have yet to be identified.

Changes in the fetal immune system after birth

Some major changes which occur in the immune system of the fetus after birth are outlined in Fig. 17.6. At term, the

FETAL LIFE

POSTNATAL LIFE

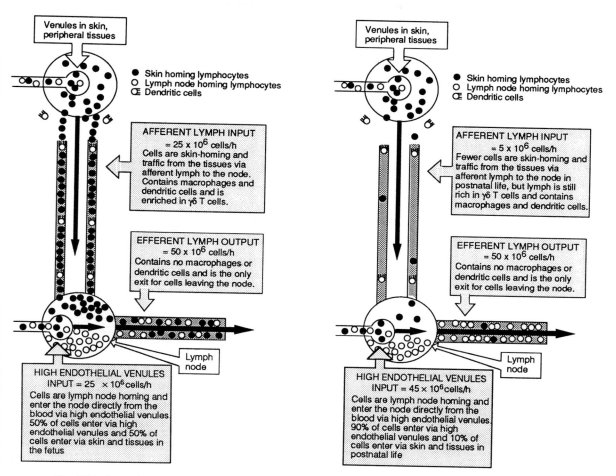

Fig. 17.5 Comparison of lymphocyte migration through fetal and postnatal skin and peripheral lymph nodes.

sheep fetus displays a highly developed immune system associated with a pool of more than 10^{10} recirculating lymphocytes, which has arisen as part of the normal growth and development of the fetus. Fetal recirculating lymphocytes are long-lived cells that continue to recirculate in the fetus from the time they first enter lymph but, within a few weeks after birth, they have been largely replaced by a pool of short-lived lymphocytes.

It is quite remarkable that the fetus develops, in such a tightly controlled manner, a large pool of lymphocytes that are continuously circulating throughout the fetal immune system; yet these lymphocytes have no known function in the sheep fetus which is protected from foreign antigens by the placenta. At birth, a process is triggered whereby, over the next 10 days, 90 per cent of these fetal lymphocytes are replaced by new lymphocytes that have been formed after birth. In this way the fetal immune system is largely replaced by a new, neonatal immune system that renews itself every 7 to 10 days. The relationship between the fetal immune system and the neonatal immune system,

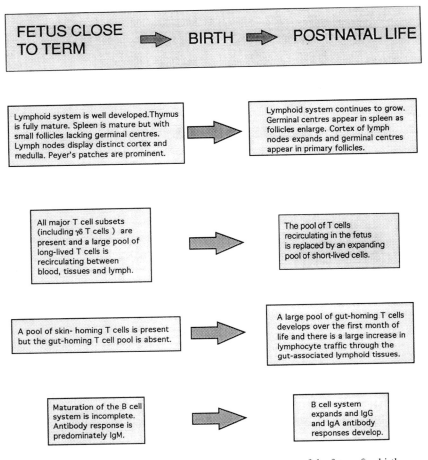

Fig. 17.6 Major changes that occur in the immune system of the fetus after birth.

i.e. the relationship between the fetal lymphocytes and those new lymphocytes making up the immune system after birth, is completely unknown. Do, for example, the fetal cells die after birth, or do they divide and so contribute to the new neonatal lymphoid system? These are not trivial questions because answers to them and to related questions will almost certainly explain how the immune system of the new-born is assembled in the face of a hostile new environment crowded with potentially pathogenic microorganisms.

Further reading

Cahill, R.N.P. and Trnka, Z. (1980). Growth and development of recirculating lymphocytes in the sheep fetus. In *Essays on the anatomy and physiology of lymphoid tissues* (ed. Z. Trnka and R.N.P. Cahill), pp. 38–49. Karger, Basel.

Cahill, R.N.P. and Trnka, Z. (1981). Recirculation and life span of lymphocytes in the sheep fetus and in post-natal lambs. In *The immune system* (ed. C.M. Steinberg and I. Lefkovits), pp. 367–74. Karger, Basel.

Cahill, R.N.P., Heron, I., Poskitt, D.C., and Trnka, Z. (1980). Lymphocyte recirculation in the sheep fetus. In *Blood cells and vessel walls: functional interactions*, CIBA Foundation Symposium, No. 71, pp. 145–66. Excerpta Medica, Amsterdam.

Cahill, R.N.P., Kimpton, W.G., Dudler, L., and Trnka, Z. (1985). Lymphopoiesis in fetal and perinatal sheep. In *Immunology of the sheep* (ed. B. Morris and M. Miyasaka), pp. 46–67. Editiones Roche, Basel.

Fahey, K.J. and Morris, B. (1974). Lymphopoiesis and immune reactivity in the fetal lamb. *Series Haematologica*, **7**, 548–67.

Haynes, B.F., Denning, S.M., Singer, K.H., and Kurtzberg, J. (1989). Ontogeny of T-cell precursors: a model for the initial stages of human T-cell development. *Immunology Today*, **10**, 87–91.

Haynes, B.F., Martin, M.E., Kay, H.H., and Kurtzberg, J. (1988). Early events in human T cell ontogeny: phenotypic characte

ization and immunohistologic localization of T cell precursors in early human fetal tissues. *Journal of Experimental Medicine*, **168**, 1061–80.

Horst, E., Meijer, C.J.L.M., Duijvestijn, A.M., Hartwig, N., Van der Harten, H.J., and Pals, S.T. (1990). The ontogeny of human lymphocyte recirculation: high endothelial cell antigen (HECA-452) and CD44 homing receptor expression in the development of the immune system. *European Journal of Immunology*, **20**, 1483–9.

Kimpton, W.G. and Cahill, R.N.P. (1985). Circulation of autologous and allogeneic lymphocytes in lambs before and after birth. In *Immunology of the sheep* (ed. B. Morris and M. Miyasaka), pp. 306–26. Editiones Roche, Basel.

Kimpton, W.G., Washington, E.A. and Cahill, R.N.P. (1989). Recirculation of lymphocyte subsets (CD5+, CD4+, CD8+, T19+, and B cells) through fetal lymph nodes. *Immunology*, **68**, 575–9.

Kimpton, W.G., Washington, E.A., and Cahill, R.N.P. (1990). Non-random migration of CD4+, CD8+ and γδ+T19+, and B cells between blood and lymph draining ileal and prescapular lymph nodes in the sheep fetus. *International Immunology*, **2**, 937–43.

Mackay, C.R., Beya, M., and Matzinger, P. (1989). γδ T cells express a unique surface molecule appearing late during thymic development. *European Journal of Immunology*, **19**, 1477–83.

Mackay, C.R., Maddox, J.F., and Brandon, M.R. (1986). Thymocyte subpopulations during early fetal development in sheep. *Journal of Immunology*, **136**, 1592–9.

Maddox, J.F., Mackay, C.R., and Brandon, M.R. (1987) Ontogeny of ovine lymphocytes. I. An immunohistological study on the development of T lymphocytes in the sheep embryo and fetal thymus. *Immunology*, **62**, 97–105.

Maddox, J.F., Mackay, C.R. and Brandon, M.R. (1987). Ontogeny of ovine lymphocytes. II. An immunohistological study on the development of T lymphocytes in the sheep fetal spleen. *Immunology*, **62**, 107–12.

Maddox, J.F., Mackay, C.R., and Brandon, M.R. (1987). Ontogeny of ovine lymphocytes. III. An immunohistological study of the development of T lymphocytes in sheep fetal lymph nodes. *Immunology*, **62**, 113–18.

Miyasaka, M. and Morris, B. (1988). The ontogeny of the lymphoid system and immune responsiveness in sheep. *Progress in Veterinary and Microbiological Immunology*, **4**, 21–55.

Morris, B. (1986). The ontogeny and comportment of lymphoid cells in fetal and neonatal sheep. *Immunological Review*, **91**, 219–33.

Pearson, L.D., Simpson-Morgan, M.W., and Morris, B. (1976). Lymphopoiesis and lymphocyte recirculation in the sheep fetus. *Journal of Experimental Medicine*, **143**, 167–86.

Reynolds, J.D. and Morris, B. (1983). The evolution and involution of Peyer's patches in fetal sheep and postnatal sheep. *European Journal of Immunology*, **13**, 627–35.

18. The structural development of the nervous system

Sandra M. Rees

The human nervous system is composed of hundreds of billions of nerve cells (~10^{11}) each with many processes that must make highly specific synaptic connections with target cells. In addition, the nervous system contains many times more neuroglial cells, which carry out many vital roles to support the function of neurones. Understanding the events that lead to the formation of this highly complex structure from a single sheet of ectodermal cells in the embryo is one of the most exciting challenges in biology today. This chapter will describe the major events in neural development, drawing on data from *in vivo* and *in vitro* experiments performed on the nervous systems of vertebrates and invertebrates. As there is a strong likelihood that major developmental processes are much the same across species, understanding how simpler brains form will ultimately help us to understand how the human brain develops.

Early embryonic events in neural development

The development of the nervous system proper begins at the end of the gastrula stage of embryogenesis when the embryo is composed of three layers of germ cells surrounding a primitive gut. The outer layer, the ectoderm, forms the skin and nervous system. The middle layer, the mesoderm, forms muscle, skeleton, connective tissue, and the cardiovascular and urogenital systems, and the inner layer, the endoderm, gives rise to the gut and many of the major organs associated with the gut. In the human, this stage is reached during the third week of embryonic development. Neurulation begins when a longitudinal sheet of ectodermal cells (neural plate) on the dorsal surface of the embryo becomes transformed into the specialized tissue from which the entire nervous system is to develop (Fig. 18.1). It has been known since the 1920s that the critical event in this process, which is called neural induction, is an interaction between the ectoderm and the underlying mesoderm. The mechanism of the interaction remains to be elucidated, but it may involve the diffusion of sub-

stances from the mesoderm to the ectoderm. Recently, for example, one of the phorbol esters (a second messenger that relays signals from the cell surface) has been suggested as a possible candidate on the basis of experiments on neural induction in amphibian tissue. As it is known that anterior mesoderm induces embryonic forebrain formation while posterior mesoderm induces the formation of spinal cord structures, each region of the mesoderm must produce highly specific factors. Initially, if cells in, for example, the forebrain–eye field of the neural plate are removed, neighbouring cells will take their place and no deficit in this region will result. As development proceeds, however, different regions of the field become committed to the formation of a specific region of the forebrain and, if removed, will result in a defect in the development of that part of the nervous system.

The margins of the neural plate gradually become raised to form the neural folds and the midline region is depressed to form the neural groove (Fig. 18.1). As there is relatively little cell division in the neural plate at this stage, the changing shape of the plate is thought to occur partly because the apical surfaces of cells decrease in size while the basal surfaces enlarge (the cells become flask-shaped) causing the neural plate to curve in on itself. Changes in the distribution and types of cell adhesion molecules (CAMs) on cell surfaces during this period might also be important in regulating cell movement. CAMs are transmembrane proteins that promote selective cell-to-cell adhesion. The continued elongation of the neural plate in the midline may also contribute to cell movement. The neural folds gradually become more pronounced and eventually meet and fuse in the midline to form the neural tube, which then separates from the overlying ectoderm. Fusion occurs first in the region of the future hindbrain and then progresses rostrally and caudally from this point. Defective closure of the neural tube is a frequent cause of congenital malformations. When closure fails at rostral levels, a condition known as anencephaly occurs in which the overall structure of the brain is grossly disturbed. When the caudal portion fails to close, a condition called spina bifida results. In this condition, the spinal cord is not confined by the vertebral column and the

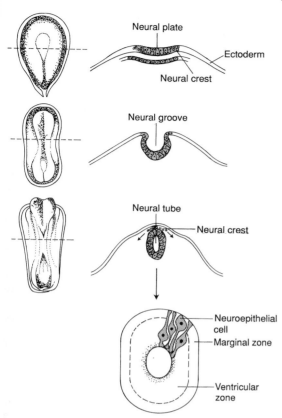

Fig. 18.1 These diagrams show an external (dorsal) view of the developing nervous system (left) and a corresponding cross-sectional view at about the middle of the future spinal cord (right). The nervous system begins as the neural plate, a flat sheet of ectodermal cells on the dorsal surface of the embryo. The plate subsequently folds into a hollow structure, the neural tube, from which the central nervous system forms. The neural crest cells give rise to the cells of the peripheral nervous system. Inset: A more detailed cross-section of the tube at the time of closure. The long process of the neuroepithelial cells span the wall of the tube across the ventricular and marginal zones. (Redrawn from Cowan, W.M. (1979). *Scientific American*, **241**, 106–17.)

functions subserved by the lumbar and sacral spinal cord are disrupted.

As will be described below, the neural tube eventually gives rise to the cells of the central nervous system (CNS), that is, the brain and spinal cord. The peripheral nervous system (PNS) originates from a distinct group of cells called the neural crest (Fig. 18.1). The neural crest is a transient structure that arises from the dorsolateral margins of the neural plate just before neural tube closure. Neural crest cells migrate extensively throughout the body and differentiate into a variety of cells including the spinal (sensory) and autonomic ganglia, Schwann cells, melanocytes, chromaffin cells of the adrenal medulla, and some skeletal and connective tissues of the face. It is not yet known how the crest cells are set apart from the other cells of the neural plate, what determines the migration routes of these cells, or what factors determine the differentiation of specific cellular phenotypes from neural crest precursor cells.

Cell generation and migration

Cell generation

At the time of closure of the neural tube there is only a single, or at most, double layer of columnar epithelial cells (neuroepithelium). Once the tube has closed, rapid cell division occurs. Assuming that the fully developed human brain contains approximately 100 billion neurones, and that neurones do not divide after birth, neurones must be generated in the developing brain at an average rate of 250 000/min. (The accepted view that, with the exception of olfactory receptors, no new neurones are formed after birth (or the immediate neonatal period in some species) in vertebrates was challenged in 1983 when Nottebohm demonstrated that some neuronal proliferation probably occurred in adult songbirds in the nucleus controlling vocalization. There is now also evidence for the generation of neurones in the adult mouse striatum. There is, however, no evidence yet for the occurrence of neurogenesis in the adult human brain.) Most of the proliferation occurs at the inner surface of the neutral tube, the ventricular zone, (Fig. 18.1), so-called because it lies adjacent to the cavity that will become the ventricular system of the brain and the central canal of the spinal cord. Initially, the neuroepithelial cells put out long processes that span the entire wall of the neural tube. The cell nucleus then migrates up the process towards the outer or marginal zone where it undergoes DNA replication. It is not clear why this migration occurs but it possibly allows the nucleus to be exposed to different cytoplasmic factors located in different regions of the cell. The nucleus then migrates back to the ventricular surface, retracts its processes, and divides to produce two daughter cells. Early in the process of cell genesis, both daughter cells will remain in the mitotic cycle. However, after several divisions the precursor cell will lose its ability to divide and will begin to migrate away from the germinal zone to its appropriate position in the developing nervous system. This is called the birth date of the cell. Cellular proliferation also occurs in additional zones, notably, in the forebrain where the subventricular zone gives rise to small neurones of the basal ganglia and some cells of the cerebral cortex, in the cerebellum where the external granule layer gives rise to granule cells and interneurones, and in the hilar zone of the dentate gyrus of the hippocampus that also produces granule cells.

Formation of specific neuronal populations

As cells begin to migrate (see below), the neural tube expands and an intermediate zone forms between the ventricular and marginal zones. By weeks 4–5 of human embryonic development, this process results in the formation of three distinct swellings or vesicles at the rostral end of the tube, in the region of the future brain. These are (1) the forebrain vesicle (prosencephalon), which gives rise to the cerebral hemispheres and diencephalon; (2) the midbrain vesicle (mesencephalon), which gives rise to the adult midbrain; and (3) the hindbrain vesicle (rhombencephalon), which gives rise to the pons, medulla, and cerebellum (Fig. 18.2). It appears that different populations of cells are withdrawn from the mitotic cycle at rigidly determined times, although it is not yet known what controls the turning on and off of the proliferative mechanisms for

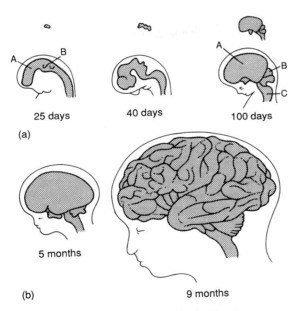

(a) 25 days   40 days   100 days

(b) 5 months   9 months

Fig. 18.2 Development of the human brain. The diagrams are about 40 per cent of life size with the exception of the lower diagrams in row (a), which have been enlarged for clarity. The cerebral hemispheres that develop from the forebrain vesicle (A) eventually overgrow the midbrain (B) and hindbrain (C). (Redrawn from Cowan, W.M. (1979). *Scientific American*, **241**, 106–17.)

specific cell types. For example, the genesis of cortical neurones starts around the fortieth embryonic day (E40) in primates and continues until approximately E100 in monkeys and possibly E125 in humans.

We have learnt a great deal about the birth date of populations of cells by pulse-labelling with tritiated thymidine. This is injected into a pregnant animal where it crosses the placenta and is actively taken up and incorporated into DNA by all cells in the synthetic phase of the mitotic cycle, during the 30–60 min that it remains in the circulation. Those cells that cease dividing after incorporation of the thymidine (and can therefore be considered to be 'born' at that time) will retain a strong radioactive signal that can be detected at a later stage when the tissue is processed for autoradiography. In cells that continue to divide the signal will be diluted.

From these studies some generalizations can be made.

1. The larger neurones (e.g. Purkinje cells in the cerebellum) are produced before smaller neurones (e.g. cerebellar granule cells).
2. In the cerebral cortex, hippocampus, and optic tectum the cells that form the deepest layers are produced first while those that are generated at successively later times form the progressively more superficial layers (e.g. in the cerebral cortex layer VI is formed before layer V, etc.).
3. There appear to be two distinct periods of cell proliferation: the first correlates with the major period of neurogenesis (10–18 weeks of gestation in the human) and the second with gliogenesis (mid-gestation to 18–24 months after birth in the human), although, as will be discussed in the next section, glial precursors and radial glial cells are present from very early in development.
4. The parts of the nervous system that appeared first in phylogeny have a tendency to appear early in ontogeny and those that arise later in evolution often arise late in ontogeny.
5. There is a ventral to dorsal progression of cell proliferation in the developing nervous system (e.g. the proliferation of motoneurones is completed before that of sensory neurones in the dorsal horn).

Cell lineage commitment

A most interesting question is whether the cells in the ventricular zone represent a homogeneous population of multipotential precursor cells capable of giving rise to neurones and neuroglia throughout cell genesis or whether the cells are committed to neuronal or glial pathways at a very early stage in development. It appears that in different parts of

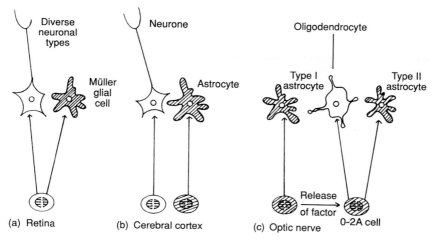

Fig. 18.3 Schema of cell lineage commitment in various regions of the central nervous system, indicating that precursor cells can exhibit a range of capabilities. (a) The retina where precursor cells exhibit multipotential capabilities, and can give rise to diverse neuronal types and Müller glial cells. (b) The cerebral cortex where neuronal and glial cell lines differentiate very early in development. (c) The optic nerve where there are two distinct glial cell lineages, one with precursor cells giving rise to astrocytes (type I) alone and the other to oligodendrocytes and astrocytes (type II).

the developing nervous system precursors exhibit a range of capabilities (Fig. 18.3). Using a recently devised method based on the ability of retroviruses (an RNA-containing virus that can transfer its genetic material into the DNA of its host's cells) to deliver a genetic marker to cells *in vivo*, studies on the developing vertebrate retina have shown that a single precursor may give rise to several diverse types of retinal neurones and also to the non-neuronal Müller glial cells. This suggests a multipotentiality in precursor cells located in the mammalian retina. A proportion of early neural crest cells also appear to be multipotential. On the other hand, results using retroviral techniques in the cerebral cortex suggest that separate cell lines for astrocytes and neurones have already differentiated early in development. These results support earlier immunohistochemical evidence that demonstrated that glial fibrillary acidic protein, a glial-specific marker, is already present in some of the dividing cells in the ventricular zone at the peak of neurogenesis, that is, very early in development. Another precursor cell type that has restricted but nevertheless bipotential differentiation capabilities has been isolated from rat optic nerve. It has been shown that the so-called 0-2A progenitor cells produce oligodendrocytes (which form the myelin sheath) or type-II astrocytes depending on conditions, at least *in vitro*. The behaviour of the 0-2A cell is thought to be regulated by growth factors produced by type-I astrocytes, the first glial cell type that is produced in the optic nerve. Therefore, present evidence suggests that, in some locations, cells retain a pluripotentiality, while in others the precursor cells follow separate pathways from very early in development.

Neuronal migration

The young neurones now have to move away from the germinal zone to their definitive location in the developing nervous system. In the early 1970s, Pasko Rakic first proposed that the processes of specialized supporting cells (radial glia), in several but not all parts of the CNS, form a scaffold along which immature neurones migrate. These glial processes, which are formed early in development, have very different lengths and trajectories in different regions of CNS. For example, in the cerebral cortex, neurones follow extremely long glial processes from the ventricular surface, through the intermediate zone, to the emerging cortical plate (the first cells to arrive in the developing cortex). In primates, a massive migration of cortical neurones occurs during mid-gestation and coincides with the rapid growth of the cerebral wall. In the cerebellum, there is a unique class of radial glia, Bergmann glia, that provide guidance for granule cell bodies migrating shorter distances from the external to the inner granular layers (Fig. 18.4). Using video microscopy, Mary Hatten and her colleagues have recently been able to observe the movement of granule cells along glial fibres in tissue culture. The cell forms a specialized junction with the glial fibre along the length of the neuronal soma and extends a motile leading process in the direction of

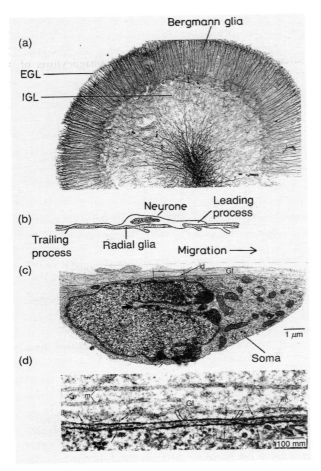

Fig. 18.4 Neuronal migration. (a) Bergmann radial glial fibres in the fetal (sheep) cerebellum stained for glial fibrillary acidic protein-like immunoreactivity. (b) This diagram illustrates how granule cells migrate along the fibres from the external granule layer (EGL) to the inner granule layer (IGL). (c) The migrating neurone forms an extensive apposition (interstitial density, id) with the glial process (GI). (d) At the interstitial density, the intercellular space is filled with fibrils (short thin arrows) and microfilaments from both the neuronal soma and glial fibres (long thin arrows) appear to be in contact with them. ((c) and (d) reprinted with permission from Gregory, W.A., Edmondson, J.C., Hatten, M.E., and Mason, C.A. (1988). *Journal of Neuroscience*, **8**, 1728–38.)

motion. This process, which extends along and enfolds the glial fibre, is highly active. Movement of the neurone appears to be generated along the neurone–glia apposition at the cell soma. In a companion electron microscope study they showed that, at the junction (or interstitial density), the intercellular space is filled with fibrillar material.

Microfilaments in both the neuronal soma and glial fibres project towards this material and appear to be in register with it. This might link or co-ordinate the cytoskeletal elements needed for locomotion (Fig. 18.4).

Surprisingly, little is known about the signals that control the movement of the neurone on to, along, and then off the glial fibre at the appropriate location. There might be a specific neurone–glia adhesion molecule to initiate the movement of the neurone from the ventricular zone on to the glial fibre. The locomotion of the neurone along the fibre probably involves other ligands (the neuronal antigen astrotactin is a possible candidate) possibly linked with the fibrillar system described above. The navigation of neurones from glial fibres into cortical layers or nuclei is likely to involve neurone–neurone adhesion ligands. This would result in the aggregation of cells of a similar kind.

As mentioned above many neurones migrate without the benefit of glial guides within the CNS and within the PNS from the neural crest. These immature neurones rely on cellular and extracellular molecules in the matrix through which they move, and the cues probably include differential adhesiveness and movement along a chemical gradient (chemotropism). These mechanisms will be discussed below in relation to axonal guidance.

Neuronal differentiation

A remarkable feature of vertebrate development is that the proliferating cells of the neural tube ultimately produce a huge variety of neuronal phenotypes. At some point during development each of these young neurones must acquire a specific identity that defines the pattern and type of connections it should form with other neurones. With cortical neurones the birth date of a neurone will predict its normal laminar specific fate; e.g. the first formed neurones form the deepest layers. However, within a layer some neurones will become projection neurones with a range of targets (e.g. corticospinal, corticobulbar); others will become interneurones. What controls this cellular differentiation? As yet we have no firm answer but the possibility is that it will involve an interplay between the cell's lineal history and environmental signals. For example, cells in layer V throughout the cerebral cortex in the rodent initially extend an axon down the pyramidal tract to the spinal cord. However, during development, visual cortical cells retract this axon collateral and develop stable connections with the tectum. So, initially, all layer V cells follow a general pattern for axon outgrowth, and only later are region-specific differences in projection sculpted out. Recent experiments have shown that, if small pieces of frontal cortex are transplanted to visual cortex in developing rats, the transplanted neurones form connections that are appropriate for their new location and withdraw their projection fibres from the pyramidal tract, a movement that would not have occurred in their original site.

Columnar organization of the cortex

In addition to being arranged in layers, the sensory and motor areas of the cerebral cortex contain columns that are perpendicularly oriented to the layers and in which cells have similar physiological properties. It has been suggested that functional columns can be subdivided into smaller, radially oriented ontogenetic columns, each centred around a single glial fibre with the cells in each layer sequentially generated from a common precursor located near the ventricle. This idea has recently been challenged, however, in experiments using retroviruses to trace neuronal lineages. Clonally related cortical neurones have been shown to enter several different radial columns, apparently by migrating along different radial glial fibres. The relationship between cell lineage and ontogenetic columns might prove to be less closely aligned than originally proposed.

Neuroglial cells

There are two major classes of glial cells in the vertebrate nervous system: macroglia (oligodendrocytes and astrocytes) and microglia. As described above, the macroglia originate from neuroepithelial cells with their production extending well into the second postnatal year in the human. In addition, they are continuously renewed by a steady rate of cell proliferation and cell death throughout life. The mechanisms that control the turnover are not well understood. Microglia, on the other hand, originate from a specific population of mononuclear leucocytes that penetrate the blood–brain barrier and enter the CNS during late fetal to early postnatal periods and transform into microglial cells.

In addition to their roles as supporting elements in the CNS, providing firmness and structure, neuroglia play many important active roles in brain development and function. Astrocytes buffer the K^+ ion concentration in the extracellular space and some take up and remove chemical transmitters released by neurones during synaptic transmission. They also produce growth factors, contribute to the formation of the blood–brain barrier, are involved in the formation of glial scars after CNS injury, and some play a role in neuronal and axonal guidance during development. The main function of oligodendrocytes is to form and maintain a myelin sheath around axons of certain cells in the CNS. Myelination of axons

begins before birth in humans and continues for several years postnatally. In the PNS myelination is carried out by Schwann cells, which originate from the neural crest. Microglia play an important role in phagocytosis of cellular debris.

The development of neuronal processes

Having migrated to the appropriate position in the nervous system, the young neurone now faces the challenging task of forming precise connections with targets that might be anything from micrometers to centimeters away. This section deals with the formation of appropriate connections between neurones.

A neurone can have many dendrites but has only a single axon, each type of process having a distinct morphology and function. The ability of axons, but not dendrites, to elongate rapidly must result from a difference in their molecular constituents. Recently, it has been shown in cultured hippocampal neurones that, preceding the onset of axonal growth, all minor neural processes have growth-associated protein (GAP-43)-like immunoreactivity. At about the time of initial axonal outgrowth, this immunoreactivity becomes highly concentrated in the axonal growth cone and, within a few days, largely disappears from the other processes. GAP-43, therefore, appears to be selectively sorted into the neurite, which will become the axon, but the means of achieving this and the part it then plays in axonal elongation are not yet known.

The initial outgrowth of the axon and its orientation appear to be genetically determined since cells that have been experimentally moved or rotated show similar growth patterns to their sister cells in the developing nervous system. However, there is not enough genetic information in a cell to carry the complete instructions for making the appropriate connections. The genome probably contains only the instructions for the general directions, the details of which we are now just beginning to understand.

The growth cone

At the tip of most growing neurites (axons and dendrites) is a distinctive structure called a growth cone (Fig. 18.5). This consists of a flattened region (lamellipodium) from which extend finger-like processes (filopodia) that are typically about 5–30 μm long and 0.1 μm in diameter, although the length, and number, of filopodia varies considerably. The growth cone was first described in 1890 by

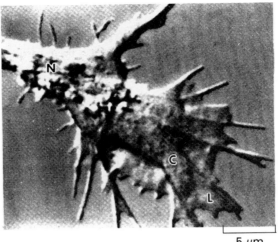

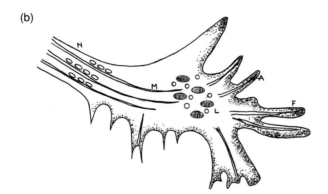

(b)

Fig. 18.5 (a) Digitally processed video micrograph of a growth cone of *Aplysia* in culture. (b) Diagram of cellular components of a growth cone. N, Consolidated neurite; L, lamellipodium; F, filopodium; M, microtubule; A, actin; C, organelle-rich central domain; shaded region, meshed actin filaments. Scale bar 5 μm. ((a) reprinted with permission from Goldberg, D. and Burmeister, D.W. (1989). *Trends in Neuroscience*, **12**, 503–6.)

the Spanish neuroanatomist Ramón y Cajal in developing chick spinal cord. Some years later, Harrison described a living growth cone in explants of frog spinal cord and made the fundamental observation that growth cones are highly motile, dynamic structures, constantly moving and exploring the environment by extension and retraction of their filopodia. When growth cones are observed *in vitro* it is noted that most filopodia ultimately regress into the growth cone but some persist when they contact and differentially adhere to particular surfaces. The molecular basis of this differential adhesiveness is of great interest to neurobiologists and will be discussed below. If the filopodium adheres to a surface that is more highly adhesive than its local surroundings, tension is generated in the filopodium and the tip of the growth cone appears to be guided in the direction of the adhesion. The mechanism of the tension-generating process is not yet understood but it is known that the growth cone is rich in actin filaments

especially in those regions involved in extending its margins. Filopodia often have core bundles of actin filaments that extend from their tips back through the lamellipodia to the organelle-rich central regions. Myosin II, the traditional type of muscle myosin, is scattered throughout the growth cone, so it is possible that interaction between these molecules generates tension.

Growth of the neurite plasma membrane occurs at the growth cone by local insertion of membrane components. These components are delivered from their site of synthesis in the neuronal cell body to the growth cone by fast axoplasmic transport directed by microtubules that extend into the central region of the cone (Fig. 18.5). The terminal ends of microtubules appear to be dynamically unstable, growing and shrinking rapidly by exchanging tubulin with a large soluble pool in the growth cone. As most neurite branches arise by division of growth cones, this means of microtubule assembly would allow tubules to be diverted into newly forming branches. Experiments in which actin filaments are disrupted with cytochalasins show that neurite extension can continue in the absence of growth cone motility (which is actin-based), presumably sustained by continuous arrival of new material by axoplasmic transport; however this growth is not directed. It seems that the growth cone's contribution to growth is to influence its direction, while axoplasmic transport provides the material.

Axonal pathfinding

As an axon grows towards its target it is probably influenced by several factors, each of varying significance according to the stage of growth of the process.

Mechanical guidance

Evidence to support the idea that axonal growth can be directed by mechanical means initially came from the observation that explanted nervous tissue will send out neurites along scratches on the surface of a culture dish. In vivo, it is possible that, in some circumstances, axons run along surface contours, interfaces, preformed neuroglial pathways, pioneer fibres, or paths of least resistance. In the developing chick hind limb, it has been reported that cells die and are phagocytosed in the vicinity of growth cones as they enter the major nerve trunk pathways in the limb. This might enable axons to travel down a pathway of reduced cell density (alternatively, growth cones might be attracted to components of necrotic tissue). Consistent with these ideas is the observation that, in certain regions, a pattern of holes or spaces is generated in the neuroepithelial matrix before axon outgrowth occurs. For example, axons growing out from the mouse retina enter open spaces in the region of the optic nerve head. In a mutant mouse in which these spaces are obliterated, axon outgrowth is impeded; thus, such neuroepithelial spaces appear to be required for normal axon outgrowth in this situation.

Electrical guidance

The early suggestion that electric fields might influence axonal growth has recently been revived. It has been demonstrated that electric fields can influence both the rate and direction of neurite extension from cultured cells with neurites orienting towards the cathode. Although evidence for physiological effects on axon outgrowth is still lacking, axon growth is clearly responsive to field strengths that are probably present at the relevant embryonic stages.

Chemotropism

The observation that bacteria and slime moulds in culture will grow towards the source of a diffusible chemical led to the suggestion that axons might be guided along gradients of a diffusible chemical being produced by the axon's target. Support for this theory has come from two experiments using the protein, nerve growth factor (NGF). In tissue culture, local application of this molecule to growing neurites of dorsal root ganglion (sensory) cells causes them to change the direction of their growth over a very brief time (approximately 30 min). In addition, an injection of NGF into the brains of neonatal rodents causes sympathetic axons to take completely aberrant pathways into the spinal cord and ascend to the site of the NGF injection.

There are several objections, however, to chemotropism playing a major role in axonal guidance on the basis of the evidence provided so far. Tissue levels of NGF, for example, are very low, several orders of magnitude below that required to cause the chemotropism described above. As the targets of sensory and sympathetic ganglion cells (the major classes of neurones known to be sensitive to NGF) are so widespread in the body, it is difficult to envisage how gradients of diffusable molecules could, by themselves, cause the highly stereotyped patterns of outgrowth apparent in normal development. Perhaps NGF and other target-derived factors serve as local attractants to axons that have reached the vicinity of those targets by other means.

Inhibitory mechanisms

Repulsion or inhibition of axonal growth in particular regions has been demonstrated to encourage their growth into more attractive areas. Inhibitory mechanisms may take the form of cellular boundaries or barriers along a pathway or may involve the production of an inhibitory molecule. For example, keratan sulfate, a glycosaminoglycan, has been shown to inhibit growth of axons across the roof plate in the embryonic rat spinal cord.

Differential adhesiveness

As stated above, growth cone filopodia seem to explore the local environment, at least in tissue culture, by seeking surfaces of preferential adhesiveness. When neurones are grown on a culture plate with a surface bearing geometric patterns of artificial materials to which neurites adhere differentially, axons follow the pattern described by the substance to which they adhere best. Although this principle probably also operates in vivo, it has been much more difficult to demonstrate. One example, however, is found in the moth wing where it has been shown that the preferred direction of growth of sensory axons is along a distal-to-proximal axis that appears to correlate with an increase in epithelial adhesiveness. Also, axons *in vivo* often fasciculate (form bundles) in a highly selective manner and this is thought to be due to specific adhesive interaction. What is the molecular basis of this selective adhesion?

Molecular basis of cellular adhesion

As a result of many years of work (1940s–1960s) on the specificity of connections in the retinotectal system in the frog, Roger Sperry proposed that nerve cells recognize one another by means of complementary surface labels. Although Sperry's original theory of chemoaffinity has been modified over the years there is considerable support for the idea that surface recognition molecules assist in guiding the axon to its target in the developing nervous system. In attempting to discover the molecular basis of this recognition, assays of cell adhesion and immunological techniques have been used. Immunological techniques have shown that neurone–neurone adhesion is mediated in part by the neural cell adhesion molecule (NCAM), a cell surface glycoprotein discovered by Edelman and others. NCAM-mediated adhesion has been postulated to occur by a homophilic binding mechanism, in which NCAM on the surface of one cell binds to NCAM on a neighbouring cell. It has been suggested that spatial and temporal variations in the amount and form of this molecule can produce differences in cell adhesivity that influence the pathway and target choice of migrating cells and growing axons.

For example, in relation to temporal variation, neural crest cells that are destined to become spinal ganglia express NCAM and adhere tightly together prior to migration. As they become motile, they no longer have detectable amounts of NCAM on their surface and instead are found in association with fibronectin (see below) in the extracellular matrix. Upon arrival at the appropriate site, the cells once again express NCAM and form a compact group of cells that differentiates into a ganglion. An example of spatial variation is found in the neural retina of the developing eye in which NCAM is most concentrated on neuroepithelium at the dorsal posterior pole, the point of origin of the optic nerve, and decreases in concentration towards more peripheral anterior regions. One can speculate that the gradient might be used to guide axons towards the optic fissure, but there is as yet no unequivocal evidence that this is so. Clearly then, these molecules can be up- or downregulated according to the needs of the cell.

Other cell surface adhesion molecules of the vertebrate nervous system that have been extensively studied are neuroglial CAM, myelin-associated glycoprotein, and N-cadherin. Each appears to be capable of mediating neurone–neurone or neurone–glia adhesion.

Fibronectin and laminin, two of the components of the extracellular matrix have been shown to promote neurite outgrowth in culture. It has been proposed that interaction between the receptor molecules on growth cones and the heparin-binding domains of these molecules might be the molecular basis for such a response. Although both of these molecules are major constituents of the extracellular matrix of the PNS they are found only in restricted areas of the CNS and then only during development.

It is probably reasonable to conclude that all of the above mechanisms play some part in axonal guidance, although none by themselves adequately explains the range of behaviour of growing axons. Perhaps mechanical guidance, electrical gradients, inhibitory effects, and differential adhesiveness might be important in leading axons to their targets, while chemotropism and molecular recognition might come into play by confirming the arrival of the axon at an appropriate target.

Formation of connections

Synaptogenesis

On reaching the target cell, axonal elongation ceases by a mechanism not yet determined, and the formation of a synapse is initiated. The sequence of events in synaptogenesis appears to be similar whether an axon forms a synapse on a muscle fibre or on a neurone. In tissue culture, growing mammalian axons have been observed by phase contrast microscopy and fixed at different stages of synapse formation. Prior to contact, no ultrastructural specializations were recognized in either the axonal growth cone or the postsynaptic cells. When one or more filopodia made contact, there was cessation of all filopodial activity for up to 30 min (Fig. 18.6). One of the filopodia eventually broadened and an area of postsynaptic specialization began to form within 6 h. In the growth cone, presynaptic dense projections developed along with the postsynaptic specializations. Synaptic vesicles appeared 18 h after first contact, and by 36 h those organelles traditionally associated with the growth cone were no longer found in the presynaptic terminal area. There may be different sequences of development of synaptic components in different parts of the nervous system. In the eutherian nervous system, synaptogenesis reaches a peak in the immediate postnatal period. As described below, some of

(a)

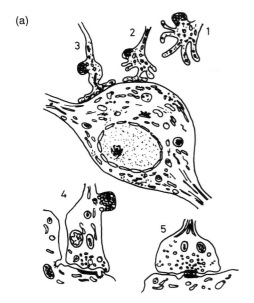

(b)

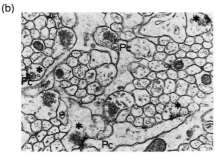

Fig. 18.6 Development of synapses on a mammalian neurone. (a) Diagrams illustrating the sequence of events (1–5) in the formation of the synapse between a growth cone arising from a spinal cord implant and a superior cervical ganglion cell in culture. (After Rees, R.P., Bunge, M.B., and Bunge, R.P. (1976). *Journal of Cell Biology*, **68**, 243–63.) (b) Electronmicrograph of synapses between granule cell axons (parallel fibres) (*) and Purkinje (PC) cell dendritic spines in late-gestation fetal cerebellum (sheep).

these synapses will be eliminated as the mature nervous system is shaped.

Specificity of connections

When an axon courses through the developing brain it does not innervate everything along its path but is highly selective as to where it forms a synapse. This selectivity can be manifest in several ways.

Specific types of neurones will invariably be attracted to make connections with each other (cell specificity). In systems where it is of the utmost importance to preserve a topographical map of the peripheral input, for example, in the visual system, axons must synapse in the target in a highly specific order (topographical specificity). In the visual system, it has recently been proposed that groups of newly generated retinal ganglion cells are temporally matched with groups of target cells. The axons of each group of retinal ganglion cells must gain access to an appropriately mature group of target cells before the period of target dependence (to be described below) begins. Such chronological determinants, either alone or in combination with long-term molecular markers, seem to be the most crucial factors for the establishment of topographic order. A third form of specificity is called synaptic site specificity where an axon synapses on a particular region of the dendritic field of a cell. These restricted sites might depend on the timing of the arrival of the particular afferent fibre and are possibly due to competition between the afferents for the available sites.

Regressive events during development

So far, cell proliferation and the growth of neural processes (i.e. progressive developmental effects) have been described. There is now considerable evidence that regressive phenomena also play an important role in determining the form of the mature nervous system.

Neuronal death

Although it was known from the beginning of the century that neuronal death occurred during development, it was not until the late 1940s, beginning with the important observations of Viktor Hamburger and Rita Levi-Montalcini on the sensory ganglia in chick embryos, that neuronal death was recognized as a major and widespread feature of normal development in both the CNS and PNS. As the death and removal of neurones by phagocytic glial cells occur very rapidly (3–7 h), the extent of cell loss is only apparent when neurones are systematically counted at the beginning and end of a specific developmental period. This might account for the phenomenon remaining undetected for so long.

Cell death has now been described for many parts of the nervous system, including the retina, various brainstem nuclei, autonomic ganglia, cranial motor nuclei, spinal motoneurone pools, and the cerebral cortex, with the

extent of death ranging from 15 to 75 per cent of the population (most typically, 50 per cent). The absence of neuronal degeneration during development has only been
reported for a few regions such as the avian pontine nuclei,
the locus ceruleus, and the red nucleus. The phase of cell
death is usually confined to a specific period that is distinctive for each neuronal population. For example, in the
chick embryo between days 5 and 9 of incubation, 40 per
cent of motoneurones die; between days 10 and 17, just
under 60 per cent of neurones in the isthmooptic nucleus,
the nucleus of origin of the centrifugal fibres to the contralateral retina, die. Young neurones appear to develop
structurally in much the same way whether they eventually
survive or die. It is of interest, however, that a specific
68kDa protein (Alz-50) found in degenerating neurones
in Alzheimer's disease has recently been identified in
neurones in the cortical plate of rats just prior to and/or
during natural neuronal death. It is yet to be determined
whether Alz-50 is a marker of cell death during normal
development.

What is the purpose of this overproduction of neurones
followed by the subsequent elimination of considerable
numbers of cells? The major purpose appears to be the
matching of the size of innervating populations to the
capacity of their targets. In addition, cell death also
corrects errors of migration (mislocation) and projection
(misprojection).

Target-dependent cell death

In most neuronal populations cell death occurs at the time
of synaptogenesis in the target tissue. This temporal coincidence, together with the demonstration that manipula-

tion of the availability of putative synaptic targets alt
the numbers of surviving neurones, led to the propo
that the target is somehow instrumental in defining
final population of innervating cells. For example, expe
ments performed on amphibian larvae have shown t
removal or addition of a limb bud could respectiv
deplete or increase the number of motor and sensory n
rones in the related innervating neuronal populations (F
18.7). In other systems it has been shown that a par
ablation of the target field results in a roughly prop
tional increase in cell death in the innervating neuro
population. No given cell, therefore, is destined to c
Clearly, cells will die if they are deprived of a target
innervate, but do cells die in the normal course of dev
opment simply because they fail to reach their targ
Experiments in several systems have shown that
majority of cells that ultimately degenerate have alrea
innervated their target tissue. For example, in the ch
isthmooptic nucleus, at day 10 of incubation the nucle
contains 22 000 cells but, by day 17, contains only 95
cells. If the enzyme horseradish peroxidase is injec
into one eye of the embryo at day 12, most cells in
contralateral isthmooptic nucleus will be labelled by r
rograde transport of the enzyme, indicating that cells d
tined to die had successfully sent axons to the target ar
It has also been shown that, when both isthmooptic nuc
are induced to innervate one eye as a result of ea
removal of the other eye, fewer neurones in the contra
eral nuclei will ultimately survive. Considering this a
other evidence, it appears that neurones compete for so
sustaining factor supplied by the target. It is not cert
whether neuronal death is caused by the limited prod
tion of such a factor or the ease of access to it. If the la

Extent of cell death

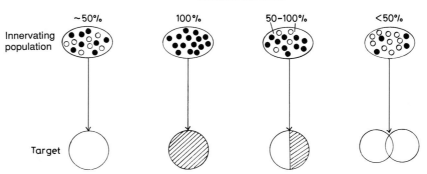

Fig. 18.7 Target-dependent cell death. In most neuronal populations approximately 50 per cent of cells die at about the time of
synaptogenesis in the target tissue. If the target is partially or totally ablated neurones are lost proportionately to the amount of
target removed. Adding additional target or providing trophic factors rescues some of the neurones that would otherwise have die

proves to be the case, as recently suggested by Oppenheim (1989), then those cells with a greater intrinsic capacity to branch and form synapses would be expected to have a competitive edge over their less well 'endowed' neighbours. It is also conceivable that the availability of trophic factors could be regulated solely by the availability of synaptic sites, with little if any need to postulate the involvement of synaptic branching. Not only does the target influence cell survival, but it has recently been shown that the target can influence cell properties such as the neurotransmitter phenotype.

At this stage the only molecule that has been shown to meet virtually all of the criteria necessary to qualify as a specific neurotrophic factor involved in regulating the survival of some classes of neurones (mammalian and avian sensory and sympathetic ganglion cells) is nerve growth factor. Many other growth factors have now been identified and shown to be important for the survival of various neuronal populations *in vitro* but their role *in vivo* has yet to be unequivocally demonstrated. For example, brain-derived neurotrophic factor enhances the survival of retinal and nodose ganglion cells, basic fibroblast growth factor increases the survival and neurite extension of hippocampal, substantia nigral, and cortical neurones, and ciliary neurotrophic factor promotes the survival of chick sympathetic ciliary ganglion cells in culture and enhances chick spinal motoneurone survival *in vivo*.

Cell death as a mechanism for eliminating mislocated or misprojecting neurones

There is evidence that neurones that have made erroneous connections are eliminated during the phase of naturally occurring cell death. For example, early in the development of the isthmooptic nucleus, a small number of neurones (~ 0.1 per cent) project erroneously to the ipsilateral rather than to the contralateral eye. There is also a population of cells that migrate past the isthmooptic nucleus to form an ectopic nucleus that projects to the contralateral isthmooptic nucleus. During the phase of cell death in which 60 per cent of the total population of cells dies, both the misprojecting and mislocated neurones can be shown to have degenerated in their entirety. At present, little is known about the mechanisms that lead to this preferential death of neurones during a phase of general cell death. In the case of mislocated cells, the lack of an appropriate afferent input might be significant in their failure to survive since it is known that the removal of an afferent input from correctly located neurones can result in their deaths. There is no definitive explanation for the selective elimination of the misprojecting neurones.

Hormone-dependent neuronal death

A third type of neuronal death during development relates directly or indirectly to the action of circulating hormones. For example, in the rat, the spinal nucleus of the bulbocavernosus (SNB), which innervates the penile and anal musculature, contains more motoneurones in the mature male than in the female. This sexual dimorphism arises primarily because motoneurone loss is greater in the female than in the male during the period of naturally occurring cell death. It is not yet certain whether androgens rescue these SNB neurones in the male by saving the muscle to which they project or by acting as direct trophic agents for neurones. Penile musculature is initially present in the female but atrophies during the first postnatal weeks. Wherever the locus of action of the sex steroids turns out to be, the result is a developmental difference in the nervous system.

Elimination of neuronal processes

Neuronal death, of course, results in the elimination of all the processes of the degenerated cell, but there is now evidence that, in addition, there is selective elimination of some of the neuronal processes of a cell without the death of that cell. Initially, evidence for the elimination of processes during development came from studies on the neuromuscular junction and it has now been convincingly shown, using physiological and anatomical techniques, to occur in the PNS and in the CNS in at least three systems: the cerebellar Purkinje cells; the visual system of the cat; and the avian cochlear nucleus. This form of process elimination involves the removal of axons and their synapses over relatively short distances and occurs in the early postnatal period appreciably later than cell death in the same neuronal system. Its purpose appears to be sharpen and refine the initial exuberant growth of synaptic connections and, as with cell death, possibly involves a competitive mechanism resulting in the death of some processes and the survival of others.

There is evidence to suggest that one factor that might give some fibres a competitive edge over others is their state of functional activity. This has been shown most strikingly in the mammalian visual system. In layer IV of the visual cortex of cats and monkeys, the inputs from the relevant layers of the lateral geniculate nucleus, which are connected with the two eyes, overlap extensively at first, but in time become progressively separated into distinct eye dominance columns or stripes. In normal animals the stripes connected with each eye are of the same width but if, during a critical period in early postnatal life, one eye is deprived of form vision by suturing the eyelids closed, the

stripes in layer IV associated with the deprived eye
become significantly reduced in width whilst those connected to the non-deprived eye are correspondingly
enlarged. This system retains a degree of plasticity for
some time so that, if the originally deprived eye is opened
and the eyelids of the other eye are sutured closed, the initially reduced eye dominance columns now increase and
those that had expanded now shrink. This suggests that the
relative absence of activity in the deprived eye somehow
places it at a competitive disadvantage.

Process elimination without cell death is not limited to
the fine-tuning of neuronal circuits; in some situations it is
also responsible for the elimination of long collateral projections during development. This was first demonstrated
in the callosal system but is now known to occur in the
corticospinal tract (see the section on 'Neuronal differentiation'). In the callosal system, using retrograde labelling of
cells with fluorescent dyes, it has been possible to show
that the initial widespread distribution of callosal fibres in,
for example, the visual cortex gradually becomes restricted
during development to the vertical meridian in the cortex.
The major factor in the restriction of the callosal projections is the selective elimination of callosal collaterals
without the death of their parent cells. It is not yet known
how widespread this process might be nor what factors are
responsible for the removal of certain branches of an axon
while others persist. It seems reasonable to suggest that a
mechanism like this in the developing cerebral cortex
would substantially reduce the amount of genetic information required to encode for all the connections which must
be made.

So it appears that, prior to maturation, neurones lose their
ability for self-sustenance and become dependent on target-
derived factors for survival. At a later stage the refining of
neuronal connections by elimination of processes might also
result from a competition for trophic factors.

Factors that influence brain development

The developing mammalian nervous system is vulnerable
to many agents and conditions including viruses, radiation,
hyperthermia, hypoxia, malnutrition, and circulating hormone levels, particularly thyroid hormone levels. The
effects that these factors have on the brain depends on the
developmental processes that are most active at the time
the insult is imposed. Much of what we know about the
vulnerability of cellular events in the developing brain
is derived from studies on experimental animals.
Extrapolation of such data to humans requires that due

regard be given to the developmental processes in progress
at the time of the insult. Although different species follow
a similar sequence of neuronal development, the stage of
brain maturation at the time of birth varies considerably
across eutherian species. At birth the human brain is
approximately 25 per cent of its adult weight, the rat brain
15 per cent, guinea-pig brain 62 per cent, and monkey
brain 72 per cent. As no new neurones are produced after
birth in the human, the increase in brain weight is
accounted for by an increase in cell size, proliferation of
axonal and dendritic processes, synaptogenesis, multiplication of glial cells, and myelination of axons. These processes will all be vulnerable to environmental influences.

Malnutrition

One of the most extensively studied of the factors that
influence brain development is malnutrition (maternal and
neonatal), particularly in rats. Several parameters are
affected, depending on the time of insult as indicated
above. There is a reduction in glial numbers, in myelination, in growth of axonal and dendritic processes, and, perhaps most importantly, a reduction in the number of
synapses per neurone. Restoration of normal nutrition
allows for the reversal of most, but not all, of these effects.

The effects of maternal malnutrition on the human fetus
are not well understood. It is known, however, that malnutrition can result in placental insufficiency causing premature birth, which carries a risk of mental retardation. When
children are subjected to chronic malnutrition from birth to
18 months of age severe enough to result in growth retardation, they have been shown to suffer from permanent
deficits in emotional, cognitive, and intellectual function.
It is possible that the structural deficits underlying these
problems are similar to those described for malnourished
rats. Nutritional therapy and enriched social and educa-
tional conditions provided during this period are successful in reversing this situation. If therapy is delayed beyond
2 years, the deficits are only partially reversed.

Thyroid hormone

Thyroid hormone plays an important role in stimulating
neural development. For example, it increases cell proliferation, myelination, and the assembly of microtubules
that are a component of axons and dendrites and essential
for their elongation. Hypothyroidism in experimental animals results in a marked reduction of brain development
particularly in the growth of neural processes and synapse
formation in the cerebral and cerebellar cortices. Children
with congenital hypothyroidism have a high incidence of
congenital mental deficiency. There is now strong evi-

ence that severe deficiency of iodine, an essential component of thyroid hormone, affects fetal brain development and is associated with endemic cretinism and mental retardation.

Hypoxia

Both chronic and acute hypoxia affect the developing brain. The extent and location of lesions produced by an acute episode of hypoxia will depend on the severity and timing of the insult. Studies of acute hypoxia in experimental animals have usually been in late gestation or in term animals. The most consistently described gross lesions occur in the white matter (myelinated axons) adjacent to the ventricles in the forebrain, but cell death in the cerebral cortex and in forebrain nuclei have also been described. Less is known about the effects of hypoxic events occurring earlier in gestation. This is now of interest since recent evidence supports the view that prenatal, (i.e. fetal) as well as intrapartum hypoxia is an important antecedent of neurological impairment in infants. Chronic hypoxia, which is usually due to placental insufficiency, can result in intrauterine growth retardation. Growth-retarded children have a higher than normal incidence of neurological deficits.

Conclusion

Thus, it can be seen that the developing brain is vulnerable to many environmental factors and its normal development depends on adequate oxygenation, nutrition, and a favourable hormonal balance both before and after birth. Postnatal sensory and cognitive inputs are also necessary for its optimal development and the full realization of an individual's potential.

Further reading

Bray, D. (1987). Growth cones: do they pull or are they pushed? *Trends in Neuroscience*, **10**, 431–4.

Cowan, M., Fawcett, J., O'Leary, D., and Stanfield, B. (1984). Regressive events in neurogenesis. *Science*, **225**, 1258–65.

Edelman, G.M. (1983). Cell adhesion molecules. *Science*, **219**, 450–7.

Goldberg, D. and Burmeister, D.W. (1989). Looking into growth cones. *Trends in Neuroscience*, **12**, 503–6.

Goldman, S.A. and Nottebohm, F. (1983). Neuronal production, migration, and differentiation in a vocal control nucleus of the adult female canary brain. *Proceedings of the National Academy of Sciences of the United States of America*, **80**, 2390–4.

Goodman, C. and Bastiani, M.J. (1984). How embryonic nerve cells recognise one another. *Scientific American*, **251**, 50–83.

Gregory, W.A., Edmondson, J.C., Hatten, M.E., and Mason, C.A. (1988). Cytology and neuro-glial opposition of migrating cerebellar granule cells *in vitro*. *Journal of Neuroscience*, **8**, 1728–38.

Hubel, D., Weisel, T., and Le Vay, S. (1977). Plasticity of ocular dominance in the monkey striate cortex. *Philosophical Transactions of the Royal Society, London (Biology)*, **278**, 377–409.

Jacobson, M. (1991). *Developmental neurobiology* (3rd edn). Plenum Press, New York.

Letourneau, P. (1983). Axonal growth and guidance. *Trends in Neuroscience*, **6**, 451–5.

Levi-Montalcini, R. and Hamburger, V. (1951). Selective growth stimulating effects of mouse sarcoma on the sensory and sympathetic nervous system of chick embryo. *Journal of Experimental Zoology*, **116**, 321–61.

McConnell, S.K. (1989). The determination of neuronal fate in the cerebral cortex. *Trends in Neuroscience*, **12**, 342–9.

Oppenheim, R.W. (1989). The neurotrophic theory and naturally occurring cell death. *Trends in Neuroscience*, **12**, 252–5.

Purves, D. and Lichtman, J.W. (1985). *Principles of neural development*. Sinauer Associates Inc, Sunderland, Massachusetts.

Raff, M. (1989). Glial cell diversification in the rat optic nerve. *Science*, **243**, 1450–5.

Rakic, P. (1972). Mode of cell migration to the superficial layers of fetal monkey neocortex. *Journal of Comparative Neurology*, **145**, 61–84.

Rees, R.P., Bunge, M.B., and Bunge, R.P. (1976). Morphological changes in the neuritic growth cone and target neuron during synaptic development in culture. *Journal of Cell Biology*, **68**, 243–63.

Rutishauser, V. and Jessell, T.M. (1988). Cell adhesion molecules in vertebrate neural development. *Physiological Reviews*, **68**, 819–57.

Sperry, R. (1963). Chemoaffinity in the orderly growth of nerve fibre patterns and connections. *Proceedings of the National Academy of Sciences of the United States of America*, **50**, 703–10.

Walsh, C. and Cepko, C. (1992). Widespread dispersion of neuronal clones across functional regions of the cerebral cortex. *Science*, **255**, 434–40.

19. Development of somatosensory systems in the fetus

Maria Fitzgerald

The role of the adult somatosensory system is to provide information to the central nervous system about stimulation of the body surface. This system begins to develop early in fetal life, and understanding somatosensory function during this period depends on a knowledge of the underlying developmental processes that are taking place in the fetal nervous system.

First, we should consider the sequence of events in the structural development of somatosensory pathways: the formation of connections between the periphery and the spinal cord or brainstem by primary sensory neurones; the development of projections from sensory neurones in the cord and brainstem to appropriate interneurones, reflex motoneurones, and higher brain structures; and the growth and maturation of thalamocortical connections.

At a cellular level, each of these stages requires the formation of precisely organized terminals and synapses; the development of appropriate transmitter substances, receptors, and second messenger systems; and the maturation of axonal and dendritic membrane biophysical properties.

In addition to knowledge of events at the anatomical and cellular levels we need to study the physiological development of the system to understand how fetal somatosensory processing is functionally organized. In particular, this allows us to address important issues such as whether the fetal somatosensory system is capable of detecting different sensory qualities (touch, temperature, pain) and how accurately it can locate a stimulus by creating a (somatotopic) map of the body surface in the central nervous system (CNS). Furthermore, we can assess the functional status of fetal cortical processing as well as local somatosensory reflexes and study the onset of descending and local control mechanisms. Finally, we may speculate as to the role and overall importance of the somatosensory system in fetal life.

Peripheral somatosensory transmission in the fetus

Structural features

In the rat, dorsal root ganglion (DRG) and trigeminal ganglion (TG) cells are born over the period embryonic (E) day 11–14 (21.5 days of gestation). In man this is complete by 6 weeks after menstruation. The large cells giving rise to myelinated afferent fibres are born before the small cells that give rise to unmyelinated fibres. The production of DRG and TG cells is followed by a period of cell death, which in chicks is complete by birth but has not been fully studied in rat and man. An important function of this is to adjust the numbers of neurones to the requirements of their target fields. In man, judging from data on motoneurone cell death, DRG cell death is complete by 25 weeks. Soon after the sensory neurones are born, their peripheral and central processes begin to grow out and contact skin, muscle, and visceral targets in the periphery and the spinal cord or brainstem centrally. An excessive number of peripheral nerve axons and dorsal roots are formed initially and a considerable number are lost postnatally independent of ganglion cell death. In the rat, facial innervation begins on day E13 and hind limb innervation takes place over days E15 to E19. In man, cutaneous innervation of the face, shoulder, axilla, and thigh have begun at 8 weeks of gestation.

Studies of hind limb innervation show the cutaneous nerves appear to reach their targets before muscle nerves do. They grow directly to their targets without sprouting or growth in aberrant directions and are thought to be attracted by a specific chemotropic factor, perhaps of epithelial origin, that diffuses from the skin. Dorsal root ganglion cells display spontaneous electrical activity during this period that may also be related to locating their appropriate targets. In both rat and man cutaneous nerve terminals initially form a dense plexus penetrating into the fetal epidermis (Fig. 19.1). This gradually withdraws and reduces in density as end organs, such as hair follicles and Meissner's corpuscles, appear and become innervated some considerable time after the initial innervation of the skin. In the rat hind limb, hair follicles are not innervated until postnatal day 7, despite being present from before birth. In the abdominal skin of the human fetus, follicles appear in the first trimester but innervation does not begin until 22–24 weeks gestation. Cutaneous innervation may have a growth stimulatory effect on the skin and may even induce the maturation of specialized target organs.

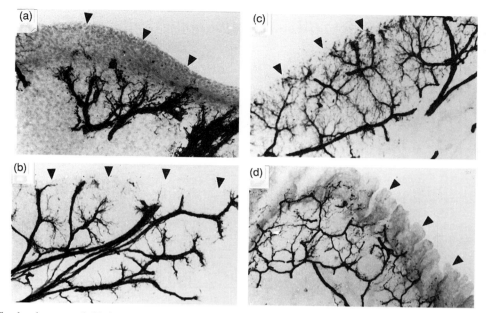

Fig. 19.1 The development of skin innervation in the fetal rat using GAP-43 immunostaining. Transverse sections from hind limb skin at embryonic (E) days, (a) E14; (b) E15; (c) E17; and (d) E19, show the progressive development of skin innervation as it penetrates the epidermis and then withdraws to form a subepidermal plexus. Arrowheads indicate the surface of the skin. × 800. (Taken with permission from Reynolds, M.R., Fitzgerald, M., and Benowitz, L.I. (1991). *Neuroscience*, **14**, 201–11.)

Functional features

In the rat hind limb, primary afferents can be activated from small well-defined peripheral receptive fields soon after skin innervation and it is possible to distinguish the rapidly and slowly adapting low-threshold afferent fibres responding to touch and pressure as well as high-threshold, polymodal nociceptor afferents responding to pinch, heat, and irritant chemicals. These physiological afferent types develop, therefore, in the absence of morphological skin receptors, demonstrating that many of the properties of cutaneous sensory nerves are intrinsic and reflect their membrane properties. Nevertheless, low-threshold afferents responding to touch and hair movement do display a considerable increase in the amplitude and frequency of response over the postnatal period and this is likely to be the main role of sensory receptor organs in the skin. Polymodal nociceptors, on the other hand, which provide information concerning painful or tissue-damaging stimuli and have unmyelinated (C) axons and unspecialized, 'free' nerve endings in the adult, are functionally mature very early in fetal life. In the hind limb of the fetal rat near term (E19–20) they are indistinguishable from adult nociceptors.

Equivalent physiological information is not available in the human but, since the anatomical data compares well with that of the rat, it is reasonable to assume that the same early maturation of C afferent nociceptors and slower maturation of afferent mechanoreceptor A fibres also takes place in man.

Chemical features

The chemical profile of developing sensory ganglion cells is also developing at this time. In the rat, thiamine monophosphatase (TMP), an enzyme found in small dorsal root ganglion (DRG) cells, develops at day E15 at lumbar levels, although it does not reach the central terminals in the cord until 24 h after birth. The endogenous substrate for this enzyme is unknown but TMP may be related to the metabolism of nucleotides, fulfilling a transmitter function in primary afferents, particularly those arising in the skin. Substance P (SP), somatostatin, and vasoactive intestinal polypeptide (VIP), which are found in small DRG cells, and calcitonin gene-related peptide (CGRP), which is in both small and large cells, all appear at days E16–17 and

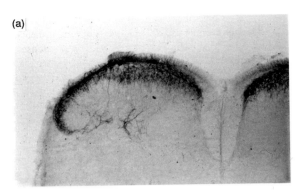
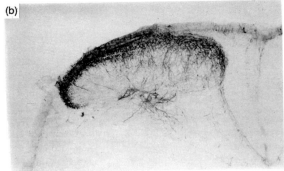

Fig. 19.2 CGRP immunostaining in primary afferent terminals in the dorsal horn of: (a) a 4-day-old rat pup; (b) a 20-day-old rat pup. The pattern of staining at the two ages is the same indicating established anatomical arrangement of terminals at postnatal day P4, but to obtain the same density of staining approximately twice the concentration was used at postnatal day 4 than at postnatal day 20. Scale bar, 100 μm. (Courtesy of M.R. Reynolds.)

in the spinal cord a day or two later. Again, the function of these neuropeptides is not fully understood but they clearly function as neurotransmitters or neuromodulators in the sensory neurones. While it is possible that their synthesis is triggered by peripheral innervation, they are expressed in neural crest cells in culture in the absence of normal targets. In the human fetal DRG these peptides appear at between 12 and 16 weeks of gestation. Despite their early appearance it takes a considerable period for levels of TMP and neuropeptides to reach adult levels. In the rat this is achieved by 10 days after birth and in the human at 30 weeks of gestation (Fig. 19.2).

Somatosensory processing in the fetal spinal cord and brainstem

Primary afferent connections

The spinal dorsal root fibres reach the lumbar spinal cord at days E11–12 in the rat and at 6.5–7 weeks of gestation in man. They do not, however, penetrate the cord for some time, but rather travel rostrocaudally up and down the outside of the grey matter for, in the case of the rat, several days. Collaterals only start to enter the cord when peripheral innervation has begun, supporting the idea that a peripheral 'signal' is required to trigger growth of primary afferents into the central nervous system. The large A fibres grow into the cord first, penetrating into the deep layers of the dorsal horn at days E15–16 in lumbar cord. The same afferents grow into the gracile nucleus in the medulla on day E18. The C fibres grow into the superficial dorsal horn (i.e. into the substantia gelatinosa) later than A fibres, at day E19. In the adult cord, primary afferent ter-

minals are arranged in a precise somatotopic pattern according to their peripheral field of innervation. This somatotopy is established very early in development as the afferent fibres grow into the spinal cord, and there is no evidence of any overlap of terminal fields arising from different peripheral nerves (Fig. 19.3).

The formation of sensory terminal arbors and formation of their synaptic connections with cells in the dorsal horn is a long drawn-out affair. In the rat fetus terminal arbors belonging to A fibres begin to form from day E18 onward in lumbar cord but C fibre synaptic boutons are not evident until a few days postnatally. The early postnatal period is one of intense synaptogenesis in the rat dorsal horn. Such a period is also observed in the human fetal cord at 17–2 weeks of gestation.

Central sensory neurones — structure and chemistry

All the dorsal horn cells have been born and have migrated into position as the primary afferent terminals grow into the cord to form their connections. Their dendritic and axonal development takes place over a very prolonged period, however, and there is good evidence to suggest that the local interneurones in substantia gelatinosa mature much later than the projection neurones whose axons will grow up the white matter to the thalamus. Certainly in the rat, enkephalin, a peptide marker of these neurones, is not expressed in the cord until birth. Other peptides, such as substance P, appear to be expressed only transiently in spinal cord cells at this time. In the human cord, neurones and their axodendritic synapses in the superficial cord are found to be mature by 25 weeks of gestation at cervical levels and therefore perhaps by 28 weeks in lumbar levels

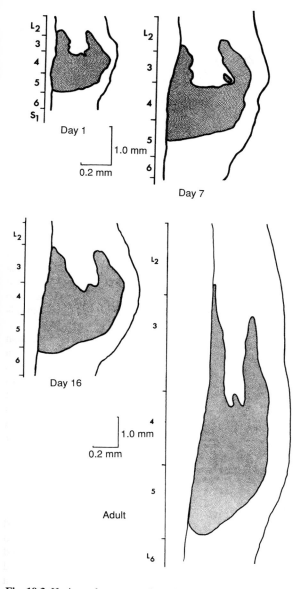

Fig. 19.3 Horizontal reconstruction of C-fibre terminal labelling (shaded) in the rat lumbar dorsal horn at postnatal days 1, 7, and 16 and in the adult. WGA–HRP (wheat germ agglutinin–horse radish peroxidase) was applied to the sciatic nerve and transganglionically transported to the C-fibre terminals in substantia gelatinosa of the dorsal horn. The map was plotted from serial histological sections. The vertical line marking the lumbar spinal segments represents the midline of the spinal cord and the shaded terminal region lies in the outline of the grey matter. (Taken with permission from Fitzgerald, M. and Swett, J. (1983). *Neuroscience Letters*, **43**, 149–54.)

Some studies have been made of the development of postsynaptic transmitter receptors in the rat dorsal horn. Substance P binding sites remain very dense and widespread over the whole grey matter until postnatal day 21 in the rat lumbar cord and then begin to decrease and become concentrated in the substantia gelatinosa. This may have important functional implications in terms of specificity of action of substance P in the neonatal dorsal horn. There is evidence that kainate and NMDA glutamate receptors may also be distributed diffusely early in development. NMDA-evoked postsynaptic potentials are augmented in immature CNS and less sensitive to Mg ions. Opiate receptors also undergo postnatal changes in rat cord. κ-receptors develop first followed by the appearance of high-affinity μ-receptor binding. The expression of GABA receptor subunits also changes with development.

Central sensory neurones — function

Dorsal horn cells in the fetal rat lumbar cord begin to respond to primary afferent inputs at day E17, but clearly defined cutaneous receptive fields responding to localized low-intensity skin stimulation cannot be recorded until days E18–19, some 2 days after the primary afferents develop functional receptive fields. This delay presumably represents the maturation of central synapses within the dorsal horn. The initial input is weak, skin stimulation producing only 1–3 spikes per stimulus, but this rapidly matures so that, by the end of fetal life, substantial cutaneous responses are established. Furthermore, where initial inputs may be weak, cells are capable of producing long-lasting firing of action potentials that outlasts the stimulus by up to 30 s. Repeated stimulation, or application of an irritant chemical to the skin, results in prolonged periods of increased sensitivity whereby responses to cutaneous stimuli are greatly augmented.

It is likely that all of the action potential responses of fetal and neonatal dorsal horn neurones to sensory stimulation in the rat are produced by A fibres. C fibres are not able to produce action potentials in dorsal horn cells for a considerable period (postnatal days 8–10 in rat lumbar cord). Before that time the effects of C-fibre stimulation are subthreshold, but by no means non-existent. Stimulation of C fibres in the fetus and neonate can cause the release of substance P and profound and long-lasting depolarization of motoneurones and sensitization of dorsal horn cells to subsequent stimuli. It can also activate expression of early-onset genes such as c-*fos* in the dorsal horn. The significance of this group of genes, many of which code for transcription factors, is unclear. The appearance of Fos protein in differentiating neurones may indicate a controlling role in the development of the adult

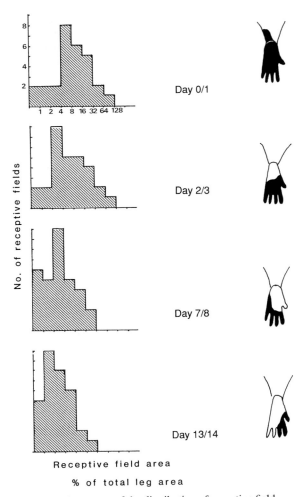

Fig. 19.4 Histograms of the distribution of receptive field areas of dorsal horn cells in neonatal rat spinal cord. The receptive fields are expressed as a percentage of total hind limb skin area (note the logarithmic scale). The histograms show a gradual decrease in receptive field area over the first 2 postnatal weeks. Typical receptive fields are illustrated beside each histogram. (Taken with permission from Fitzgerald, M. (1985). *Journal of Physiology*, **364**, 1–18.)

phenotype of a cell, whereas in more mature neurones c-*fos* induction is thought to be a candidate for coupling neuronal excitation to long-term adaptative modification of transcription. Activation of c-*fos* expression in neonatal spinal cord suggests functional C-fibre inputs but the effects are subtle and prolonged and the fast millisecond responses that are present in the mature cord are absent.

A further feature of the developing rat dorsal horn is that, once the synaptic input from the primary afferents has

been established, it is clear that there is a period during which receptive fields are larger than in the adult (Fig. 19.4). This means that a given cell can be excited from a larger area of skin and therefore has a much greater chance of being excited by external cutaneous stimuli. It also means that, while there is a somatotopic map in the immature cord, it is blurred compared to that in the adult. One possible reason for this is the delayed maturation of local inhibitory neurones that normally act in the adult to suppress or 'silence' the input from a wide range of primary afferents and therefore 'focus' the receptive fields of dorsal horn cells.

Fetal somatosensory reflexes

Few of the above studies in developing rats can be directly compared to studies in the human but useful data can be gained from examination of cutaneous evoked reflexes. Such reflex movements in response to skin stimulation begin at day E15.5–16 in the rat and at 8–10 weeks of gestation in the human fetus. In both species, sensitivity appears first in the lips and gradually moves caudally to involve the lower limbs at days E18.5–19 in the rat fetus and at 14 weeks of gestation in the human. These times correlate well with the physiological and anatomical studies of cutaneous afferent connections within the cord, but such behavioural studies on young fetuses are open to criticism since 'spontaneous' movements that occur in the apparent absence of any external stimulus are prominent in young fetuses and can easily be mistaken for evoked reflex movements.

Physiological studies show that the first reflex connections in the spinal cord are polysynaptic, presumably from cutaneous A fibres and perhaps group Ib and II muscle afferents. These are followed by the formation of the monosynaptic Ia connections. Both are very augmented in the neonatal rat and human infant when compared to such reflexes in the adult. C-fibre evoked reflexes develop some considerable time later at postnatal days 8–10 in the rat coinciding with the appearance of C-fibre evoked spikes in the spinal cord (see the section on 'Central sensory neurones — function').

Thalamic and cortical development of somatosensory processing in the fetus

Structure

In the rat, thalamic neurones are born at day E14. The human thalamus is detectable at 22–23 days of gestation and all the major nuclear subdivisions identifiable by midterm. Lemniscal fibres, the axons arising from the gracile and cuneate nuclei in the brainstem reach the thalamus by the end of fetal life in the rat but take several days to distribute themselves throughout the thalamic nuclei. In the particular case of the vibrissal afferents in the rat, these can be seen to arrange themselves postnatally into discrete clusters in the thalamus at the same time as the thalamic cells also organize themselves into groups relating to each whisker.

The birth of rat spinothalamic cells is complete by day E14, just at the time that thalamic neurones are generated. Dendritic and axonal growth of these projection cells is very active in the late fetal period but little is known of their arrival in the thalamus.

Rat neocortical cells are generated over days E16–21, depending on their final laminar position. Neurones of layer IV are generated on day E18 and thalamocortical afferents begin to arrive in this layer at postnatal day 3. At birth, thalamocortical projections have reached the cortex but remain in the 'subplate' white matter beneath the developing cortical plate and penetrate it on the first postnatal day. Neurones of layer IV are generated on day E18 and thalamocortical afferents begin to arrive in this layer at postnatal day 3. These afferents are organized in such a way as to confer a somatotopic map on to the developing cortex.

In the human fetus, the cortical plate develops at approximately 11 weeks of gestation and the cerebral cortical layers emerge over the whole fetal period and into postnatal life. Layer IV is generally thought to be generated at 24–26 weeks of gestation and thalamocortical afferents begin to penetrate the cortical plate at 26–34 weeks of gestation. Before that time (between 17 and 25 weeks of gestation) thalamic afferents form temporary synapses with subplate cortical neurones while apparently 'waiting' for cortical pyramidal cells to differentiate.

Function

Electrophysiological recording of single cells in the rat somatosensory cortex on the seventh postnatal day shows their synaptic responses to be immature with long latencies and low-frequency responses. Columnar organization is present, but receptive fields are considerably larger than in the adult. A further feature is cyclical firing of action potentials, with periods of increased and decreased firing rate in both spontaneous and evoked responses.

By postnatal days 10–12 the somatosensory evoked potential recorded from the cortical surface in the rat is mature in pattern. Before that time, a predominantly negative wave is recorded. A similar pattern is recorded in human infants, where the potential is slow, long-latency,

and essentially negative until approximately 30 weeks post menstruation, at which time an early positive wave appears. The pattern continues to mature through infant life.

Descending control systems in the fetus

In the adult, nuclei in the brainstem, particularly nucleus raphe magnus (NRM) and nucleus reticularis paragigantocellularis (NRPG), project via the dorsolateral funiculus to the dorsal horn of the spinal cord and have a profound inhibitory influence on the responses of sensory neurones. This and other such descending pathways have an important function in controlling and processing of sensory inputs as at a spinal cord level by higher centres. In the rat, the projections of these axons descending from the brainstem appear to reach the cord early in fetal life and are well-established at birth. However the main neurotransmitters contained in these axons, 5-hydroxytryptamine (5-HT) and noradrenaline, develop much later, and the adult pattern and levels of 5-HT are not achieved until postnatal day 21 in lumbar cord. Noradrenaline levels mature somewhat earlier at postnatal day 14. This delayed neurochemical maturation, long after anatomical growth, may explain the lack of functional influence of these descending axons on spinal sensory processing before postnatal day 10 in the rat. Electrophysiological recording of single cells in the neonatal dorsal horn shows that stimulation of the dorsolateral funiculus (DLF) has no effect and that the first signs of inhibition appear at postnatal day 10, reaching adult levels at postnatal day 19. As yet, it is not clear if there is a similar absence of descending inhibition in the human fetal and neonatal cord, but the observed exaggerated sensory reflexes suggest that there is.

Another major descending pathway is the projection from the cortex to the thalamus. Corticothalamic afferents appear in the rat thalamus at birth but do not invade the grey matter for several days, i.e. well after the lemniscal afferents have invaded the thalamus. Over the next week the cortical afferents distribute themselves throughout the thalamus but not in the discrete clusters seen in lemniscal afferents.

Some generalizations on the fetal somatosensory system

1. The somatosensory system develops very early in fetal life. It matures in a rostrocaudal direction, starting

with the trigeminal system followed by cervical through to the sacral spinal cord.

2. Initially, the somatosensory system develops at the local segmental level, forming functional connections with reflex motoneurones. Projections to the thalamus and cortex develop in late fetal and postnatal life. Early fetal somatosensation is therefore restricted largely to the levels of the spinal cord and brainstem, and higher CNS levels need only be considered later.

3. Fetal sensory receptors are capable of distinguishing between noxious and innocuous stimuli and displaying both rapidly and slowly adapting responses. Nociceptors mature very early; mechanoreceptors need longer to acquire high-frequency responses.

4. Receptive fields of central somatosensory cells are large, which effectively reduces the accuracy of locating a stimulus but increases the chance of exciting a given cell.

5. Inhibitory processes appear to develop after excitatory ones. In particular, the late development of descending inhibition from the brainstem to the spinal cord means that fetal sensory reflexes are low-threshold and highly exaggerated.

The role and importance of the somatosensory system in fetal life

The somatosensory system becomes functional extremely early in fetal life. At 8–10 weeks of gestation a human fetus is capable of a reflex response to stimulation of the body surface. And yet one might consider that under normal conditions the fetus lives in a virtually 'sensation-free' environment, so why should such a system be required so early in development?

One possibility is that the systems that are most urgently required in immediate postnatal life and that therefore need to be relatively mature at birth are the ones that develop first. Thus, location of the mother's nipple requires functional sensory inputs from the infant's mouth and facial region and that, indeed, is the area that displays the first sensitivity to touch *in utero*. Consistent with this proposal is the early development of the olfactory system, also needed for feeding postnatally and the later development of the auditory and visual systems. It does not, however, explain why the somatosensory system should begin to function in the first trimester.

A further possibility is that somatosensory input is actually a necessary requirement for the subsequent normal development of the nervous system. In other words, sensory inputs or 'feedback' are needed if the developing CNS is to form the correct connections. From 7.5 weeks of gestation onwards the human fetus displays a variety of movements that become increasingly vigorous and complex. Thus the fetus is capable of considerable self-stimulation as, for instance, limbs touch the trunk and digits touch the face. The resulting sensory input may well provide important information in developing organized reflex connections within the spinal cord and brainstem, strengthening appropriate synapses, and allowing inappropriate ones to weaken.

Further reading

Armstrong-James, M. (1975). The functional status and columnar organization of single cells responding to cutaneous stimulation in neonatal rat somatosensory cortex S1. *Journal of Physiology*, **246**, 501–38.

Bradley, R.M. and Mistretta, C.M. (1975). Fetal sensory receptors. *Physiological Reviews*, **55**, 352–82.

Davies, A. and Lumsden, A. (1990). Ontogeny of the somatosensory system: origins and early development of primary sensory neurons. *Annual Reviews of Neuroscience*, **13**, 61–73.

Fitzgerald, M. (1987). Spontaneous and evoked activity of fetal primary afferents in vivo. *Nature*, **326**, 603–5.

Fitzgerald, M. (1987). Cutaneous primary afferent properties in the hindlimb of the neonatal rat. *Journal of Physiology*, **383**, 79–92.

Fitzgerald, M. (1987). The prenatal growth of fine diameter primary afferents into the spinal cord — a transganglionic study. *Journal of Comparative Neurology*, **261**, 98–104.

Fitzgerald, M. (1991). A physiological study of the prenatal development of cutaneous sensory inputs to dorsal horn cells in the rat. *Journal of Physiology*, **432**, 473–82.

Fitzgerald, M. (1991). The development of pain mechanisms. *British Medical Bulletin*, **47**, 667–75.

Fitzgerald, M. (1991). The development of descending brainstem control of spinal cord sensory processing. In *The fetal and neonatal brainstem: developmental and clinical issues* (ed. M. Hanson), pp. 127–36. Cambridge University Press.

Fitzgerald, M., Shaw, A., and McIntosh, N. (1988). Postnatal development of the cutaneous flexor reflex: a comparative study of preterm infants and newborn rat pups. *Developmental Medicine and Child Neurology*, **30**, 520–6.

Jonakait, G.M., Ni, L., Walker, P.D., and Hart, R.P. (1991). Development of substance P (SP) containing cells in the developing nervous system: consequences of neurotransmitter colocalization. *Progress in Neurobiology*, **36**, 1–21.

Killackey, H.P., Jacquin, M.F., and Rhoades, R.W. (1990). Development of somatosensory system structures. In *Development of sensory systems in mammals* (ed. J.R. Coleman). Wiley, New York.

Klimach, V.J. and Cooke, R.W.I. (1988). Maturation of the neonatal somatosensory evoked response in preterm infants. *Developmental Medicine and Child Neurology*, **30**, 208–14.

Marti, E., Gibson, S.J., Polak, J.M., Facer, P., Springall, D.R., van Aswega, G., Aitchison, M., and Koltzenberg, M. (1987).

Ontogeny of peptide and amine containing neurons in motor, sensory and autonomic regions of rat and human spinal cord. *Journal of Comparative Neurology*, **266**, 332–59.

Saito, K. (1979). Development of spinal reflexes in the rat fetus studied in vitro. *Journal of Physiology*, **294**, 581–94.

Windle, W.F. and Baxter, R.E. (1936). Development of reflex mechanisms in the spinal cord of albino rat embryos.

Correlations between structure and function and comparison with the cat and with the chick. *Journal of Comparative Neurology*, **63**, 189–209.

Wise, S.P. and Jones, E.G. (1978). Developmental studies of thalamocortical and commissural connections in the rat somatic sensory cortex. *Journal of Comparative Neurology*, **178**, 187–208.

20. Development of auditory and visual systems in the fetus

David R. Moore and Glen Jeffery

Every parent knows that the senses of hearing and sight are highly developed in the new-born human. Following birth there are further improvements in the optics and accommodation of the eye, in the transmission of sound through the middle ear, and in the processing of both auditory and visual stimuli within the central nervous system. However, these postnatal improvements are only the final scene of an act that began during the third trimester of pregnancy, when sensory function started. The earlier stages of this development took place much earlier, between fetal weeks 3 and 24. During this period, cells of the primordial ear and eye were formed from ectoderm, migrated and differentiated, connected with other cells, and began to send messages to the brain.

This chapter deals primarily with the early stages of visual development and with the later stages of auditory development in the fetus. The presentational bias is a reflection of intrinsic differences between the two modalities, the relative amount of knowledge in the two systems, and the personal interests of the authors. For more detailed and complete reviews of auditory and visual development the reader is referred to the 'Further reading' at the end of the chapter. Most of the data underpinning this chapter were derived from experimental anatomical and physiological studies in rodents, cats, and monkeys. As elaborated below, these studies have revealed common timetables for the development of sensory systems that, with certain additional assumptions, allow an increasingly detailed picture of development in the human fetus. Throughout the chapter, dates will be provided (*in italics*) when the relevant stage of human maturation is known in relation to developmental mechanisms revealed in animals.

The mature auditory system

The auditory system consists of the outer, middle, and inner ears; the auditory nerve; and the various nuclei and fibre tracts of the central auditory system (Fig. 20.1). The spectrum of incident sound is transformed by the outer and middle ears. They selectively amplify the midfre-quencies, leading to the characteristic shape of the mature audiogram. In the cochlea of the inner ear, sound produces a travelling wave of displacement along the basilar membrane. Maximum displacement occurs at the basal end of the cochlea for high-frequency sounds, and at the apical end for low-frequency sounds. Basilar membrane displacement is converted into neural activity by the deflection of minute hairs, the stereocilia of the inner hair cells. Stereocilia displacement opens ion channels, and the resulting depolarization is the generator for synaptic transmission between the hair cells and the auditory nerve fibres. Outer hair cells are thought to sharpen the mechanical tuning of the cochlea through an active feedback mechanism. Auditory-nerve fibre responses to tones are characterized by sharp frequency tuning and, for low frequencies, by a tendency to discharge at a particular phase of the stimulus waveform.

In the brain, nerve impulses are carried to the cochlear nucleus for further processing. The cochlear nucleus has two representations of the frequency organization established in the cochlea (tonotopicity). In the ventral cochlear nucleus (VCN) many neurones respond to sound in the same way as do auditory nerve fibres. These VCN neurones project medially to provide bilateral innervation of neurones in the superior olivary complex (SOC). Neurones in the dorsal cochlear nucleus (DCN) are excited directly by auditory nerve input, and inhibited indirectly by interneurones from the VCN. DCN neurones project primarily to the contralateral midbrain (inferior colliculus, and nuclei of the lateral lemniscus, NLL). The inferior colliculus also receives input from the SOC nuclei, the NLL, and the contralateral inferior colliculus. Neurones in the inferior colliculus provide the dominant ascending projection to the auditory thalamus (the medial geniculate nucleus, MGN) and lateral inferior colliculus neurones also innervate the superior colliculus. The final major path in this anatomically complex system is a parallel one from the MGN to several areas of the auditory cortex. The known physiology of the higher auditory centres is similar to that of the DCN, except that binaural interaction first appears in the SOC and is maintained in the majority of neurones at higher levels.

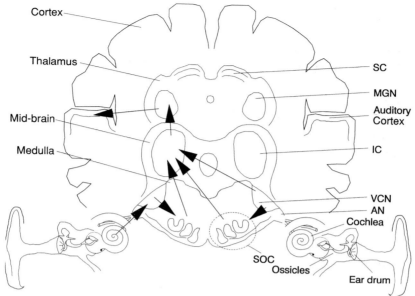

Fig. 20.1 Organization of the auditory system. Each ear provides input to the brain via the auditory nerve (AN). Major ascending neural pathways from the ventral cochlear nucleus (VCN) are indicated by arrows. Sections through the medulla (superior olivary complex, SOC), and through the dorsal midbrain (superior and inferior colliculi, SC, IC, respectively) and thalamus (medial geniculate nucleus, MGN), are shown superimposed and are not to scale.

The mature visual system

The retina has three cellular layers. The outer nuclear layer contains the photoreceptor nuclei. Photoreceptors can be subdivided into rods and cones. Cones respond at relatively high light levels and are used for colour vision, while rods respond at lower levels and do not give a sense of colour. The inner nuclear layer contains the horizontal, bipolar, and amacrine cell populations, which provide vertical and horizontal pathways for visual information, while the ganglion cell layer contains three main projection cell types that give rise to the retinofugal pathway. Ganglion cells can be divided into subtypes on the basis of their morphology, physiological properties, and function. The density of each cell type in each layer varies across the retina, being higher in central regions. The peak ganglion cell density is located laterally in the retina, along a line that divides ganglion cells into different populations depending on the side of the brain to which they project.

Ganglion cell axons traverse the retina and leave the eye via the optic disc to form the optic nerve (Fig. 20.2). The nerves fuse at the optic chiasm, where ganglion cells located nasal to the region of highest ganglion cell density cross to form part of the optic tract, while those located temporal to this region remain uncrossed. The

crossed projection is larger than the uncrossed projection. The majority of these projections terminate in the lateral geniculate nucleus (LGN), which relays the visual information to the cortex, and in the superior colliculus. In each structure the projections form orderly retinotopic maps that have both monocular and binocular segments. In the binocular segments, the projections from each eye terminate in different regions. However, they are aligned so that the two representations of the visual fields, one from each eye, are in register. The projections to the LGN and to the superior colliculus subserve different visual functions. The LGN is a relay nucleus that conveys visual information to the visual cortex. This pathway is concerned with high spatial resolution and colour, which are necessary for the identification of objects.

The visual cortex has six main layers. The projection from the LGN enters layer 4 and information is processed in the cortical layers with regard to specific features such as orientation, depth, and location in the visual field. There is a very large projection from the cortex back to the LGN, the function of which remains to be explored. The function of the retinal projection to the superior colliculus is probably more concerned with identifying the location of objects and directing visual attention, perhaps in relation to the other main senses, hearing and touch.

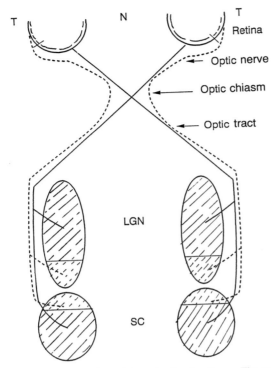

Fig. 20.2 Representation of the retinofugal pathway. The two eyes are at the top. The temporal (T) aspect of the retinae are to the outside and the nasal (N) aspect is towards the midline. The retinae are divided, as shown by lines, into a temporal side, projecting ipsilaterally through the chiasm (dashed line), and a nasal side projecting contralaterally (continuous line). The two projections terminate in separate regions of the lateral geniculate nucleus (LGN) and superior colliculus (SC).

Development of hearing

Studies of auditory system development have been motivated by two general principles: (1) that the attainment of mature auditory function is an essential part of normal communication; (2) that the system has unique advantages for addressing broader issues in developmental neurobiology. The common link in these principles is the role of genetic and environmental influences on the normal and abnormal development of the system, and we present in this section a synthesis of present knowledge of those influences.

Early development of the cochlea

The auditory and vestibular apparatus begin as a thickening of the ectoderm (otic placode) in the region of the hindbrain (*3 weeks*). The thickening invaginates, separates from the ectoderm (to form the otocyst), and develops three ducts. These ducts are the sacculus, utricle, and endolymphatic duct. The cochlear duct forms by elongation of an outgrowth from the sacculus. Throughout this earliest period of development there are inductive interactions, first between the placode and the surrounding tissue (neural crest, neural tube, neuroectoderm, and mesoderm), and subsequently between the otocyst and the rhombencephalon. The spiral ganglion neurones (somata of the auditory nerve) are also derived from placodal tissue, and send distal processes into the otocyst at a very early stage (*4 weeks*), possibly as they migrate from the lumen of the otocyst.

Three general principles of organ of Corti maturation have been identified. First, there is a basal-to-apical gradient of morphological differentiation. Second, there is a simultaneous maturation of all structures at a given cochlear location. Finally, there is a slight precedence of inner hair cell over outer hair cell development. Recent studies have confirmed these generalizations and extended them by showing, for example, that efferent as well as afferent auditory nerve terminals arrive early in the cochlea and wait until just before the onset of auditory function before making synapses with the hair cells. The onset of auditory function (*25 weeks*) ends the early stage of auditory system development.

Development of the place principle

As outlined above, the most fundamental aspect of auditory system organization is the orderly representation of sound frequency in the cochlea and in the central auditory system. This organization is known as the place principle (Fig. 20.3(a)). In maturity, the audible spectrum is represented logarithmically, so that each octave occupies an approximately equal length or depth of tissue. However, at the onset of auditory function, only a very limited range of low-frequency sounds produce physiological or behavioural responses. These functional data present a paradox, since the basal end of the cochlea, representing high-frequency sounds in the adult, achieves structural maturation before the apical end (*30 weeks*). However, recent results have confirmed an earlier suggestion that the region of the cochlea transducing sounds of a given frequency shifts towards the apical end during normal development. Thus, in newly functional cochleas, low-frequency sounds produce maximum stimulation of the structurally mature basal end, whereas high-frequency sounds are ineffective. Since the major neural connections between the cochlea and the brain, and between the vari-

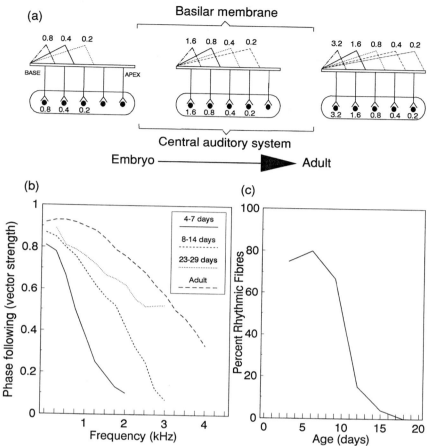

Fig. 20.3 Development of auditory system function. (a) Development of the place principle. Pure tones of the frequencies indicated (in kHz) produce maximum displacement of the basilar membrane at varying distances along the cochlea and, through topographic connections, excitation of neurones in central auditory nuclei. Initially the basal half of the cochlea responds to low-frequency sounds. Gradually, the apex begins responding to low frequencies and the base becomes sensitive to high frequencies. (Model based on chicken auditory system development, from Rubel, E.W. (1984). (*Annual Review of Physiology*, **46**, 213–29.) Development of (b) stimulus phase following and (c) regular discharge patterns by cat auditory nerve fibres. In neonatal animals (probably corresponding to late fetal humans) auditory neurones respond sluggishly and in short burst ('rhythmic' responses) to acoustic stimuli. Immature phase following is characterized by low 'vector strengths' (VS). VS of 1 and 0 represent, respectively, perfect and random phase following. (Taken with permission from Walsh E.J. and Romand, R. (1992). In *Development of auditory and vestibular systems 2* (ed. R. Romand), pp. 161–219. Elsevier, Amsterdam.)

ous brain nuclei, are established in the correct topographic order prior (*5–12 weeks*) to auditory function, the tonotopic organization of the central auditory system develops in parallel with that of the cochlea. During the period following the onset of hearing there is a gradual expansion of both the cochlear and neural representation of high-frequency sounds, and a contraction of the representation of low frequencies. In humans, the full audible spectrum may not be represented until after the time of birth.

Development of timing and binaural interaction

At high sound levels the discharges of auditory nerve fibres saturate, leading to a 'spread of excitation', particularly across the fibre population representing low frequencies, and a blurring of the place principle. Frequency resolution at high levels is thought to be enhanced by the timing information carried by auditory nerve fibres responding to low-frequency sounds. Although low frequencies are the first to stimulate the developing system (see above), the initial responses of auditory nerve fibres (and other neurones in the central pathway) are immature in several respects. They are insensitive to low sound levels, they are broadly tuned to sound frequency (probably due, in part, to the later development of the outer hair cell contribution to tuning), they are poorly responsive to variations in sound phase (Fig. 20.3(b)), and they have rhythmic discharge patterns (Fig. 20.3(c)). The development of timing properties, like that of the place principle, occurs gradually during the days and weeks following the onset of hearing. However, unlike the place principle, the development of timing has reasonably well understood anatomical bases. Within the nervous system, the most important of these is probably the myelination of nerves (*onset at 20–44 weeks*), leading to faster axon conduction and allowing better time resolution of impulses. Another major contribution is the expansion and elaboration of synapses, leading to a high fidelity of transmission between neurones. Regressive events, such as 'programmed' cell death and neuronal process elimination (see below), may also contribute to a refinement in central auditory coding.

Binaural interaction is dependent on the timing and level of neural activity in the auditory periphery, and is therefore also subject to developmental change. Although at least some neurones sensitive to interaural time and level differences are present soon after the onset of hearing, the proportion and precision of sensitive neurones are less than in adults. Binaural properties of neurones mature in parallel with those of the periphery ('monaural' response properties). However, both binaural and monaural response maturation appear to be more retarded at higher than at lower levels of the system, and there is increasing evidence that binaural properties are more sensitive than monaural properties to alterations in peripheral input during development.

Hearing in the human fetus

Most mammals do not begin to hear until after birth. In those species, auditory development proceeds and may be studied through the presentation of sound to the ear(s). In humans, hearing is thought normally to begin *in utero* (*25 weeks*), since ultrasound images of fetal behavioural responses to sound have been obtained, and it is possible at birth to elicit electrical responses to sound from the auditory brainstems of very premature infants. The sound environment of the fetus has several special properties. These have been likened to being submerged in a swimming pool, but this simple analogy obscures other import-

ant properties of the transmission of sound to the fetal cochlea. The fetal middle ear contains fluid. For normal airborne sound, middle ear fluid attenuates (as in 'glue ear'). In the fetus it may actually aid sound transmission, since the primary purpose of the ossicular chain is to provide an impedance match between air in the outer ear and fluid in the cochlea. If the outer ear is fluid-filled, the chain becomes redundant. On the other hand, fluid-borne sound delivered in phase to both the oval and round windows of the cochlea will produce energy cancellation, thus reducing or eliminating basilar membrane motion. In addition, for sounds external to the mother, the impedance mismatch between air and tissue is not eliminated, but is moved out from the middle ear to the surface of the mother's body. Finally, there is a substantial level of internally generated, low-frequency intrauterine noise. This will be well transmitted to the cochlea and will mask the reception of externally generated sound.

One way of measuring the contribution of these factors to hearing is to examine the physiological output of fetal animal cochleae in response to external sound. The cochlear microphonic response has recently been compared between fetal and new-born sheep. The results indicated a 15-dB (at low frequencies, 125 and 250 Hz) to 45-dB (at 2 kHz) attenuation in the fetuses, although calibration problems (e.g. cochlear microphonic current shunt through the middle ear fluid) make the quantification of this work difficult.

Nevertheless, it seems likely that, through a combination of sound attenuation to the inner ear and auditory system immaturity, the hearing of external sounds by the human fetus *in utero* is extremely poor, unless those sounds are intense and of low frequency.

Modification of fetal hearing

At least half of all congenital hearing defects are caused by hereditary factors. Molecular genetics is beginning to make a contribution to the understanding of hereditary deafness, primarily through research on mutant mice and through genetic screening of carrier human families. The coming decade holds the promise of substantial progress in this field. Diseases of the fetal ear, primarily produced by ototoxicity and infection, are better understood at present. Although the primary site of action of these factors is usually in the cochlea, a great deal of evidence shows that cochlear malfunction during developmental sensitive periods can disrupt brain development. In addition, the cochlea is hypersensitive to chemical or noise trauma during development, and this exacerbates the potential for further neurological sequelae. Finally, there is evidence that the sensitive periods, particularly those for binaural/spatial hearing, may extend beyond the time of birth. In the postnatal environment, auditory experience is thought to play a role in the final shaping of the system.

Development of vision

The eye provides a window to the only part of the central nervous system that can be viewed directly, the retina. This sheet of neural tissue and the visual pathway of which it is the source, have been the subject of considerable interest, not only because of the biological advantage that a complex visual system affords, but also because it is employed in many aesthetic judgements. The accessibility of the system and the ease with which its function can be tested have made it a popular part of the central nervous system for studies concerned with the mechanisms that regulate its development.

This section will concentrate on the early development of the retina and its projections. In a wide range of mammals there is a common sequence of developmental events that gives rise to the mature retinofugal pathway. These events even have a common timetable if expressed as a proportion of the period between conception and eye opening. Consequently, they will be described here without direct reference to a particular animal model. This section will be predominantly anatomical in orientation, because it is unlikely that visual function exists prior to birth. In addition to the obvious lack of intrauterine light, the optics of the eye do not start to clear until relatively late in fetal development. After birth, the eye remains hypermetropic and the lens does not take on its adult shape until the end of the first postnatal year. This section will also be predominantly confined to the development of the retina and its connections, rather than to the cortex, since many of the factors that determine the development of the visual cortex rely upon the ability to see. In the human, functional development of the visual cortex is a postnatal event.

Development of the retina

In the early embryo, two stalks emerge from the region that will develop into the future forebrain. These become invaginated to form cuplike structures (*4 weeks*), the inner layer of which develops into the retina. At this stage, the retina consists of a homogeneous population of ventricular cells, elongated in parallel, and moving between the superficial and the deep layers, depending upon their position in the cycle of cell division. Mitosis occurs along the outer edge of the retina. A newly formed cell becomes elongated until it reaches the vitreal surface. DNA duplication then takes place, after which the cell contracts towards the ventricular edge where the cycle is repeated. At any time, different cells are at different stages of this cycle. Studies that have traced cell lineage have shown that there is not a deterministic pattern of cell production in the retina, in that a mitotic cell is capable of giving rise to progeny that have widely differing morphologies. The cycle continues at a progressively slower rate as cell number increases.

The processes of cell division and differentiation overlap. Because many different cell types are produced simultaneously, it is unlikely that temporal factors alone are significant in determining cell type. Retinal neurogenesis takes place in two phases, with ganglion cells, displaced and orthoptic amacrine cells, horizontal cells, and cones born in the first phase. Glia in the ganglion cell layer, orthoptic amacrine cells, bipolar and horizontal cells, Müller glia, and rods are generated in the second phase. Differential addition of cells generated in the second phase helps establish density gradients in cell populations generated in the first phase.

Ganglion cells are the first cells to differentiate at any given eccentricity. The three types are generated with a rough centre-to-periphery gradient, in distinct but overlapping waves. However, at this stage there is no gradient in ganglion cell density across the retina and the cells' morphology is relatively simple (Fig. 20.4). Ganglion cell

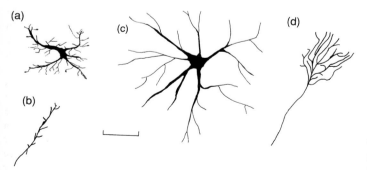

(a)

(b)

(c)

(d)

Fig. 20.4 The morphology of retinal ganglion cells and their terminal processes during maturation in the ferret: (a) an immature ganglion cell in the retinal periphery; (b) a ganglion cell axon terminal at the same developmental stage in the lateral geniculate nucleus. These cells have been labelled on the day of birth, which in this species occurs before the main period of ganglion cell death. (c) The morphology of a ganglion cell and (d) the morphology of a terminal process are both shown on the tenth postnatal day, after the main period of ganglion cell death is over. Scale bar, 50 *μ*m.

division ceases first in the central retina, close to or at the region which will develop into the area of highest cell density. The cessation of division then spreads out towards the retinal periphery. While this is occurring other cell types start to differentiate. The amacrine cells are the next to differentiate, again in a central-to-peripheral manner, followed closely by the horizontal cells, the bipolar cells, and the photoreceptors (*15–20 weeks*). The final sculpting of the density gradients in these populations occurs as a consequence of differential retinal growth that extends beyond the fetal period.

Development of the optic nerve

The developmental mechanisms that give rise to the retinofugal pathway are complex and vary along its length. The optic nerve initially appears as a stalk (*6 weeks*). Developing axons that form the intraorbital segment of the optic nerve appear to be organized in a roughly retinotopic manner. The majority of new fibres are added around the periphery, reflecting the pattern of ganglion cell neurogenesis. In the intracranial segment of the nerve, the pattern changes to one in which the younger axons express a preference for locations relatively close to the ventral surface. This change in fibre order is related to a change in the glial environment. Intraorbitally, glial cells are distributed in an interfascicular manner, surrounding groups of axons. Intracranially, the glial cells occupy a periventricular position with radial processes extending to the pial surface.

Development of the optic chiasm and optic tract

The factors outlined above are important in the development of the optic chiasm. However, simple spatial relationships alone can not determine whether developing fibres do or do not cross the midline, because, when axons reorganize in the intracranial segment of the nerve, fibres destined to form the crossed and the uncrossed chiasmatic pathway become mixed. Interactions between fibres from the two eyes and between fibres and midline glia play a significant role in the formation of the chiasm. The relative pattern of fibre distribution from one eye to each optic tract is disrupted in a time-dependent manner if the other eye is removed during development. Also, evidence that developing fibres are responsive to the chiasmatic environment comes from studies of their growth cones (the leading edges of the developing axons). In their course along the optic nerve and tract, growth cones are relatively streamlined. In the chiasmatic region they become relatively flat and extend processes into their local environ-

ment. This is not seen again until the growth cones reach the LGN.

In the optic tract, fibres from both eyes adopt a configuration that reflects their relative time genesis. This chronotopic representation develops because relatively young fibres express a preference for growing along the pial surface, and in doing so displace relatively older axons to deeper regions of the tract.

Refinement of projections

As with other parts of the nervous system, many more retinal cells are generated than are found in the adult. The period of neurogenesis is followed by a wave of cell death, which probably occurs in two phases reflecting the two phases of retinal neurogenesis. This establishes the adult number of retinal neurones. The periods of neurogenesis and cell death probably overlap. However, it is estimated that approximately two to three times as many ganglion cells are generated compared to the number found in the adult. During the period of overproduction, the morphology of their cell bodies and terminals remains relatively simple. However, they project more widely at this stage than they do when mature. Before the period of cell death, projections from each eye to the LGN and superior colliculus are retinotopically organized, but they are not segregated into separate territories. At this stage, ganglion cell projections can also be identified in structures that do not receive a retinal input at maturity. The refinement of these projections is coincident with the period of cell death and the time during which distinct morphological types appear (Fig. 20.4). Once normal cell numbers are established, differential retinal growth reinforces gradients in retinal cell density initiated by differential cell addition in the second phase of neurogenesis.

It has been proposed that the main role of cell death is to refine projections by removing errors. It is assumed that cell death occurs because of competitive interactions between axons, particularly between those from each eye, for terminal space and/or trophic factors in the LGN and superior colliculus. Evidence can be found both in favour of, and against this proposition. If one eye is removed prior to cell death, the uncrossed projection from the remaining eye does not retract, and cell death is reduced. But the number of cells that are saved is relatively small compared to the number that normally die, and to the amount of space and/or trophic factors made available in the target regions. Further, as the projections are retinotopically organized prior to cell death, the number of errors in the system is estimated to be relatively small. The temporal coincidence of cell death with axon segregation implies a functional role, although the nature of that role remains unclear. Because the projections are mixed at a stage when they are retinotopically organized, the adult pattern must develop due to relative shifts in the two projections. As each shifts in a different direction, it is unlikely that there is a single set of landmarks in the LGN and superior colliculus used to form the adult pattern.

The mechanisms that drive the segregation of the projections into their separate territories are unknown. One possible explanation is that this process results from a comparison of the patterns of neuronal activity in this projections from each eye. If activity is blocked in the retinofugal pathways, the projections fail to segregate fully in the LGN. However, action potential activity is probably not the only factor regulating the maturation of the projections. Unfortunately, many of the other experiments that have been undertaken to determine these mechanisms have done so by damaging the system early in development and assessing the effects of such damage at maturity (e.g. eye removal or retinal lesions). Because these experiments create pathological conditions, they probably only provide a crude tool with which to investigate a complex series of interactions and relationships on which the normal development of this system depends.

The development of the LGN

Relatively little is known about the development of the LGN in relation to what we know about retinal develop-ment. Ganglion cells and LGN neurones are generated over similar time periods. There are no obvious spatial gradients in the patterns of neurogenesis in the LGN, but there is a period of cell death that succeeds the period of cell generation. After the adult number of LGN neurones is established, the nucleus changes considerably in size and geometry through rotation. This explains why the optic radiation, the projection from the LGN to the visual cortex, is twisted. The cytoarchitectural lamination of the LGN does not take place until relatively late, after the patterns of innervation are established and the nucleus has adopted a relatively mature configuration.

The sequence of events that gives rise to the development of the retinofugal pathway is shown in Fig. 20.5. While the sequence of events is similar in all mammals that have been studied, the length of this sequence varies depending on the species.

Cortical development

There is a general pattern of cortical development that occurs prior to innervation and that appears to be common to all cortical areas, independent of the modality that is to be subserved. In fact, sensory projections can be redirected such that auditory information can replace visual information in what would otherwise be visual cortex and vice versa. Consequently, cortical regions do not appear to be pre-specified for a particular modality.

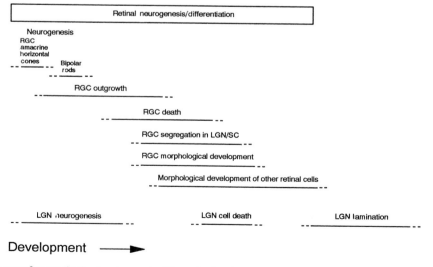

Fig. 20.5 The sequence of events in the development of the retinofugal pathway. This pattern is relatively consistent between mammals, including primates. In humans, most or all of these events occur prenatally. RGC, retinal ganglion cells; LGN, lateral geniculate nucleus; SC, superior colliculus.

Cell generation in the cortex takes place in an inside-out manner with cells destined to reside in the deep layers being formed before those destined to reside in the superficial layers. The developing cortical plate is connected to the zones of cell proliferation by radial glia fibres. These serve as guides along which neuronal migration takes place and assist in the establishment of the columnar organization of the mature cortex. Within each radial column, newly arrived cells adopt the most superficial position — adjacent to the developing cortical plate in a region that will become the first cortical layer. In doing so they migrate past cells generated earlier that are now located relatively deeper.

The development of the mature pattern of cortical innervation is dependent on a population of neurones in the cortical subplate. Projections from the thalamus are probably guided to the subplate region by the axons of these cells. Here they wait before entering the cortex. If the subplate neurones are removed, the axons from the thalamus do not invade appropriate regions. When thalamic axons enter the cortex their organization and terminal morphology are relatively immature. The maturation of the adult cortical configuration is predominantly a postnatal event and, at least in the visual system, depends largely upon sensory experience rather than on intrinsic factors.

Further reading

Coleman, J.R. (ed.) (1990). *Development of sensory systems in mammals*. Wiley, New York.

Finlay, B.L. and Sengelaub, D.R. (ed.) (1989). *Development of the vertebrate retina*. Plenum Press, New York.

King, A.J. and Moore, D.R. (1991). Plasticity of auditory maps in the brain. *Trends in Neuroscience*, **14**, 31–7.

Lam, D.M. and Shatz, C.J. (ed.) (1991). *Development of the visual system*. MIT Press, Cambridge, Massachusetts.

Lund, R.D. (1978). *Development and plasticity of the brain: an introduction*. Oxford University Press, New York.

Romand, R. (ed.) (1992). *Development of auditory and vestibular systems 2*. Elsevier, Amsterdam.

Rubel, E.W. (1978). Ontogeny of structure and function in the vertebrate auditory system. In *Handbook of sensory physiology*, Vol. IX (ed. M. Jacobson), pp. 135–237. Springer-Verlag, Berlin.

Rubel, E.W. (1984). Ontogeny of auditory system function. *Annual Review of Physiology*, **46**, 213–29.

Shatz, C.J. and Stretavan, D.W. (1986). Interactions between retinal ganglion cells during development of the mammalian visual system. *Annual Review of Neuroscience*, **9**, 171–207.

21. Development of the autonomic nervous system, including adreno-chromaffin tissue

David W. Walker

Gross structure of the autonomic nervous system

What is the autonomic nervous system? In 1921 Langley defined it as '... all the nerve cells and nerve fibres by means of which efferent impulses pass to the tissues other than multinucleate striated muscle.'

This definition thus includes the innervation of visceral, vascular, and other smooth muscle, cardiac muscle, and secretory cells, and permits one to identify the anatomical distribution of the autonomic nervous system as:

1 running within cranial nerves III, VIII, IX, X, and XI;
2 the thoracic, lumbar, and sacral spinal outflows, which pass via the ventral spinal roots.

The anatomical and functional subdivision of this entire system into the sympathetic and parasympathetic nervous systems was first delineated by Gaskell in 1916 and Langley in 1921, and their idea has required only minor modification since. A distinguishing feature of the autonomic nervous system is that all the efferent pathways consist of a two-neurone chain with a ganglionic synapse outside the neuraxis (i.e. brain and spinal cord). The preganglionic parasympathetic neurones have long fibres and terminate with few collaterals close to, or even within, the organ they supply; their postganglionic fibres are therefore short and tend to have a restricted field of distribution. In contrast, preganglionic sympathetic neurones are short tracts with several or many collaterals terminating within the prevertebral ganglia or the paravertebral ganglion chains; the postganglionic sympathetic neurones typically have long fibres and the field of distribution from each ganglion is usually more widespread than those arising from parasympathetic ganglia.

There has been some difficulty in classifying the intramural nerve plexuses that lie within the walls of the abdominal organs, especially the gut. Langley had considered the plexuses of Auerbach and Meissner to be separate from the autonomic nervous system, although later workers came to consider them to be cholinergic in nature and part of the parasympathetic nervous system. More recent work has established the unique properties of this neural system, which is neither adrenergic nor cholinergic in nature but may be purinergic (i.e. the neurones release ATP, adenosine, or a related substance). This system has been called the 'enteric' division of the autonomic nervous system.

The older descriptions of the autonomic nervous system tended to emphasize the motor components of autonomic function, although autonomic reflexes, especially those involving the heart and vasculature, were clearly understood. In the 1930s and 1940s Cannon and Rosenblueth established experimentally the concept of autonomic homeostasis, concluding a major conceptual theme first delineated by Claude Bernard in the nineteenth century. There is now clear anatomical evidence for afferent fibres from the viscera and vasculature within autonomic pathways, and visceral sensation and pain are an undeniable reality. Some workers regard sensory pathways as anatomically and functionally autonomic if they carry afferent impulses from the viscera and blood vessels, whether they travel entirely, in part, or even not at all within autonomic motor pathways. Afferent neurones from other regions are not autonomic, even if they elicit autonomic motor responses, because of their central interconnections (e.g. olfactory and trigeminal afferent pathways). The anatomical distribution and physiological significance of afferent fibres from the viscera and vasculature are still poorly understood.

Fine structure

Postganglionic fibres enter smooth muscle in bundles of 50–100 axons and first split into bundles of 10–20 axons that pass between the muscle bundles, and then into smaller bundles of 3–5 axons that run longitudinally between the individual muscle fibres. The axons may end within the Schwann cell covering of the nerve bundle or may leave it to run naked between the cells and end in a splay of nerve twigs, each about 100 μm long. The axons produce a complicated collateral branching and form a plexus or network in which the fibres of several axons may overlap.

Two types of neuromuscular contact have been distinguished. In one, typical of the mammalian gut and uterus, the Schwann cell covering of the axon bundle is interrupted at intervals and the individual nerve fibres are exposed to the muscle fibre or tissue cell across a gap of 100–200 nm; different fibres of the axon bundle may contact the same muscle fibre at successive openings. In the second type, single axons come into close (20 nm) contact as they pass along the muscle fibres, and they often appear to indent the profile of the muscle cell. Some tissues receive their innervation exclusively by bundles of axons in close apposition, or by single axons in close apposition (vas deferens, seminal vesicles), but most smooth muscle receives a mixture of the two.

Blood vessels have another, unique pattern of innervation. Axons do not enter the muscular coat but form a plexus around the vessel at the medio-adventitial border with nerve and muscle fibres separated by 100–200 nm. There are some exceptions to this (e.g. carotid artery of sheep, saphenous artery of the rabbit, and cutaneous veins), where the nerve fibres penetrate the muscle layer.

In general, parasympathetic postganglionic neurones release acetylcholine and sympathetic postganglionic neurones release noradrenaline. The transmission between pre- and postganglionic neurones in all autonomic ganglia is cholinergic, and excitatory and inhibitory actions due to acetylcholine have been identified. However, postganglionic sympathetic tracts that are entirely cholinergic in nature are known (e.g. innervation of sweat glands, and vasodilatation of the hind limbs in dogs and cats) and many sympathetic adrenergic nerve tracts contain cholinergic fibres. The opposite is also true, i.e. parasympathetic tracts that contain sympathetic fibres, as in the vagus nerve. Many of these apparent anomalies are now considered to be evolutionary relics of a more primitive organization of the autonomic nervous system.

Whereas the parasympathetic nervous system consists only of the pre- and postganglionic nerve cells, the sympathetic nervous system is often thought of as one component of a larger 'sympathoadrenal system.' As such, this system includes not only the two-neurone pathways of the sympathetic nervous system itself, but also the chromaffin tissue system of the adrenal medulla, paraganglia, and other widely distributed cells that store catecholamines and have a 'chromaffin' reaction. The extraadrenal chromaffin cells are widely distributed as cellular agglomerations and single cells, and they are found within sympathetic ganglia, heart muscle, the carotid and aortic chemoreceptors, blood vessels, and in association with nerve trunks. Concentrations of chromaffin cells are found near the thoracic aorta (the 'organ of Zuckerandl') and in the peritoneal region. Like the cells of the adrenal medulla some receive a cholinergic innervation, and some produce short fibres that appear to contact other neurones or the cells of the embedding tissue, as in the heart. The chromaffin cells of the adrenal medulla may contain adrenaline, noradrenaline, or both, as well as a number of neuropeptides, and the extraadrenal chromaffin cells may also contain dopamine. In ganglia, chromaffin cells may act as interneurones. In the carotid body they may influence the sensitivity of the chemoreceptor cells, while in other tissues the catecholamines may be released in response to a variety of situations, such as hypoxia.

In addition to this heterogeneity within the sympathoadrenal system, two populations of sympathetic postganglionic fibres have been identified — 'long' and 'short' adrenergic neurones. Long adrenergic axons are those of the paravertebral and prevertebral ganglia, while the short axons arise from the pelvic plexus and innervate the rectum and pelvic organs. Short axons, in common with the medullary and extramedullary chromaffin cells, are largely unaffected by the actions of the neurotoxin 6-hydroxy-dopamine, by immunosympathectomy, or by long-term treatment with guanethidine. Ultrastructurally, these 'short' neurones have a distinct arrangement of the granular vesicles within their terminals, and unlike 'long' neurones they are sensitive to the influence of steroid hormones. Thus, the short adrenergic neurone and the ubiquitous chromaffin cell have common biochemical and ultrastructural characteristics and appear to be related although the significance of this is unclear. In lower vertebrates diffusely distributed chromaffin cells are the most important source of catecholamines, and they may therefore be the evolutionary precursor of the adrenergic nervous system, which first appeared as a 'short' neurone type.

Autonomic responses have also been described that are due to the release of a transmitter substance that is neither acetylcholine nor noradrenaline. The inhibitory intramural nerves of the mammalian gut release a purine compound, perhaps ATP. Other non-cholinergic, non-adrenergic responses have been described, such as vasodilatation in skeletal muscle and skin and some excitatory responses of the mammalian urinary bladder and intestine. It has been suggested that histamine, substance P, or prostaglandin may mediate these responses. Immunocytochemical techniques have shown that a number of peptides are localized in autonomic nerves, including substance P, vasoactive intestinal polypeptide (VIP) neuropeptide Y (NPY), calc tonin gene-related peptide (CGRP), cholecystokinin (CCK), neurotensin, galanin, and several opioid peptides including met- and leu-enkephalin, dynorphin, and β-endorphin. These peptides may themselves be transmitters, but the current opinion is that they are co-released with noradrenaline and acetylcholine and that they act to modify the post synaptic actions of these classical transmitters, i.e. they act as 'neuromodulators'. Several of these peptides appear early in development. For example, the density of VIP and CGRP nerve plexuses reach a peak at birth in the mesenteric, carotid, and renal arteries of the guinea-pig, whereas full development of the noradrenergic plexus does not occur until 4 weeks of postnatal age. Burnstock has suggested that neuropeptides in autonomic ganglia and postganglionic fibres may have a trophic role during development.

Embryology and development of the autonomic nervous system

The preganglionic fibres of the sympathetic and parasympathetic nervous systems arise from cells in the general visceral efferent columns of the neural tube. In the thoracic, lumbar, and sacral regions of the spinal cord in mammals (including man) these fibres issue forth in the ventral spinal roots, whereas in lower vertebrates preganglionic fibres may have both ventral and dorsal root outflows. In the cranial region the preganglionic fibres also arise from cells of the general visceral efferent columns and leave the neuraxis in the oculomotor, facial, glossopharyngeal, and vagus nerves; they thus have both a dorsal and ventral root outflow.

The embryological origin of the postganglionic autonomic neurones has been somewhat controversial. Two principal ideas had been distilled by 1900 — one that postganglionic autonomic neurones, like their preganglionic counterparts, developed from cells of the neural tube itself, while an opposing view was that the postganglionic neurone was a derivative of the neural crest. Albert Kuntz, extending the ideas of Cajal and Froriep, was the principal proponent of the first view, summarizing his arguments in a final book in 1953. He had observed that, in many vertebrate species, the primordia of sympathetic ganglia were composed of large and deep-staining cells arranged in continuous columns on the dorsolateral aspects of the aorta, and that cells of 'identical' appearance could be observed to pass along both the dorsal and ventral roots from the neural tube and spinal ganglia. However, Streeter pointed out in 1912 that aggregations of cells are present on either side of the aorta before the spinal nerve roots have developed and sent projections to this site, and he believed that

the bilateral and segmented cell groups were the primordia of the paravertebral sympathetic ganglia and, further, that they were derived from the neural crest.

Experimental evidence for Streeter's views was provided by Hammond and colleagues in the early 1950s. Using glass needles and miniature forceps they removed the neural crest or neural tube from chick embryos of 7–28 somites (24–48 h postfertilization) and showed that:

(1) the sympathetic trunks and ganglia are derived from the neural crest tissue in the lower thoracic and upper abdominal region;

(2) the intrinsic ganglionated nerve plexuses within the heart, lungs, and alimentary canal are derived from neural crest contributions to the hindbrain and upper spinal cord, and the autonomic innervation of the pelvic organs have their origin in cells from the sacral regions of the neural crest;

(3) the neural tube gives rise to preganglionic, but not postganglionic, autonomic fibres.

The 'enteric' nervous system within the gut musculature has been shown to arise from neural crest cells in the upper thoracic region, from the vagal ganglion, and also from neural crest cells in the lumbosacral region. The vagal and cranial contributions are quantitatively the most important, supplying cells that migrate the whole length of the gut.

Hammond's experiments have been confirmed and extended more recently by Le Douarin who used the elegant technique of transplanting neural crest cells from quail embryos into chick embryos. The nuclei of quail cells are always distinguishable from those of the chick, and these 'foreign' cells appear to move and differentiate as if they are host cells. Their ultimate fate in terms of physical migration and differentiation can be followed with great precision, and a 'fate' map of cells arising from different parts of the nervous system can therefore be drawn up. Such a map for all the components of the autonomic nervous system has been composed from Le Douarin's studies. Figure 21.1 shows the parts of the neural crest that provide the cells of the intramural plexuses of the gut, the sympathetic ganglia, the adrenal gland, and the ciliary and superior cervical ganglia.

The following general description of the formation of the autonomic nervous system can therefore be given. Cells originating from the neural crest migrate along the primordia of the spinal ganglia to form bilateral columns of cells ventral to the notochord and just lateral to the aorta. These appear first in the thoracic and lower cervical region and extend in both a cranial and caudal direction. These columns, which are the primordia of the paravertebral sympathetic chains, have a primitive segmentation in

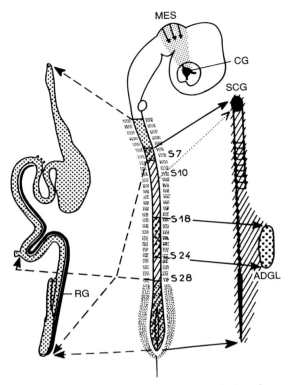

Fig. 21.1 Origin of the enteric ganglia, sympathetic and ciliary ganglia from the neural crest. The left-hand side shows the parasympathetic innervation of the gut. The right-hand side shows the sympathetic chain and the adrenal medulla. The sympathetic chain arises with somites 5–28 contributing to the superior cervical ganglion, and somites 18–24 to the adrenal medulla. The vagal innervation of the gut arises principally from somites 1–7, and additional innervation arises from somites beyond no. 28. The ciliary ganglion arises from the mesencephalic neural crest. ADGL, adrenal gland; SCG, superior cervical ganglion; S, somite; RG, ganglion of Remak; CG, ciliary ganglion; MES, mesencephalic crest. (Reproduced with permission from Le Douarin, N.M. and Cochard, P.L. (1983). *In Somatic and autonomic nerve-muscle interactions*. **Research Monographs in Cell and Tissue Physiology**, Vol. 8 (ed. G. Burnstock, R. O'Brien, and E. Vrbova), pp. 1–33. Elsevier, Amsterdam.)

register with the myotomes, but owing to the curvature of the embryo the segments may be in such proximity, especially in the cranial and sacral regions, as to appear as continuous columns of loosely aggregated cells. Secondary segmentation may occur later, particularly in the thoracic and lumbar regions. Other groups of cells continue to move ventrally beyond these columns in the upper thoracic and abdominal regions to form the primordia of the pre-

vertebral ganglia and plexuses. A similar migration also occurs from the lower lumbar regions of the paravertebral chains into the pelvic region, where they form the pelvic plexus in association with parasympathetic fibres originating directly from the sacral spinal cord. It is apparent that the preganglionic outflow from the neural tube occurs after the approximate positions of the autonomic ganglia are established.

The vagal plexuses in the thorax and abdomen are derived from cells originating from both the hindbrain and the vagal ganglia within the cranium. These cells extend in a tract to innervate the heart, lungs, stomach, and upper intestine.

The challenge of determining how cells navigate when migrating from the neural crest to their final target position continues to exercise embryologists and has been reviewed extensively elsewhere. An important associated question is whether the ultimate fate of the cells is already determined by the position they initially occupy along the axis of the neural crest, or whether their differentiation is determined by the route of migration and the position they finally occupy in the body. It is clear that the cells from somites 1–7, which infiltrate the primordial gut, differentiate into cholinergic, peptidergic, and serotonergic neurones largely to the exclusion of the adrenergic cell type, whereas cells from somite 5 downwards give rise to the sympathetic structures. What determines this differentiation.?

Le Douarin's group again used the heterotopic grafting method in which quail cells were transplanted into the chick neural crest. When cells from the vagal neural crest of the quail were introduced into the adrenomedullary region of the chick neural crest, these cells *became* adrenomedullary in phenotype. Conversely, when quail cells from the adrenomedullary region of the neural crest were placed in the 'vagal' region of the chick neural crest, the cells colonized the gut and were transformed into cholinergic ganglia and peptidergic neurones. These experiments show that neural crest cells are 'pluripotent' and that the phenotype expressed is determined by signal received during migration and/or by the target tissue in which the cell is ultimately embedded. Some of the signal that may be involved are discussed in the section 'Trophic and growth factors'.

It has been suggested that there is a primitive stem cell giving rise to all the cell types within the autonomic nervous system. As described previously, the postganglionic component of the sympathetic nervous system consists of at least three cell types, all derived from the primary sympathicoblast of the neural crest — the 'short' and 'long' adrenergic neurones and the 'chromaffin' tissue of the adrenal medulla; paraganglia; and the scattered 'small intensely fluorescent' (SIF) cells within the heart, smooth muscle, and many ganglia. Many of the secretory cells of the intestinal mucosa (amine precursor uptake and decarboxylating cells — APUD cells) are also derived from the same neural crest element. In contrast, the postganglionic tracts of the parasympathetic nervous system appear to consist only of fully differentiated neurones and, although groups of cells have been described as 'parasympathetic paraganglia', the significance and function of these have received little attention.

Adrenal medulla and chromaffin tissue

As shown in Fig. 21.1 the adrenal medulla has its origins in neural crest cells originating from somites 18–24, but little is known of what attracts these precursor cells to the developing cortical mass. Although the pattern of adrenal growth differs between species, this generally reflects variations in steroidogenic capacity of the cortex; in contrast, the volume of the medulla increases steadily during embryonic and fetal development. The primitive sympatho-chromaffin cells, migrating from nearby sympathetic ganglia, invade the loose mass of cortical cells medially and occupy the centre of the gland. At quite an early stage, primitive sympathetic cells, which appear to be capable of differentiating into either neurones or chromaffin cells, become aggregated as clusters of phaeochromoblasts that then acquire the capacity to synthesize and store catecholamines in intracellular granules; these cells are the definitive phaeochromocytes. Towards the end of fetal development or soon after birth, depending on species, different types of granules are present in the cells in the middle of gland compared to the cells located more peripherally near the cortex. The granules within the medullary cells of this juxtacortical rim have long been thought to contain mainly adrenaline, and this has been confirmed recently by immunocytological studies that show that the adrenaline-forming enzyme phenylethanolamine-methyl transferase (PNMT) is largely restricted to this region of the gland.

Isolated chromaffin cells are found dispersed through thoracic and abdominal organs throughout life, but several large agglomerations of chromaffin tissue arise in fetal life. These 'extraadrenal' chromaffin cells originate, like the adrenal medullary cells, from the neural crest. In some species (e.g. rabbit) there is a structural continuity between the intra- and extraadrenal chromaffin tissue but, more commonly, they appear as scattered encapsulated tissue masses near the aorta in the abdominal and thoracic cavities (Fig. 21.2). The largest of these chromaffin cell

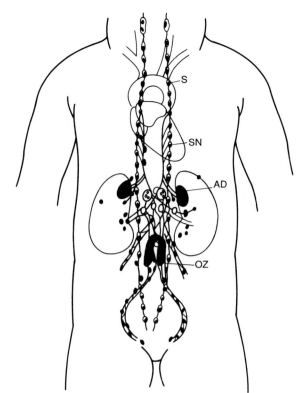

Fig. 21.2 Sympathetic chains and extraadrenal chromaffin tissue in the human. The largest discrete body of extraadrenal chromaffin tissue is the Organ of Zuckerkandl (OZ) adjacent to the aorta at the level of the inferior mesenteric artery. AD, adrenal medulla; S, paravertebral, sympathetic chain; SN, splanchnic nerve. (Reproduced with permission from Coupland R.E. (1980). In *Biogenic amines in development* (ed. H. Parvez and S. Parvez), pp. 3–27. Elsevier, Amsterdam.)

masses, known as the Organ of Zuckerkandl, occurs in the abdominal cavity near the inferior mesenteric artery. In most species, groups of chromaffin cells, or 'paraganglia', grow during fetal life and regress after birth, although remnants may persist into childhood. In some, these pre-aortal paraganglia are sparsely innervated by cholinergic nerves (e.g. guinea-pig and squirrel monkey), but in others there appears to be little or no innervation of the major clumps of chromaffin cells. More recently, met-enkephalin positive nerve fibres have been identified in paraganglia of the guinea-pig. The paraganglia synthesize and store noradrenaline, but not adrenaline, although exposure of chromaffin cells to corticosteroids induces the formation of adrenaline, presumably by induction of the enzyme

PNMT. Neuropeptides such as neuropeptide Y, galanin, met-enkephalin, calcitonin gene-related peptide (CGRP), vasoactive intestinal peptide (VIP), and cholecystokinin (CCK) have also been shown to be present.

It is not clear if extraadrenal chromaffin tissue has a physiological role in the fetus and new-born. In rabbit fetuses severe anoxia has been shown to deplete the noradrenaline content of the paraganglia, whereas little or no change occurs in the adrenal medulla. Human fetal chromaffin tissue also responds to hypoxia by releasing noradrenaline. These observations suggest either that: (1) both extraadrenal and adrenal chromaffin tissues release noradrenaline during hypoxia but synthesis and recovery of stores occurs earlier and faster in the adrenal medulla; or (2) the contents of extraadrenal chromaffin tissue are released more readily during hypoxia (or during other stresses), and thereby act as a buffer reserve for the adrenal medulla.

In the rabbit fetus, the relatively large volume of extraadrenal chromaffin tissue could conceivably function as an 'additional' adrenal medulla for the supply of noradrenaline to the circulation before birth and in the early neonatal period. In other species the small total size of the extraadrenal chromaffin cell mass makes it unlikely that these cells could contribute significantly to the circulating pool of catecholamines. If the neuropeptides also stored within the tissue are released during stress situations, more complex effects on the circulation could be expected to occur.

Growth and trophic factors

As in other parts of the nervous system, the precursor cells of the autonomic nervous system are produced overabundantly and there is a period of time when some cells are selected to survive and others die. It is now generally accepted that immature neurones compete with each other to make contacts with target tissue, and that survival and trophic factors are produced that are taken up by nerve cells and then either permit survival or promote further growth and development. A number of such neurotrophic factors have been identified and characterized, among them nerve growth factor (NGF), acidic and basic fibroblast growth factor, ciliary neuronotrophic growth factor (CTNF), and brain-derived growth factor. These 'permissive' trophic factors should be distinguished from another class of substances that influence phenotype; i.e. the particular combination of transmitter and neuropeptide synthesized and released by the neuron. These 'instructive' growth factors are discussed in the section 'Neuronal plasticity'.

In tissue culture, adrenergic axons grow preferentially towards explants of tissues, which in the adult are densely

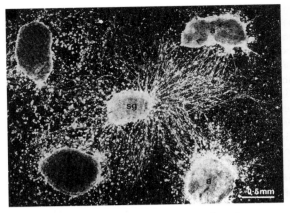

Fig. 21.3 Growth of nerve fibres from an explant of a sympathetic ganglion (sg) after 5 days in culture toward explants of lung (1), and atrium (a). Note the greater growth of fibres towards the atrial explants. (Reproduced with permission from Chamley, J., Campbell, G., and Burnstock, G. (1973). *Developmental Biology*, **33**, 344–61.)

innervated (Fig. 21.3), and it has long been assumed that tissues synthesize a chemotactic substance that attracts growing axons and causes functional connections to be made between the neuron and target tissue. The first of the neurotrophic substances that promoted growth of autonomic neurones to be isolated was given the name 'nerve growth factor' (NGF), a protein with a molecular weight of approximately 44 000. It is present in higher amounts in developing as compared to adult tissues. NGF stimulates the development and survival of sensory neurones, and some (but not all) peripheral sympathetic neurones. In the brain, neurones synthesizing acetylcholine respond to NGF, whereas catecholaminergic neurones are relatively insensitive. The actions of NGF are more easily demonstrated with embryonic or neonatal neurones than with adult tissues. Even during embryonic life there are subtle changes in the ability of neuronal tissue to respond to NGF. Thus, the classical target tissue for NGF, the chick embryonic superior cervical ganglion, only responds well to NGF during specific periods of embryonic life.

The most compelling evidence that NGF is involved in the development of the peripheral nervous system is that NGF antiserum causes widespread and almost permanent destruction of postganglionic sympathetic neurones when administered to immature animals. This effect is much smaller when the antibody is administered to adult animals. NGF, when administered to young animals, causes hypertrophy of adrenergic cells and promotes outgrowth of fibres from them. NGF is probably the trophic and chemotactic agent produced by the tissues that normally receive sympathetic innervation. Sympathetic nerves fibres have a membrane binding site of high affinity at which NGF is internalized and then transported retrogradely to the cell body. Tissues maintained in culture do produce an NGF-like substance. In some instances this substance has been shown to have some of the biochemical characteristics of β-NGF, the biologically active moiety of the high-molecular-weight form of the protein, which can be isolated in high amounts from the salivary glands of the mouse. However, it is difficult to detect NGF in all tissues that receive sympathetic innervation, and it may be that, once a tissue has been invaded by a certain number of nerve fibres, either NGF production is suppressed or NGF is removed by retrograde transport within the sympathetic axons so that only insignificant quantities remain in the tissues. Thus, tissues that are yet to receive their nerve supply or that have been denervated respond by increased production of NGF. The mechanisms that normally regulate NGF production remain to be elucidated.

NGF may also be produced by the placenta. A high-molecular-weight (150 kDa) protein with NGF-like activity on the chick dorsal root ganglion assay has been isolated from the human placenta and amnion, and blood sampled from the umbilical cord also contained some NGF-like activity. In this context, if an NGF-like substance is produced by the placenta it is curious that the extraabdominal vessels in the umbilical cord and the blood vessels on the fetal side of the placenta are not innervated. This raises the possibility that 'inhibitory' nerve growth factors may be present in some tissues that prevent extension of nerve fibres from one territory to the next.

Although the action of NGF on sympathetic and sensory ganglia has received the most attention, this substance may have other actions relevant to development. During fetal life considerable quantities of catecholamines are synthesized and stored in the extraadrenal chromaffin tissue. These regress after birth for reasons that are not understood, but it is possible that they are maintained *in utero* by the presence of the NGF-like substance of placental origin. Chromaffin cells are responsive to NGF in the immature animal where the substance promotes the development of biochemical and morphological features characteristic of sympathetic neurones. Injection of NGF into fetal and new-born rats stimulates the development of neurone-like cells in the adrenal medulla. The presence of glucocorticoids effectively inhibit these 'neurone'-like responses of chromaffin tissue, and instead enhance other actions of NGF, such as the induction of the enzyme tyrosine hydroxylase. Thus, the adjacency of the cortical and medullary cells in the adrenal gland may have the important consequence of allowing the adrenal cells to respond maximally to NGF without causing a 'phenotype' transi-

tion of chromaffin cells from the adrenal to the neural cell type.

NGF is not the only neurotrophic factor relevant to the autonomic nervous system. Other factors with a different spectrum of activity and whose effects are not reversible with NGF antiserum have been identified. The most carefully characterized has been ciliary ganglion survival factor, isolated from the ciliary body and lens of the 15-day chick embryo and from the ox heart. This substance appears to be an acidic protein with a higher molecular weight than NGF. It is unaffected by antiserum to murine NGF, and may be the principal growth factor for parasympathetic cholinergic nerves. A specific sensory neurone survival factor has also been described recently. It is a small basic protein obtained from porcine brain extracts that can enhance the survival of sensory neurones isolated from 10-day chick embryos.

Neuronal plasticity

In contrast to the 'permissive' neurotrophic factors that permit survival and growth of autonomic neurones, other diffusible factors are present in target tissue that determine aspects of neuronal phenotype, i.e. they appear to 'instruct' the neurone as to the type of neurotransmitter and the complementary neuropeptides that will be synthesized and then stored within the terminals. These 'instructive' factors appear to be highly specific and localized. For instance, sweat glands in the glabrous skin of the foot in rats are normally innervated by cholinergic sympathetic nerve fibres, and the hairy skin areas receive noradrenergic sympathetic innervation. The cholinergic innervation of the sweat glands develops postnatally from a noradrenergic prototype. If glabrous skin is transplanted to an area on the chest wall that normally receives the noradrenergic innervation typical of hairy skin, the neurones that innervate the implant change to the cholinergic phenotype over a period of several weeks. These findings demonstrate that target tissues are able to induce appropriate neurotransmitter traits in the neurones that innervate them. Such 'plasticity' has also been demonstrated in cell culture. Conditioned medium taken from cultures of heart muscle has been shown to contain not only a factor which induces expression of cholinergic traits in adrenergic nerves, such as choline acetyltransferase activity, but also other factors that induce expression of neuropeptides such as substance P, somatostatin, vasoactive intestinal peptide, enkephalins, and cholecystokinin. Subtle changes in the culture conditions produce variations in the expression of these traits of phenotype (see Table 21.1). For example, co-culture of sympathetic neurones with non-neuronal ganglion cells

Table 21.1 Effects of altering culture conditions on neurotransmitter and neuropeptide expression in dissociated superior cervical ganglion cells from neonatal rats*

	Tyrosine hydroxylase	Choline acetyltransferase	Substance P	Somatostatin
Serum-free medium	=	↓	↓	↓
+ non-neuronal cells	↓	↑	↑	↓
Membrane depolarization	↑	↓	↓	↓
Increased neuronal density	↓	↑	↑	↓

= No change; ↓, decrease; ↑, increase.
*From Kessler, J.A. (1985) *Neuroscience*, **15**, 827–39.

increases substance P and choline acetyltransferase activities, whereas the absence of these non-neuronal cells abolishes substance P and choline acetyltransferase expression but increases somatostatin and tyrosine hydroxylase activities. Increased neuronal density during culture increases substance P and choline acetyltransferase, but decreases tyrosine hydroxylase and somatostatin activities. Membrane depolarization, which may mimic nerve activity *in vivo*, increases tyrosine hydroxylase but suppresses cholinergic and peptide expression. Thus, neurotransmitter and neuropeptide expression is sensitive to many factors in the cellular environment during development. Cholinergic traits seem frequently to be linked to substance P expression, whereas noradrenergic traits appear to be coupled to somatostatin expression, but independent regulation of each of these traits of phenotype also appears likely.

Autonomic receptors

Embryonic and fetal tissues of many species (chick, mouse, rat, guinea-pig, sheep, and man) respond to autonomic neurotransmitters and the analogues of these agonists. A number of different tissues (skeletal muscle, heart, ileum, vascular smooth muscle, vas deferens) and various experimental situations have been used (*in ovo*, *in vivo*, and tissue culture). Autonomic 'receptors' may be genetically coded, fixed entities that are assembled by the cell independently of extrinsic innervation. Therefore, it is not surprising to find receptors in tissues which never receive an innervation, e.g. chick and mammalian amnion, placental vascular smooth muscle, and the experimentally non-innervated chick heart.

However, chemical sensitivity is an intrinsic and fundamental property of all cells. Most cells have a surface negative charge, so that many substances probably act by altering this charge density and the conformation of membrane macromolecules. Many cells, including central and peripheral neurones, are not uniquely sensitive to the hormone or neurotransmitter that is their primary physiological messenger. Tissues may respond to a wide range of natural agonists (neurotransmitters or hormones) and this appears to be particularly true during development and at any time that new membranes are being formed at a high rate, e.g. during repair after injury.

Receptor binding capacity may appear before cellular biochemical responses can be recorded reliably. For example, in the embryonic chick, α-bungarotoxin binds to nicotinic receptors on cultured sympathicoblasts and skeletal muscle, and cardiomyocytes bind the specific muscarinic antagonist 3-quinyclidinyl benzylate (3-QNB) from at least 70 h of incubation age. The binding of 3-QNB increases with embryonic age but there is no alteration in the rate of increase at the time of vagal invasion of the myocardium (days 4 and 5) or at the time when vagal transmission can first be demonstrated (days 10–12). Thus there is no evidence from binding studies for a nerve mediated change of receptor characteristics in this tissue.

Desensitization occurs more readily in developing tissues. 'Adaptation', 'tachyphylaxis', or 'desensitization' to acetylcholine have been shown to occur in the heart of the chick, fetal rat, and fetal mouse heart in culture. Tachyphylaxis (adaptation) to the pressor effects of noradrenaline and adrenaline occurred in the chick embryo but not in older chicks and hens, even when very high doses were used. The mechanism of desensitization is uncertain, but in the chick heart it appears to involve the acetylcholine receptor specifically because it is not prevented by treatment of the tissue with physostigmine or propranolol and because it occurs simultaneously to a number of muscarinic agonists, but not to K^+.

At present it is not entirely certain if the receptors that bind the radiolabelled ligands have all of the same properties in immature and mature tissues. It is not known if receptors are always present in apparent excess ('receptor reserve') with respect to obtaining a maximum physiological response, and, also, whether the receptor protein is always present on the cell surface as a macroaggregate.

Not only do embryonic tissues bind agonists early in development, but agonist–receptor interactions produce intracellular responses before the cellular machinery is capable of producing a fully developed cellular response, such as increased contractility in heart muscle. Noradrenaline stimulates the adenyl cyclase/cyclic AMP system in homogenates of human fetal hearts at 6 weeks of gestation, and in several organs of the fetal rat. In the rat these effects arise before glucagon has the property of stimulating adenyl cyclase activity, and before catecholamines produce distinct chronotropic responses. This suggests that responses of the intracellular apparatus associated with cyclic AMP, Ca^{2+} fluxes, and the contractile system take a longer time to develop.

Receptor plasticity

The interconvertibility of α and β-adrenoceptors, and of H_1 and H_2 histamine receptors, shown to occur at different temperatures and metabolic states in the adult frog and mammalian heart, may also occur in embryonic or fetal hearts and be a normal feature of development. The interconvertibility of adrenoceptor subtypes is influenced by thyroid hormone, and the thyroid gland has been implicated in the regulation of receptor numbers in heart cells. Triiodothyronine renders cultured fetal mouse hearts more sensitive to the chronotropic effects of catecholamines. It is not known if there is a differentiation of the major receptor subtypes from a common type during development, perhaps controlled by neural or hormonal maturation of the fetus, or whether each receptor subtype is specifically determined by the genome.

There are qualitative changes with age in the response of some tissues to catecholamines that suggests that modifications of receptor characteristics take place during development. β-adrenoceptor activity of rat aortic smooth muscle decreases with age, although the time-scale involved is longer than is generally considered for developmental purposes (birth to 90 days). α-adrenergic responses of rat tracheal smooth muscle is at first weak and then increases with age and at roughly the same time that the β-adrenoceptor response has been shown to diminish. In the coronary circulation of the fetal human and fetal cat noradrenaline produces an α-mediated vaso-constriction, whereas in juvenile postnatal cats (500–850 g) a β vasodilatation occurs. The adult coronary arteries of many species show both α and β responses. In the 13-day fetal rat, noradrenaline, but not adrenaline has a positive chronotropic effect upon the isolated heart, whereas at 21 days of gestation the reverse is observed; adrenaline, but not noradrenaline, accelerates the heart. In the chick, isoprenaline produced a propranolol-sensitive vasodilatation and hypotension on day 3 and did not accelerate the heart; α-adrenergic vasoconstriction did not occur until day 6. Thus, there are organ-specific changes in receptor responses during development that are probably related to maturational events within each tissue. However, there are great difficulties in studying very immature tissues. For instance, most arteries in the 4 day-old chick embryo consist only of endothelium. Also, the question of tone in embryonic and fetal smooth muscle, which must influence the response to excitatory and inhibitory drugs, has received scant attention. It is possible that some of the qualitative differences in response to autonomic agonists may not involve the receptors directly, but may depend upon the degree to which the tissue or organ is actually developed.

Functional development of autonomic neurones

There are obvious differences in the maturity of different animals at birth (e.g. rat pups versus lamb or foal) and probably no two animal species are born at the same point in their ontogenetic development. In addition, autonomic innervation of the various effector systems proceeds at different rates within the organs of each species. The heart, gut, and vasculature are usually innervated earlier and more completely compared to the iris, pineal gland, vas deferens, and brown fat.

Although the tissue content of noradrenaline and acetylcholine and their related enzymes increases with development, this is not a reliable guide as to whether functional autonomic transmission is possible, or whether tonic nervous control is actually present. In the chick heart, the autonomic innervation is essentially established by day 6 of development, and catecholamine levels are relatively high by day 8. However, reflex changes in heart rate in response to somatic stimulation do not appear until day 17, and the reflex bradycardic response to an increase of blood pressure does not occur until 3 months after hatching. Conversely, the hearts of new-born rabbits and lambs have relatively low concentrations of catecholamines and histologically there is poor development of the terminal plexus even though neurally mediated changes of heart rate can be shown to occur at this time. However, despite differences in the timing and pattern of development within and between species, developmental processes in the autonomic nervous system are probably similar in all.

Growing autonomic neurones are smooth and their terminal regions lack the varicosities characteristic of mature neurones for some time after contact with the effector tissue has been established. The intraneuronal granules, which represent the storage pool of transmitter, together with the enzymes of synthesis and a complex of binding proteins, are absent in the very immature neurone. They appear first in the cell body where they probably originate from the Golgi apparatus. When they first appear in the terminal region of developing axons it is uncertain whether they are granules that have been transported down the axon, or whether they are granules formed *de novo* from the endoplasmic reticulum. Granules appear in the periphery before there is evidence of functional neuromuscular transmission. Free unbound acetylcholine probably predominates in the terminals of newly developed neuromuscular junctions; bound acetylcholine and synaptic vesicles appear at the same time in explants of rat spinal cord and this is also close to the time that spontaneous electrical activity appears at the synapse.

Immature sympathetic neurones have the ability, albeit diminished, to take up and store noradrenaline from the extracellular space. Intraneuronal levels of catecholamines remain low at first, and the number of storage granules are few, so that much of the catecholamines so acquired may be lost or metabolized until the intraneuronal storage mechanisms are fully developed. The uptake, storage, and retention of choline and acetylcholine by developing cholinergic autonomic neurones has received less attention, but may not be different in principle. The appearance of cholinesterases in developing organisms has been studied extensively. Cholinesterases have a wide and ubiquitous distribution and frequently their ontogenetic appearance has little obvious relationship with neural development and the onset of neural function. However, like noradrenaline and some of the enzymes of catecholamine synthesis, neural acetylcholinesterase first appears in the cell body of the neurone and is later found in peripheral regions.

There is a delay of at least several days, or even weeks (depending on species), between the time at which autonomic neurones first reach the effector tissues and at which neuromuscular transmission begins. This is also true of regenerating nerve terminals in the adult and of transmission through developing autonomic ganglia, since preganglionic fibres appear to form synapses on ganglion cells some time before evidence of transmission can be obtained.

The reason for this delay in the onset of functional transmission is unclear. Insufficient transmitter available for release is unlikely to be the cause because in adult tissues it is possible to deplete neuronal stores of noradrenaline by up to 90 per cent before transmission fails. Also, it is clear that, when nerve transmission does begin in the heart, trachea, and intestine of new-born rabbits the endogenous levels of noradrenaline are still very low and therefore little of the eventual intraneural store of transmitter is required. There may be a critical development of the arrangement of the various intraneuronal pools of transmitters within the neurone before neurotransmission can occur. For instance, reserpine has a limited capacity to deplete noradrenaline in the new-born rat heart, and this increases over several weeks of postnatal life in this species, whereas in the kitten reserpine produces a more complete and prolonged depletion of catecholamines. The developmental changes in transmitter storage are a poorly understood aspect of autonomic development, and certainly there are differences between species.

Whilst developing nerve fibres (both somatic and autonomic) may be able to sustain and propagate impulses, they are unable to conduct impulses at high frequencies. This limitation might be due to a conduction block brought about by the intraneuronal accumulation of Na^+ as a result of a low capacity of the membrane pump to remove Na^+ from the axoplasm in immature neurones, which would favour a high rate of entry of Na^+ during the depolarization. However, there seem to be few measurements of conduction in autonomic nerves at this particular time of development. In the adult frog a proportion of the reinnervating fibres growing back into muscle do not produce an excitatory postsynaptic potential (EPSP). These 'non-transmitting' terminals do contain transmitter as shown by the effect of focal electrical stimulation, and they exert the usual trophic influences upon the muscle sensitivity to acetylcholine. It is possible, therefore, that a non-transmission stage is characteristic of the development of both new and regenerating nerve terminals, involving incomplete development of the action potential mechanism in the region of the endplate.

As discussed above, it is clear that postsynaptic receptors are present in tissues because embryonic and fetal tissues respond to the application of acetylcholine and noradrenaline. It is not known if the arrival of autonomic nerve terminals induces a rearrangement and concentration of receptors towards the region of the neuromuscular junction like that which occurs in skeletal muscle. The development of 'functionally important' α-adrenoceptors in the close vicinity of the nerve fibres has been proposed as an explanation for the increase in sensitivity that occurs postnatally in the rat portal vein in response to extrinsic nerve stimulation and to exogenously applied noradrenaline. However, the existence of localized areas of high sensitivity in smooth muscle adjacent to the neuromuscular synapse is controversial.

It is probable that most cholinergic neurones become functional before adrenergic pathways begin to transmit nerve impulses. Cholinergic responses of the chick atrium to field stimulation occur on the twelfth day of incubation, whereas the adrenergic responses do not appear until days 18 to 21. Parasympathetic motor responses occur in the intestine of new-born rats and mice and in fetal rabbits several days before adrenergic inhibitory responses can be obtained. In the fetal rabbit, the non-adrenergic, non-cholinergic inhibitory nerve supply to the gut becomes functional several days after cholinergic responses can be obtained and some time before the classical adrenergic inhibitory responses appear. Stimulation of the lumbar sympathetic chain in dogs produces cholinergic vasodilatation in the hind limbs at 18 days of postnatal age, whereas the normal adrenergic vasoconstriction does not occur until 11 weeks of age.

Thus, it appears that cholinergic nerves develop functionally before adrenergic fibres possess that capacity, irrespective of whether the cholinergic fibre runs in anatomically sympathetic or parasympathetic pathways. Non-adrenergic sympathetic transmission also seems to precede adrenergic transmission. Thus the 'long' adrenergic axon appears to be the last to develop, both phylogenetically and ontogenetically.

Development of autonomic control

Plasma concentrations of catecholamines

Plasma concentrations of noradrenaline in the fetus are usually < 1 ng/mL and adrenaline concentrations are very much lower, generally < 0.05 ng/mL. This is because the adrenal medulla, which is the major source of this catecholamine in the body, does not synthesize a significant amount of adrenaline until late in gestation when cortisol causes the induction of the enzyme PNMT, which converts noradrenaline to adrenaline. PNMT activity is largely restricted to a rim of medullary tissue where there is intimate interdigitation of medullary and cortical tissue and where, presumably, the local cortisol concentrations are high. Because cortisol synthesis is itself dependent on the maturation of the hypothalamo-pituitary axis, surgical removal of the pituitary in fetal sheep results in the failure of this PNMT activity to develop in the adrenal medulla.

The presence of noradrenaline in fetal plasma suggests that there is constant secretion from either the adrenal medulla or sympathetic nerve endings. The placenta contains high concentrations of monoamine oxidases and catechol-O-methyl transferase, which account for about half the clearance rate of noradrenaline from the fetal circulation and prevent the passage of catecholamines across the placenta from the maternal circulation. Virtually all the adrenaline in fetal plasma comes from the adrenal medulla as shown by the fact that destruction of this part of the gland reduced the plasma adrenaline to unmeasurably low concentrations, but reduced plasma noradrenaline concentrations by only 40 per cent. It appears that much of the noradrenaline in plasma comes from sympathetic nerve endings, i.e. that fraction of the released transmitter that is not taken up by the nerve endings and 'spills' over into the circulation. Destruction of sympathetic nerves by neurotoxins such as guanethidine or 6-hydroxydopamine significantly reduces plasma concentrations of noradrenaline. Noradrenaline concentrations can still be measured in fetal sheep after combined adrenal demedullation and chemical sympathectomy, and this presumably comes from those sympathetic nerve endings that are resistant to neurotoxins such as guanethidine (i.e. the 'short' axon neurones), or from paraganglia that may undergo a compensatory hypertrophy after removal of the other, principal, sources of catecholamine synthesis in the body.

Plasma catecholamine concentrations in fetal animals are increased by hypoxia, asphyxia, haemorrhage, and hypoglycaemia. In relatively mature species, such as sheep, these stimuli produce a reflex increase in sympathetic nerve activity, and a neurally mediated increase of secretion from the adrenal medulla. In less mature species, such as rabbit and rats, and early in gestation in sheep, hypoxia and asphyxia directly stimulate release of noradrenaline from the adrenal medulla. This so-called 'direct' response of the adrenal medulla is lost some time after the development of a functional splanchnic nerve supply. The changes occurring within the medulla that make it refractory to low oxygen tensions after the splanchnic nerves become functional have not been fully described. It may be that the onset of activity in the splanchnic nerve causes a small hyperpolarization of the resting membrane potential of adrenal chromaffin cells so that depolarization to the level critical for the exocytotic release of catecholamine granules is only brought about by the focal release of acetylcholine from the splanchnic nerve terminals.

In altricial species, such as rats and mice, hypoxic conditions also cause depletion of the catecholamine content of the extraadrenal chromaffin tissue (paraganglia), whereas little or no change occurs in adrenal medullary catecholamine content. From these observations it has been suggested that the extraadrenal stores are specifically adapted to provide the fetus and new-born with a source of catecholamine to respond to stress. It is possible that release actually occurs from both the adrenal and extra-adrenal stores, but that the adrenal has a greater capacity to quickly resynthesize and store catecholamine after a stressful episode, such as hypoxia or asphyxia.

Cardiovascular regulation and autonomic reflexes

Changes in the heart rate and blood pressure of the fetus and new-born, together with the effects of autonomic blocking drugs, have been used to indicate the level of development of autonomic control of cardiovascular function. In general, the ontogenetic decline of heart rate is due to an increasing vagal tone acting on the heart. However, there are exceptions. In the rat there is a postnatal increase of heart rate that is sympathetic in origin since it does not occur after pre-treatment of the rat pup with reserpine or anti-serum to NGF. In new-born dogs, sympathetic mechanisms seem to be primarily responsible for much of the control over blood pressure and heart rate since propranolol causes cardiac slowing and a fall in cardiac contractility, and section of the vagi does not increase the resting heart rate.

Resting arterial pressure rises and heart rate falls during prenatal and postnatal development in most species. Studies in the larger fetal animals, particularly sheep, have shown that autonomic cardiovascular responses can be evoked in early gestation by a number of experimental manoeuvres, such as hypoxia, occlusion of the umbilical cord, and stimulation of the aortic chemoreceptors by cyanide. Baroreflexes can be evoked in the rabbit and dog at birth and in the sheep fetus by about 0.6 gestation. In sheep, the carotid arterial baroreceptors and chemoreceptors are active and responsive to pressure and blood gas changes from about 0.6 of gestation, and reflex changes of heart rate and blood pressure occur in response to asphyxia, hypoxia, and haemorrhage from about this time. There is a progressive decrease in sympathetic cardiac activity and increase of vagal cardiac activity with development, so that there is a steady decrease of fetal heart rate as gestation proceeds. Sympathetic activity to other regions of the circulation probably increases, accounting for some of the increase of total peripheral resistance in the fetus and new-born. These changes suggest an underlying maturation of the brainstem and hypothalamic pathways that control regional activity of the autonomic nervous system, but this is yet to be studied in detail. The emergence of control by higher brain centres is shown by the fact that, when differentiated EEG activity is established in fetal sheep at about 125 days gestation, small but consistent changes of regional cerebral and peripheral blood flows occur in association with the changes between the EEG states.

Much attention has been focused on the cardiovascular responses to hypoxia in fetal and new-born animals. Hypoxia of acute onset is associated with a redistribution of the cardiac output so that blood flow to the heart and brain is increased and flows to the lungs, gut, kidney, and muscles of the body are decreased. The decrease of blood flow is due to increased sympathetic vasoconstrictor activity, reflexly activated by the aortic and carotid chemoreceptors. This response is important initially, but more prolonged vasoconstriction may be also mediated by the increase of plasma catecholamines of adrenal origin or by the increased secretion of vasopressin. This is shown by studies in which pulmonary and femoral blood flows were measured continually with implanted flowmeters, and the decrease of blood flow following onset of hypoxia proceeded more *slowly* in carotid-sinus-denervated compared to intact fetal sheep. These observations suggest that, as in the adult, the sympathetic nervous system initiates rapid changes of regional blood flows in the fetus that may then be maintained and further modified by circulating vasoactive substances. Neuropeptides such as NPY and opioid peptides, which are also released from sympathetic nerve endings, may contribute to the more prolonged changes of regional vascular resistances, but clear evidence for this is not yet to hand.

The vasoconstriction produced in vascular beds such as skeletal muscle and the gut provides the extra blood that flows to the heart and brain. The increased blood flow to these organs is therefore largely passive and the result of the increase of systemic arterial pressure (although part of the increase of coronary blood flow is mediated by β adrenoceptors, perhaps stimulated by the increase of plasma catecholamines, rather than active neurogenic vasodilatation).

Metabolic responses

Besides its facilitated transport across the placenta, glucose concentrations in the fetal blood are partly controlled by insulin and glucagon secretion from the fetal pancreas which in turn is controlled by plasma concentrations of catecholamines. Fetal blood glucose is increased by infusion of physiological doses of noradrenaline or adrenaline and this response is blocked by pre-treatment with α adrenoceptor antagonists. Catecholamines inhibit the release of insulin and increase the release of glucagon, adrenaline being more potent than noradrenaline. The hyperglycaemia that follows infusion of catecholamines in fetal sheep is also partly due to inhibition of gluco uptake by muscle and release of glucose from the fetal liver.

Many of the metabolic effects produced by catecholamines in the fetus are like those that normally occur at birth, particularly the increase of plasma glucagon and decrease of insulin. It is probable that the surge of catecholamine release that occurs at birth (which, as mentioned above, consists of the release of both noradrenaline and adrenaline from the adrenal medulla), mediates those endocrine responses that aid in the transition from intrauterine life (where exogenous glucose is supplied by placental transfer) to postnatal life where glucose must be supplied principally from endogenous stores. It is not surprising to find, therefore, that the use of adrenergic drugs in pregnancy can interfere with these sympatho-adrenal mediated changes in metabolic state at birth. For instance, the use of β-agonists such as ritrodrine to prevent premature labour is associated not only with neonatal hypotension and ileus, but also with hypoglycaemia, hyperinsulinaemia, and depletion of hepatic glycogen stores.

Catecholamines are also important regulators of plasma free fatty acid concentrations in the fetus, and contribute to their increase immediately at birth. There is a close correlation between the increase of plasma free fatty acids and adrenaline concentrations, again suggesting that adrenomedullary release of adrenaline probably mediates the lipolytic responses in the fetus. In fetal sheep adrenaline is more effective than noradrenaline in increasing the plasma concentration of free fatty acids.

Brown fat and thermoregulation

The regulation of brown fat is dealt with in Chapter 31. Brown fat is innervated by two sets of adrenergic fibres — one set innervating the blood vessels within the tissue and the other set investing the individual adipocytes with a dense network of fibres. Most of the noradrenaline content of brown fat is thought to be associated with the innervation of the adipocytes. The thermogenic and lipolytic responses of brown fat are produced by reflex activation of this postganglionic innervation, although circulating catecholamines also have effects as shown by the thermogenic response that follows injection of noradrenaline into the circulation.

The thermogenic response to cold is present in the newborn of many species at birth (e.g. lamb), or soon after birth (e.g. rat), but in some species (e.g. hamster) it is delayed for several weeks. This delay is probably not due to immature sympathetic responses but is the result of a delay in the production of the essential protein thermogenin, which is responsible for the production of metabolic

heat (see Chapter 31). Circulating catecholamines may have a role in the synthesis of thermogenin (a process known as recruitment) because high plasma concentrations of catecholamines lead to an increase in the thermogenic capacity of brown fat. In species in which brown fat is well developed before birth, thermogenic responses do not occur *in utero* when the sympathetic nervous system is activated — e.g. by asphyxia or by cooling the fetus. This is thought to be due to the presence of inhibitory substances in the fetal circulation that prevent the action of noradrenaline on the adipocytes. Prostaglandin E_2, and adenosine are present in relatively high concentrations in the fetal circulation and may act as such inhibitors of thermogenesis before birth.

Birth and the lungs

The liquid that is secreted by the alveolar epithelium during fetal life is quickly absorbed at birth. In fact, this resorption appears to begin at around the time of the onset of labour, and is, in part, a response to the surge of adrenaline in fetal blood at this time. There is an increase in both the sensitivity of the lung to this resorptive action of adrenaline, and an increase in the effectiveness of the β-receptor mediated response. Catecholamines also increase the release of surfactant into the alveolar space. Thus, the release of catecholamines from the adrenal medulla during labour and delivery promotes increased release of surfactant and causes resorption of liquid so that the alveolar spaces are made ready for gas exchange. In new-borns in whom these processes have not occurred fully, either because there was insufficient release of catecholamines (e.g. when the baby is delivered by Caesarean section) or because the action of the catecholamines was blocked (e.g. by β-blockers, such as propranolol), some lung liquid may remain so that gaseous ventilation is inadequate; this has been termed 'wet lung' and the poor ventilatory state, a 'respiratory distress syndrome'.

The successful establishment of ventilation at birth requires not only the entry of air into the lung, but also dilatation of the pulmonary blood vessels. Pulmonary blood flow in the fetus is approximately 7 per cent of the cardiac output and, by adult standards, the vasculature is constricted. The low PO_2 of fetal blood may be responsible for this; in other words there may be a chronic state of hypoxic vasoconstriction analogous to the response to acute hypoxia that occurs in the adult. Further decrease of PO_2, as during asphyxia or hypoxia, leads to greater pulmonary vasoconstriction in the fetus. It can be shown that electrical stimulation of the vagus and thoracic sympathetic nerves in fetal sheep produces an increase and decrease, respectively, of pulmonary blood flow, and reflex hypoxic constriction of the pulmonary circulation can be demonstrated from at least 0.68 of term in the sheep fetus. However, denervation of the aortic and carotid chemoreceptors only reduces and does not abolish pulmonary vasoconstriction during fetal hypoxia. The slow, but considerable vasoconstriction that persists is probably caused by blood-borne catecholamines released from the adrenal medulla.

It might be expected that the high concentrations of catecholamines that are released during labour and birth could interfere with the pulmonary vasodilatation that must occur at the onset of gaseous ventilation. Again, it may be that endogenous inhibitory substances in fetal blood counteract the vasoconstricting effects of catecholamines, or that the sudden rise of PO_2 in the alveolar space and consequent release of the hypoxic vasoconstriction is so great that neither circulating nor neurally released catecholamines have significant effects on the lung vasculature.

Conclusion

In most species some sympatho-adrenal and parasympathetic function can be demonstrated before birth, and it is likely that co-ordinated responses to stress *in utero* depend upon autonomic mechanisms. Birth requires that there is a rapid and major adjustment of many body systems to cope with the change from intrauterine life to life in the external world, and the adrenal gland appears to play a crucial role in this transition, particularly in those species born in an immature state. Even those species that are relatively mature at birth (such as sheep) rely upon adrenal mechanisms, and can survive quite well after chemical destruction of the sympathetic nervous system, but not after removal of the adrenal gland or destruction of the adrenal medulla.

While much is known about the development of the peripheral autonomic nervous system and the adrenal gland, little is known about the development of the centres and pathways in the brain and spinal cord that control them. The increase of autonomic activity may be important, not only for regulation of whole physiological systems

(i.e. the reflex regulation of the circulation, thermoregulation, etc.), but it may also have important effects on tissue growth and differentiation. It is well known that activity-dependent induction of enzyme systems occurs between the pre- and postganglionic neurones in both sympathetic and cholinergic nerves. The postganglionic release of transmitter is likely to influence receptor numbers and their distribution in the target organ. These events may, in turn, lead to modulation of cellular growth and differentiation within the organ. For instance, in new-born rats it has been shown that cell replication and RNA-dependent growth in the adrenal gland, heart, and kidney is influenced by the onset of sympathetic nerve activity. That is, the autonomic nervous system provides signals that determine maturational events at the cellular level. This role in development is in addition to the control of whole organ systems, which is normally considered the principal function of the autonomic nervous system.

Further reading

Bartlett, P.F. and Murphy, M. (1991). Nerve growth factors. *Today's Life Science*, **3**, 12–20.

Black, I.B. (1978). Regulation of autonomic development. *Annual Reviews of Neuroscience*, **1**, 183–214.

Coupland, R.E. (1980). The development and fate of catecholamine secreting endocrine cells. In *Biogenic amines in development* (ed. H. Parvez and S. Parvez), pp. 3–27. Elsevier, Amsterdam.

Le Douarin, N.M. and Cochard, P. (1983). Embryonic development of the autonomic nervous system. In *Somatic and autonomic nerve–muscle interactions*, Research Monographs in Cell and Tissue Physiology, Vol. 8. (ed. G. Burnstock, R. O'Brien, and E. Vrbova), pp. 1–33. Elsevier, Amsterdam.

Mott, J.C. and Walker, D.W. (1983). Neural and endocrine regulation of the circulation in the fetus and newborn. In *Handbook of Physiology*, Section 2. The cardiovascular system, Vol. III. Peripheral circulation and organ blood flow, Part 2 (ed. J.T. Shepherd and F.N. Abbond). pp. 837–82. American Physiological Society, Bethesda, Maryland.

Slotkin, T.A. (1988). Adrenomedullary catecholamine release in the fetus and newborn: secretory mechanisms and their role in stress and survival. *Journal of Developmental Physiology*, **10**, 1–16.

22. Development of motor functions in the fetus

Iain C. Bruce and John A. Rawson

In man, the motor functions that have matured at birth are quite limited, the voluntary control of movements developing and maturing during the prolonged period prior to sexual maturity. However, many involuntary and reflex functions develop according to as yet poorly understood timetables in order to ensure the survival of the neonate in the extrauterine environment; for example, the neural circuits underlying respiration, sucking, swallowing, postural reactions, and responses to noxious stimuli are functional at the end of fetal life. Furthermore, the development of motor control circuits *in utero* forms substrates on which postnatal

experience can build systems for the co-ordinated control of both automatic and voluntary movements. Thus, any disturbance to the harmonious assembly of a substrate may result in a cascade of deficits as 'higher' systems are prevented from making the appropriate connections at the optimal times. In this chapter we therefore adopt a hierarchical approach (Fig. 22.1), recognizing that the 'lower' systems in the brainstem and spinal cord are well-developed at birth (indicated in Fig. 22.1 by cross-hatching), while the 'higher' systems in the cerebellum and motor cortex are quite immature, and the 'highest' areas, involving pre-

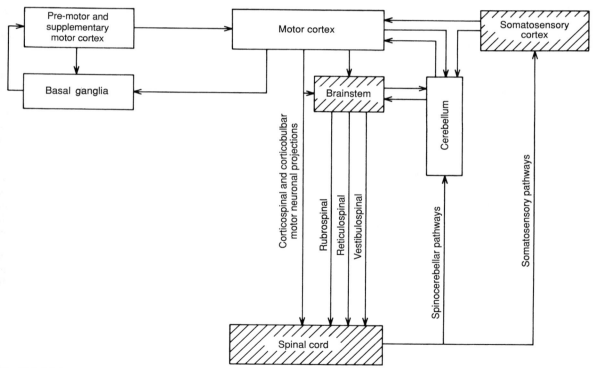

Fig. 22.1 Main components and interconnections of the motor system. The structures that mature earliest during development are indicated by cross-hatching. In primates, the motor and somatosensory cortical areas are localized around the central sulcus with the motor region being located in the precentral gyrus of the frontal lobe and the somatosensory region in the postcentral gyrus of the parietal lobe. Pre-motor cortex lies anterior to the motor area and supplementary motor cortex is located along the medial aspect of the frontal lobe.

motor and association cortices and the basal ganglia, do not become fully functional until infancy is long past. As the focus of this chapter is on the human, the stages of development refer to this species unless otherwise stated.

Spinal cord and brainstem

Motor functions in the adult

The brainstem and spinal cord contain the executive elements of the motor system, the motoneurones and the neural networks that underlie simple protective and postural reflexes as well as largely automatic functions such as swallowing, breathing, and locomotion.

Motoneurones can be broadly divided into three categories; large α motoneurones that innervate fast-twitch, fatiguable fibres to generate large forces at high thresholds; small α motoneurones that innervate slow-twitch, fatigue-resistant fibres to generate smaller forces at low threshold, and γ motoneurones that set the sensitivity of the muscle spindles. Recruitment of α motoneurones according to size provides the basis for the control of force, while the activity of γ motoneurones can modulate the operation of various feedback pathways during movement.

While the simple reflexes operate automatically at segmental levels, the networks and the motoneurones are, to varying degrees, under the control of higher centres as in the initiation of swallowing, the modulation of breathing required for speech, and the initiation and modification of locomotion according to environmental requirements.

Several motor behaviours exhibited by the fetus are absent in the normal adult, for example, the rooting reflex (turning the head towards a perioral stimulus) and the grasp reflex (finger flexion in response to palmar stimulation). Some of these 'primitive reflexes' are of evident survival value for the neonate (as in the rooting reflex for suckling), while others may be of little functional significance, merely representing evolutionary relics. For example, the grasp and Babinski reflexes may have had great survival value in our primate ancestors in whom the neonate was transported by clinging to the mother's pelt, but appear to have little functional significance in our relatively hairless species. However, since these reflexes often reappear in the degenerative diseases of old age, it seems that the circuits subserving them are operational in the fetus but are suppressed as control from higher centres develops during postnatal life.

Functional development in utero

Fetal movements have been observed, through ultrasound, at 2 months as apparently random twitches that may originate from spontaneous activity of the developing muscle cells, or may be generated by motoneurones as they begin to establish endplates on the muscle fibres. Throughout the rest of gestation, fetal movements exhibit a phasic character, as the motoneurones can only maintain action potential activity for brief periods. Immature motoneurones in animal preparations do, in fact, show high excitability presumably because of their small size and consequent high input resistance that results in large postsynaptic potentials. However, factors such as the density and distribution of synapses, the effectiveness of recurrent inhibition, and the capacity of immature glia to maintain the ionic milieu in the face of repetitive activity are all likely to affect the firing properties of fetal motoneurones. While the motor units are established, development appears to proceed in a craniocaudal sequence, beginning with the face and proceeding to the upper and then the lower limbs.

Assuming homology with animal models, motoneurones early in development are uniform in size, α and γ motoneurones being morphologically indistinguishable. At this stage, recruitment by size cannot permit force control and conduction velocities are low. As the motor units differentiate, and the processes of programmed cell death are completed (see Chapter 18), those motoneurones destined to form fast motor units increase in size, ultimately resulting in the population distribution characteristic of the motor neurone pool for a given muscle. The factors that determine this characteristic are presently unknown. By the end of gestation the muscle fibres and their motoneurones are still differentiating and conduction velocities remain significantly lower than in the adult.

Segmental reflexes, involving one or a few synapses from primary afferents to motoneurones, are the major functional circuits of the fetus. As with observed motor activity, reflex movements first appear in the face and proceed caudally. Along with the appearance of simple reflexes, such as withdrawal in response to an irritating stimulus, more complex movements resembling simple motor programmes also make their appearance early in the second trimester. These include the tonic neck reflexes, the rooting reflex, and the grasp reflex. However, when stimuli are delivered at this stage, responses are poorly organized and radiate from the point of stimulation to evoke generalized muscle activity, suggesting that the pathways from sensory afferents to motoneurones are wide-ranging and the inhibitory interneurones are not sufficiently developed to focus input in appropriate ways. Experiments in immature animals have revealed long, multisegmental branches from primary afferent fibres, projections that are normally ineffective in the adult. In addition, electromyographic recordings fail to show reciprocal inhibition between antagonist muscles at these early stages. It is in the third trimester that relatively well-organized automatisms appear, along with reciprocal inhibition and focused segmental reflexes that activate only those muscles that are functionally related to the stimulus.

Respiratory movements are first observed at about 11 weeks and progressively become more frequent. Thus, the motoneurones to the intercostal muscles and diaphragm, along with the circuits creating the pattern generator in the brainstem, show relatively advanced development.

Locomotor movements also appear early, attesting to the genetically programmed assembly of the segmental circuitry that gives rise to the pattern of sequential and reciprocal excitation and inhibition of limb muscles in locomotion. Careful study of this sequence in neonates indicates a digitigrade pattern in the distal lower limb resembling that seen in quadrupeds, presumably another relic of our evolutionary heritage. This feature is expressed as activation of the ankle extensor muscles prior to contact of the foot with the ground during 'reflex stepping' in the neonate, in contrast to the ankle flexion that ensures heel contact in adult walking. Again, the simple pattern generator laid down by the genes will be superseded when higher centres establish their connections with the segmental circuitry.

As illustrated in Fig. 22.1 spinal cord interneurones and motoneurones are the targets of major descending system for the control of posture and movement. Consequently, any developmental failure that affects the growth of these neurones may be expected to have wide-ranging effects. Indeed, failure of the presumptive spinal cord to form during the first gestational month usually leads to spontaneous abortion. At subsequent stages, defects in closure of the caudal spinal cord (dysraphism) are frequently associated with the loss of descending (voluntary) control of the lower limbs.

The early development *in utero* of ascending systems from receptors in skin, muscles, and joints is also indicated in Fig. 22.1. The significance of such afferent activity in the fetus, and the sequelae of abnormalities in afferent pathways for subsequent development, are as yet unknown. However, the appropriate development of such pathways is clearly essential for the fetal development of reflexes and receptor-influenced 'programmes' such as locomotion and breathing.

Cerebellum

Structure and motor functions in the adult

The main features of the mature cerebellum are outlined in Fig. 22.2(a). The somata of the distinctive Purkinje cells,

which are the output neurones of the cerebellar cortex, are arranged in a monolayer that spans the entire cortical sheet. Their dendrites project into the molecular layer and their axons descend in the white matter and terminate mainly in the intracerebellar nuclei which transmit cerebellar output to the brainstem and thalamus. Purkinje cells are contacted indirectly by mossy fibre afferents. These terminate on the granule cells, which send axons upwards into the molecular layer where they branch in a T-shaped fashion to become the parallel fibres, each of which synapses on the dendrites of many Purkinje cells. Climbing fibres provide the other major input to the Purkinje cells. These fibres originate solely in the inferior olive and terminate directly on the Purkinje cells in a 1:1 relationship.

Control of ongoing movement

It has long been recognized that the cerebellum exerts an important influence in the motor system, which ensures that movements are carried out smoothly, rapidly, and accurately. A full repertoire of voluntary movements can still be initiated in the absence of the cerebellum but there are marked irregularities and inaccuracies in the timing, force, and co-ordination of muscle activities. Such observations and neuroanatomical considerations have led to the widely held view that the cerebellum participates in motor performance by correcting errors in the execution of movements. As can be seen from Fig. 22.1 it receives input from both motor and sensory systems and sends its output to the motor cortex and motor centres in the brainstem. The motor input is believed to provide a copy of command signals that are being transmitted to the motor apparatus and the sensory input to provide information about the status of the body parts. It is though that the cerebellum uses these inputs to detect the occurrence of any mismatch between intended and actual movement and then computes corrective signals that are transmitted back to the motor cortex or brainstem.

The cerebellum and motor learning

Several theories of cerebellar function have proposed that the cerebellum is important for motor learning and is a site at which motor memory traces are stored. In essence, these theories suggest that memory traces representing new motor skills are established by permanent changes in the efficacy of transmission at the parallel fibre–Purkinje cell synapses and that these changes are brought about by concurrent activity in the climbing fibres. These ideas have provided a considerable impetus to cerebellar research and there is now convincing evidence that, under

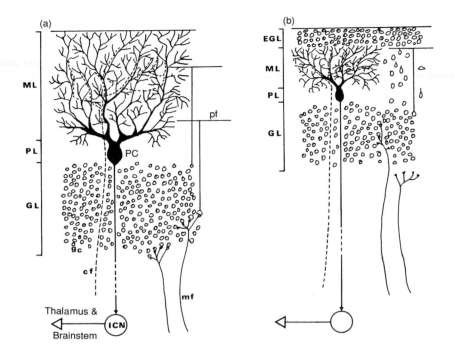

Fig. 22.2 Main structural features of cerebellum. (a) Adult; (b) at birth. **ML**, molecular layer; **PL**, Purkinje cell layer; **GL**, granular layer; **EGL**, external granular layer; **PC**, Purkinje cell; **gc**, granule cell; **cf**, climbing fibre; **mf**, mossy fibre; **pf**, parallel fibre; **ICN**, intracerebellar nuclei. Part (b) was compiled using data from the vermis as this area has been systematically studied. The indications are that other regions, particularly the hemispheres are likely to be even less well developed at birth. For clarity, cortical interneurones are not shown and only the projections of a few granule cells are indicated. Diagrams are on approximately the same scale.

experimental conditions, climbing fibres can indeed induce modifications in parallel fibre transmission that can last at least for many minutes.

Further evidence that seems to implicate the cerebellum in motor learning/memory comes from findings that animals with cerebellar lesions fail to adapt their motor performance to changing circumstances and apparently lose the ability to acquire and express classical conditioned reflexes. However, the memory theory of cerebellar function is by no means universally accepted among cerebellar researchers. Critics of the theory have pointed out, for example, that long-term changes in parallel fibre transmission may not be induced under physiological conditions or that the cerebellum may only be an essential link in the motor learning pathway rather than the site at which memory traces are stored. On balance though, available evidence does indicate a role for the cerebellum in the acquisition of motor skills but whether it facilitates or enables motor learning or actually stores memory traces remains to be clarified.

Structural development in utero

The cerebellum originates from bilateral expansions of the alar lamina of the metencephalon. These thicken, arch over the fourth ventricle, and fuse in the midline to form the cerebellar plate by the eighth week. Initially the plate consists of a neuroepithelial, a mantle, and a marginal layer. The Purkinje cells and neurones destined to form the intracerebellar nuclei evolve from the mantle layer and their production is complete by the thirteenth week. The Purkinje cells then start to migrate outwards into the marginal layer where they will ultimately become organized into a layer one-cell thick. Simultaneously, a stream of undifferentiated neuroepithelial cells migrate around the mantle layer into the marginal layer above the Purkinje cells. These cells form the external germinal or granular layer in which they start to proliferate. The overwhelming majority will subsequently differentiate into granule cells and the remainder will form the inhibitory interneurones of the cerebellar cortex. In about the twentieth week the external granular layer starts to release cells which migrate inwards. The granule cells migrate past the developing monolayer of Purkinje cells to form the internal (or definitive) granular layer. All the granule cells are destined to join the internal layer and the external one will eventually disappear. As can be seen from Fig. 22.2(b), the cerebellar cortex is still immature at birth. The external granular layer is still present and indeed its cells will continue to proliferate for some time. Purkinje cell differentiation continues for 12–15 months or longer after birth, and the external granular layer gradually disappears over a similar time period as its cells migrate inwards and receive their mossy fibre contacts. Full cerebellar function is therefore unlikely to be attained until these processes are complete some time during childhood.

Functional development in utero

There is little definitive information about cerebellar function in utero but the limited physiological evidence that is available suggests that activity may only start towards late-gestation and is unlikely to influence movements that appear during fetal life (see Table 22.1). Different regions of the cerebellum do, however, mature at different times and in a sequence that suggests that the parts concerned with the control of eye movements and posture are the first to begin to function. These are the archi- and palaeocerebellar regions (flocculonodular lobe and vermis, respectively) where myelination begins in the fifth to sixth month and is well established by birth. Recordings from the vermis of fetal sheep in utero have shown that the Purkinje cells display vigorous impulse activity and modulations in firing rate at a stage where the development of the cortex corresponds to that at 35–36 weeks gestation in man. If one extrapolates these findings in the sheep to man, it may be tentatively assumed that the vermis is active at term and capable of influencing developing postural mechanisms very early in neonatal life. The massive neocerebellum

(cerebellar hemisphere), which is intimately connected with the cerebral cortex and corticospinal tract in the control of skilled movements, matures last. Myelination begins in the seventh or eighth month and continues after birth. Neocerebellar influences thus evolve later in postnatal life and presumably continue to develop into childhood as corticospinal control matures (see below) and complex motor skills are acquired.

Functional sequelae of abnormal developments

Congenital defects of the cerebellum do not appear to produce symptoms at birth unless there are associated cerebral or brainstem anomalies. This is probably due to immaturity of both the cerebellum and the motor functions it controls. As the nervous system matures, cerebellar symptoms usually develop beginning with hypotonia, truncal ataxia, and disturbed eye movements in about the third to fourth month then limb ataxia during late infancy or early childhood. The onset of crawling, standing, walking, and speech is markedly delayed in these patients even though they may have no paralysis. Delayed acquisition of postural control has also been demonstrated following cerebellar lesions in neonatal animals and it appears that the cerebellum participates in the development of motor skills as well as in their execution. Again, as with the involvement of the cerebellum in motor learning, it is not clear how it participates in developmental processes.

The range of symptoms due to defects sustained in utero is, however, very variable and some patients do not develop obvious motor disturbances or the symptoms may largely subside with time. Indeed, there are case reports of asymptomatic cerebellar ageneses which remained undetected throughout life, indicating that other parts of the nervous system can compensate well for the absence of cerebellar influences. Although clinical reports have long hinted that an intact cerebral cortex is essential for good compensation, the role of the cortex has only just begun to be investigated in detail and the results of recent studies show that the primary somatosensory area is of special importance, particularly the region which is known as area 2. Adult animals compensate quite well for unilateral lesions of the intracerebellar nuclei but cerebellar symptoms reappear if area 2 of the cerebral cortex is subsequently lesioned. Moreover, no compensation occurs if this part of the cortex and cerebellar nuclei are lesioned at the same time, implying that area 2 is essential for maintaining the recovery of function as well as establishing it. A particularly interesting finding in one of the studies was that the density of synaptic input from area 2 to motor cortex increases substantially while compensation is being

Table 22.1 Development of fetal movements

Gestational Age (Weeks)	Comments
7–8	Movements first detectable
10–12	Respiratory movements begin
12–13	Elements of rooting reflex expressed
	First components of grasp reflex appear
16	Withdrawal reflex of foot expressed
	Gastrocnemius stretch reflex reliably elicited
28	Stepping movements begin

acquired. It seems as though this additional somatosensory input to motor cortex makes up for the loss of signals via the damaged cerebello-thalamo-cortical pathway. No doubt future experiments will begin to examine the neurophysiological mechanisms of this important compensatory process.

Motor cortex and basal ganglia

Motor functions in the adult

The role of the motor cortex in the voluntary movement of adults has been intensively studied since Fritsch and Hitzig first reported that discrete contralateral movements could be elicited by focal electrical stimulation of the cortex. Over the ensuring century, the motor cortex has come to be viewed as a 'summing point' at which neural commands elaborated in the pre-motor cortex–basal ganglia–cerebellar loops are integrated with somatosensory inputs from skin, muscles, and joints, and the resulting instructions are relayed to the motoneurones by way of the corticobulbar and corticospinal tracts.

In primates, section of the corticospinal tract alone produces a relatively mild deficit: the loss of independent finger movements results in the hand being used in a way reminiscent of the neonate in which all digits move together. Lesions at the level of the motor cortex or internal capsule have more profound effects as they presumably also interrupt corticobulbar projections. The result is the complete absence of voluntary control over the affected areas and hyperreflexia (spasticity).

The exact roles played by the basal ganglia in the generation of behaviour remain enigmatic, the present consensus being that they are crucial for voluntary or willed acts, and for cognition and intellectual function. Like the cerebellum, the motor regions of the basal ganglia participate in a feedback loop, receiving topographically organized inputs from all areas of the cerebral cortex and sending their processed output via the thalamus to the premotor and motor cortices. Current thinking attributes to the basal ganglia the role of processing information for the control of voluntary movements and postures. Recent experiments have revealed that basal ganglia neuronal activity is more related to the goal of a movement (direction, amplitude, and force) irrespective of the specific pattern of muscle activity driving the movement (in contrast to cerebellar functions).

Pathological changes in motor regions of the basal ganglia are associated with uncontrollable motor acts (as in Huntington's chorea), the absence or difficulty in initiating movements (as in Parkinson's disease), and the assumption of bizarre, fixed postures (as in dystonia musculorum deformans).

In summary, the motor cortex and basal ganglia of adults are essential to the normal generation of voluntary movements, with specialized functions in the fine control of the fingers and face.

Structural development in utero (Table 22.2)

As is true for the entire cerebral neocortex, the cortical plate from which the motor cortex will evolve is generated by cells migrating from their origin at the ventricular surface during the first 8 weeks of gestation. At this time, the axons trailing behind them and growing towards their targets begin to form the pathways from the cortex. The somata of the presumptive cortical cells migrate, guided by a structural 'scaffolding' of glial processes, to the cortical surface where they form an unlaminated plate and meet the afferent axons of cells destined to form the nuclei of the thalamus and other projection areas.

From this time onwards, the neurones destined to form the efferent pathways from area 4 develop their input

Table 22.2 Development of motor cortex

Gestational Age (Weeks)	Comments
7	Afferent fibres appear in internal capsule
8	First appearance of cells to form cortical plate
12	Corticospinal tract fibres enter medulla
16	Decussation of corticospinal tract at pyramids complete
	Precentral gyrus becomes distinguishable
20	Corticospinal tract enters lumbar cord
	Layer V distinguishable
	Betz cells with short dendrites, few dendritic spines
28	Layers I, V, and VI distinguishable
30	Corticospinal tract reaches full extent
32	All cortical layers distinguishable, including IV
	Corticospinal tract myelination begins in medulla
	Increasing dendrite length and spine density of Betz cells
35	Continued growth of all layers
	Betz cell dendrites extend to layer II
	Sparse myelination in peduncles and cord

and output functions, co-ordinated by as yet unknown mechanisms.

1. Once a soma has reached the end of its migratory path (i.e. near the surface but above the somata which arrived earlier) dendritic growth is an integral component of establishing contact with afferents from subcortical and cortical sources. Until such contacts have been established, motor cortical neurones cannot respond to sensory inputs or 'instructions' from higher centres.
2. At the same time, axonal growth proceeds, possibly by following 'pioneer' axons along pathways laid out in the structure of the brain and/or by growing in response to molecular signals generated in appropriate target cell groups at specific times. Until these processes are complete, the output of the motor cortex cannot produce functional effects on behaviour.

The identifying features of the mature motor cortex and its connections are the absence of layer IV (agranular cortex) and the presence in layer V of giant Betz cells whose axons contribute to the corticospinal tract. Since the presumptive motor cortex develops all six layers during fetal life and the loss of granule cells from layer IV occurs postnatally, attention will be focused on the development of the Betz cells.

Following the appearance of the cortical plate, the axons of the Betz cells grow through the cerebral peduncle and medulla to form the decussation and enter the cervical spinal cord during the fourteenth week. By about week 28, corticospinal tract axons are found in the caudal spinal cord. However, the immaturity of this tract is reflected in the late onset of myelination and the small diameters of the axons. At the time when the growing terminals enter the caudal spinal cord, myelination has not yet reached medullary levels. In the neonate, the area of the corticospinal tract is about 15 per cent that of the adult, the thickest fibres having only reached about 10 per cent of their adult diameter, and myelination is sparse in the cervical cord and absent in caudal segments. On the basis of such observations, the functional role of motor cortical output during even the latest states of fetal development is likely to be minimal. This tract continues to mature during postnatal life, achieving adult-like structure between the third and seventh years.

Insights from animals studies

Anatomical studies in non-human primates have shown that, while the terminals of corticospinal axons are present in the spinal cord late in fetal development, their distribution is restricted to the dorsal horn. It is only in the eighth month of postnatal life that the growth of these axons shows the adult-like invasion of the ventral horn where effective synapses may be established on the proximal dendrites of motoneurones. This coincides with the time at which relatively independent finger movements first appear. Such late maturation of functional transmission from motor cortex to motoneurones has also been demonstrated in the cat. While great caution must be exercised in making cross-species comparisons, it is generally found that the sequence in which functions develop tends to be preserved across species while the time relative to birth at which a given function develops is widely variable. Given this caveat, it may be assumed that, in the human, effective connections between corticospinal tract neurones and motoneurones are not established until the end of the first year of postnatal life, when relatively independent finger movements are first observed.

Considerably less is known about the development of the basal ganglia. In rhesus monkeys, the genesis of synapses in the caudate and putamen begins in the second trimester, reaching a peak just before birth (when the major cell types become distinguishable) and continuing at a slower rate for a month thereafter. Thus, there seems little reason to suppose that the basal ganglia make significant contributions to fetal motor activity. As they are intimately connected with the motor areas of frontal cortex that have not reached a significant functional capacity at term, it may be supposed that these structures are functionally ineffective during fetal development. Indeed, the motor functions associated with the basal ganglia in adults are conspicuously absent from the fetal behavioural repertoire, namely, the initiation and control of goal-directed voluntary motor acts.

Pathology in utero

Like the cerebellum, the motor cortex and basal ganglia may be too immature to affect motor activity in the fetus but abnormalities in their development during fetal life may have striking consequences for the postnatal development of voluntary motor control. It seems likely that the abnormalities observed in children with some variants of the cerebral palsies, with inherited disorders such a dystonia musculorum deformans or Wilson's disease, and who were exposed to toxins may have begun *in utero* although the symptoms are not generally expressed (or recognized) until months or years after birth. Facts are scarce and the problem of how to determine the long-term effects of a specific lesion occurring at a specific time is a seemingly intractable but challenging one.

Conclusions

During fetal life, limited but crucial motor functions develop, presumably as a result of the orderly expression of genetic instructions operating according to a 'molecular programme'. As motoneurones in the brainstem and spinal cord grow, the axons compete for their targets, the intra- and extrafusal muscle fibres, and those that fail that to make contact are eliminated. Primary afferents from skin, muscle, and joints contact local motoneurones and interneurones to establish fast, protective reflexes. Subsequently, interneurones in the cord and brainstem link up to form circuits that generate stereotyped movements (i.e. reflex stepping). These primary circuits form substrates or targets on which projections from late-maturing higher areas terminate. These areas ultimately control motor function whose postnatal development is not so much under genetic control but rather is dependent upon experience, for example, manual dexterity and speech.

Any processes that interfere with the orderly assembly of these primary circuits in the fetus are likely to have widespread effects. Such processes include abnormal genes or expression of normal genes at inappropriate times, hostile fetal environment as induced by maternal abnormalities or trauma, and disturbance of the vascular supply to the growing nervous system. The effects of such factors may be determined not only by the location of the insult to the central nervous system but also by the time of occurrence of the insult in relation to unfolding developmental sequences. For example, assume that, in the adult, 'nucleus X' projects to 'nucleus Y' and that this connection is essential for a specific behaviour. If 'nucleus Y' fails to develop, when the axons of 'nucleus X' enter their target area, they may make inappropriate connections with the nearby nuclei 'W' and 'Z', or they may fall victim to cell death, thus eliminating 'nucleus X' and creating further abnormalities in its growing afferent pathways. The formation of inappropriate connections is likely to generate abnormal motor activity such as mirror movements in cases of cerebral palsy where the corticospinal tract from one hemisphere terminates bilaterally in the spinal cord, while the failure to establish connections may lead to the persistence of fetal behaviours and/or various poorly understood 'compensatory' processes.

For obvious and practical reasons, the study of fetal motor behaviour has, in the past, been severely limited. In recent years, however, with the advent of benign, non-invasive techniques such as magnetic resonance imaging, the ideal of observing functional activity in the developing fetal brain is being realized, and the sequelae of abnormal anatomy may soon be more clearly understood.

Further reading

Albin, R.L., Young, A.B., and Penney, J.B. (1989). The functional anatomy of basal ganglia disorders. *Trends in Neurosciences*, **12**, 366–75.

Brand, S. and Rakic, P. (1984). Cytodifferentiation and synaptogenesis in the neostriatum of fetal and neonatal rhesus monkeys. *Anatomy and Embryology*, **169**, 21–34.

Bruce, I.C. and Tatton, W.G. (1980). Sequential output–input maturation of kitten motor cortex. *Experimental Brain Research*, **39**, 411–19.

Chugani, H.T. and Phelps, M.E. (1986). Maturation changes in cerebral function in infants determined by 18FDG positron emission tomography. *Science*, **231**, 840–3.

De Vries, J.I.P., Visser, G.H.A., and Prechtl, H.F.R. (1982). The emergence of fetal behaviour. I. Qualitative aspects. *Early Human Development*, **7**, 301–22.

Dow, R.S. and Moruzzi, G. (1958). *The physiology and pathology of the cerebellum*. University of Minnesota Press, Minneapolis.

Forssberg, H. (1985). Ontogeny of human locomotor control. I. Infant stepping, supported locomotion and transition to independent locomotion. *Experimental Brain Research*, **57**, 480–93.

Harding, R., Rawson, J.A., Griffiths, P.A., and Thorburn, G.D. (1984). The influence of acute hypoxia and sleep states on the electrical activity of the cerebellum in the sheep fetus. *Electroencephalography and Clinical Neurophysiology*, **157** 166–73.

Humphrey, T. (1960). The development of the pyramidal tracts in human fetuses, correlated with cortical differentiation. In *Structure and function of the cerebral cortex* (ed. D.B. Tower and J.P. Schade), pp. 93–103. Elsevier, Amsterdam.

Ito, M. (1984). *The cerebellum and neural control*. Raven Press, New York.

Keller, A., Arissian, K., and Asanuma, H. (1990). Formation of new synapses in cat motor cortex following lesions of the deep cerebellar nuclei. *Experimental Brain Research*, **80**, 23–33.

Koh, T.H.H.G., and Eyre, J.A. (1988). Maturation of corticospinal tracts assessed by electromagnetic stimulation of the motor cortex. *Archives of Disease in Childhood*, **63**, 1347–52.

Macchi, G. and Bentivoglio, M. (1987). Agenesis or hypoplasia of cerebellar structures. In *Handbook of clinical neurology*, Vol 61. *Malformations* (ed. N.C. Myrianthopoulos), pp. 175–96 Elsevier, New York.

Mackel, R. (1987). The role of the monkey sensory cortex in the recovery from cerebellar injury. *Experimental Brain Research*, **66**, 638–52.

Marin-Padilla, M. (1970). Prenatal and early postnatal ontogenesis of the human motor cortex: A Golgi study. I. The sequential development of the cortical layers. *Brain Research*, **23**, 167–83.

Petrosini, L. and Molinari, M.T. (1990). Hemicerebellectomy and motor behaviour in rats. I. Development of motor function after neonatal lesion. *Experimental Brain Research*, **82**, 472–82.

Rakic, P. and Goldman-Rakic, P.S. (1982). *Development and modifiability of the cerebral cortex*. MIT Press, Cambridge, Massachusetts.

Rakic, P. and Sidman, R.L. (1970). Histogenesis of cortical layers in human cerebellum, particularly the lamina dissecans. *Journal of Comparative Neurology*, **139**, 473–500.

Sarnat, H.B. and Acala, H. (1980). Human cerebellar hypoplasia. A syndrome of diverse causes. *Archives of Neurology*, **37**, 300–5.

Schulte, F.J. (1981). Development neurophysiology. In *Scientific foundations of paediatrics* (ed. J.A. Davis and J. Dobbing), pp. 785–829. Heinemann, London.

Zecevic, N. and Rakic, P. (1976). Differentiation of Purkinje cells and their relationship to other components of developing cerebellar cortex in man. *Journal of Comparative Neurology*, **167**, 27–48.

23. Development of skeletal muscle and its innervation

Uwe Proske

This chapter is a brief account of the currently recognized sequence of events in the development of skeletal muscle and its innervation. Given the space limitations, it has not been possible to provide detailed accounts of the structure and physiology of adult skeletal muscle, the sliding filament hypothesis, excitation–contraction coupling, and neuromuscular transmission. These topics have been adequately dealt with in other textbooks and reviews. However a little more detail has been provided on muscle receptors since this represents a somewhat specialized subject.

While much of the work on the development of muscle has been carried out on chickens, because of the ready availability of the various stages *in ovo*, there is now a growing body of information about the development of mammalian muscle, a large part of it coming from work using tissue culture preparations. The species used have included rat, mouse, and guinea-pig. For mammals, whenever interspecific comparisons are made, it is important to remember the large differences in gestational periods. Here it is of interest to reflect on the fact that in different animals the time of birth is imposed somewhat arbitrarily on the developmental process. In altricial animals like the cat, the new-born is still quite immature while in precocious animals like sheep the new-born is able to stand and, if necessary, run alongside the mother to escape predators from soon after birth.

As to 'Further reading', accounts of the early development of muscle have been written by Konigsberg (1986), Fishman (1986), and Schmalbruch (1986). Recent accounts of the development of the different fibre types can be found in Sanes (1987) and Hoh (1991). The innervation of skeletal muscle has been comprehensively reviewed by Jansen and Fladby (1990). Reviews of the development of muscle receptors can be found in Barker (1974), Proske (1981), and Gregory and Proske (1988).

Early development

During the first half of the last century it was proposed that muscle fibres were formed by the fusion of mononuclear cells. This theory was subsequently displaced by the proposal that a muscle precursor cell could undergo repeated division of its nucleus, without each time passing through the intermediate stage of cytoplasm formation. Once nuclear multiplication had reached a critical point, the fibre was thought to split longitudinally to form two daughter fibres, within each of which the process would then be repeated. Foci of replication in the muscle were thought to be the Weissman bundles, which were later identified as the muscle spindles. Muscle fibres were therefore thought to be a plasmodium, like some microorganisms, and not a syncytium, the result of cellular fusion. Two facts argue against the amitotic division of myoblast nuclei: first, they are always diploid, that is, they have pairs of homologous chromosomes; and second, they do not synthesize DNA (deoxyribose nucleic acid). These two facts, plus the observation that in maturing muscle fibres no mitotic figures were ever seen, led to the conclusion that the main process involved in muscle development was the fusion of myoblasts with assembly into myotubes and myocytes.

The emergence of somites and limb buds in the embryo involves cells whose destiny has already been determined. Here the principle emerges that a cell's genealogy, that is, its ancestral origins, determines whether, for example, it becomes a muscle precursor or a cartilage precursor. During the process of differentiation and transcription, cell–cell contacts are likely to be important as well as position effects and the extracellular microenvironment, but the overriding determinant will remain the cell's inherited genetic traits.

The muscle precursor cells migrate from the somite before the bulging out of the limb bud. Once in the limb bud they congregate in dorsal and ventral pre-muscle masses. Repeated cleavage of these masses produces the characteristic muscle pattern of each limb segment. Connective tissue cells are responsible for the spatial organization of the muscles. While muscle precursor cells are derived from somitic mesoderm, connective tissue, cartilage, and bone are derived from somatopleural mesoderm.

The developmental process is asynchronous; that is, the stages of development reached by different muscles at an point in time are not the same. One of the first signs of myogenesis is the adoption of a bipolar spindle shape by myoblasts (see Fig. 23.1). This is accompanied by cessation of DNA synthesis. Adjacent myoblasts then fuse with one another to form myotubes (Fig. 23.1). At this stage the surface of the myotube appears enveloped by a multinucleate syncytium comprising other fused myoblasts from which pseudopodial-like projections bulge into the myotube membrane. Once fusion has occurred, the process of cytodifferentiation begins with the formation of myofilaments. Initially, after myoblast fusion, the nuclei are centrally placed and the first myofibrils develop along the periphery. As the myofibrils proliferate they gradually push the nuclei to a more peripheral location. Adoption of a subsarcolemmal position by the nuclei signals the transition from myotube to myocyte (Fig. 23.1). Normal muscle synthesis leads to the formation of parallel running, nonbranching muscle fibres.

Alongside the initial or primary myotubes are longitudinally orientated myoblasts that subsequently fuse to form secondary myotubes. As the secondary myotubes are formed they align themselves around each primary myotube to give the muscle a rosette-like appearance in cross-section. Initially the primary myotubes and their associated myoblasts and secondary myotubes are coupled by gap-junctions and share a common basal lamina.

The primary myotubes are the first to become innervated, the secondary myotubes being coupled to them by the gap-junctions. The clusters of myotubes then gradually break up and a separate basal lamina is laid down around each presumptive muscle fibre. The gap-junctions disappear. A sheet of endomysium is laid down around each fibre and becomes part of the sarcolemma. From this stage on, the number of muscle fibres remains constant and each fibre grows in length by incorporation of new myoblasts accompanied by the addition of sarcomeres to the existing myofibrils. The increase in fibre girth is the result of the addition of myofilaments to the existing myofibrils, leading to eventual longitudinal splitting of myofibrils.

Some reserve myoblasts continue to remain in association with muscle fibres for the rest of their lives. These are the satellite cells, which are in contact with the muscle fibre and are enclosed within the same basal lamina (Fig. 23.1). About one satellite cell is found for every 20 myonuclei. During muscle regeneration one of the daughter cells of a satellite cell mitosis becomes a myoblast capable of fusing with other myoblasts or with the existing muscle fibre. Other daughter cells remain mitotically active and are the source of more myoblasts. The number of satellite cells remains more or less constant during development, except during the final stages of maturation when the proportion of myonuclei increases.

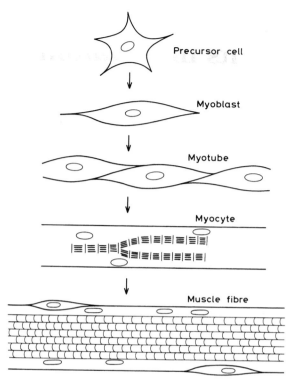

Fig. 23.1 Sequence of events in the development of muscle fibres. The precursor cell adopts an elongated, spindle shape to become a myoblast. Myoblasts fuse end-to-end to form a myotube. The myofilaments begin to form and occupy the central space, pushing the nuclei to a peripheral location. These are the features of the myocyte. Further growth in girth from proliferation of myofilaments and the incorporation of sarcomeres at each end characterize the maturing muscle fibre. Notice the satellite cells lying in close proximity to the muscle membrane.

The processes of organization and differentiation of developing muscle fibres may involve the protein synthesized by fibroblasts, fibronectin. This protein may be involved in, amongst other things, the binding of cytoplasmic actin and the establishment of transmembrane connections. Formation of myotubes in an ordered arrangement may also be controlled by a fibronectin-containing matrix of connective tissue cells. A thin sheet of fibronectin surrounds each mature muscle fibre. This may prevent lateral fusion of myotubes and therefore branching of fibres. A conspicuous feature of dystrophic muscle is the extensive branching of muscle fibres.

Myofibrillogenesis

Only the postmitotic myoblast and its fusion product, the myotube, have the capacity to transcribe the entire complex of myofibrillar genes. Within 10 h of its birth a single postmitotic myoblast can initiate transcription of its myofibrillar genes, translate the transcripts into monomers of the muscle-specific isoform of actomyosin, polymerize them into thick and thin filaments, and, finally, assemble them into sarcomeres.

All eukaryotic cells contain a cytoplasmic fibrillar network, a cytoskeleton that serves to organize intracellular organelles. Once myoblasts have stopped migrating, a reorganization of their cytoskeleton occurs. Actin monomers and oligomers and microfilaments become bundled into a cable-like structure called stress fibres (Fig. 23.2). This process is accompanied by a rearrangement of the membrane attachment sites for the microfilaments. Proteins integral to the cell membrane distribute in a linear, periodic array in register with the array of dense bodies found at intervals along microfilaments. These dense bodies may be precursor Z bands. The alignment of membrane proteins and microfilaments represents the first step in the formation of sarcomeres.

Associated with the dense bodies are non-muscle-specific contractile proteins. Once muscle-specific proteins begin to be expressed by the cell, a co-polymerization of muscle and non-muscle contractile proteins may occur along the microfilament cables. Further growth of the future myofibril occurs by the addition of other sarcomeric proteins to its ends and sides. The inward growth of myofibrils probably leads to their splitting away from the sarcolemma, but attachment would be retained in the form of intermediate filaments (Fig. 23.2). Soluble proteins present in the sarcoplasm provide the framework for new isoforms of actomyosin to be incorporated into the pre-existing myofibrils.

The polymerization of myosin, triggered by an increase in monomer concentration beyond a critical level, together with other myosin binding proteins leads to the formation of thick filaments — a three-stranded arrangement of myosin polymers from which the eventual double hexagonal lattice is constructed. The co-ordinated synthesis and accumulation of troponin, tropomyosin, and α-actin leads to the formation of thin filaments. The association of thin filaments with cross-bridges projecting from thick filaments produces the final lattice of six thin filaments surrounding each thick filament. M-band linkages between thick filaments may serve to stabilize the lattice and help with its alignment. Adjacent sarcomeres are held in register by filamentous protein bridges between myofibrils. It remains unknown when the elastic proteins, titin and nebulin, are incorporated into the sarcomere. These two recently discovered proteins are thought to be concerned with anchoring thick filaments to the nearby Z lattice. They, rather than the sarcolemma, are now thought to be the main source of passive tension in the muscle.

Muscle fibre types

Skeletal muscle can be broadly classified into two types, red and white. Red muscle has a slow contraction speed; white muscle is fast. An important additional criterion used by physiologists is that of fatiguability. It is now generally agreed that there are three kinds of muscle: FF (fast, fatiguable: type IIB); FR (fast, fatigue-resistant: type IIA); and S (slow: type I). Fatigue-resistant fibres are red in

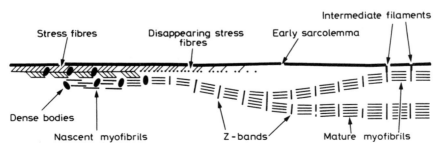

Fig. 23.2 Diagram illustrating the formation of myofibrils in a developing muscle fibre. The assembly of nascent myofibrils occurs in close association with stress fibres, which lie in orderly arrangement along the developing sarcolemma. The dense bodies may be precursor Z lines. As myofibrils increase in girth by the addition of actin and myosin filaments they separate from membrane structures. Some attachment may be retained by means of intermediate filaments. (Redrawn in part from Dlugosz, A.A., Antin, P.B., Nachmias, V.T., and Holzer, H. (1984). *Journal of Cell Biology*, **99**, 2268–78.)

colour because they use oxidative metabolism as their source of energy, which requires large numbers of mitochondria and a rich blood supply. Fatiguable fibres depend on the breakdown of intracellular glycogen for their energy supply, so they are less vascularized with fewer mitochondria, giving them their white appearance.

A topic that has received much attention during the last 30 years concerns the processes by which these different fibre types are established during development. Until recently, it was thought that the nerve played a prominent role, either through the influence of substances produced in the cell body and secreted at the nerve terminal or through the pattern of action potentials carried by the axon. Thus motoneurones supplying slow muscle typically maintain a low level of tonic activity while motoneurones innervating fast muscle are silent most of the time, firing only occasionally and then in brief, high-frequency bursts. An experiment that appeared to point directly at the nerve as the determinant of muscle fibre type was that of cross-reunion where the nerve of a slow muscle was forced to grow into a fast muscle and vice versa. Both in young animals and in adults this led to at least partial conversion of muscle properties, the slow muscle becoming faster and the fast muscle slower. Other experiments involving transection of the spinal cord and chronic stimulation led to similar conclusions.

The biochemical basis for the difference in contraction speed between fast and slow muscle lies in the rate of splitting of ATP (adenosine triphosphate) by myosin. Actin exists in all muscles as the α-isoform. Myosin, on the other hand, is distinct for different fibre types and its rate of ATP hydrolysis determines the contraction speed. Apart from the adult types of myosin isoforms, embryonic and neonatal isoforms exist as well. Incidentally, the nerve cross-reunion experiments succeeded in converting the slow and fast myosin isoforms as part of the process of transformation of muscle properties.

The first evidence of nerve-independent differentiation of muscle fibre types came from experiments on chickens where it was shown by histochemical methods that the first myotubes formed at a time when the nerve had not yet reached the muscle, and were already of at least two distinct kinds, fast-like and slow-like. It appears that, while the development of primary myotubes proceeds quite normally in the absence of a nerve supply, development of secondary myotubes requires the presence of nerves in the muscle. Tissue culture studies in the chicken have shown that, using a classification based on myosin isoforms, primary myotubes are of three types — fast-like, slow-like, and a fast–slow mixture. Secondary myotubes, on the other hand, were all of the one kind, although distinct from primary myotubes. In other words, in the chicken there

appear to be four distinct lineages of developing fibre types. How these different types eventually give rise to the adult pattern is still being resolved.

In mammalian muscle there are at least two types of primary myotube, slow and fast. Slow primary myotubes express embryonic and adult slow myosins but not neonatal myosin. The embryonic myosin is subsequently replaced by adult slow myosin. Fast primary myotubes initially express embryonic, neonatal, and adult slow myosins, that is, in most respects they are like the slow primaries. However, as the fibres mature under the influence of the nerve, these isoforms are all eventually replaced by fast myosin. There are two types of secondary myotube. Both initially express embryonic and neonatal myosins, but these are replaced by either adult fast or slow myosin.

At birth all mammalian muscles are slow-oxidative and fatigue-resistant, that is, they exhibit physiological characteristics resembling adult type I fibres. All subsequent conversion is neurally determined.

How the nerve may influence the development of muscle fibres can be studied by denervation experiments. When adult slow fibres are derived from slow primary myotubes, denervation of the muscle does not change their properties. When, under the influence of the nerve, slow primary myotubes have been converted to adult fast fibres, nerve section leads to reconversion into the slow type. However it must also be taken into account whether the adult fibres arose from primary or secondary myotubes. Thus denervation of adult rat muscle, which is phenotypically slow, leads to the expression of fast myosin in only half of its fibres. Half of the fibres in soleus derive from primary myotubes and half from secondary myotubes. To conclude, to be able to reliably predict a muscle fibre's response to denervation requires knowledge of its lineage.

The innervation

Axonal outgrowth

The limb muscles are innervated by motoneurones from a number of pools spread over several spinal segments. There is a rough correspondence between the rostrocaudal location of a pool within the spinal cord and the muscle its axons innervate. Medially placed pools innervate the ventral muscle mass; lateral pools supply the dorsal muscle mass.

The process of innervation of limb muscles proceeds in an orderly fashion, axons growing out towards their appropriate target muscle. There appear to be two kinds of cues

provided by the extracellular matrix to which the axons
adhere. One, a non-specific cue, provides a preferred tissue
substrate for the outgrowing axons. This gives the axons
their direction for growth. A second, specific cue directs
an individual axon towards its target muscle. The exact
nature of this guidance cue is not known. However,
laminin, a component of the extracellular matrix, may be
involved in generating the non-specific cue. In addition,
substances released by the target muscle are involved in
directing the axons towards their destination.

Formation of neuromuscular junctions

By the time of myoblast fusion, outgrowing axon terminals have already reached the muscle primordium. The
first functional synapses are formed within hours of the
appearance of myotubes. The recognition process by the
nerve terminal of the muscle fibre as an appropriate target
involves the protein neural cell adhesion molecule (N-
CAM). It is located in the plasma membrane of myoblasts
and early myotubes as well as in the motoneurones themselves. This protein promotes the contact between the
motoneurone and the myotube. After innervation has
occurred N-CAM expression in the extrajunctional regions
of the myofibre is suppressed. There are probably other
cues involved in the selection of its target by an outgrowing axon. These are reflected in the segmental innervation
of the muscle and the preference shown by the growing
axons for particular muscle fibres. The protein s-laminin is
thought to be involved here.

The outgrowing growth cones have the cellular machinery for the spontaneous and activity-evoked release of
acetylcholine (ACh), the transmitter at the skeletal neuromuscular junction. Efficiency of the release process
improves after contact is made with the muscle. In the
muscle membrane the ACh receptors combine with ACh to
effect the postsynaptic conductance change that will ultimately trigger an action potential in the muscle fibre. In
newly formed myotubes ACh receptors are evenly distributed across the membrane surface (Fig. 23.3). As development proceeds, the receptors begin to aggregate in
randomly distributed clusters. The arriving motoneurone
terminals make contact with the myotube membrane, not
necessarily at the site of a receptor cluster. At the contact
point new clusters begin to form while extrajunctional
receptors become fewer, the result of entrapment by the
axon terminal and suppression of the synthesis of extrajunctional receptors. The new receptors formed under the nerve
terminal are the result of a selective transcription of ACh
receptor–messenger RNA (ribose nucleic acid) in the subsynaptic nucleii.

A well known feature of the pattern of innervation of
adult muscle is that the nerve endplates are located in a

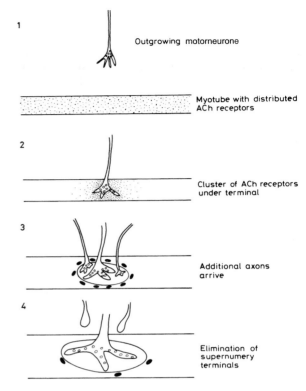

Fig. 23.3 Diagram to illustrate aspects of the development of
neuromuscular junctions. 1. The growth cone of the
outgrowing motor neurone has the capacity to release ACh.
The myotube membrane which the axon is approaching has
ACh receptors distributed uniformly across it. 2. At the contact
point, clusters of ACh receptors form while extrajunctional
receptors become fewer. 3. Additional axons innervate the
newly formed endplate. It is hypothesized that release by nerve
terminals of ACh triggers the production of proteolytic
enzymes that begin to digest the nerve terminal membranes.
4. This leads to withdrawal of some terminals leaving one axon
to innervate the muscle fibre. (Redrawn in part from O'Brien,
R.A.D., Östberg, A.J.C., and Vrbova, G. (1978). *Journal of
Physiology*, **282**, 571–82.)

clearly identifiable band across the muscle. Early
myotubes are receptive to outgrowing motor axons all over
their surface. How is the adult endplate pattern achieved?
Early myotubes are quite short and they elongate by growing at both ends. Thus, a motoneurone that makes contact
anywhere along an early myotube will eventually finish up
somewhere near the middle of the adult muscle fibre.

After the initial contact is made between a motor axon
and the muscle fibre, later arriving axons are restricted to
making their contact at the site of the first-formed junction
(Fig. 23.3). This is because the extrajunctional region has
become unreceptive, partly because of activity-dependent

suppression of receptivity and partly from the reduced synthesis of extrajunctional receptors.

Once the new subsynaptic receptor aggregates have formed under the nerve ending, further development of the neuromuscular junction proceeds more slowly. As the nerve terminals grow in size and branch to increase their contact area there is a corresponding expansion of the sub-synaptic receptor aggregates. At the same time the ACh receptors are stabilized by a 20–30 fold increase in their metabolic half-life and by an anchoring action from a membrane-bound protein that attaches to the receptor.

Within a few days of formation of the first nerve–muscle junctions, ACh esterase is deposited in the subsynaptic membrane. This enzyme is important for the rapid breakdown of ACh and therefore ensures that there is not unduly long transmitter action. An additional change occurs in the kinetics of the membrane channels opened by the coupling between ACh and its postsynaptic receptors. Extrajunctional receptors and early junctional receptors have a channel opening time of 4 ms. The receptors synthesized later during development have a channel opening time of only 1 ms.

There is some evidence from experiments on adult muscle that suggests that muscle activity is required for the normal developmental processes. Nerve section prevents the formation of normal receptor aggregates, but, if, at the same time, the muscle is made to contract at regular intervals by direct stimulation, a near-normal development of receptor clusters will occur. In paralysed muscles ACh esterase fails to appear, but it is present in normal amounts if the muscle is stimulated. A similar situation exists for the shortening of the membrane channel opening time

although here, in the presence of an inactive nerve, some slow conversion does occur.

A compound associated with the junctional basal lamina, 'agrin', which is produced by motoneurones, enhances the aggregation of ACh receptors and the production of ACh esterase and so may be involved in the process of junctional development. Another compound produced by the motor neurone and a candidate for a nerve–muscle trophic factor is calcitonin gene-related peptide (CGRP). The presence of this peptide triggers an increase in the rate of ACh receptor synthesis.

Motoneurone death

Soon after the outgrowing motor axons make their initial contact with the muscle the first major degenerative event in muscle development begins. Motoneurones begin to die; this occurs in the rat at 5–10 days postconception. Eventually up to 50 per cent of the cells will degenerate. In the rat the process continues over about 10 days and is complete by the time of birth (20 days, Fig. 23.4). A great deal of discussion has centred around the possible significance of this event. Death during development of part of the population appears to be a common property of neurones in the central nervous system. This process was first thought to be a mechanism by which incorrect connections were eliminated. Since, as already mentioned, motoneurones are accurately guided to the appropriate muscle, any process of elimination of incorrect connections would have to be restricted to within a muscle.

Activity in the muscle appears to be important for determining how many motoneurones will die. If, early during

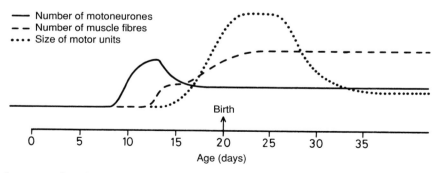

Fig. 23.4 Graphical representation of the changes, with time, in the number of muscle fibres, motoneurones, and the size of motor units during muscle development. Data are based on events in the rat. The vertical scale is purely arbitrary. (In fact there are likely to be 100 times as many muscle fibres as motoneurones.) During the period before birth motoneurone numbers increase and then decline as about half of them die. At the time of onset of cell death, muscle fibres become innervated. Initially, there is a period of hyperinnervation followed by synapse elimination several days after birth. The onset of synapse elimination begins at a time when muscle fibre numbers have reached their adult values. (Modified and redrawn from Jansen, J.K.S. and Fladby, T. (1990). *Progress in Neurobiology*, **34**, 39–90).

development, a limb is amputated, all of the motor neurones destined to innervate that limb will eventually die. On
the other hand, total paralysis of the embryo during the
critical motoneurone-death period virtually prevents the
loss of any neurones. This finding emphasizes that even
those motoneurones normally destined to die are inherently equipped for prolonged survival. Curarization after
limb amputation does not prevent death of the relevant
neurones. The current hypothesis is that paralysis of a
muscle leads to overproduction by the inactive muscle of a
substance that permits more neurones to survive.

Synapse elimination

The final event in the establishment of a mature innervation pattern is the elimination of all but one of the axons
supplying a single nerve terminal. Early during development several neurones send axon terminals to a single endplate. In the new-born rat all muscle fibres show
polyneuronal innervation, the number of terminals per
muscle fibre reaching a peak soon after birth (Fig. 23.4).
Extrajunctional sites are not occupied because of the progressive suppression of extrajunctional receptors (see
above). The elimination process begins in the rat at 5–10
days after birth and takes about 10 days to complete. After
degeneration of terminals belonging to dying motoneurones, the remaining supernumary terminals are removed
by one of two mechanisms. A motoneurone has an upper
limit to the number of terminals it can support for any
length of time, given the normal metabolic requirements of
those terminals. That, in turn, sets an upper limit to the
number of muscle fibres innervated. All additional terminals established during early development are eventually withdrawn.

A second, more specific mechanism is a competitive
interaction between the terminals of the different axons
innervating a muscle fibre. Within a muscle, particularly
one of mixed, slow–fast fibre composition, there is probably a continuous ongoing exchange of terminals between
inappropriate and functionally appropriate fibres. This
exchange goes on before the elimination process begins
and it eventually ensures that the contacts made by a motoneurone are all on muscle fibres which are of the same
kind.

The competitive vigour of a terminal is inversely related
to the size of the motor unit, that is, the number of muscle
fibres innervated. A specific mechanism has been proposed
for the competitive elimination process. It is suggested that
the release of ACh causes the production and liberation of
proteolytic enzymes, either from the muscle fibre itself or
from non-muscle cells adjacent to the nerve terminal. The
enzymes diffuse into the proximity of the endplate and

begin to digest the nerve terminal membrane. Terminals
must therefore be continuously replaced if they are to
maintain contact with the muscle. Small motoneurones
will have inherently less capacity for replacement, so their
terminals are eliminated first. Small motoneurones are also
tonically active, so they will tend to release more transmitter, again leading to stimulation of the elimination process.
The combined activity of several terminals at one endplate
will lead to elimination of terminals by digestion (Fig.
23.3). The net production of proteolytic enzyme will fall
as terminals are eliminated, to a point where the motoneurone can provide sufficient replacement to maintain the
integrity of the remaining ending. The end result is small
motoneurones that innervate a smaller number of muscle
fibres than do large motoneurones.

Finally, the existence of growth factor has been postulated, motoneurone growth factor (cf. nerve growth
factor). This factor is thought to be produced in small
amounts, inversely related to the level of activity. In addition, calcitonin gene-related peptide may be involved in
inhibiting neurone terminal growth.

Muscle receptors — muscle spindles

The nerves of skeletal muscles contain afferent as well as
motor fibres. The afferent fibres, with their cell bodies in
the dorsal root ganglion, cover a wide range of sizes. They
include the very largest of the myelinated nerve fibres in
the body, classified as group Ia and Ib, which supply
muscle spindles and tendon organs, respectively. Then
there are the group II fibres that supply the secondary endings of muscle spindles. Finally there are fibres within the
small myelinated and unmyelinated range, whose terminals within the muscle are free-branching and which are
thought to subserve the senses of muscle pressure and
pain. Nothing is known about the development of muscle
pressure receptors and pain receptors so this account will
be restricted to muscle spindles and tendon organs.

Before discussing the development of muscle receptors
in detail it is of interest to consider some recent observations on the re-innervation of adult muscle and the implications these may have for development. Cutting a muscle
nerve and then letting it re-innervate the muscle leads to
disruption of receptor properties accompanied by an overall slowing of afferent axon conduction velocities. Based
on the spinal synaptic action of muscle afferents, it has
been argued that the re-innervation process is essentially
random, so that afferents may innervate inappropriate targets, for example, spindle afferents supplying tendon

organs and vice versa. Some afferent fibres appear unable to find any target and remain non-functional. This result suggests that, during development, there must be some kind of matching process between the outgrowing axon, its central connections, and the peripheral target.

In a second series of experiments the properties of outgrowing axonal sprouts of regenerating muscle nerves were studied and found to contain a majority of axons that responded with a slowly adapting discharge to mechanical displacement of the cuff into which they had been grown. A similar experiment on skin nerves showed that the majority of axons were rapidly adapting. In a related experiment a muscle nerve that had been guided to innervate skin contained more slowly adapting afferents than a skin-nerve-innervating muscle. The conclusion from both of these experiments is that not all of the response properties of mechanoreceptors are attributable to the influence of the target tissue. Part of the response appears to be inherent to the axon terminals. It raises the possibility that, during development, outgrowing axons are predestined to become slowly or rapidly adapting receptors even before they reach their intended targets.

Adult muscle spindles

Muscle spindles are unique amongst the various kinds of sensory receptors with which the body is endowed in that they represent an example of a receptor with an in-built sensitivity control via the fusimotor system. Spindles consist of a number of structurally highly specialized muscle fibres, the intrafusal fibres, that are enclosed in a fluid-filled capsule which is expanded in its middle to give the structure its spindle-like shape. This bundle of intrafusal fibres lies alongside the ordinary or extrafusal muscle fibres. Intrafusal fibres are of two types, nuclear bag and nuclear chain. Nuclear bag fibres are larger both in length and diameter. Their ends project well beyond the ends of the capsule. The name nuclear bag is derived from the lack of myofibrillar material and an accumulation of nuclei in the middle of the capsular segment of these fibres. There are several minor differences in the structure and innervation of bag fibres that allows them to be subdivided into two kinds, bag$_1$ and bag$_2$. Nuclear chain fibres are shorter and thinner than bag fibres. The ends of chain fibres attach to the inside of the capsular wall. Chain fibres are not entirely devoid of myofibrillar material in their equatorial region and here the nuclei are arranged as a chain. A typical spindle has two bag fibres (one each of bag$_1$ and bag$_2$) and four chain fibres.

Afferent fibres penetrate the spindle capsule and make spiral terminations around the intrafusal fibres. The large Ia fibre sends branches to the centre of the nucleated region of all intrafusal fibres. Terminals of the primary ending have an annulospiral form. A smaller axon, the group II axon, innervates predominantly nuclear chain fibres and its endings lie to one side of the primary ending. The terminals are less obviously spiral in shape (flower spray appearance) and are referred to as secondary endings. Minor terminals of the group II axon lie on bag fibres, especially the bag$_2$ fibre. The intrafusal fibres receive at least two kinds of motor terminals, plate endings on bag$_1$ fibres and trail endings on both bag$_2$ and chain fibres.

Muscle spindles are stretch receptors, and respond to muscle stretch with trains of impulses that signal both size of the stretch and its rate. Larger, faster stretches generate higher rates of impulse discharge. The motor axons that supply spindles can be separated on functional grounds into two kinds, static and dynamic. Static fusimotor axons, which terminate on intrafusal fibres as trail endings, powerfully increase the level of firing of the spindle at a particular length, when stimulated, but do not very much alter the spindle's length or rate sensitivities. Dynamic fusimotor axons, which terminate as plate endings on bag$_1$ fibres, are, when stimulated, somewhat weaker in their excitatory action on the spindle while the muscle is being held still but greatly increase the spindle's sensitivity to movements. They are therefore dynamic in action. Secondary endings of spindles also respond to muscle stretch but they have a low rate sensitivity and are excited only by static axons.

Some spindles are innervated by branches of motor axons that also supply extrafusal muscle fibres. These skeletofusimotor or β axons are also of two types, static and dynamic. Again, axons with a dynamic action innervate bag$_1$ fibres, while axons with a static action innervate bag$_2$ and chain fibres.

Development of structure

Myogenesis of intrafusal fibres follows essentially the same sequence of events as for extrafusal fibres (Fig. 23.5). The assembly of myoblasts into myotubes begins at the time of arrival in the muscle of the first axons. Motor and sensory axons arrive at more or less the same time. As the axons grow into the muscle they make contact with primary myotubes. It appears to be a matter of chance whether a primary myotube is first contacted by a Ia afferent fibre or an α motoneurone. Group II afferents and γ motor neurones, by virtue of their smaller size, lag behind the bigger axons in their outward growth.

The majority of primary myotubes of developing skeletal muscle contain a myosin heavy-chain isozyme resembling that of adult slow-twitch muscle. A small fraction of primary myotubes stains for the slow-tonic isozyme. Slow-tonic muscle fibres, common in reptiles and birds, are thought to be found in mammals only amongst the intrafusal fibres of muscle spindles and in extraocular muscles. An as yet unanswered question is whether the primary myotubes, containing the slow-tonic isozyme and destined, therefore, to become intrafusal fibres, arise from a distinct lineage, or whether the expression of this isozyme is the result of the influence of the afferent innervation.

Further muscle spindle development is determined by the Ia nerve fibre. The axon induces the aggregation of nuclei in the region of myotube beneath its terminals. Soon, simple motor terminals become visible as well as a thin capsule of perineural epithelium (Fig. 23.5). This first developing intrafusal fibre is destined to become the bag_2 fibre. Once the capsule is present there begins production of additional intrafusal fibres by proliferation and fusion of intracapsular myoblasts. The next intrafusal fibre to differentiate is the bag_1 fibre. Its myotube separates and acquires terminals from the Ia axon. All of the subsequently formed myotubes are shorter and thinner. They will eventually become the nuclear chain fibres. As is the case for extrafusal fibres, the intrafusal fibres develop in close apposition to one another and interdigitate at their surfaces. In the cat, assembly of the full complement of intrafusal fibres is complete at birth. At this time secondary sensory endings as well as γ motoneurones have made functional contacts.

The Ia afferent axon is not only an important influence in the development of the muscle spindle, but in the early stages ensures its very survival. In the rat, if the limb is denervated just before initiation of spindle formation, or a few days later, the muscle develops without any muscle spindles. Denervation in 14-day-old rat pups, however, does not lead to spindle disintegration and the spindles are subsequently re-innervated successfully. Survival of immature spindles appears to depend predominantly on the sensory innervation since de-efferentation during the critical period does not lead to spindle disintegration. Neural activity does not appear to be essential for spindle development

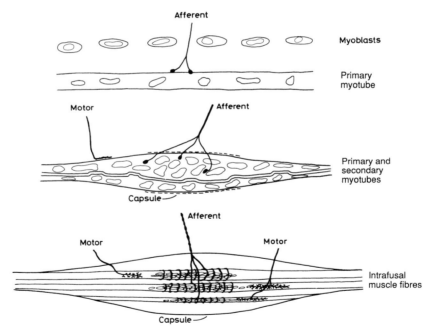

Fig. 23.5 Development of the muscle spindle. Myoblasts, probably already predestined to become intrafusal fibres, fuse to form myotubes. The myotube receives first an afferent innervation and subsequently a motor innervation. Both sensory and motor axons are involved in the further differentiation of the spindle. More intrafusal fibres form inside the rudimentary capsule by the alignment of myoblasts alongside existing myotubes. Adjacent myotubes are coupled by tight junctions. Under the influence of the innervation the intrafusal fibres acquire their special features. Nuclear bag fibres develop first followed by nuclear chain fibres. Gradually the spiral terminals of the afferent axons are formed. The capsular wall thickens and a network of elastic filaments becomes apparent around and between the intrafusal fibres. (Redrawn in part from Barker, D. (1974). In *Handbook of sensory physiology*, Vol. 3, Part 2 (ed. C.C. Hunt), pp. 1–190. Springer, Berlin.)

since, in paralysed developing muscles, spindles are still present, although exhibiting some abnormal morphological features.

Development of function

Spindles in new-born kittens have no resting discharge at normal muscle lengths. They will respond to muscle stretch but only with a brief burst of impulses at low frequency. Gradually, over the first 3 weeks of life, responses begin to assume a more adult-like pattern, although overall firing rates remain low until at 4 weeks all essential features of the adult response can be recognized. In the new-born, not only are the firing rates low but peak rates during ramp stretch show some form of saturation. This is not due simply to saturation of the impulse generator or to a limitation in the rate of impulse propagation by the immature axon since kitten spindles are able to maintain relatively high rates of discharge in response to longitudinal vibration of the muscle. One suggested explanation is that the mechanical properties of intrafusal fibres are not yet fully developed. Perhaps the network of elastic fibres seen in association with adult intrafusal fibres and which provides a splinting action, particularly for bag_2 fibres, is not yet present.

A recent structure–function correlation study has shown that, in new-born kittens, the morphology of the endings on intrafusal fibres of the Ia afferent axon consists of a terminal network of fine branches that run longitudinally along the fibre, with no sign of an annulospiral shape. Spiral endings develop during the first 3 weeks after birth and a typical terminal, encircling the intrafusal fibres, is present at 4 weeks. These changes in terminal morphology cannot be correlated in any simple way with changes in stretch responsiveness of the afferents. A 1-week-old animal has spindles in its soleus muscle that, apart from their lower overall rate of discharge, show stretch responses not unlike those in the adult. Yet at this stage there is no sign of a spiral shape to the afferent terminals. This leads to the conclusion that the annulospiral shape is not essential for stretch receptor-like behaviour and poses the question, what is the significance of the Ia terminal morphology in adult spindles?

Muscle spindles in the cat, as also in rat and sheep, already have a motor innervation *in utero*. A functioning fusimotor innervation can be demonstrated in the new-born kitten; fusimotor effects can be identified as static or dynamic by 7 days of age.

Extrafusal fibres pass through a stage of hyperinnervation in their development (see earlier). Although evidence from the spindle motor innervation remains fragmentary there is a suggestion that intrafusal fibres pass through a similar stage. In a histological study of motor terminal morphology it was found that, in new-born kittens, many of the innervation patterns seen did not occur in the adult, suggesting that some kind of remodelling goes on during development in which certain terminal types are shed. This appears to include the intrafusal terminals of axons innervating extrafusal fibres (β axons).

During early life the intrafusal fibres destined to become the nuclear bag fibres pass through a stage in their development where their length and girth is about the same as that of extrafusal fibres. Adult intrafusal fibres are about one-fifth of the diameter and one-third of the length of extrafusal fibres. The similar dimensions of intrafusal and extrafusal fibres in the kitten provided the structural basis for the claim that intrafusal fibres in kittens, when contracting in response to nerve stimulation, develop tension that can be measured at the muscle tendon. Adult intrafusal fibres, even when a number of them is contracting together, synchronously, do not develop measurable tension. A recent detailed re-examination of this problem in new-born kittens has shown that, here too, intrafusal fibres do not develop measurable tension.

Muscle receptors — tendon organs

Adult tendon organs

Tendon organs are encapsulated nerve endings located at the ends of muscle fibres, at their point of attachment to the tendon. Muscle spindles lying alongside muscle fibres are thought to provide the central nervous system with information about muscle length. Tendon organs are ideally sited to monitor muscle tension.

There are tendon organs in nearly all skeletal muscles. Typically, they are slightly less numerous than muscle spindles. Each tendon organ consists of a loose connective tissue capsule that encloses the branching terminals of a Ib axon. The axon, on penetrating the capsule, branches repeatedly, each of the branches making complex terminals on the thin tendon strands to which one or more muscle fibres are attached (Fig. 23.6). Often the tendon strands and their attached nerve endings are intertwined with one another. Individual strands may divide and refuse. From 4 to 25 muscle fibres insert on tendon strands lying within the capsule. Most of these strands will have nerve endings on them; some may not. Because the muscle fibres of a motor unit occupy a large territory in the muscle, fibres from different motor units are thoroughly intermixed so that adjacent muscle fibres rarely belong to the same motor unit. This means that muscle fibres insert-

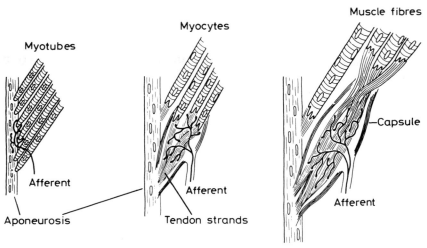

Fig. 23.6 Diagrammatic representation of changes seen in developing rat tendon organs. An afferent fibre approaches the muscle–tendon junction and sends branches to the points of insertion of adjacent myotubes. Gradually, the structure becomes encapsulated, the myocytes receding from the tendon proper but remaining in contact with it through intertwining tendon strands. Finally the capsule encloses only the tendon strands and afferent terminals. The muscle fibre ends have receded to an extracapsular position. (Redrawn from Zelena, J. and Soukup, T. (1977). *Journal of Neurocytology*, **6**, 171–94.)

ing into the tendon organ normally come from different motor units.

Although tendon organs are tension receptors, they will respond to passive muscle stretch provided that tension during the stretch reaches high enough levels. However, tendon organs are much more sensitive to active muscle contraction. A single motor unit may powerfully excite a tendon organ. Typically, tendon organs ignore the contraction of most parts of the muscle but are sensitive to contraction of a small select number of motor units. It is presumed that muscle fibres from these motor units insert on tendon strands that lie within the receptor capsule and on which some of the sensory terminals are located. The conclusion is that each tendon organ is concerned mainly with monitoring the tension in a small group of motor units. Calculations based on the known numbers of tendon organs and motor units in a muscle have led to the conclusion that all motor units are associated with at least one tendon organ, three on average. The picture that emerges is that tendon organs are regional tension sensors and do not accurately monitor global tension levels (i.e. from the entire muscle) so that the earlier notion of tendon organs as overload monitors is no longer tenable.

Development

In the rat, tendon organs form late during development. The first sign of a developing tendon organ is a sensory

axon lying near the muscle–tendon boundary. In this area the ends of myotubes are seen, beginning to form myotendinous junctions (Fig. 23.6). At birth the tendon organ has become elongated as a result of proliferation of Schwann cells and fibroblasts lying between the developing tendon and the ends of the myotubes. Soon the capsule is formed. During this period the ends of the muscle fibres gradually recede from the tendon and are no longer in direct contact with the nerve terminal, but attach to the tendon strands (Fig. 23.6). Subsequently, septal cells appear and begin to divide the tendon organ into a number of compartments, the Ib axon becomes myelinated, and the whole end-organ grows further in length and girth.

Very little is known about the response properties of developing tendon organs. In new-born kittens tendon organs are already selectively sensitive to contraction of particular motor units. However, discharge frequencies are low. As for muscle spindles, tendon organs of developing muscle are unable, initially, to maintain a tonic discharge. In response to rapid muscle stretch they fire a burst of impulses during the length change, demonstrating their rate sensitivity, but they are unable to maintain any activity during the hold-phase of the stretch. Measurements of the relationship between tendon organ discharge and muscle tension show an approximately linear relationship, at least over the lower end of the tension range, but the slope of the relation for tendon organs of new-born kittens is lower than that for the adult receptors.

Further reading

Barker, D. (1974). The morphology of muscle receptors. In *Handbook of sensory physiology*, Vol. 3, Part 2 (ed. C.C. Hunt), pp. 1–190. Springer, Berlin.

Fishman, D.A. (1986). Myofibrillogenesis and the morphogenesis of skeletal muscle. In *Myology* (ed. A.G. Engel and B.Q. Banker), pp. 5–37. McGraw Hill, New York.

Gregory, J.E. and Proske, U. (1988). Development of proprioception, In *Handbook of human growth and developmental biology*, Vol. 1, Part B (ed. E. Meisami and P.S. Timiras), pp. 95–108. CRC Press, Boca Raton, Florida.

Hoh, J.F.Y. (1991). Myogenic regulation of mammalian skeletal muscle fibres. *News in Physiological Sciences*, **6**, 1–6.

Jansen, J.K.S. and Fladby, T. (1990). The perinatal reorganisation of the innervation of skeletal muscle in mammals. *Progress in Neurobiology*, **34**, 39–90.

Konigsberg, I.R. (1986). The embryonic origin of muscle. In *Myology* (ed. A.G. Engel and B.Q. Banker), pp. 39–71. McGraw Hill, New York.

Proske, U. (1981). The Golgi tendon organ. *International Review of Physiology*, **25**, 127–71.

Sanes, J.R. (1987). Cell lineage and the origin of muscle fibre types. *Trends in Neurosciences*, **10**, 219–21.

Schmalbruch, H. (1986). Muscle regeneration: fetal myogenesis in a new setting. *Bibliotheca anatomica* (Basel) **29**, 126–52.

Zelena, J. and Soukup, T. (1977). The development of Golgi tendon organs. *Journal of Neurocytology*, **6**, 171–94.

24. Ontogeny of behavioural states in the fetus
Bryan S.Richardson

Studies in human adults more than 30 years ago first demonstrated that sleep occurs in two distinct phases, namely rapid eye movement sleep (REMS) and slow-wave sleep (SWS). Since that time much study has been directed at the nature of these so-called 'sleep states' including their physiological correlates, control mechanisms, possible function, and their use as an expression of neurodysfunction with apparent maldevelopment. Sleep states have been shown to exist in apparently healthy new-born infants, although they are not evident in preterm infants born much before 36 weeks gestation (term ~ 40 weeks). While sleep states have also been well described in most mammals studied to date, REM sleep appears rudimentary in birds and non-existent in reptiles with SWS sleep predominating, thus suggesting an evolutionary need for the REM state. More recently, the use of chronic catheterization techniques in fetal animals, primarily sheep, and the use of high-resolution ultrasound equipment for the study of the human fetus, have firmly established the existence of activity or behavioural states *in utero* that have similarities to postnatal sleep states. Moreover, a developmental process has become apparent whereby the REM state predominates in early life and may have functional importance for growth and developmental over the perinatal period.

Sleep states

Sleep states are recognized by temporal patterns in various electrophysiological and behavioural parameters that are relatively stable and repeat themselves in time. Sleep state recognition and scoring is largely based on three criteria, namely: (1) neocortical EEG pattern; (2) eye movements (representing phasic motor activity); and (3) chin or neck electromyographic (EMG) recordings (representing tonic postural motor activity) and, usually, characteristic postures. As noted, sleep is classified into two distinct types:

1. SWS (slow wave sleep) — high-voltage, slow-wave electroencephalogram (EEG), eye movements absent, postural muscle tone present;
2. REMS (rapid eye movement sleep) — low-voltage, fast activity EEG, rapid eye movements, postural muscle atonia.

SWS, also called NREMS (non-REMS) is usually further divided into stages I, II, III, and IV in humans, or light and deep SWS in laboratory animals, depending on the presence of specific EEG patterns that are unique to SWS. At the transition to REMS, EEG patterns are similar to those at waking, but their coupling with postural muscle atonia is used to recognize REMS. Rapid eye movement sleep, also called paradoxical sleep, may be further divided into tonic REMS and phasic REMS, the latter identified by a relative profusion of rapid eye movements, twitches of other muscles, and phasic electrical events recorded in the nervous system. In addition to the parameters noted above, the onset of sleep and changes in state are characterized by discrete changes in many other physiological systems.

Behavioural states

Human neonate

The importance of the 'sleep state' concept to the assessment of neurological function has led to extensive studies in young infants. Numerous classification scales have been used, depending on the particular parameters chosen, to characterize the different states. One set of state definitions gaining wide acceptance is that introduced by Prechtl based on easily observed parameters which are continuously evident. The use of terms with a physiological connotation is avoided and, instead, an arbitrary classification scale is used.

State 1 — eyes closed, regular respiration, no movements;
State 2 — eyes closed, irregular respiration, no gross movements;
State 3 — eyes open, no gross movements;
State 4 — eyes open, gross movements, no crying;
State 5 — eyes open or closed, crying.

Although physiological parameters that are not directly observable were not included in the definitions of states, in many instances recognizable patterns of these variables occur as consistent concomitants of the states. Thus state 1 corresponds to SWS or NREMS, with absent eye movements and a high-voltage EEG pattern. State 2 is REMS or paradoxical sleep, with rapid eye movements and a low-voltage EEG. States 3 to 5 are awake states, with increasing degrees of arousal and body movements. Studies concerned with infants may also use the terms 'quiet sleep' and 'active sleep', reflecting the behavioural difference between the state, with SWS or NREMS characterized by quiescence and REMS by phasic muscular movements, particularly in neonates. Due to the reliance on state-related differences in behaviour in the neonate and the usual inclusion of periods of wakefulness, the concept of 'sleep states' gives rise to that of 'behavioural states'.

Fetal animals

Studies in unanaesthetized fetal animals near term, with chronic placement of electrodes and pressure sensors, have identified the equivalents of behavioural states with similarities to those described after birth. Most of these studies have involved the use of the chronic fetal sheep preparation, now the primary animal model for the study of fetal physiology (Fig. 24.1). Behavioural state classification is based for the most part upon electrophysiological recordings (electrocorticogram, electrooculogram, and nuchal muscle EMG) rather than behavioural observations. The following states can be recognized:

1. NREMS — high-voltage, slow-wave electrocortical activity, eye movements absent, nuchal muscle tone present or absent;
2. REMS — low-voltage electrocortical activity, rapid eye movements, nuchal muscle atonia;

3. Awake — low-voltage electrocortical activity, eye movements present or absent, nuchal muscle tone present.

The NREM and REMS behavioural states described for the ovine fetus are directly comparable to these same states seen after birth and thus are seen to represent *in utero* sleep state activity. The third behavioural state (wakefulness) can be characterized using criteria used to identify postnatal wakefulness. Direct observations of exteriorized sheep fetuses, behavioural responses to external stimuli, and the different effect of evoked potentials in skeletal muscles provide further support for the existence of a fetal awake state. However, although the behavioural state criteria for such a state are directly comparable to those used after birth, whether or not the fetus attains a level of behavioural arousal such as to be deemed awake remains open to question.

As in the adult and neonate, additional physiological parameters are recorded as consistent concomitants to these fetal behavioural states, although species differences may be evident. In the ovine fetus, rapid irregular breathing movements occur only during the REMS state. They are not continuous, however, as approximately one-third of REM time is not associated with breathing movements identified in recordings of tracheal pressure. Similarly, fetal heart rate is usually higher during the NREMS behavioural state. In fetal sheep episodes of repeated swallowing (resembling postnatal feeding episodes) and bladder contractions (indicative of bladder emptying) are also

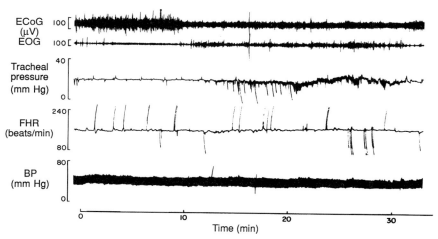

Fig. 24.1 A chart recording of behavioural parameters from a chronically catheterized ovine fetus at 132 days of gestation. Electroocular activity (EOG) and breathing movements (shown as fluctuations in tracheal pressure) normally occur during episodes of low-voltage electrocortical activity (ECoG) and are absent during episodes of high-voltage ECoG. FHR, fetal heart rate; BP, blood pressure.

influenced by behavioural state. Near term, both of these fetal activities mainly occur during the low-voltage electrocortical activity state in association with eye movements, breathing movements, and increased nuchal muscle activity, suggesting a state of heightened activity resembling wakefulness. Detection of these concomitant physiological parameters, as well as individual state criteria, provide evidence for the existence of behavioural states in the fetal rhesus monkey and fetal guinea-pig.

Human fetus

Body movement and heart rate patterns in the human fetus near term reveal a cyclicity which, by reference to postnatal patterns, appears to have a state-related basis (Fig. 24.2). Initial studies used fetal body movements and heart rate patterns as scoring criteria for behavioural state assessment. With improvement in real-time ultrasound imaging the scoring of fetal eye movements also became possible. One set of state definitions widely used is that introduced by Nijhuis based on parameters observed with ultrasound and the simultaneous recording of fetal heart rate patterns. Following the strategy used for the new-born infant by Prechtl, four distinct behavioural states were recognized from the stability of the association of parameters over prolonged periods, and by the simultaneous changes

in these parameters at state transitions. In order to avoid an interpretative terminology, an arbitrary classification scale was used, with the letter F (for fetal) added to indicate that the criteria used to define the behavioural states of the fetus differ from those of the neonate.

State 1F — quiescence (occasional brief, gross body movements), eye movements absent, fetal heart rate stable with a narrow oscillation bandwidth;

State 2F — frequent gross body movements, eye movements continually present, fetal heart rate with a wider oscillation bandwidth and frequent accelerations during body movements;

State 3F — no gross body movements, eye movements continually present, fetal heart rate stable but with wider oscillation bandwidth than for state 1F;

State 4F — frequent and vigorous gross body movements, eye movements continually present (when observable), fetal heart rate unstable with large and long-lasting accelerations.

When eye movements, body movements, and heart rate patterns are compared, states 1F and 2F are directly comparable to state 1 (quiet sleep or NREM sleep) and 2 (active sleep or REM sleep), respectively, in the new-born. However, comparisons between fetal states 3 and 4 and postnatal awake states are less clear, as state 3F is seldom

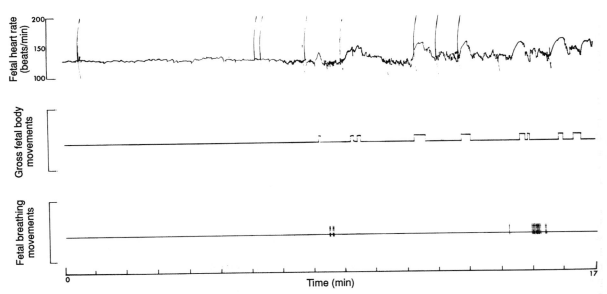

Fig. 24.2 A chart recording of behavioural parameters from a healthy human fetus near term. For the first 8 min there are no gross body movements and the heart rate pattern demonstrates no accelerations. From 8 to 17 min a normal episode of gross body movements accompanied by accelerations in the heart rate occurs. (Taken with permission from Richardson, B.S., Campbell, K., Carmichael, L., and Patrick, J. (1981). *American Journal of Obstetrics and Gynecology*, **139**, 344–52).

seen and the frequency of body movements for state 4F is somewhat less than that seen in state 4 after birth. While the behavioural response of the human fetus to external vibroacoustic stimulation supports the existence of a fetal awake state, it is also possible that states 2F and 4F represent different levels of activity occurring in the same state rather than truly different states.

Additional behavioural parameters are evident as state-related concomitants, although they are not continuously present and thus unsuitable for use as state-defining criteria. Fetal breathing movements are much more regular during state 1F than during 2F, while their incidence is increased in 2F. Fetal micturition detected by ultrasound recognition of bladder emptying, while inhibited during episodes of low heart rate variation suggestive of state 1F, appears facilitated by a change to an episode of high heart rate variation with rapid eye movements and fetal breathing movements suggestive of 2F or possibly 4F. Conversely, regular mouthing movements, as in the neonate, only occur during periods of quiescence and can be considered a concomitant for state 1F. The recording of fetal EEG during labour, although impractical as a routine scoring criterion, shows patterns that resemble those in the neonate, further supporting the existence of behavioural states in the human fetus.

Ontogeny of behavioural states

The study of the temporal relationship between behavioural and physiological parameters has firmly established, not only the existence of behavioural states, but also developmental changes beginning at different times before birth and continuing through postnatal life.

Human neonate

The human neonate born at or near term exhibits well developed behavioural states from birth. In the immediate new-born period, approximately 70 per cent of the time is spent sleeping with two-thirds of total sleep as REM sleep and one-third as NREM sleep. This high ratio of REM sleep to NREM sleep diminishes progressively with maturation especially over the first year of life and largely involves the substitution of wakefulness for REM sleep with little actual change in the amount of NREM sleep. By 1 year of age the percentage of NREM sleep is greater than the percentage of REM sleep, a reversal of the proportional relationship at birth, although the REM state still occupies approximately 40 per cent of night-time sleep. In old age, however, REM sleep diminishes to less than 20 per cent of total daily sleep.

Observations of healthy preterm infants have demonstrated that organized behavioural state cycles appear at about 36 weeks of postconceptional age (gestational age at birth plus postnatal age). Before this, behavioural states are less organized as cycles of rest and activity, regular and irregular respiration, and epochs with and without eye movements alternate more or less independently. Although there are periods of congruency between state parameters that fit the state definitions, changes in individual parameters do not occur simultaneously. From about 30–32 weeks onward, however, a certain bias in the relationship between state parameters appears to exist, with the earliest association being that between the EEG pattern and regular respiration. The eye movement cycles become synchronized later while somatic movements appear to be the last parameter to come into phase with the other variables. The developmental course of sleep–wake patterns in full-term and 'healthy' premature infants is similar when compared on a conceptional age basis, suggesting a biologically, rather than environmentally, mediated course.

Fetal animals

The ovine fetus exhibits well-differentiated electrocorticographic (ECoG) patterns from approximately 120 days of gestation onward with a temporal relationship to episodic muscle and breathing activity indicative of behavioural states (term ~ 145 days). As in the human neonate, there is initially a high proportion of time in the REM state (> 50 per cent), with approximately 40 per cent of time in NREMS and only brief periods of apparent wakefulness. Quantitative changes in the distribution of these behavioural states then continue through gestation. There is a progressive decrease in the incidence of the REM state to approximately 40 per cent of total recording time by term, mainly due to an increase in periods of wakefulness, although NREMS is also increased. Postnatally there is a marked fall-off in the incidence of REM sleep to less than 10 per cent, again primarily due to an increase in time spent awake.

Prior to the establishment of well defined behavioural states, the ovine fetus also has a cycling of behavioural parameters as seen in the human neonate. Episodic breathing movements and ocular activity appear before electrocortical differentiation and are initially associated with increased, rather than decreased, activity in the neck muscles. The subsequent phasic relationship to electrocortical activity occurs over a few days once initiated and involves the appearance of high-voltage ECoG activity superimposed on existing low-voltage ECoG activity. Maturational changes in the waveform characteristics of the electrocorticogram then continue throughout the rest of gestation primarily involving the low-voltage, fast-activity periods. There is, however, little change in the power spectrum of the high-voltage ECoG activity once it becomes established.

Observations on the maturation of electrocortical patterns *in utero* or of behavioural state activity after birth provide insights into the development of sleep–wake states in other species. The organization of electrocortical activity in the guinea-pig fetus is similar to that in the ovine fetus, with the onset of cyclic variation occurring prenatally and with progressive maturation in electrocortical waveforms thereafter. The new-born guinea-pig, lamb and monkey all demonstrate the ECoG aspects of NREMS, REMS, and wakefulness at the time of birth. Conversely, the adult-like ECoG aspects of sleep and wakefulness are not fully developed until several days postnatally in the rat, rabbit, and cat. Although comparison between the immature of several species of animals indicates considerable differences in the rate of development of sleep–wake patterns, all demonstrate a similarly high proportion of the REM state during the establishment of well-defined behavioural states.

Human fetus

The human fetus first demonstrates clearly defined behavioural states from approximately 36 to 38 weeks of gestation with stable alignment of behavioural parameters and synchrony of changes at state transitions. At this time state 2F (or REMS) predominates, occurring approximately 40 per cent of the time (see Table 24.1). State 1F (or NREMS) occurs approximately 25 per cent of the time while state 4F (or wakefulness) occurs but briefly. Behavioural states cannot be identified approximately 20 per cent of the time. With advancing gestation the incidence of NREMS and wakefulness increases, that of the REMS state remains little changed, while the percentage of time with no identifiable states decreases. This time course of behavioural state development *in utero* is qualitatively similar to that observed in healthy premature neonates. Moreover, although the NREM and REM states are observed more frequently in the fetus at term (~ 80 per cent) than in the neonate during early infancy (~ 70 per cent), the ratio of these two types of sleep remains similar.

Prior to 36 weeks of gestation, periods of 'activity' and 'quiescence' are evident and, although 'coincidence' of behavioural parameters may mimic states, they lack stability in temporal relationship and synchrony of changes at transition of states (Table 24.1). This linkage in behavioural parameters can be demonstrated at as early as 25 to 30 weeks of gestation with a progressive increase in coincidence within state 1F from approximately 5 per cent

Table 24.1 Coincidence of human fetal states 1F and 2F as percentages of total recording time

	Coincidence of states 1F and 2F (%) at gestational age (weeks)			
	25–30[*]	32[†]	36[†]	40[†]
1F	6	15	20	36
2F	17	46	38	42
No coincidence	77	29	22	9

[*]Data (mean values) from Drogtrop, A.P., Ubels, R., and Nijhuis, J.G. (1990). *Early Human Development*, **23**, 67–73.
[†]Data (median values) from Nijhuis, J.G., Prechtl, H.F.R., Martin, C.B., and Bots, R.S.G.M. (1982). *Early Human Development*, **6**, 177–95.

to 20 per cent of recording time by 36 weeks, and in state 2F from approximately 15 per cent to approximately 40 per cent of the recording time by 36 weeks of gestation. There is a corresponding decrease in the amount of time with no 'coincidence' of behavioural parameters from about 80 to 20 per cent. The duration of the so-called 'activity-quiescent' cycle also shows a maturational change, with a progressive increase from about 20 min at 28 weeks gestation to about 60 min by term; this duration at term is similar in length to the REMS–NREMS cycle of the new-born infant.

Clinical implications

Fetal health

As behavioural state activity characterizes the healthy fetus, disturbances in the ontogeny or organization of these states may be a potential indicator for central nervous system dysfunction and/or of adaptive change to an adverse fetal environment. The appearance of well-defined behavioural states is delayed in the intrauterine growth-restricted (IUGR) human fetus and the 4F or active (awake) state is less evident that in the healthy fetus. Assessment of behavioural states may also indicate delayed development of brain function in the fetuses of diabetic pregnancies. However, in the ovine fetus, the incidence of the low-voltage ECoG state (or REMS state) is decreased during short-term hypoxaemia reflecting, in part, a protective change to a metabolic state with lower oxidative needs. The behavioural state alterations of the IUGR human fetus may similarly reflect a long-term metabolic compensatory response to mild hypoxaemia rather than neurodevelopmental delay.

The characteristic behavioural state activity of the healthy fetus also provides the basis for the biophysical assessment of fetal well-being, whereby body movements, breathing movements, and heart rate pattern analyses are used in the monitoring of high-risk pregnancies. In the ovine fetus, body and breathing movements have been shown to decrease with acute hypoxaemia. These biophysical alterations may be protective, as decreases in energy expenditure and thus oxygen requirements would occur, depending on the extent of the decrease in neuromuscular activity. In the human fetus the assessment of behavioural parameters is used in the biophysical profile of the fetus, with the monitoring of five fetal variables including body movements, breathing movements, and heart rate pattern. The predictive value for a normal perinatal outcome is similar among these dynamic fetal variables and is about 95 per cent accurate. Conversely, decreased or absent fetal body movements, prolonged fetal apnoea, and fetal heart rate tracings that fail to show accelerations in association with fetal body movements help identify the compromised fetus in need of delivery. However, the inherent cyclical nature of these behavioural parameters and their developmental changes within the behavioural state framework must be considered when establishing biological norms against which pathological change is measured.

Fetal growth and development

The abundance of *in utero* behavioural activity in the healthy fetus and the poor outcome associated with its restriction in animal experimentation and in the human fetus support an important role for behavioural activity in fetal growth and development. Fetal breathing movements appear necessary for normal development of the lung as fetal sheep and rabbits in which these movements were abolished by high spinal transection demonstrate diminished lung growth. The neural activity of fetal motility and its motor effects may contribute to the development of muscles, joints, and even the fine structure of the central nervous system itself.

The high proportion of REM sleep or behavioural-like activity during the perinatal period would also indicate that the immature being has a high requirement for such states and that the REM state itself might play an additional role in fetal/neonatal growth and development. This led Roffwarg to speculate many years ago that 'the REMS mechanism serves as an endogenous source of stimulation, furnishing great quantities of functional excitation to higher centres. Such stimulation would be particularly crucial during the periods *in utero* and shortly after birth, before appreciable exogenous stimulation is available to the central nervous system.' The concept of a functional role for the REM state mechanism in brain development is further supported by the comparison of the anatomical and electrophysiological maturation of the brain across species. Whereas the sheep and guinea-pig may be classified as prenatal brain developers from a neuro-anatomical standpoint, the rat is a postnatal brain developer, with the peak in brain growth velocity occurring after birth. Similarly, the sheep and guinea-pig, as prenatal brain developers, have relatively mature electrocortical patterns at birth, which are qualitatively similar to their adult counterparts, while the rat, as a postnatal brain developer, has a poorly differentiated ECoG. The anatomical development of the brain thus appears to correlate with its electrophysiological development as indicated by behavioural state maturation.

Cerebral metabolism has been studied in the ovine fetus in relation to behavioural state development; a significant increase has been measured during the low-voltage ECoG state or REM state when compared to the high-voltage ECoG state or NREM state (Fig. 24.3). Blood-flow velocity waveform studies in the human fetus using Doppler ultrasound suggest a similar relationship given the close correlation between cerebral blood flow and cerebral metabolism. The increased metabolic rate of the brain during the REM state may reflect increased neuronal activity or synthetic processes and supports an important role

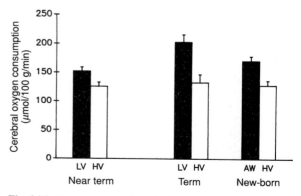

Fig. 24.3 Measurements of cerebral oxygen consumption in the ovine fetus near term (~ 130 days gestation), at term (~ 140 days gestation), and at 24 h after birth. Cerebral oxidative metabolism shows no significant change with age. A coupling to behavioural state is evident as oxygen consumption significantly increases during both the perinatal low-voltage (LV) ECoG state and the awake (AW) state at 24 h. HV, High voltage electrocortical state.

for the increased incidence of this state during the accelerated growth and development of the brain during the perinatal period. It is possible that alterations in behavioural state activity and associated changes in cerebral metabolism during this critical period may provide a mechanism whereby pathological factors give rise to abnormal brain growth and development.

Further reading

Martin, C.B. (1981). Behavioural states in the human fetus. *Journal of Reproductive Medicine*, **26**, 425–32.

McGinty, D.J. and Drucker-Colin, R.R. (1982). Sleep mechanisms: biology and control of REM sleep. *International Review of Neurobiology*, **23**, 391–436.

Nijhuis, J.G. (1986). Behavioural states: concomitants, clinical implications and the assessment of the condition of the nervous system. *European Journal of Obstetrics and Gynaecology, Reproductive Biology*, **21**, 301–8.

Nijhuis, J.G. Prechtl, H.F.R., Martin, C.B., and Bots, R.S.G.M. (1982). Are there behavioural states in the human fetus? *Early Human Development*, **6**, 177–95.

Pillai, M. and James, D. (1990). Are the behavioural states of the newborn comparable to those of the fetus? *Early Human Development*, **22**, 39–49.

Prechtl, H.F.R. (1974). The behavioural states of the newborn infant (a review). *Brain Research*, **76**, 185–212.

Prechtl, H.F.R. (1985). Ultrasound studies of human fetal behaviour. *Early Human Development*, **12**, 91–8.

Richardson, B.S. (1989). Fetal adaptive responses to asphyxia. *Clinics in Perinatology*, **16**, 595–611.

Richardson, B.S. (1989). Fetal breathing movements. In *Fetal monitoring* (ed. J. A.D. Spencer), pp. 64–72. Castle House Publication Ltd., Tunbridge Wells, Kent.

Richardson, B.S. (1992). The effect of behavioural state on fetal metabolism and blood flow circulation. *Seminars in Perinatology*, **16**, 227–33.

Roffwarg, H.P., Muzio, J.N., and Dement, W.C. (1966). Ontogenetic development of the human sleep dream cycle. *Science*, **152**, 604–19.

Szeto, H.H. (1992). Behavioural states and their ontogeny: animal studies. *Seminars in Perinatology*, **16**, 211–16.

25. Ontogenesis of biological rhythms
I. Caroline McMillen

The fetus spends gestation sequestered in the darkness of the womb where it is not directly exposed to the range of environmental time cues that normally signal the sequence of day and night to the adult. It is known, however, that during late gestation there are a range of physiological parameters in the fetus, such as gross body movements, breathing activity, heart rate, and certain hormone concentrations, that show a predictable variation within each 24-h period. This chapter will discuss the possible mechanisms that underlie the ontogenesis and co-ordination of such daily fetal rhythms. Such daily rhythms may simply reflect the physiological and neuroendocrine responses of the fetus to certain stimuli (e.g. maternal feeding) that occur at the same time each day. A further possibility is that rhythms in fetal plasma hormone concentrations are entirely derived from the mother as a consequence of placental transfer. One final possibility is that, in some cases, fetal behavioural and hormonal rhythms represent the outcome of an interaction between the circadian-rhythm-generating systems of the mother and fetus. The exact 24-h periodicity of prenatal rhythms would imply that the activity of the fetal circadian pacemaker is entrained to a 24-h period by maternal behavioural or hormonal signals that act as time cues. The focus of this chapter is the role of the maternal and fetal circadian systems in the generation of the daily fetal rhythms in melatonin, prolactin, and cortisol concentrations and in behavioural patterns including electrocortical state and the incidence of breathing movements.

Rhythms, pacemakers, and entrainment

In the adult mammal, many body functions, including temperature, sleep and wakefulness, electrolyte excretion, and the plasma concentrations of hormones such as melatonin and cortisol, show a regular daily rhythm. These daily rhythms are defined as circadian (circa — about; dies — day) rhythms because they fulfil certain criteria.

The first of these criteria is that a circadian rhythm is endogenously generated by an internal 'body clock' or circadian pacemaker. In the absence of any external information about time of day, endogenously generated circadian rhythms free-run, that is, they repeat with a periodicity of around, but not exactly, 24 h. External time cues or 'zeitgeber' (time-givers) are required to entrain a free running circadian rhythm to an exact 24-h schedule. The most important zeitgeber is the 24-h cycle of light and dark; others include external variations in temperature, social cues, and food availability.

Extensive evidence suggests that the major circadian pacemakers in the mammalian brain are the bilateral suprachiasmatic nuclei (SCN) in the anterior hypothalamus. Lesions in these nuclei result in the loss of circadian rhythmicity and, after postnatal ablation of the rat SCN, behavioural and hormonal circadian rhythms fail to develop. The metabolic and electrophysiological activities of neurones in the SCN are high during the day and low at night. Entrainment of SCN activity by the light and dark cycle is thought to be mediated by a neural pathway that projects from the retina to the SCN — the retinohypothalamic tract. Neurones in the SCN project to and receive projections from the hypothalamus and the output signals from the SCN are presumably processed within the hypothalamus and ultimately drive a wide range of hormonal and behavioural circadian rhythms.

Development of circadian pacemaker function

A series of studies in the rat have shown that the development of circadian activity in the hypothalamic SCN occurs before birth. Initial experiments used the timing of overt rhythms during the postnatal period to infer the state of the circadian pacemaker during fetal life. Subsequently, labelling with $[^{14}C]2$ deoxyglucose and autoradiography were used to demonstrate that there was a circadian variation in metabolic activity in the fetal rat SCN from as early as day 19 of gestation. It has also been shown that there is a circadian rhythm in vasopressin mRNA levels in the SCN at around day 21 of gestation. A major finding of these studies was that the metabolic activity rhythm in the fetal SCN was synchronous with the circadian time of the blinded dam and was not affected by changes in the external lighting. Furthermore, studies in rats and hamsters have shown that the maternal SCN are essential for entrainment of the fetal SCN, because destruction of the SCN in the pregnant animals disrupts the circadian timing of the fetal hypothalamic clock and the expression of postnatal rhythms. The mechanism of maternal-fetal entrainment is not known. In the rat, maternal pinealectomy or the separate extirpation of the maternal pituitary, thyroid, adrenals, or gonads did not abolish the maternal synchronization of fetal rhythmicity.

In the rat and mouse, the dam continues to influence the circadian rhythm generating system in the pup during postnatal life, until the innervation of the pup SCN by the retinohypothalamic tract is complete and direct photic entrainment can occur. In a precocial species, the spiny mouse, in which the retinohypothalamic innervation of the SCN is complete on the day of birth, there is prenatal, but no postnatal, influence of the mother on the circadian rhythmicity of the pup.

While it is clear that the SCN are functioning before birth in species such as the rat and spiny mouse, the nature and role of the output signals from the circadian pacemaker in these species are less clear. In the rat the circadian activity of the fetal SCN is expressed at a time in gestation when these nuclei are still immature and lack synapses. Therefore the prenatal significance of maternal entrainment of the metabolic activity of fetal SCN neurones in this species is not clear.

In contrast to the rat, in which hormonal and behavioural rhythms develop postnatally, there is evidence in the more precocial species, such as man and sheep, that daily rhythms are present in several fetal behavioural and hormonal activities. In these latter species, however, there are only limited data available on the development of fetal SCN neurones. One study reported that there is a prominent circadian rhythm of vasopressin in the cerebrospinal fluid of the late-gestation fetal sheep and also found that exposure of the pregnant ewe to constant light disrupted this rhythm. While this evidence suggests that SCN neurones do oscillate in a circadian pattern before birth in this precocial species, more extensive studies on the pattern of SCN synaptogenesis and activity in the human and sheep are required. The rest of this chapter will explore the evidence that the maternal and fetal circadian timing systems may have a role in the control of specific daily fetal rhythms, in particular, those in the fetal plasma concentrations of cortisol, melatonin, and prolactin and in the incidence of fetal breathing movements.

Cortisol

Maternal plasma oestriol concentrations provide a useful measure of fetal adreno-placental function in the human.

In late pregnancy, maternal plasma concentrations of cortisol and oestriol vary inversely during a 24-h period. This relationship is consistent with the hypothesis that the circadian rhythm in maternally derived glucocorticoids acts via negative feedback at the fetal pituitary to generate an inverse rhythm in fetal adrenal steroidogenesis and oestriol production. This is supported by the finding that exogenous glucocorticoids suppress the maternal cortisol rhythm and abolish the diurnal rhythm in maternal oestriol in late pregnancy. Thus it appears that there is no evidence for a rhythm in fetal adrenal function that is endogenously generated by the fetal SCN in the human.

The emergence of the cortisol rhythm in the human new-born has been investigated in a series of studies, and the earliest age at which an entrained 24-h cortisol rhythm emerges is at around 3 months of age. It is not clear whether the postnatal delay in the emergence of this hormonal rhythm is a consequence of immaturity of function of the retinohypothalamic tract, or of an immaturity of the circadian pacemaker itself. The coincidence of the timing of the emergence of the circadian rhythms in melatonin and cortisol in the human infant suggests that the maturation of a common central control pathway is the limiting factor in the expression of hormonal rhythmicity.

Melatonin

In the adult, the presence of a daily melatonin rhythm is considered to be a marker of a functional circadian rhythm generating system. Secretion of melatonin from the pineal gland is under the direct control of the retinohypothalamic tract and the SCN. From the SCN, synaptic connections are made within the hypothalamus and descending projections from the hypothalamus pass via the intermediolateral column of the spinal cord to the superior cervical ganglia. Postganglionic noradrenergic fibres from the SCG pass through the tentorium cerebelli and supply the pineal gland via the paired nervi conarii. The synthesis and secretion of melatonin from pinealocytes is controlled by these noradrenergic fibres that are stimulated by darkness and inhibited by light.

In the human, there is no information about the presence of a circadian melatonin rhythm *in utero* or about the capacity of the human fetal pineal gland to synthesize and secrete melatonin. Our current understanding of the source and role of the melatonin rhythm in prenatal life is based on experiments in the pregnant ewe and sheep fetus in late gestation.

It has been reported that there is a clear daily rhythm in the plasma concentrations of melatonin in both the ewe and fetus from as early as 114 days of gestation (term, 147 $\pm$ 3 days gestation) and that melatonin crosses the sheep placenta in the maternal–fetal direction. It has also been shown that, irrespective of the daily time of onset of darkness, maternal and fetal plasma concentrations of melatonin are highest during the hours of darkness. As pinealectomy of the pregnant ewe completely abolishes the daily rhythm in maternal and fetal melatonin concentrations it appears that increased secretion of melatonin by the maternal pineal gland during the hours of darkness generates the increase in melatonin concentrations in the fetal circulation. Therefore, the plasma melatonin rhythm measured in the late-gestation fetal sheep is not an endogenously generated fetal rhythm but rather a hormonal rhythm imposed on the fetus by the ewe. The emergence of an autonomous circadian melatonin rhythm does not occur until after birth in both the sheep and human. The postnatal delay in the emergence of this hormonal rhythm may be a consequence of a limited synthetic capacity of the new-born pineal gland or it could be a result of a delay in the entrainment of the circadian pacemaker to the 24-h light–dark cycle.

Role of the prenatal melatonin rhythm

In the non-pregnant ewe, the duration of the nocturnal surge of plasma melatonin is virtually identical to the period of darkness and mediates the effect of changing day length on the reproductive system of this seasonal breeder. There is evidence that the maternal melatonin rhythm provides the fetal sheep with information about the length of the photoperiod. It has been shown that fetal and maternal plasma prolactin concentrations increase at times of long day length and decrease at times of short day length. The changes in fetal prolactin secretion with photoperiod are mediated by the varying duration of the nocturnal increase in melatonin concentrations. Whilst this phenomenon is of importance in a seasonal breeder such as the sheep, it is only of limited interest in the human. Of greater relevance to the human is the possible role of maternal melatonin acting as an 'internal zeitgeber' to influence the activity of the fetal circadian pacemaker. Exogenous melatonin entrains free-running activity rhythms in the adult rat and also inhibits the metabolic activity of neurones in the adult rat SCN. Although it has been found that removal of the maternal pineal gland in the rat does not abolish the maternal entrainment of fetal SCN activity, melatonin binding sites have been identified in the human fetal SCN. It is therefore possible that melatonin may entrain daily fetal rhythms via an action at the fetal SCN. In this context it is interesting that there is an apparent relationship in fetal sheep between the daily rhythm in the incidence of breathing movements in the sheep and the prevailing light–dark and melatonin cycles.

Fetal breathing movements

The ovine and human fetus have periods of fetal breathing activity that are episodic in nature in late gestation and that, in the sheep, are associated with episodes of low-voltage electrocortical activity. Daily rhythms in the incidence of fetal breathing movements have been reported in both the human and sheep. It is important to note that in the human the incidence of fetal breathing movements is increased during the second and third hour following maternal meals and that this increase follows the normal increase in maternal blood glucose concentrations that occurs in the postprandial period. It therefore appears that maternal carbohydrate intake is the primary factor responsible for increased fetal breathing activity following meals. In addition to the increase in fetal breathing activity that occurs after meals, there is a prolonged increase in the incidence of fetal breathing activity in the late evening and early hours of the morning when the maternal blood glucose concentrations are stable. The time of the daily peak in the incidence of fetal breathing activity shifts from the late evening at 24–26 weeks gestation to 02.00–05.00 h at 26–28 weeks and to 04.00–07.00 h at term. The mechanisms underlying these changes in the 24-h profile of fetal breathing activity are unknown and it is not clear whether this daily rhythm is a circadian rhythm generated by an endogenous fetal circadian pacemaker.

In the fetal sheep, there is an increase in the hourly incidence of breathing movements at the time of daily feeding and also in the early evening hours preceding and extending into the hours of darkness. In an experiment in which ewes were exposed to a 12-h period of darkness from either 11 a.m. or 7 p.m. it was found that the daily peak in the incidence of fetal breathing movements occurred at the onset of darkness independently of the time of onset of darkness. This suggests that the fetus is receiving information about the external light and dark cycle and can alter its pattern of breathing activity accordingly. There are also differences in the 24-h profiles of fetal breathing activity between fetuses of intact and pinealectomized ewes maintained under a 12-h light:12-h dark regime with dark onset at 19.00 h. In pinealectomized ewes, in contrast to intact ewes, there is no increase in the incidence of fetal breathing movements between 17.00 and 21.00 h. This suggests that the high incidence of fetal breathing activity that normally occurs at the onset of darkness in intact ewes is dependent on the cyclical changes in maternal and fetal melatonin concentrations. The mechanism by which melatonin modulates the pattern of fetal breathing activity is unknown. One possibility is that melatonin acts on the fetal brain via its metabolite, N-acetyl 5-formylkynurenamine. This metabolite has been shown to inhibit prostaglandin synthesis and it is known that administration of prostaglandin synthesis inhibitors can increase the incidence of fetal breathing movements. It is also possible that the rhythmic melatonin signal entrains the pattern of activity of the neurones in the fetal SCN. This area awaits the outcome of experiments in which the fetal suprachiasmatic nuclei are lesioned and the 24-h profiles of breathing activity recorded.

Prolactin

There is a significant diurnal rhythm in plasma prolactin concentrations in both the pregnant ewe and fetal sheep in late gestation. As in the non-pregnant ewe and ram, there is an increase in maternal prolactin concentrations at the time of onset of darkness in experiments in which darkness begins at around 7 p.m. Fetal plasma concentrations of prolactin increase in parallel between 7 and 11 p.m. which is interesting as prolactin does not cross the placenta. After pinealectomy of the pregnant ewe the daily fetal and maternal prolactin rhythms persist, suggesting that the ontogenesis and presence of this hormonal rhythm in the sheep is relatively independent of the pineal gland. The relative roles of the light–dark cycle and melatonin in the control of the diurnal rhythm of prolactin in the ewe and her fetus need to be explored further.

Whereas the 24-h prolactin rhythm in the sheep is related to the light–dark cycle, in the human plasma prolactin concentrations show a marked diurnal variation primarily related to the pattern of sleep and wakefulness. Prolactin concentrations in the adult human are highest during the later part of sleep and decrease on waking. The 24-h prolactin rhythm is present in women during the last trimester of pregnancy and it has been reported that there is a diurnal variation in prolactin concentrations in umbilical venous blood during delivery in term infants. The source of this variation is unclear and the development of the neural control of the daily prolactin rhythm requires further investigation.

Summary

While there are daily rhythms present in a range of fetal hormonal and behavioural parameters during late gestation, our understanding of the mechanisms which underly the generation of these daily patterns is incomplete. An evening peak in gross body movements, breathing activity, and heart rate variability has been described in the human fetus but the source and role of these fetal rhythms are unknown. In contrast, there is evidence that the daily rhythms in fetal adrenal function and fetal melatonin concentrations exist as a passive consequence of maternal rhythmicity. A key issue is whether changes in the daily profile of maternal melatonin release provide the fetus with information about the external environment. There is clear evidence from work in seasonal breeders that maternal melatonin can convey information to the fetus about the length of the external photoperiod. In non-seasonal breeders, it is possible that maternal melatonin conveys information about the time of day to the fetus acting as a circadian signal to enable the fetus to entrain to changes in the external environment. One possible site of action of melatonin may be the fetal circadian pacemakers but it is not yet clear in species such as the sheep and human whether the fetal circadian pacemakers generate any of the daily fetal hormonal or behavioural rhythms that have been observed.

An understanding of the functional status of the circadian pacemakers before birth is important if we are to understand the relative roles of external environmental cues (such as the light–dark cycle) and infant maturity in the ontogenesis of circadian rhythmicity after birth. If the fetal circadian pacemakers do oscillate before birth then the postnatal delay in the emergence of entrained circadian rhythms such as the sleep–wake cycle would be a consequence of the time taken for re-entrainment of these pacemakers to occur. If the circadian oscillators are not functional *in utero* then the postnatal delay in the emergence of entrained rhythms would be a consequence of maturation and entrainment of the circadian pacemakers in the new-born infant. An understanding of circadian rhythm development is important if we are to understand the impact of the intensive care nursery in which there are no external cues as to the passage of day and night on the physiological development of the preterm infant. A better understanding of developmental chronobiology may therefore lead to an improved definition of the optimal environment for neonatal intensive care.

Further reading

McMillen, I.C. and Nowak, R. (1989). The pre- and postnatal development of hormonal circadian rhythms. *Ballière's Clinical Endocrinology and Metabolism*, **3**, 707–21.

McMillen, I.C. and Walker, D.W. (1991). Effects of different lighting regimes on daily hormonal and behavioural rhythms in the pregnant ewe and sheep fetus. *Journal of Physiology*, **442**, 465–76.

Patrick, J., Challis, J.R.G., and Campbell, K. (1980). Circadian rhythms in maternal plasma cortisol and estriol concentrations at 30 to 31, 34 to 35 and 38 to 39 weeks' gestational age. *American Journal of Obstetrics and Gynaecology*, **136**, 325–34.

Reppert, S.M. and Schwartz, W.J. (1983). Maternal coordination of a fetal biological clock in utero. *Science*, **220**, 969–71.

26. Neuroendocrinology of the fetus
Peter Gluckman

After birth, the hypothalamic-pituitary axis regulates growth, reproduction, and metabolic homeostasis via the secretion of growth hormone (GH), prolactin (PRL), thyrotrophin (TSH), adrenocorticotrophin (ACTH), luteinizing hormone (LH), and follicle-stimulating hormone (FSH). The secretion of each of these hormones is regulated in a precise and distinct manner by hypothalamic neurohormones. The neurones that secrete these factors have their cell bodies within distinct hypothalamic nuclei and their axon terminals lie close to the primary plexus of the hypophyseal portal circulation. Neurohormones are secreted into the portal circulation and are distributed by the secondary plexus within the pituitary gland to either stimulate or inhibit adenohypophyseal hormone synthesis and release. The secretion of the hypothalamic neurohormones is, in turn, influenced by endocrine feedback loops and by neurotransmitter systems that modulate environmental and other influences. The action of neurohormones may also be influenced by endocrine factors — for example oestrogen exerts part of its negative feedback influence on LH secretion by inhibiting the ability of gonadotrophin-releasing hormone (GnRH) to stimulate LH release.

In fetal life, homeostasis is to some extent passive (e.g. thermoregulation) and determined by the maternal environment. Fetal growth must be tightly linked to nutrient supply as, in contrast to postnatal growth, there are no strategies available to the fetus to overcome limitations in substrate availability. The growth of the fetus in late gestation is regulated by factors that differ from those influencing postnatal growth. After birth growth is primarily determined by genetic factors and is not normally constrained, whereas fetal growth is limited by nutrient availability. Thus it is not surprising that there are basic differences in the mechanisms of the endocrine control of growth before and after birth. The organism must be capable of immediate and independent homeostasis from the time of birth and thus the neuroendocrine system must achieve a high level of maturation by birth. The individual components of the neuroendocrine unit are mature by mid-gestation, but differing strategies are employed to determine the activity of each hormonal axis appropriate to the conditions.

While the focus of this chapter will be the human fetus, it is not possible to discuss this topic without extensive reference to the experimental animal, particularly to the chronically instrumented fetal sheep. Because of the different rates of neural maturation between species, details of timing are not easily extrapolated between species and are only given where there are human data.

Morphogenesis and functional development of the hypothalamic-pituitary unit

The hypothalamus is the earliest forebrain structure to differentiate. In the human fetus the major hypothalamic nuclei differentiate between 6 and 12 weeks postconception. The neurohormones and neurotransmitters can be detected in the fetal hypothalamus between 8 and 16 weeks of gestation.

The embryonic origin of the adenohypophysis remains controversial. The most widely held view is that Rathke's pouch originates as a evagination from the primitive buccal cavity — the rostral limb becoming the anterior lobe and that behind the enclosed cleft, the intermediate lobe. Rathke's pouch comes into apposition with an evagination of the ventral diencephalon, the neurohypophyseal bud, that will form the median eminence and the posterior lobe. The alternative view is that Rathke's pouch is also of neural plate origin and comes secondarily to lie in contact with the stomatodeum. Rathke's pouch first forms at 4 weeks postconception and by 6 weeks is in contact with the developing neurohypophysis. By 8 weeks it is separated from the oral cavity and by 12 weeks has a largely mature anatomical form. Small groups of cells (rests) originating along the line of separation of the pituitary from the oral cavity may form a pharyngeal pituitary unit that can be quite prominent in fetal life.

The differentiated pituicytes develop from a common precursor. Hormone-specific cell types develop soon after Rathke's pouch comes into contact with the neural plate: corticotrophs are present by 7 weeks. By 11 weeks, somatotrophs, lactotrophs, thyrotrophs, and gonadotrophs are visible. It is unclear as to whether neural influences are necessary for development of the differentiated hormone-secreting cells. Evidence from the rat suggests that the adenohypophysis is dependent on neural, intrapituitary, and possibly extrapituitary factors for normal differentiation. For example, luteinizing hormone-releasing hormone (LHRH) may be essential for gonadotroph differentiation and the α-glycoprotein subunit produced by the gonadotrophs may, in turn, influence mammotroph development. Recently, at least one gene product responsible for regulating this development has been identified. *Pit-1* (also termed *GHF-1*) is a pituitary-specific transcriptional activation factor that determines differentiation of GH-, PRL-, and TSH-secreting cells and later is essential for the function of lactotrophs and somatotrophs. A defect in the expression of *pit-1* would explain GH deficiency in association with PRL and TSH deficiency as seen in the Snell dwarf rat and in rare forms of congenital hypopituitarism in man.

The development of the GH- and PRL-secreting cells is of interest. While the mature adenohypophysis has distinct somatotrophs and lactotrophs there is, at least in mid-gestation, a population of cells that secrete both GH and PRL. These somatomammotrophs may be regulated in a different manner and have a distinct secretory pattern. However, they coexist with both types of individual secreting cells and their significance is not yet certain.

As the primitive pituitary is forming, a dense network of capillaries surrounds it and will come to form the hypophyseal portal system. A capillary plexus coalesces on the ventral surface of the hypothalamus and axon terminals staining for neurohormones terminate in apposition with this 'mantle plexus' by 12–16 weeks. Secondarily, capillary loops invade the hypothalamus to form the true primary plexus. However, this mantle plexus, which is present before 12 weeks, may well be functional as similar arrangements persist throughout life in some reptiles. The major hypophyseal venous trunks are functional by 11.5 weeks. It has been suggested that at earlier ages hypothalamic factors might reach the developing pituitary anlage by diffusion.

Between 7 and 14 weeks the pituitary hormones can be detected in fetal plasma, although *in vitro* their secretion can be detected earlier. Hypothalamic neurohormones are detected at about this time. *In vitro* and limited *in vivo* data demonstrate that the capacity for neurohormonal control of pituitary hormone secretion is present by this age. The patterns of secretion in plasma from animals and as given in limited human data clearly show distinct patterns of active regulation of each of the adenohypophyseal hormones in the mid- and late-gestation fetus.

In considering the fetal neuroendocrine system a further complicating factor is the potential role of placental factors. The placenta will allow the passage of some neurohormones (e.g. TRF). However, except when given therapeutically in pharmacological doses to the mother (e.g. TRF), circulating neurohormones in the fetal circulation are probably not of biological significance. More recently, it has been recognized that the placenta can synthesize a wide range of neurohormones including thyrotrophin-releasing factor (TRF), GnRH, somatostatin (SRIF), GH-releasing factor (GRF), and corticotrophin-releasing factor (CRF). However, there is no convincing evidence that they regulate fetal pituitary function; they are more likely to have intraplacental function, perhaps influencing other endocrine secretion in a paracrine manner. There is, for example, some evidence that placental GnRH may affect placental human chorionic gonadotrophin (hCG) production. The potential interaction between pituitary hormones and their placental homologues must also be considered. For example it may be that the placental variant of human growth hormone (hGH-v) and placental lactogen (hPL) suppress maternal pituitary GH secretion. It is generally accepted that hCG in the fetal circulation affects the fetal gonadal axis prior to the secretion of fetal pituitary gonadotrophins and that this is the stimulus for fetal testosterone production in the male at 7–8 weeks, which is essential in sexual differentiation.

The somatotrophic axis

The mature somatotrophic axis

Growth hormone (GH) is the primary regulator of postnatal skeletal growth. It also has a major role in the regulation of whole body metabolism by promoting protein deposition, reducing protein degradation, promoting lipolysis and glycogen deposition, and by antagonizing some actions of insulin. GH is secreted by somatotrophs in a characteristic pulsatile fashion. These pulses are generated by the interplay of two hypothalamic peptides, GRF and SRIF. Growth hormone-releasing factor (GRF) is synthesized by neurones with cell bodies in the medial-basal hypothalamus and stimulates GH release via G_s (stimulatory G protein)-linked receptors. SRIF inhibits GH release via G_i (inhibitory G protein)-linked receptors. SRIF also inhibits TSH release. Both SRIF and GRF are also found within the gastrointestinal tract, but peripheral SRIF and GRF do not normally influence pituitary secretion. Both neurohormones, in turn, are influenced by other neuropeptides and neurotransmitters, some of which (e.g. dopamine) may also act directly on the somatotroph. The major stimuli to GH release include stress, food, exercise, and, in man, slow-wave sleep.

Growth hormone acts through membrane receptors. These somatogenic receptors are related to the lactogenic

(PRL) receptor and, more remotely, to some cytokine receptors. However, the second messenger system is not known. Somatogenic receptors are found in many tissues including the growth plate, erythroid precursors, liver, fat, and muscle. GH exerts its anabolic and somatogenic actions via stimulation of secretion of the insulin-like growth factors (IGFs). IGF-I appears to be the primary mediator of GH action. GH acts on the liver to cause the release of IGF-I into the circulation. However, the relative importance of this endocrine mode of secretion compared to direct stimulation of IGF-I production within tissues such as the growth plate remains debatable. The biological role of the closely related IGF-II remains to be resolved. Both IGF-I and II appear to have their numerous effects via the type I IGF receptor which is closely homologous to the insulin receptor. Both also have insulin-like effects. These may be mediated through insulin receptors but depend on IGF-I being present in concentrations in excess of the binding capacity of the IGF binding proteins; this may not be physiological. There are at least six different IGF binding proteins that, in part, prolong the circulating half-life of the IGFs. They may also act to modify the biological role of the IGFs and may help target IGF action within tissues. The significance of the type II receptor that only binds IGF-II is not known.

Growth hormone secretion in the fetus

Growth hormone is present in human fetal blood by 10 weeks. Fetal blood sampling confirms that concentrations in the fetal circulation are 10- to 20-fold higher than post-natal concentrations by 16 weeks. GH levels peak at 20 to 24 weeks of gestation but remain high through fetal life. In umbilical cord blood at term they are two to three times adult concentrations. In infants born at term they fall to adult values within a few days after birth, whereas in premature infants they are somewhat elevated for several weeks after birth (Fig. 26.1). The high fetal levels are due to increased secretion, not to reduced clearance. Experimental studies, primarily in the sheep, show that fetal GH release is pulsatile in fashion implying a complex form of regulation.

Regulation of fetal growth hormone secretion

Studies in the sheep fetus, supported by studies in human anencephaly (where the pituitary is small and the hypothalamus is only variably present as a diencephalic remnant), show that fetal GH secretion is dependent on the fetal hypothalamus. Although the placenta can synthesize GRF,

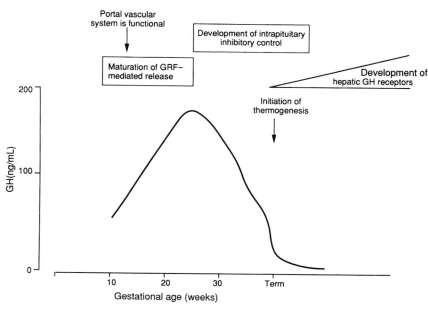

Fig. 26.1 Plasma growth hormone (GH) concentrations in the human fetus and neonate. The major determinants of this pattern of secretion are indicated. GRF, Growth hormone-releasing factor.

there is good evidence that placental factors do not directly stimulate fetal GH release. The favoured hypothesis to explain the high levels of GH in the fetal circulation had been that there was immaturity of hypothalamic control; however, recent experimental studies suggest that the primary cause lies within the somatotroph itself. Both GRF and SRIF can appropriately modulate GH release in the fetal sheep *in utero*, in the human fetal pituitary *in vitro*, and in the human neonate. The response to GRF appears to be exaggerated in the fetus, whereas the response to SRIF and other inhibitors is blunted. These observations suggest that stimulatory pathways predominate within the fetal somatotroph. Studies in the sheep fetus, supported by *in vitro* studies in several species including man, of the interaction between GRF and SRIF add weight to this hypothesis. Whereas SRIF is effective in suppressing the stimulatory effect of GRF postnatally, this mechanism is defective in the fetus. This suggests that part of the reason for the high fetal GH concentrations is an immaturity of intracellular mechanisms within the somatotroph whereby G_i-mediated processes cannot override G_s-dependent stimulation of GH release. The gradual decline in GH secretion in late gestation, which in part is accelerated by the prenatal rise in glucocorticoids, may reflect gradual maturation of this mechanism. This may also be reflected in the gradual decline in GH secretion postnatally in premature infants. A second factor that may contribute to the high fetal GH concentrations may be defective feedback. In the mature somatotrophic axis there is evidence for negative feedback in part mediated by IGF-I. IGF-I levels are relatively low in fetal life and are largely independent of GH control. Further, because of the defective intrapituitary mechanisms, exogenous IGF-I is relatively ineffective in suppressing fetal GH release (Fig. 26.2).

In term infants and particularly in the sheep there is a rapid decline in GH secretion in the first hours after birth. This has been shown to be linked to the onset of non-shivering thermogenesis and the marked rise in plasma free fatty acids (FFA). FFA inhibit GH release by blocking the action of GRF. In some species and in the premature infant, the rise in FFA is blunted or absent and it is of interest that in these species the fall in GH secretion at birth is slower.

There is some evidence that integrated control of GH release remains immature in the human neonate. For example, the response to hypoglycaemia may be paradoxically suppressive and slow-wave sleep-associated GH release may take some months to develop. While these observations suggest some degree of immaturity of neural control of GH release, it has not been possible to identify abnormalities of a specific neurotransmitter system. In the fetal sheep a broad range of neurotransmitter ago-

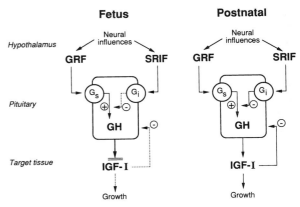

Fig. 26.2 A schematic summary of the differences in the regulation of growth hormone between fetal and postnatal life. The dotted lines represent parts of the axis less efficient in fetal life. GRF, Growth hormone-releasing factor; SRIF, somatostatin; GH, growth hormone; G_s, stimulatory G protein; G_i, inhibitory G protein; IGF-I, insulin-like growth factor-I.

nists, including opiate, serotoninergic, GABAergic, dopaminergic, and adrenergic agents, appropriately affect fetal GH release.

The functions of the somatotrophic axis in the perinatal period

Despite the high concentrations of GH in the fetal circulation, any effects of GH on fetal development are relatively subtle. However, in contrast to past dogma that fetal GH had no somatogenic role, it is now clear that GH deficiency is associated with definite growth abnormalities. For example, in human infants with gene defects leading to a defective GH receptor (Laron dwarfs) or with deletions of the GH gene, there is a slight but definite reduction in length at birth and relative obesity. Hypoglycaemia, due to reduced hepatic glycogenesis, is also observed. Similarly, in experimental animals the effects of intrauterine GH deficiency are difficult to define. Assessment of the role of GH *in utero* may be confounded in part by as yet unknown interactions between GH and other related hormones such as placental lactogen or other placental GH/PRL-like peptides.

The major reason for the relatively minor effect of GH deficiency *in utero* is receptor immaturity. In the liver the gene for the GH receptor is not expressed until late in gestation and then only in low levels. This is confirmed by binding studies that show that there is a marked postnatal increase in hepatic GH binding. This immaturity leads to relative resistance to GH in terms of its classical endocrine functions. However, the gene for the GH receptor has recently been shown to be expressed in other fetal tissues including the brain, the gastrointestinal tract, and developing bone. The functional significance of these potential receptors remains to be elucidated. GH may play a role in promoting tissue differentiation earlier in gestation; for example, it stimulates the differentiation of preadipocytes *in vitro*. GH has also been shown to affect pancreatic islet cell replication, possibly through the local production of IGFs.

The absence of GH receptors in tissues associated with growth until close to term may create a mechanism by which growth *in utero* can be largely removed from hypothalamic control and be directly regulated by nutrient availability. This can be considered in teleological terms as ensuring that fetal growth is separated from genetic control and thus reduce the risk of fetal overgrowth leading to dystocia. Postnatally, the pattern of GH release is determined, at least in part, genetically.

Because of immaturity of the hepatic GH receptor there is less dependence of fetal plasma IGF-I secretion on GH than is observed after birth. However fetal IGF-I does appear a strong candidate as a regulator of fetal growth. Fetal growth is tightly linked to substrate supply across the placenta, and fetal plasma IGF-I is rapidly responsive to rises and falls in plasma glucose. There is increasing experimental evidence that elevation of fetal plasma IGF-I promotes fetal anabolism. Studies in a variety of species have shown that plasma IGF-I levels in the fetus and at term correlate with fetal size.

Prolactin

The mature lactotrophic axis

Prolactin is secreted by specific cells called lactotrophs. In contrast to other adenohypophyseal hormones where the dominant hypothalamic control is stimulatory, PRL is under tonic inhibitory control by the hypothalamus. The major prolactin inhibitory factor is dopamine secreted into the portal circulation by neurones originating in the arcuate nucleus of the hypothalamus. Other hypothalamic factors also affect PRL secretion including TRF which is stimulatory. Oestradiol is the other major endocrine influence on PRL secretion with actions at both the hypothalamic and pituitary level. Within the lactotroph, oestrogen reduces the inhibitory effect of dopamine. Prolactin release is stimulated by stress and has both a diurnal and seasonal rhythm in many mammals. In fish and amphibia, it plays a regulatory role in fluid and electrolyte balance with actions on water and salt excretion. In mammals PRL can modulate reproductive function, but its major role would appear to be in the regulation of the onset of lactopoiesis. PRL is evolutionarily related to GH and binds to a lactogenic receptor that is structurally related to the somatogenic receptor.

Prolactin secretion in the fetus

Fetal PRL levels are very low prior to 25 weeks and then rise progressively to very high concentrations late in gestation. They are high at birth and decline over several weeks after birth. It has been suggested that the principal determinant of the pattern of PRL secretion in the perinatal period is oestrogen of placental origin. There is a close temporal relationship between the rise in PRL levels in the second half of pregnancy and rising oestrogen levels in the fetal circulation. Infusion of oestrogen into the sheep fetus prematurely accelerates PRL secretion. The dopaminergic system is mature *in utero*: agents that stimulate hypothalamic dopamine release and dopaminergic agonists inhibit PRL secretion; dopaminergic antagonists stimulate PRL secretion *in utero*. The latter demonstrates the presence of tonic dopaminergic inhibition and suggests that the high levels of PRL *in utero* are not due to immaturity of this system. Experimental studies suggest that other aspects of the neurotransmitter control of PRL secretion are also mature by late gestation. Prolactin secretion in the sheep fetus shows a diurnal rhythm (see Chapter 25).

The function of prolactin in the fetus

The functional significance of the high PRL levels in the perinatal circulation is poorly understood. There is evidence that the high PRL levels determine the relatively high total body water content of the fetus and neonate. In experimental animals, the use of dopaminergic agonists to suppress PRL rapidly at birth accelerates the loss of body water. In infants with respiratory distress syndrome, cord-blood PRL levels are lower and in fetal sheep it has been shown that PRL is synergistic with glucocorticoids and thyroid hormones in promoting lung maturation. As TRF can cross the placenta and stimulate both thyroid hormone release and PRL secretion, clinical trials have commenced using TRF and glucocorticoids administered to the mother to promote fetal lung maturation.

Amniotic prolactin

In humans, amniotic fluid PRL levels are very high. This amniotic PRL is of decidual origin and is regulated independently of fetal pituitary PRL by local decidual factors. It has been suggested that it plays a role in the regulation of amniotic fluid volume and osmolality.

Gonadotrophins

By mid-gestation gonadotrophin secretion into the fetal circulation shows many characteristics not again seen until puberty. Plasma concentrations of LH and FSH are similar to those at puberty and the pattern of secretion is pulsatile. Sexual dimorphism is observed with males having lower levels of FSH and LH, presumably reflecting testicular secretion of both testosterone and inhibin. Fetal orchidectomy is associated with increased gonadotrophin release. Exogenous GnRH can stimulate LH and FSH release in the mid-gestation fetus. Administration of GnRH superagonists inhibit gonadotrophin release due to downregulation of the GnRH receptor. Later in gestation, gonadotrophin secretion declines. In more precocial species such as the sheep, gonadotrophin secretion is almost entirely suppressed in late gestation; in less precocial species such as man there is a progressive inhibition, which continues over several postnatal months. Several mechanisms contribute to the late-gestation inhibition of the gonadotrophic axis. First, oestrogen receptors develop at about mid-gestation in the hypothalamus and placental oestrogen secretion begins to increase; this allows progressive feedback inhibition of gonadotrophin release. Second, there appears to be a change in the sensitivity of the negative feedback loop such that it becomes more efficient later in gestation. This changing sensitivity is reduced at the onset of puberty and in part explains its initiation. Third, there may be intrinsic neural mechanisms that progressively mature and inhibit gonadotrophin release. Tonic opioid-mediated inhibition of gonadotrophin release is demonstrable in the late-gestation sheep and pig fetus.

As discussed in Chapter 27 the gonadotrophic axis is functional in fetal life. Early in gestation the testis under hCG stimulation produces testosterone, essential for normal male differentiation. Later in gestation fetal pituitary gonadotrophins maintain gonadal function; in the male this is necessary for phallic growth and in the female for primary folliculogenesis.

Thyrotrophic axis

While the thyrotrophic axis is quiescent prior to mid-gestation, there is a marked increase in activity in the second half of gestation. There is a marked increase in hypothalamic TRF content and in TSH release. Late in gestation fetal TSH secretion is relatively high. The hypothalamic control of fetal TSH release appears functional by this stage in that exogenous TRF can stimulate its release as can hypothermia induced in fetal lambs.

However, the negative feedback loop by which thyroid hormones inhibit TSH release appears to be relatively inefficient. This may reflect immaturity of thyroid hormone receptors in the pituitary or immaturity of thyroid-dependent transcription factors, or may possibly be due, in part, to altered thyroid hormone metabolism. However, negative feedback is not absent; fetal thyroidectomy does lead to very high TSH levels in the fetus. There is evidence that negative feedback continues to mature for some time after birth.

At birth there is a rapid rise in circulating TSH levels that is dependent on cutaneous cooling. This plays a role in regulating thermogenesis. The thyroid axis is active *in utero*. Fetal hypothyroidism due to hypothalamic, pituitary, or thyroid disorder is associated with delayed maturation of a number of organ systems including the brain and lung. For further discussion of the perinatal thyroid axis see Chapter 29.

Adrenocorticotrophic axis

ACTH is derived from a larger precursor molecule, proopiomelanocortin (POMC). While ACTH is the principal bioactive product of POMC, a number of other molecules including β lipotrophin, β endorphin, α melanocyte-stimulating hormone (α MSH), and met-enkephalin are derived from POMC. POMC-secreting cells are found both in the anterior and intermediate pituitary lobes. The processing of POMC differs between these lobes with ACTH and β lipotrophin being the principal anterior lobe products and α MSH and the other smaller peptides being largely of intermediate origin. Adenohypophyseal ACTH is largely under regulation by CRF and other neuropeptides of hypothalamic origin, whereas intermediate lobe POMC-containing cells are largely regulated by dopaminergic innervation, which provides tonic inhibition. The anterior lobe is the dominant component of the pituitary POMC system, but in fetal life the intermediate lobe is relatively hyperplastic. There are reports of relatively greater secretion of other POMC products into the fetal circulation but their functional significance is unclear.

ACTH release from the fetal pituitary can be demonstrated from 8 weeks and corticotrophin-releasing factor (CRF) appears in the fetal hypothalamus by 12 weeks. Most studies of the maturation of the control of ACTH have been performed in the sheep fetus. There is a progressive increase in cortisol secretion in the last quarter of gestation associated with the induction of enzymes in the lung, placenta, liver, etc. This rise in cortisol release is dependent, in part, on a rise in ACTH release. This is associated with a reduction in the sensitivity of negative feedback by adrenal steroids. Nevertheless, fetal adrenalectomy leads to a further rise in ACTH release. Earlier the secretion of ACTH is readily inhibited by low concentrations of glucocorticoids. A further reason for the late-gestation rise in plasma cortisol is that adrenal sensitivity to ACTH increases late in gestation.

Fetal ACTH secretion is regulated primarily by hypothalamic CRF. Lesions of the hypothalamic paraventricular nucleus, which is the site of CRF cell bodies, leads to a loss of the prepartum rise in ACTH and an abolition of the release of ACTH in response to stimuli such as hypoxia and hypotension. Such responses can normally be seen throughout the third trimester in the fetal sheep. However, fetal ACTH release is not entirely abolished perhaps reflecting the contribution of arginine vasopressin (AVP) and other potential regulators. There is considerable evidence suggesting that, in fetal life, AVP is synergistic with CRF in the regulation of fetal ACTH release. While the placenta synthesizes CRF, there is no evidence implicating it in the regulation of fetal pituitary secretion. There may be a role for placental prostaglandin E_2, which can stimulate ACTH release directly.

The adrenal axis is essential to normal fetal maturation and to homeostasis in the new-born and is discussed in detail in Chapter 28.

Concluding remarks

The development of the fetus and its survival, both *in utero* and during the adaptation to independent existence, require a high level of physiological and biochemical integration. Thus, it is not surprising that key elements in the neuroendocrine axis mature early in fetal life and that, by mid-gestation, all components have become differentiated. However, the function of the system *in utero* is modulated by alterations in the sensitivity of feedback loops to allow for escalating secretion of hormones critical for maturation of organs such as the lung that are needed for independent existence (e.g. ACTH and TSH) or, alternatively, to inhibit their secretion once their requirements for fetal organogenesis has passed (e.g. gonadotrophins). The other strategy used to modulate activity is changes in receptor number/affinity. This increases, in the case of ACTH, to provide for escalating cortisol production prior to delivery. However, in the somatotrophic axis, receptor immaturity provides a route for alternate control of fetal growth appropriate to intrauterine existence. However, the presence of receptors in other tissues may allow GH to exert actions *in utero* that are not prominent after birth. The changing profiles of circulating hypophyseal hormones *in utero* are a consequence of these changes in feedback

mechanisms, changes in hypothalamic influences, and the influence of other endocrine factors such as placental oestrogens.

Further reading

Altman, J. and Bayer, S.R. (1978). Development of the diencephalon in the rat: Autoradiographic study of the time of origin and settling patterns of neurons of the hypothalamus. *Journal of Comparative Neurology*, **182**, 945–72.

Asa, S.L., Kovacs, K., Laszlo, F.A., Domokos, I., and Ezrin, C. (1986). Human fetal adenohypophysis: histologic and immunocytochemical analysis. *Neuroendocrinology*, **43**(3), 308–16.

Barnard, R., Haynes, K.M., Werther, G.A., and Waters, M.J. (1988). The ontogeny of growth hormone receptors in the rabbit tibia. *Endocrinology*, **122**, 2562–9.

Begeot, M., Hemming, F.J., Martinat, N., Dubois, M.P., and Dubois, P.M. (1983). Gonadotropin releasing hormone (GnRH) stimulates immunoreactive lactotrope differentiation. *Endocrinology*, **112**, 2224–6.

Blanchard, M.M., Goodyer, C.G., Charrier, J., and Barenton, B. (1988). In vitro regulation of growth hormone (GH) release from ovine pituitary cells during fetal and neonatal development: effects of GH-releasing factor, somatostatin, and insulin-like growth factor-I. *Endocrinology*, **122**, 2114–20.

Bugnon, C., Fellmann, D., and Bloch, B. (1978). Immunocytochemical study of the ontogenesis of the hypothalamic-somatostatin-containing neurons in the human fetus. *Metabolism*, **27** (9 suppl.), 1161–5.

Garcia-Aragon, J., Lobie, P.E., Muscat, G.E.O., Gobius, K.S., Nerstedt, G., and Waters, M.J. (1992). Prenatal expression of the growth hormone (GH) receptor/binding protein in the rat: a role for GH in embryonic and fetal development? *Development*, **114**, 869–76.

Gluckman, P.D. (1982). The hypothalamic–pituitary unit: the maturation of the neuroendocrine system in the fetus. In *Clinical neuroendocrinology* (ed. L. Martini and G.M. Besser), pp. 1–30. Academic Press, New York.

Gluckman, P.D. (1983). The fetal neuroendocrine axis. In *Fetal endocrinology and metabolism, current topics in experimental endocrinology* (ed. L. Martini and V. James), pp. 1–42. Academic Press, New York.

Gluckman P.D. (1986). Hormones and fetal growth. *Oxford Reviews in Reproductive Biology* **8**, 1–60.

Gluckman, P.D. and Bassett, N.S. (1989). The development of hypothalamic function in the perinatal period. In *Handbook of human growth and developmental biology*, Vol. II, Part A (ed. E. Meisami and P. Timeras), pp. 3–20. CRC Press, Boca Raton, Florida.

Gluckman, P.D., Grumbach, M.M., and Kaplan, S.L. (1980). The human fetal hypothalamus and pituitary gland. In *Maternal–fetal endocrinology* (ed. D. Tulchinsky and K.J. Ryan), pp. 196–232. Saunders, Philadelphia.

Gluckman, P.D., Grumbach, M.M., and Kaplan, S.L. (1981). The neuroendocrine regulation and function of growth hormone and prolactin in the mammalian fetus. *Endocrine Reviews*, **2**, 363–95.

Goodyer, C.G. (1989). Development of the anterior pituitary. In *Handbook of human growth and developmental biology* (ed. A.E. Meisami and P.S. Timiras), pp. 21–49. CRC Press, Boca Raton, Florida.

Hemming, F.J., Aubert, M.L., and Dubois, P.M. (1988). Differentiation of fetal rat somatotropes in vitro: effects of cortisol, 3,4,3'-triiodothyronine and glucagon, a light microscopic and radioimmunological study. *Endocrinology*, **123**, 1230–6.

Shiino, M., Ishikawa, H., and Rennels, E.G. (1978). Accumulation of secretory granules in pituitary clonal cells derived from the epithelium of Rathke's pouch. *Cell Tissue Research*, **186**, 53–7.

Takor-Takor, T. and Pearse, A.G.E. (1975). Neuroectodermal origin of avian hypothalamo-hypophyseal complex: the role of the ventral neural ridge. *Journal of Embryology and Experimental Morphology*, **34**, 311–25.

Thlivers, J.A. and Currie, R.W. (1980). Observations on the hypothalamo–hypophyseal portal vasculature in the developing human fetus. *American Journal of Anatomy*, **157**, 441–4.

Watanabe, Y.G. and Daikoku, S. (1976). Immunohistochemical study on adenohypophysial primordia in organ culture. *Cell Tissue Research*, **166**, 407–12.

27. The ontogeny of the pituitary-gonadal axis and sexual differentiation in the human fetus

Jeremy S.D. Winter

Sexual reproduction permits each parent to contribute equal (haploid) amounts of genetic material to the offspring, a mechanism that provides for rapid gene assortment and therefore increased opportunity for favourable combinations of genes to be selected in a fluctuating environment. In many lower vertebrates sex is not irreversibly assigned; some fish change sex under appropriate environmental conditions, while in amphibians and reptiles sex determination may be influenced during embryogenesis by external signals such as temperature or sex steroids. In some birds removal of the single ovary causes the second gonad to develop as a testis.

In nematodes and *Drosophila* sex determination depends on a chromosomal signal, the ratio of X chromosomes to sets of autosomes. In higher animals parents are of different sexes, maintained separate and distinct by the actions of unpaired sex chromosomes that contain the genetic information for sex determination. One sex (in mammals XX or female) is homogametic while the other (XY or male) is heterogametic. The Y chromosome contains a master control gene that directs the embryo away from an otherwise predetermined female pattern of differentiation. Not all male characteristics need to be regulated by genes on the Y chromosome. Rather, there is a hierarchical cascade of regulatory genes — a single Y-linked gene initiates differentiation of a testis; in turn, testicular hormones can regulate genes on any chromosome, a strategy that permits any gene product to be sexually dimorphic.

Normal sex determination and sexual differentiation

The appearance of a fertile adult with appropriate secondary sexual characteristics and psychosexual orientation represents the outcome of a logical and ordered sequence that begins at conception with the establishment of genetic sex (Fig. 27.1). The presence or absence of a specific Y-linked testis-determining (TDF) gene defines gonadal sex, causing the embryonic gonad to become either a testis or an ovary. The subsequent differentiation of the internal and external genitalia follows a common paradigm: the indifferent common primordia show an innate tendency to feminize unless male development is imposed by secretions from the fetal testis. At birth the external genitalia provide the basis by which society assigns both legal sex and gender role. Later during puberty, testicular or ovarian sex steroids induce secondary sexual characteristics that serve both to reinforce psychosexual identity and to signal adult reproductive capability.

Genetic sex: properties of the sex chromosomes

The unpaired sex chromosomes probably arose from an ancestral pair of autosomes through translocation of sex-specific genes plus some mechanism to discourage crossing over or gene exchange between the sex chromosomes during meiosis.

The Y chromosome

The Y chromosome is the most specialized mammalian chromosome, being almost exclusively involved in the control of sex determination and fertility. By conventional cytogenetic techniques it contains a euchromatic portion, involving the short arm and the proximal long arm, and a quinacrine-fluorescing heterochromatic portion involving the centromere and the rest of the long arm. The occasional absence of this entire quinacrine-positive region in normal fertile males suggests that it does not contain functional genes, being composed mainly of tandem DNA repeats.

During male meiosis X/Y genetic exchange normally occurs only within the so-called pseudoautosomal region at the distal end of the short arm. The testis-determining gene (termed SRY in humans) is located on the distal short arm immediately adjacent to this pseudoautosomal region (Fig. 27.2). A highly conserved gene has been identified here, which is expressed only in testis and encodes a

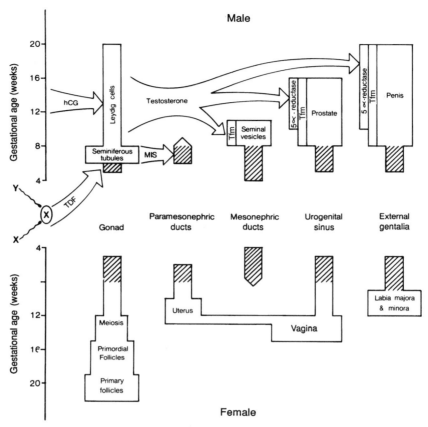

Fig. 27.1 The major determinants of sexual differentiation, showing the cascade of genetic and endocrine effects and their timing in relation to fetal age in the male (top) and female (bottom). The broad arrows show the sequential effects of the Y-linked testis-determining factor (TDF), human chorionic gonadotrophin (hCG), Müllerian-inhibiting substance (MIS), and testosterone in directing male differentiation. Tfm indicates the target cell androgen receptor. The cross-hatched segments refer to a sexually undifferentiated structure. (Taken with permission from Winter, J.S.D. (1986). In *Endocrinology and metabolism* (2nd ed.) (ed.) P. Felig, J.D. Baxter, A.E. Broadus, and L.A. Frohman), p. 991. McGraw-Hill, New York.)

protein with a potential DNA-binding domain. Loss of this gene causes XY gonadal dysgenesis, while its inappropriate exchange to the X chromosome during male meiosis results in an XX male. While other genetic factors may be required for spermatogenesis (as evidenced by the failure of maturation of germ cells in XX males), it does appear that the SRY gene is necessary and sufficient for testicular somatic development.

There is considerable circumstantial evidence for the existence of other functional genes on the human Y chromosome. Some of these may be involved in the regulation of spermatogenesis and of stature. Others appear to suppress the appearance of somatic features of Turner's syndrome and to influence predisposition to gonadoblastoma in the dysgenetic gonads of this disorder. The H–Y antigen

is a male-specific minor transplantation antigen whose expression is controlled by a gene on the proximal long arm of the Y chromosome.

The X chromosome

The X chromosome is a large metacentric chromosome that contains over 200 mapped genetic loci involved in the function of every system in the body, including sexual differentiation of both males and females. The gene for the androgen receptor, essential in the male for both genital differentiation and secondary sexual characteristics, is located in band Xq11–Xq12. Males with X chromosomal deletions or inversions may show abnormalities ranging from oligospermia to genital ambiguity, evidence of the

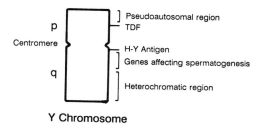

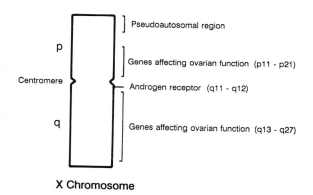

Fig. 27.2 Mapping of genes involved in sexual differentiation on the Y chromosome (top) and the X chromosome (bottom).

critical but still incompletely defined role that this chromosome plays in male reproductive function.

A segment on the long arm of the X chromosome (Xq13–Xq27) is involved in normal ovarian differentiation; females with structural abnormalities in this region present with accelerated follicular atresia, gonadal dysgenesis, and infertility. Some deletions of the short arm (Xp) cause short stature, but more severe abnormalities such as monosomy (45, X) give rise to the full Turner's syndrome with gonadal dysgenesis, short stature, and various somatic abnormalities. It would appear, therefore, that genes on both arms of the X chromosome are necessary for normal female growth and development.

The X chromosome contains a large number of unpaired genes not present on the Y chromosome, defects of which cause disease in hemizygous males or females with monosomy X. Since normal 46, XX females obtain twice as many of these genes as normal males, a dosage difference with effects similar to those of aneuploidy might be expected were there not some mechanism for dosage compensation. This compensation is effected by random inactivation of one X chromosome in each cell of the female early in development, so that the expressed dosage of X-linked genes in males and females is equivalent. This X

inactivation occurs in the early blastocyst stage. However both X chromosomes become active again in the primary oocytes of the developing ovary, where they appear to be essential for normal germ cell survival, since monosomy X results in accelerated follicular atresia. Recent evidence indicates that, even in somatic cells, X inactivation is incomplete, so that females normally express some gene loci, such as the Xg^a red blood cell antigen and steroid sulfatase, in double dosage.

Gonadal sex

Although genetic sex is established at conception, for the first few weeks of fetal life the gonadal anlagen remain similar in both sexes. This indifferent gonad appears at about 7–10 days in the mesenchyme at the ventral edge of the cranial medial part of the mesonephros, and is invaded in turn by both primordial germ cells and cells originating from the mesonephros and the coelomic epithelium. The germ cells arise in embryonic ectoderm; from here they migrate via the yolk sac and dorsal mesentery to reach the gonad at around days 35–45, where they proliferate rapidly. The mesonephros contributes dense cell cords, which in the male form the rete testis and epididymis and in the female the rete ovarii. It appears that mesonephric cells may give rise to testicular Sertoli cells and ovarian granulosa cells, while Leydig cells and ovarian theca and stroma arise from mesenchyme. These common origins emphasize that, although the testis and ovary differentiate into dissimilar organs, they continue to show numerous homologies in their steroidogenic and gametogenic functions.

Testicular differentiation

The earliest sign of male differentiation is the appearance at about 6–7 weeks within the indifferent gonad of large numbers of Sertoli cells, and their later organization into seminiferous tubules. Subsequently, germ cells (gonocytes and spermatogonia) undergo mitotic division within the testicular cords as these grow and eventually canalize, but germ cells play no role in the initiation of testicular differentiation. In contrast to the female, male meiosis does not begin until puberty.

At 8 weeks typical Leydig cells, equipped with membrane receptors for the gonadotrophic hormones, human chorionic gonadotrophin and luteinizing hormone, appear in the interstitium between the testicular cords; by 14 weeks these make up more than half the volume of the testis. These fetal Leydig cells have the typical ultrastructural appearance of steroid-secreting cells, and contain abundant amounts of steroidogenic enzymes. After the fourth month of pregnancy the number of fetal Leydig cells decreases, and only a few are visible at term. The appearance and later decline of Leydig cells closely follows the rise and fall of fetal circulating hCG levels. The final stage of testis differentiation is descent through the inguinal canal to reach the scrotum at about 34 weeks of gestation.

Ovarian differentiation

In contrast to testis development, which is independent of the presence of germ cells, differentiation of the ovary is largely defined in terms of germ cell maturation. At 7–9 weeks these oogonia are found undergoing mitotic division in clumps scattered throughout the ovarian parenchyma. The first definitive sign of ovarian development is the onset of meiotic prophase in primary oocytes at around 10–11 weeks. The first meiotic division is arrested at the diplotene stage, and is not completed until ovulation occurs some 12 to 40 years later. At about 13 weeks of gestation oocytes become surrounded by a single layer of follicular cells. After 22–25 weeks, multilayered and even a few antral follicles can be seen; the new-born ovary still contains mainly primary and primordial (single-layer) follicles, although small numbers of more mature and atretic forms occur. Small amounts of sex steroids can be produced by theca and interstitial cells, which are homologous with Leydig cells, but their nature and significance for oocyte maturation remain unclear.

The most remarkable phenomenon of ovarian development during the latter half of pregnancy and indeed during postnatal life is a decline in total germ cell population due to atresia. At midpregnancy the ovaries contain up to 7 million oogonia, oocytes, and degenerating cells; by term they contain about 2 million germ cells, but only about 400 of these will survive to ovulate during a woman's reproductive life. Preservation of a viable follicle requires an oocyte equipped with two active X chromosomes. Females with X chromosomal abnormalities such as monosomy show accelerated rates of follicular atresia until eventually they are left with streak gonads containing no germ cells. The syndrome of autosomal recessive XX gonadal dysgenesis suggests that other genes are involved in oocyte survival.

Genital duct differentiation

The mesonephric, or Wolffian, ducts appear at about 4 weeks of gestation; they then grow caudally, become canalized, and open into the cloaca. The mesonephric nephrons degenerate once the definitive kidney (metanephros) forms, but their ducts become incorporated into the genital system. The paramesonephric, or

Müllerian, ducts originate at about 6 weeks of gestation as solid cords of coelomic epithelial cells lateral to the mesonephric ducts that grow caudally and cross the latter ventrally to reach the urogenital sinus. These paired cords become canalized; caudally, where they meet in the midline, they fuse to form a single uterovaginal canal. Thus by 7 weeks each fetus is equipped with the primordia of both male and female internal genitalia.

In female fetuses the mesonephric ducts degenerate relatively early, leaving only remnants as the epoophoron, paraoophoron, and Gartner's duct. The partially fused paramesonephric ducts provide the paired Fallopian tubes, the midline uterine primordium, and, together with tissue contributed by the urogenital sinus, the vaginal primordium. Development of these structures does not depend upon any hormonal or other influences from the ovary; rather it is contingent upon prior appearance of the mesonephros. Although the female internal genitalia differentiate somewhat later than those of the male, they appear to be functionally committed to becoming female by about 8 weeks of gestation.

The development of male internal genitalia involves two separate processes: regression of the paramesonephric ducts and stabilization of the mesonephric ducts. These are mediated by two secretions of the fetal testis — Müllerian inhibiting substance (MIS, also called anti-Müllerian hormone) from the Sertoli cells, and testosterone from the fetal Leydig cells.

The role of MIS

Regression of the paramesonephric ducts begins at about 8 weeks of gestation, initially adjacent to the caudal pole of the testis and then extending cranially and caudally. Close juxtaposition of the duct to the ipsilateral testis is necessary, but even large amounts of testosterone applied locally cannot duplicate this effect. Rather, paramesonephric regression is induced by the local paracrine action of MIS, a dimeric glycoprotein (140 kDa) secreted by Sertoli cells. This factor possesses interesting structural homologies with both TGFβ and the β chain of inhibin. It is encoded by a gene on the short arm of chromosome 19, and is expressed at around 7 weeks of gestation at the time of Sertoli cell differentiation. Secretion of MIS appears to be enhanced by cyclic AMP, but is not regulated by known hormonal factors. The mechanism of its paracrine action on the paramesonephric duct is not clear, but may involve inhibition of membrane protein phosphorylation.

MIS is still detectable after birth, although circulating levels are lower after the onset of puberty. The function of MIS apart from paramesonephric duct involution is not known. Exposure of fetal ovaries to MIS causes inhibition of germ cell replication, suppression of aromatase activity, and the appearance of seminiferous tubule-like structures, evidence that it may also play a role in normal testis development. Sertoli cells not only secrete MIS, but also have MIS receptors; MIS might act in an autocrine fashion to induce further Sertoli cell differentiation, and thus could eventually influence germ cell proliferation, Leydig cell function, and seminiferous tubule formation. Finally, there is evidence that MIS may be responsible for caudal migration of the testis to the entrance of the inguinal canal, although a direct effect on gubernacular cells has not been demonstrated.

Circulating MIS is not detected in females before birth, but some is secreted by postnatal granulosa cells. There is some evidence that in the female MIS may act as an inhibitor of oocyte meiosis.

The role of testosterone

Differentiation of the paired mesonephric ducts to form vasa efferentia, epididymis, vasa deferentia, ejaculatory ducts, and seminal vesicles is dependent upon local high concentrations of testosterone from the ipsilateral testis. Although androgen receptors are present in ductal cells, they lack 5α-reductase for conversion of testosterone to the more active dihydrotestosterone until after differentiation is complete. That this effect of testosterone is a local paracrine effect rather than a blood-borne endocrine effect is underscored by the failure of circulating androgens of adrenal origin to prevent mesonephric duct regression in females with virilizing congenital adrenal hyperplasia.

Differentiation of the urogenital sinus and external genitalia

For the first 2 months of fetal development the external genitalia are identical in both sexes. They consist of a midline phallic tubercle and lateral labioscrotal swellings. Behind these lies the urogenital sinus, into which open both the ureters and the internal genital ducts.

In females the phallus remains small, the anogenital distance is short, and the labioscrotal swellings and urethral groove do not fuse in the midline. The labioscrotal swellings become the labia majora, the separate genital folds the labia minora, and the phallic part of the urogenital sinus the vestibule. The urogenital sinus is divided by downgrowth of a vaginal plate that approaches the perineum to provide separate urethral and uterovaginal access to the exterior. This female development of the external genitalia is independent of any ovarian influence.

In the male fetus the labioscrotal swellings and urethral folds fuse in the midline to form the scrotum and a penis enclosing the phallic urethra. This process, which is a response to circulating testicular androgen, begins at 9 weeks and is complete by 14 weeks of gestation. The urogenital sinus develops into the male urethra and the prostate. During later gestation the penis continues to grow more rapidly than the female clitoris. During this same period the testes descend into the scrotum, a process that is at least partly androgen-dependent.

Differentiation of the hypothalamic-pituitary-gonadal axis

Fetal gonadal sex steroids

By the time Leydig cells are visible at 6–7 weeks of fetal age, the testis can synthesize testosterone, using circulating low-density lipoprotein cholesterol as the preferred substrate. Testicular concentrations of testosterone rise to a peak at 12–14 weeks of gestation and then decline, a pattern that correlates closely with the growth and involution of fetal Leydig cells and with testicular mRNA levels for key steroidogenic enzymes. Although fetal ovaries also contain enzymes for synthesis of androgens and oestrogens, there is no evidence for significant secretion of these steroids, at least during early pregnancy when sexual differentiation is proceeding.

The initial local or paracrine action of fetal testosterone is to promote development of the ipsilateral mesonephric duct and growth of the testis itself. Soon, however, male fetal serum testosterone concentrations begin to rise, reaching peak levels of 7 to 21 nmol/L at 16 weeks of gestation, values that are in the adult male range. After mid-pregnancy male serum testosterone levels gradually decline, but they remain higher than female values. Levels of sex steroid-binding globulin (TeBG) are only one-twentieth those in maternal serum; although albumin serves as a low-affinity carrier of many steroids, its concentration is also low in fetal serum. The net effect is to increase the fraction of circulating testosterone that is unbound and presumably active.

The developing fetus is also exposed to high concentrations of steroids from the placenta, notably oestrogen and progesterone, and from the adrenal cortex, in particular dehydroepiandrosterone sulfate. There is no sex difference in levels of these steroids; presumably they play no direct role in sexual differentiation, but they may influence the process through effects on pituitary function, gonadal steroidogenesis, serum protein binding of steroids, or target cell receptor occupancy.

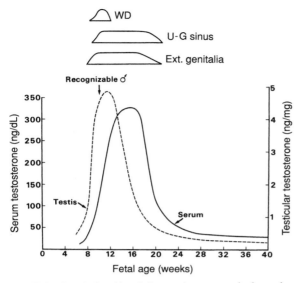

Fig. 27.3 The relationship of changes in mean testicular and serum testosterone concentrations to the time of development of the mesonephric ducts (WD) and virilization of the urogenital sinus (U–G) and external genitalia in human male fetuses. (Taken with permission from Winter, J.S.D., Faiman, C., and Reyes, F.I., (1977). In *Morphogenesis and malformation of the genital system* (ed. R.J. Blandau and D. Bergsma), p. 47. Alan R. Liss, New York.)

Action of sex steroids

The events of male genital development, in relation to the changing patterns of fetal testis testosterone production, are summarized in Fig. 27.3. Testosterone appears first in high concentrations in the testis itself where it induces masculinization of the ipsilateral mesonephric duct. Subsequently, circulating testosterone levels become sufficiently high to virilize the external genitalia.

The androgen receptor

All these target cells contain specific nuclear receptors that bind androgen with high affinity; this steroid–receptor complex then binds to chromatin acceptor sites to initiate transcription, which leads eventually to synthesis of androgen-induced proteins. There is evidence that the actual target for these androgen effects is the mesenchyme of the genital anlagen, which in turn induces specific morphological and functional changes in the overlying epithelium. The gene (Tfm) for the androgen receptor has been cloned and localized to band Xq11–Xq12 of the X chromosome.

5α-reductase

The androgen receptor has a higher binding affinity for dihydrotestosterone than for testosterone itself, so intracellular conversion to DHT will significantly amplify the androgen effect. The necessary microsomal enzyme, termed 5α-reductase, is found in the urogenital sinus and external genitalia, but is not expressed in the mesonephric ducts until their development is already well advanced. The gene for 5α-reductase is autosomal; thus both sexes can respond to androgen and the direction of genital differentiation is determined solely by testicular production of testosterone. Defects in testosterone synthesis, 5α-reductase activity, or androgen receptor function will each prevent normal masculinization of a genetic male. Conversely, exposure of a genetic female to high levels of circulating androgen will cause varying degrees of external genital virilization.

Control of fetal gonadal function

Chorionic gonadotrophin

There is considerable circumstantial evidence that Leydig cell testosterone production during the period of genital differentiation is regulated by chorionic gonadotrophin, although other factors may be involved in the initiation of testicular steroidogenesis. In man, human chorionic gonadotrophin (hCG) appears in the maternal circulation immediately after implantation; levels rise to a peak at 10 weeks and then decline, although placental production of hCG continues until term. The pattern in the fetal circulation of either sex is similar, but fetal hCG values are only about one-thirtieth of those in maternal serum.

Fetal Leydig cells, but not ovarian cells, have specific membrane luteinizing hormone–hCG receptors, the number of which increases to a maximum at 15–18 weeks and then declines. Acting via intracellular cyclic AMP, hCG increases the number of low-density lipoprotein (LDL) receptors, enhances *de novo* cholesterol synthesis, and stimulates mitochondrial cholesterol side-chain cleavage, all of which serve to increase the pool of pregnenolone available for androgen biosynthesis.

Pituitary gonadotrophins

Maternal gonadotrophins do not cross the placenta, but after 9 weeks the fetal pituitary produces significant amounts of follicle-stimulating hormone (FSH) and LH. Although the fetal pituitary can synthesize glycoprotein hormone α-subunit by 4–5 weeks, it seems likely that gonadotrophin-releasing hormone (GnRH)-mediated

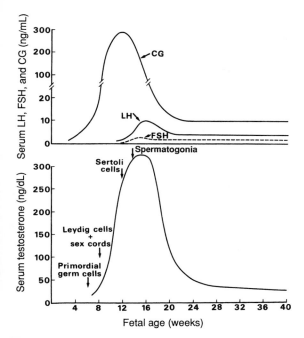

Fig. 27.4 A schematic summary of the temporal relationship between serum concentrations of hCG, LH, FSH, serum testosterone concentrations, and development of the human fetal testis. (Taken with permission from Winter, J.S.D., Faiman, C., and Reyes, F.I., (1977). In *Morphogenesis and malformation of the genital system* (ed. R.J. Blandau and D. Bergsma), p. 53. Alan R. Liss, New York.)

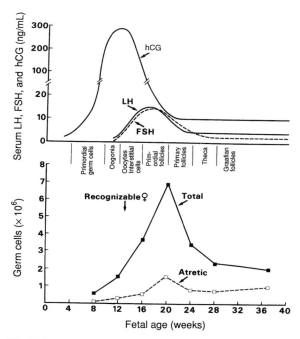

Fig. 27.5 A schematic summary of the temporal relationships between serum concentrations of hCG, LH, and FSH, morphological development of the fetal ovary, and ovarian total content of germ cells. (Taken with permission from Winter, J.S.D., Faiman, C., and Reyes, F.I. (1977). In *Morphogenesis and malformation of the genital system* (ed. R.J. Blandau and D. Bergsma), p. 54. Alan R. Liss, New York.)

hypothalamic stimulation is necessary for production of hormone-specific β-subunit and thus of intact FSH and LH.

Both gonadotrophins appear in the fetal circulation at about 11–12 weeks and rise to peak concentrations at mid-pregnancy. Serum FSH concentrations are much higher in female fetuses, in whom adult castrate values occur, than in male fetuses, in whom values are in the normal adult range. This sex difference demonstrates that a negative feedback mechanism mediated by testicular androgen is already operative. After 28 weeks, serum gonadotrophin levels decline in both sexes, a phenomenon that probably reflects maturation of placental oestrogen-mediated feedback systems. Thus the decline in Leydig cell testosterone production in the latter half of pregnancy appears to result from the combined effects of declining hCG, FSH, and LH stimulation, as well as direct inhibition of androgen biosynthesis by circulating placental oestrogens.

The patterns of serum hCG, FSH, and LH concentrations in the male fetus, in relation to changes in testicular morphology and function, are summarized in Fig. 27.4. While hCG clearly drives testosterone synthesis during

genital differentiation, pituitary gonadotrophins become relatively more significant in later gestation when they influence phallic growth and testicular descent. The testes of hypogonadotrophic fetuses show not only reduced numbers of Leydig cells, but also reduced numbers of spermatogonia, evidence that FSH and/or LH may play a role in germ cell maturation.

The relationships between serum gonadotrophins and ovarian development are summarized in Fig. 27.5. Ovarian development proceeds normally in apituitary fetuses until about 32 weeks. After this, ovarian size declines and accelerated follicular atresia occurs. This evidence of inadequate granulosa cell differentiation coincides with the normal appearance of FSH receptors in ovarian granulosa cells.

There is considerable evidence that fetal pituitary gonadotrophin secretion is regulated by hypothalamic GnRH. Neurones expressing pro-GnRH mRNA appear in the nasal brain by 5 weeks fetal age and migrate to the basal hypothalamus. By 9–11 weeks GnRH-secreting neurones can be identified in the pericommissural, preoptic, lamina terminalis, and perimamillary regions; their axons end in the median eminence apposed to a capillary plexus connecting to the anterior pituitary, although a well-defined hypothalamic-hypophyseal portal system appears later. Between 11 and 14 weeks of gestation there is a 14-fold rise in hypothalamic GnRH content that parallels the increase in pituitary gonadotrophin secretion.

By mid-gestation the fetus demonstrates pulsatile GnRH release; both *in vitro* and *in vivo* pituitary gonadotrophin secretion is enhanced by GnRH administration and blocked by specific GnRH antagonists. Sex steroids have profound effects upon this process; thus, fetal orchidectomy increases FSH and LH levels, an effect that can be blocked by administration of testosterone. As one would expect, ovariectomy has no effect on gonadotrophin secretion in the female fetus.

Less is known about central nervous system regulation of fetal hypothalamic GnRH release. Certainly by mid-gestation hypothalamic nuclei can produce significant amounts of potential regulators such as dopamine, serotonin, noradrenaline, and endorphins, and there is some evidence that endogenous opiates exert a tonic inhibiting influence on GnRH release at this time. In addition, the placenta produces some GnRH; its relevance for regulation of pituitary function is unclear, but it may influence placental hCG synthesis.

Other factors

In addition to gonadotrophins, various other factors of placental or pituitary origin, such as growth hormone, may influence fetal gonadal growth and differentiation. The fetal gonad shows significant expression of insulin-like growth factor II (IGF-II); other local growth factors such as IGF-I, fibroblast growth factor, epidermal growth factor, and transforming growth factor β (TGFβ) influence fetal development and may well play a role in the developing gonad. Inhibin and activin are structurally related dimeric glycoproteins produced by the gonads that appear to modulate various intragonadal processes as well as pituitary FSH secretion. Inhibin secretion is stimulated by both FSH and hCG; it appears earlier and in greater abundance in the testis than the ovary, and may play a role in the decline of fetal FSH secretion in the latter half of gestation. Inhibin is also produced by the placenta, where it may be involved in suppression of hCG secretion in late gestation.

Perinatal adaptation of the reproductive endocrine system

At term the pituitary-gonadal axis of the human infant is structurally complete and capable of relatively mature function, but it is profoundly influenced by placental hormones. Serum hCG levels range from 20 to 9000 international units (IU)/L (median 50 IU/L). Circulating levels of placental steroids such as oestrone (33–147 nmol/L), oestradiol (7–60 nmol/L), oestriol (44–550 nmol/L), 17-hydroxyprogesterone (30–212 nmol/L), and progesterone (360–1520 nmol/L) are extremely high; these are cleared from the neonatal circulation by 48 h, but hCG and various steroid sulfates are cleared more slowly. The only hormonal sex difference is in serum levels of LH and testosterone, which are slightly higher in males at term than in females.

At the end of the first week of postnatal life, a remarkable phenomenon occurs in infants of both sexes that can only be interpreted in neuroendocrine terms as the apparent onset of puberty. Released from the inhibitory influence of placental steroids, pituitary gonadotrophin levels rise rapidly. This secretion is pulsatile and clearly driven by the hypothalamic GnRH pulse generator. In male infants serum FSH and LH levels peak at around 1 month of age and then decline to prepubertal levels by 4 months. This is accompanied by a parallel rise in serum testosterone concentrations to a peak at 1–3 months of age. Although levels of sex hormone-binding globulin also increase, values of free testosterone are distinctly raised in normal male infants; the only apparent clinical response to this phenomenon is testicular and phallic growth, plus occasional acne. However this neonatal activation of the pituitary–gonadal axis may influence the timing and course of subsequent sexual development, since blockade with a GnRH agonist in infancy causes a significant delay in the later onset of puberty.

In females or agonadal infants serum FSH and LH levels continue to rise until 3 or 4 months of age, and do not decline to their prepubertal nadir until about 3 years. The ovarian response to this gonadotrophin surge is not as immediate as that of the testis. One can observe an increase in the number of ovarian antral follicles, and variable but unsustained increases in serum oestradiol levels. The major clinical response to these events is the common appearance of transient breast enlargement in girls during the first 2 years of life.

The most interesting aspect of this perinatal adaptation is the subsequent decline in pituitary gonadotrophin secretion, which occurs in both sexes and is independent of any gonadal feedback influences. Instead, it reflects maturation of a central mechanism for inhibition of the hypothalamic GnRH pulse generator. This prepubertal suppression of gonadotrophin secretion is the penultimate stage of sexual differentiation. The final phase is puberty, with the appearance of sexually dimorphic secondary sex characteristics and acquisition of full adult reproductive function.

Gender role and psychosexual differentiation

Gender role encompasses a person's psychosexual self-identity, legal and social gender designation, certain aspects of dress and social comportment, and erotic responsiveness to various cues. For years a nature-versus-nurture controversy has raged regarding the potential significance of prenatal sex steroid exposure as an organizer of these postnatal phenomena. At one extreme is the concept that each infant is psychosexually neutral at birth, and gender identity is permanently imposed during the first few years of life by social and environmental influences. Numerous reports of persons raised in a sex discordant with their genetic or gonadal sex certainly testify to the importance of such influences, although it seems simplistic to assume that their impact is limited to infancy. More probably, gender role continues to be reinforced throughout childhood and puberty by both external influences and awareness of one's own genital and secondary sex characteristics. The demonstration of spontaneous and successful gender role reversal in some male pseudohermaphrodites following virilization at puberty attests to the impact of these sex hormone-mediated effects even in later life.

The opposite view holds that postnatal behaviour and intellectual function, to the extent that they are sexually dimorphic, are programmed by sex steroids during a critical prenatal phase of neural differentiation. In its most extreme form this theory proposes that subtle deficiencies of androgens in genetically male fetuses permit female organization of the brain and predispose to later homosexual behaviour. Conversely, excess androgen in a female might induce masculine behaviour, homosexuality, or anovulatory infertility. It must be emphasized that this model invokes not just the usual direct activational effect of androgens, such as that upon aggressive behaviour and libido in adults, but rather an organizational effect on neuronal growth and synaptogenesis that is permanent and expressed for years after the actual hormone exposure. Such organizational effects can be readily demonstrated in female rodents, in whom brief exposure to sex steroids at a critical time can permanently impose male-type acyclic gonadotrophin secretion and copulatory behaviour in late adult life. One can observe sexually dimorphic synaptogenesis in the preoptic area of rat brain, which is regulated by the level of sex steroids during prenatal development, but it is not at all clear to what extent these phenomena have a counterpart in human brain development. The developing human brain does, however, express not only androgen receptors but also 5α-reductase and aro-matase, suggesting that it can transform circulating androgen to products that could have greater or lesser biological activity. Prenatal exposure of female primates to testosterone in amounts sufficient to masculinize the genitalia can lead to increased postnatal aggressiveness and aberrant adult sexual behaviour, but does not prevent normal ovulatory cycles. Early reports of higher IQ scores in androgen-exposed human infants have not been confirmed. Prenatal exposure of females to the synthetic oestrogen, diethylstilbestrol, may slightly increase the risk of later homosexuality; however, the fact that the majority of such individuals are exclusively heterosexual suggests that many variables, some as yet unidentified, act to determine sexual orientation.

A reasonable viewpoint would suggest that prenatal androgen may play some role in the differentiation of human psychosexual behavior, but is certainly not the decisive determinant. Successful differentiation of gender role appears to depend primarily upon appropriate assignment of the sex of rearing plus continued unambiguous reinforcement by social interaction, concordant genital appearance, and appropriate secondary sexual characteristics.

Summary

This chapter focuses on human sexual differentiation, but it is important to recognize that in this process, as in all developmental processes, there is considerable species diversity. Thus in man sex is genetically determined, depending on the presence or absence of a single Y-linked gene, which then initiates a cascade of sex-specific developmental processes. In bees, however, a haplodiploid system is seen, in which males develop from unfertilized eggs and females from fertilized eggs. In various reptiles, amphibians, and nematodes sex, or rather the expression of sex-determining genes, is determined by environmental conditions rather than genotype. In chickens sex reversal can be accomplished by blockade of enzymatic conversion of testosterone to oestradiol. Finally there are numerous examples of species-specific sexually dimorphic characteristics, ranging from brain nuclei to body size, that may be either sex-hormone- or gene-dependent. From an evolutionary perspective these represent components of specific strategies to maximize reproductive success and to provide optimal conditions for gene assortment and selection. The challenge for the physiologist is to be able to discriminate between evolutionary 'noise' and those central themes of sexual development that transcend species differences.

Further reading

Asa, S.L., Kovacs, K., Horwath, E., Losinski, N.E., Laszlo, F.A., Domokos, I., and Halliday, W.C. (1988). Human fetal adenohypophysis: electron microscopic and ultrastructural immunocytochemical analysis. *Neuroendocrinology*, **48**, 423–31.

Austin, C.R. and Edwards, R.G. (1981). *Mechanism of sex differentiation in animals and man*. Academic Press, London.

Baker, M.L., Metcalfe, S.A., and Hutson, J.M. (1990). Serum levels of Mullerian inhibiting substance in boys from birth to 18 years, as determined by enzyme immunoassay. *Journal of Clinical Endocrinology and Metabolism*, **70**, 11–15.

Behringer, R.R., Cate, R.L., Froelick, G.J., Palmiter, R.D., and Brinster, R.L. (1990). Abnormal sexual development in transgenic mice chronically expressing mullerian inhibiting substance. *Nature*, **346**, 167–70.

Blandau, R.J. and Bergsma, D. (1977). *Morphogenesis and malformation of the genital system*. Alan R. Liss, New York.

Burgoyne, P.S. (1988). Role of mammalian Y chromosome in sex determination. *Philosophical Transactions of the Royal Society of London*, **B322**, 63–72.

Delemarre-van de Waal, H.A., Plant, T.M., van Rees, G.P., and Schoemaker, J. (1989). *Control of the onset of puberty III*. Excerpta Medica, Amsterdam.

Donahoe, P.K., Cate, R.L., MacLaughlin, D.T., Epstein, J., Fuller, A.F., Takahashi, M., Coughlin, J.P., Ninfa, E.G., and Tayler, L.A. (1987). Mullerian inhibiting substance: gene structure and mechanism of action of a fetal regressor. *Recent Progress in Hormone Research*, **43**, 431–67.

Gorski, R.A. (1985). Sexual differentiation of the brain: possible mechanisms and implications. *Canadian Journal of Physiology and Pharmacology*, **63**, 577–94.

Gubbay, J., Collignon, J., Koopman, P., Capel, B., Economou, A., Munsterberg, A., Vivian, V., Goodfellow, P., and Lovell-Badge, R. (1990). A gene mapping to the sex-determining region of the mouse Y chromosome is a member of a novel family of embryonically expressed genes. *Nature*, **346**, 245–50.

Kurilo, L.F. (1981). Oogenesis in antenatal development in man. *Human Genetics*, **57**, 86–92.

Lyon, M.F. (1988). X-chromosome inactivation and the location and expression of X-linked genes. *American Journal of Human Genetics*, **42**, 8–16.

Muller, U. (1989). Molecular biology of the human Y chromosome. In *Evolutionary mechanisms in sex determination* (ed. S.S. Wachtel), pp. 91–8. CRC Press, Boca Raton, Florida.

Rabinovici, J. and Jaffe, R.B. (1990). Development and regulation of growth and differentiated function in human and subhuman primate fetal gonads. *Endocrine Reviews*, **11**, 532–57.

Reyes, F.I., Winter, J.S.D., and Faiman, C. (1989). Endocrinology of the fetal testis. In *The testis* (2nd edn) (ed. H. Burger and D. deKretser), pp. 119–42. Raven Press, New York.

Ronnekleiv, O.K. and Resko, J.A. (1990). Ontogeny of gonadotropin-releasing hormone-containing neurons in early fetal development of rhesus macaques. *Endocrinology*, **126**, 498–511.

Rundlett, S.E., Wu, X-P., and Miesfield, R.L. (1990). Functional characterizations of the androgen receptor confirm the molecular basis of androgen action is transcriptional regulation. *Molecular Endocrinology*, **4**, 708–14.

Sinclair, A.H., Berta, P., Palmer, M.S., Hawkins, J.R., Griffiths, B.L., Smith, M.J., Foster, J.W., Frischauf, A.M., Lovell-Badge, R., and Goodfellow, P.N. (1990). A gene from the human sex-determining region encodes a protein with homology to a conserved DNA-binding motif. *Nature*, **346**, 240–4.

28. The function of the fetal pituitary–adrenal system

Charles E. Wood

The fetal hypothalamus–pituitary–adrenal axis is uniquely important. The activity of this endocrine axis directs much of fetal visceral and pulmonary development and is important in the initiation of parturition. The function of this endocrine axis varies among species. For example, increased fetal adrenocorticotrophic hormone (ACTH) and cortisol secretion are required for the initiation of parturition in the sheep fetus; the role of these hormones in the initiation of parturition in the human is less clear. On the other hand, fetal plasma glucocorticoids are thought to be essential for stimulation of fetal visceral development in both the sheep and the human fetus.

This chapter will review the development of fetal hypothalamus–pituitary–adrenal function. It is not practical to present a detailed comparison of the function of this endocrine axis in all mammalian species studied. Therefore, the focus of the chapter will be on the sheep. The sheep is the most extensively studied model of fetal development because it is possible to chronically catheterize fetal sheep and collect blood samples for endocrine studies. The non-human primate fetus has also been studied *in utero*, although comparatively fewer data have been obtained from this species. The human fetus has provided even less data, although understanding human development and parturition is the ultimate goal of most of the research in sheep and non-human primates.

The components of the fetal hypothalamus–pituitary–adrenal axis

The arrangement of the hypothalamus–pituitary–adrenal axis differs somewhat between primate and non-primate species. In the sheep fetus, corticotrophin-releasing factor (CRF) and arginine vasopressin (AVP), secreted into the portal blood perfusing the median eminence, stimulate ACTH secretion by the anterior pituitary gland (Fig. 28.1). ACTH, in turn, stimulates biosynthesis and release of glucocorticoids from the adrenal cortex. Cortisol, the major glucocorticoid in both the fetal and adult sheep, stimulates 17α-hydroxylase and 17,20 lyase activities in the placenta,

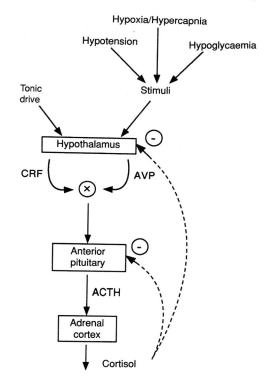

Fig. 28.1 An overview of the ovine fetal hypothalamus–pituitary–adrenal axis. –, Inhibition; ×, a multiplication effect.

which increase placental production of oestrogen and decrease placental production of progesterone. Zonation of the ovine fetal adrenal cortex is similar to that of the adult gland, in that there is no distinct 'fetal zone'.

In the primate, increases in portal plasma concentrations of CRF and AVP stimulate ACTH secretion, and ACTH stimulates adrenal steroidogenesis (Fig. 28.2). However, the hypothalamus–pituitary–adrenal axis in the primate fetus differs from that in the sheep fetus because the scheme of adrenal steroidogenesis is different. Two distinct zones can be discerned histologically in the primate (and human) adrenal cortex. The majority of the adrenal

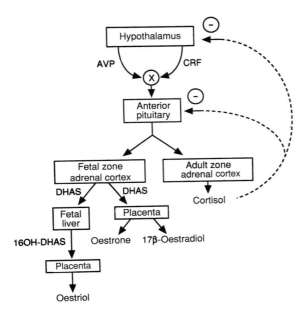

Fig. 28.2 An overview of the primate fetal hypothalamus–pituitary–adrenal axis. Symbols as in Fig. 28.1.

cortex is comprised of the so-called 'fetal zone', and only a small portion functions as the 'adult' or 'definitive' zone. The fetal zone lacks 3β-hydroxysteroid dehydrogenase activity and therefore secretes mainly dehydroepiandrosterone sulfate (DHAS); the adult zone secretes glucocorticoids and mineralocorticoids. Steroidogenesis is stimulated in both zones of the fetal adrenal by ACTH. The fetal zone also responds to human chorionic gonadotrophin (hCG), and it has been proposed that hCG secreted by the placenta stimulates fetal DHAS production early in gestation.

In the primate and human, ACTH increases production of DHAS from the fetal zone and cortisol from the adult zone. Steroidogenesis proceeds in both the fetal and adult zone by synthesis *de novo* from low-density lipid (LDL)-cholesterol. DHAS is hydroxylated by the fetal liver, to 16-hydroxy-dehydroepiandrosterone sulfate (16-OH-DHAS), and converted to oestriol by the placenta. DHAS can be converted by the placenta to adrenostenedione and then to oestrone or to testosterone. Testosterone is converted to 17β-oestradiol. The fetal adrenal gland and placenta which interact to effect oestrogen synthesis, have together been termed the 'feto-placental' unit. Glucocorticoids do not induce 17α-hydroxylase and 17,20 lyase activities in the placental of these species. However, through modulation of the activity of the feto-placental unit, circulating concentrations of oestrogen are raised by

increases in fetal ACTH secretion and lowered by reductions in fetal ACTH secretion.

The ontogeny of the fetal hypothalamus–pituitary–adrenal axis

Because the fetal adrenal axis is responsible for the initiation of parturition in the sheep, many studies have been performed to investigate the normal development of the axis *in utero*. Serial measurements of fetal plasma cortisol concentrations in chronically catheterized fetal sheep have revealed that, throughout much of normal fetal life, plasma cortisol concentrations are lower than maternal plasma cortisol concentrations. Late in gestation, fetal cortisol concentrations begin to increase, and continue to increase in a semilogarithmic fashion until parturition (Fig. 28.3). Like cortisol, fetal plasma ACTH concentration is low throughout much of gestation. Fetal plasma ACTH concentration also increases semilogarithmically starting either simultaneous with, or after, the increase in fetal plasma cortisol (Fig. 28.3). In many studies, it is apparent that there is not a clear correlation between fetal plasma ACTH and cortisol concentrations. Other factors are therefore important in elevating cortisol secretion rate.

One factor that produces an apparent dissociation of the ontogenetic changes in fetal plasma ACTH and cortisol concentrations is increased fetal plasma transcortin

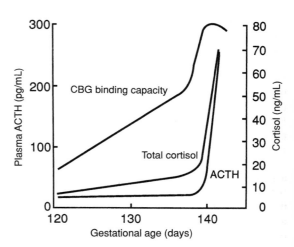

Fig. 28.3 The ontogeny of ovine fetal plasma adrenocorticotrophic hormone (ACTH) and total cortisol (both bound and unbound) concentrations and the binding capacity of corticosteroid-binding globulin (CBG).

(corticosteroid binding globulin, CBG) concentration (Fig. 28.3). Because most assays measure the total plasma concentration of cortisol, increases in circulating CBG concentrations would increase total measured concentrations. While increased CBG does contribute to the elevated total fetal cortisol, estimation of free plasma cortisol suggests that the free concentrations are increased as well.

Another factor that produces an apparent dissociation of ACTH and cortisol is increased adrenal sensitivity to ACTH. Injection or infusion of ACTH into the fetal circulation has demonstrated that the fetal adrenal becomes more sensitive to ACTH as the fetus matures. This is caused, in part, by the growth of the fetal adrenal which accelerates at the end of the gestation, and in part by an increase in sensitivity to ACTH at the cellular level. ACTH binding and adenylate cyclase activity are increased. The mechanisms controlling these changes in fetal adrenal cortical function are not fully understood. However, it is likely that circulating ACTH augments adrenal maturation. Adrenal sensitivity is known to be dependent, in part, upon adrenal secretion of cortisol because infusion of metyrapone (an 11-hydroxylase inhibitor) reduces the ACTH-dependent augmentation of adrenal sensitivity to ACTH. Other factors may be important, because fetal adrenal cortical cells appear to increase in sensitivity *in vitro*. This suggests that circulating factor(s) maintain low adrenal sensitivity, and that the increased adrenal sensitivity at the end of gestation results from reduction of plasma levels of these factors.

It is possible that plasma ACTH and cortisol concentrations are dissociated because of changes in the relative bioactivity of circulating ACTH in late gestation. In most studies of the fetal hypothalamus–pituitary–adrenal axis, fetal plasma ACTH concentration is measured by radioimmunoassay. Depending upon the cross-reactivity of the antiserum used and the method of extraction of ACTH from plasma, the values for plasma ACTH concentration represent a function of the sum of various proopiomelanocortin (POMC)-related peptides. Several groups of investigators have found that, in addition to ACTH, larger-molecular-weight POMC-derived peptides circulate in fetal plasma. It is possible that these other forms of 'immunoreactive ACTH' either add to or modulate the action of ACTH at the adrenal cortex. In one study of 'immunoreactive ACTH' and biologically active ACTH, the 'b/i ratio' (the ratio of biological activity to immunological activity) during stress increased at the end of gestation. The change in b/i ratio most probably reflects secretion of different proportions of large- and small-molecular-weight forms of immunoreactive ACTH.

Relatively little is known about the maturation of the fetal pituitary. The anterior pituitary contains two sub-populations of corticotrophs. The number of 'fetal' corticotrophs is reduced and the number of 'adult' corticotrophs is increased near term. Injection of CRF, AVP, or CRF plus AVP into the fetal circulation at various gestational ages demonstrates that the fetus secretes ACTH in response to these releasing factors, and that the responsiveness of the corticotrophs is reduced in late gestation. The reason for this reduced responsiveness is unclear at the present time.

Evidence suggests that the ontogenetic increases in fetal plasma ACTH concentration are produced by an increasing drive to ACTH secretion (Fig. 28.1). In adrenalectomized fetal sheep, which do not have increasing fetal plasma cortisol concentration and therefore have a constant negative feedback signal, the ontogenetic increases in fetal plasma ACTH are exaggerated. It is possible that the drive to ACTH secretion represents neuronal maturation within the fetal central nervous system. It is not clear, however, whether this drive to ACTH is a hypothalamic drive to ACTH secretion mediated by increasing portal plasma concentrations of CRF and AVP, or whether this drive represents increasing concentrations of hormones or other substances released from the placenta into the fetal circulation. Some investigators have suggested that there is a direct release of ACTH or CRF by placenta into fetal plasma. However, studies in chronically catheterized fetal sheep demonstrate that the rise in fetal plasma ACTH observed in that species at the end of gestation does not originate from the placenta and is not the result of CRF released from the placenta. It has been proposed that prostaglandin E_2 (PGE$_2$) secreted by the placenta at the end of gestation might be an important stimulus for increased secretion of ACTH and cortisol in the fetus. In the sheep, high levels of PGE$_2$ stimulate the secretion of ACTH and cortisol. However, increases in arterial plasma PGE$_2$ concentrations that mimic the ontogenetic rise occurring spontaneously do not stimulate ACTH secretion from the fetal pituitary. Recent evidence suggests that PGE$_2$, generated locally within the cerebral vasculature, might be a stimulus to pituitary ACTH secretion, and that circulating arterial plasma PGE$_2$ might release ACTH immunoreactivity from an extrapituitary site.

It is possible that ACTH is secreted from sources other than the fetal pituitary. ACTH-like immunoreactivity has been identified in placental tissue from both humans and sheep. The mRNA for proopiomelanocortin has been found in human placental tissue, but has not been identified in the placenta of the sheep. Placental tissue has a low concentration of ACTH immunoreactivity relative to that in the anterior pituitary. The concentration of ACTH immunoreactivity in human placenta has been reported to be 1–20 ng/g tissue wet weight. Therefore, the total content of ACTH immunoreactivity in human placenta is approximately 0.5–10 μg ACTH. This total content is low compared to that of the human anterior pituitary (approximately 500 μg ACTH). In the human, the placental processing of the proopiomelanocortin is more similar to the processing in the intermediate lobe than to that in the anterior lobe of the pituitary. That is, the placenta produces proportionately more α melanocyte-stimulating hormone (αMSH) and proportionately less 1–39 ACTH. Placental tissue from humans releases ACTH *in vitro*. The ovine placenta also contains ACTH immunoreactivity; however, chronic catheterization studies in sheep have not demonstrated any release of ACTH into the fetal circulation under either basal or hypoxia-stimulated conditions. It is possible that placental ACTH is released during uterine contractions, contributing to the fetal plasma ACTH concentration during labour. It is also possible that placental ACTH serves a paracrine function within the placenta, and that it does not contribute appreciably to circulating concentrations of ACTH in the fetus.

It is also possible that the placenta secretes CRF into fetal plasma. Ovine fetal plasma concentrations of CRF are normally low (5–10 pg/mL), but increase near the time of spontaneous parturition (30–80 pg/mL). The fetal plasma concentrations of CRF in the sheep fetus just prior to parturition are comparable to those in human umbilical cord blood at the time of birth. The content of CRF in ovine placental tissue is low but measurable. Chronic catheterization studies in the sheep have not demonstrated any consistent release from the placenta into fetal blood under either basal or hypoxia-stimulated conditions. However, infusion of adrenaline into the maternal uterine artery, a manoeuvre that reduces maternal placental blood flow and thereby produces placental ischaemia (perhaps similar to labour contractions) promotes the release of CRF into the fetal blood. While it is not clear that CRF is synthesized in the ovine placenta, and it seems that the normal ontogenetic rises in fetal plasma CRF and ACTH are not derived from placental secretion of CRF, it is possible that CRF might be released from the placenta during labour. It is not known at the present time whether this release of CRF represents wash-out of CRF destined for a paracrine action within the placenta or whether it ultimately stimulates fetal pituitary ACTH secretion.

Maternal plasma concentrations of CRF in the human are quite high and are thought to result from secretion by the placenta and the presence of a CRF binding protein which circulates in maternal plasma. The function of the high concentrations of CRF in maternal plasma is not known, since the bound CRF is biologically inactive. Maternal plasma concentrations of CRF in the sheep are low, in contrast to these in the human, and do not increase

as a function of fetal gestational age. It is possible that much of the difference between maternal plasma CRF concentrations in the sheep and human might be explained by the lack of binding protein for CRF in ovine plasma and the lower placental CRF concentration in ovine placenta.

It is likely that ACTH and CRF do serve a paracrine function in the placenta. The human placenta secretes ACTH *in vitro*. The rate of ACTH secretion by the placenta is stimulated by CRF, PGE_2, and prostaglandin $F_{2\alpha}$ ($PGF_{2\alpha}$). Glucocorticoids increase the rate of CRF secretion by human amnion. Interestingly, the rate of secretion of PGE_2 and $PGF_{2\alpha}$ are increased by CRF, an effect that is partially blocked by administration of anti-ACTH antiserum. It has therefore been proposed that there is a positive feedback loop within the placenta, such that CRF stimulates the production of prostaglandins and the increased levels of prostaglandins increase the rate of secretion of CRF.

Fetal adrenal gland and fetal development

Much of the maturation of the fetal viscera is accelerated by glucocorticoids from the fetal adrenal. Organ systems known to be affected by glucocorticoids include the lung, gastrointestinal tract, liver, and kidneys. One of the most studied effects of glucocorticoids in the fetus is that on the lung. Cortisol stimulates the production of all the components of pulmonary surfactant. Fetuses born prematurely are not exposed to the high plasma concentrations of glucocorticoids at the end of fetal life and are therefore at a higher risk for respiratory distress, or hyaline membrane disease. This is primarily a lack of pulmonary surfactant, producing distress because of the inability of the neonate to ventilate its lung effectively. Because steroids readily pass between the maternal and fetal circulations, it is possible to infuse synthetic glucocorticoids into the maternal circulation to stimulate pulmonary maturation in fetuses at risk for premature delivery. Infants treated in this manner *in utero* have a lower incidence of the respiratory distress syndrome (RDS) than those not treated with glucocorticoids (see Chapter 11).

Fetal adrenals and parturition

Prematurity is an important cause of morbidity and mortality in human infants. A thorough understanding of the mechanism of the initiation and completion of parturition in the human would be of enormous and obvious clinical importance. In sheep, destruction of the fetal pituitary gland or removal of the fetal adrenal glands delayed or blocked parturition. Conversely, infusion of ACTH or glucocorticoid into the fetal circulation stimulated premature parturition. These observations established a central role of the fetal hypothalamus–pituitary–adrenal axis in the initiation of parturition.

The ontogenetic increase in fetal plasma cortisol concentration at the end of gestation stimulates the production of 17α-hydroxylase and 17,20 lyase activities by the placenta (Fig. 28.4). The increased enzyme activity increases the production of oestrogen and decreases the production of progesterone by the placenta. The increased oestrogen/progesterone ratio stimulates increased activity of the uterine muscle (increasing the frequency of contractions) and increases the sensitivity of the uterine muscle to oxytocin.

The human and primate placentae do not respond to corticosteroid with the changes in 17α-hydroxylase and 17,20 lyase activities. However, fetal ACTH stimulates DHAS production by the fetal zone of the fetal adrenal and the DHAS, in turn, is ultimately converted to oestrogens. ACTH in the primate fetus, therefore, increases oestrogen production via a mechanism that is different from that in the sheep. While the link between ACTH and oestrogen biosynthesis is well-established, the role of fetal ACTH in the initiation of parturition in the primate or human is not firmly established. The events leading to parturition are more complicated. For example, the absence of the fetal brain (anencephaly) increases the variability in the timing of parturition but does not greatly alter the average length of gestation.

The mechanism of the increase in fetal plasma cortisol that stimulates parturition in the sheep is only partially understood. Ultimately, it will be important to understand the mechanism of the ontogenetic increase in fetal plasma ACTH. For example, it will be necessary to determine whether all of the immunoreactive ACTH in fetal plasma originates in the fetal pituitary. It will also be necessary to determine whether the increase in fetal ACTH is caused ultimately by fetal neuronal maturation, relatively independently of other hormones or circulating factors. It is possible that some of the ACTH in the last days or hours of fetal life prior to delivery is secreted from an extrapituitary source and that the extrapituitary ACTH provides the stimulus for parturition.

In summary, the human and primate fetuses appear to be somewhat different to the sheep fetus in the zonation of fetal adrenal, the stimulation of placental oestrogen production via a mechanism not related to induction of placental steroidogenic enzyme activity, and in that control of parturition by the human or primate fetal pituitary is not obligatory. However, it is possible that the two species

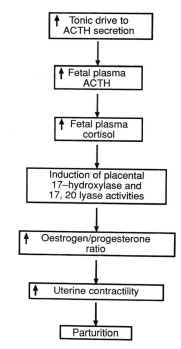

Fig. 28.4 A simplified scheme of events leading to parturition in the fetal sheep.

share one or several components of the scheme controlling parturition. Hopefully, a more complete understanding of the control of parturition in the sheep will improve our understanding of the process in the human.

Fetal stress

'Stress' is a term that was first adapted to physiological systems by Hans Selye, to describe what appeared to be non-specific noxious stimuli. The responses to stress were characterized by what we now know to be the results of hypersecretion of ACTH and glucocorticoid secretion, including the appearance of peptic ulcers, adrenal hypertrophy, and thymic involution. Yates later suggested that the term 'stress' be used to describe any condition that increases plasma glucocorticoid concentration above that expected for that time of day (i.e. above any expected circadian increases in corticosteroids). Fetal 'stress' is therefore probably best defined as a noxious stimulus that elevates fetal plasma ACTH concentration above that expected at a particular gestational age.

The criterion of elevation in plasma cortisol does not always hold true in the fetus, because adrenal sensitivity to

ACTH varies as a function of gestational age. For example, the adrenal sensitivity to ACTH is high at approximately 60 days of gestation in the fetal sheep (term is approximately 147 days in this species), but quite low between 100 and 120 days. After 120 days of gestation, the sensitivity of the adrenal increases dramatically until parturition occurs. Therefore, fetal 'stress' at 120 days of gestation is marked by increases in fetal plasma ACTH but little increase in fetal plasma cortisol concentration, and fetal 'stress' at 140 days is marked by increases in both ACTH and cortisol.

Repeated fetal stress 'matures' the fetal hypothalamus–pituitary–adrenal axis and hastens labour and delivery. For example, repeated periods of hypoxia in chronically catheterized fetal sheep stimulate increases in fetal plasma cortisol concentration of increasing magnitude. This is most probably due to the maturation of the fetal adrenal cortex in response to the repeated elevations in fetal ACTH concentration. The final result of the repeated hypoxia is premature parturition. Indeed, ovine fetal stress caused by any stimulus causes premature parturition if the stress causes elevations in fetal cortisol concentration. Most commonly in the laboratory setting fetal surgery for chronic catheterization or instrumentation often advances the timing of parturition.

The role of the hypothalamus–pituitary–adrenal axis in the survival of stress

Maintenance of circulating concentrations of glucocorticoids are required for maintenance of vascular reactivity to vasoconstrictors for maintenance of normal blood volume in adult animals. Increases in plasma glucocorticoid concentrations after various 'stresses' are homeostatic. For example, increases in plasma cortisol concentration in dogs after haemorrhage are required for full restitution of plasma proteins.

The role of the hypothalamus–pituitary–adrenal axis in maintaining basal cardiovascular function and in homeostatic responses to 'stress' in the fetus might be somewhat different than in the adult. In the period of fetal life during which fetal adrenal sensitivity is low (approximately 100–120 days of gestation), most of the circulating glucocorticoids in fetal plasma originate in the maternal adrenal gland and are transferred to the fetal circulation across the placenta. After the activation of the fetal hypothalamus–pituitary–adrenal axis at approximately 125–130 days of gestation, a greater and increasing proportion of fetal plasma cortisol originates from the fetal adrenal. Therefore, because the fetus benefits from a constant infusion of cortisol from its mother, adrenalectomized fetuses do not demonstrate any signs of adrenal insufficiency.

Adrenalectomy of the fetus does not impair blood volume or lead to vascular collapse as it would in the adult and does not impair the rate of blood volume or plasma protein restitution in the fetus. The relative independence of blood volume regulation from the hypothalamus–pituitary–adrenal axis might result from the ready supply of fluid available for support of fetal blood volume and the relative high permeability of the fetal capillaries to solutes.

Although the interactions between the hypothalamus–pituitary–adrenal axis and the cardiovascular system have been well studied in the sheep fetus, relatively little is known about the possible role of cortisol in survival of other stresses in the fetus. It is possible, however, that fetal cortisol responses to stress are less important in homeostasis than in the adult because of the influence and ready supply of maternal cortisol.

Mechanisms controlling fetal responses to stress

Several stimuli have been identified that increase fetal plasma ACTH concentration above 'basal' levels at any gestational age. These stimuli include haemorrhage, hypotension, hypercapnic and isocapnic hypoxia, and hypoglycaemia. It is assumed that, in the fetus as in the adult, the ACTH responses to these various stimuli are mediated by specific neural pathways. However, these pathways are less well identified in the fetus than in the adult. Evidence indicates that ACTH secretion is stimulated by activation of central chemoreceptors, and by arterial baroreceptors. The role of other baroreceptors (in the heart and elsewhere) in the control of ACTH secretion is unknown at the present time. It is assumed that ACTH responses to hypoglycaemia are mediated by a glucose-sensitive system in the central nervous system.

The magnitude of the ACTH response to stress is a function of the type of stress, the intensity of the stress, and the circulating concentrations of glucocorticoids before the stress. In the adult animal and human, negative feedback regulation of ACTH secretion by corticosteroids has been characterized as occurring in three separate time domains. Fast feedback inhibits ACTH secretion within seconds of an increase in plasma corticosteroid concentration. This mode of negative feedback has been best characterized in adult rats and dogs, but does not appear to function in either fetal or adult sheep. Intermediate feedback can be demonstrated after approximately 30–45 min, enough time to allow for binding of the steroid hormone to the cytoplasmic receptor, transformation of the receptor, transcription of mRNA, and translation of the mRNA into protein. This mode of negative feedback is an important regulator of plasma ACTH concentration in fetal animals.

Slow negative feedback is observed after increases in plasma corticosteroids for more than 6–8 hours, and is characterized by a reduction in POMC synthesis and therefore a reduction in ACTH content in pituitary corticotrophs. This mode of negative feedback has not been well characterized in fetal animals, and may function differently than in the adult, since the fetal pituitary contains a heterogeneous population of corticotrophs.

The sensitivity of the fetal hypothalamus–pituitary–adrenal axis to inhibition by cortisol negative feedback is influenced by the gestational age of the fetus. Between approximately 120 and 130 days of ovine gestation, increases in fetal plasma cortisol of only 1.6 ng/mL (i.e. approximately 5 per cent of the increase in fetal plasma cortisol observed after severe stress) completely inhibit fetal ACTH responses to hypotension. Within the last 10 days before parturition, the sensitivity of fetal ACTH to cortisol is greatly diminished. Increases in fetal plasma cortisol to maximal physiological concentrations do not inhibit fetal ACTH responses to hypotension. Therefore, fetal ACTH responses to hypotension (and perhaps to other stresses as well) are not inhibited by the high endogenous plasma cortisol concentrations immediately prior to parturition. The mechanism of this reduction in negative feedback sensitivity is not known at the present time.

Cortisol-induced inhibition of ACTH secretion is a function of the concentration of cortisol in plasma that is not bound to plasma proteins (i.e. corticosteroid binding globulin or albumin). For example, the sensitivity of hypotension-stimulated ACTH secretion to cortisol negative feedback is quite different in the 117–131 day fetal sheep compared to the adult sheep, if the comparison is made on the basis of total plasma cortisol concentrations. If the comparison is made on the basis of the unbound plasma cortisol, however, the apparent sensitivity is equal at the two ages. On the other hand, changes in the plasma binding of cortisol do not account for the reduction in negative feedback sensitivity in the near-term fetus.

The inhibition of fetal ACTH secretion by corticosteroids is a function of the input to the fetal hypothalamus (the type of stimulus) as well as gestational age and the timing of the administration of corticosteroid relative to the measurement of ACTH. The ontogenetic increase in fetal plasma ACTH is under negative feedback control at the end of gestation when hypotension-stimulated ACTH secretion is not. This might reflect the involvement of different neural pathways in the ontogenetic or 'tonic' drive to fetal ACTH secretion compared to the pathways involved in the ACTH responses to hypotension, and differential inhibition of transmission in these pathways by corticosteroid-sensitive feedback elements within the fetal central nervous system. This differential regulation might

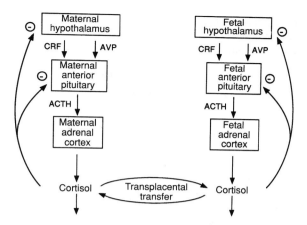

Fig. 28.5 Communication of the maternal and fetal adrenocortical axes.

also reflect the stimulation of ACTH responses to various inputs by differing proportions of CRF and AVP, and a possible alteration in negative feedback sensitivity at the corticotroph by differing proportions of CRF and AVP.

Differential sensitivity of the hypothalamus–pituitary–adrenal axis to negative feedback inhibition has been demonstrated several times in the adult animal. In adult dogs, the sensitivity of basal ACTH secretion to negative feedback inhibition is lower than that of hypoglycaemia-stimulated ACTH secretion. In rats and dogs, some intense stimuli to ACTH secretion (such as haemorrhage producing severe hypotension or laparotomy with intestinal traction in anaesthetized animals) have been termed 'steroid-insensitive' because the ACTH or corticosteroid responses to these stimuli cannot be inhibited with high physiological doses of corticosteroids.

The maternal hypothalamus–pituitary–adrenal axis interacts with the fetal hypothalamus–pituitary–adrenal axis via transplacental transfer of cortisol to the fetal circulation (Fig. 28.5). The inhibition of fetal ACTH responses to hypotension can be demonstrated by infusion of cortisol into the fetal circulation directly, by infusion of cortisol into the maternal circulation, or by infusion of ACTH into the maternal circulation to mimic a response to maternal stress. It is, therefore, apparent that the small increases in fetal plasma cortisol, similar to those that can be measured after fetal stress, regulate the activity of the fetal hypothalamus–pituitary–adrenal axis. The responsiveness of fetal ACTH to stimulation is therefore influenced by the past activity of the maternal hypothalamus–pituitary–adrenal axis: maternal stress inhibits subsequent fetal responses to stress.

Conclusion

The hypothalamus–pituitary–adrenal axis is perhaps the most actively investigated endocrine axis of the fetus. Much of fetal well-being hinges on the proper function of this axis. Control of fetal corticosteroid concentrations at appropriate levels is required for much of normal fetal visceral development and the proper timing of parturition relative to fetal development. If the fetus is born prior to the production of pulmonary surfactant, for example, the risk of respiratory distress in the new-born is increased

While the fetal sheep has been the animal model most frequently used in the study of this endocrine axis, we know that substantial differences in the function of the axis exist between the sheep and the human. However, we also know that there are similarities. The challenge for future investigations will be to more fully understand the mechanisms controlling this axis in the sheep, and to identify the essential similarities in the primate and human fetus. Ultimately, a more complete understanding of these control mechanisms in the sheep might provide the key to understanding normal parturition and prematurity in the human.

Future reading

Carr, B.R. and Simpson, E.R. (1981). Lipoprotein utilization and cholesterol synthesis by the human fetal adrenal gland. *Endocrine Reviews*, **2**, 306–26.

Challis, J.R.G. and Brooks, A.N. (1989). Maturation and activation of hypothalamus–pituitary–adrenal function in fetal sheep. *Endocrine Reviews*, **10**, 182–204.

Keller-Wood, M. and Dallman, M.F. (1984). Corticosteroid inhibition of ACTH secretion. *Endocrine Reviews*, **5**, 1–22.

Liggins, G.C. (1976) Adrenocortical-related maturational events in the fetus. *American Journal of Obstetrics and Gynecology*, **126**, 931–41.

Liggins, G.C., Fairclough, R.J., Grieves, S.A., Kendall, J.Z., and Konx, B.S. (1973). The mechanism of initiation of parturition in the ewe. *Recent Progress in Hormone Research*, **29**, 111–59.

Mulchahey, J.J., DiBlasio, A.M., Martin, M.C., Blumenfield, Z., and Jaffe, R.B. (1987). Hormone production and peptide regulation of the human fetal pituitary gland. *Endocrine Reviews*, **8**, 406–25.

Novy, M.J. and Walsh, S.W. (1981). Regulation of fetoplacental steroidogenesis in rhesus macaques. In *Fetal endocrinology* (ed. M.J. Novy and J.A. Resko), p. 65. Academic Press, New York.

Rice, G.E. and Thorburn, G.D. (1989). The gestational development of ovine placental prostaglandin-synthesizing capacity. In *Advances in Fetal Physiology: reviews in honor of G.C. Liggins*, (ed. P.D. Gluckman, B.M. Johnston, and P.W. Nathanielsz), pp. 387–407. Perinatology Press, Ithaca, New York.

Selye, H. (1936). A syndrome produced by diverse nocuous agents. *Nature*, **138**, 32.

Yates, F.E. and Maran, J.W. (1974). Stimulation and inhibition of adrenocorticotropin release. In *Handbook of physiology*. Vol. 4. *The pituitary gland and its neuroendocrine control*, part 2 (ed. E. Knobil and W.H. Sawyer), pp. 367–404. American Physiological Society, Washington, DC.

29. Development of the fetal thyroid system

Delbert A. Fisher and Daniel H. Polk

The development of the thyroid system in the human fetus can be divided into three phases that roughly correlate temporally with the three classic trimesters of pregnancy. Phase I is the early period of embryogenesis, which, for the thyroid system, includes embryogenesis of the hypothalamus, the pituitary gland, and the thyroid gland. Phase II, during the second trimester of pregnancy, is a period of continuing fetal growth and relatively quiescent thyroid function manifest by largely autonomous pituitary and thyroid action. Phase III, including the third trimester and the neonatal period, is the phase of maturation of hypothalamus–pituitary–thyroid interaction and control. Our understanding of thyroid system development during the past 3 decades has arisen from limited human studies and more extensive investigations in two animal models, the fetal-neonatal rat and the fetal sheep. The events of thyroid development are roughly comparable in the three species as indicated in Fig. 29.1, which shows the relative timing of phases. The rat is an altricial species, born hypothyroid and poikelothermic. In contrast, the human and sheep are precocial species with relatively mature thyroid function and homeothermic metabolism at birth. In the rat it is possible to study much of thyroid system development in the neonatal period in the absence of the placenta. The chronically catheterized sheep fetus provides a good precocial model of human thyroid development.

The similarities in the patterns of thyroid maturation in the three species have provided important insights and understanding. However, there are important differences among the species which must be considered in the interpretation of experimental results. First, there are differences in placental structure in the three species. The sheep has an epitheliochorial placenta in which the maternal and fetal circulations are separated by six tissue layers (maternal endothelium, maternal connective tissue, maternal endometrial syncytium, cytotrophoblast, fetal basement

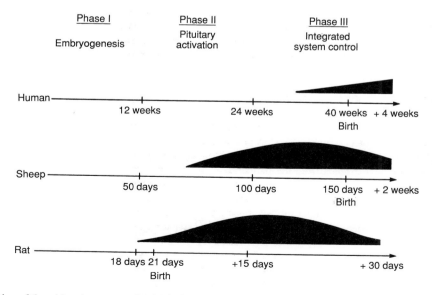

Fig. 29.1 The timing of thyroid system maturation in the human, sheep, and rat. The timing in weeks (for humans), days (for sheep), and days (for rats) is shown below. The periods or portions of periods of hormone-dependent brain maturation are indicated by the darkened areas.

membrane, and fetal endothelium). The rat has a haemotri-chorial placenta with four tissue layers separating maternal and fetal blood (chorionic epithelium, syncytiotrophoblast, cytotrophoblast, and fetal endothelium). The human placenta is haemomonochorial with three tissue layers between the maternal and fetal circulations (syncytiotrophoblast, fetal basement membrane, and fetal endothelium). Thus, differences in placental permeability to various substances (including drugs and hormones might be expected among the three species.

All three placental types express an iodothyronine inner ring monodeiodinase enzyme that degrades thyroxine (T_4) to inactive 3,3′,5′ (reverse) triiodothyronine (rT_3) and deiodinates active 3,5,3′triiodothyronine (T_3) to inactive 3,3′diiodothyronine. As a result of these structural and enzymatic characteristics, the placentae of all three species are relatively impermeable to thyroid hormones and there is a marked maternal-to-fetal gradient of free T_4 and free T_3 across the placental barriers. Moreover, all three placental types are impermeable to the circulating thyroid hormone-binding proteins and to thyroid-stimulating hormone (TSH) from the pituitary gland. The human and rat placentae are permeable to the tripeptide thyrotrophin-releasing hormone (TRH). In contrast the sheep placenta is impermeable to TRH. There are quantitative differences in placental iodothyronine permeability in the three species. The sheep placenta is less permeable to maternal-fetal iodothyronine transfer than the rat or human placentae, presumably due to its structural complexity; there is little or no placental transfer of thyroid hormone in the sheep. In contrast, limited amounts of thyroid hormones do cross the placental barrier in the rat and human.

The three species also differ with regard to the timing of thyroid-hormone-dependent brain maturation relative to intrauterine development. In the infant rat, the major period of thyroid dependency of brain maturation extends from birth, or slightly before, to about 40 days of postnatal age. In the sheep, thyroid-dependent brain maturation extends from 70 to 90 days of gestation to several postnatal weeks. In the human infant the period of thyroid-dependent brain development extends from the perinatal period to about 2 years of postnatal age (Fig. 29.1). Much of our detailed information regarding the effects of thyroid hormone on brain maturation and function has been obtained from the neonatal rat because of its relative availability.

Phase of embryogenesis

Embryogenesis of the hypothalamus–pituitary thyroid system has been characterized in some detail in all three species. The present brief review will focus on the human system. As indicated, the hypothalamic and pituitary structures are quite well developed by 12 weeks, and the hypothalamic nuclei are apparent by 16 weeks of gestation. TRH is identifiable in low concentrations in hypothalamic tissues between 10 and 20 weeks; there is little information in the human fetus thereafter. Human fetal pituitary TSH concentrations are detectable at 8–10 weeks, remain low from 16 to 18 weeks, and then increase progressively to term.

The pituitary portal blood vascular system develops initially as a network of capillaries within the proliferating anterior pituitary mesenchymal tissue around Rathke's pouch. Definitive pituitary portal vessels are present by 12–17 weeks. A superficial system of capillaries also develops in association with the diencephalon and matures progressively to form the rostral (primary or hypothalamic) plexus of the portal vascular system. This primary plexus is first visible by 16–19 weeks. There is a progressive maturation of the plexus thereafter with increasing tortuosity and development of capillary tufts that progressively increase in volume and penetrate hypothalamic tissue. This vascular maturation as well as histological maturation of the hypothalamus is largely complete by 30–35 weeks.

The thyroid gland of the human embryo is derived from a midline outpouching of the endoderm of the floor of the primitive buccal cavity. Lateral extensions (ultimobranchial anlagen) from the fourth pharyngeal pouches contribute the thyroid interstitial calcitonin-secreting (C) cells. The thyroid develops as a flasklike vesicle with the narrow neck attached to the buccal cavity. The vesicle enlarges, becomes bilobed, separates from the buccal cavity, and migrates caudally. By the end of the seventh week the thyroid weighs 1–2 mg and has assumed its definitive position in the anterior lower neck. Histologically, the gland develops in three stages; precolloid, early colloid, and follicle. The factors controlling thyroid gland development are not well understood. TSH does not appear to be necessary for thyroid gland embryogenesis. Growth factors, acting via autocrine or paracrine routes, may be involved.

Phase II maturation

During the midperiod of human gestation, embryogenesis is largely complete and organ system functions can be demonstrated. However, endocrine and neuroendocrine control systems remain quite immature. With regard to thyroid function, the human placenta provides iodide substrate and produces significant amounts of TRH. TRH i also synthesized by the fetal pancreas and other tissues. Fetal serum, in contrast to adult serum, has little or no TRH degrading enzyme activity so that the circulating half-life of TRH is relatively prolonged; thus the serum concentrations of TRH in the mid-gestation fetus are quite high. This circulating TRH is largely extrahypothalamic in origin; TRH concentrations in hypothalamic tissue remain relatively low until phase III maturation.

Fetal serum T_4 and TSH concentrations remain low between 15 and 18 weeks of gestation. At this time, serum TSH levels begin to increase. Early studies of human fetal serum TSH levels from the umbilical cord in premature deliveries suggested a rapid increase in concentration, with a peak level of 10–15 mU/L at 22–24 weeks and a slow decline thereafter to a value approximating 10 mU/L at term. Recent data from *in utero* fetal cord blood sampling suggest a progressive increase in fetal serum TSH between 18 and 20 weeks and term. Figure 29.2 summarizes TRH levels in extrahypothalamic (shown as pancreas) and hypothalamic tissues and TSH levels in pituitary and serum of the human fetus. Fetal pituitary TSH content and fetal serum TSH concentrations begin to increase at mid-gestation and, because hypothalamic TRH levels are quite low, it is possible that extrahypothalamic TRH may be stimulating pituitary TSH release at this time. Alternatively, autonomous pituitary function may be involved. In addition, the progressively increasing pituitary and serum TSH concentrations correlate temporally with the maturation of the pituitary portal vascular system so that hypothalamic TRH may also be playing a role at this time (Fig. 29.2).

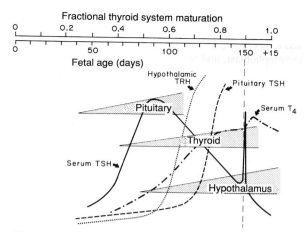

Fig. 29.3 Timing of maturation of hypothalamic TRH, serum and pituitary TSH, and serum T_4 concentrations in the fetal and neonatal sheep. The periods of maturation of pituitary, thyroid, and hypothalamic function are indicated by the hatched areas. See text for details.

More complete data are available for the fetal sheep model shown in Fig. 29.3. At mid-gestation in this species, gut and placental TRH concentrations greatly exceed hypothalamic TRH concentrations. Available data indicate a mid-gestation peak in serum TSH with later increases in hypothalamic TRH and pituitary TSH concentrations. The progressive increase in fetal serum T_4 concentrations after mid-gestation resembles that in the human fetus. These

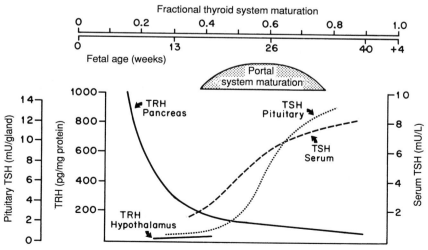

Fig. 29.2 Timing of changes in pancreatic TRH, early hypothalamic TRH, and pituitary and plasma TSH concentrations in the human fetus. The period of maturation of the pituitary portal vascular system is shown as the hatched area. See text for details.

data also suggest an early pituitary phase of control, perhaps modulated, at least in part, by extrahypothalamic TRH. This phase is followed by progressive maturation of hypothalamic control of pituitary TSH secretion.

Phase III maturation

The final phase of thyroid system development includes several simultaneous maturational events. These include: the maturation of hypothalamus–pituitary neuroendocrine control of TSH secretion; the maturation of thyroid gland responsiveness, the maturation of thyroid hormone metabolism; and the maturation of thyroid receptor systems and actions.

Maturation of neuroendocrine control

The events of pituitary and hypothalamic maturation with regard to thyroid function have been studied to varying degrees in rat, sheep, and man. Available information is most complete in the developing rat and, where species comparisons are available, there is good agreement. The approximate timing of the events of hypothalamus–pituitary functional maturation in the rat is summarized in Table 29.1. As indicated in Table 29.1, the capacity for

pituitary TSH synthesis is present early, and pituitary TRH and nuclear T_3 receptors are present in near-adult concentrations in the new-born rat; serum TSH responses to TRH and T_3 are present at this time. The pituitary enzyme, iodothyronine 5′monodeiodinase, which catalyses intrapituitary conversion of T_4 to T_3 is also present at birth in the rat. T_3 inhibition of TRH receptor binding has been observed in the 10-day-old rat. A TRH effect on glycosylation of TSH *in vivo* was not observed at 5 days but was clearly present at 56 days in the rat. Inhibition of pituitary T_3 receptor binding by propylthiouracil-induced hypothyroidism is observed at 14 days after birth (about 0.7 of thyroid system maturation). These data indicate that the pituitary cellular mechanisms for modulation of TSH synthesis and release are largely mature by 0.4–0.6 of thyroid system maturation in the rat. This corresponds to the mid or late second trimester of human development.

In contrast, hypothalamic maturation is relatively delayed. Hypothalamic control of pituitary thyrotroph cell function is quite complex. The major factors include the modulation of TRH synthesis and release into the pituitary portal vascular system (to stimulate pituitary TSH synthesis and release) and hypothalamic production of somatostatin and dopamine, which tend to inhibit pituitary TSH release. One of the factors influencing hypothalamic TRH production is body temperature. The effect of temperature

Table 29.1 The timing of events of hypothalamic pituitary functional maturation in the rat

Event	Time of appearance*
Maturation of pituitary thyrotroph function	
TSH synthesis present	< 0.3
TRH receptor present	< 0.4
TSH response to TRH	< 0.4
T_3 receptor present	< 0.4
T_3 inhibition of TSH synthesis and release	< 0.4
T_4 5′monodeiodinase activity present	< 0.4
T_3 inhibition of pituitary TRH binding	< 0.6
TRH effect on TSH glycosylation	< 0.6
T_3 stimulation of T_3 receptor binding	< 0.7
Maturation of hypothalamic TSH control	
TRH synthesis	< 0.4
TRH stimulation of TSH	0.4
T_3 inhibition of TRH synthesis	0.5
TSH response to cold	1.0
Somatostatin inhibition of basal TSH secretion	> 1.0

*Fractional proportion of thyroid system maturation time: complete maturation time = 1.0. Figures following the symbol '<' indicate the earliest time at which studies were conducted to assess the maturation event.

can be assessed by measuring TSH secretion in response to cold exposure. This response is observed at 3–4 weeks of life in the rat (i.e. at the completion of thyroid system development). Somatostatin inhibition of TRH-stimulated TSH secretion is present at 3–4 days in the neonatal rat; however, there is no effect of somatostatin antiserum on basal TSH levels in the postnatal rat until 10 days of age. Dopamine receptor blockade has no effect on serum TSH levels in rats during the first 40 days of postnatal life. T_3 inhibition of hypothalamic TRH synthesis is observed in the rat by 0.5 of gestation. These data are summarized in Table 29.1. Examination of Table 29.1 indicates that pituitary thyrotrophic function matures earlier than the more complex integrative hypothalamic events that modulate pituitary TSH secretion. This is true in all three species and is illustrated for the fetal sheep in Fig. 29.3.

Maturation of thyroid gland responsiveness

There is a progressive maturation of fetal thyroid gland responsiveness to TSH during the last half of human or ovine gestation. This has been best quantified in the sheep where a progressive increase in the serum T_4 response to TRH-stimulated endogenous TSH release has been observed during the third trimester of gestation. Whether this involves TSH receptor and/or TSH postreceptor responsiveness is not yet clear.

The mature thyroid follicular cell can modify iodine uptake relative to blood iodine levels despite variations in serum TSH concentrations. The immature thyroid gland lacks this autoregulatory mechanism. Thyroid follicular cell autoregulation of iodide transport develops after 36–40 weeks in the human fetus and at 18–20 days in the neonatal rat. The cellular mechanisms modulating this autoregulation are not yet known. It has been shown that the failure of the immature thyroid follicular cell to exhibit autoregulation relates to the absence or reduced iodination of an 8–10 kDa protein within the follicular cell. The period of maturation of thyroid gland responsiveness to TSH and iodide is shown in Fig. 29.3.

Maturation of thyroid hormone metabolism

The structures of the thyroid hormones and various analogues are shown in Fig. 29.4. Of the analogues shown in Fig. 29.4, only thyroxine, a tetraiodothyronine (T_4), and 3,5,3'triiodothyronine (T_3) are biologically active. T_3 has three to four times the metabolic potency of T_4 as it binds to nuclear thyroid hormone receptors with 10-fold greater affinity than T_4. T_4 is the major thyroid secretory product, and the first step in metabolism of tetraiodothyronine or T_4 is monodeiodination to a triiodothyronine, either bioactive T_3 or inactive reverse T_3 (rT_3). The monodeiodination is mediated by at least three types of particulate cellular iodothyronine monodeiodinase (MDI) enzymes. Type I MDI demonstrates a high affinity for T_4, is inhibited by propylthiouracil (PTU), and stimulated by thyroid hormone; it catalyses monodeiodination of T_4 to T_3 and rT_3 to $3,3'T_2$. The preferred substrate for type I MDI is rT_3. Type II MDI is a low-affinity enzyme, insensitive to PTU and inhibited by thyroid hormone; it catalyses T_4 conversion to T_3; the preferred substrate is T_4. The type III MDI catalyses conversion of T_4 to rT_3. The maturation of tissue MDI activities has been most comprehensively studied in the sheep. The patterns of type I, type II, and type III MDI activities in ovine fetal tissues are summarized in Fig. 29.5. As indicated in Fig. 29.5, the predominant MDI activity by 130 days of gestation in the fetal sheep is type III activity in liver with significant levels of type III activity in kidney and placenta (not shown). There is also type I activity in fetal liver and significant type II activities in fetal brain and brown adipose tissues.

The age-related changes in rT_3 and T_3 concentrations in human fetal serum are shown in Fig. 29.6. Reverse T_3 is the predominant product of T_4 metabolism and levels increase progressively during the last half of gestation. Serum T_3 concentrations increase only at term. The major pathway of T_4 metabolism in the fetus is type III MDI-mediated monodeiodination to inactive rT_3 by fetal liver and perhaps kidney and placenta. Reverse T_3 is probably degraded to $3,3'T_2$ by type I MDI in fetal liver. Although T_4 is actively secreted by the fetal thyroid, its effects are largely neutralized due to rapid conversion to inactive rT_3. Brain and brown adipose tissue can convert T_4 to T_3; this T_3, in contrast to that produced in liver and other tissues, remains sequestered locally, thus serving as the source of T_3 for local thyroid effects. The triiodothyronines are progressively monodeiodinated to a T_2, and T_1 and eventually T_0 by the several MDI activities (see Fig. 29.4). Thyroid metabolism in the fetus seems designed to provide thyroid hormone preferentially to selected tissues, such as brain and brown adipose tissue, while avoiding exposure of most tissues to active T_3.

There are other thyroid hormone metabolic pathways including glucuronide and sulfate conjugations and progressive cleavage of the alanine side chain via decarboxylation and transamination. Thyroid hormones in the adult are excreted as glucuronide and sulfate conjugates in urine and faeces and there is significant recycling of iodothyronines via conjugate hydrolysis and reabsorption. These pathways are more significant in sheep than in humans. However, there is no information regarding maturation of these pathways in the fetus of any species.

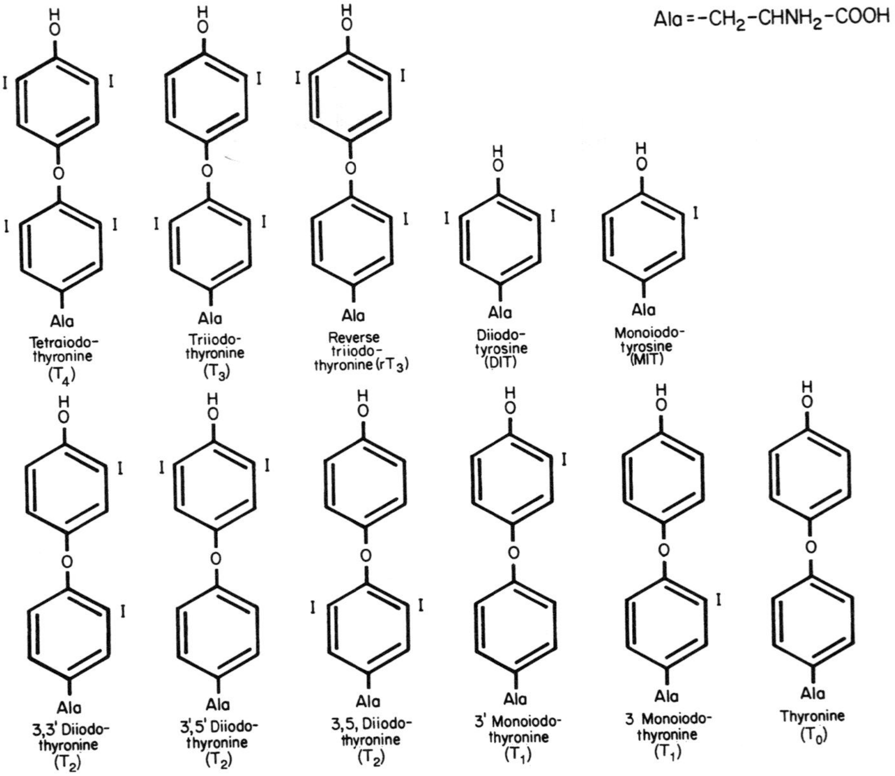

$Ala = -CH_2-CHNH_2-COOH$

Fig. 29.4 Structure of thyroid hormones and selected analogues. Thyroid hormones are synthesized in the thyroid gland as tyrosine residues within the large 660-kDa thyroglobulin molecule. Iodination of the tyrosine residues is mediated by the thyroidal peroxidase enzyme complex producing both monoiodotyrosine (MIT) and diiodotyrosine (DIT) residues. Coupling of MIT and DIT, which are appropriately located sterically within thyroglobulin, is catalysed by peroxidase to form tetraiodothyronine (T_4) or triiodothyronine (T_3). Both are biologically active, but T_3 has three to four times the biological activity of T_4 since it binds with 10-fold higher affinity to thyroid nuclear receptors. Reverse T_3 (rT_3) is biologically inactive as are the several diiodo and monoiodothyronines (T_2s and T_1s). The alanine side chain is shown as Ala. See text for further discussion.

Appearance of thyroid hormone receptors and actions

Thyroid hormone receptors have been identified in plasma membrane preparations, in inner mitochondrial membranes, and within cell nuclei. The significance of plasma membrane and mitochondrial receptors is not clear. It is generally accepted that the important actions of thyroid hormones are mediated via the nuclear receptors. Thyroid nuclear receptors are members of the steroid receptor superfamily that function as DNA-binding transcriptional control factors. Two thyroid hormone receptor genes (TR-α and TR-β) have been characterized in humans and rats. In humans, these genes, localized to chromosomes 3 and 17, generate at least five receptor mRNA species coding for five receptor-related proteins. Of these, the TR-α_1 and TR-β_1 transcripts code for T_3-binding receptor distributed in many tissues; expression of TR-β_2 appears unique to the pituitary gland. Two additional TR-α mRNA species code for non-T_3 binding, DNA-binding proteins of unknown significance.

Studies of T_3 nuclear receptor maturation have been conducted in rats, sheep, and humans, but no comprehensive data are available. These studies have measured T_3-binding by radioreceptor assays that measure both TR-α_1 and TR-β_1 receptor species. In general, the results indicate the appearance of T_3 nuclear receptors in fetal pituitary and brain tissues during the second trimester when T_3 binding (per mg DNA) approximates adult levels. Nuclear T_3 binding has been observed in other fetal tissues including lung, heart, and liver. In the rat and sheep, studies in developing liver show a progressive increase in

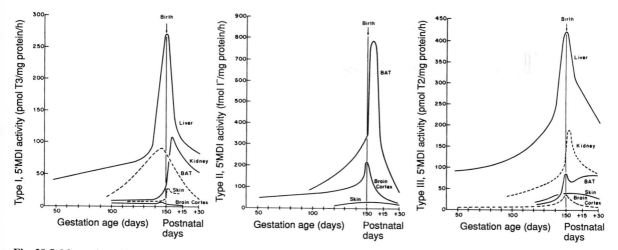

Fig. 29.5 Maturation of iodothyronine monodeiodinase enzyme activities (MDI) in fetal sheep tissues. Type I MDI activities are shown in the left panel in pmol of T_3 generated from T_4 per mg protein per hour. Type II activities are summarized in the middle panel as fmol of iodide generated from radioiodine-labelled rT_3 per mg protein per hour. Type III activities are shown in the right panel as pmol of T_2 generated from T_3 per mg protein per hour. See text for details.

T_3-binding capacity during the final phases of thyroid system development (during the latter third of ovine fetal development and during the neonatal period in rats). Thus, two general patterns of receptor maturation are

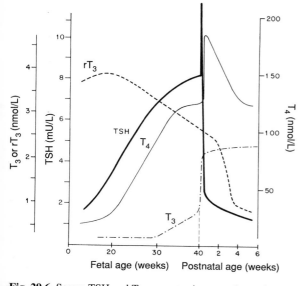

Fig. 29.6 Serum TSH and T_4 concentrations are shown in the latter half of gestation and neonatal period in the human fetus and new-born. Serum 3, 3′, 5′ triiodothyronine (reverse T_3 or rT_3) and 3,5,3′ triidothyronine (T_3) concentrations are also shown. See text for details.

observed: an early pattern in brain and pituitary tissues followed by a late pattern in liver and probably most other tissues.

The time of appearance of thyroid hormone effects has also been studied in the rat, sheep, and human. The most comprehensive data are available from rats and sheep. Data for the developing rat are summarized in Table 29.2. As indicated, there is a marked variation in the time of appearance of thyroid hormone actions in various tissues and some variation of timing within tissues (e.g. liver malic enzyme synthesis versus liver EGF receptor binding). Some of this variation reflects the timing of appearance of T_3 nuclear receptor binding (e.g. brain versus liver). Other variations must reflect other mechanisms such as nuclear receptor–transcription factor interactions, DNA methylation, and translation events.

The pattern of ontogenesis of the actions of thyroid hormone in the human fetus and new-born is derived largely from clinical observations. As in the rat, the ability of thyroid hormones to modulate serum TSH levels appears early. Increased serum TSH levels have been observed in response to hypothyroidism in 22–24 week premature infants (0.5 of thyroid system development). Delayed bone maturation, equivalent to 6–10 weeks of gestation, is observed in one-third to one-half of hypothyroid new-borns suggesting an effect of T_3 on bone maturation by 0.7 of thyroid system development. However, other than the serum TSH concentration and bone maturation, the athyroid human neonate appears normal at birth both biochemically and physically. Moreover, if treatment is begun

Table 29.2. Maturation of thyroid hormone actions in the rat

Effect[*]	Time of appearance[†]
Inhibit pituitary TSH release	0.3
Stimulate brain maturation	0.4
Stimulate pituitary GH synthesis	0.4
Increase skin and eye EGF levels	0.4
Increase skin EGF receptor binding	0.4
Inhibit hypothalamic TRH secretion	0.45
Increase cardiac β-adrenoceptor binding	0.5
Increase hepatic malic enzyme synthesis	0.5
Increase tissue thermogenesis	0.6
Increase kidney EGF levels	0.6
Increase carcass growth	0.6
Increase lung β-adrenoceptor binding	0.7
Increase salivary gland EGF synthesis	0.7
Increase liver EGF receptor synthesis	0.8

[*]GH, Growth hormone; EGF, epidermal growth factor.

[†]Fractional proportion of thyroid system maturation time: complete maturation time = 1.0. The time of appearance is approximate.

within 30 days of birth, growth and development are normal. The classic signs and symptoms of cretinism only appear in untreated infants during the first 3 to 6 months after birth. This suggests either that thyroid hormone has limited effects on the developing human fetus until the perinatal period or that the small amounts of maternal thyroid hormone crossing the human placenta (producing levels at term of 30–70 nmol/L T_4 in serum) are protective, particularly in those fetal tissues with the capacity for local T_4 to T_3 conversion.

Adaptation to extrauterine life

Parturition involves the abrupt fetal transition from the protected intrauterine milieu to the relatively cold extrauterine environment, and separation from the uterine nutrient supply. Both the adrenal cortex and the autonomic nervous system are essential to successful, immediate extrauterine adaptation of the precocial fetus. Thyroid hormones also play a critical role. In most mammals a prenatal cortisol surge initiates a series of metabolic alterations that adapt the fetus for extrauterine survival. Among the effects of the prenatal increase in serum cortisol is stimulation of hepatic type I iodothyronine 5′MDI activity augmenting the conversion of T_4 to T_3 and increasing fetal serum T_3 concentrations. This increase for the human fetus is shown in Fig. 29.6.

Parturition evokes a dramatic increase in catecholamine release, and serum catecholamine concentrations peak in the first minutes after birth. The catecholamine surge is important in the cardiovascular and metabolic adaptation of the neonate. In addition, catecholamines stimulate thermogenesis in fetal brown adipose tissue (BAT), which limits the hypothermia of extrauterine exposure and mediates neonatal homeothermy. BAT is the major site of endogenous chemical (non-shivering) thermogenesis in the mammalian new-born and is especially prominent in precocial species such as the sheep and human. The largest masses of BAT envelop the kidneys and adrenal glands, while smaller amounts surround the blood vessels of the mediastinum and neck. BAT tissue peaks in mass and metabolic activity at the time of birth in precocial species and gradually decreases in volume during the early weeks and months of life.

BAT tissue is rich in mitochondria and contains a unique 32-kDa protein (thermogenin) that uncouples mitochondrial energy production from phosphorylation of ADP to ATP. This uncoupling enhances the generation of heat. Thermogenin levels in BAT are thyroxine-dependent and BAT type II MDI activity provides for optimal local conversion of T_4 to T_3. Type II MDI activity in BAT increases markedly in the neonatal period (Fig. 29.5), perhaps due to catecholamine stimulation.

There is also a dramatic increase in serum TSH concentration in the early minutes after birth in precocial mammals (Fig. 29.6). This TSH surge is stimulated by abrupt cooling of the neonate in the extrauterine environment. The human TSH surge peaks at 30 min after birth and evokes increased T_4 and T_3 secretion by the thyroid gland.

Thyroidectomy of the fetal sheep several days prior to delivery is associated with low and unchanging serum T_4 and T_3 levels in the neonatal period and severe neonatal hypothermia with markedly inhibited thermogenesis in BAT *in vitro*. Thyroidectomy a few hours before birth is associated with normal but unchanging neonatal T_4 levels, failure of elevation of serum T_3 levels during the first 2–4 h of life, and normal BAT thermogenesis. These data indicate that normal fetal thyroid function with normal serum T_4 levels is necessary for normal neonatal BAT thermogenesis and that the early neonatal increase in serum T_3 levels is thyroid-gland-dependent.

Re-equilibration of TSH levels to the normal extrauterine range from the relatively increased intrauterine range (Fig. 29.6) presumably relates to the increase in prevailing serum T_3 levels after birth. The early increase in serum T_3 levels persists indefinitely whereas the neonatal TSH surge is transient. The persistence of relatively increased T_3 levels is due to the permanent increase in T_4 to T_3 conversion conditioned by the increased tissue type I MDI levels in the postnatal period (Figs 29.5 and 29.6). Continuing maturation of negative feedback control of TSH secretion by thyroid hormones probably also contributes to re-equilibration of serum TSH. Reverse T_3 production in neonatal tissues progressively decreases during the early weeks of life, and the high serum rT_3 concentrations characteristic of fetal life gradually decline to characteristic adult levels. The elevation of serum T_3 levels in the neonatal period presumably conditions the gradual increase in non-shivering thermogenesis by non-BAT tissues as well as other late appearing thyroid hormone actions (Table 29.2), since BAT tissue gradually disappears during the early weeks of postnatal life.

Clinical correlations

The complexity of the development of the thyroid system and the delayed maturation of thyroid control during the latter third of gestation in the human fetus lead to a variety of alterations of thyroid function in premature infants. Fetal thyroid function at mid-gestation is characterized by hypothalamic immaturity and pituitary autonomy. Thus the 20 to 25 week premature human infant typically presents with a state of hypothalamic hypothyroidism characterized by low serum free T_4 and TSH concentrations and a normal serum TSH response to TRH. This period of transient hypothyroxinaemia gradually resolves and serum T_4 and free T_4 concentrations progressively increase in the extrauterine environment just as they would have *in utero*. Treatment of such infants is not required.

Other functional and transient thyroid disorders are observed with lesser frequency. These include transient hypothyroxinaemia in euthyroid neonates and delayed maturation of negative feedback control of TSH secretion by thyroid hormones in infants born with hypothyroidism. Neonatal hypothyroxinaemia may last several months in the absence of other manifestations of thyroid dysfunction. The mechanism is not known. A few infants born with congenital hypothyroidism will manifest a delayed increase in serum TSH levels in spite of their low serum free T_4 and free T_3 concentrations. The mechanism for the delay also remains obscure.

Congenital hypothyroidism (sometimes called cretinism) can occur endemically or sporadically. Endemic cretinism is caused by iodine deficiency or iodine deficiency plus endemic goitrogens. This is the most common cause of congenital hypothyroidism and preventable mental deficiency worldwide. The disorder can be prevented by dietary iodine supplementation. However, political, economic, and logistic barriers have prevented effective supplementation in many parts of the world.

Sporadic congenital hypothyroidism is usually caused by defects in thyroid gland embryogenesis (thyroid dysgenesis). Abnormalities include agenesis, hypogenesis, and ectopic glands. Such defects account for 80–90 per cent of infants with sporadic congenital hypothyroidism. Defects in hypothalamus–pituitary embryogenesis, hereditary defects in TRH or TSH synthesis, genetic abnormalities in thyroid TSH receptors, defective thyroid thyroglobulin synthesis, or genetic abnormalities in the thyroid peroxidase enzyme complex, the thyroid iodotyrosine deiodinase system, or the thyroid cell membrane iodide transport system also can lead to congenital hypothyroidism. These disorders account for 10–20 per cent of cases of sporadic congenital hypothyroidism. In rare instances TSH receptor antibodies acquired transplacentally from mothers with autoimmune thyroid disease can produce transient congenital hypothyroidism. Screening programmes have been developed throughout most of the industrialized world to detect elevated serum TSH levels in new-born infants as a biochemical marker for congenital, primary (thyroidal) hypothyroidism. Confirmation of hypothyroidism in infants with elevated TSH levels and early treatment has normalized brain development and childhood IQ values in most affected infants.

Finally, in rare instances, maternal TSH receptor antibodies that stimulate thyroid function have been associated with intrauterine or neonatal hyperthyroidism. This neonatal Graves' disease may be severe and life-threatening, requiring meticulous management with drugs that block thyroid hormone production, drugs that inhibit tissue T_4 to T_3 conversion, and/or drugs that block selective thyroid hormone actions.

Further reading

Fisher, D.A. (1989). The thyroid gland. In *Clinical paediatric endocrinology* (ed. C.D. Brook), pp. 309–37. Blackwell Scientific Publications, London.

Fisher, D.A. (1989). Maturation of thyroid hormone actions. In *Research in congenital hypothyroidism* (ed. F. Delange, D.A. Fisher, and D. Glinoer), pp. 61–77. Plenum, New York.

Fisher, D.A. (1989). Development of fetal thyroid system control, I. In *Iodine and the brain* (ed. G.R. Delong, J. Robbins, and P.G. Condliffe), pp. 167–76. Plenum, New York.

Fisher, D.A. and Klein, A.H. (1981). Thyroid development and disorders of thyroid function in the newborn. *New England Journal of Medicine*, **304**, 702–12.

Fisher, D.A. and Polk, D.A. (1989). Development of the thyroid. *Baillière's Clinical Endocrinology and Metabolism*, **3**, 627–57.

Fisher, D.A., Dussault, J.H., Sack, J., and Chopra, I.J. (1977). Ontogenesis of hypothalamic pituitary thyroid function and metabolism in man, sheep and rat. *Recent Progress in Hormone Research*, **33**, 59–116.

Morreale de Escobar, G. and Escobar del Rey, F. (1980). Brain damage and thyroid hormone. In *Neonatal thyroid screening* (ed. G.N. Burrow and J.H. Dussault), pp. 25–50. Raven Press, New York.

Morreale de Escobar, G., Calvo, R., Obregon, M.J., and Escobar del Rey, F. (1990). Contribution of maternal thyroxine to fetal thyroxine pools in normal rats near term. *Endocrinology*, **126**, 2765–7.

Roti, E. (1988). Regulation of thyroid-stimulating hormone (TSH) secretion in the fetus and neonate. *Journal of Endocrinological Investigation*, **11**, 145–58.

Roti, E., Gnudi, A., and Braverman, L.E. (1981). The placental transport, synthesis and metabolism of hormones and drugs which affect thyroid function. *Endocrine Reviews*, **4**, 131–49.

Taylor, T., Gyres, P., and Burgunder, J.M. (1990). Thyroid hormone regulation of TRH mRNA levels in rat paraventricular nucleus of the hypothalamus changes during ontogeny. *Neuroendocrinology*, **52**, 262–7.

30. Calcium and mineral metabolism in the fetus

Oussama Itani and Reginald C. Tsang

Calcium and mineral metabolism in the fetus represents a delicate and complex biological process comprising a number of intricate and interrelated components. Fetal mineral homeostasis depends on the availability of mineral substrates that are actively acquired from the mother against a concentration gradient, and interactions with the calciotrophic hormones, namely parathyroid hormone (PTH), calcitonin (CT), and 1,25 dihydroxyvitamin D $[1,25(OH)_2D]$.

Mineral physiology

Body mineral content

Calcium is the fifth most abundant inorganic element in the human body. The adult human body contains about 1200 g of calcium (19 g of Ca/kg body weight). The total body Ca content in a full-term new-born is approximately 28 g, almost all of which (99 per cent) resides in bone

where it serves a structural function. The remainder resides in body fluids and serves a crucial role in a multitude of physiological processes involving muscular contraction, neurotransmission, membrane transport, enzyme reactions, hormone secretion, and blood coagulation. In the term infant, the skeletal system contains 8 g of Ca/kg body weight. Calcium accretion during the latter half of gestation is shown in Fig. 30.1.

The adult human body contains about 1000 g of phosphorus mostly in the skeleton. Total body P in the term new-born infant is 16 g (4.5 g of P/kg body weight), 80 per cent of which is concentrated in the hydroxyapatite crystal lattice of bone, 9 per cent in the skeletal muscle, and the remainder in the viscera and extracellular fluid. Normally, calcium and phosphorus are deposited in bone in a ratio of 2.1:1. Phosphorus accretion during the latter half of human gestation is shown in Fig. 30.2.

Magnesium is the fourth most abundant cation and the second most abundant intracellular cation within the body. Total body Mg content in the new-born amounts to 0.8 g

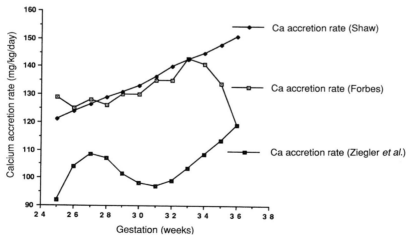

Fig. 30.1 Calcium accretion in the human fetus during the latter half of gestation. Data are from independent studies. In general the rate of calcium accretion in the fetus increases with gestation, most remarkably in the third trimester. (Data adapted from Ziegler, E., O'Donnel, A., Nelson, S., and Fomon, S. (1976). *Growth*, **40**, 329–41; Forbes, G. (1976). *Pediatrics*, **57**, 976–7; Shaw, J. (1973). *Pediatric Clinics of North America*, **20**, 333–58.)

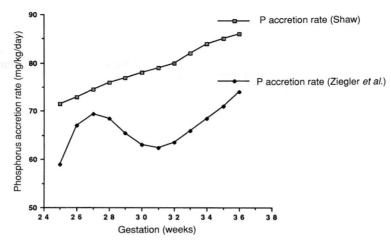

Fig. 30.2 Phosphorus accretion in the human fetus during late gestation. The rate of phosphorus accretion in the human fetus increases with advancing gestational age. (Data are from two studies: Ziegler, E., O'Donnel, A., Nelson, S., and Fomon, S. (1976). *Growth*, **40**, 329–41; Shaw, J. (1973). *Pediatric Clinics of North America*, **20**, 333–58.)

(0.22 g of Mg/kg body weight), most of which (50–60 per cent) is concentrated in bone tissue as an integral component of the hydroxyapatite lattice (30–40 per cent) and as an exchangeable fraction (15–20 per cent) adsorbed to apatite and in equilibrium with the extracellular fluid compartment. About 20 per cent of total body Mg is concentrated in muscle, and another 20 per cent is in the intracellular compartment of blood cells and other body tissues. Changes in total body Mg content are largely reflected in changes in skeletal Mg and to a lesser extent in serum Mg concentrations. Magnesium has a significant regulatory role in a multitude of biological processes involved in the storage, transfer, and production of energy. Further, Mg plays a significant role in calcium and bone homeostasis. Mg cations may contribute to the control of the onset of bone matrix and osteoid matrix vesicle-induced mineralization. Magnesium accretion during the latter half of human gestation is shown in Fig. 30.3.

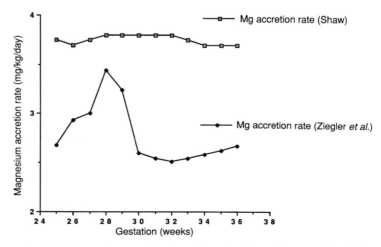

Fig. 30.3 Magnesium accretion in the human fetus as a function of gestational age. (Data are adapted from two studies: Ziegler, E., O'Donnel, A., Nelson, S., and Fomon, S. (1976). *Growth*, **40**, 329–41; Shaw, J. (1973). *Pediatric Clinics of North America*, **20**, 333–58.)

Serum mineral concentration

Most of the body minerals are concentrated in the tissues, primarily in bone (99 per cent of total body calcium; 80 per cent of total body phosphorus; and 60 per cent of total body magnesium). Less than 1 per cent of the total body content of each of these minerals is in the circulation.

In the circulation, calcium exists in three forms: 45 per cent of total serum calcium is there as the biologically active ionized calcium; 45 per cent is protein-bound mainly to albumin; and 10 per cent is complexed to anions (phosphate, lactate, citrate, bicarbonate, sulfate). Total serum calcium concentration is routinely assayed by atomic absorption spectrophotometry. However, determination of serum ionized calcium concentration is more desirable because it is a better physiological indicator of Ca homeostasis.

Serum phosphorus concentration is high at birth and decreases from infancy (4–7 mg/dL) to adulthood (2.7–4.5 mg/dL). About 10 per cent of total serum P is protein-bound, 50 per cent exists as free phosphate ions, and the remainder is complexed to Ca, Mg, and Na salts.

Total serum Mg concentration in infancy and early childhood is 2.2 ± 0.3 mg/dL. Ionized Mg is the fraction that is important for biochemical processes. Because Mg is predominantly an intracellular cation, serum Mg concentrations, as determined by atomic absorption spectrophotometry, may not reflect total body Mg content. Recently, it has been demonstrated that mononuclear blood cell Mg might be a better predictor of intracellular Mg and total body Mg status than the concentration of Mg in plasma or erythrocytes. Normally, Mg exchange occurs between the extracellular fluid compartment and bone in response to alterations in serum Mg concentration. About one-third of bone Mg content is on the surface of hydroxyapatite crystals and may be freely available for exchange. Magnesium homeostasis is only partly understood. It does not appear to be primarily controlled through hormonal mechanisms, and plasma concentrations seem to be under renal control.

The calciotrophic hormones

Parathyroid hormone

Parathyroid hormone (PTH) is an 84-amino-acid polypeptide (molecular weight of 9500) synthesized in the parathyroid glands, which develop early in gestation from the third and fourth branchial pouches. Production of PTH, demonstrated histologically by the presence of electron-dense secretory granules, first appears at the 6 to 7 cm long stage (6–10 g weight) of embryological development, and gradually increases during gestation. In rats, the parathyroid glands are well differentiated by day 17 of gestation. Human fetal parathyroid glands elaborate PTH from as early as 10 weeks of gestation; immunoreactive-PTH staining cells were demonstrated early in fetal development using antibodies specific to the carboxyl terminal PTH.

Secretion

The PTH gene is located on the short arm of chromosome 11. It codes for pre-pro-PTH, which undergoes two enzymatic cleavages before it yields PTH. Serum ionized calcium (iCa) concentration is the main determinant of PTH secretion: a drop in serum iCa concentration stimulates PTH secretion while a rise in serum iCa concentration suppresses it. However, other ions and hormones influence PTH secretion by the parathyroid glands: for instance, a rise in serum 1,25(OH)$_2$D decreases PTH secretion. An acute drop in serum Mg concentration stimulates PTH secretion but to a much smaller extent (10-fold less) compared to the effect of acute hypocalcaemia. Chronic hypomagnesaemia impairs PTH secretion and causes blunting of PTH action at target organs. Magnesium ions are essential for adenylate-cyclase-mediated secretion of secretory granules from the parathyroid chief cells. Therefore, magnesium deficiency may cause secondary hypocalcaemia. Hypermagnesaemia also suppresses PTH secretion. Aluminium inhibits PTH secretion *in vitro*.

Actions

Full biological activity resides in the amino-terminal 1–34 peptide, the middle and carboxy-terminal sequence (35–85 amino acids) being biologically inert although immunologically highly reactive. PTH regulates serum concentrations of Ca and P by modulating the activity of specific cells in bone and kidney, generating intracellular cyclic adenosine monophosphate (cAMP). Known actions of PTH include:

1. Stimulation of the release of calcium and phosphorus from bone into the circulation. The most obvious histological action of PTH is an increase in osteoclast number and activity. However, osteoclasts do not carry receptors for PTH in contrast to osteoblasts. It is likely that osteoclastic bone resorption is mediated by PTH–osteoblast–osteoclast interaction mechanism. PTH has a synergistic effect with 1,25(OH)$_2$D in stimulating bone resorption.
2. Enhancement of fractional reabsorption of calcium from the glomerular filtrate.

3. Stimulation of renal 1-α hydroxylase activity to increase 1,25(OH)$_2$D synthesis, which leads to increased intestinal Ca and P absorption.

4. Action on renal tubules to decrease reabsorption of P resulting in significant phosphaturia and an overall decrease in serum phosphorus concentration, despite PTH-induced bone resorption and release of bone phosphorus into the circulation.

Recent evidence indicates that PTH has dual effects on bone homeostasis: anabolic (increase in bone formation) and catabolic (increase in bone resorption). Intermittent and continuous PTH administration increases bone formation independently from resorption effects. However, the increase in bone mass and ^{47}Ca accretion rate were consistent with intermittent PTH administration (which mimics the physiological pulsatile secretion of PTH) but not with continuous PTH administration.

Parathyroid hormone-related peptide (PTHRP)

Parathroid hormone-related peptide (PTHRP) is a recently characterized peptide that was first demonstrated by cytochemical bioassay and discovered to have a major role in the aetiology of hypercalcaemia of malignancy. Parathyroid hormone and PTHRP genes are members of the same gene family, and the amino terminal of PTHRP has a sequence homology in eight amino acids with the PTH amino terminal. It is also equipotent to that of PTH when assessed by cytochemical bioassay and *in situ* biochemistry. PTHRP is produced by normal keratinocytes and has been demonstrated by *in vivo* histochemical studies of the skin and by detection of PTHRP in the medium of keratinocytes cultured *in vitro*. Squamous cells, which originate in keratinocytes, are a common source of PTHRP and have a common association with hypercalcaemia of malignancy.

Several studies suggest that this peptide may have a significant physiological role. One of the most common production sites of PTHRP is lactating breast tissue. For unknown reasons, PTHRP is present in large quantities in milk. Parathyroid hormone-like peptides have been detected in cultures of fetal rat long bones. Using a radioimmunoassay for human PTHRP it has been determined that plasma PTHRP concentrations, in normal adult subjects, ranged from less than 2 to 5 pmol/L.

PTHRP and bone

Infusion of PTHRP causes an elevation in serum 1,25-dihydroxy-vitamin D concentration and an increase in bone formation parameters. In cultures of fetal rat long bones, PTHRP causes a rise in cAMP concentration and bone resorption which is qualitatively and quantitatively similar to that induced by PTH. It is possible that PTH and PTHRP act on the same receptor.

PTHRP and pregnancy

PTHRP is produced in the fetal parathyroid glands and placenta and may be responsible for stimulation of placental calcium transport. PTHRP gene is expressed in the rat myometrium, with a major peak in PTHRP mRNA expression occurring in the 48 h immediately preceding parturition. A similar peak was found in tissue extracts by biological and immunological assays, but PTHRP could not be detected in the peripheral circulation or in uterine vein plasma during late gestation. PTHRP mRNA has been demonstrated in the myometrium by *in situ* hybridization histochemistry. The rise in myometrial PTHRP mRNA in late gestation was dependent on intrauterine occupancy; the myometrium of non-gravid uterine horns had greatly reduced or absent PTHRP mRNA. It is possible that the expression of the PTHRP gene in the myometrium is dependent on local factors. PTHRP has been demonstrated in human fetal tissues, serum, and amniotic fluid.

PTHRP and lactation

PTHRP is expressed in lactating rat mammary glands after suckling, as a result of rises in serum prolactin concentration rather than suckling *per se*. PTHRP mRNA and PTH-like bioactivity has been found in the breast tissue of lactating mice. The precise role of PTHRP in breast milk remains unclear at present. It is speculated that PTHRP may play a role in stimulating calcium transport into breast milk.

Calcitonin

Calcitonin (CT) is a 32-amino-acid polypeptide with a molecular weight of 3.5 kDa. It is secreted by the parafollicular C cells of the thyroid gland, which develop embryologically from a neural crest origin. Secretion of CT is affected by several factors. Serum Ca concentration is the major regulator of CT secretion. Plasma CT concentration increases when serum ionized Ca concentration rises, and declines when it falls. Secretion of CT is also induced by gastrin, cholecystokinin, and glucagon. The physiological role of these secretagogues in CT regulation is not very clear.

Vitamin D has a direct regulatory (inhibitory) effect on CT gene expression and CT secretion; receptors for 1,25(OH)$_2$D have been demonstrated on parafollicular C cells.

Actions

The best recognized physiological effect of CT is to counteract the action of PTH at several organ sites in the human body. In bone, receptors specific for CT have been demonstrated on osteoclasts. CT antagonizes PTH-mediated bone resorption by suppressing osteoclastic activity. Consequently, CT decreases the flux of Ca and P from bone into the circulation. It is controversial whether CT has a stimulatory effect on bone osteoblasts.

In the kidney, CT may enhance Ca, Mg, and P excretion. CT also acts on vitamin D metabolism; it enhances $1,25(OH)_2D$ production by proximal renal tubules, and increases fetal renal synthesis of $1,25(OH)_2D$. The overall effects of CT are to decrease serum Ca and P concentrations.

Calcitonin gene-related peptide (CGRP)

CGRP is another product of the translation of the calcitonin gene. CGRP is a 37-amino-acid peptide that has little amino acid sequence homology with calcitonin. Receptors for this peptide are primarily located in neural tissues, and biosynthesis of CGRP is confined primarily to neural tissues where it acts. It is likely that this peptide's prime role is in neurotransmission. It is not known to date whether CGRP shares calciotrophic activity with calcitonin.

Vitamin D metabolism

The human body obtains vitamin D from two major sources : (1) endogenously from the skin where vitamin D synthesis occurs under the effect of ultraviolet light; and (2) exogenously from dietary vitamin D_2 (derived from plant sterols) and D_3 from animal or synthetic origin (Fig. 30.4). Normally, at least 90 per cent of the vitamin D requirement is provided by endogenous photosynthesis in the skin which amounts to 2.5–10 µg/day (100–400 international units (IU)/day). The synthesis of vitamin D comprises a sequence of biochemical reactions in several organs that start with cholesterol and end up with the formation of 1,25 dihydroxyvitamin D, the most active vitamin D metabolite. Under the effect of small intestinal mucosal dehydroge-

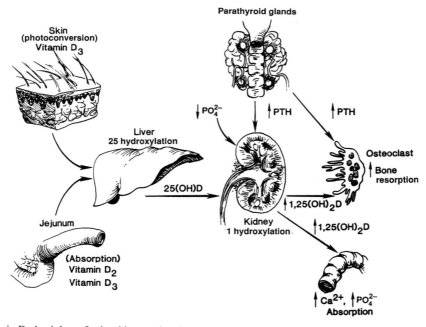

Fig. 30.4 Vitamin D physiology. In the skin, provitamin D_3 (7-dehydrocholesterol) is converted under the effect of ultraviolet radiation to previtamin D_3 which undergoes thermal isomerization to vitamin D_3. Alternatively, vitamin D_2 and vitamin D_3 are obtained from the diet and absorbed mainly in the jejunum. In the circulation, vitamin D is bound to vitamin D binding protein which carries it to the liver where it is hydroxylated to 25 hydroxyvitamin D (25 OHD). The latter is carried to the kidney where it is hydroxylated to 1,25 dihydroxyvitamin D [$1,25(OH)_2D$]. PTH and $1,25(OH)_2D$ act synergistically to enhance osteoclastic bone resorption. $1,25(OH)_2D$, the most potent vitamin D metabolite, enhances calcium and phosphorus absorption from the small intestines. Renal 1 α-hydroxylation (i.e. $1,25(OH)_2D$ synthesis) is stimulated by PTH and hypophosphataemia.

nase, dietary cholesterol is converted into 7-dehydrocholes-terol, which is then transported to the Malpighian layer of the skin. Ultraviolet light (of wavelengths 290–320 nm) penetrates the skin to break the C9–C10 bond of 7-dehy-drocholesterol (provitamin D_3) to form pre-vitamin D_3. Pre-vitamin D_3 undergoes several reactions: it may be pho-toisomerized to lumisterol and tachysterol or converted by a temperature-dependent isomerization to cholecalciferol (vitamin D_3). Cholecalciferol is then released in the circula-tion where it is bound to vitamin D binding protein and transported to the liver. In the liver, cholecalciferol under-goes 25-hydroxylation to yield 25 hydroxy-vitamin D (25OHD), which is released into the circulation once again before reaching the kidney. In the kidney mitochondria, 25OHD undergoes 1-α-hydroxylation to produce 1,25-dihydroxyvitamin D or 24-hydroxylation to form 24,25-dihydroxyvitamin D [24,25(OH)$_2$D]. The role of 24,25(OH)$_2$D in mineral and vitamin D homeostasis is not very well known. The activity of renal 1-α hydroxylase is stimulated by PTH and hypophosphataemia and by periods of high calcium demand such as growth, pregnancy, or low calcium intake, and may be regulated by calcitonin, 1,25(OH)$_2$D, and other vitamin D metabolites. Hypomagnesaemia may suppress PTH secretion and induce end-organ resistance to the effect of PTH, in partic-ular renal resistance with a secondary drop in renal synthe-sis of 1,25(OH)$_2$D. Magnesium supplementation restores parathyroid and calcitriol responses to low calcium diet. A summary of vitamin D physiology is given in Fig. 30.4.

Actions

1,25(OH)$_2$D is the major active vitamin D metabolite. It acts on several organ systems where specific vitamin D receptors are located : gut, bone, kidney, parathyroid glands, and immune system. It increases calcium and phosphorus absorption from the small intestine and the kidney tubules, mobilizes Ca and P from bone, and induces the synthesis of a specific calcium-binding protein, calbindin D. Calbindin D_{9k} is a water-soluble protein with a high affinity for calcium. Intestinal calbindin D_{9k} is an excellent marker for calcitriol action, and its concentration correlates with the rate of intestinal calcium absorption. It enhances calcium transport through facilitated diffusion; uptake of calcium at the brush border or luminal surface (through either a channel- or carrier-mediated mechanism) is enhanced by binding of calcium to Ca binding protein.

Calcitriol may have a synergistic effect with PTH in stimulating osteoclastic bone resorption.

Calcitriol has a regulatory effect on PTH and CT gene transcription. It decreases PTH mRNA and CT mRNA concentrations *in vitro* and *in vivo* in rats. In humans with

secondary hyperparathyroidism, intravenous 1,25(OH)$_2$D administration leads to marked reduction in serum PTH concentration. Oral administration of calcitriol to children with hypophosphataemic rickets and secondary hyper-parathyroidism also has an inhibitory effect on PTH secre-tion. Calcitriol upregulates its own receptor in the parathyroid gland's chief cells: *in vivo* and *in vitro* 1,25(OH)$_2$D administration increases the concentration of vitamin D receptor mRNA in the parathyroid gland.

Perinatal mineral homeostasis

Maternal physiology

Pregnancy is associated with a number of physiological adaptations that provide an optimal milieu for the growth of the fetus and preserve maternal homeostasis at the same time. The extracellular fluid volume expands, and renal blood flow and glomerular filtration rate increase. Serum total calcium concentration decreases gradually throughout pregnancy, reaching a nadir by the mid-third of gestation and rising slightly thereafter. This pattern parallels that of serum albumin concentration, while serum ionized Ca con-centration remains constant. A recent longitudinal study demonstrated a slight but significant drop in serum iCa after 30 weeks of gestation and a similar decrease post-partum. Serum P concentration also declines during pregnancy and parallels that of serum Ca.

Studies applying traditional radioimmunoassays (target-ing the amino terminal) demonstrated that serum PTH con-centration may rise during pregnancy and a state of 'physiological hyperparathyroidism' appears to set in, pos-sibly as a response to loss of calcium from mother to fetus across the placenta. However, studies using immunoradio-metric assay (IRMA) of the intact PTH molecule reported either a gestational drop, no change, or a rise in serum PTH concentration. In a longitudinal study from Denmark, serum PTH rose significantly from 18 to 36 weeks of ges-tation and continued rising for 12 months postpartum. A recent longitudinal study demonstrated that serum PTH declined toward the middle of pregnancy and increased thereafter. Serum CT concentration does not change significantly with gestation.

Maternal serum 25OHD concentration is little affected by pregnancy. It is mainly influenced by dietary vitamin D intake and ultraviolet light exposure. Serum vitamin D binding-protein concentration doubles during pregnancy. Serum 1,25(OH)$_2$D concentration is high early in preg-nancy and rises with advancing gestation. This is explained partly because of a rise in vitamin D binding-protein concentration, and possibly from rises in serum

PTH concentration, and is thought to increase the synthesis of intestinal calcium binding proteins and the intestinal absorption of calcium and phosphorus to meet the mineral requirements of the developing fetus.

Further, placental tissues are capable of 1-α-hydroxylation of 25OHD and may significantly contribute to the rise in serum $1,25(OH)_2D$ concentration in pregnancy. This mechanism is supported by the fact that patients with pseudohypoparathyroidism who previously required vitamin D therapy were able to maintain normal serum $1,25(OH)_2D$ and Ca concentrations during pregnancy.

Early in pregnancy, serum osteocalcin concentration is low and declines further towards the middle of pregnancy. It increases in late pregnancy and is correlated significantly with serum PTH concentration. PTH may influence the synthesis of osteocalcin indirectly by increasing osteoclast-mediated bone resorption. $1,25(OH)_2D$ directly increases the synthesis of osteocalcin. It is possible that during pregnancy the stimulatory effect of increasing $1,25(OH)_2D$ concentration on osteocalcin synthesis may be overridden by the inhibitory effect of decreasing PTH concentration particularly during mid-pregnancy. Although there is concern that pregnancy *per se* is stressful to maternal bone and theoretically may deplete maternal calcium stores, few studies have examined this issue. Although one study suggested that as much as 10 g of net bone loss may occur during normal pregnancy, this has not been supported by other clinical studies. Further, there is increasing awareness of the clinical benefits of optimal calcium and magnesium homeostasis in the perinatal period. Calcium supplementation during pregnancy has been associated with lower incidence of prematurity and pregnancy-induced hypertension, possibly related to a relaxing effect of calcium on vascular smooth-muscle fibres. Similar clinical trials of magnesium supplementation during pregnancy showed a decreased incidence of pregnancy-induced hypertension and prematurity. The effects of these gestational dietary interventions on fetal bone and mineral homeostasis have not been evaluated yet.

Fetal mineral physiology

The fetus relies on maternal resources to acquire minerals that are indispensable for optimal fetal mineral and bone homeostasis. In the rat, radiocalcium studies demonstrated that up to 92 per cent of fetal calcium is acquired from the maternal diet. In human pregnancy, stable calcium isotope studies demonstrated a doubling of intestinal calcium absorption from the end of the second trimester to term. There is little transfer of minerals to the fetus early in gestation in rats and humans. Fetal acquisition of these minerals increases exponentially during gestation, and most remarkably during the third trimester (Figs 30.1–30.3). There is enough evidence showing that minerals, primarily calcium, phosphorus, and magnesium, are actively transported across the placenta from maternal to fetal circulation against a concentration gradient. Several studies have demonstrated that cord-blood Ca concentration at birth in humans exceeds the corresponding maternal blood concentration throughout the third trimester of pregnancy. Serum total calcium concentration in the fetus as estimated by cord-blood analysis is 5.5 mg/dL in the second trimester and approaches 11 mg/dL by the end of the third trimester. It is possible that the rising fetal serum calcium concentration during pregnancy, in parallel with fetal calcium accretion, will facilitate deposition of calcium into rapidly growing and ossifying bone.

Similarly, it has been shown that cord-blood P and Mg concentrations at birth are higher than corresponding maternal blood concentrations. In contrast to serum Ca concentration, which increases with increasing gestational age, cord-blood serum P concentration decreases from about 14 mg/dL in the second trimester to about 6 mg/dL at term. Yet, total body phosphorus content increases directly with increasing gestational age. The drop in fetal serum phosphorus may be caused by the continuous incorporation of inorganic phosphorus into fetal bone, most remarkably in the last trimester of pregnancy.

The chemical analysis of dead fetuses at different gestational ages for Ca, Mg, and P made possible a rough estimation of the fetal mineral accretion rate during pregnancy. As shown in Figs. 30.1–30.3, the fetal accretion rates for these minerals increase exponentially during the last trimester of pregnancy (24–38 weeks). The peak mineral accretion rate is reached at 34 to 36 weeks of gestation and amounts to 117 mg of Ca, 74 mg of P, and 2.7 mg of Mg per kg fetal weight per day. In fact, more than two-thirds of fetal body Ca content is acquired during the third trimester of gestation at a rate of up to 150 mg/kg/day.

The placenta plays a major role in regulating the active transfer of minerals from mother to fetus. Calcium transport across the placenta to the fetus may involve an ATPase-dependent active transport. Calcium binding proteins (CaBP, calbindins) may play a significant role in perinatal mineral homeostasis. These proteins are primarily synthesized by calcium transporting tissues such as intestine, kidney, and placenta. Calbindins have also been isolated from mouse and mammalian kidneys, mammalian placenta and yolk sac, rat uterus and growth cartilage and osteoblasts, brain, pancreas, lung, and avian egg shell gland ('uterus') of the egg-laying hen. Two major subclasses of calbindins have been described: calbindin D_{9k}

and calbindin D_{28k}. The former has a molecular weight of 9 kDa and is characteristic of mammals; the latter has a molecular weight of 28 kDa and is characteristic of avian and other species. Synthesis of intestinal and renal calbindins has been reported to be vitamin-D dependent, and their concentrations rise in response to rising blood concentrations of $1,25(OH)_2D$ in pregnancy. The concentration of these peptides is tightly regulated by vitamin D status. Vitamin D deficiency reduces their concentrations while vitamin D supplementation increases their concentrations. A high-calcium diet suppresses PTH and $1,25(OH)_2D$ synthesis, and reduces intestinal Ca binding protein concentrations and calcium absorption. In rat and chick brain, calbindin D_{28k} has been reported to be unresponsive to vitamin D metabolites. From studies in the rat and mouse it has been found that intestinal calbindin D_{9k} concentration increases progressively with advancing gestation, reaching a peak concentration in late gestation at a time when calcium transfer to the fetus and fetal bone mineralization are at their highest rate. Also, maximal intestinal concentrations of calbindin D_{9k} parallel those of maternal plasma $1,25(OH)_2D$ concentrations and maternal intestinal calcium absorption.

A placental calcium binding protein has been isolated. It is similar but not identical to the intestinal calcium binding protein. Placental calcium binding protein concentration increases with gestation in parallel with the increasing transplacental calcium transfer to the fetus. Theoretically, in analogy with intestinal Ca binding protein, placental Ca binding protein, which is also $1,25(OH)_2D$-dependent, may play a role in the regulation of transplacental calcium transfer to the fetus, particularly because its concentration increases late in gestation in parallel with $1,25(OH)_2D$. Calbindins have been localized in the rat uterus (D_{9k}) and the shell gland of the egg-laying hen (D_{28k}). However, in these tissues calbindins are not vitamin D-dependent. Instead their concentrations increase in response to oestradiol administration. Although the presence of calbindins in uterine epithelial cells has been suggested to have a role in the transfer of calcium to the fetus, this concept needs further investigation.

Fetal bone physiology is not completely understood. We alluded above to the fact that most of the body's minerals are stored in bone, which serves both as a mechanical support organ for the whole body and as a metabolic reservoir for these minerals. Osteocalcin is a major component of non-collagenous bone proteins and a sensitive indicator of osteoblastic activity. Immunocytochemical localization of osteocalcin in human fetal bones has been demonstrated as early as at 12 weeks of development. These studies demonstrated that osteocalcin is most actively synthesized by osteoblasts early in fetal life, and is then deposited on collagen fibres of the osteoid and bone matrix by 17 weeks of gestation. Localization of calbindins in bone and cartilage cells suggests that they may have a significant role in bone formation and ossification. Calbindins have been immunocytochemically localized to chondrocytes of the growth plate in rats and chick, and rat and human osteoblasts. Calbindin D_{9k} has been localized in the cytoplasm of maturing rat chondrocytes, in the extracellular lateral edges of longitudinal septa, where mineralization of cartilage is initiated and where matrix vesicles are preferentially located. From these findings it is suggested that this protein may be involved in cartilage ossification. The results of several immunocytochemical studies suggest that calbindins play a role in the movement of intracellular calcium in the chondrocyte and in the movement of calcium toward extracellular sites of calcification in the growth plate.

Perinatal homeostasis

The placenta transfers calcium ions from mother to fetus against a concentration gradient resulting in relative fetal hypercalcaemia, and higher cord-blood calcium concentration than in maternal blood. In the rat, calcium fluxes are bidirectional. Passive transfer accounts for the majority of placental calcium movement; however, an additional active component results in a net calcium movement favouring calcium deposition in the fetus. It is unclear whether bidirectional fluxes of calcium across the placenta exist in the human, or whether transport is entirely unidirectional from mother to fetus. The exact molecular mechanisms involved in placental mineral transport still need elucidation. Evidence supports an active energy-dependent process of transplacental Ca transfer to the fetus. The intrinsic placental calcium binding protein (CaBP, calbindin) in the rat and mouse is vitamin D-dependent. Calcium-dependent adenosine triphosphatase systems have been demonstrated in animal as well as in human placental tissue and calcium ions have been localized in placental plasma membrane vesicles and mitochondria within the fetal capillary endothelium. It is likely that calbindin may play a significant role in the active transport of calcium transplacentally to the fetus.

Phosphorus and magnesium are also actively transported across the placenta to the fetus although the exact molecular mechanisms involved are still unclear.

Perinatal PTH homeostasis

Theoretically, since PTH does not cross the placenta, the relative fetal hypercalcaemia should suppress the fetal parathyroid glands. Paradoxically, fetal PTH secretion is not suppressed. Cord-blood PTH concentrations are lower than, or similar to paired maternal samples. The inconsistency in PTH assay results stems from the use of different PTH antisera; assays may detect the amino- or carboxy-terminal or midmolecule of the PTH molecule. A possible explanation for the non-suppressible PTH secretion, despite relative fetal hypercalcaemia, is that the negative feedback system regulating PTH secretion by calcium concentration operates with a higher setpoint in the fetus in such a way that suppression of PTH secretion in the fetus requires higher serum calcium concentrations than after birth. Although serum PTH concentrations by radioimmunoassay are often lower in the fetus than in the mother, PTH-like bioactivity, measured by a sensitive cytochemical assay, is higher in the fetus than in the mother.

From recent data it has been suggested that PTH or PTH-related peptide (PTHRP) may play a significant role in placental transport of calcium. PTHRP is a recently characterized peptide that may play a major role in humoral hypercalcaemia of malignancy. Evidence shows that transcription and splicing of the PTHRP gene give rise to three peptides consisting of 139, 141, and 173 amino acids, respectively. The 13 amino-terminal amino acids of each of these peptides are 70 per cent homologous with the corresponding region of PTH. Unlike PTH, whose synthesis in normal subjects is restricted to the parathyroid glands, PTHRP messenger RNA is widely distributed in normal tissues, including the skin, thyroid, bone marrow, hypothalamus, pituitary, parathyroid, adrenal cortex, adrenal medulla, and stomach. In thyroparathyroidectomized (with thyroxine replacement) pregnant sheep, fetal parathyroidectomy results in an acute drop in fetal serum calcium concentration and a reversal of the materno–fetal calcium gradient; the lambs are born rachitic (i.e. with rickets). Interestingly, parathyroid extracts, but not immunoreactive PTH infusions, normalized the materno–fetal calcium gradient. From these experiments it was suggested that PTHRP may have a putative role in maintaining the transplacental materno–fetal calcium gradient. It is interesting to note that PTHRP (1–34) is ineffective in increasing calcium transport across the sheep placenta, whereas PTHRP (1–84), PTHRP (1–108), and PTHRP (1–141) are effective, suggesting that specific receptors for PTHRP bind to the protein at a site distal to amino acid 34.

Perinatal calcitonin homeostasis

In the human fetus, the thyroid C cells appear to be well developed by 14 weeks of gestation. However, the role of CT in fetal mineral and bone homeostasis is not very well understood. Maternal serum calcitonin may be elevated or unchanged during pregnancy. Calcitonin does not cross the placenta, and, as with PTH, fetal CT may function independently from that of the mother. Evidence suggesting fetal production of CT is indirectly inferred from the observations that cord-blood CT concentrations are significantly higher than corresponding maternal values in all trimesters, and that arterial cord-blood CT is higher than venous cord-blood concentration. The human fetal thyroid contains a larger number of C cells than the adult thyroid. Immunocytochemical evidence for CT in the thyroid gland has been demonstrated and confirmed by direct radioimmunoassay in human and rat fetuses. Because cord-blood CT concentrations are elevated at birth in term and preterm infants and presumably during the last trimester of gestation, it is speculated that elevated fetal blood CT concentration may play a significant role in promoting fetal bone mineralization and modulation of placental transfer of calcium. In rats, thyroidectomy of the pregnant rat results in lower fetal bone density. Calcitonin possibly plays a role in protecting bone mineralization in the mother as well as in the fetus.

Perinatal vitamin D physiology

The role of vitamin D metabolites in fetal mineral homeostasis is not completely understood. There is enough evidence that vitamin D metabolites are necessary for optimal fetal and maternal bone mineralization. As described later, vitamin D deficiency results in osteomalacia and rickets in the mother and the infant, respectively, and can be prevented by dietary vitamin D supplementation. The fetus is totally dependent on maternal supplies of vitamin D. In humans, a supplementation of 400 IU of vitamin D per day results in normal serum vitamin D concentration in the pregnant woman. Normally, the human new-born has undetectable plasma vitamin D concentration. Placental transfer of radiolabelled vitamin D has been assessed in the rat and the sheep. These studies demonstrated limited placental transfer of vitamin D from maternal to fetal circulation: less than 10 per cent of vitamin D activity was present in rat fetus in the form of unchanged vitamin D (5 per cent in the sheep). The major vitamin D metabolite that crosses the placenta is 25 hydroxyvitamin D (25OHD) in all species studied.

In most species studied, fetal or cord plasma 25OHD concentrations are lower than, and correlate positively with, corresponding maternal plasma concentrations supporting the thesis that fetal plasma 25OHD concentrations are dependent on maternal 25OHD status. Although pregnancy in animals (sheep, rats, and rabbits) is associated with a drop in plasma 25OHD concentrations, presumably because of increased utilization of 25OHD for synthesis of

1,25(OH)$_2$D, this effect was not apparent in human pregnancy.

In animals and humans, cord plasma 24,25(OH)$_2$D concentrations are lower than corresponding maternal values. It is not known whether fetal 24,25(OH)$_2$D comes from endogenous fetal synthesis or is of maternal origin. Feto-placental tissues have the enzymatic system to synthesize 24,25(OH)$_2$D *in vitro*, and there is evidence that feto-placental synthesis of this compound may occur *in vivo*, at least in rats. However, the role of this metabolite in human fetal mineral and bone homeostasis is not known.

The role of 1,25(OH)$_2$D in the fetal handling of calcium and other minerals is at present controversial. Serum calcitriol concentrations are elevated during pregnancy possibly in order to increase maternal intestinal absorption of calcium to meet the requirements during pregnancy for fetal bone mineralization. There is no general agreement regarding transplacental passage of maternal 1,25(OH)$_2$D. A recent longitudinal study has shed some light on perinatal vitamin D homeostasis; the four common vitamin D metabolites (25OHD, 1,25(OH)$_2$D, 24,25(OH)$_2$D, and 25,26(OH)$_2$D) were determined simultaneously in mothers and infants from delivery to several months. At delivery, total vitamin D metabolites in maternal and fetal plasma were closely correlated, maternal concentrations being higher. Free unbound vitamin D metabolite concentrations were higher in fetal than in maternal plasma, except for free 1,25(OH)$_2$D concentrations which were equal. The results of this study suggest that transplacental materno–fetal transfer of 25OHD, 24,25(OH)$_2$D and 25,26(OH)$_2$D is effected by an active transport system and that of 1,25(OH)$_2$D by a passive transport system.

Disorders of perinatal mineral homeostasis

Fetal mineral homeostasis is closely linked to that of the mother. In the pregnant woman and the fetus there is an intimate and delicate relationship amongst the calciotrophic hormones PTH, CT, calcitriol, and possibly PTHRP and the minerals Ca, P, and Mg. Any perturbation of maternal homeostatic mineral balance may affect that of the fetus and may have metabolic sequelae in the fetus manifesting in the neonatal period and infancy.

Disorders of calcium homeostasis

A wide variety of factors can cause significant disturbances in maternal and fetal calcium homeostasis.

Maternal hypocalcaemia

Maternal hypocalcaemia results in fetal hypocalcaemia, which stimulates the fetal parathyroid glands to synthesize and secrete more PTH. PTH does not appear to cross the placenta in either direction. Causes of maternal hypocalcaemia are listed in Table 30.1. Impaired secretion of PTH because of hypoparathyroidism or magnesium depletion results in maternal hypocalcaemia. Hypocalcaemia may be a manifesting feature of abnormal vitamin D metabolism; in particular, maternal vitamin D deficiency may be caused by inadequate dietary intake, insufficient sunlight exposure, or malabsorption. Defective 25-hydroxylase activity may be associated with maternal liver disease and may result in hypocalcaemia, low serum 25OHD concentration, and rickets.

Defective 1 α-hydroxylase activity may be caused by renal or parathyroid gland diseases. For example, renal hypophosphataemic rickets is a familial disorder characterized by decreased reabsorption of phosphate in the

Table 30.1 Causes of maternal hypocalcaemia

Impaired secretion of parathyroid hormone

Hypoparathyroidism
Magnesium deficiency

Abnormal vitamin D metabolism

Vitamin D deficiency
  Inadequate intake
  Insufficient sunlight exposure
  Malabsorption
Defective 25-hydroxylase activity
  Hepatic disease
Defective 1 α-hydroxylase activity
  Renal disease
    Chronic renal failure
    Hypophosphataemic rickets
    Vitamin D-dependent rickets type I
  Parathyroid-related disease
    Hypoparathyroidism
    Pseudohypoparathyroidism
Defective response to 1, 25-dihydroxyvitamin D
  Vitamin D-dependent rickets type II

Miscellaneous

Tumour-induced osteomalacia
Anticonvulsant therapy
Acute pancreatitis
Phosphate therapy
Chemotherapy

proximal renal tubules resulting in hypophosphataemia and rickets. It often has an X-linked dominant mode of inheritance. Serum Ca concentration is either normal or low. Serum PTH concentration is either normal or increased. Serum 1,25(OH)$_2$D concentration may be normal or low.

Chronic renal failure is associated with hyperphosphataemia, hypocalcaemia, and depressed 1 α-hydroxylase activity with low serum 1,25(OH)$_2$D which results in secondary hyperparathyroidism.

Vitamin D-dependent rickets type I is a familial disorder with an autosomal recessive inheritance pattern. It is characterized by depressed renal 1 α-hydroxylase activity, hypocalcaemia, low serum 1,25(OH)$_2$D concentration, and elevated serum PTH concentration. Vitamin D-dependent rickets type II is a rare disease characterized by defective target organ response to 1,25(OH)$_2$D which is attributed to a defect in the cellular and nuclear uptake of 1,25(OH)$_2$D. Receptors for 1,25(OH)$_2$D are either absent or defective and bind abnormally to DNA (deoxyribonucleic acid) molecules because of the expression of a point mutation in the gene coding for these receptors. Patients with this disease have elevated serum 1,25(OH)$_2$D concentration and associated alopecia.

Maternal hypoparathyroidism is associated with defective renal 1 α-hydroxylase activity, hypocalcaemia, and low serum 1,25(OH)$_2$D concentration but the impact on the fetus is a transient hyperparathyroidism state with hypercalcaemia.

Tumours of mesenchymal origin (e.g. sarcomas, hemangiomas, giant-cell tumours of bone, carcinoma of the breast) and the epidermal naevus syndrome may cause osteomalacia and hypocalcaemia, which resolve after tumour ablation. Most patients have marked hypophosphataemia because of excessive renal phosphate wasting and reduced proximal tubular reabsorption. The pathogenesis of the disease involves the production of a phosphaturic substance by tumour cells that reduces renal tubular phosphate reabsorption and results in phosphate depletion. Hypocalcaemia may complicate the course of acute pancreatitis, phosphate therapy, or chemotherapy.

Maternal hypercalcaemia

Maternal hypercalcaemia causes an increase in transplacental calcium transfer from mother to fetus resulting in fetal hypercalcaemia which suppresses the fetal parathyroid glands resulting in neonatal hypocalcaemia, once the fetus is delivered and the excessive maternal calcium input is abruptly halted. Causes of maternal hypercalcaemia are listed in Table 30.2. Primary hyperparathyroidism is a major cause of hypercalcaemia. In 80–85 per cent of cases it is caused by parathyroid adenoma; in 15–20 per cent it

Table 30.2 Causes of maternal hypercalcaemia

Endocrine

Hyperparathyroidism
Hyperthyroidism
Adrenal insufficiency
Phaeochromocytoma
VIPoma syndrome
Williams's syndrome
Familial hypocalciuric hypercalcaemia
Hypophosphatasia

Drugs

Thiazides
Lithium
Vitamin A intoxication
Vitamin D intoxication
Milk–alkali syndrome

Others

Abnormal vitamin D metabolism
    Sarcoidosis
    Tuberculosis
Malignancy
Immobilization

results from parathyroid hyperplasia; and rarely (in 3 per cent) it is caused by parathyroid carcinoma. Hyperthyroidism is associated with increased bone turn-over and hypercalcaemia Maternal hypercalcaemia may occur in patients with untreated hypoadrenalism. It is attributed to diminished renal excretion of calcium and volume depletion or reduced secretion of corticosteroids. Hypercalcaemia and hyperparathyroidism are sometimes associated with phaeochromocytoma or multiple endocrine neoplasia type II syndrome. Pancreatic VIPoma tumours secrete vasoactive intestinal peptide (VIP), which causes severe diarrhoea. About 50 per cent of the cases have hypercalcaemia, possibly because of VIP-induced bone resorption.

Hypercalcaemia may be caused by drugs. Thiazides act directly to increase calcium release from the skeleton and promote renal tubular reabsorption of calcium. Chronic lithium intake may cause hyperparathyoidism and mild hypercalcaemia and hypermagnesaemia. Milk–alkali syndrome and hypercalcaemia may be caused by excessive intake of calcium and phosphate in the form of milk and absorbable alkali, sodium carbonate, sodium bicarbonate, or calcium carbonate, usually for the treatment of peptic ulcer disease. Bone resorption and hypercalcaemia may be caused by vitamin D or vitamin A intoxication.

Hypercalcaemia may be a feature of granulomatous diseases such as tuberculosis and sarcoidosis which are associated with abnormal vitamin D metabolism and increased circulating concentrations of $1,25(OH)_2D$ in blood. Hypercalcaemia may be associated with malignant tumours. The mechanism of hypercalcaemia in malignancy may be due to production of PTHRP (humoral hypercalcaemia of malignancy), $1,25(OH)_2D$ in lymphoma, lymphotoxin in multiple myeloma, or other substances that cause bone resorption.

Immobilization causes bone demineralization and hypercalcaemia.

Fetal and neonatal hypocalcaemia

Causes of fetal and neonatal hypocalcaemia are listed in Table 30.3. Parathyroid gland adenoma is the most common cause of maternal hyperparathyroidism and hypercalcaemia which suppresses the fetal parathyroid glands resulting in transient neonatal, or congenital, hypoparathyroidism with or without rickets. At least 50 per cent of infants born to hyperparathyroid mothers present with hypercalcaemia tetany. However, the fact that some of these hypercalcaemia infants had normal or elevated serum PTH concentrations, suggests that other factors may be involved in the pathogenesis of hypercalcaemia (total serum calcium concentration < 7mg/dL or serum ionized calcium concentration < 4.4mg/dL) in these infants. Likely explanations may involve hypomagnesaemia, defective calcitriol synthesis, or end-organ resistance to PTH.

Primary hypoparathyroidism may present as neonatal hypercalcaemia and may be sporadic or familial. Primary hypoparathyroidism is a chronic disease that requires supplementation with calcium and vitamin D metabolites to maintain eucalcaemia and prevent osteopenia.

Table 30.3 Causes of fetal and neonatal hypocalcaemia

Maternal-related

Hyperparathyroidism
Diabetes mellitus
Anticonvulsant therapy
Malabsorption
Abnormal vitamin D metabolism (see Table 30.1)

Fetal-related

Prematurity
Hypoparathyroidism
Hypomagnesaemia
Asphyxia neonatorum

Prematurity is the major cause of neonatal hypocalcaemia which may develop in a large proportion (30–90 per cent) of preterm infants. The incidence of hypocalcaemia correlates inversely with gestational age and birth weight. Its cause is uncertain: earlier reports provided data consistent with neonatal hypoparathyroidism but recent reports of bioassayed PTH appear to indicate a significant rise in circulating PTH in response to hypocalcaemia in preterm infants.

About 30 per cent of infants with birth asphyxia may develop hypocalcaemia in the neonatal period. Infants with intrauterine growth retardation may develop hypocalcaemia only if they are born prematurely or have experienced birth asphyxia. Hypercalcitoninaemia is suggested to have a significant role in the development of neonatal hypocalcaemia. The results of a recent study showed no correlation between neonatal serum calcium, or the postnatal fall in serum calcium concentrations, and serum CT concentrations in normal term infants and infants of diabetic mothers.

Mothers with insulin-dependent diabetes mellitus have excessive urinary magnesium losses especially if euglycaemia is not maintained. Consequently these mothers — and theoretically their fetuses — may be magnesium-depleted. Hypomagnesaemia impairs PTH secretion and may explain the transient neonatal hypoparathyroidism and hypocalcaemia that develop in about 50 per cent of infants of diabetic mothers.

Neonatal, and presumably fetal, hypocalcaemia and rickets have been described in infants born to mothers with abnormal vitamin D homeostasis (discussed earlier): vitamin D deficiency osteomalacia may result from inadequate maternal dietary intake and sunlight exposure, or intestinal malabsorption. There is often an association with increased maternal parity and lower socio-economic status.

Hypocalcaemia has been described in infants, and presumably in fetuses, born to mothers with gestational exposure to anticonvulsant therapy. The mechanism may be related to the effect of phenobarbital or phenytoin in enhancing accelerated metabolism of 25OHD by hepatic microsomal P450 oxidase activity. Hypocalcaemia has also been described in infants born to hypercalcaemic women with familial hypocalciuric hypercalcaemia. The mechanism of hypocalcaemia in these infants is possibly related to fetal parathyroid suppression by long-lasting maternal hypercalcaemia.

Fetal and neonatal hypercalcaemia

Causes of fetal and neonatal hypercalcaemia are listed in Table 30.4. Fetal and neonatal hypercalcaemia (total serum Ca concentration > 11 mg/dL or serum ionized Ca

Table 30.4 Causes of fetal and neonatal hypercalcaemia

Maternal-related

Excessive vitamin A and D intake
Thyrotoxicosis
Hypoparathyroidism
Pseudohypoparathyroidism

Fetal/neonatal-related

Endocrine

Primary hyperparathyroidism
Congenital hypothyroidism

Renal

Familial hypocalciuric hypercalcaemia
Diuretics: thiazides
Bartter's syndrome

Others

Idiopathic infantile hypercalcaemia
Williams's syndrome
Infantile hypophosphatasia
Congenital nephroblastoma
Subcutaneous fat necrosis

concentration > 5.8 mg/dL) can be the result of prolonged maternal hypocalcaemia, which can be due to a multitude of causes (Table 30.1). Maternal hypoparathyroidism is most commonly due to inadvertent parathyroidectomy at thyroid surgery; less commonly, it may be idiopathic. Mechanistically, maternal hypocalcaemia results in a decrease in placental calcium supply to the fetus despite a physiologically intact active transport system for calcium. As a result, fetal hypocalcaemia stimulates the fetal parathyroid glands to normalize fetal serum Ca concentration. Consequently, a variable degree of congenital transient hyperparathyroidism results. At least 10 cases of transient secondary hyperparathyroidism have been reported. These infants are described to have a spectrum of manifestations that depend on the severity of the maternal homeostatic disturbance: at one end of the spectrum are normal infants; at the other end are infants with severe hyperparathyroidism with elevated cord-blood PTH concentration, severe congenital bone demineralization, and intrauterine fractures with clinical and radiological bowing of the long bones. Fortunately, the manifestations of secondary hyperparathyroidism improve rapidly after birth once these infants are provided with adequate amounts of Ca, P, and vitamin D.

Primary hyperparathyroidism in the new-born is an important cause of hypercalcaemia which may occur sporadically or have a familial basis. This entity may be inherited in an autosomal dominant or recessive manner. Neonatal hyperparathyroidism may be associated with familial hypocalciuric hypercalcaemia which is inherited in an autosomal dominant manner with high penetrance at all ages and which is characterized by a variable degree of hypercalcaemia associated with decreased renal clearance of calcium. Clinically, the spectrum of the disease ranges from the asymptomatic heterozygous person with mild hypercalcaemia detected by blood testing, to the homozygous subject with neonatal hyperparathyroidism and severe life-threatening hypercalcaemia. Recently, a self-limited form of neonatal hyperparathyroidism and hypercalcaemia has been reported in a family of three siblings, in association with hypercalciuria, and renal tubular acidosis.

Infants with congenital agoitrous hypothyroidism have also been described to be hypercalcaemic. These infants possibly have congenital absence or dysfunction of the parafollicular calcitonin-producing C cells; therefore they cannot mount a CT response to a Ca load.

Excessive maternal ingestion of vitamin D or vitamin A metabolites may result in bone resorption, and maternal and neonatal hypercalcaemia. The same outcome may occur in infants of mothers who have prolonged hypercalcaemia because of thyrotoxicosis or chronic diuresis with thiazides. It is possible that the fetal parathyroid gland threshold for PTH secretion becomes higher, or the glands themselves become 'tolerant' to chronic maternal and fetal hypercalcaemia.

Neonatal hypercalcaemia is a basic feature of the severe infantile form of hypophosphatasia, a rare disorder characterized by severe bone demineralization most pronounced at growth plates, low serum alkaline phosphatase concentrations, and elevated urine phosphoethanolamine concentrations. The severe form of the disease is inherited in an autosomal recessive manner. Because of alkaline phosphatase deficiency, bone mineralization is severely depressed and most of body's calcium is transferred to the circulation resulting in severe hypercalcaemia, hypercalciuria, and nephrocalcinosis. Severely affected infants may die *in utero* or shortly after birth because of impaired skeletal support of the thorax and skull.

A subgroup of Bartter's syndrome, which is characterized by hypokalaemic alkalosis, hyperreninaemia, and hyperaldosteronism, may be associated with hypercalcaemia and prostaglandin-mediated hypercalciuria. The mechanism of hypercalcaemia in these infants is attributed to increased intestinal Ca absorption enhanced by elevated serum $1,25(OH)_2D$ concentration, the serum 25OHD concentration being normal. It is though that increased

$1,25(OH)_2D$ synthesis is mediated by prostaglandin E_2 (PGE_2), which has been demonstrated to stimulate renal 1-α-hydroxylase activity *in vivo* in thyroparathyroidectomized rats and *in vitro* in chick and rat kidney cells. In patients with this syndrome, serum $1,25(OH)_2D$ concentrations correlate significantly with urinary PGE_2 concentrations. Therapy with prostaglandin synthesis inhibitors such as indomethacin lowers serum calcitriol concentration and urinary PGE_2 and calcium concentration and improves the associated nephrocalcinosis.

Idiopathic hypercalcaemia may occur in infancy. The basic defect in this disease is still obscure. A subgroup of infants with idiopathic hypercalcaemia have 'Williams's syndrome' which is characterized by elfin facies, short stature, mild mental retardation, hyperacusis, and congenital heart disease (commonly supravalvular aortic stenosis). The hypercalcaemia in these infants is associated with hypersensitivity to the effect of vitamin D metabolites manifesting as increased intestinal absorption of calcium. Defects in vitamin D or calcitonin metabolism have been suggested.

New-borns who had difficult vaginal deliveries may develop subcutaneous fat necrosis over areas subjected to increased pressure such as the shoulders, back, upper arms, and outer thighs. These infants may develop transient hypercalcaemia days or weeks later, probably because of calcium mobilization from necrosed tissues.

Rarely, hypercalcaemia may be an associated feature with congenital renal tumours, particularly mesoblastic nephroma. The cause of hypercalcaemia in these neoplasms is due to the production of ectopic PTH or prostaglandin E by tumour cells. In most cases, extirpation of the tumour corrects the hypercalcaemia and recurrence is rare.

Disorders of phosphate homeostasis

Hypophosphataemia

Moderate hypophosphataemia is defined as a serum inorganic phosphorus concentration between 2.5 and 1 mg/dL, and is usually asymptomatic. Severe hypophosphataemia is defined as a serum inorganic phosphorus concentration below 1.0 mg/dL.

Hypophosphataemia may be caused by: (1) decreased intestinal absorption of phosphate; (2) increased urine losses of phosphate, and an endogenous shift of inorganic phosphorus from extracellular to intracellular fluid compartments. Causes of hypophosphataemia are listed in Table 30.5.

Hyperphosphataemia

Hyperphosphataemia is most often the result of decreased renal excretion of phosphate anions as encountered in

Table 30.5 Causes of maternal and neonatal hypophosphataemia

Endocrine

Hyperparathyroidism
Diabetic ketoacidosis
Vitamin D disorders
    Vitamin D deficiency
    Vitamin D resistant rickets type I
    Vitamin D resistant rickets type II

Renal (tubular disorders)

Primary
    Fanconi syndrome
    Renal tubular acidosis
    Familial hypophosphataemia
Secondary
    Diuresis
    Glycosuria
    Hypomagnesaemia
    Hypercalciuria
    Volume expansion of ECF
    Gout

Gastrointestinal disease

Decreased intake
Decreased absorption
    Malabsorption
    Antacids

Others

Malignancy: mesenchymal tumours
Metabolic acidosis
    Keto acidosis
    Lactic acidosis
Respiratory alkalosis
Thermal burns
Gram-negative sepsis

acute or chronic renal failure particularly when the glomerular filtration rate is reduced to less than 25 per cent. Hyperphosphataemia could also be the result of increased body phosphate load from phosphate-containing laxatives and enemas, blood transfusions, hyperalimentation, and is also a major component of the tumour lysis syndrome secondary to cell lysis by cytotoxic therapy, and tissue injuries (hyperthermia, hypoxia, or crush injuries) resulting in rhabdomyolysis and haemolysis. Decreased renal tubular reabsorption of phosphate is responsible for hyperphosphataemia seen in hypoparathyroidism, hyper-

Table 30.6 Causes of maternal and neonatal hyperphosphataemia

---

Endocrine

Parathyroid disorders
    Idiopathic hypoparathyroidism
    Transient hypoparathyroidism
    Pseudohypoparathyroidism
Hyperthyroidism
Growth hormone excess
Vitamin D intoxication

Renal (decreased glomerular filtration)

Acute renal failure
Chronic renal failure

Increased load

Parenteral
    Hyperalimentation
    Blood transfusions
Enemas
Tumour lysis syndrome
Malignant hyperthermia
Rhabdomyolsis

Others

Postmenopause

---

thyroidism, hypogonadism, and growth hormone excess (Table 30.6).

Disorders of magnesium homeostasis

The importance of optimal Mg homeostasis in the perinatal period is becoming recognized. Pregnancy is associated with increased renal excretion of magnesium and a drop in serum Mg concentration, which may become more exaggerated if maternal dietary Mg intake is insufficient for optimal Mg homeostasis and fetal requirements. Recent clinical studies demonstrated the apparent benefits of maternal magnesium supplementation in reducing the incidence of preterm labour and allowing greater fetal growth. In one study, maternal Mg intake of 300 mg/day was associated with optimal birthweight, length, and head circumference.

Fetal and neonatal hypomagnesaemia

Fetal and neonatal hypomagnesaemia (serum Mg concentration < 1.5 mg/dL) is most commonly due to depletion of maternal body Mg stores, which is associated with decreased transplacental Mg supply to the fetus. Causes of maternal hypomagnesaemia are listed in Table 30.7. Aetiological factors in congenital hypomagnesaemia are listed in Table 30.8 and include endocrine, familial, and hereditary disorders.

Mothers with intestinal malabsorption were described in the literature as giving birth to infants who develop hypomagnesaemia. Maternal hypomagnesaemia may be caused by reduced Mg intake as in malnutrition or alcoholism, or increased intestinal Mg losses as in chronic diarrhoea and malabsorption because of ulcerative colitis, Crohn's disease, coeliac sprue, or gluten enteropathy. Maternal hypomagnesaemia may result from excessive renal Mg losses because of acute tubular necrosis or chronic renal disease,

Table 30.7 Causes of maternal hypomagnesaemia

Decreased intake

Enteral or parenteral
Malnutrition
Alcoholism

Gastrointestinal disease

Chronic diarrhoea
    Chronic ulcerative colitis
    Crohn's disease
    Laxative abuse
    Villous adenoma
Malabsorption
Short bowel syndrome
Gluten enteropathy
Tropical sprue
Familial magnesium malabsorption

Renal

Renal tubular acidosis
Acute tubular acidosis (diuretic phase)
Chronic pyelonephritis
Chronic glomerulonephritis
Familial and sporadic renal magnesium loss
Diuretics (furosemide, thiazides, ethacrynic acid)
Antibiotics (gentamycin, tobramycin, amphotericin B)
Cyclosporin

Endocrine

Hyperaldosteronism
Hyperthyroidism
Hypercalcaemia
Hyperparathyroidism
Uncontrolled diabetes mellitus

Table 30.8 Causes of fetal and neonatal hypomagne-saemia

Maternal-related

Malabsorption
Hyperparathyroidism
Insulin-dependent diabetes mellitus

Fetal-related

Endocrine

Hypoparathyroidism
Hyperaldosteronism
Hypercalcaemia

Gastrointestinal

Primary Mg malabsorption
Short bowel syndrome
Hepatobiliary
    Biliary atresia
    Neonatal hepatitis

Renal

Decreased tubular reabsorption (primary)
Decreased tubular reabsorption (drug-induced)
    Diuretics: furosemide, thiazides
    Antibiotics: gentamicin, amphotericin
Renal tubular acidosis
Acute tubular necrosis

Others

Birth asphyxia
Small for gestational age
DiGeorge's syndrome
Zellweger's syndrome

---

as in renal tubular acidosis, chronic pyelonephritis, or glomerulonephritis, or the use of diuretics and antibiotics. As described earlier, mothers with uncontrolled diabetes mellitus are often hypomagnesaemic presumably because of hyperglycaemia-induced hypermagnesuria (osmotic diuresis) and deficient insulin-dependent cellular uptake of Mg. Infants of diabetic mothers may be secondarily magnesium-depleted and may develop hypomagnesaemia, hypocalcaemia, and hypoparathyroidism. As described earlier, Mg is an intracellular cation and serum Mg concentration may not reflect total body Mg content. Therefore, tissue Mg deficiency may coexist with normomagnesaemia.

Familial hypomagnesaemia is a well known disease entity characterized by primary intestinal malabsorption or decreased renal tubular reabsorption of Mg. It is usually recognized in the neonatal period and predominantly occurs in males. The primary defect in renal tubular re-absorption of Mg may occur as an isolated or familial disease. Hypomagnesaemia may cause hypocalcaemia either by decreased Mg-dependent adenylate cyclase-mediated secretion of PTH, or end-organ resistance to PTH, or by decreased heteroionic exchange of Ca for Mg at the bone surface. Secondary hypocalcaemia responds to magnesium supplementation. Hypercalcaemia of any cause (hyperparathyroidism, vitamin D intoxication) or hypophosphataemia may cause excessive Mg losses in urine and hypomagnesaemia.

Renal tubulopathy and excessive Mg losses may be associated with administration of aminoglycosides and amphotericin. Also, the use of diuretics causes renal Ca and Mg loss, potentially resulting in negative magnesium balance and secondary hyperparathyroidism. The polyuric phase of acute renal failure or acute tubular necrosis is associated with hypomagnesuria and may result in hypomagnesaemia and other electrolyte disturbances.

Maternal hypomagnesaemia may have a significant role in the aetiology of pre-eclampsia with secondary placental insufficiency and possibly impaired Mg transfer to the fetus. Because most fetal Mg accretion occurs in the third trimester, preterm infants are at risk of developing hypomagnesaemia if not supplied with enough Mg in their diet. Small-for-gestational-age infants frequently develop hypomagnesaemia. A proportion of these infants are born to hypomagnesaemic pre-eclamptic mothers who may have placental insufficiency. Hyperaldosteronism is associated with excessive losses of magnesium in urine and feces and may result in hypomagnesaemia and negative magnesium balance.

Maternal hyperparathyroidism and hyperthyroidism may be associated with negative Mg balance in the mother and fetus because of excessive Mg losses in urine. Maternal hyperparathyroidism predisposes to transient neonatal hypoparathyroidism manifesting as hypocalcaemia and hypomagnesaemia.

Neonatal hypoparathyroidism may be associated with neonatal hypomagnesaemia. Possible mechanisms involved are: decreased PTH-induced Mg release from bone; increased renal tubular Mg loss; and hyperphosphataemia. Persistent congenital hypoparathyroidism may occur as an isolated or familial X-linked recessive or autosomal dominant entity as part of the DiGeorge syndrome (thymic aplasia and T-lymphocyte immunodeficiency, and aortic arch anomalies), the Zellweger syndrome (cerebrohepato-renal syndrome), or as an autoimmune disorder with mucocutaneous candidiasis.

Neonatal cholestasis due to congenital biliary atresia or severe neonatal hepatitis may cause excessive intestinal

Table 30.9 Causes of fetal and neonatal hypermagnesaemia

Maternal-related

MgSO$_4$ administration for tocolysis and pre-eclampsia

Fetal-related

Decreased renal excretion
    Oliguric renal failure
    Asphyxia neonatorum

Mg losses and magnesium depletion. It is possible that decreased metabolism of aldosterone due to hepatocellular damage results in secondary hyperaldosteronism which causes hypermagnesuria. Excessive Mg losses and hypomagnesaemia may occur in infants with short bowel syndrome following intestinal resection. Hypomagnesaemia has been described in new-borns with birth asphyxia but the pathogenesis is not clear.

Fetal and neonatal hypermagnesaemia

The most common cause of fetal and neonatal hypermagnesaemia (serum Mg concentration > 2.5mg/dL) is maternal hypermagnesaemia due to therapeutic magnesium sulfate administration for pre-eclampsia and preterm labour (Table 30.9). Theoretically, high maternal serum Mg concentrations may lead to transplacental Mg transfer to the fetus that exceeds the physiological rate of Mg transport resulting in fetal and neonatal hypermagnesaemia. Decreased Mg excretion because of oliguric renal failure and impaired glomerular filtration may result in elevated serum Mg concentrations. Hypermagnesaemia (serum Mg concentration > 2.5 mg/dL) blocks transmission at the neuromuscular junction and motor endplate and antagonizes the effects of calcium, thereby decreasing skeletal and smooth muscle tone. Hypermagnesaemic new-borns present with a variable degree of flaccidity and hypotension and respiratory depression. Because of depressed muscular tone, hypermagnesaemic neonates may develop the meconium-plug syndrome. In preterm new-borns, immature renal function and delayed renal excretion of Mg predispose to more pronounced prolonged hypermagnesaemia.

Serum Ca concentrations may be normal or elevated in infants with hypermagnesaemia possibly due to heteroionic exchange of Ca for Mg at the blood–bone interface. Recently, congenital bone demineralization and rickets have been described in infants born to mothers who were treated with magnesium sulfate for preterm labour for a long period (up to 4 weeks). Presumably, excess Mg enters the skeleton and displaces Ca out into the circulation, thus distorting the mineral proportion (Ca, P, Mg) of the crystal lattice and resulting in bone demineralization.

Disorders of vitamin D homeostasis

Vitamin D deficiency

Vitamin D deficiency in pregnant women may have drastic effects on bone mineralization of the fetus. Low dietary vitamin D intake, as in vegetarian women, and insufficient exposure to sunlight predispose to vitamin D deficiency manifesting as low serum concentration of 25OHD and calcium, with or without rickets in the mother and the baby. Congenital rickets, though rare, has been described in infants of mothers with vitamin D deficiency, severe intestinal malabsorption, or malnutrition. Vitamin D supplementation during pregnancy improves fetal skeletal growth in the rat.

Vitamin D toxicity

Excessive dietary intake of vitamin D metabolites has been reported to cause hypercalcaemia in both the mother and the new-born. Vitamin D metabolites have a long half-life. Transplacental transfer of excessive amounts of vitamin D or 25OHD from mother to fetus may be responsible for prolonged hypercalcaemia in the neonate. In pregnant rabbits, excessive vitamin D administration resulted in offspring with supravalvular aortic stenosis and craniofacial and dental anomalies. It is unclear how these findings may relate to the idiopathic hypercalcaemia syndrome of human infancy, in which hypersensitivity to vitamin D has been implicated and which includes craniofacial anomalies and supravalvular aortic stenosis.

Further reading

Abbas, S.K., Pickard, D.W., Illingworth, D., Storer, J., Purdie, D.W., Moniz, C., Dixit, M., Caple, I.W., Ebeling, P.R., Rodda, C.P., Martin, T.J., and Care, A.D. (1990). Measurement of parathyroid hormone-related protein in extracts of fetal parathyroid glands and placental membranes. *Journal of Endocrinology*, **124**, 319–25.

Alfrey, A.C. (1992). Normal and abnormal magnesium metabolism. In *Renal and electrolyte disorders* (4th edn) (ed. R. Schrier), pp. 371–404. Little Brown, Boston.

Brehier, A. and Thomasset, M. (1990). Stimulation of calbindin-D9K (CaBP9K) gene expression by calcium and 1,25-dihydroxycholecalciferol in fetal rat duodenal organ culture. *Endocrinology*, **127**, 580–7.

Bruns, M.E., Boass, A., and Toverud, S.U. (1987). Regulation by dietary calcium of vitamin D-dependent calcium-binding protein and active calcium transport in the small intestine of lactating rats. *Endocrinology*, **121**(1) 278–83.

Bruns, M.E., Kleeman, E., Mills, S.E., Bruns, S.E., and Herr, J.C. (1985). Immunochemical localization of vitamin D-dependent calcium binding protein in mouse placenta and yolk sac. *Anatomical Record*, **213**, 514–17.

Bruns, M., Overpeck, J.G., Smith, G.C., Hirsch, G.N., Mills, S.E., and Bruns, D.E. (1988). Vitamin D-dependent calcium binding protein in rat uterus: differential effects of estrogen, tamoxifen, progesterone, and pregnancy on accumulation and cellular localization. *Endocrinology*, **122**, 2371–8.

Christakos, S., Gabrielides, C., and Rhoten, W.B. (1989). Vitamin D-dependent calcium binding proteins: chemistry, distribution, functional considerations, and molecular biology. *Endocrine Reviews*, **10**, 3–26.

Coe, F. and Favus, M.J. (ed.) (1992). *Disorders of bone and mineral metabolism*. Raven Press, New York.

Corradino, R.A. and Fullmer, C.S. (1991). Positive cotranscriptional regulation of intestinal Calbindin-D28K gene expression by 1,25-dihydroxyvitamin D3 and glucocorticoids. *Endocrinology*, **128**, 944–50.

Davis, O.K., Hawkins, D.S., Rubin, L.P., Posillico, J.T., Brown, E.M., and Schiff, I. (1988). Serum parathyroid hormone (PTH) in pregnant women determined by an immunoradiometric assay for intact PTH. *Journal of Clinical Endocrinology and Metabolism*, **67**, 850–2.

Delvin, E.E., Richard, P., Pothier, P., and Ménard, D. (1990). Presence and binding characteristics of calcitriol receptors in human fetal gut. *Federation of European Biochemical Societies*, **262**, 55–7.

Drinkwater, B.L and Chestnut, C.H. (1991). Bone density changes during pregnancy and lactation in active women: a longitudinal study. *Bone and Mineral*, **14**, 153–60.

Favus, M.J. (ed.) (1990). *Primer on the metabolic bone diseases and disorders of mineral metabolism*. The American Society for Bone and Mineral Research, Kelseyville.

Forbes, G. (1976). Calcium accumulation by the human fetus. *Pediatrics*, **57**, 976–7.

Gross, M. and Kumar, R.C. (1990). Physiology and biochemistry of vitamin D-dependent calcium binding proteins. *American Journal of Physiology*, **259**, F195–F209.

Halbert, K. and Tsang, R. (1992). Neonatal calcium, phosphorus, and magnesium homeostasis. In *Fetal and neonatal physiology*, Vol. 2 (ed. R. Polin and W. Fox), pp. 1745–60. Saunders, Philadelphia.

Hock, J.M. and Gera, I. (1992). Effects of continuous and intermittent administration and inhibition of resorption on the anabolic response of bone to parathyroid hormone. *Journal of Bone and Mineral Research*, **7**, 65–72.

Itani, O. and Tsang, R.C. (1991). Calcium, phosphorus and magnesium in the newborn: pathophysiology and management. In *Neonatal nutrition and metabolism* (ed. W. Hay), pp. 171–202, Mosby, St Louis.

Khosla, S., Johansen, K.L., Ory, S.J., O'Brien, P.C., and Kao, P.C. (1990). Parathyroid hormone-related peptide in lactation and umbilical cord. *Mayo Clinic Proceedings*, **65**, 1408–14.

Koo, W. and Tsang, R.C. (1988). Calcium, magnesium and phosphorus. In *Nutrition in infancy* (ed. R. Tsang), pp. 175–89. Hanley and Belfus, Philadelphia.

Loveridge, N., Dean, V., Goltzman, D., and Hendy, G.N. (1991). Bioactivity of parathyroid hormone and parathyroid hormone-like peptide: agonist and antagonist activities of amino-terminal fragments as assessed by the cytochemical bioassay and in situ biochemistry. *Endocrinology*, **128**, 1938–46.

Martin, T.J., and Ebeling, P.R. (1990). A novel parathyroid hormone-related protein: role in pathology and physiology. In *Molecular and cellular regulation of calcium and phosphate metabolism*, pp. 1–37. Alan R. Liss, New York.

Marya, R.K., Saini, A.S., and Jaswal, T.S. (1991). Effect of vitamin D supplementation during pregnancy on the neonatal skeletal growth in the rat. *Annals of Nutrition and Metabolism*, **35**, 208–12.

Mimouni, F. and Tsang, R.C. (1992). Pathophysiology of neonatal hypocalcemia. In *Fetal and neonatal physiology*, Vol. 2 (ed. R. Polin and W. Fox), pp. 1761–6. Saunders, Philadelphia.

Moseley, J.M., Hayman, J.A., Danks, J.A., Alcorn, D., Grill, V., Southby, J., and Horton, M.A. (1991). Immunohistochemical detection of parathyroid hormone-related protein in human fetal epithelia. *Journal of Clinical Endocrinology and Metabolism*, **73**, 478–84.

Mughal, M. and Tsang, R.C. (1992). Calcium, phosphorus, and magnesium transport across the placenta. In *Fetal and neonatal physiology*, Vol. 2 (ed. R. Polin and W. Fox), pp. 1735–44. Saunders, Philadelphia.

Ohta, T., Mori., M., Ogawa, K., Matsuyama, T., and Ishii, S. (1989). Immunocytochemical localization of BGP in human bones in various developmental stages and pathological conditions. *Virchow's Archiv A, Pathological Anatomy and Histopathology*, **415**, 459–66.

Popovtzer, M.M., Knochel, J.P., and Kumar, R. (1992). Disorders of calcium, phosphorus, vitamin D, and parathyroid hormone activity. In *Renal and electrolyte disorders* (4th edn) (ed. R. Schrier), pp. 287–369. Little Brown, Boston.

Purdie, D.W., Aaron, J.E., and Selby, P.L. (1988). Bone histology and mineral homeostasis in human pregnancy. *British Journal of Obstetrics and Gynaecology*, **95**, 849–54.

Quamme, G.A. and Dirks, J.H. (1987). Magnesium metabolism. In *Clinical disorders of fluid and electrolyte metabolism* (4th edn), (ed. M. Maxwell, C. Kleeman, and R. Narins), p. 297. McGraw-Hill, New York.

Rasmussen, N., Frohlich, A., Hornnes, P.J., and Hegedüs, L. (1990). Serum ionized calcium and intact parathyroid hormone levels during pregnancy and postpartum. *British Journal of Obstetrics and Gynaecology*, **97**, 857–62.

Repke, J.T. (1991). Calcium, magnesium, and zinc supplementation and perinatal outcome. *Clinical Obstetrics and Gynecology*, **34**, 262–7.

Ross, R. (1992). Calcium regulating hormones. In *Fetal and neonatal physiology*, Vol. 2 (ed. R. Polin and W. Fox), pp. 1698–1734. Saunders, Philadelphia.

Rubin, L.P., Posillico, J.T., Anast, C.S., and Brown, E.M. (1991). Circulating levels of biologically active and immunoreactive intact parathyroid hormone in human newborns. *Pediatric Research*, **29**, 202–7.

Saggese, G.I., Baroncelli, G.I., Bertelloni, S., and Cipolloni, C. (1991). Intact-parathyroid hormone levels during pregnancy, in healthy term neonates and in hypocalcemic preterm infants. *Acta Paediatrica Scandinavica*, **80**, 36–41.

Seki, K., Makimura, N., Mitsui, C., Hirata, J., and Nagata, I. (1991). Calcium-regulating hormones and osteocalcin levels during pregnancy: a longitudinal study. *American Journal of Obstetrics and Gynecology*, **164**, 1248–52.

Shaw, J. (1973). Parenteral nutrition in the management of sick low birth-weight infants. *Pediatric Clinics of North America*, **20**, 333–58.

Sibai, B.M., Villar, M.A., and Bray, E. (1989). Magnesium supplementation during pregnancy: a double blind, randomized controlled clinical trial. *American Journal of Obstetrics and Gynecology*, **161**, 115.

Spätling, L., Disch, G., and Classen, H.G. (1989). Magnesium in pregnant women and the newborn. *Magnesium Research*, **2**, 271–80.

Villar, J. and Repke, J. (1990). Calcium supplementation during pregnancy may reduce preterm delivery in high risk populations. *American Journal of Obstetrics and Gynecology*, **163**, 1124.

Wallach, S. (1990). Effects of magnesium on skeletal metabolism. *Magnesium Trace Elements*, **9**, 1–14.

Walters, M.R., Bruns, M.E., Carter, R.M., and Riggle, P.C. (1991). Vitamin D independence of small calcium-binding proteins in nonclassical target tissues. *American Journal of Physiology*, **260**, E794–E800.

Ziegler, E.E., O'Donnell, A.M., Nelson, S.E., and Fomon, S.J. (1976). Body composition of the reference fetus. *Growth*, **40**, 329–41.

31. Development of thermogenesis in the fetus

Barbara Cannon and Jan Nedergaard

Passive versus active (thermoregulatory) thermogenesis

What is thermogenesis?

All life has as its by-product heat production. In a broad sense this follows from the fact that life is a fight against entropy and disorder and that only by adding a constant input of energy is it possible to counteract this increasing disorder. Thus, the 'basal' heat production (or the basal metabolism, which is essentially the same) of an organism can be interpreted as the running cost of maintenance. The heat released does not have a purpose in itself, nor should it really be considered an effect of metabolism. Rather, the metabolic rate is adjusted so that the input of energy equalizes that which is dissipated as heat.

Although it is generally not possible to divide steady-state processes into consecutive steps, the processes may more easily be understood in such an analysis. Thus, as seen in Fig. 31.1(a), we may have a certain degree of organization in a living organism. This includes, for instance, a non-uniform distribution of molecules in different fractions of this organism. As time passes, these molecules will tend to distribute uniformly (Fig. 31.1(b)) and the energy gained from this redistribution is released as heat. We then have to put more energy into the system in order to restore the uneven distribution desired (Fig. 31.1(c), and the energy input needed is therefore governed fully by the rate of disorganization of the previous 'ordered' system. Thus, the metabolic rate observed equals the heat released as the order of the system is being disorganized.

As can thus be understood from the above, the amount of energy input into an organism and the amount of heat released are virtually identical, and we can therefore express metabolism and heat production by the same units. The units used can therefore be the classical calories (cal) (per unit of time) or the presently preferred joules (J) (per unit of time). However, as joules per second has its own unit, the preferred unit to use in discussion of heat production or metabolism is the watt (W). As the amount of energy released when a given amount of oxygen is used

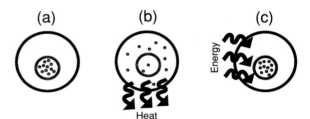

Fig. 31.1 The energetic basis for the so-called basal heat production.

for combustion does not vary much with different substrates being combusted (metabolized, 'eaten'), heat production can also be expressed in mL O_2 combusted per unit of time.

Factors that influence the basal rate of heat production

All living organisms, including the fetus, are heat producers, i.e. they are thermogenic. However, the rate of thermogenesis in different organisms is not similar, but is governed by some simple principles, primarily body temperature and body weight.

Temperature

As the metabolism of the organism obeys the laws of chemistry and physics, the sum of the chemical reactions (which we refer to as metabolism) is dependent upon temperature. This can be understood in terms of the kinetic energy of the molecules interacting in a given process: in order for an interaction to occur, the kinetic energy of the reactants has to exceed a certain value, which we refer to as the activation energy or the Arrhenius energy. The fraction of molecules that have a kinetic energy in excess of the activation energy is increased with increasing temperature (as temperature is simply an expression for the mean kinetic energy of molecules). Within a temperature range compatible with life, the rate of chemical reactions (and

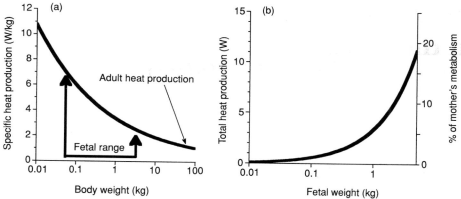

Fig. 31.2 (a) Specific and (b) total heat production in the fetus. The figures show theoretical values, based on the so-called law of metabolic reduction, stating that basal metabolism is 3.3 watts × (kg body weight)$^{0.75}$. (b) The mother's body weight is set to be 60 kg.

thus the heat production) increases by a factor of 2–3 for each 10°C increase in temperature (or by about 7–12 per cent per °C).

As the fetus is normally in constant thermal surroundings of about 38°C (in the human), the rate of heat production rarely alters due to the temperature effect. However, as the temperature of the fetus is fixed in relation to that of the mother, any increase of her body temperature is transferred to the fetus and results in a higher rate of heat production in the fetus.

Body weight

Of course, with increasing mass of living material, the total heat production increases. It has, however, been realized for a long time that the rate of heat production (the metabolic rate) in living organisms is not simply proportional to the weight of the organism — i.e. the heat production of a 10 times larger organism is not 10 times larger. It *is* larger — but not as much larger as expected. This phenomenon has important consequences for the understanding of the development of the heat production of the fetus.

When the specific rate of basal heat production (metabolism) (i.e. W/kg body weight) of a wide variety of mammalian species of different sizes is plotted as a function of the body weight, it is observed that this specific rate of heat production is reduced principally with increasing body size, as seen in Fig. 31.2(a). The metabolic rate is 3.3 W for a body size of 1 kg and is otherwise proportional to body weight raised to the power of 0.25. (This means (as seen in Fig. 31.2(a)) that the specific rate is nearly halved for each 10-fold increase in body weight). As the weight of fetus is much less than one-tenth that of the

mother, the specific thermogenic rate of the fetus, based on this relationship, could be expected to be much higher than that of the mother (who has a specific thermogenic rate of about 1 W/kg body weight).

The cause of the increase in specific metabolic rate with decreasing body size is still not understood. It is, however, still not uncommon to see the heat loss from the organism discussed as an explanation for this phenomenon, with reference to the surface/volume ratio increasing with smaller size. It must, therefore, be pointed out that this explanation does not hold. This can be understood, for example, from the fact that even 'cold-blooded' (poikilothermic) organisms show the same relationship, without having any temperature gradient between the body and the surrounding. Thus, the metabolism of the fetus cannot be expected to be lowered in spite of the lack of temperature gradient between the fetus and its environment.

It may be argued that the fetus cannot be considered a 'free-living' organism and therefore does not conform to this kind of relationship. There is no simple way to investigate this question, at least not in relation to the human fetus, but available evidence would tend to indicate that the fetus is indeed in this respect behaving as an independent organism, with the expected high metabolism. From which stage during embryonic development this occurs is not known, and the extension of the specific metabolic rate to the very small 'organism weight' of the fertilized egg may be doubtful; on the other hand, from this point onwards the growth of the prospective child is governed by forces endogenous to the new organism, and there is, therefore, good reason to think that metabolism is regulated as if the embryo or fetus were free-living.

Analysed in this way, the extra weight of the fetus cannot metabolically be considered as just some tissue added to the mother but rather as an independent and small (and parasiting) organism. This is illustrated in Fig. 31.2(b) where it is shown how the calculated heat production increases with fetal body weight; it is also seen how, due to this high endogenous metabolism, the basal heat production of the fetus in the last phases of gestation amounts to a significant increase in the metabolism of the mother. It may be mentioned that the metabolism of very premature infants has been found to be as expected from the above relationship.

Thermoregulatory thermogenesis

The heat production discussed above is that which is a by-product of life processes and which must always occur in a living organism. This heat production should be contrasted with that which can be analysed as occurring from processes initiated with the apparent purpose of producing heat — i.e. thermoregulatory thermogenesis. Such processes can only be envisaged if the organism endeavours to obtain or remain at a specified temperature higher than that of the surroundings. As we have no reason to believe that the fetus would ever strive to reach a temperature higher than that of the mother, the situation of the fetus may seem to be clear-cut: never would the fetus under physiological conditions experience situations in which it would need to call such thermoregulatory thermogenic functions into action.

However, the fetus has to prepare itself for the dramatic changes that await it after birth, and during fetal life normal development must ensure that the new-born child is prepared to encounter the thermically changing and challenging conditions with which it will be faced at birth.

Developmental stage at birth

To understand the development of thermoregulatory thermogenesis, different animal systems have been investigated. It is, however, essential to keep in mind that the conclusions from such experiments cannot directly be applied to the human fetus. An important issue to clarify is that the fetal to new-born transition occurs at different stages of development in different types of organisms. Roughly, different species may be categorized into three types, based on the developmental stage reached by the young at the time of the fetal to new-born transition: the precocial, the altricial, and the immature groups.

The precocial new-borns are often one or few in a litter, but they are well developed in all respects: their eyes are open, they are furred, and they can move around. These animals have a capacity for thermoregulatory thermogen-

esis that is already fully developed at birth. The typical experimental example is the lamb.

The altricial new-borns are often many in a litter. They are blind, naked, and cannot really move around; they stay in some form of nest and are often even warmed by the mother. In these animals, the capacity for thermoregulatory thermogenesis is not fully developed at birth but is recruited as an effect of the cold experienced by the pups after birth. The typical experimental example is the rat.

The immature new-borns are apparently 'cold-blooded' at birth and do not react to cold with a recruitment of thermoregulatory thermogenesis. Only after a period of postnatal development are the mechanisms that allow for the recruitment of these processes developed, and the new-borns then behave as altricial new-borns. The typical experimental example is the hamster.

The classification of man into one of these groups is not self-evident but the human new-born is probably somewhere between being a typical altricial and a typical precocial new-born. If results concerning fetal thermogenic development are transferred from experimental animals to man, the risk is thus that they yield an impression of a fetus more developed than is really the case if the lamb is the model, and too undeveloped if the rat is the model. As the lamb has generally been the experimental model used for physiological studies, and the rat that for biochemical studies, the results are not always compatible.

The question of how to classify the human fetus within this scale may be of more than academic interest. This is because, if the human infant is considered altricial, the normal development of thermoregulatory thermogenesis is not completely finished until after the new-born has been exposed to cold after birth. Following this interpretation, there may be reason to discuss the routine use of incubators after birth for small infants and perhaps to limit the use of incubators to those infants by whom it is critically needed.

Development of central thermoregulatory control mechanisms

The ability to control body temperature in mammals is based on an integrated system as illustrated in Fig. 31.3. In general, the functional system includes an input from some kind of monitoring system(s), a central regulatory system, and peripheral effector systems.

In mammals, the main body temperature monitoring system and the central control system are both found in the hypothalamic area and are sensitive to the temperature of

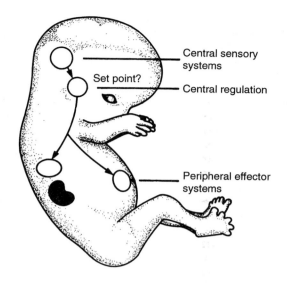

Fig. 31.3 Organization of the thermoregulatory system in the fetus.

the cerebrospinal fluid in the third ventricle. There are no investigations allowing for a direct approach to the question of when during fetal development the central control system matures, i.e. when the developing fetus changes from being a true 'cold-blooded' creature to one that has a desired body temperature (a so-called 'setpoint') and that, by invoking peripheral mechanisms, tries to defend this temperature. Clearly, it must mature earlier than that point at which a thermogenic response to cooling can be invoked, as this also presupposes maturation of the peripheral response mechanisms.

If we consider fetal maturation as a reflection of evolution ('ontogeny recapitulates phylogeny'), it would be predicted that the central control mechanism would mature relatively early. Although we normally consider only mammals (and birds) to be homeothermic (i.e. able to maintain a given body temperature), it is now realized that many species of animals that are generally considered to be poikilotherms ('cold-blooded') also show behavioural thermal adjustments in order to keep a stable body temperature. Thus, they possess a central thermoregulatory control mechanism. (What they lack is an endogenous metabolic capacity to achieve this goal.)

The present ability to keep even extremely preterm infants alive would theoretically allow for a more detailed investigation of this question concerning the maturation of the central thermoregulatory control mechanism. In reality, what has to be monitored are their responses; i.e. do

thermoregulatory responses occur even in these preterm new-borns? Very small changes in environmental temperature would be sufficient to evoke such responses if they exist. Concerning the defence against high body temperatures, investigations have shown that even infants of a gestational age of less than 30 weeks (28 weeks) are able to respond adequately to an increased temperature with posture changes or vasodilatation. No similar studies have been performed concerning defence against cold but it is unlikely that the central control system would be one-sided. Although these observations do not in themselves demonstrate that these systems are operative even when the fetus is *in-utero*, this would seem most likely. Thus, there is good reason to assume that the central thermoregulatory mechanism has matured at least from week 28 of gestation in humans.

Development of peripheral thermogenic mechanisms

As seen in Fig. 31.3, the function of the central thermoregulatory system would be to evoke thermogenic responses in the fetal body. However, there is no reason to think that this would ever happen during fetal life. Thus, what can be discussed is the nature and the maturation of the potential thermogenic mechanisms and whether these mechanisms can be artificially stimulated while the fetus is still *in utero*.

There are two kinds of thermogenic mechanisms: shivering and non-shivering thermogenesis.

Shivering thermogenesis is a thermogenesis that utilizes for thermogenic purposes the heat released as a by-product of muscle contraction. The efficiency of muscular work is quite low, about 25 per cent; this means that for every joule of work being done by a muscle on the environment, 4 joules of energy are used, and 3 of these are released as heat. This is undoubtedly the major acute thermogenic process called upon in cold-exposed adult man, but, in new-borns, this mechanism plays a minor role. As the basis for this heat production is the maturation of muscle innervation, the mechanism as such may be considered to be developed from an early fetal age, but the capacity is low (as vigorous muscle work cannot be evoked in small infants). When fetal lambs are cold-exposed *in utero*, some shivering may occur, but again the capacity of this process is low.

Non-shivering thermogenesis thus refers to all other kinds of thermogenesis, i.e. where the heat produced is not a by-product of muscular work. By this definition, the basal heat production is really a 'non-shivering thermogenesis', but this expression should normally be reserved for a thermoregulatory non-shivering thermogenesis, i.e. any reaction other than muscle contraction evoked with the purpose of producing heat. Three types of reactions are normally discussed in this context: futile cycles, thyroid thermogenesis, and brown adipose tissue thermogenesis. There is presently no doubt that it is the maturation of brown adipose tissue thermogenesis that is most relevant in connection with fetal development.

Futile cycles (or substrate cycles) refer to processes in which two metabolic processes simultaneously occur, one utilizing ATP for phosphorylation of a substance and the other dephosphorylating this substance. Such a process would be, for example, the synthesis of fructose 1,6-bisphosphate from fructose 6-phosphate and its subsequent dephosphorylation back to the original monophosphate. Although there are many examples both theoretically and experimentally of such processes, their thermogenic significance is probably limited, and there has been no indication that they are recruited during cold stress. They will therefore not be further discussed here.

Thyroid thermogenesis refers to such purposive thermogenic processes that are stimulated by thyroid hormone. There is a connection between the thyroid status and the level of basal metabolism (heat production) of an organism, but there is no evidence that thermoregulatory thermogenesis is directly regulated by thyroid hormones. It has been discussed earlier whether a stimulation of the $Na^+K^+ATPase$ could have a specific thermogenic purpose. Although some evidence has been presented for this hypothesis, it would seem that the thermogenic function of this ion pump in normal cells (except kidney and nerves) is minor and that the high values obtained earlier were due to the cells studied not being intact.

Brown adipose tissue thermogenesis

Brown adipose tissue development and evolution

It is not so long ago that brown adipose tissue was considered to be an esoteric tissue, thought to be found only in a few mammalian species (such as marmots) which were true hibernators. The tissue had an unknown function and was though to be rudimentary in some way. Only in recent years has research revealed that this tissue is indeed a major mammalian accomplishment and that it is found in all new-born mammals (with the apparent exception of the domestic pig). Indeed, it may be assumed that, for the evolution of homeothermic mammals, the development of a

few fundamentally 'new' organs was essential. Of course, these organs include the mammalian uterus and the mammae, but the development of brown adipose tissue as a new organ gave the evolving mammals new possibilities: to keep active even when the surrounding temperature was low. It is also likely that the evolution of brown adipose tissue was essential because the mammalian organism became organized to function well only at 37°C so that the mammalian new-born needed to be able to protect its body temperature.

In larger mammals such as man, the functional role of brown adipose tissue thermogenesis is probably limited or even non-existent in adulthood. There is, however, no doubt that, in the new-born infant, brown fat is the major source of themoregulatory thermogenesis. This is a relatively new insight, and the quantitatively prominent role of brown adipose tissue for postnatal thermogenesis is not generally emphasized in common medical textbooks.

The function of brown adipose tissue

The major — if not the only — function of brown adipose tissue is to produce heat. In many ways, the tissue resembles white adipose tissue, but it has, in addition, some specific features, mainly based on the existence in the tissue of the unique uncoupling protein thermogenin. The tissue may also be active as a part of the endocrine system.

The adipocyte character of the brown fat cell

As seen in Fig. 31.4(a) the fully developed brown fat cell has many of the characteristics of a white fat cell. The cell is filled with triglyceride droplets that are formed from fatty acids taken up from the circulation through the action of the enzyme lipoprotein lipase. The breakdown of the triglycerides is under neurotransmitter control by noradrenaline which is released from the sympathetic nerves innervating the tissue. The noradrenaline acts via adrenoceptors that are mainly of the β_3-type, although α_1-adrenoreceptors are also found. The β-adrenergic stimulation gives rise to cyclic AMP (cAMP), which activates a protein kinase, which in its turn activates a so-called hormone-sensitive lipase. Through the action of this enzyme, free fatty acids are released within the cell. These may be released to the circulation, but a significant fraction of the fatty acids are transferred to the mitochondria where they are combusted, and heat is released.

The uncoupling protein thermogenin

The catabolism of fatty acids in the mitochondria proceeds metabolically in brown-fat mitochondria just as in other

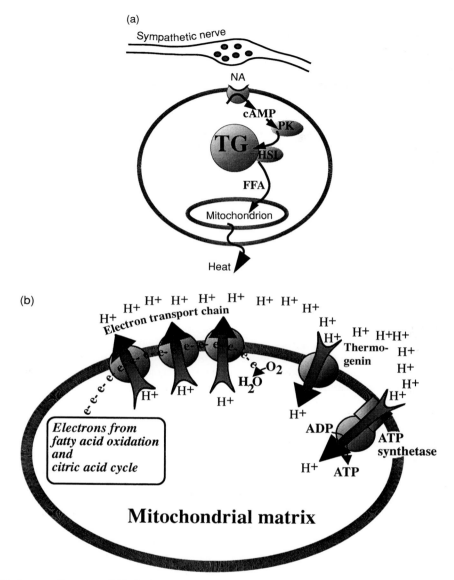

Fig. 31.4 (a) The brown adipocyte and (b) the brown fat mitochondrion. NA, Noradrenaline; PK, protein kinase; TG, triglyceride; HSL, hormone-sensitive lipase; FFA, free fatty acid.

fatty-acid-metabolizing mitochondria (Fig. 31.4(b)). The fatty acids are broken down to two-carbon units of acetyl-coenzyme A and are finally fully oxidized (combusted) to yield CO_2 and water. The chemical energy thus released is, as in other mitochondria, initially stored in the form of a transmembranous H^+ electrochemical gradient — i.e. the electron transport chain converts the energy obtained from the transfer of electrons from the substrate to

O_2 to a proton gradient, by pumping protons out of the mitochondria.

In all other mitochondria, and also in unstimulated brown-fat mitochondria, the protons are allowed to re-enter the mitochondrion through the ATP–synthetase complex, and the energy is finally used to drive the synthesis of ATP. In brown-fat mitochondria, however, there is an alternative pathway: entry through the uncoupling protein thermogenin.

Thermogenin is a regulated carrier (or channel) for protons, and the re-entry of protons through thermogenin is not coupled to any energy-conserving process (in contrast to the case in ATP–synthetase). Therefore the whole energy stored in the proton electrochemical gradient (which originally was the energy released from the combustion of the fatty acids) is released as heat, as the protons re-enter.

The activity of thermogenin is, as stated, regulated. Physiologically, this is a necessity, as a constantly high proton conductance would lead to an incessant heat production at the highest enzymatically possible rate. The activity of thermogenin is increased after noradrenaline stimulation of the cell; the mechanism behind this is still not fully settled but probably involves a stimulatory action of the free fatty acids (or some derivative); the fatty acids thus molecularly function as activators of their own combustion.

Deiodination of thyroxine

Besides the direct thermogenic action of brown adipose tissue, the tissue may also indirectly affect thermogenesis. It has been found that the tissue has a high activity of the enzyme thyroxine deiodinase, which transfers the metabolically inactive thyroxine (T_4) to the metabolically active T_3. This is apparently necessary for the full functioning of the tissue itself, but it is likely that T_3 may also leave the brown fat cells and be transferred with the circulation to other cells, where it may affect the basal metabolic rate. At least in the fetal rat, the activity of thyroxine deiodinase is high shortly before birth and this may be involved in the final maturation of the thermogenic system in these animals.

Development of brown adipose tissue

Brown adipose tissue develops (just like white adipose tissue) from mesodermal cells. There is reason to believe that these two cell types, although being closely related, are already at the mesodermal stages predetermined to develop into one or the other of the two adipose tissues. In accordance with this, undifferentiated cells isolated from brown or white adipose tissue that are allowed to differentiate in culture under identical conditions develop into cells that are distinctively different from each other.

In the developing human fetus, brown adipose tissue can be distinguished morphologically from the twentieth gestational week. After this time, it increases more than in parity with fetal somatic growth. The fractional amount of brown fat increases with increasing fetal weight from about 0.5 per cent to more than 1 per cent of body weight,

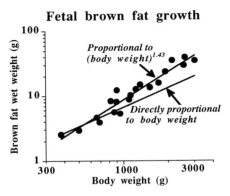

Fig. 31.5 Development of brown adipose tissue during fetal life. (The graphs calculated from data presented by Merklin, R.J. (1974). *Anatomical Record*, **178**, 637–46.)

and in Fig. 31.5 it can be seen that the total amount of brown adipose tissue increases in proportion to the fetal weight raised to the power of 1.43. Thus, as fetal development progresses, brown adipose tissue growth becomes a priority.

The different depots of brown adipose tissue do not grow in parallel. Brown adipose tissue is found in the human fetus in more than 12 different locations, including major interscapular, periaortic, and perirenal depots. The relative fraction of total brown fat found in the perirenal depots increases with increasing fetal weight. The growth consists of accumulation of both lipid and protein (enzymes, etc.)

In the human fetus, the lipid accumulated probably consists primarily of fatty acids that have passed the placenta (this is not the case in all animals). Some ketone bodies may also pass the placenta and be taken up by the tissue, and some *de novo* fatty acid synthesis may occur. In the fetus there is no reason to believe that lipoprotein lipase is involved in the accumulation of lipid in the tissue. It is clear that in the human fetus, a significant amount of fatty acid is accumulated in the tissue before birth. Therefore, in contrast to, for example, the altricial new-born rat, the somewhat more precocial new-born human infant carries a supply of substrate for thermogenesis at birth and can initiate thermogenesis before receiving the first postnatal meal.

Fetal development of the thermogenic apparatus (requisite enzymes) is, however, even more important for the acquisition of thermogenic potential. Unfortunately, very little is known about this important aspect of fetal development and, whereas the presence of the uncoupling protein thermogenin in human infant brown adipose tissue has been verified, the emergence of this protein during fetal development has not yet been studied. As adequate immunological methods for detection of thermogenin are now available, it should be possible to follow this important parameter during fetal growth.

The recruitment process

An understanding of the fetal development of the thermogenic potential of brown adipose tissue and especially the acquisition of thermogenin may be discussed in the light of what is known about these processes in other physiological conditions.

In non-stimulated, non-fetal mammals, the brown adipose tissue is in a state that can best be referred to as atrophied. In particular, very low levels of thermogenin are found. However, if a continuous thermogenic response is required, the tissue becomes recruited: a series of different processes are initiated, but among the most remarkable is the induction of an increased transcription of the thermogenin gene, leading to an increased thermogenin mRNA level and an increased thermogenin synthesis. All evidence suggests that these effects are due to the stimulation of the cells with noradrenaline — thus, the same agent that directly controls thermogenesis recruits the tissue. Similarly, when brown fat cells differentiate in culture, they spontaneously exhibit only a very low level of thermogenin. Only when the cells are stimulated with noradrenaline does the thermogenin level increase.

In altricial species (such as the rat), the fetal level of thermogenin is also very low, and not until the new-born becomes successively exposed to a cold environment does thermogenin gene expression increase. Thus, in all cases summarized above, the recruitment of brown adipose tissue and especially the increase in thermogenic capacity is due to a continual stimulation of the tissue with noradrenaline. As a necessary consequence, this stimulation must also evoke the thermogenic function as such. From this point of analysis, the case of the fetal development in species like the sheep, and most probably also the human, are very difficult to understand.

In precocial species like the sheep brown adipose tissue is apparently fully recruited at the moment of birth. This is physiologically meaningful because their new-borns are expected to be independent from birth. However, this means that the regulation of brown adipose tissue thermogenic recruitment in these species in some aspects must be fundamentally different from that in other animal models studied.

In this connection the following points can be made.

1. The recruitment process in these types of fetuses could be assumed to proceed due to the same processes that govern recruitment in other conditions: a constant noradrenaline stimulation. However, this would necessarily also stimulate thermogenesis in the fetus, a situation that is physiologically meaningless, that would present a heavy thermal load on the mother, and for which there is no experimental evidence. Further, this process necessitates the activation of the sympathetic nervous supply to the tissue. Although it seems clear that adrenergic innervation has reached the tissue during fetal life, there is no physiological reason for these nerves to be stimulated: as concluded above, the central thermoregulatory centres are developed at this time and there is no reason to assume that the central control systems would send signals to the tissue to produce heat as long as the surrounding temperature is 38°C or more.

2. Some of these analytical problems could perhaps be solved if a fetal inhibitor of thermogenesis (and perhaps of other processes) were active. Indeed, there is some evidence that such an inhibitor is present and is active. Thus, fetal sheep that still have a placental circulation cannot be provoked into non-shivering thermogenesis, but if the connection with the placenta is blocked (while ventilating the lungs), thermogenesis can be evoked. Interesting as these observations are, they only partly explain the problems discussed above. Thus, if the inhibitor is working by counteracting the effect of noradrenaline, this would indeed abolish thermogenesis, but, as recruitment is apparently stimulated via the same intracellular pathways, recruitment would also be inhibited *in utero*. The other problem mentioned above — that there is no reason to assume that the sympathetic nerves are stimulated *in utero* — also remains unsolved.

3. It can alternatively be assumed that fetal thermogenic recruitment does not proceed via the noradrenaline pathway explored in the other systems studied. Although there is experimental evidence that may support this notion (brown adipose tissue apparently develops in the fetus even when the sympathetic nervous system is chemically destroyed), the consequences of such an assumption are also difficult to reconcile with observations. Thus, if the tissue can become spontaneously recruited without adrenergic stimulation, why does it not remain stimulated after birth (where experimental evidence clearly indicates that brown adipose tissue becomes atrophied)?

4. The only acceptable hypothesis would therefore be that a specific fetal stimulatory substance exists that does not proceed in its action via the noradrenaline/cAMP pathway but that is still able to induce thermogenin gene expression and the recruitment process in general. This substance would thus be something like a placental activator of recruitment. Unfortunately, there is absolutely no experimental evidence for the existence of such a compound.

Thus, the most challenging feature concerning fetal thermogenic development remains the understanding of the regulation of the fetal recruitment process. Although not directly demonstrated, the human fetus would seem to undergo such a recruitment event (although further postnatal recruitment also occurs in humans), and this process has to be completed before birth.

The modern improvements in techniques that allow premature babies to survive began with the introduction of incubators. Since then it has been understood that the prevention of hypothermia is indeed an essential point in rescuing premature babies. This understanding is thus a consequence of knowledge about the development of a thermogenic mechanism in the fetus.

Further reading

Cannon, B. and Nedergaard, J. (1985). The biochemistry of an inefficient tissue: Brown adipose tissue. *Essays in Biochemistry*, **20**, 110–64.

Clausen, T., van Hardeveld, C., and Everts, M.E. (1991). The significance of cation transport in the control of energy metabolism and thermogenesis. *Physiological Reviews*, **71**, 733–74.

Gluckman, P.D., Gunn, T.R., Johnston, B.M., Power, G.C., and Ball, K.T. (1988). Maturation of thermoregulatory and thermogenic mechanisms in fetal sheep. In *The endocrine control of the fetus* (ed. W. Künzel and A. Jensen), pp. 300–5. Springer, Berlin.

Gunn, T.R., and Gluckman, P.D. (1989). The endocrine control of the onset of thermogenesis at birth. *Ballière's Clinical Endocrinology and Metabolism*, **3**, 869–86.

Hull, D. (1988). Thermal control in very immature infants. *British Medical Bulletin*, **44**, 971–83.

Nedergaard, J., Connolly, E., and Cannon, B. (1986). Brown adipose tissue in the mammalian neonate. In *Brown adipose tissue* (ed. P. Trayhurn and D.G. Nicholls), pp. 153–213. Edward Arnold, London.

Power, G.C. (1989). Biology of temperature: the mammalian fetus. *Journal of Developmental Physiology*, **12**, 295–304.

Schönbaum, E. and Lomax, P. (ed.) (1990). *Thermoregulation: physiology and biochemistry*. Pergamon Press, New York.

Sinclair, J.C. (ed.) (1978). *Temperature regulation and energy metabolism in the newborn*. Grune and Stratton, New York.

Trayhurn, P. and Nicholls, D.G. (1986). *Brown adipose tissue*. Edward Arnold, London.

32. Maternal adaptation to pregnancy
Alan D. Bocking

The physiological alterations that take place in the mother during pregnancy are profound and vital for the successful completion of gestation. Many of these adaptations are hormonally mediated and others are due to the effect of the expanding gravid uterus. This chapter will review some of the changes that occur within the maternal cardiovascular, renal, haematological, respiratory, gastrointestinal, and endocrine and metabolic systems as well as the skin in order to promote and allow the normal growth and development of the fetus.

Cardiovascular system

Blood volume

Total circulating blood volume in the mother is increased during pregnancy by approximately 45 per cent reflecting an increase in both red blood cell mass and plasma volume. Since the increase in plasma volume is greater than the increase in red cell mass, however, there is a relative decrease in haemoglobin content and haematocrit giving rise to what has been called the relative or 'physiological' anaemia of pregnancy.

Plasma volume begins to rise at as early as 6 to 8 weeks of gestation and has increased by approximately 45 per cent by 30 weeks of gestation in human pregnancy. In multiple gestations, the increase in blood volume is even greater than that for singletons. A similar increase in plasma volume has been noted over the last 30 days of gestation in pregnant sheep. It is of interest that the failure of plasma volume to increase during human pregnancy is associated with an increased incidence of pregnancy complications such as hypertension and/or intrauterine growth restriction. In pregnant sheep, when the normal increase in plasma volume with advancing gestation is prevented over the last 30 days of gestation, there is a fall in fetal PO_2 in association with a decrease in uterine blood flow, highlighting the importance of the physiological increase in plasma volume during pregnancy.

It has been suggested that the effects of pregnancy on maternal plasma volume and blood pressure are mediated by the increase in circulating oestrogen concentrations. Support for this hypothesis is provided by studies in non-

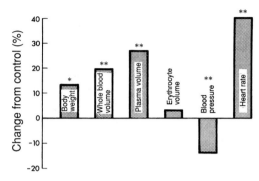

Fig. 32.1 Percentage change from control values of physiological variables in non-pregnant sheep following 3 weeks of infusion with 17β-oestradiol. *, P<0.05, **P<0.01. (Reprinted with permission from Ueda, S., Fortune, V., Bull, B.S., Valenzuela, G.J., and Longo, L.D. (1986). *American Journal of Obstetrics and Gynecology*, **155**, 195–201.)

pregnant sheep that were infused with 17B-oestradiol for 3 weeks with a subsequent 27 per cent increase in plasma volume (Fig. 32.1). Acute injections of oestrogens into the uterine artery of non-pregnant sheep have also been shown to cause a significant increase in uterine blood flow. The observed increase in red blood cell mass during pregnancy occurs as a result of an increase in red cell production, not prolongation of the life-span of the cells.

Cardiac output

In addition to the increase in blood volume, pregnancy is accompanied by an increase in maternal cardiac output of approximately 30–40 per cent from early in the first trimester (Fig. 32.2). This increase in cardiac output occurs initially as a result of an increase in stroke volume and subsequently of an increase in heart rate as well. The increase in stroke volume parallels the increase in blood volume previously described. Systemic vascular resistance in healthy pregnant women at term, measured using invasive techniques, has been shown to be decreased by 20 per cent in the same women when compared to the non-pregnant state (at approximately 12 weeks postpartum) as shown in Fig. 32.3.

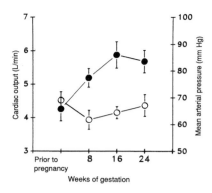

Fig. 32.2 Cardiac output (filled circles) and mean arterial pressure (open circles) for eight healthy women prior to and during early pregnancy. (Reprinted with permission from Capeless, E.L. and Clapp, J.F. (1989). *American Journal of Obstetrics and Gynecology*, **161**, 1449–53.)

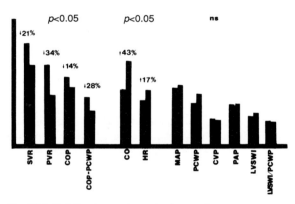

Fig. 32.3 Relative haemodynamic changes in pregnancy: non-pregnant versus late phase of third trimester. SVR, systemic vascular resistance; PVR, pulmonary vascular resistance; COP, colloid osmotic pressure; PCWP, pulmonary capillary wedge pressure; CO, cardiac output; HR, heart rate; MAP, mean arterial pressure; CVP, central venous pressure; PAP, pulmonary artery pressure; LVSWI, left ventricular stroke work index; ns, not significant. (Reprinted with permission from Clark, S.L., *et al.* (1989). *American Journal of Obstetrics and Gynecology*, **161**, 1439–42).

Recent studies have shown that as much as 50 per cent of the increase in cardiac output observed during pregnancy is present by 8 weeks of gestation as compared to the preconceptual period. This suggests that a considerable portion of the increase in cardiac output that occurs in association with a fall in systemic vascular resistance early

in pregnancy is due to endocrine alterations. Pulmonary vascular resistance is also decreased during pregnancy, whereas pulmonary capillary wedge pressure and central venous pressure are unchanged.

Measurements of haemodynamic alterations in pregnancy are very dependent on maternal position as studies carried out with the subject in the supine position are associated with a decrease in central venous return secondary to compression of the vena cava by the gravid uterus. This in turn gives rise to a fall in stroke volume and therefore cardiac output. It is of utmost importance, then, in interpreting studies of maternal cardiovascular physiology during pregnancy, that the position of the subject during the study be known and accounted for. It is not only unwise but often impossible for pregnant women, particularly during the latter part of gestation, to lie on their back for a prolonged period of time.

Maternal heart rate increases during the first 16 weeks of gestation and then plateaus at approximately 11 per cent more than preconceptual rates. Mean arterial blood pressure decreases initially during the first trimester and then returns to prepregnancy values. There is little information regarding the regional distribution of blood flow during human pregnancy. However, in pregnant sheep at term, uterine blood flow constitutes approximately 15 per cent of cardiac output, compared to 2 per cent during the non-pregnant state. The mammary gland also receives a significantly greater proportion of cardiac output during pregnancy as compared to the non-pregnant state (2 versus 0.2 per cent).

Renal system

As a consequence of the increase in cardiac output, renal blood flow increases by approximately 40 per cent by 20 weeks of gestation, with a further 10 per cent increase during the latter half of gestation. Glomerular filtration rate increases in a parallel fashion to renal blood flow.

It is quite common for pregnant women to have glucose in their urine (glycosuria) as a consequence of the increase in glucose filtered by the glomerulus exceeding the amount that can be reabsorbed by the renal tubule. Overall, tubular reabsorption of glucose is unchanged in pregnancy and, therefore, glycosuria occurs in those pregnant women with a relatively low tubular maximum threshold for glucose.

In contrast to glucose, there is a large increase in the amount of sodium that is reabsorbed by the renal tubule during pregnancy. Plasma sodium concentrations decrease slightly during pregnancy as a consequence of a relatively greater reabsorption of water compared to sodium. There are many competing influences on sodium and water balance during pregnancy. Factors, such as increased glomerular filtration, increased circulating concentrations of progesterone, increased antidiuretic hormone concentrations, decreased plasma albumin concentrations, and decreased systemic vascular resistance, all promote sodium excretion. In contrast, increased circulating concentrations of aldosterone, oestrogens, cortisol, placental lactogen, and prolactin give rise to increased sodium reabsorption. Clearly, the net effect of all these influences is, however, to decrease sodium excretion relative to the amount that is filtered by the glomerulus. Positional changes during pregnancy will also give rise to significant changes in sodium homeostasis with sodium excretion being decreased in the supine position along with the fall in cardiac output and presumably renal blood flow.

The increase in plasma concentrations of aldosterone begins very early during pregnancy. There are also increases in circulating concentrations of renin substrate and angiotensin I and II. Renin substrate production by the liver is known to be increased by oestrogen, providing further evidence that the rise in plasma renin activity observed from as early as 8 weeks of gestation is mediated by an increase in circulating oestrogen concentrations. Pregnancy is also characterized by a decrease in vascular responsiveness to exogenously infused angiotensin. In women who subsequently develop pregnancy-induced hypertension or pre-eclampsia, the refractoriness to infused angiotensin is observed to decrease with advancing gestation.

Haematology

The physiological anaemia of pregnancy has already been discussed in relation to the relatively greater increase in plasma volume than in red cell mass. In contrast to red blood cells, the number of white blood cells increases from an average of 5000 cells/cm^3 to 12 000 cells/cm^3 in late gestation. Maternal platelet counts may be diminished during normal pregnancy with a recent study demonstrating that as many as 8 per cent of healthy pregnant women in labour may have thrombocytopenia, defined as a platelet count of less than 150×10^9/L. It is important for clinicians to be aware of these normal physiological changes when interpreting haematological indices during pregnancy.

Pregnancy is also characterized by an increase in a number of coagulation factors and therefore thrombin activity. There is no compensatory rise in antithrombin III which is the main inhibitor of thrombin. Fibrinolytic activity is reduced due to a decrease in plasminogen activators despite a rise in overall plasminogen concentration.

Fibrinogen concentration is also increased, as a consequence of the rise in circulating oestrogen concentrations. The overall effect of these changes is to make blood relatively hypercoaguable during pregnancy; therefore, pregnant women are at increased risk for various coagulation disorders such as venous thromboembolism.

Immune system

The effect of pregnancy on the maternal immune system has been a subject of great interest and controversy. The reason for this is the fascinating ability of the fetus to survive essentially as an 'allograft' within the maternal environment. It was originally suggested that one possible explanation for this was a generalized suppression of the maternal immune response. Recent studies, however, have refuted this hypothesis. It is now known that circulating concentrations, as well as function, of B cells are maintained during pregnancy. Normal concentrations of IgG, IgM, IgA, and complement are present in the sera of pregnant women, indicative of systemic antibody-mediated immunity being no different from that during the non-pregnant state. In addition, total T-cell numbers and percentages of their subsets in the peripheral blood, as measured using specific monoclonal antibodies or flow cytometry in pregnant women, are either increased or unchanged compared to those in non-pregnant women. Although there are conflicting reports regarding systemic T-cell function during pregnancy, the majority of evidence suggests that maternal cell-mediated immunity is also unchanged during pregnancy.

More recently, attention has been directed at determining the immune events occurring at the maternal-fetal interface as it is apparent that these are the mechanisms most probably responsible for the ability of the fetus to survive within the maternal organism. A variety of maternal antibodies directed against fetal epitopes (or antigenic determinants), have been detected including leucocytotoxic antibodies, anti-FcR antibodies, and antibodies to oncofetal and trophoblastic antigens. These have been collectively termed 'blocking' antibodies as they may protect the fetus by preventing further immune processing by maternal cells. The exact importance of these antibodies in the maintenance of pregnancy is yet to be confirmed.

It is important to recognize that the primary fetal tissue in contact with maternal cells in the human is the extravillous cytotrophoblast. These cells have recently been shown to express an immunoreactive HLA antigen, a non-classical, truncated, non-polymorphic class I antigen referred to as HLA-G. There is, however, an overall low density of potential immunogens on the surface of the extravillous cytotrophoblast which may limit the antigenic challenge to the maternal immune system. It is interesting that the syncytiotrophoblast cells do not express any HLA antigens and that, in addition, they are covered by a highly charged sialomucin coating that provides protection from immune surveillance. It is known that trophoblast cells have a reduced susceptibility to specific and non-specific cell lysis.

An important consideration in local immune responses at the feto-placental-maternal interface is the presence of immunoregulatory substances such as progesterone, corticosteroids, eicosanoids, α-fetoprotein, and others. Some of these compounds are present in large concentrations systemically in the mother and may exert their effect locally while others are present in high concentrations within the placenta and decidua and may have an effect on the immune response in a paracrine fashion. In addition to the reduced degree of trophoblast cell antigenicity and the presence of immunoregulators, it is known that there are natural suppressor cells present in the decidual bed which may in turn affect the number of natural killer cells locally.

In summary, during pregnancy, there appears to be very little change in the maternal systemic immune system. In contrast, there are very important immune mechanisms present within the decidua and placenta that are critical to the maintenance of normal pregnancy.

Respiratory system

The alterations that occur within the respiratory system during pregnancy can be placed within two broad categories: those occurring due to the mechanical effects of the enlarging uterus and those that are secondary to the endocrine changes. The increasing abdominal size with advancing gestation gives rise to an elevation in the resting position of the diaphragm but does not alter diaphragmatic function.

The major effect of pregnancy on respiratory function is one of hyperventilation. This occurs largely due to an increase in tidal volume with only a slight increase in respiratory rate. The increase in tidal volume begins during the first trimester and by term the increase is approximately 50 per cent. There is also a 20 per cent decrease, during the latter third of pregnancy, in functional residual capacity (FRC), which is the lung volume at rest after a normal respiration. Expiratory reserve volume and residual volume also decrease progressively with advancing gestation, whereas vital capacity remains unchanged. The cause of the increase in tidal volume during pregnancy is most probably the increase in circulating progesterone concen-

trations. Progesterone is a well-known respiratory stimulant acting, possibly, by increasing the sensitivity of the respiratory centre to fluctuations in PCO_2. The role of oestrogens in mediating the respiratory changes during pregnancy is less certain.

As a consequence of the increased ventilation during pregnancy, maternal alveolar and arterial PCO_2 are consistently lower than in non-pregnant women, giving rise to a respiratory alkalosis and a compensatory metabolic acidosis. As a consequence of the decrease in PCO_2, alveolar and arterial PO_2 are increased during pregnancy by approximately 5 mmHg. Large-airway function, as determined using the forced expiratory volume in 1 s (FEV_1), is not altered during pregnancy. The effect of pregnancy on small-airway function is less certain with conflicting results from different investigators, probably reflecting the fact that this aspect of pulmonary function is more difficult to study. The diffusing capacity of the lung is altered only slightly during pregnancy with an initial increase during the first trimester followed by a decrease during the remainder of pregnancy.

Urinary tract

Pregnancy is characterized by a dilatation of the urinary collecting system including the renal calices, pelves, and ureters. Dilatation is generally not present until 20 weeks of gestation and is present in virtually all pregnant women by term. There are two components to the observed uretero-pelvic dilatation. The first of these is the effect of increased circulating concentrations of progesterone, which may have an inhibitory effect on ureteral smooth muscle function although there is some evidence indicating that ureteral contractile pressure, frequency, and tone are not altered in human pregnancy. The second component is that of a relative obstruction to drainage of the ureters above the pelvic brim that is probably due to compression by the enlarging uterus. Further evidence to support this suggestion is the fact that the relative obstruction is greater on the right side than the left, in keeping with the known dextrorotation of the uterus during human pregnancy.

As a consequence of these alterations in the urinary collecting system, there is an increase in the 'dead space' or amount of urine remaining in the ureters and bladder at any given time during pregnancy, and this predisposes pregnant women to urinary tract infections. It is also important for clinicians to be aware of the normal changes during pregnancy when interpreting radiological or ultrasonic investigations of the urinary tract.

Gastrointestinal tract

Biliary tract

Pregnancy is characterized by a decrease in contractility of the gall bladder as a result of the inhibitory effect of progesterone on baseline smooth muscle contractility as well as a decrease in cholecystokinin-induced contraction of the gall bladder. The result of this change in biliary contractility is increased biliary stasis or the amount of bile remaining in the gall bladder, thereby predisposing pregnant women to the formation of bile stones. In addition, non-pregnant women who have been exposed to increased oestrogen concentrations have a decrease in biliary in the proportion of chenodeoxycholic acid which regulates the amount of cholesterol that can be incorporated into bile. The effect of this alteration in bile salt metabolism is to decrease the bile salt pool causing a further predisposition to the formation of bile stones. Whether this same alteration occurs in pregnant women is unknown, although it is highly likely given the large increase in circulating oestrogen concentrations with pregnancy. It has been well documented using ultrasound that there is an increase in the incidence of gall stones or cholelithiasis during pregnancy as a consequence of these changes.

Bowel

The generalized inhibitory effect of progesterone on gastrointestinal smooth muscle contractility is also evident in the lower oesophageal sphincter. The administration of oestrogen and progesterone to animals giving rise to a condition known as pseudopregnancy has been shown to cause a decrease in lower oesophageal sphincter pressure in keeping with observations in humans. The major manifestation of this is a high incidence of 'heartburn' in pregnant women, which can generally be relieved by antacids and/or maintained elevation of the head during sleep. Gastric emptying time is also reduced during pregnancy and, particularly, during labour. This has important clinical implications in that extra precautions are required to prevent the aspiration of gastric contents in pregnant women who are undergoing induction of general anaesthesia.

As a result of the enlarging uterus with advancing gestation, the small and large bowels are displaced upwards with traditional abdominal landmarks being altered. For example, the appendix, which is normally present in the right lower quadrant of the abdomen in the non-pregnant state, is displaced to the right upper quadrant during late pregnancy. A change in the position of the appendix can, on occasion, cause a delay in the diagnosis of appendicitis since the pain may not be present in the usual location.

Liver

Overall liver size remains unchanged during pregnancy as do the serological tests of liver function with the exception of alkaline phosphatase. Alkaline phosphatase is an enzyme that is produced by placenta as well as the liver, and therefore its concentration in blood may increase as much as two fold or more during pregnancy. Placental alkaline phosphatase may be differentiated from that from hepatic sources by gel electrophoresis as well as by its stability to heating.

Serum total protein and albumin concentrations decrease during pregnancy as a consequence of the previously described rise in plasma volume. In contrast, a number of proteins that are produced by the liver, such as fibrinogen, globulins, and caeruloplasmin, increase during pregnancy, probably as a result of the increased circulating concentrations of oestrogen. Changes in plasma concentrations of fibrinogen, albumin, and globulins similar to those occurring in human pregnancy have been induced in non-pregnant sheep exposed to high concentrations of 17β-oestradiol for 3 weeks, adding further support to the suggestion that these alterations are mediated through oestrogens.

Endocrine changes

Pituitary

During pregnancy, the pituitary gland increases in size by approximately two to threefold due primarily to hyperplasia and hypertrophy of the lactotrophs. This is associated with a progressive rise in maternal plasma concentrations of prolactin with advancing gestation. The increase in number and size of lactotrophs is most probably a result of the increased circulating concentrations of oestrogen and progesterone. The effect of increasing concentrations of prolactin appears to be the preparation of the mammary gland for lactation.

Maternal serum concentrations of the pituitary gonadotrophins, follicle-stimulating hormone (FSH), and luteinizing hormone (LH) are lower during pregnancy than in the non-pregnant condition, probably as a result of increased negative feedback exerted by the increased concentrations of oestrogen and progesterone. Concentrations of thyroid-stimulating hormone (TSH) are unchanged during pregnancy in keeping with the lack of a change in the normal negative feedback relationship between thyroid hormones and TSH. Baseline concentrations of growth hormone (GH) are also unchanged, although the pituitary responsiveness to stimulation tests of GH release, such as insulin-induced hypoglycaemia, appears to be reduced during pregnancy. This alteration in GH release during pregnancy is possibly related to the increased concentrations of progesterone, cortisone, and human placental lactogen (hPL).

It is of interest that plasma ACTH concentrations are elevated during pregnancy despite a threefold increase in plasma cortisol concentrations. This suggests that there may be other factors that regulate ACTH production by the pituitary or, alternatively, that there may be other sources of ACTH during pregnancy such as the placenta, which is known to produce a variety of peptides. Plasma concentrations of corticotrophin-releasing-hormone (CRH) also increase progressively during the latter two-thirds of pregnancy and fall immediately after birth suggesting that the major source of this peptide is the placenta.

Plasma concentrations of vasopressin, which is produced by the posterior pituitary, are unchanged during pregnancy. This is of particular interest since, as previously discussed, there is a 5–10 per cent decrease in plasma osmolality during normal pregnancy as compared to the non-pregnant state, in addition to a decrease in plasma sodium concentrations. This would indicate that, during normal pregnancy, there is a downwards shift in the osmotic threshold for vasopressin release.

Oxytocin is also synthesized and released from the posterior pituitary during pregnancy and is important in the process of parturition. Oxytocin is also important in regulating uterine contractility following delivery and in lactation. The primary action of oxytocin in lactation is to cause contraction of the myoepithelial cells within the mammary gland as well as contraction of the smooth muscle in the mammary duct leading to ejection of milk from the gland. This is mediated through a neurohumoral reflex with neural impulses in response to suckling, being transmitted through the intercostal nerves and dorsal roots of the spinal cord, and terminating in the supraoptic and paraventricular nuclei of the hypothalamus. This then leads to the release of oxytocin in a pulsatile fashion.

Adrenal function

Plasma cortisol concentrations increase approximately threefold during pregnancy largely due to an increase in plasma cortisol-binding-globulin (CBG) produced by the liver. The plasma concentration of free cortisol also increases during pregnancy, although this is due to a decrease in clearance and metabolism with no increase in production. The diurnal rhythm in plasma cortisol concentrations normally present in non-pregnant individuals persists during pregnancy. The changes in CBG and cortisol clearance and metabolism observed during normal pregnancy may be reproduced in non-pregnant women by administering oestrogens over 1 to 2 weeks implicating oestrogens once again in this normal physiological adaptation to pregnancy.

As discussed previously, plasma renin activity and angiotensin and aldosterone concentrations all increase during normal pregnancy. Concentrations of deoxycorticosterone (DOC), a mineralocorticoid produced by the adrenal cortex, increase markedly during pregnancy, although its role in regulating fluid balance remains uncertain. Plasma concentrations of testosterone increase significantly due to the increase in sex hormone-binding globulin (SHBG). SHBG is increased due to the increase in oestrogen concentrations during pregnancy and this increase in SHBG gives rise to an increase in bound testosterone but a decrease in free testosterone.

Thyroid function

The major change in thyroid hormone physiology during pregnancy is an increase in thyroid-binding globulin (TBG) produced by the liver in response to the elevation in oestrogen concentrations. This gives rise to an increase in total thyroxine (T_4) concentrations with a minimal increase in free T_4. Although the basal metabolic rate increases during pregnancy by approximately 25 per cent, this is not due to changes in thyroid hormone function but is a result of the addition of the fetus, placenta, and increased myometrial mass to the maternal organism.

Parathyroid function

The absorption of calcium across the gastrointestinal tract of the mother is increased during pregnancy in order to meet the significant requirements of the fetus for its skeletal development. This increase in maternal calcium absorption is mediated by an increase in parathyroid hormone (PTH) secretion as well as an increase in 1,25-dihydroxyvitamin D ($1,25(OH)_2D$) synthesis. Total serum calcium concentrations decrease in the mother due to the decrease in serum albumin concentrations. Circulating calcitonin concentrations appear to be unchanged during normal pregnancy.

Carbohydrate metabolism

The major alterations in carbohydrate metabolism during early pregnancy are directed at augmenting tissue storage of nutrients as a preparatory phase for later fetal growth. Maternal concentrations of oestrogen and progesterone increase, giving rise to pancreatic B-cell hyperplasia and increased insulin secretion. As a consequence, tissue glycogen deposition and peripheral glucose utilization are increased, while hepatic glucose production and plasma glucose concentrations decrease.

With advancing gestation, there is a progressive rise in circulating concentrations of human placental chorionic somatomammotrophin (hCS), or human placental lactogen as well as prolactin, cortisol, and glucagon. The net effect of this change in hormonal milieu is to promote a state of insulin resistance and to increase hepatic production of glucose by decreasing stores of hepatic glycogen. Actual circulating concentrations of insulin are increased by hCS and fluctuations in insulin concentrations increase around meals with advancing gestation. In addition, although fasting plasma glucose concentrations decrease with advancing gestation, oscillations in glucose concentrations around the time of meals are accentuated.

There are similar increases in the magnitude of oscillations in the utilization and production of most fuels (glucose, free fatty acids, and triglyceride) during the transition between the fed and fasted state. Freinkel and colleagues have popularized the terminology of accelerated starvation alternating with facilitated anabolism to characterize the metabolic fluctuations that occur with meals in later gestation. The purpose of these changes is to ensure a constant supply of nutrients to the fetus and placenta through their rapid mobilization and ready transfer, when fetal growth is maximal. The importance of these metabolic adjustments in regulating fetal growth has been described in the 'modified Pedersen hypothesis' (Fig. 32.4).

Fasting plasma concentrations of amino acids tend to be lower in late gestation compared to those in non-pregnant women, although, in contrast to most other fuels, postprandial increases tend to be greater in the non-pregnant state than in the pregnant. The explanation for this variation among metabolic fuels is unknown.

Skin

There are a number of adaptations that take place within the skin during pregnancy including changes in pigmentation and connective tissue, development of vascular abnormalities, and alterations in hair growth. Increases in pigmentation occur in approximately 90 per cent of pregnant women and are usually located diffusely in the face (melasma or chloasma), or locally in the areola of the breast, axillae, perineal skin, genitalia, anal region, or inner thighs. Another common area in which hyperpigmentation occurs is the linea nigra, which is a darkened streak along the lower midline of the abdominal wall. The mechanism for this hyperpigmentation is uncertain as there is conflicting evidence as to the concentrations of melanocyte-stimulating hormone (MSH) during pregnancy and the time course of its increase (if present) in relation to the development of hyperpigmentation.

Pregnancy is also associated with dilatation and proliferation of small blood vessels within the skin. These can either be in the form of telangiectasiae or spider angiomas. Palmar erythema may also occur and, like the vascular proliferative lesions, are probably due to the increase in circulating oestrogen concentrations. Similar changes can be seen in the skin of non-pregnant individuals with liver disease in which the metabolism of oestrogens is impaired.

Striae or 'stretch marks' may also occur during pregnancy particularly over the abdomen, breasts, thighs, buttocks, groin, and axilla. Striae are the result of linear tears in the dermal collagen and manifest themselves as pink or purple atrophic bands. The aetiology of striae is unknown although they are thought to be hormonally mediated.

It is known that oestrogen decreases the rate of hair growth and lengthens the duration of time which the hair follicles spend in the growing phase. During late pregnancy, the percentage of follicles in the resting phase is approximately 5–10 per cent, which represents a decrease from 10–15 per cent in the non-pregnant state. Following delivery, the percentage of follicles in the resting phase increases to 35 per cent with a resultant overall loss of hair that peaks at 3 to 4 months postpartum. Hirsutism or increased hair growth may uncommonly occur during the latter part of pregnancy, presumably as a result of increased production of androgens by the placenta or ovary.

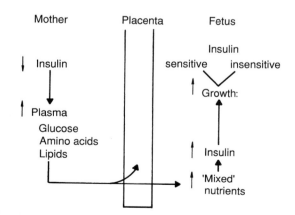

Fig. 32.4 Fetal development according to the modified Pedersen hypothesis. (Reprinted with permission from Freinkel, N. (1980). *Diabetes*, **29**, 1023–35.)

Further reading

Brinkman, III, C.R. (1989). Biologic adaptation to pregnancy. In *Maternal fetal medicine: principles and practice* (ed. R.K. Creasy and R. Resnik), pp. 734–45. Sunders, Philadelphia.

Capeless, E.L. and Clapp, J.F. (1989). Cardiovascular changes in early phase of pregnancy. *American Journal of Obstetrics and Gynecology*, **161**, 1449–53.

Clark, S.L., Cotton, D.B., Lee, W., Bishop, C., Hill, T., Southwick, J., Pivarnik, J., Spillman, T., DeVore, G.R., Phelan, J., Hankins, G.D.V., Benedetti, T.J., and Tolley, D. (1989). Central hemodynamic assessment of normal term pregnancy. *American Journal of Obstetrics and Gynecology*, **161**, 1439–42.

Colbern, G.T. and Main, E.K. (1991). Immunology of the maternal–placental interface in normal pregnancy. *Seminars in Perinatology*, **15**, 196–205.

Dunlop, W. (1979). Renal physiology in pregnancy. *Postgraduate Medical Journal*, **55**, 329–32.

Feely, J. (1979). The physiology of thyroid function in pregnancy. *Postgraduate Medical Journal*, **55**, 336–9.

Feinberg, B.B. and Gonik, B. (1991). General precepts of the immunology of pregnancy. *Clinical Obstetrics and Gynecology*, **34**, 3–16.

Freinkel, N. (1980). Of pregnancy and progeny. *Diabetes*, **29**, 1023–35.

Gant, N.F., Daley, G.L., Chand, S., Whalley, P.J., and MacDonald, P.C. (1973). A study of angiotensin II pressor response throughout primigravid pregnancy. *Journal of Clinical Investigation*, **52**, 2682–9.

Hollingsworth, D. (1983). Alterations of maternal metabolism in normal and diabetic pregnancies: differences in insulin-dependent, non-insulin-dependent, and gestational diabetes. *American Journal of Obstetrics and Gynecology*, **146**, 417–29.

Hollingsworth, D.R. (1989). Endocrine disorders of pregnancy. In *Maternal fetal medicine: principles and practice* (ed. R.K. Creasy and R. Resnik), pp. 989–1031. Saunders, Philadelphia.

Howie, P.W. (1979). Blood clotting and fibrinolysis in pregnancy. *Postgraduate Medical Journal*, **55**, 362–6.

Kalkhoff, R., Ahmed, H.K., and Kim, H.-J. (1978). Carbohydrate and lipid metabolism during normal pregnancy: relationship to gestational hormone action. *Seminars in Perinatology*, **2**, 291–307.

Longo, L.D. (1983). Maternal blood volume and cardiac output during pregnancy: a hypothesis of endocrinologic control. *American Journal of Physiology*, **245**, R720–R729.

Lund, C.J. and Donovan, J.C. (1967). Blood volume during pregnancy. *American Journal of Obstetrics and Gynecology*, **98**, 393–403.

Metcalfe, J. and Parer, J.T. (1966). Cardiovascular changes during pregnancy in ewes. *American Journal of Physiology*, **210**, 821–5.

Milne, J.A. (1979). The respiratory response to pregnancy. *Postgraduate Medical Journal*, **55**, 318–24.

Rosenfeld, C.R. (1977). Distribution of cardiac output in ovine pregnancy. *American Journal of Physiology*, 232, H231–H235.

Seymour, C.A. and Chadwick, V.S. (1979). Liver and gastrointestinal function in pregnancy. *Postgraduate Medical Journal*, **55**, 343–52.

Ueda, S., Fortune, V., Bull, B.S., Valenzuela, G.J., and Longo, L.D. (1986). Estrogen effects on plasma volume, arterial blood pressure, interstitial space, plasma proteins, and blood viscosity in sheep. *American Journal of Obstetrics and Gynecology*, **155**, 195–201.

Ueland, K., Novy, M.J., Peterson, E.N., and Metcalfe, J. (1969). Maternal cardiovascular dynamics. *American Journal of Obstetrics and Gynecology*, **104**, 856–64.

Weinberger, S.E., Weiss, S.T., Cohen, W.R., Weiss, J.W., and Johnson, T. (1980). Pregnancy and the lung. *American Review of Respiratory Disease*, **121**, 559–81.

Wong, R.C. and Ellis, C.N. (1984). Physiologic skin changes in pregnancy. *Journal of the American Academy of Dermatology*, **10**, 929–44.

33. Myometrial activity during pregnancy and parturition

Graham Jenkin and Peter W. Nathanielsz

The mechanisms involved in the control of uterine activity are important for normal fertility, including fertilization, the maintenance of pregnancy, and the initiation of labour, and are relevant to clinical problems such as dysmenorrhoea, infertility, miscarriage, and pre- and post-term labour. It is also becoming clear that uterine activity during pregnancy can have a profound effect on the well-being of the developing fetus.

Co-ordinated contraction of the entire myometrium requires that many of the smooth muscle cells composing the myometrium contract at approximately the same time. This can occur as a result of the syncytial nature of the myometrium; that is, pathways of low resistance between the smooth muscle cells allow the flow of current, generated in a localized area, to spread. The ease with which the spread of current can occur, and hence the extent of co-ordinated contraction, depends on many factors that interact to result in the initiation of organized uterine contraction. Such activity during pregnancy eventually results in the initiation of parturition. The recording of myometrial activity in the conscious pregnant animal allows us to assess the progress of the pregnancy towards this goal. Evacuation of the uterine contents presents a different challenge to monotocous and polytocous species. In polytocous species, the uterus must contract to expel each fetus in turn while not compromising the flow of blood to the remaining fetuses. The monotocous species most studied are sheep and non-human primates such as the rhesus monkey.

Myometrial function has also been extensively studied in isolated tissue as well as *in vivo*. Recently, *in vitro* studies have been refined and attention has been paid to the need to distinguish between the activities of the individual circular and longitudinal muscle layers of the myometrium. In addition, it is now clear that *in vitro* studies may provide only partial information, since the *in vitro* environment lacks many factors that are present in the whole animal. For example, it has been shown that even the wash-out of steroids from uterine muscle, as it is studied in the organ bath, may alter myometrial function. A clear understanding of the regulation of myometrial activity thus requires assessment of systems that operate at the whole animal level as well as at the cellular level. Studies on myometrial activity in the whole animal are discussed in this chapter while intracellular events leading to uterine contraction are discussed in the following chapter.

Anatomical considerations

The uterus is composed of three functionally distinct layers, the inner secretory endometrium, the smooth muscle component (the myometrium) and a thick outer serosa consisting predominantly of connective tissue. Total uterine mass increases approximately 20-fold during gestation, this increase being mainly due to hypertrophy (enlargement) and hyperplasia (increased numbers) of the muscle cells. The force-generating potential of these cells does not, however, vary at different times of pregnancy since maximal stimulation generally produces the same force per unit area at any time of pregnancy. The myometrium is subdivided into an inner layer in which the muscle fibres are aligned in a circular fashion and an outer layer consisting of longitudinally aligned muscle fibres. The two layers are separated by loose connective tissue containing blood vessels. Adrenergic nerve axons penetrate the uterine wall with the vascular supply and nerve bundles located in the loose connective tissue separating the two myometrial layers. Axons branching from these nerve bundles innervate both the endometrium and myometrium. Adrenergic axons within the myometrium run parallel to the direction of the muscle fibres. During pregnancy, however, there is a loss of adrenergic innervation to the smooth muscle of the uterus.

Measurement of uterine muscle activity

Information on uterine muscle activity during pregnancy and at parturition is generally based on recordings of either electrical or mechanical activity. In live animals, uterine electromyographic (EMG) activity is recorded by means of electrodes implanted directly on or in the myometrium or by the recording of abdominal surface potentials. Recordings of mechanical activity are usually obtained by means of open-ended or balloon catheters placed directly within the uterus and connected to a pressure transducer or by microtransducers placed at the tip of the catheter. Such methods detect the pressure changes that occur beneath the myometrium as a result of its contractile activity.

Recording of intrauterine pressure (IUP) by direct measurement with open-ended catheters provides an index of uterine muscle activity and is relatively simple to perform. It has been widely employed in animal research as well as in human pregnancy. Although the frequency response is normally satisfactory, there is, however, a small time lag between electrical activity within the myometrium and the resulting recording of pressure change. Furthermore, it is not possible to record activity reliably using this method after rupture of the membranes during pregnancy. To overcome this disadvantage, microtransducer catheters have been developed consisting of semiconducting silicon strain gauges placed at the tip of a catheter. They provide identical results to those obtained by fluid-filled catheters but the lag time is considerably reduced.

Extracellular EMG electrodes have been used extensively in animal studies to detect electrical activity in uterine muscle *in vivo*. On a time scale of minutes or hours, episodes of EMG activity are closely correlated with changes in IUP. EMG activity, however, demonstrates only the capacity for myometrium to contract and is thus not a direct indication that the linkage between EMG activity and the contractile process is complete. For example, bursts of EMG activity may be relatively localized and not sufficient to generate sufficient wall tension to produce a rise in IUP. Attempts to extend recording of uterine EMG activity via the recording of skin surface potentials have not generally been successful, due mainly to the low signal-to-noise ratio obtained, and have thus had limited success in assessing the progress of labour in humans. Similarly, although ultrasound devices have also been used to monitor uterine activity near term, the attendant disadvantages have limited their use in clinical practice. Finally, a simple non-invasive measure of uterine activity, using the guard-ring tocodynanometer, has been employed in clinical practice since its introduction in 1957. Such devices measure intrauterine pressure changes during spontaneous uterine contractions by means of a pressure-sensitive device, surrounded by a guard ring that is firmly applied to the abdomen by means of a belt. These devices have been shown to be effective provided that they are strapped to the abdomen sufficiently firmly to flatten the underlying uterine wall.

Electrophysiological and mechanical properties of myometrium have been characterized *in vitro* as well as *in vivo*. Transmembrane events have been recorded using intracellular microelectrodes while tension has been simultaneously measured in the tissue. Furthermore, the contractile characteristics of isolated segments of longitudinal or circular muscle can be observed in a tissue bath using 'muscle strip' preparations. Using such preparations, it has been demonstrated that the pattern of electrical and mechanical activity of entire or separated circular and longitudinal myometrium differs considerably from that observed in the intact animal indicating that endogenous factors, not present in the *in vitro* situation, may have an influence on contractile activity *in vivo*.

Contractile activity patterns

In all studies conducted in horses, cows, pigs, goats, guinea-pigs, rats, rabbits, sheep, monkeys and humans, different patterns of uterine activity are observed at parturition compared to those observed throughout the majority of pregnancy. Using both uterine EMG and IUP recording, the characteristics of these patterns of activity have been studies most extensively in the sheep and rhesus monkey. Throughout the majority of pregnancy, apart from the last 6 to 18 h before parturition, the uterus is relatively quiescent apart from recurring contractile events of low amplitude (3–10 mmHg) and low frequency (0.5 to 3/h), which have a duration of 3–15 min. In view of their long duration and low amplitude, this form of contractility has been called by some a uterine 'contracture' to distinguish it from the type of activity typically associated with labour which is characterized by high-frequency, high-amplitude 'contractions' with a relatively short duration. Throughout this chapter the two types of uterine activity will be referred to by the terms 'contractures' and 'contractions' respectively. The nature of individual uterine 'contractures' (Fig. 33.1) and contractions (Fig. 33.2) in the non-human primate is very similar to those observed in the sheep. Indeed, it is very difficult to distinguish a recording of a few hours of myometrial activity in the pregnant sheep that is not in labour from a recording at the same stage of pregnancy in a monkey.

Characterization of the state of uterine contractility can be made in several ways. The frequency of contractile episodes or the area under the curve of the contraction record can be measured in an attempt to integrate the total amount of muscle activity observed. Integrating the area under the curves shown in the raw data in Figs 33.1(a), (b) and 33.2 (a), (b) does not tell us the qualitative nature of the myometrial activity. In these figures spectral analysis

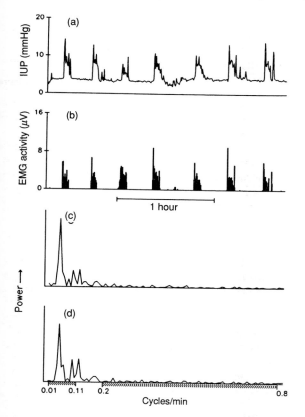

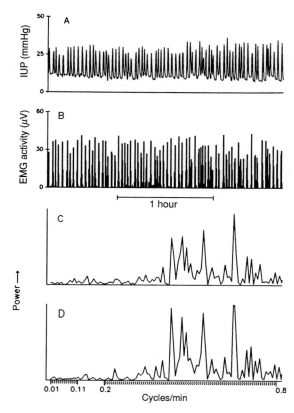

Fig. 33.1 Myometrial activity during the hours of daylight in a pregnant monkey. (a) Raw data for intrauterine pressure (IUP) when only uterine 'contractures' were present. (b) Integrated myometrial electromyogram (EMG) activity simultaneously recorded with (a). (c) Power spectral analysis of intrauterine pressure recording in (a). (d) Power spectral analysis of integrated EMG recording. In (c) and (d), the window from 0.01 to 0.11 cycles/min corresponds to the majority of 'contracture' activity and the window from 0.2 to 0.8 cycles/min corresponds to uterine contraction activity.

Fig. 33.2 Myometrial activity during the hours of darkness in the pregnant monkey shown in Fig. 33.1. (a) Raw data for intrauterine pressure (IUP) when contractions only were present. (b) Integrated myometrial electromyogram (EMG) activity simultaneously recorded with (a). (c) Power spectral analysis of intrauterine pressure recording in (a). (d) Power spectral analysis of integrated EMG recording. In (c) and (d), the window for 0.01 to 0.11 cycles/min corresponds to the majority of 'contracture' activity and the window from 0.2 to 0.8 cycles/min corresponds to contraction activity.

has been applied to the contractility and EMG data. It can be seen that the frequency patterns are very different. Uterine contractures occur at a frequency of less than one every 10 min or so, whereas uterine contractions occur at frequencies greater than once every 5 min.

The clear distinction between uterine contractures and uterine contractions, as seen in Figs 33.1 and 33.2, and the association of contractions with labour and delivery provides a means of monitoring myometrial activity and making conclusions about how soon labour is likely to occur If the uterus is in the 'contractures' mode it is not about to deliver its contents. However, when the uterus is

in the 'contractions' mode, labour and delivery may be imminent.

Co-ordination of uterine activity

Electrical processes

Since individual smooth muscle cells of the myometrium are small (200–600 μm long and 3–10 μm wide), synchronization of activity between cells is essential for the co-ordinated uterine contractions necessary for successful labour. The concept of the myometrium as an electrical syncytium at the time of co-ordinated uterine activity associated with parturition is generally accepted. Evidence for the presence of, and localization of, a pacemaker site for the propagation of impulses has, however, proved elusive. Detailed experiments on the cardiac syncytium, together with sophisticated computer-modelling of the data, have provided considerable insights into the propagation of activity throughout the heart. On the one hand, the coupling between the cells in the generator region and the bulk of the tissue must be restricted to allow the build-up of sufficient current to fire an active response, that is, an action potential. On the other hand, the coupling between the cells of the generator region and the bulk of the tissue must be sufficiently large to allow the current generated by the action potential to spread out to invade adjacent areas.

Although it is usually suggested that contractility is propagated through the uterus from the fundus to the cervix, there is little experimental evidence to support this proposal. Studies in conscious animals fitted with fine extracellular EMG recording electrodes sewn into the myometrium suggest that activity is likely to be generated in localized regions of the uterus, but that the generator regions do not remain static and appear to move around the uterus. The mechanisms involved in the spread of electrical activity from the generator regions in the uterus have received scant attention. Most studies indicate that propagation may occur more readily in the longitudinal orientation rather than circumferentially but there appears to be no consistent direction of propagation during pregnancy in all species studied.

Some studies in sheep and humans indicate that the direction of propagation alters during parturition. Uterine activity in the active phase of labour usually, but not always, starts from the uppermost part of the body of the uterus, above the level of the entrance of the oviducts (fundus), and progresses towards the cervical end (procervical). Using abdominal surface potential measurements it has also been shown that, in the human, the front of relaxation moves in the opposite direction to the contraction. A change in the electrical resistance of myometrial cells to allow propagation of membrane currents has been observed at term and may well account for the marked increase in the ability of activity to spread throughout the entire uterus during delivery.

Intrinsic factors

Studies of uterine activity in non-pregnant and pregnant ewes have led investigators to propose that uterine contractures are not likely due to neural stimulation but to an intrinsic activity of the myometrium *per se*. This conclusion has been reached as a result of studies using isolated and autotransplanted segments of myometrium. These studies have demonstrated that uterine smooth muscle *in vitro* and *in vivo* has an inherent rhythmicity that is not dependent on adrenergic innervation and that 'pacemaker' sites are not limited to the tips of the uterine horns, as was previously proposed, but are present at multiple sites in the myometrium of the body of the uterus (Fig. 33.3). During labour, however, circulating factors (such as prostaglandins and oxytocin) can be shown to have major effects on uterine activity in autotransplanted, as well as entire, myometrium, the increase in frequency and amplitude of bursts of activity now being in synchrony.

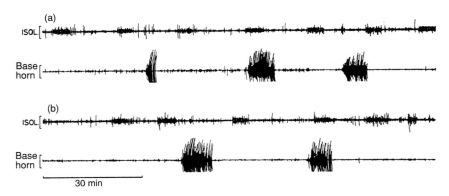

Fig. 33.3 Myometrial electromyogram (EMG) activity recorded from two ewes ((a) and (b)) at 120 days of gestation (term ~ 145 days) from the base of the pregnant horn (Base horn) and from tissue removed from the tubal extremity of one uterine horn and transplanted into a pocket of abdominal fat (ISOL). Calibrations: isolated tissue 40 μV; uterus 40 μV. (Adapted from Sigger, J.N., Harding, R., and Jenkin, J. (1984). *Journal of Reproduction and Fertility*, **70**, 103–14.)

Hormonal factors

Endogenous hormonal factors reaching the myometrium either from the maternal or fetal circulation (endocrine factors), directly from uterine tissues (paracrine factors), or from within myometrial cells (autocrine factors) are known to play a significant role in the control of uterine activity. In 1956 Csapo proposed that progesterone was the major factor responsible for inhibiting uterine activity during pregnancy. He demonstrated that, when the uterus is dominated by the influence of progesterone, co-ordinated uterine activity is reduced. Progesterone does not, however, abolish the spontaneous 'contractures' which occur throughout the latter part of pregnancy and administration of progesterone to women will not usually prevent uterine activity associated with premature or term labour. Other hormones that can inhibit uterine activity are metabolites of progesterone, prostaglandin I_2, and relaxin. Whether these hormones play a physiological role in the inhibition of uterine activity is presently unclear.

Hormonal factors that have been shown to stimulate uterine activity in a variety of animal models are oestrogen, prostaglandins $F_{2\alpha}$ and E_2, and oxytocin. Acute increases in the circulating concentration of these hormones, either before or during uterine contractions or contractures, have not, however, been demonstrated. Experiments using *in vivo* autotransplanted myometrial tissue in sheep suggest that suppression of uterine activity during gestation is due to the systemic action of hormones such as progesterone. Uterine contractures may, however, be caused by imposition of paracrine or autocrine mechanisms in the myometrial cells that overcome the inhibitory effects of progesterone, thus explaining the lack of change in concentrations of systemic hormones during these events.

In many situations the balance between uterine contractures and contractions appears to depend on the level of oestrogen stimulation of the uterus and thus a change from progesterone to oestrogen dominance. Oestrogen, in turn, has been shown to increase uterine responsiveness to other agonists and to increase the ability of the uterus to propagate activity. Such actions of oestrogens could be manifested either by a direct action on the uterus to increase agonist binding, intracellular calcium, potassium conductance, or gap-junction formation between myometrial cells or, since the effects of oestrogen usually take several hours to become apparent, indirectly by increasing contractile protein synthesis, agonist receptor formation, prostaglandin $F_{2\alpha}$ or E_2, or oxytocin production. Recent observations that both prostaglandins and oxytocin are synthesized in increasing amounts in uterine tissues during late pregnancy and at parturition also indicate that autocrine or paracrine, as well as endocrine, factors may be involved in the switch from uterine contractures to contractions at term.

Myometrial activity in the non-pregnant uterus

In non-pregnant, cycling sheep uterine activity, as measured by EMG activity, is unpatterned and infrequent during the greater part of the luteal phase of the cycle. The amplitude and frequency of activity increases 2–3 days before oestrus and remains elevated during the time that progesterone concentrations are basal. Studies in ovariectomized sheep and monkeys indicate that uterine contractures are brought about by low levels of oestrogen and that, as levels of oestrogen increase, the activity switches to the contraction type of activity. Thus, in the non-pregnant sheep and monkey with their ovaries intact, uterine contractures occur throughout the majority of the oestrous cycle and are replaced by 'contractions' at ovulation when oestrogen levels are elevated. Exogenous oxytocin is also able to induce an increase in EMG activity similar to that observed at oestrus in the ewe. Short-duration pulses of endogenous oxytocin have been observed at the time of oestrus in this species. It is thus possible that oxytocin may also affect uterine activity at oestrus. As immunization of ewes against oxytocin will result in a significant decrease in fertilization rate, the increase in activity at oestrus may be involved in assisting the transport of sperm and egg at the time of fertilization. As in pregnant animals, directional propagation of uterine activity at oestrus has not been convincingly demonstrated.

Patterns of myometrial activity throughout pregnancy

Early pregnancy

Myometrial activity has been recorded in pregnant sheep at as early as 13 days of pregnancy (term is 145–150 days in this species). In these early stages of pregnancy the bursts of spikes are irregular and do not occur widely over the uterus. By 40 days of pregnancy they are becoming better co-ordinated and, as pregnancy progresses, they increase in amplitude. It is likely that uterine stretch plays an important role in the increase in amplitude since the force with which smooth muscle contracts is increased by stretch. In both monkeys and sheep, low-amplitude, low-frequency uterine contractures are the characteristic form of myometrial activity by mid-pregnancy.

Mid-pregnancy

In mid-pregnancy the characteristic form of myometrial activity recorded either with EMG electrodes or an IUP catheter is of the contracture type in all species studied to date. In sheep the appearance of uterine contractures varies from animal to animal both in extent and in frequency, but the highest frequency is approximately 3–4/h and the lowest frequency is one approximately every 2 h. The uterine contracture pattern recorded at mid-pregnancy in pregnant rhesus monkeys or baboons is similar to that of the sheep but the frequency is generally greater than in sheep. We do not know the cause of these differences or the effects that different patterns of these 'contractures' will have on the developing fetus. It is, however, known that uterine contractures are associated with small decreases in fetal arterial PO_2, a reduction in fetal breathing movements, and increases in fetal intracranial and intrathoracic pressure. Increasing the frequency of uterine contractures by the administration of single or repeated pulses of oxytocin will also increase fetal plasma ACTH and cortisol concentrations and alter the pattern of development of the fetal electroencephalogram.

Late pregnancy

In sheep, little indication of the approach of parturition can be observed from the pattern of uterine contractures until the last 6 to 18 h of pregnancy. At this time the duration of these contractile events decreases progressively and their amplitude increases. The switch from 'contractures' to 'contractions' is evenly progressive and occurs only once on the day of delivery.

In contrast to the sheep, in non-human primates the switch from uterine contractures to uterine contractions is sudden and dramatic and occurs on several nights before the onset of labour (Fig. 33.4). On the first night on which

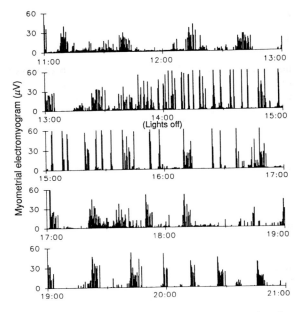

Fig. 33.4 Switch from uterine contractures to contractions in a pregnant baboon. Each vertical line of recording represents 2 h of summed EMG activity data and the record is continuous, beginning at 09.47 h.

the switch occurs, contractions may last 2 to 3 h before myometrial contractility reverts to the 'contracture' pattern. The next night, the switch occurs again. On this second night uterine contractions are larger and last longer before reverting again to 'contractures'. This pattern is repeated with uterine contractions recurring and intensifying on several nights until eventually parturition occurs (Fig. 33.5). Thus, it is possible to anticipate delivery over several nights in the pregnant monkey that has EMG electrodes implanted in the myometrium.

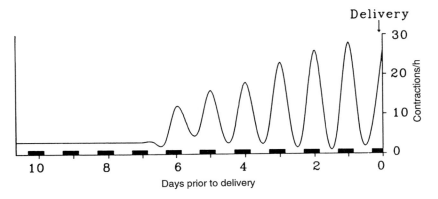

Fig. 33.5 Diagrammatic representation of myometrial electromyogram (EMG) contraction-type activity in the pregnant monkey during the last third of gestation, in the immediate preparturient period, and at delivery. Periods of darkness are indicated by black bars. (Reproduced, with permission, from Nathanielsz, P.W. (1992) as in Further reading.)

Circadian rhythms

The 24-h patterns of recurring switching from uterine contractures to uterine contractions in the last days of gestation in the monkey and baboon have recently been shown to be true circadian rhythms. A circadian rhythm is an endogenous rhythm that is maintained in the absence of external cues and, as the name suggests, the rhythm recurs about every 24 h. When a pregnant monkey is kept in a constant low-level light environment and every effort made to avoid giving her daily cues, uterine contractures still switch to contractions as term approaches at roughly 24-h intervals (Fig. 33.6). Thus this rhythm is truly circadian. The factors that control the periodicity of the switch are not yet known. However, it is known that pregnant non-human primates, like pregnant women, have a pronounced rhythmicity of plasma concentrations of several key steroid hormones circulating in their blood. These changes in hormones such as oestrogen and progesterone may alter the population of receptors on endometrial and myometrial cells as well as change the patterns of synthesis of uterotonic agents such as oxytocin. Maternal plasma oxytocin concentrations have been shown to vary with a 24-h rhythm in late pregnancy in the rhesus monkey. In studies using pregnant monkeys it has been demonstrated that the responsiveness of the myometrium to oxytocin is greater in the early hours of darkness than in daylight hours. Furthermore, the increase in nocturnal uterine activity is associated with elevated endogenous maternal oxytocin concentrations.

Parturition

It is clear that the basis for uterine contractility lies in the electrical and chemical changes that take place within individual muscle cells and the propagation of the action potentials that arise from pacemaker regions throughout the myometrium. It is also generally agreed that propagation is greatest during parturition when co-ordinated high-amplitude contractions take place. Measurements of the length constant (λ, a measure of the degree of current

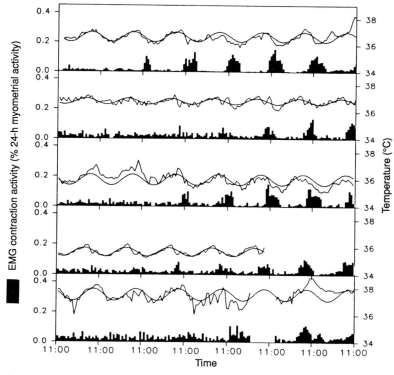

Fig. 33.6 Intraabdominal temperature (continuous lines) and uterine contractions recorded as integrated myometrial electromyograms (EMG) activity (black bars) over the last 7 days of pregnancy in five rhesus monkeys maintained in continuous 5-Lux low-level light from 61–77 days of gestation until delivery at approximately 160 days of gestation. Maternal body temperature is shown as the raw data and as the best fit cosinor curve.

spread between muscle cells) during gestation in the sheep myometrium indicate that there is a large increase in the degree of current spread between muscle cells at the time of parturition. This increase in cell-to-cell coupling may be the physiological correlate of the increase in the number of myometrial gap junctions that also occurs at this time, thus providing the means whereby ionic and molecular activities are synchronized during parturition. The increase in gap-junctions occurs whether labour takes place at term or before full term and represents an increase in both the number and size of these junctions and thus an increase in total junctional area. Preventing the formation of these junctions by manipulation of the hormonal environment will delay or inhibit labour.

After delivery

Labour is divided for clinical purposes into three stages. The third stage begins after the fetus is delivered and ends after delivery of the placenta. During this period uterine contractions continue, but, within a few hours of final evacuation of the uterus, myometrial activity is minimal and of the contractures type.

Preterm labour

Attempts to identify those pregnant women who will subsequently have a preterm delivery have been singularly unsuccessful. Increased antenatal uterine activity, combined with cervical dilatation and effacement, has been observed in some studies of patients subsequently having preterm labour, while other studies have failed to demonstrate such changes, or have reported increased uterine activity in patients who were subsequently delivered at term. Similarly disappointing results have been obtained in studies carried out to determine if there are changes in hormone concentrations in the maternal circulation before the onset of preterm labour. Abnormal concentrations of oestrogens and progesterone have been proposed as a mechanism for the onset of labour before term. In many studies, however, it has not been possible to define whether concentrations that were normal before the onset of preterm labour had become abnormal as a consequence of labour itself.

Studies of concentrations of plasma cortisol, vasopressin, oxytocin, and of oxytocin receptor concentrations in myometrium of patients in preterm labour have also produced disappointing results. Elevated hormone concentrations and receptor density have been observed. None of the studies was, however, carried out on a prospective basis and it is not possible to determine whether the observed changes were responsible for, or a consequence of, premature labour.

Prostaglandins play a central role in the initiation of labour in all species studied. One would, thus, expect their concentrations to be elevated in preterm labour. The plasma concentration of the circulating metabolite of prostaglandin $F_{2\alpha}$, PGFM, is higher in women during preterm labour than in non-labouring women of similar gestational age. The mechanisms responsible for the observed increases in prostaglandin activity have not, however, been defined.

Despite this, stimulatory prostaglandins have been shown to act on the secretory membrane receptor to stimulate phosphatidylinositol turnover leading to calcium mobilization. They can also mobilize calcium from the sarcoplasmic reticulum, increase transmembrane calcium flux, and increase electrical propagation by induction of gap-junction formation. It will become clear from the following chapter that prevention of the formation of or closure of gap-junctions, and thus prevention of calcium movements, is essential for the prevention of preterm labour. Inhibition of prostaglandin synthesis should, therefore, have the potential of providing a means of inhibiting premature labour.

Prevention of preterm labour

Since premature birth remains the major factor responsible for neonatal mortality and neurodevelopmental handicaps, a plethora of treatments, including prostaglandin inhibition, have been advocated to inhibit uterine activity associated with premature delivery. Such treatments for the inhibition of, and prevention of (prophylaxis), preterm labour have been classified into six major groupings (Table 33.1).

Non-pharmacological interventions such as bed rest and sedation appear to owe their success to a placebo response of 20–50 per cent in cases of threatened premature labour, reflecting the inherent difficulty of diagnosing true preterm labour (see above). Treatments such as cervical cerclage (placement of a suture around the cervix in women who have had a previous preterm delivery or who have 'cervical incompetence') and sedation may even be harmful to the fetus as well as inducing, rather than inhibiting, labour. There is a sound physiological basis for the use of β-adrenergic agents, and this is the only group of pharmacological agents approved by the United States Food and Drug Administration for inhibition of preterm labour. β_2-adrenoceptors predominate in the uterus and β-adrenergic agonists act on these receptors, activate adenyl cyclase, and, thus, increase intracellular concentrations of cyclic

Table 33.1 Prevention of preterm labour

Non-pharmacological intervention	Calcium channel blocking agents
Antenatal home	Nicardipine
visits by midwives	Nifedipine
Bed rest	Verapamil
Cervical cerclage	
Hydration	
Sedation	
β-Adrenergic agents	**Prostaglandin synthetase inhibitors**
Isoxsuprine	
Ritodrine	Fenoprofen
Salbutamol	Indomethacin
Terbutaline	Naproxen
Other agents	**Anti-oxytocic agents**
Aminophylline	Alcohol
Diazoxide	Atosiban
Magnesium sulfate	
Progestogens	
Theophylline	

AMP. A reduction of intracellular calcium concentration ensues and leads to a decrease in uterine contractility. In theory, therefore, such treatments should be effective in inhibiting uterine activity. Recent large-scale surveys of the use of β-adrenergic agents in the prevention of preterm labour have, however, shown no significant effect. Intravenous administration of β-mimetics can stimulate β-adrenoceptors in a number of tissues besides the uterus, most notably in cardiac muscle, thereby leading to significant side-effects such as hypertension, tachycardia, and arrhythmia. Furthermore, the inhibitory effect of β-mimetics is transitory in nature and, although various treatment regimes have been used to try to overcome this problem, there has been little progress in the effectiveness of such treatments since these drugs were first introduced to clinical practice more than 20 years ago.

Magnesium sulfate treatment is thought to act due to magnesium competing with calcium for entry into the sarcoplasmic reticulum, as well as inhibiting activation of the actin–myosin unit; it has fewer side-effects than other tocolytic agents. This agent has generally been used when β-mimetics have failed to prevent premature labour activity and has the advantage that its effects can be reversed by treatment with calcium gluconate. Other agents that have their effect by preventing passage of extracellular

calcium into smooth muscle cells via voltage-dependent channels have been shown to suppress both prostaglandin- and oxytocin-induced uterine activity. Drugs such as nifedipine and verapamil may also suppress the release of calcium from intracellular stores and increase calcium extrusion from smooth muscle cells. In view of the importance of calcium in the control of uterine muscle activity, one would expect such drugs to represent a powerful group of tocolytic agents. Relaxation of the myometrium by these drugs is, however, non-specific and can result in peripheral vasodilatation, as well as having direct effects on conduction in cardiac muscle.

The prostaglandin inhibitor that has been most widely used in clinical practice for the prevention of premature labour is indomethacin, an inhibitor of cyclooxygenase, a key enzyme in the synthesis of prostaglandins. Unfortunately, such drugs are able to pass across the placenta and thus inhibit fetal, as well as maternal, prostaglandin formation. Such inhibition can lead to the inhibition of fetal prostaglandin E_2 production, which is essential for the maintenance of the patency of the ductus arteriosus in the fetus. Premature closure of the ductus arteriosus during indomethacin infusion to the mother can compromise fetal well-being and lead to primary pulmonary hypertension in the newborn. Furthermore, it is now evident that prostaglandin E_2 plays an important role in enabling the fetus to survive periods of stress, such as hypoxia. The use of indomethacin is not, therefore, recommended as the first choice of treatment of premature labour.

Most of the other tocolytic agents listed in Table 33.1 have not undergone properly randomized clinical trials and, although many are claimed to delay delivery, they are not effective at decreasing the incidence of preterm delivery. In fact, some obstetricians have suggested that there has never been an obstetric trial that showed that tocolytic drugs help in the *prevention* of preterm labour and that, since such treatments may also be harmful to both the fetus and the mother, they should be used with caution.

A final group of agents listed in Table 33.1 that shows some promise for the future are antioxytocic drugs. Before the development of β-mimetic drugs for the treatment of preterm labour, alcohol was extensively used, despite the drawback of its inherent side-effects. It was probably the first tocolytic agent used for this purpose and has now been shown to act by inhibiting the secretion of oxytocin and vasopressin from the posterior pituitary. Very recently, more direct means of inhibiting oxytocin-induced uterine activity have been developed. Oxytocin analogues, such as Atosiban, bind to uterine oxytocin receptors but prevent translocation of the signal to initiate prostaglandin release and prostaglandin-stimulated uterine activity. The concentration of myometrial oxytocin receptors and plasma oxytocin concentrations, in both term and preterm spontaneous labour, are higher than those in women at term and not in labour. The uterus is also more sensitive to oxytocin in women who deliver prematurely. It has, therefore, been suggested that oxytocin may be the cause of uterine hyperactivity associated with preterm labour. In preliminary trials in Scandinavia and in a recent multicentre trial in the USA, Atosiban has been shown to be effective at inhibiting premature labour and does not have any detrimental effects on the mother or its fetus. Recent animal studies have shown that, although it inhibits oxytocin-induced prostaglandin $F_{2\alpha}$ release, it has no effect on fetal prostaglandin E_2 release and, unlike prostaglandin synthetase inhibitors, does not, therefore, have the potential of compromising the fetal circulation and the ability of the fetus to withstand hypoxic stress.

Further reading

Carsten, M.E. and Miller, J.D. (ed.) (1990). *Uterine function: molecular and cellular aspect.* Plenum Press, New York.

Garfield, R.E. (ed.) (1990). *Uterine contractility, mechanisms of control.* Sereno Symposia USA, Norwell, Massachusetts.

Nathanielsz, P.W. (1992). *Life before birth and a time to be born.* Promethean Press, Ithaca, New York.

Wynn, R.M. and Jollie, W.P. (ed.) (1989). *Biology of the uterus.* Plenum Press, New York.

Yu, V.Y.H. and Wood, E.C. (ed.) (1987). *Prematurity.* Churchill Livingstone, Edinburgh.

34. Regulation of contraction in uterine smooth muscle

Helena C. Parkington and Harold A. Coleman

The contractile state of the muscle component of the gravid uterus is important for the well-being of the fetus and its delivery at term. Relative quiescence must be ensured during pregnancy, whereas co-ordinated, forceful contractions are required for successful labour. This chapter will consider some of the mechanisms controlling contraction in uterine smooth muscle and how co-ordinated contractions can be achieved.

Contraction in uterine smooth muscle

Initiation of contraction

Uterine smooth muscle cells are fusiform in shape, 200–600 μm long and 3–10 μm in diameter at the centrally placed, single nucleus. The proteins actin and myosin are the key players in contraction in uterine smooth muscle, as in all muscle. These two proteins are distributed throughout the cytoplasm in smooth muscle. Actin is a filamentous protein formed by the polymerization of identical globular subunits. One end of each actin filament may be attached to the plasma membrane at a region that appears dense in electron microscopy. Alternatively, an actin filament may be fixed to a 'dense body' in the cytoplasm, to which many actin filaments are invariably attached. The other end of each actin filament is free to interact with myosin. Myosin is also filamentous, formed by the bundling together of the long legs of L-shaped proteins. The short legs of the L's protrude from the filament and are called cross-bridges. At the free tip of the cross-bridge is a region that possesses ATPase activity; that is, it can cleave a phosphorus from ATP to release energy essential to power contraction. During contraction, the cross-bridges attach to actin filaments and move in a ratchet fashion, pulling the actin filaments along. The sliding of two actin filaments that are attached at different dense bodies in the cytoplasm or in the plasma membrane along each end of one myosin filament, draws the two regions together and shortens the cell. If the cell cannot shorten, tension is developed.

There is wide acceptance of the view that calcium-dependent phosphorylation of myosin is pivotal in the initiation of contraction in smooth muscle. Consistent with this, in uterine smooth muscle the increased concentration of phosphorylated myosin has been shown to precede contraction. An important step involves the binding of calcium to the cytoplasmic protein, calmodulin. The calcium/calmodulin complex binds to and activates the enzyme myosin light-chain kinase (MLCK). MLCK then phosphorylates the cross-bridge, which unmasks the ATPase region and which, in the presence of actin, releases energy. Actin and myosin interact, the cross-bridge rotates, actin is drawn along, and contraction results.

Maintenance of contraction

While the necessity of myosin phosphorylation for the *initiation* of contraction of smooth muscle is well established, its absolute requirement for the *maintenance* of contraction has been questioned. The concentration of phosphorylated myosin increases early in the response of uterine smooth muscle to stimulants, such as oxytocin and prostaglandin $F_{2\alpha}$. However, the level of phosphorylated myosin appears to decline with time, while contraction is well maintained. In an attempt to explain this it has been suggested that, if the phosphorus on myosin is removed while the actin and myosin are bonded, then unbonding is slowed. It was said that the actin and myosin were 'latched'. This would allow for the maintenance of contraction in the face of low concentrations of phosphorylated myosin. This hypothesis is also consistent with the observations that smooth muscle can maintain contraction very economically, that is, with relatively low expenditure of energy.

An alternative hypothesis to explain the maintenance of contraction in the presence of lower levels of phosphorylated myosin suggested that regulation of actin may be involved. Activation of the ATPase of myosin, and hence the release of the energy for contraction, requires actin. The hypothesis envisages that the ability of actin to activate the ATPase is blunted by the presence, in association with actin, of one or other of two proteins called caldesmon and calponin. High concentrations of calcium/calmodulin have been shown to remove caldesmon from actin in uterine smooth muscle and this could explain the maintenance of contraction in the face of a decline in the levels of phosphorylated myosin.

Relaxation

Agents that increase the concentration of cyclic AMP (cAMP) within uterine smooth muscle are potent relaxants. Drugs that activate β-adrenoceptors, e.g. ritodrine, and prostacyclin, prostaglandin E_2 and relaxin, all increase the production of cAMP as a result of stimulating the membrane-bound enzyme adenylate cyclase. Theophylline and similar drugs increase cAMP by decreasing the breakdown of cAMP. cAMP activates a kinase that phosphorylates and inactivates MLCK.

Phosphatases remove the phosphorus from the myosin and this is necessary for the termination of contraction. However, it will be appreciated from the discussion above that dephosphorylation of myosin will not inevitably result in immediate relaxation. In the event of a decrease in the rate of myosin phosphorylation, or an increase in the activity of phosphatases, smooth muscle inevitably relaxes. Control of contraction via regulation of phosphatase activity is presently under close scrutiny, but firm evidence to support this avenue of control in uterine smooth muscle is lacking at present.

Role of calcium in uterine contraction

Cytoplasmic calcium

It will be apparent from the foregoing that the regulation of cytoplasmic calcium is crucial to the initiation of contraction in uterine smooth muscle. The development of calcium-sensitive dyes in recent years has made it possible to measure cytoplasmic free calcium directly. The concentration of free calcium within the cytoplasm of uterine smooth muscle at rest is in the range 70–130 nM. At this concentration the levels of calcium/calmodulin complex are low, there is insufficient activated MLCK, only around 5 per cent of myosin is phosphorylated, and the uterus is quiescent. If contraction is to occur, the level of cytoplasmic free calcium must increase.

The concentration of free calcium in the cytoplasm results from a balance between calcium movement into the cytoplasm, calcium removal, and the buffering of calcium. Calcium may enter the cell from the extracellular fluid, where its concentration is in the mM range. The concentration of calcium within the endoplasmic reticulum, an organelle present in all smooth muscle cells, is also in the mM range, and release of this internal store can also

increase calcium levels in the cytoplasm. Calcium is extruded from the cell by pumps and ion exchangers, and is also moved out of the cytoplasm back into the endoplasmic reticulum by the actions of pumps.

While the requirement for calcium in the initiation of contraction is accepted, its role in the maintenance of contraction is less clearly understood. It can be appreciated from Fig. 34.1 that exposure of uterine smooth muscle to prostaglandin $F_{2\alpha}$ results in a substantial increase in cytoplasmic calcium early in the response, but calcium declines with time (Fig. 34.1(a)). Tension develops rapidly and is maintained, or even continues to develop slowly, while cytoplasmic calcium declines. It may be that at least some of the decline in cytoplasmic calcium illustrated in Fig. 34.1 reflects calcium buffering, as it binds to intracellular components such as calmodulin, since the calcium-sensitive dyes detect only calcium that is *free*. In this eventuality, the low concentration of free calcium, but

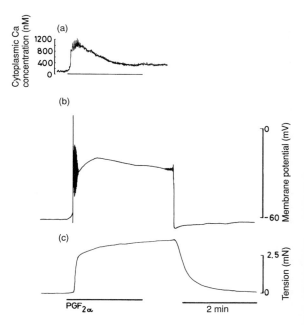

Fig. 34.1 Cytoplasmic calcium concentration, membrane potential and tension development in the longitudinal smooth muscle of guinea-pig uterus in mid-pregnancy in response to prostaglandin $F_{2\alpha}$ ($PGF_{2\alpha}$). (a) When $PGF_{2\alpha}$ was applied for min there was an initial rise in cytoplasmic calcium that peaked and then declined to a sustained level of around 300 nM. (b) $PGF_{2\alpha}$ caused depolarization to threshold for the initiation of a spike and sustained depolarization. (c) Tension development, measured simultaneously with membrane potential (b), was rapid and sustained until the membrane repolarized to resting levels.

high concentration of calcium/calmodulin, would facilitate contraction. Alternatively, the initial peaks, in both cytoplasmic calcium and phosphorylated myosin, may represent gross overkill in relation to requirements for the amplitude of uterine contraction. However, it is likely that the peaks are important in ensuring a fast rate of contraction.

Calcium influx

Ion channels

The most important route of calcium influx into uterine smooth muscle is through specific ion channels in the plasma membrane. These channels are large proteins that span the membrane several times. Aspects of the structure of the channels have been determined by techniques including amino-acid sequencing, X-ray crystallography, and nuclear magnetic resonance (NMR). Such studies indicate that these proteins contain a number of subunits that are likely to be clustered together. Under some conditions the membrane-spanning regions are positioned in such a way that a 'tunnel' exists within the cluster. When in this configuration the channel is defined as being in the open state. When a calcium channel opens, calcium moves into the cell, driven by the 10^4 concentration gradient for calcium that exists at rest between the external environment and the cytoplasm. The patch-clamp technique has been used to study the control of current flow through ion channels and indicates that open channels conduct in the order of 10^6 ions per second.

A conformational change in the three-dimensional structure of the channel can result in the closing of the tunnel, and this is defined as the closed state of the channel. In this state no ions can pass through the channel. Under resting conditions, channels are usually closed. During activation (see below) there is an increased probability of the channel leaving the closed state and entering the open state, in which case ions can pass through the channel. From the open state channels can revert back to being shut. However, some channels can alternatively enter a third, 'inactivated' state. In the inactivated state channels do not allow the passage of ions. The difference between the inactivated state and the closed state is that inactivated channels must be returned back to resting conditions before they can open again. In contrast, channels in the closed state, although not conducting ions, are ready and available to be opened.

Channels also possess a property that is called selectivity, which, as the name implies, enables a particular channel protein to be selective in relation to which ion passes through its open tunnel to the relative exclusion of other ions. Selectivity is determined by the physical dimensions

of the tunnel, and by the chemical and electrical properties of the tunnel wall. It can be appreciated that a small tunnel will not allow the passage of a large ion. For most channels, an ion has to lose its shell of hydration and bind to sites located within the wall of the tunnel. The ion then jumps from one binding site to another, driven by thermal energy and the electrical and concentration gradients, until it reaches the other side of the channel. There, it regains water molecules to become hydrated again. Since the binding sites in different channels can bind some ions and not others, they can impose the property of selectivity on a channel such that one type of channel can be selective for one species of ion (e.g. calcium) while another type of channel can be selective for another ion (e.g. potassium). Gradients exist across the cell membrane for several important ions. In uterine smooth muscle the sodium gradient is around 15-fold, outside to inside, with the extracellular sodium concentration being around 145 mM and the intracellular concentration around 10 mM. In contrast, the extracellular potassium concentration is low, being about 6 mM, while the intracellular concentration is high at approximately 155 mM. This results in a potassium gradient of some 26-fold, inside to outside. The property of selectivity means that these cations cannot move rapidly through an open calcium channel, even though they are some 100-fold more numerous than calcium ions.

Voltage-operated calcium channels

Voltage-operated channels, as their name implies, are sensitive to the potential difference across the plasma membrane. The probability of opening of voltage-operated calcium channels is increased when the plasma membrane is depolarized, that is when the potential is changed from the resting value of around –60 mV to –45 mV. Voltage-operated calcium channels have been studied in detail and several distinctly individual calcium channels have been identified. One of these has been identified in uterine smooth muscle. It is highly selective for calcium and conducts in the order of 10^3 calcium ions for every sodium ion. Once it has been activated, it remains in the open state for a relatively long period of time and only slowly enters the inactivated state. If the membrane remains depolarized for a prolonged period, a large proportion of the calcium channels can eventually become inactivated. The consequent smaller number of open channels will then result in a decrease in the amount of calcium entering the cell through these channels.

The resting membrane potential of uterine smooth muscle cells during pregnancy ranges between –65 and –55 mV, at which levels the probability of opening of voltage-operated calcium channels is low. Depolarizing the membrane from −65 to around −45 mV increases the probability of opening of these channels some 10-fold and calcium flows across the membrane. The rapid influx of calcium can be recorded as a spike action potential.The probability of opening increases approximately 100-fold during maximal activation at the peak of the action potential, and cytoplasmic concentrations of the ion of up to 1 μM may be attained. Thus, the membrane potential plays an important role in determining the contractile state of uterine smooth muscle. For example, if the membrane can be prevented from depolarizing, the calcium channels will not be activated and a contraction is less likely to occur. The calcium current that flows during the spike action potential can be recorded with extracellular, electromyographical (EMG) electrodes.

Voltage-clamp studies have shown that oxytocin increases the total calcium current that flows into the smooth muscle cells of rat uterus. This has been attributed to recruitment of a greater number of these channels opening in response to depolarization.

Dihydropyridine drugs, such as nifedipine, bind to a region on the external face of this voltage-operated calcium channel and prevent the flow of calcium ions. Hence these drugs are called calcium channel blockers. The dihydropyridines are very effective in suppressing spontaneous contractions in uterine smooth muscle. However, this type of calcium channel is present in the membranes of most other smooth muscles, and the success of these drugs in controlling contraction in arterial smooth muscle has led to their widespread use in treating hypertension. It is hardly surprising that their possible usefulness in preventing preterm labour has been thwarted by the occurrence of unacceptable side-effects, most notably on the heart and blood vessels.

Potassium channels

The principal channels that are open at rest in uterine smooth muscle are those that conduct potassium ions. Potassium has a tendency to leave the cell through these channels and leave behind in the cytoplasm a small excess of negatively charged proteins. As a consequence, the potential difference across the smooth muscle cell membrane is negative on the inside. The value of this potential at rest is around −65 to −55 mV.

A variety of different types of potassium channels have been reported to occur in uterine smooth muscle. The most significant feature of these potassium channels is that most of them are activated by depolarization: potassium ions flow down their electrochemical gradient to leave the cell, resulting in the membrane potential moving in the more negative direction. Thus, a major function of potassium channels is to repolarize the membrane from the peak of the action potential. The various types of potassium channel differ in their rates of activation, inactivation, and their sensitivity to the value of the membrane potential, and hence the various types of potassium channel differ in their contribution to the resting potential.

One class of potassium channel that occurs in the uterine smooth muscle is activated by depolarization of the membrane, and also by the binding of calcium ions to the cytoplasmic face of the channel. These characteristics mean that it will be strongly activated by both the depolarization of the upstroke of an action potential, and also by the calcium that enters during the action potential. It is important in restoring the smooth muscle cell to resting conditions following the upstroke of the action potential.

Another type of potassium channel has been reported to occur in uterine smooth muscle. This channel activates quickly when the membrane is depolarized. It appears to be most significant in mid-pregnancy when its effects are to shorten the duration of the action potential and to reduce the amplitude of the action potential. Both of these effects would be expected to result in smaller contractions.

A class of potassium channel that is maintained in the closed state by cytoplasmic ATP has been identified in uterine smooth muscle. A fall in cytoplasmic ATP, such as might occur during strong uterine contraction that both used up ATP to supply energy and also restricted the blood supply and nutrients, could result in activation of these channels. Hyperpolarization of the membrane would ensue and this would tend to close voltage-operated calcium channels and facilitate relaxation. These channels are opened by relaxants that are used clinically to treat hypertension (e.g. pinacidil) and are blocked by the oral hypoglycaemic drugs, tolbutamide or glibenclamide.

Oestrogen induces the formation of mRNA which codes for the production of a potassium channel whose opening requires a prolonged period of depolarization; that is, it is slowly-activating. The mRNA, induced in oestrogen-treated rat uterus, stimulates the production of these channels when injected into *Xenopus* oocytes.

Action potentials in uterine smooth muscle

Depolarization increases the probability of opening of voltage-operated calcium channels. Calcium then enters the cell and causes further depolarization until the threshold voltage for the initiation of a spike action potential is reached. Threshold is defined as the potential at which calcium influx is sufficiently high to first balance, and then overtake the resting tendency for potassium to leave the cell. At this potential there is a rapid, and runaway influx of calcium that gives rise to the rapidly rising upstroke of the spike action potential (see Fig. 34.6(b)).

The spike can overshoot 0 mV, up to +20 mV. The equilibrium potential for calcium, that is, the potential that would be required to prevent calcium influx along the 10^4 concentration gradient, is approximately +130 mV. Thus, if everything else remained unchanged, the peak of the spike should overshoot to +130 mV. Changes in several important parameters limit the peak of the spike, decrease calcium influx, and terminate the action potential.

1. The depolarization, up to the approximate maximum of +20 mV, that occurs during the upstroke of the spike drives the voltage-operated calcium channels into the inactivated, and hence non-conducting state.
2. The very calcium that enters through the voltage-operated calcium channels hastens the rate of this inactivation.
3. The calcium that has entered also increases the probability of opening of the calcium-activated potassium channel, discussed above. Opening of these channels repolarizes the membrane back towards the resting level. Repolarization decreases the probability of opening of the calcium channels, that is, effectively closing them.
4. The depolarization during the upstroke increases the probability of opening of yet a second class of potassium channel, the voltage-operated potassium channel. Again, this repolarizes the membrane, and hence closes the calcium channels.

These mechanisms serve to bring about repolarization and to terminate the spike action potential. Thus, the spike is self-limiting and brief, with a duration at half maximal amplitude of around 20 ms. Nevertheless, during this period, significant calcium can enter the cells to give rise to considerable contraction (refer to Fig. 34.6(b)).

The action potential that occurs in uterine smooth muscle does not necessarily take the form of a spike. In the circular muscle layer of the uterus of rats, the longitudinal layer of guinea-pigs, and some uterine smooth muscle cells of women, the action potential is complex. An initial spike is followed by a sustained period of depolarization that is of variable duration, and can last up to tens of seconds (refer to Fig. 34.6(a)). The duration of the contraction accompanying these complex action potentials parallels the duration of the sustained depolarization. Thus, complex action potentials give rise to prolonged contraction, demonstrating the manner in which uterine contractile activity is largely determined by the underlying electrical activity. Prolonged, complex action potentials may be terminated as a result of the opening of the slowly-activating potassium channels that have been expressed and studied in *Xenopus* oocytes (discussed above). Alternatively, strong, tetanic contractions, which occur in the uterus

during labour, may reduce the levels of ATP in the cytoplasm of the smooth muscle cells sufficiently to allow the activation of the potassium channels that are blocked by the oral hypoglycaemic drugs. Opening of these channels would lead to repolarization, closing of voltage-operated calcium channels, and a decrease in calcium influx.

The form of the action potential not only differs according to the muscle layer, but may also vary as a function of gestation. Complex action potentials are recorded in the circular layer of rat uterus in mid-pregnancy. As pregnancy progresses, the plateau becomes less pronounced, whereas the spikes become more significant. In contrast, in the longitudinal layer of guinea-pig uterus, complex action potentials persist throughout gestation.

Receptor-operated channels

Channels whose opening is absolutely dependent upon the interaction between a hormone and its receptor are called receptor-operated channels. Most of the receptor-operated channels that have been described so far are poorly selective in that they readily conduct more than one type of ion. Receptor-operated channels have been most thoroughly studied in non-uterine smooth muscle and in these tissues the channels conduct mainly sodium and potassium ions. The principal result of activating these channels is depolarization. The calcium permeability of receptor-operated channels is highly variable between different tissues and between different agonist receptors. For example, a channel that is activated by extracellular ATP, and has been described in detail for arterial smooth muscle, has a permeability ratio of three calcium ions to one sodium ion. Activation of this channel by extracellular application of ATP can result in significant calcium influx, raising the cytoplasmic calcium concentration to 500 nM. A cation channel that is activated by extracellular ATP occurs in uterine smooth muscle, but its calcium permeability has not been determined.

More recently, a channel that has a relatively high permeability for calcium and that is activated by prostaglandins and oxytocin has been implicated in the contractile response of uterine smooth muscle (Fig. 34.1). Initially, cytoplasmic calcium rises from approximately 100 nM to exceed 1 μM, and then declines to around 300 nM, and is sustained at this level. There appears to be little inactivation of the response over tens of minutes and the accompanying contraction is large. The detailed characteristics of these channels remain to be elucidated.

Receptor-operated channels are not blocked by the dihydropyridines or other clinically used drugs that block voltage-operated calcium channels. Attempts to develop drugs that would be useful in blocking receptor-operated channels have so far met with little success.

Stretch-activated channels

The last few years have witnessed an explosion in the literature describing stretch-sensitive ion channels in a wide variety of tissues, and most recently in uterine smooth muscle. Stretching the plasma membrane in the vicinity of these channels in uterine smooth muscle results in an increase in the probability of their opening. They are poorly selective for cations, conducting both sodium and potassium, and hence their opening gives rise to depolarization. Rapid stretch results in depolarization that can be sufficiently large to reach threshold for the initiation of spike action potentials and contraction (see Fig. 34.5).

Relationship between voltage- and receptor-operated mechanisms

An important function of poorly selective cation channels is their ability to depolarize the plasma membrane. Depolarization results from the predominance of sodium influx through these channels. This depolarization, brought about by interactions between circulating or locally produced hormones such as oxytocin and prostaglandins and their receptors, is an important physiological mechanism in that it leads to an increase in the probability of opening of voltage-operated calcium channels, action potentials, significant calcium influx, and contraction.

The dihydropyridines are very effective in eliminating spontaneous contractions in uterine smooth muscle because the underlying calcium influx is purely through voltage-operated calcium channels. However, when uterine activity is stimulated by hormones, such as prostaglandins or oxytocin, the dihydropyridines may be ineffective as blockers of the contraction. Spike action potentials, resulting from calcium influx through voltage-operated calcium channels, are all-or-none events in uterine smooth muscle. If the level of depolarization evoked by low concentrations of hormone reaches threshold for the initiation of spikes, then a large, dihydropyridine-sensitive contraction will inevitably follow (Fig. 34.2). In contrast, receptor-operated calcium influx is graded. Thus, in the presence of dihydropy-

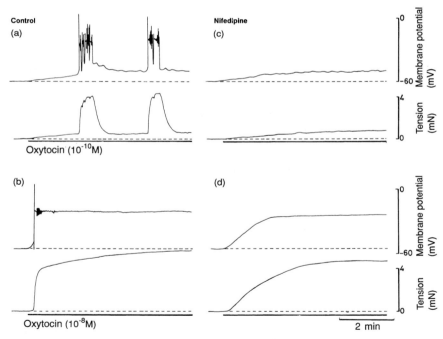

Fig. 34.2 The effect of the dihydropyridine, nifedipine, on the membrane potential and tension evoked by oxytocin. (a) Oxytocin (10^{-10}M) caused by depolarization to threshold for the initiation of action potentials, each of which was accompanied by a large phasic contraction. (b) Oxytocin (10^{-8}M) caused rapid depolarization, a spike, followed by sustained depolarization and a large sustained contraction. (c) In the presence of nifedipine, oxytocin (10^{-10}M) evoked only depolarization. No action potentials occurred. Contraction was small. (d) In the presence of nifedipine, oxytocin (10^{-8}M) caused only the sustained depolarization, without the spike. The accompanying contraction was substantial but relatively slow in onset.

ridines to block voltage-operated calcium entry, calcium influx and hence contraction are only small in response to low concentrations of hormone. However, substantial contraction can be achieved in response to large concentrations of hormone, in the presence of dihydropyridines (Fig. 34.2).

Release of calcium from internal stores

Interactions between many hormones and their receptors lead to mobilization of phospholipids in the plasma membrane. This involves a GTP-binding protein (a G protein) and its activation of phospholipase C which interacts with phosphatidyl inositol, resulting in the release into the cytoplasm of inositol 1,4,5,trisphosphate (IP_3) (Fig. 34.3).

IP_3 interacts with a protein that is embedded in the membrane of the endoplasmic reticulum. This protein is both a receptor for IP_3 and a calcium channel. Interaction between IP_3 and the receptor/channel increases the probability of opening of the channel leading to a substantial amount of calcium flooding into the cytoplasm down its 10^4-fold concentration gradient. In uterine smooth muscle the concentration of calcium in the cytoplasm may reach 200 nM as a result. This response is invariably transient, and cytoplasmic calcium returns to resting levels within 1 min. The accompanying contraction is also transient, with a duration of 1–3 min.

Interaction between prostaglandin $F_{2\alpha}$ and oxytocin with their receptors on uterine smooth muscle leads to the production of IP_3 and the release of calcium from the endoplasmic reticulum. In the uterus of guinea-pigs at mid-pregnancy the contraction that results from the release of stored calcium alone can be up to 50 per cent of the maximum contraction possible. The amplitude of the contraction attributable to the release of stored calcium declines to around 10 per cent of maximum at term. This may reflect a decline in the extent of the store, a failure of the store to fill, or a failure in the release process. A recent study has suggested that a decline in the functioning of the G protein, or an impairment of its ability to stimulate phospholipase C may be involved. In any event, the apparent decline in the prominence of the endoplasmic reticulum in supplying calcium for contraction at term places increasing importance on calcium influx from the extracellular fluid in initiating the contractions required for successful delivery.

Calcium extrusion

Extrusion pumps

Energy-requiring ATPases that are capable of moving calcium against considerable electrochemical gradients are present in the plasma membrane and in the membrane of the endoplasmic reticulum. The pumps in these two locations are distinctly different from one another and hence the two can be regulated separately. The activities of both of these pumps are stimulated when cytoplasmic calcium rises. Thus, during sustained calcium influx and contraction, the pumps continue to translocate calcium. There is evidence to suggest that the activities of both pumps are stimulated by β-adrenoceptor agonist drugs. This would explain, in part, the ability of these drugs to bring about relaxation of the uterus in the face of preterm labour. In contrast, oxytocin and prostaglandin $F_{2\alpha}$ both inhibit the binding of calcium to the plasma membrane pump, and decrease the pump's activity. These actions would facilitate calcium retention in the cytoplasm.

The calcium pump in the endoplasmic reticulum is blocked by cyclopiazonic acid and thapsigargin. These agents prevent filling of the store, which becomes depleted as a consequence. As yet, thapsigargin and cyclopiazonic acid are only used as research tools but they could provide the basis for a search for suitable therapeutic agents with similar actions.

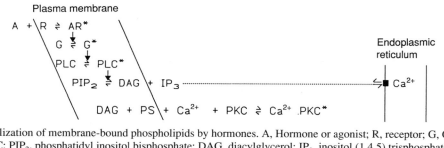

Fig. 34.3 Mobilization of membrane-bound phospholipids by hormones. A, Hormone or agonist; R, receptor; G, G-protein; PLC, phospholipase C; PIP_2, phosphatidyl inositol bisphosphate; DAG, diacylglycerol; IP_3, inositol (1,4,5) trisphosphate; PS, phosphatidyl serine; Ca^{2+}, calcium; PKC, protein kinase C; *, activated molecules.

Calcium exchange

A protein that is capable of exchanging calcium for sodium across the cell membrane is known to exist in uterine and other smooth muscle. At rest, sodium ions leak into the cytoplasm, driven by the considerable electrical and chemical gradient that exists across the plasma membrane for these ions. This electrochemical gradient is maintained by the energy-consuming, $Na^+ K^+$ ATPase pump. The inward leak of sodium, down its electrochemical gradient, 'powers' the countermovement of calcium out of the cell by exchange. The importance of this exchanger as a mechanism for removing calcium from the cytoplasm in uterine smooth muscle can be demonstrated when the chemical gradient for sodium is removed. Lowering the concentration of sodium in the external medium results in a large contraction that is sensitive to the concentration of external calcium. The exchanger is a large membrane-spanning protein and hence is potentially vulnerable to pharmacological manipulation. However, no drug has yet been identified that stimulates or blocks this exchanger.

Sensitivity of the contractile apparatus to calcium

It has recently been discovered that exposure of smooth muscle to the α-toxin produced by *Staphylococcus aureus* leads to the formation of holes in the plasma membrane that are 2 nm in diameter, while leaving intact hormone receptors and membrane-bound proteins that are associated with second messenger systems. The holes produced by staphylococcal α-toxin are small enough to ensure that important cytoplasmic constituents are not lost from the cells, yet they allow small molecules and ions such as calcium to enter the cell with ease. As a consequence, it is possible to experimentally control, or 'clamp', the cytoplasmic calcium concentration. The experimental preparations resulting from treatment of smooth muscle with α-toxin made it possible to study in detail the relationships between cytoplasmic calcium concentrations and events initiated as a result of hormone/receptor interaction. A startling realization emerged. Hormone/receptor interaction dramatically altered the relationship between cytoplasmic free calcium and contraction, and this appeared to be achieved as a result of protein phosphorylation.

Spasmogenic hormones, that is, hormones that increase or evoke contraction, such as acetylcholine, noradrenaline, or the prostaglandins, shifted the calcium/contraction curve to the left. This meant that, for a given concentration of cytoplasmic calcium, a larger contraction was obtained in the presence of a hormone than in its absence. Such an increase in the sensitivity of the contractile apparatus to calcium in the presence of a hormone was associated with an increase in protein phosphorylation. The identity of the proteins that are phosphorylated has not been established, nor has the kinase(s) that is likely to be responsible. Protein kinase C is a possible candidate. This kinase is activated by diacylglycerol (DAG), which is the by-product remaining in the plasma membrane following cleavage of IP_3 from phosphatidyl inositol (see Fig. 34.3). In fact, DAG may even be produced in the absence of IP_3, as a result of cleavage of other phospholipids, such as phosphatidylcholine. The calcium that enters the cell through voltage-operated calcium channels, or that is released from internal stores early in the responses to hormones, will bind to calmodulin, and calmodulin-dependent protein kinase II will be activated. This kinase only requires calcium/calmodulin during the initial stages of its activation, as a small quantity of activated kinase autoactivates more molecules in snowball fashion. The ability of protein kinase II to phosphorylate myosin at the same site as MLCK makes it a candidate for the maintenance of 'calcium-independent' contraction.

The process of DAG formation, activation of protein kinase C, and the subsequent protein phosphorylation leading to an increase in the sensitivity of the contractile apparatus to calcium is likely to result in a slow build-up in the contractile response. This is in contrast with the fast contraction associated with calcium influx through channels or following release from internal stores.

Agents that relax smooth muscle, such as β-adrenoceptor agonists and the nitrovasodilators, have also been shown to influence the sensitivity of the contractile apparatus of smooth muscle to calcium. These agents decrease the sensitivity via a mechanism involving activation of protein kinases A and G by cAMP and cGMP, respectively. These kinases phosphorylate MLCK resulting in its inactivation and a decrease in contraction (see above). Relaxin stimulates cAMP production in uterine smooth muscle and is thus likely to decrease the sensitivity of the contractile apparatus for calcium.

The ability to maintain uterine relaxation decreases with time during prolonged or frequent application of β-mimetics such as ritodrine. This results from a decrease in the number of β-adrenoceptors. This is primarily due to a decrease in the mRNA responsible for the synthesis of new β-adrenoceptors. The inability of β-mimetics to sustain relaxation has also been demonstrated in women. A recent study by the Canadian Preterm Labor Investigators Group has shown that ritodrine is only effective in delaying labour for up to 48 h.

Interactions between uterine smooth muscle cells

Structural aspects

The individual smooth muscle cells that make up the wall of the muscle layer of the uterus are 3–5 μm in diameter and 200–300 μm long in the non-pregnant state, and 5–10 μm in diameter by 500–600 μm long at term. The cells are held together by virtue of the arrangement of the cytoskeleton and the external connective tissue. Filamentous molecules, including the contractile protein actin, cross and link throughout the cytoplasm and are anchored to the plasma membrane in thickened regions, recognizably dense in electron microscopy due to the high concentration of proteins. These proteins include α-actinin, also present in the Z bands of skeletal muscle, and vinculin. A family of fibrous, extracellular molecules, the fibronectins, also attach to the plasma membrane in these dense regions. The fibronectins, in turn, attach to collagen fibres. Collagen fibres attached to one smooth muscle cell will invariably attach to another cell. This arrangement provides support for the individual smooth muscle cells and ensures that the smooth muscle structure of the uterine wall is not pulled apart during contraction.

The smooth muscle cells of the uterine wall are arranged into bundles, and the longitudinal axes of the cells are oriented along the bundles. In the uterus, bundles branch and fuse, analogous to the situation in the heart. In those species in which outer longitudinal and inner circular smooth muscle layers can be distinguished, the long axes of the smooth muscle cells are oriented in the ovarian-cervical axis or around the uterus, respectively. Interconnection between the layers is achieved by the existence of bundles of muscle that connect with bundles in both layers. In this way it is possible for activity generated in one layer to spread to the other.

Gap-junctions

The smooth muscle cells of the uterus are interconnected in an even more intimate way. Cytoplasmic continuity is achieved between neighbouring cells via gap-junctions. The fundamental unit of the gap-junction is the connexon, a cluster of six similar proteins that are embedded in the plasma membrane. Connexons have similarities with ion channels in that a tunnel may exist, nestled within the cluster, and it can be open or closed. The principal difference between a connexon and an ion channel is the size of the tunnel. For example, highly selective sodium and potassium channels have a nominal diameter of 0.3–0.5 nm,

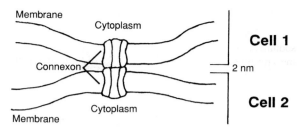

Fig. 34.4 Schematic representation of a gap-junction. Hexameric proteins, called connexons, embedded in the plasma membranes of two smooth muscle cells, fuse to form a gap-junction. The behaviour of the protein complex is similar to that of an ion channel in that it can be open and conducting, or closed and not conducting. The open gap-junction allows passage of molecules of approximately 1-kDa mass. The plasma membranes of the two cells involved are separated by a distance of 2 nm at the junction.

while the connexon is permeable to molecules that have a nominal diameter of 1.3 nm. A gap-junction results when two connexons from adjacent smooth muscle cells fuse, spanning a gap of approximately 2 nm between the membranes of the two cells (Fig. 34.4). When a gap-junction is open and conducting, it allows not only ions, but molecules up to a nominal diameter of 1.3 nm and a molecular mass of around 1 kDa to pass from one smooth muscle cell to the next.

The consequences for contraction of the movement of large molecules between smooth muscle cells is completely unknown. In contrast, the importance of gap-junctions in allowing freedom of movement of ions between cells, and consequently the spread of current, is well recognized. The existence of gap-junctions means that the potential difference across the plasma membrane is similar in many cells. Thus, uterine smooth muscle is an electrical syncytium. Current injected through a microelectrode into one cell flows rapidly into neighbouring cells along the lower resistances of the gap-junctions and is only slowly lost out across the plasma membrane. When the resistance between cells is lower, current generated locally spreads further within the syncytium. This situation is achieved when the number of open gap-junctions between cells is increased. Such an occurrence is observed just prior to labour.

Histological examination has revealed a rapid increase in the number of gap-junctions at term, and electrophysiological studies have demonstrated a five- to seven-fold increase in the distance travelled by current in strips of uterus at this time. These changes would be highly conducive to the production of co-ordinated and forceful contractions at a time when such activity is necessary. It has been suggested that agents such as prostaglandin $F_{2\alpha}$ and oxytocin facilitate the spread of activity via a mechanism involving gap-junctions. Conversely, agents that increase cAMP, such as β-mimetics, prostacyclin, and relaxin, appear to close gap-junctions and thus facilitate a decrease in organ contraction.

When a smooth muscle cell is damaged, excessive amounts of calcium enter into the cytoplasm and this causes gap-junctions to close. This isolates the damaged cell from its neighbours and prevents the entire syncytium from 'bleeding' through a damaged cell.

Stretch

The growth of the uterus during pregnancy and its rapid involution postpartum is nothing short of spectacular. It has been recognized for more than 30 years that acute stretch of smooth muscle can lead to contraction. From the results of an elegant series of experiments, Takeda suggested that synchrony of electrical and mechanical activity throughout the uterus in rabbits could be achieved when areas of contraction stretched the neighbouring regions. In these experiments, the uterus was cut circumferentially to form two sacs, oviductal and cervical, which were then connected via a fluid-filled tube. Activity in both halves remained synchronized, provided that continuity of the fluid in the two sacs was maintained.

The possibility has since been considered that the pull resulting from contraction in one region of the uterus might be sufficient to activate stretch-sensitive channels in an adjacent region, and hence induce propagation of contraction. To test this, membrane potential and tension were recorded simultaneously in two separate strips of uterus that were tied together with silk. One of the strips was stimulated with an electric shock. The action potentials and contraction in the stimulated strip were followed by depolarization to threshold for the initiation of action potentials and contraction in the strip to which it was attached only by silk (Fig. 34.5). This, together with the recording of stretch-activated ion channels at the single-channel level, provide a modern verification of Takeda's interpretation some 25 years ago. The elements of the cytoskeleton are attached to, or in the close vicinity of stretch-sensitive channels and this relationship between the channel, the cytoskeleton, and the external connective tissue elements is consistent with a role for these channels in stretch-induced propagation of contraction.

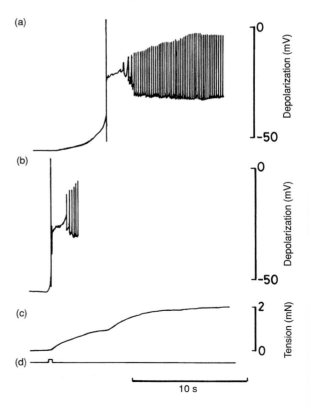

Fig. 34.5 Depolarization and contraction induced by stretch in uterine smooth muscle. Two strips of longitudinal muscle, obtained from guinea-pigs during late gestations, were tied together by a length of silk. One strip (b) was stimulated electrically (d). The resulting contraction (c) caused depolarization and action potentials in the other strip (a) and an additional component of contraction (c). (a) Depolarization to threshold for the initiation of an action potential was evoked in the unstimulated strip following contraction of the stimulated strip. (b) Action potentials were evoked as a result of electrical stimulation. (c) The contraction was biphasic. The initial component arose from the electrically evoked action potential activity in one strip (b). Stretch-induced contraction in the second strip (a) added to the latter part of the response. (d) The stimulating pulse.

Spontaneous contractions

Pacemaker potentials

Contractions may occur spontaneously, that is, in the absence of stimulation, when strips of uterine smooth muscle are placed in warmed, oxygenated physiological saline. Recordings from individual smooth muscle cells with a microelectrode reveal that each contraction is preceded by a single action potential or is associated with a burst of action potentials. When recordings are made with a microelectrode from different regions of a spontaneously contracting strip of uterus, an area can usually be found in which the smooth muscle cells generate pacemaker potentials. Pacemakers in uterine smooth muscle can take two forms. Slow progressive depolarization, similar to the pacemaker potentials that occur in the heart, eventually reach −45 to −40 mV, the threshold for the firing of action potentials, and contraction ensues (Fig. 34.6(a)). The slow depolarization is likely to result from a progressive increase in the probability of opening of a class of poorly selective cation channels. The additional depolarization that occurs during the action potentials, together with a calcium-induced opening of potassium channels (refer to the section 'Action potentials in uterine smooth muscle'), are responsible for the precipitous repolarization of the membrane to around −65 mV. Then the process starts all over again. This gives rise to rhythmic, usually regular, spontaneous contractions. Blockade of the voltage-operated calcium channels prevents the firing of actions potentials, abolishes the contractions, and prevents the repolarization of the membrane. As a result, the membrane potential becomes stable at −45 to −40 mV in the presence of these blockers.

Slow waves

More discrete episodes of depolarization, 5–10 mV in amplitude, 1–2 s in duration and superimposed on a steady membrane potential, have also been reported in uterine smooth muscle and have been called 'slow waves'. If the resting membrane potential is sufficiently depolarized, or if the slow waves are sufficiently large, threshold may be reached for the initiation of action potentials, and hence contraction (Fig. 34.6(b)). The detailed mechanisms underlying slow waves have not been elucidated but they cease when the concentration of sodium in the external medium is lowered. Calcium channel blockers do not affect slow waves, although action potentials and contractions no longer occur. In muscle preparations displaying slow waves, the membrane potential remains at the resting level and regular slow waves continue in the presence of calcium channel blockers.

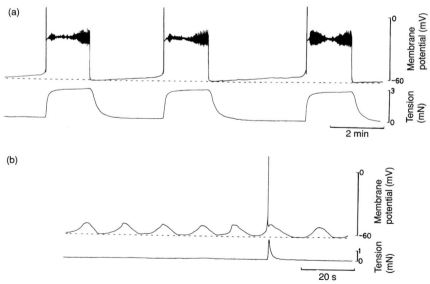

Fig. 34.6 Spontaneous changes in membrane potential and tension in uterine smooth muscle. (a) Progressive depolarizing pacemaker potentials gave rise to action potentials and contractions in a longitudinal strip of uterine muscle. (b) When the amplitude of a slow wave was sufficiently large to reach threshold for the initiation of a spike action potential in a circular strip of uterine muscle, a contraction occurred.

Sites of pacemaker initiation

Pacemaker activity always arises in the same region of the healthy, intact heart (i.e. the sinoatrial node), although all cardiac muscle cells are capable of pacemaking, as illustrated when they are isolated from the faster node region. In contrast, the generation of pacemaker activity in the uterus typically moves around the organ. This has been demonstrated in intact animals in which arrays of recording electrodes have been distributed over the uterus, both close together and far apart. The electrode in which electrical activity first appears changes continuously. Even in small strips of isolated uterine smooth muscle, the site of initiation of pacemaker activity can be observed to change location within hours. Thus, all smooth muscle cells in the uterus appear to be capable of generating pacemaker potentials.

The gap-junction coupling between the smooth muscle cells of the relatively small, discrete area in which pacemaker activity is initiated is likely to be different to the gap-junction coupling between smooth muscle cells in the non-pacemaking areas. This would allow the build-up of sufficient depolarization, with minimum loss, to fire an action potential. The large current during the action potential would then spread and propagate further throughout the muscle of the uterus.

Effects of steroids

Hormones such as oxytocin and the prostaglandins interact with receptors that are embedded in the plasma membrane to activate processes such as second messenger enzymes and ion channels that are already present and available. In contrast, interaction between the steroid hormones, oestrogen and progesterone, and their receptors in the cytoplasm results in transcription of specific genes to produce messenger RNA, and this leads to the synthesis of the very enzymes and channel proteins that are activated by agents that act acutely, such as oxytocin, prostaglandins, and β-adrenoceptor agonists. Thus, the principal role of the steroids appears to be more long-term, to establish the machinery that is necessary for the acute regulation of contraction.

Oestrogen

Oestrogen is the principal female steroid. It is largely responsible for the maintenance of the female reproductive tract, and is important for uterine growth during pregnancy. Oestrogen treatment of ovariectomized rats leads to hyperpolarization of uterine smooth muscle cells, from a resting membrane potential of around −40 mV following ovariectomy, to −60 mV in response to oestrogen. This is likely to be due to an effect on the genome involving transcription of the DNA and the incorporation into the plasma membrane of a potassium channel. The messenger RNA for this channel has been isolated and the channel induced has been studied in detail following transfection of the mRNA into *Xenopus* oocytes.

The role of oestrogen in regulating the level of the resting membrane potential during pregnancy is unclear. Again, most of the work has involved rats and this makes interpretation easier. The concentration of oestrogen in the blood of rats during pregnancy is very low, and a surge occurs in the 24–48 h prior to delivery. Yet several thorough studies have shown that the resting membrane potential of rat uterine smooth muscle decreases from around −60 mV during pregnancy to −35 mV at term. Taken together with the mRNA/potassium channel results, described in the paragraph above, it has to be concluded that the concentration of oestrogen in the circulation of rats cannot explain the changes in the resting membrane potential observed at term. It is likely that factors peculiar to pregnancy, other than oestrogen, are more important.

Progesterone

Progesterone was dubbed the hormone of pregnancy by Csapo. This arose from the substantial evidence that he and his collaborators had amassed that it was responsible for maintaining the relative quiescence of the uterus during pregnancy. Although the fact of relative uterine quiescence has long been accepted, the mechanisms responsible are unfolding only slowly. Of relevance to this chapter, progesterone has a profound effect in decreasing the spread of current through the uterine syncytium, and under its influence the number of gap-junctions is small.

It is likely that progesterone and oestrogen are responsible for determining some of the basic properties of uterine smooth muscle, for example, the concentration of the contractile proteins actin and myosin and the nature and density of various ion channels and hormone receptors. The nature and density of ion channels determine the value of the resting membrane potential, and hence the proximity to threshold for the initiation of action potentials and consequently, the ease with which contractions can be evoked. This parameter also determines the occurrence or not, and the nature of, pacemaker potentials and hence the level of spontaneous contractile activity. An increase in the density of voltage-operated calcium channels facilitates greater calcium influx, and lowers threshold. In contrast, an increase in the relative density of potassium channels makes the generation of action potentials more difficult. An increase in receptor-operated channels increases the effectiveness of the relevant hormones in bringing about depolarization and/or calcium influx. The nature and density of hormone receptors dictate the susceptibility to regulation of the fundamental properties inherent to the muscle.

Uterine contractions *in vitro*: a summary

Calcium influx through voltage-operated calcium channels during the action potential is pivotal in providing the calcium that is required for the initiation of contractions in uterine smooth muscle. The depolarization of the membrane that is necessary to activate these channels is provided by pacemaker activity, or results from hormone/receptor interaction. A burst of action potentials usually occurs, perhaps sustained by the pacemaker. Such bursts are always associated with contraction. Eventually, additional potassium channels are activated, by the depolarization and elevated cytoplasmic calcium (see the section on 'Action potentials in uterine smooth muscle'). The calcium channels enter the inactivated state, also caused by the depolarization and elevated cytoplasmic calcium. Together, these events lead to the re-establishment of the dominance of potassium efflux over calcium influx, and repolarization occurs.

The stimulatory effects of hormones, and their activation of poorly selective, receptor-operated channels can result in membrane depolarization, activation of bursts of action potentials, and muscle contraction in the absence of pacemakers. While action potentials are perhaps the most important mechanism of intracellular calcium delivery in uterine smooth muscle, when voltage-operated calcium channels are blocked, receptor-operated calcium influx and release of calcium from internal stores can still cause contraction. In addition, hormones can also suppress calcium removal from the cytoplasm.

In contrast, relaxation can be facilitated by agents that increase the proportion of potassium channels in the open state. This makes the inside of the smooth muscle cell more negative. In order for contraction to occur, calcium influx must be stimulated to an even greater extent in order to overcome the resulting increase in the tendency for potassium to leave. Thus, it is more difficult for pacemakers and stimulators to reach threshold for the initiation of spikes. Inhibitors may also act to decrease calcium entry through channels. Relaxation is facilitated as a result of stimulating calcium extrusion from the cytoplasm.

Regulation of contraction is not the sole prerogative of events involving ion channels. It is becoming increasingly

clear that the sensitivity of the contractile apparatus for calcium plays an important role in controlling contraction. Thus, the contraction evoked by stimulatory agents may be enhanced as a result of an increase in the sensitivity of the contractile apparatus, and relaxation may be achieved by decreasing its sensitivity.

Finally, co-ordinated contraction of the uterus requires that activity spreads throughout the entire organ. This is made possible by the existence of limited cytoplasmic continuity between the cells. Current generated in a discrete area can radiate out along the lower resistance pathways between cells. Agents that facilitate the formation of such pathways are stimulatory. Closure or suppression of these pathways leads to restriction of activity to a localized area and inhibition of organ contraction, without any assumption of inhibition of contraction in the individual smooth muscle cells.

Uterine contractility *in vivo*

Recording uterine contractility *in vivo*

What is known of the detailed mechanisms regulating contraction in uterine smooth muscle has, of necessity, been obtained from isolated tissue or individual cells. In intact animals the uterus is subjected to myriad influences, both circulating and local, only some of which are known. Thus, it is important to study uterine activity in intact animals in order to determine which of the various mechanisms identified as capable of inducing contractile activity *in vitro* are the most functionally significant *in vivo*. In addition, observations *in vivo* can and should be utilized to help provide focus for studies *in vitro*.

Uterine activity *in vivo* has been studied using a variety of methods but the most useful have involved catheters or balloons placed in the uterus to record changes in pressure resulting from contraction of the muscle. Recording electrodes attached to the muscle detect electromyographic (EMG) activity resulting from the currents that flow during spike action potentials (Fig. 34.7). There are several points of caution that must be borne in mind in relating the occurrence of contractions to recordings of EMG activity alone.

1. It is important to keep in mind the fact that EMG electrodes are usually only capable of recording relatively rapid changes in current. Slow or sustained current flow, that is, when the rate of change is negligible, will go undetected. This arises from the necessity of using high amplification of the small signal and the consequent use of filters to remove slow changes which would otherwise make meaningful recording impossible.

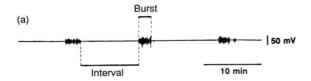

(a)

Burst

Interval

| 50 mV

10 min

(b)

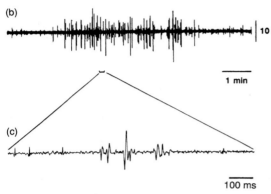

10 mV

1 min

(c)

100 ms

Fig. 34.7 Electromyographic (EMG) activity recorded from uterine smooth muscle of an intact pregnant guinea-pig. (a) A pair of electrodes was attached to the muscle of the guinea-pig uterus under general anaesthetic and EMG activity was recorded 48 h after recovery. During the onset of labour, bursts of EMG spikes occur at regular intervals. (b) On a faster chart speed the spikes within a burst are clear. (c) With a very fast chart speed the components of the individual spikes can be seen. The spikes are biphasic as expected.

2. Pharmacological agents, hormones, stretch, and other stimuli are quite capable of eliciting *changes* in current flow that can be very slow, and yet may be associated with substantial increases in calcium entry and hence considerable contraction. The associated electrical activity is unlikely to be recorded by EMG electrodes.

3. Hormones can induce changes, both increases and decreases, in the sensitivity of the contractile apparatus for calcium. Such effects are not associated with any change in current flow across the plasma membrane and will not be associated with EMG activity *per se*, although any associated phenomena that culminate in action potential discharge will be recorded.

4. Release of calcium from stores in the endoplasmic reticulum is also, of itself, independent of changes in membrane currents and will not be reflected in the EMG activity recorded.

From these considerations it follows that substantial contraction is, in theory, possible in the face of no, or only a weak change in EMG activity. Also for these reasons, it is unwise to associate, in a cause-and-effect way, changes in the amplitude of EMG responses with the amplitude of contraction. An increase in the amplitude of the EMG response will result from an increase in the current flowing in the vicinity of the electrodes and is likely to reflect an increase in the number of smooth muscle cells that are firing action potentials; thus, such an increase better reflects spread of activity than a stronger contraction by the individual smooth muscle cells. However, it must be said that co-ordinated contraction resulting from better spread of activity is likely to result in a larger change in intrauterine pressure during contraction.

Quiescence during pregnancy

Uterine contractions occur throughout pregnancy in horses, cows, guinea-pigs, rats, rabbits, sheep, monkeys, and humans. However, the contractions are of low amplitude and this is likely to result from a lack of co-ordinated spread of activity. This interpretation is strengthened by the observation that EMG activity evoked electrically by stimulating electrodes only travels a short distance before dying away. Bursts of EMG activity, 5–10 min in duration, that are associated with these weak contractions, occur at a rate of 2–5/h in several species. The regulation of this phenomenon has been studied in detail in conscious ewes. When an array of EMG electrodes was placed at various locations over the uterus, and also on isolated segments of uterine muscle explanted to an omental fold, activity occurred in the uterus and in the explants at similar rates but not at the same time. This suggested that the bursts of activity were not evoked by the release into the circulation of a spasmogen. When segments of uterine smooth muscle were isolated and studied *in vitro*, they were quiescent for the first 30–60 min, and then spontaneous contractions commenced that had a frequency of several per minute which is much higher than that which occurred in the animal *in vivo*. In addition, this fast rate of spontaneous activity persisted for many hours. It is thought that the muscle has the capacity to contract spontaneously at a high frequency, a property that may well be determined by steroids. However, this activity is suppressed by circulating factor(s) *in vivo*. The identity of this suppressant is unknown, but possible candidates include relaxin, prostacyclin, and perhaps progesterone, although inclusion of progesterone in the tissue bath *in vitro* did not restore relative quiescence.

Uterine activity during labour

At term, the ability of EMG activity to travel through uterine muscle increases some fivefold, and the EMG spikes increase in amplitude. This is consistent with the observations that the density of gap-junctions increases at term. The pattern of EMG activity during the approach to labour was also studied in sheep with explants of uterine muscle, fitted with EMG electrodes, and placed in an omental fold. As labour approached, the duration of the EMG bursts decreased and their frequency increased. These changes culminated in parturition. EMG bursts in muscle explanted to the omentum became synchronous with those in the body of the uterus. Isolated strips of muscle remained quiescent for up to 6 h *in vitro*. However, activity was easy to evoke and exposure to prostaglandins or oxytocin initiated bursts of action potentials and contraction whose duration and frequency *in vitro* were remarkably similar to those which occurred in the animal *in vivo*.

Concluding remarks

Much has been learned of the regulation of contraction in uterine smooth muscle from detailed study of strips of tissue, single cells, subcellular organelles and components, and even the isolated receptor and ion channel proteins. The recent development of fluorescent probes to measure a variety of cytoplasmic components, including calcium, and the technique of controlling cytoplasmic calcium using staphylococcal α-toxin are rapidly opening up new possibilities for further understanding of the mechanisms controlling uterine contraction. However, these exciting techniques, and the glamour of patch-clamping and molecular biology, should not blind us to the absolute necessity of returning frequently to considering uterine contractions in intact organisms. The behaviour of uterine activity *in vivo* should be used to assess the relative importance of the various regulatory mechanisms revealed *in vitro*, and to focus detailed research efforts.

Further reading

Aidley, D. (1991). Ionic channels in cell membranes. *New Scientist*, **47**, 1–4.

Amédée, T., Mironneau, C., and Mironneau, J. (1987). The calcium channel current of pregnant rat single myometrial cells in short-term culture. *Journal of Physiology*, **392**, 253–72.

Blennerhassett, M.G. and Garfield, R.E. (1991). Effect of gap junction number and permeability on intercellular coupling in rat myometrium. *American Journal of Physiology*, **261**, C1001–C1009.

Canadian Preterm Labor Investigators Group (1992). Treatment of preterm labor with the beta-adrenergic agonist ritodrine. *New England Journal of Medicine*, **327**, 308–12.

Carsten, M.E. and Miller, J.D. (1990). Calcium control mechanisms in the myometrial cell and the role of the phosphoinositide cycle. In *Uterine function: molecular and cellular aspects* (ed. M.E. Carsten and J.D. Miller), pp. 121–67. Plenum Press, New York.

Ducsay, C.A. (1990). Calcium channels: role in myometrial contractility and pharmacological applications of calcium entry blockers. In *Uterine function: molecular and cellular aspects* (ed. M.E. Carsten and J.D. Miller), pp. 169–94. Plenum Press, New York.

Garfield, R.E., Thilander, G., Blennerhassett, M.G., and Sakae, N. (1992). Are gap junctions necessary for cell-to-cell coupling of smooth muscle?: an update. *Canadian Journal of Physiology and Pharmacology*, **70**, 481–90.

Hartshorne, D.J. and Kawamura, T. (1992). Regulation of contraction in smooth muscle. *News in Physiological Sciences*, **7**, 59–66.

de Lanerolle, P. and Paul, R.J. (1991). Myosin phosphorylation/dephosphorylation and regulation of airway smooth muscle. *American Journal of Physiology*, **261**, L1–L14.

Mackenzie, L.W., Word, R.A., Casey, M.L., and Stull, J.T. (1990). Myosin light chain phosphorylation in human myometrial smooth muscle cells. *American Journal of Physiology*, **258**, C92–C98.

Neher, E. and Sakman, B. (1992). The patch clamp technique. *Scientific American*, **266**, 28–35.

Parkington, H.C. and Coleman, H.A. (1990). The role of membrane potential in the control of uterine motility. In *Uterine function: molecular and cellular aspects* (ed. M.E. Carsten and J.D. Miller), pp. 195–248. Plenum Press, New York.

Takeda, H. (1965). Generation and propagation of uterine activity *in situ*. *Fertility and Sterility*, **16**, 113–19.

Wray, S. (1990). The effects of metabolic inhibition on uterine metabolism and intracellular pH in the rat. *Journal of Physiology*, **423**, 411–23.

35. Eicosanoid biosynthesis and its regulation during human pregnancy and parturition

Murray D. Mitchell

Products of arachidonic acid metabolism (eicosanoids, e.g. some prostaglandins and leukotrienes) have important roles in the maintenance of pregnancy, the mechanism(s) of parturition, and various diseases of pregnancy such as pregnancy-induced hypertension. The wealth of literature describing these relationships dictates that this chapter be focused rather than global in nature. Only studies of human tissues will be discussed despite the limitations in experimental design that are imposed on such studies. Emphasis will be placed on studies of the cyclooxygenase and, to a lesser extent, lipoxygenase pathways of arachidonic acid metabolism, and the epoxygenase pathway and catabolic pathways will be mentioned only briefly. It should be noted also that many eicosanoids are not derived from arachidonic acid and will not be discussed here.

In order to develop concepts of regulation, one section has focused on the amnion since this tissue has received most attention for such studies. The significance of the products of the pathways described in the physiological and pathophysiological events of pregnancy will be described primarily in a section in which an attempt is made to delineate those regulatory mechanisms that are considered most significant in, or specific to, pregnancy and parturition.

Pathways of arachidonic acid metabolism

In order to understand the mechanisms that regulate eicosanoid production it is necessary to have an appreciation of the biosynthetic and catabolic pathways involved. A brief outline of such information is provided in this section. Arachidonic acid (all *cis*-5,8,11,14-eicosatetraenoic acid) is a carbon-20 polyunsaturated fatty acid that is the precursor of what is presently the most important group of eicosanoids (derivatives of carbon-20 polyunsaturated fatty acids). Formation of these eicosanoids requires that arachidonic acid be in the non-esterified form. Arachidonic acid is a typical polyunsaturated fatty acid in that it is present in cells predominantly in an esterified form, usually in

the sn2 position of a glycerophospholipid (Fig. 35.1). Hence, the liberation from glycerophospholipids of arachidonic acid is a key event in the biosynthesis of eicosanoids and, indeed, it is thought to be the rate-limiting step in this process. The release of arachidonic acid from glycerophospholipids is accomplished either directly by the action of phospholipase A_2, or indirectly by the action of phospholipase C; the latter mechanism requires the subsequent actions of diacylglycerol and monoacylglycerol lipases (Fig. 35.2).

Arachidonic acid liberated by these reactions can be metabolized by way of at least three major pathways (Fig. 35.2). The most recently discovered pathway is that designated as the epoxygenase pathway. In this pathway, arachidonic acid is metabolized by way of cytochrome P450 linked monooxygenase enzymes into biologically active epoxides, and thence to various vicinal diols and tri-

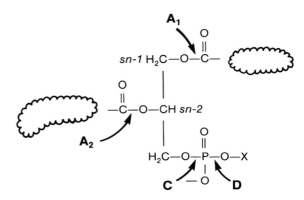

Fig. 35.1 The structure of a generalized glycerophospholipid. The various alcohols that may be found in phosphomonoester linkage are depicted as X. The sites of hydrolysis catalysed by the various phospholipases are indicated by the letters A_1, A_2, C, and D. A saturated fatty acid is typically esterified at the sn-1 position and a polyunsaturated fatty acid commonly esterified at the *sn*-2 position. The fatty acids are represented by 'clouds'.

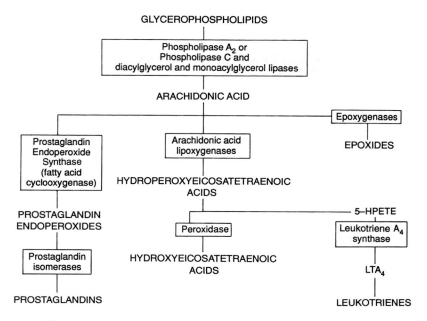

Fig. 35.2 A simplified schematic representation of the enzymatic pathways of arachidonic acid metabolism.

hydroxy acids. This metabolic pathway was originally described in liver and kidney but is present in other tissues including placental tissues. Information concerning this pathway, its products, and the possible roles of such products in pregnancy is still limited. Thus, as mentioned in the introduction, attention will be focused on the other two major pathways of arachidonic acid metabolism, i.e. the cyclooxygenase and lipoxygenase pathways.

An outline of the enzymes involved in arachidonic acid metabolism is presented in Fig. 35.2 and the major products of such metabolism are given in Fig. 35.3. The formation of prostaglandins and thromboxanes (cyclooxygenase

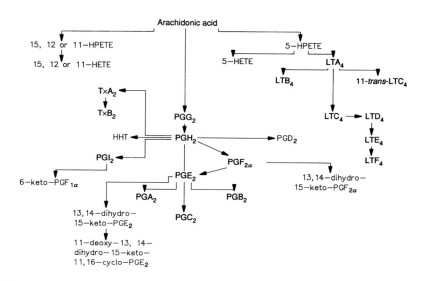

Fig. 35.3 A simplified outline of the pathways and products of arachidonic acid metabolism.

pathway) begins with the action of prostaglandin endoperoxide synthase, synonymous with fatty acid cyclooxygenase, on arachidonic acid. There is an inherent peroxidase activity in the holoenzyme, such that the prostaglandin G_2 (15-hydroperoxy derivative) formed is rapidly converted to prostaglandin H_2 (15-hydroxy derivative). Although these substances are short-lived intermediates, they do possess intrinsic biological activity. It is the formation of these endoperoxide intermediates that is inhibited by non-steroidal anti-inflammatory agents such as indomethacin. Thus, the biosynthesis of all prostaglandins and thromboxanes is inhibited by such drugs. The endoperoxide intermediates are metabolized further to prostaglandins and thromboxanes by the actions of various isomerases. Evidence for a specific reductase to form prostaglandin $F_{2\alpha}$ is still limited, and it is likely that this conversion is often non-enzymatic. The enzymes of prostaglandin and thromboxane biosynthesis are located in the microsomal fraction of the cell, with the exception of prostaglandin D_2 (11-keto-) isomerase, which is a cytosolic enzyme.

The first step in the catabolism of prostaglandins is catalysed by 15-hydroxyprostaglandin dehydrogenase (PGDH), which is a cytosolic enzyme. The 15-keto derivatives so formed are substantially biologically inactive and are rapidly converted to the 13,14-dihydro-15-keto derivatives that are the major circulating forms. A major site of such metabolism is the lung; almost all biologically active prostaglandins are metabolized during one passage through the lungs. Such metabolism occurs firstly by uptake into pulmonary cells and secondly by the action of PGDH. Elimination of biologically active prostaglandins is completed by a series of β- and ω-oxidations that result in the formation of a wide variety of products that are excreted in the urine. It should be noted that exceptions exist to the preceding description of prostaglandin catabolism. For instance, prostaglandin D_2 is an extremely poor substrate for PGDH and prostacyclin may not be metabolized completely by the lungs, since it is not a good substrate for the uptake mechanisms that transport prostaglandins into the pulmonary cells for subsequent metabolism.

Lipoxygenase enzymes are found in several subcellular fractions and catalyse the formation of hydroperoxyeicosatetraenoic acids (HPETEs) from arachidonic acid (Figs 35.2 and 35.3). These derivatives are biologically active and are converted rapidly to their hydroxyeicosatetraenoic acid (HETE) derivatives, which also have biological activity. Other pathways of metabolism exist for the HPETEs of which the leukotriene (LT) pathway is presently considered the most important. This pathway derives from the 5-lipoxygenase pathway and the key step is the conversion of 5-HPETE to LTA_4. Thereafter, LTA_4 is metabolized either by addition of water at C12, leading to the opening of the epoxide at C6, and the formation of LTB_4, or by nucleophilic opening of the epoxide at C6 by the sulfhydryl group of glutathione and the formation of LTC_4. The latter compound may be metabolized further by the sequential elimination of glutamic acid and glycine to form LTD_4 and LTE_4. Conversion of LTE_4 to LTF_4 occurs by addition of a γ-glutamyl residue to the amino group. The further metabolism of LTs is complex and has been described in detail elsewhere. It should be noted that lipoxins, which are products derived from 15-HPETE, have also been shown to have many important biological properties.

Concentrations of eicosanoids in the mother and fetus

Eicosanoids in the maternal circulation

It is difficult to obtain reliable measurements of concentrations of primary prostaglandins in peripheral plasma because, not only do platelets have the ability to synthesize prostaglandins, but also concentrations of prostaglandins in the peripheral circulation are extremely low due to highly active metabolism by the lungs. These problems can be overcome, to some extent, in experimental animals by the use of chronically implanted vascular catheters in the venous drainage of an organ of interest, e.g. the uterus. In sheep, for instance, there is considerable evidence that prostaglandin concentrations in utero-ovarian venous effluent are raised significantly during the 24 h before delivery. Similar data are now available for many other experimental animals. There is considerable evidence that these prostaglandins are the cause of the onset of labour.

The use of chronic catheterization techniques is obviously excluded in studies of human pregnancy and thus two approaches have been adopted for the evaluation of eicosanoid production when only the peripheral circulation is available for sampling. The first approach adopted was the improvement of assay techniques such that peripheral plasma concentrations of prostaglandin E_2 (PGE_2) and prostaglandin $F_{2\alpha}$ ($PGF_{2\alpha}$) could be measured accurately and be consistent with levels that have been reported by the use of gas chromatography–mass spectrometry techniques. Data obtained using such methods indicate that peripheral plasma concentrations of PGE_2 and $PGF_{2\alpha}$ do not change significantly in women during labour. This finding is consistent with the known high capacity of the lungs to metabolize prostaglandins. Indeed, there is no conclusive proof for any human disease in which circulating levels of primary prostaglandins are elevated.

The second method adopted was the measurement of circulating metabolites of prostaglandins. This approach was suggested because platelets cannot metabolize prostaglandins and circulating levels of metabolites are 10- to 30-fold greater than those of the primary prostaglandins. Hence, generation of prostaglandins during sampling and assay sensitivity were no longer major problems. The major circulating metabolite of $PGF_{2\alpha}$ is 13,14-dihydro-15-keto-prostaglandin $F_{2\alpha}$ (PGFM). The plasma concentrations of PGFM change little during gestation with the possibility of a small rise during the final month of pregnancy; a significant increase in PGFM concentrations does, however, occur during labour. It appears, therefore, that during labour there is an increasing rate of production of $PGF_{2\alpha}$, presumably by tissues within the uterus. This $PGF_{2\alpha}$ is subsequently metabolized efficiently by the lungs such that only the levels of a metabolite can be detected in increased amounts in the peripheral circulation.

Since concentrations of both PGE_2 and $PGF_{2\alpha}$ are raised in amniotic fluid during labour and circulating levels of PGFM increase concomitantly, it might be expected that levels of the comparable major circulating metabolite of PGE_2 would be raised in the peripheral circulation. Unfortunately, 13,14-dihydro-15-keto-PGE_2 is unstable in aqueous media, and this caused many problems in developing radioimmunoassays for its measurement. The problem was overcome when two groups independently established that metabolites of PGE_2 in the peripheral circulation could be converted to the same final product, which is 11-deoxy-13,14-dihydro-15-keto-cyclo-prostaglandin E_2 (PGEM-II). Surprisingly, however, the circulating concentrations of this metabolite are not elevated to any great extent during labour. This finding raised the possibility that the PGE_2 being secreted towards the maternal circulation is metabolized to some other compound. Indeed, it has been found that PGE_2 can be converted to $PGF_{2\alpha}$ by enzymes in the decidua vera. This finding is of importance because PGE_2 is approximately 10-fold more potent than $PGF_{2\alpha}$ in its actions on uterine contractility and cervical softening and effacement. Thus, this conversion is a protective mechanism. It has also been found that prostaglandin D_2 (PGD_2) can be converted to $PGF_{2\alpha}$ by intrauterine tissues of women at term of pregnancy.

Measurements of PGD_2 in plasma are difficult due to its ability to bind to plasma proteins in a covalent manner. Such measurements would be of some considerable interest, however, as PGD_2 is not a good substrate for PGDH and, thus may be a circulating hormone. It has been demonstrated that circulating levels of PGD_2 are similar to those of PGE_2 and $PGF_{2\alpha}$. In sheep, PGD_2 is a dilator of the uterine vasculature whilst being a mild constrictor of the peripheral vasculature. Hence, this prostaglandin may have therapeutic potential in pathological conditions such as pregnancy-induced hypertension.

Obtaining an index of thromboxane A_2 (TxA_2) production by measurements of the degradation product thromboxane B_2 (TxB_2) in the peripheral circulation is also difficult. TxA_2 is produced in large amounts by platelets in response to many stimuli. TxA_2 also combines with plasma proteins in a covalent manner and, thus, it is difficult to draw meaningful conclusions from measurements of TxB_2 in the peripheral circulation. Measurements have been made, however, and to date they are suggestive of little change in TxB_2 concentrations in peripheral plasma throughout gestation or even in labour. The question of TxA_2 production in pregnancy is not trivial since TxA_2 can contract the uterus and, by way of its vasoconstrictive and proaggregatory activities, can play a major role in utero-placental pathology.

There is also considerable controversy concerning the measurement of 6-keto-prostaglandin $F_{1\alpha}$ (6-keto-$PGF_{1\alpha}$), the degradation product of prostacyclin. Initially measurements using radioimmunoassay techniques and gas chromatography–mass spectrometry provided similar results, with peripheral plasma concentrations of approximately 120 to 150 pg/mL. Subsequently, it has been reported that the circulating concentrations of 6-keto $PGF_{1\alpha}$ are much closer to those of PGE_2 and $PGF_{2\alpha}$ (i.e. 2–5 pg/mL). Once again, early studies provided evidence that suggested that little change occurred in the peripheral plasma concentrations of 6-keto $PGF_{1\alpha}$ throughout pregnancy and indeed during labour. Evidence from studies using a highly sensitive technique of mass spectrometry with negative ion chemical ionization detection suggest, however, that there is indeed a significant increase in the plasma concentrations of 6-keto $PGF_{1\alpha}$ in late pregnancy with the possibility of a further elevation in labour.

In sheep, there is evidence that both utero-ovarian venous and cervical venous plasma concentrations of 6-keto $PGF_{1\alpha}$ increase during the 24 to 48 h before delivery. Intrauterine production of prostacyclin is of some importance since in sheep it has been shown that prostacyclin is highly potent in its ability to inhibit uterine contractions. In women, also, inhibitory effects on uterine contractions have been found, although the data are more inconsistent. There is, however, evidence from studies of experimental animals showing that prostacyclin can potentiate the effects of contractile agents such as oxytocin. At present no data are available concerning the plasma concentrations of lipoxygenase products of arachidonic acid metabolism.

Eicosanoids in the fetal circulation

There is considerable evidence that prostaglandins play many physiological roles within the fetus. In particular, it is well established that PGE_2 is the major factor in maintaining the patency of the ductus arteriosus during intrauterine life. The ductus arteriosus is patent throughout intrauterine life, and the lungs receive a relatively small proportion of cardiac output. Thus, this is a situation when circulating prostaglandin concentrations are likely to be high and there is the possibility that prostaglandins may act as circulating hormones in the fetus. Concentrations of PGE_2 and $PGF_{2\alpha}$ have been measured in the circulation of fetal sheep during late pregnancy and labour. Although a slight increase in $PGF_{2\alpha}$ concentrations was shown, there was also a highly significant increase in PGE_2 concentrations during labour. Increased PGE_2 production with labour is of significance, because it is known to inhibit fetal breathing movements, and such movements occur with a decreased frequency during labour. It has been suggested that the absence of fetal breathing movements in women with preterm labour is a predictor of 'true' preterm labour and that delivery will then eventuate. The concentrations of 6-keto $PGF_{1\alpha}$ in ovine fetal plasma are also increased significantly during labour. The physiological significance of this increase in prostacyclin biosynthesis is unknown.

It has been shown that circulating concentrations of prostaglandins in the human fetus are extremely high during the first half of pregnancy. At term, umbilical plasma concentrations of prostaglandins have been described. Concentrations of PGE_2, $PGF_{2\alpha}$, and PGFM in umbilical plasma were all found to be raised above those found in the maternal circulation. Moreover, a significant arteriovenous difference across the umbilical circulation was demonstrated for PGE_2 with venous levels being raised. Thus, it was suggested that the placenta is a major source of the PGE_2 circulating in the fetus. The arteriovenous difference across the umbilical circulation for PGE_2 was found whether labour had occurred or not. It should be noted that this evidence is consistent with findings that circulating PGE_2 concentrations decrease rapidly in the neonatal period. Hence, the placenta may contribute (by way of PGE_2) to the patency *in utero* of the ductus arteriosus; physical separation of the new-born from the placenta in itself will thus contribute to the postnatal closure of this vessel. Obviously, the increased blood flow through the lungs and the concomitant increase in clearance of prostaglandins in the postnatal period play a major role in this mechanism.

The concentrations of 6-keto $PGF_{1\alpha}$ and TxB_2 in umbilical plasma have also been determined. Neither of these prostanoids has a significant arteriovenous difference across the umbilical circulation. Umbilical plasma concentrations of these prostanoids exceed or are similar to circulating concentrations in the mother, and the mode of delivery does not influence measured levels. Measurements of circulating concentrations of arachidonate lipoxygenase metabolites in the fetus have yet to be reported.

Eicosanoids in amniotic fluid

The presence of prostaglandins in amniotic fluid was reported first by Sultan Karim and shortly thereafter he also observed that labour was associated with greatly increased concentrations of prostaglandins in the amniotic fluid of women at term of pregnancy. Although the identity of the specific prostaglandins in amniotic fluid has been questioned, the general precept that concentrations of prostaglandins in amniotic fluid are elevated greatly during labour has been amply confirmed. Amniotic fluid has proven to be a popular fluid in which to measure prostaglandins since there is essentially no biosynthesis or metabolism of prostaglandins in this fluid. Indeed, in numerous studies it has been shown that concentrations of PGE_2 and of $PGF_{2\alpha}$ increase slowly from mid-pregnancy until term and increase sharply with the onset of labour and continue to increase in parallel with cervical dilatation (Fig. 35.4). Since, during labour, amniotic fluid levels of PGFM exhibit changes similar to those of $PGF_{2\alpha}$, it is assumed that the increased $PGF_{2\alpha}$ (and PGE_2) concentrations reflect enhanced rates of prostaglandin biosynthesis rather than reduced rates of catabolism. This assumption has been proved correct since it has been demonstrated that the activities of prostaglandin catabolizing enzymes within intrauterine tissues do not change during labour.

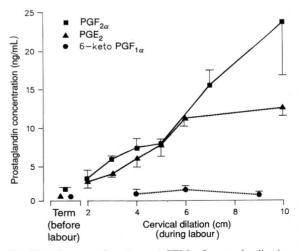

Fig. 35.4 Concentrations (mean ± SEM) of prostaglandins in human amniotic fluid during late pregnancy and labour at term.

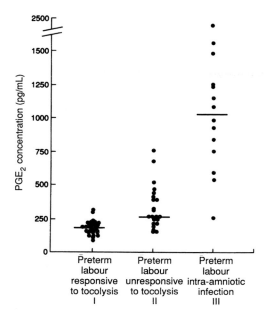

Fig. 35.5 Concentrations of prostaglandin E_2 (PGE_2) in amniotic fluid of women in preterm labour. The line represents the median value of each of the study groups. Significant differences in the distribution of amniotic fluid concentrations of PGE_2 were found among all three study groups. (Taken with permission from Romero, R., Wu, Y.K., Mazor, M. Hobbins, J.C., Mitchell, M.D. (1988). Amniotic fluid prostaglandin E_2 in preterm labour. *Prostaglandins, Leukotrienes and Essential Fatty Acids*, **24**, 141–5.)

Raised levels of prostaglandins have been found during preterm labour although it is possible that, in association with preterm labour, the increase starts a little later and does not approach the full magnitude of that observed with term labour. Recent studies have indicated that the subset of women in preterm labour associated with intrauterine infection have elevated amniotic fluid levels of prostaglandins (Fig. 35.5). This effect is prominent whether membranes are intact or ruptured. It can be seen quite clearly that, early in the course of preterm labour, amniotic fluid concentrations of prostaglandins are elevated only modestly in the absence of infection. In earlier studies these groups were not separated and thus what was reported was a composite picture of prostaglandin levels in amniotic fluid during preterm labour.

Both 6-keto $PGF_{1\alpha}$ and TxB_2, the major hydrolysis products of prostacyclin (PGI_2) and TxA_2 respectively, have been detected in amniotic fluid. Although the concentrations of these products are elevated in labour, there is no trend for levels to rise further as labour progresses (e.g. Fig. 35.4). Hence, it has been suggested that in labour

there is a redistribution in the flow through different prostanoid pathways that favours PGE_2 and $PGF_{2\alpha}$. A similar suggestion has been made on the basis of detailed investigations in pregnant sheep. Hence, labour is characterized by an increasing ratio in the rates of formation of prostaglandins that contract the uterus versus those that are inhibitory or have no action on uterine contractility.

The mechanism(s) responsible for increased prostaglandin concentrations in amniotic fluid during labour are functional prior to term since preterm labour and therapeutic abortions are associated with elevated levels of amniotic fluid prostaglandins. Such mechanisms are highly responsive since prostaglandin concentrations in amniotic fluid can increase within minutes. For instance, vaginal and cervical manipulation can raise amniotic fluid prostaglandin levels within 5 min. It is interesting to note also that a significant proportion of PGE_2 applied to the vagina can reach the amniotic sac after a lag period of a few hours without being degraded. This finding was quite unexpected since it was considered highly unlikely that PGE_2 could diffuse so far without being metabolized or cleared. Thus, some degree of caution is warranted whenever prostaglandins are applied to the vagina or cervix. Whether the prostaglandins in amniotic fluid play any part in the mechanisms of the onset and progression of labour is unknown. Nevertheless, the concentrations of prostaglandins in amniotic fluid are quite an accurate index of important changes during labour, since abnormally low levels of prostaglandin are found in the amniotic fluid of women with clinically delayed labour who are destined to require oxytocin treatment.

The possibility exists that in labour there is a shift in arachidonic acid metabolism away from the lipoxygenase pathways to the cyclooxygenase pathway. Although there may indeed be a relative shift in these pathways, as has been suggested, it is not absolute. This is likely because it has been demonstrated that amniotic fluid concentrations of LTB_4, 15-HETE, and 12-HETE are raised to varying extents during labour at term and preterm in women. Hence, in labour there is an increase in the rate of formation of lipoxygenase pathway products as well as cyclooxygenase pathway products of arachidonic acid metabolism. This is true even allowing for the possibility that 15-HETE in amniotic fluid may originate in a tissue where it is a product of the cyclooxygenase pathway.

Eicosanoids in urine

Measurements of the primary prostaglandins E_2 and $F_{2\alpha}$ in urine reflect production by only the kidney. The major urinary metabolites of PGE_2 and $PGF_{2\alpha}$ in men and women have been determined and methods for their measurement have also been established. It has been found that the concentrations of urinary metabolites of prostaglandins rise during pregnancy and particularly during labour. Thus it appears that there is an increasing rate of secretion of prostaglandins, presumably from a uterine source, throughout pregnancy. The reason why this increasing rate of production is not reflected in measurements of plasma concentrations of metabolites of prostaglandins is uncertain. Recently, there have also been measurements of a major urinary metabolite of prostacyclin. The production rate of this prostacyclin metabolite was again found to be increased with gestation although the effect of labour was not determined. Interestingly, the rate of production of this urinary metabolite of prostacyclin was significantly lower in patients with pregnancy-induced hypertension. There appears to be a consensus of opinion that pregnancy-induced hypertension (or pre-eclampsia) is associated with a deficiency in prostacyclin biosynthesis within the feto-placental unit. A local deficiency in such a potent vasodilator and anti-aggregatory agent could account for many of the features of this pathological condition.

Little is known of the concentrations of prostaglandins in the urine of the fetus. In one study the concentrations of several prostaglandins in fetal urine at term of pregnancy were determined. Fetal urine in this case was obtained as the first voided urine of the new-born which was thus formed *in utero*. Significant concentrations of PGE_2, $PGF_{2\alpha}$, PGFM, and 6-keto $PGF_{1\alpha}$ were found. Thus, it seems that prostaglandins in fetal urine do contribute to the prostaglandins that are found in amniotic fluid. Increased concentrations of these prostaglandins occur during labour although the increase observed is too small to account for a substantial fraction of the increased levels of prostaglandins in amniotic fluid during labour.

Eicosanoid production by intrauterine tissues

The measurement of rates of production of prostanoids (prostaglandins and thromboxanes) by tissues has been difficult. This is the case because trauma (e.g. in obtaining the tissue) is a major stimulus to the production of prostaglandins and, therefore, abnormally high rates of prostaglandin formation may be observed for some time after removal of tissue. Moreover, it should be noted that prostaglandins are not stored in tissues, but are secreted immediately after synthesis. Hence, it is difficult to interpret measurements of prostaglandins or other eicosanoids in tissues. Rather, we must use methods for the measurement of prostaglandin production under dynamic conditions. Several methods have been used successfully to

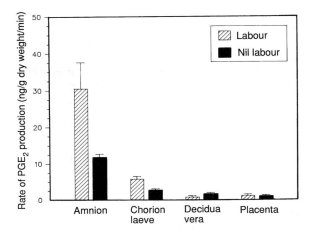

Fig. 35.6 Rates of production (mean ± SEM) of prostaglandin E_2 by intrauterine tissues obtained before and after spontaneous labour and delivery in women.

obtain such measurements. These methods include the use of cells separated from tissues and maintained either in monolayer culture or in suspension; alternatively, whole tissue fragments have been maintained in organ culture. The most widely used technique is the incubation of a homogenate of a tissue, or a subcellular fraction thereof, with precursor (radiolabelled or otherwise). Another popular method has been that of tissue superfusion. This technique permits the prostaglandins produced by the trauma of excision to be washed away and basal production rates of prostaglandins may then be established.

Using such a method of superfusion, the rates of production of a variety of prostanoids by intrauterine tissues obtained either before or after labour have been established (e.g Fig. 35.6). Surprisingly, it has been found that amnion, the inner fetal membrane, is a major site of the production of prostaglandins and specifically of PGE_2. Moreover, the rate of production of PGE_2 by amnion is increased during labour. These findings have been confirmed and extended by several groups, one of which also found evidence for increased production of prostaglandins during labour by chorion laeve and decidua vera.

In sheep there are increased rates of production of prostaglandins not only by fetal membranes but also by maternal cotyledon and myometrium. Evidence has been presented recently that the myometrium of women at term has increased activities of prostaglandin endoperoxide synthase and prostacyclin synthase. Furthermore, it has been demonstrated that the production of prostacyclin in myometrium is not confined to vascular cells, but is also found in smooth muscle cells. As mentioned earlier, although prostacyclin inhibits uterine activity in sheep, evidence for its inhibitory activity in women is at present more equivocal.

A key factor in the normal progression of labour and parturition is the softening and effacement of the uterine cervix. Cervical tissue can produce prostaglandins at high rates. In both human and ovine pregnancy there is evidence that the uterine cervix produces increased quantities of prostaglandins during the period of cervical softening, effacement and dilatation. PGE_2 is used clinically for cervical ripening prior to the induction of labour and abortion, and is a highly effective and successful treatment. Thus, the concept has been put forward that cervical ripening is a function of prostaglandins formed and acting locally. The exact mechanisms by which prostaglandins and particularly PGE_2 influence cervical ripening are unknown. However, it has been shown that prostaglandins can enhance collagenolytic activity in the cervix and also have significant effects on the proteoglycan composition of the cervix.

What of changes in eicosanoid production by intrauterine tissues taken after preterm labour? A series of studies from a British group have provided answers to this question. They have found that PGE_2 production by amniotic cells obtained after preterm labour is significantly lower than in cells from pregnancies ending in term labour. No such difference existed when similar comparisons were made of PGE_2 production by chorionic and decidual cells obtained from the same groupings. In subsequent studies it was suggested that amnion and choriodecidua obtained after preterm labour associated with chorioamnionitis produced greater amounts of prostaglandins than the same tissues obtained after preterm labour in the absence of infection. Further studies have shown that rates of production of leukotrienes C_4, D_4 and E_4 by intrauterine tissues are unaffected by preterm labour although production of leukotriene B_4 by placental tissue is increased during preterm labour associated with chorioamnionitis. Thus, only in the presence of infection have tissues obtained after preterm labour been shown to have increased rates of eicosanoid production.

Regulation of the biosynthesis of eicosanoids within the uterus

Defined substances and mechanisms

The relative abundance of arachidonic acid in uterine tissues is higher than in many other tissues; it accounts for 7–20 per cent of the non-esterified fatty acids. Concentrations of non-esterified arachidonic acid in amniotic fluid increase severalfold during labour. This increase

exceeds those of other fatty acids and indicates a degree of specificity in the mobilization of arachidonic acid. It has been shown that there is a significantly lower arachidonic acid content in fetal membranes (amnion and chorion) from women in labour compared with women not in labour. Subsequently, a more comprehensive study compared the arachidonic acid contents of amnion (and chorion) from women not in labour and women in early labour (cervix less than 4 cm dilated). The results of this study indicated that, before labour, arachidonic acid constitutes 14 per cent of the total fatty acids in amnion and, after labour, this value decreases to 10 per cent. Separation of the total glycerophospholipids of amnion into individual classes revealed a significant change with early labour in the arachidonic acid content of (diacyl)phosphatidylethanolamine and phosphatidylinositol. Similar changes have been observed in chorion, although the magnitudes of the changes are less in this tissue. It has been calculated that this release of arachidonic acid from fetal membranes is sufficient to account for all the prostaglandins formed during the onset of labour.

How is the release of arachidonic acid from these glycerophospholipids controlled? Regulation of the activities of phospholipases A_2 and C is an obvious method. In amnion there is a phospholipase A_2 that is calcium-dependent and has a substrate preference for phosphatidylethanolamine that contains arachidonic acid. This is consistent with the release of arachidonic acid from phosphatidylethanolamine in amnion during labour. There is also a phosphatidylinositol-specific phospholipase C in amnion that is calcium-dependent. Moreover, diacylglycerol and monoacylglycerol lipase activities have been detected in amnion. This is consistent with the release of arachidonic acid from phosphatidylinositol during early labour. The simplest mechanism for arachidonic acid release (and hence prostaglandin generation) in labour would be increased activities of the enzymes described above. Assay of the specific activities of these enzymes, under optimal conditions, at term before and after labour, however, has not revealed a significant change in the specific activities of the enzymes in uterine tissues.

The optimal conditions used in the assay of phospholipase activities has included concentrations of Ca^{2+} that are far greater than those found in the cytosol of amnion cells under normal conditions. Thus, it is possible that, *in vivo*, changes in the activities of phospholipases could occur that account for the mobilization of arachidonic acid. Such changes could result from the action of a regulatory factor that either raises the cytosolic concentration of Ca^{2+} in amnion cells or increases the sensitivity of the phospholipases to Ca^{2+}. It has been suggested that both basal and stimulated prostaglandin production by amnion cells are

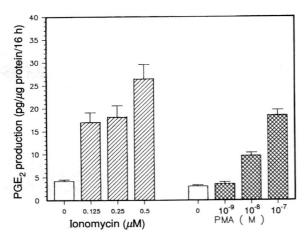

Fig. 35.7 The effects of various concentrations of ionomycin (a calcium ionophore) and of phorbol 12-myristate 13-acetate (PMA, an activator of protein kinase C) on prostaglandin production (mean ± SEM, n=4) by human amnion cells.

extremely dependent upon extracellular calcium. In these studies, not only a calcium channel blocker, but also trifluoperazine (a calmodulin antagonist) had significant inhibitory effects on prostaglandin biosynthesis. These findings have been confirmed using the calcium channel blocker nifedipine and further have shown that raising intracellular calcium concentrations by the use of calcium ionophores will enhance amnion prostaglandin production (Fig. 35.7).

Several substances have been identified in amniotic fluid that could influence intracellular Ca^{2+} concentrations in the amnion. For instance, mean concentrations of 1,25-dihydroxyvitamin D_3 in amniotic fluid are increased during late gestation and this hormone is known to affect Ca^{2+} fluxes. Another intriguing substance that has the properties of a calcium ionophore and is found in amnion cell culture media and in amniotic fluid is LTB_4. Intriguingly, LTB_4 may actually be synthesized in amnion and, hence, may have an autocrine action. Perhaps the most likely candidate for the role of Ca^{2+} regulator in amnion at term is platelet-activating factor (PAF; 1-O-alkyl-2-acetyl-Sn-glycero-3-phosphocholine). PAF is known to cause a rapid increase in the cytosolic concentration of Ca^{2+} in platelets; this results in activation of phospholipases, release of arachidonic acid, and generation of prostaglandins. It has also been suggested that PAF may induce the biosynthesis of lipoxygenase products of arachidonic acid metabolism, which may act as mediators of its actions. PAF has been identified in the amniotic fluid of women in labour at term, but is not found in the liquor

of women at term but not in labour. Moreover, the concentration of PAF in amnion tissue from women in labour at term is two to threefold greater than that of women at term but not in labour. Finally, PAF acts on fresh amnion tissue to increase the rate of prostaglandin biosynthesis substantially. Hence, at present PAF best fulfils the requirements for a regulator of cytosolic Ca^{2+} concentrations in amnion at term that may result in the cascade of events of parturition.

It should be noted that Ca^{2+} may also have effects on arachidonic acid metabolism that are independent of actions on phospholipase activities. For instance it may act via the phospholipid-requiring, Ca^{2+}-dependent protein kinase known as protein kinase C. Protein kinase C has been implicated in the control of thromboxane A_2 biosynthesis in platelets by way of an action on arachidonic acid mobilization. Furthermore, amnion tissue has been shown to have protein kinase C activity. The possibility that protein kinase C may play a part in the release of arachidonic acid for eicosanoid biosynthesis at term was strengthened by the findings that phorbol esters (which activate protein kinase C) stimulate PGE_2 production in human amnion cells (Fig. 35.7).

It has been suggested, however, that phorbol esters may also have actions that cause induction of cyclooxygenase protein. The effects of protein kinase C on prostaglandin biosynthesis may be of more importance than originally thought. Recent studies suggest that activation of protein kinase C may be critical for the stimulatory actions of several substances on prostaglandin biosynthesis in amnion and other cell types. There is some disagreement, however, over the degree of involvement of protein kinase C in such stimulatory actions with respect to amnion. The results of studies in which inhibitors of protein kinase C action were used suggest that protein kinase C activity is necessary for the stimulatory actions of epidermal growth factor (EGF) and oxytocin on amnion prostaglandin biosynthesis. Results from studies in which protein kinase C is inactivated by prolonged exposure to phorbol esters have not indicated any effect on stimulatory actions of these substances. More recent studies using the latter technique have suggested that some degree of the stimulatory action of EGF and interleukin 1β (IL-1β) on amnion prostaglandin biosynthesis is dependent upon protein kinase C activation. This is an area of significant importance for the future and perhaps the specificity of the agents and treatments used may need to be enhanced for a clearer picture to emerge.

There is an ever-increasing number of hormones that may have actions to increase the mobilization of arachidonic acid, possibly via activation of the phosphatidylinositol cycle. This effect can be of varying importance in the known actions of this group of hormones. Examples of these hormones include oxytocin, bradykinin, vasopressin, angiotensin II, catecholamines, and possibly even PGE_2. Conversely, it has been suggested that relaxin and glucocorticosteroids have inhibitory actions on phospholipase activities. In the case of glucocorticosteroids this inhibitory action is thought to be mediated by way of induction of the biosynthesis of antiphospholipase proteins now known as lipocortins. These proteins have been sequenced and the potential for their production by uterine tissues has been established.

The exact mechanism of action of lipocortins is not clear. One hypothesis states that lipocortins work by binding directly to phospholipase A_2 and inhibiting interaction with its substrate. Others believe lipocortins work by binding to and sequestering the substrate itself. The literature varies as to whether these proteins are secreted or cytosolic molecules. It has been reported that a lipocortin is present in endometrial cell-conditioned medium and that the levels of this lipocortin are increased by treatment with dexamethasone. This increase correlates with a decrease in the production of $PGF_{2\alpha}$ by the endometrial cells. Extracts of amnion and chorion laeve obtained at elective Caesarean section stain positive for lipocortins I and II in Western blot assays. Dexamethasone (10^{-5} to 10^{-7} M) treatment of amnion induces lipocortin formation and paradoxically also induces an increase in PGE_2 production. This effect may or may not be receptor-mediated and could involve induction of cyclooxygenase protein. Recent studies demonstrate that intrauterine tissues may synthesize at least five lipocortins. The physiological relevance of these substances remains ill-defined.

Non-characterized substances

Substances that have actions on phospholipase activities but that have yet to be fully characterized have been reported to be present within the uterus. Amniotic fluid contains a substance that can stimulate fibroblast phospholipase A_2 activity. Moreover, this action can be enhanced by the presence of soluble constituents of meconium from the fetus. Information on the characteristics of these substances remains limited, but they have obvious potential for a role in the regulation of arachidonic acid mobilization by amnion as well as possible roles in pathologies such as meconium aspiration syndromes. Recently, two substances have been found in amniotic fluid that inhibit phospholipase activities of endometrial cells. These substances have apparent molecular weights of 150–165 and 70–80 kDa. Chorion has been shown to be the major source of the smaller-molecular-weight inhibitory substance; the activity of this substance, named

gravidin, has been demonstrated to be reduced in chorions obtained after the onset of labour. The reduction in gravidin activity with labour could constitute the equivalent of the release of a brake on arachidonic acid mobilization. Although chorion is the major source of gravidin and is a source of lipocortins, it has also been shown to be a source of activity that stimulates protaglandin biosynthesis by amnion. Chorion produces a range of prostaglandins and has a 'normal' response to glucocorticoids in that prostaglandin biosynthesis is inhibited in a concentration-related manner with 50 per cent inhibition occurring at approximately 10^{-9} M. One report has suggested that prostaglandin output from chorion is increased with labour although this has not been observed in several other reports.

An endogenous inhibitor of prostaglandin synthase (EIPS) has been shown to be present in amniotic fluid. Work with bovine seminal vesicle (BSV) prostaglandin synthase shows that EIPS activity in amniotic fluid decreases towards term and particularly in labour. Fractionation has revealed that EIPS has a molecular weight of more than 30 kDa. Prostaglandin synthase inhibiting activity has also been described in ovine allantoic fluid. This inhibitory activity differs from EIPS in that it is a small-molecular-weight molecule that is not heat-labile and is insensitive to protease treatment. A similar inhibitory activity has been reported to be present in bovine placental tissue.

The source of EIPS in human amniotic fluid is, as yet, uncertain. EIPS-like activity has been suggested to be present in amnion at term before the onset of labour. This activity was reported to be completely absent in amnion obtained after labour although it has recently been shown to be present in amnion and extravillous trophoblast throughout gestation. The possibility has also been evaluated that amnion is a source of prostaglandin-inhibitory substances. In these experiments, conditioned media (i.e. culture media removed from cells maintained *in vitro*) from amnion obtained at elective Caesarean section or after spontaneous labour were placed on endometrial cells in culture. The conditoned media from amnion obtained after spontaneous labour had no effect on endometrial prostaglandin production, but conditioned media from amnion obtained at Caesarian section inhibited the production of both PGE and PGF by human endometrial cells. These studies did not determine whether the amnion-conditioned media acted at the level of substrate release or conversion to prostaglandin and it is not clear whether this activity is the same as the EIPS activity discussed above.

Amnion production of PGE_2 has been shown to be increased with labour. The production of lipoxygenase products of arachidonic acid metabolism by amnion does not seem to increase significantly with labour although more sensitive methods will be required to rule out this possibility. Nevertheless, with data presently available it seems reasonable to state that the ratio of cyclooxygenase to lipoxygenase products increases with labour.

The association between intra-amniotic infection and preterm labour

There are many reports of a relatively strong association between intra-amniotic infections and the incidence of preterm labour. In recent studies that have been reviewed in detail the mean rate of positive amniotic fluid cultures in samples obtained by amniocentesis from women in preterm labour with intact membranes was 16.1 per cent. The group of women with positive amniotic fluid cultures, when compared with those having negative cultures, were more frequently refractory to tocolysis (i.e. pharmacological inhibition of uterine contractions); in addition, they were at greater risk of spontaneous rupture of membranes and subsequently developing chorioamnionitis. It should be noted that intrauterine infection is also strongly associated with premature rupture of membranes and indeed the prevalence of intra-amniotic infection is higher in women with premature rupture of membranes than in women with preterm labour and intact membranes. Moreover, women admitted with preterm labour and premature rupture of membranes have a significantly greater incidence of intra-amniotic infection than those admitted with premature rupture of membranes but not in labour. Histological evidence of chorioamnionitis is correlated with intrauterine infection but should not be taken as direct evidence of this condition. The relationship between histological chorioamnionitis and preterm delivery has been reviewed extensively and, despite many variations between results of published studies, it can be observed that there is a higher incidence of chorioamnionitis in women delivering preterm. It is noteworthy that women in preterm labour who have failed tocolysis have a greatly increased incidence of histological chorioamnionitis than women who respond to tocolysis. Multiple organisms have been isolated from the amniotic fluid of women with preterm labour. Although some of these organisms are found relatively frequently under these conditions, none has been proven to be a unique marker for, or to have a causal role in, the initiation of preterm labour. In summary, there is a strong association between intra-amniotic infection and preterm labour although the precise aetiological factors and causal mechanisms remain to be determined.

Regulation of the biosynthesis of eicosanoids associated with preterm labour

Potential mechanisms by which intra-amniotic infections lead to preterm labour

Prostaglandins are considered to play a significant part in the mechanism(s) of the onset and progression of labour both at term and preterm. Raised levels of prostaglandins have been found during preterm labour although, as mentioned previously, it is possible that the increase starts a little later with preterm labour and does not approach the full magnitude of that observed with term labour. Recent studies have indicated that the subset of women in preterm labour associated with intrauterine infection have elevated amniotic fluid levels of prostaglandins. This effect is prominent whether membranes are intact or ruptured. It can be seen quite clearly that early in the course of preterm labour amniotic fluid concentrations of prostaglandins are elevated only modestly in the absence of infection. An intriguing question arises from consideration of these results. Does the modest change in amniotic fluid concentrations in the absence of infection reflect a change in intrauterine prostaglandin production sufficient to account for the increased uterine activity? If the answer is no then we must search for new contractile agents (e.g. endothelins) to evaluate as potential initiators of preterm labour of unexplained aetiology. As yet there is less reason to doubt the importance of prostaglandins in the mechanism(s) of preterm labour associated with intrauterine infection.

This section will evaluate briefly the possible mechanisms by which bacteria may enhance prostaglandin production within the uterus and thus cause preterm labour. A simplified outline of such mechanisms is provided in Fig. 35.8. The simplest theory would be that the bacteria

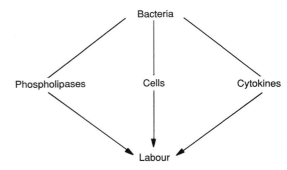

Fig. 35.8 Pathways by which intra-amniotic infection could induce preterm labour.

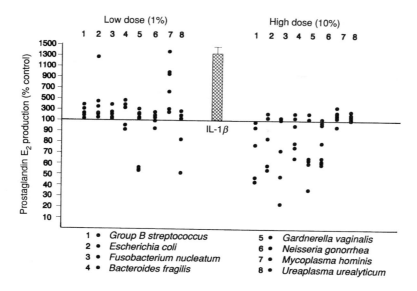

Fig. 35.9 The effects of bacterial products and interleukin-1β (IL-1β) on prostaglandin E$_2$ (PGE$_2$) production by human amnion cells. At low doses of bacterial products PGE$_2$ production tended to be increased whereas at high doses it tended to be inhibited. IL-1β consistently stimulated PGE$_2$ production. Significant stimulation of amnion PGE$_2$ production was observed with *Group B streptococcus*, *Escherichia coli*, *Fusobacterium nucleatum*, and *Mycoplasma*. (Taken with permission from Mitchell, M.D., Romero, R.J., Avila, C., Foster, J.T., and Edwin, S.S. (1991). Prostaglandin production by amnion and decidual cells in response to bacterial products. *Prostaglandins, Leukotrienes and Essential Fatty Acids*, **42**, 167–9.)

colonizing the amniotic fluid are a direct source of prostaglandins. No study, however, has yet been able to demonstrate the production of prostaglandins by such bacteria. Alternatively, it has been shown that some bacteria produce the enzyme phospholipase A$_2$, which could act on amnion or other intrauterine tissues to release arachidonic acid from glycerophospholipid stores of cellular membranes. Thus the bacteria would cause the release of substrate for prostaglandin biosynthesis by these tissues. It should be noted that substrate availability is widely considered to be the rate-limiting step in prostaglandin biosynthesis. This latter theory is unlikely to account completely for the mechanism of infection-driven preterm labour. For instance, it has been shown that intracellular or extracellular products from *E. coli* and *S. faecalis* induce an increase in amnion cell PGE$_2$ biosynthesis in a manner similar to that of purified phospholipase A$_2$. In this study, however, bacterial products seemed capable of stimulating PGE$_2$ production to an extent greater than that of phospholipase A$_2$ alone. Furthermore, it has been demonstrated that endotoxin (devoid of phospholipase A$_2$ activity) can stimulate amnion PGE$_2$ biosynthesis.

The complexity of the effects of bacterial products on amnion arachidonic acid metabolism was demonstrated in recent studies showing that such products can cause an increase in the production of both lipoxygenase and cyclooxygenase pathway metabolites. A disproportionate increase in the production of cyclooxygenase-pathway products was noted in the study. Another recent study emphasized that there is great complexity in the direct actions of bacterial products on intrauterine prostaglandin production. The results of this study evaluating the actions of multiple doses of secretory products from several bacteria that are commonly associated with intrauterine infections are presented diagrammatically in Fig. 35.9. It can be seen that there is wide variation between types of bacteria, both quantitatively and qualitatively, in their effects on amnion and decidual prostaglandin production. Moreover, there is a trend for a biphasic dose–response relationship to the bacterial products, with low doses enhancing prostaglandin production (as found by others) and high doses actually inhibiting prostaglandin production. It has been speculated that inhibition of amnion (and decidual) PGE$_2$ production by high doses of bacterial products may contribute to the poor course of labour observed under these conditions. It is obvious that the direct effects of bacteria on intrauterine prostaglandin production are complex and deserving of more detailed investigation with respect to the individual organisms, their relative concentrations, and the interplay between them.

Another potential mechanism for infection-driven preterm labour is mediation via cytokines. The theory has been advanced that the host response to infections, specifically generation of cytokines, may be fundamental to this mechanism. Cytokines are important mediators of inflammatory responses and can stimulate prostaglandin biosynthesis in several cell types (see next section). There is considerable evidence for increased cytokine generation during preterm labour associated with intrauterine infection as reflected by elevated amniotic fluid levels of IL-1β, tumour necrosis factor, and IL-6.

Cytokine regulation of prostaglandin biosynthesis

The cytokines, IL-1 and tumour necrosis factor, are known to stimulate prostaglandin biosynthesis in a range of cell types. It has been demonstrated that both of these cytokines enhance the rate of production of PGE$_2$ by human amnion cells (Fig. 35.10) and human decidual cells. In addition, in recent studies it has been found that IL-6 treatment of human amnion and decidual cells also results in increased prostaglandin production. The concentrations of these cytokines required for such stimulatory actions are within the ranges measured in amniotic fluid of women with intra-amniotic infections.

The mechanism(s) whereby cytokines stimulate prostaglandin biosynthesis have been the subject of many studies although most have addressed the actions of IL-1. There is evidence that IL-1 can induce fatty acid cyclooxygenase protein and, recently, it has been shown that it may

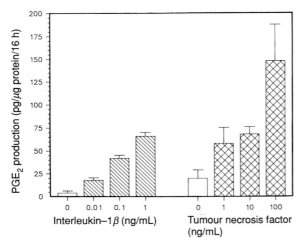

Fig. 35.10 The effects of various concentrations of IL-1β and tumour necrosis factor on prostaglandin production (mean ± SEM, $n = 4$) by human amnion cells.

also, on occasion, have actions on substrate (arachidonic acid) availability. Similarly, tumour necrosis factor has been suggested to directly stimulate arachidonic acid release in human neutrophils. Recent findings suggest that IL-1 actions on amnion are independent of exogenous substrate and are most likely directed at induction of fatty acid cyclooxygenase protein. It has recently been observed that the actions of substances that stimulate prostaglandin production may, at times, be dependent upon protein kinase C activity. It now seems unlikely that protein kinase C activation is necessary for IL-1 stimulation of amnion PGE$_2$ biosynthesis. Nevertheless, a proportion (approximately half) of the stimulatory action of IL-1β does seem to require protein kinase C activation. Clearly, the major action of IL-1β in stimulating prostaglandin production is directed at the induction of fatty acid cyclooxygenase protein and, at present, other effects appear secondary. It is of interest, however, that the time courses of these different mechanisms are quite different and this may prove to be of some consequence to the development of uterine contractions that proceed to delivery or are responsive to tocolytic therapy. Other mechanisms linking cytokines to preterm labour or abortion obviously exist. For instance, both IL-1 and tumour necrosis factor can cause placental injury in rats. Nevertheless, the actions on prostaglandin biosynthesis are likely to be amongst the most significant. Many actions of cytokines are dependent upon activation of protein kinase C or cAMP-dependent protein kinase A and thus activation of prostaglandin biosynthesis via these mediators may be a secondary event.

IL-8 or neutrophil-activating protein 1 is a 72-amino-acid protein that is produced in response to inflammation by a number of cell types such as monocytes, fibroblasts, and endothelial cells. It has actions as a chemoattractant and activator of neutrophils. In other cell types IL-8 has been shown to enhance arachidonate 5-lipoxygenase activity although not the release of cellular arachidonate. These results emphasize the often observed interplay between cytokines and reinforce how difficult it may be to isolate any one specific substance or action that could be considered of key importance among the many interacting events of preterm labour. Nevertheless, IL-8 must be regarded as an addition to the cytokines that may play a significant part in the mechanism of (preterm) labour.

There is also evidence favouring an inhibitory action of cytokines on prostaglandin production. For instance, it has been shown that IL-4 may inhibit prostaglandin production by monocytes, although its action may have been more directed at suppressing the generation of other stimulatory cytokines. In recent years the early pregnancy factor(s), ovine trophoblastic protein and bovine trophoblastic protein, have been isolated and shown to suppress prostaglandin production particularly in response to oxytocin. These factors are now known to have high homology with the cytokine family of α-interferons. Interestingly, interferon-α can enhance β-human chorionic gonadotrophin expression and thus have progestational activity. The concept has been proposed that pregnancy is maintained by tonic suppression of prostaglandin biosynthesis within the uterus and that the onset of labour requires at least partial withdrawal of such inhibition. Whether cytokines could play a part in this type of hypothesis remains uncertain. Nevertheless, it is tempting to consider the proposition that labour may commence by virtue of a redirection of cytokine secretion or action away from inhibitory (IL-4, α-interferons) towards stimulatory (IL-1, IL-6, tumour necrosis factor) cytokines.

Potential role of mediators of immune responses in preterm labour

If we accept that altered immune function can lead to preterm labour via cytokine secretion or other mechanisms then what immunomodulatory substances can be identified that may be of importance? As a massive number of characterized and non-characterized substances exist, only a few examples of potentially important substances will be mentioned.

One major modulator of immune function is PGE_2 itself, although there has been uncertainty concerning the mechanism of its immunosuppressive activity. It is clear that PGE_2 inhibits IL-2 production but conflicting results have been reported regarding actions on IL-2 receptor expression and induction of suppressor T lymphocytes. Another action of PGE_2 is to downregulate transferrin receptor expression. It is interesting to note that PGE_2 can not only inhibit both proliferation and production of granulocyte–macrophage colony-stimulating factor (GM-CSF) by murine T_H clones stimulated with antigen (AG) or anti CD3 (a cell surface marker) antibody but can also act synergistically with IL-2 for the induction of GM-CSF in some T_H1 clones.

Products of arachidonic acid metabolism by way of the lipoxygenase pathway are also known to alter immune function. LTB_4 and 15-HPETE both exhibit immunoregulatory activity and there are significant changes in amniotic fluid concentrations of these eicosanoids with preterm labour (Fig. 35.11). Initially, it was shown that LTB_4 enhanced the proliferation of mitogen-stimulated purified suppressor cytotoxic T lymphocytes; similar experiments defined the enhanced proliferation of the subset of cells as being OKT8 positive and a population that is inhibited as OKT4 positive (OKT4 and OKT8 being monoclonal antibodies). It has been shown also that 15-HPETE exhibits similar activities to LTB_4. Furthermore, LTB_4 may act synergistically with IL-2 in promoting T-cell proliferation.

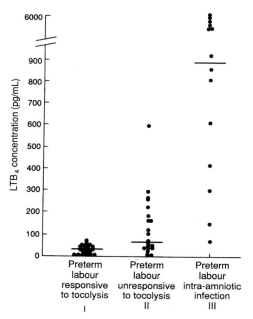

Fig. 35.11 Concentrations of leukotriene B_4 (LTB_4) in amniotic fluid of women in preterm labour. Horizontal bars represent the median values. Significant differences in the distribution of amniotic fluid concentrations of LTB_4 were found among all three study groups. (Taken with permission from Romero, R., Wu, Y.K., Mazor, M., Oyarzun, E., Hobbins, J.C., and Mitchell, M.D. (1989). Amniotic fluid arachidonate lipoxygenase metabolites in preterm labor. *Prostaglandins, Leukotrienes and Essential Fatty Acids*, **36**, 69–75.)

The cytokines IL-2 and IL-4 play critical parts in the development of cell-mediated immunity and humoral immunity respectively. IL-2 and IL-4 production may also be modulated by steroid hormones elaborated within the feto-placental unit. In recent studies it has been demonstrated that dehydroepiandrosterone can stimulate IL-2 production by helper T cells. Glucocorticoids can have the opposite effect and also may enhance IL-4 secretion. The antigestational agent RU486, which can have actions on both the progesterone and glucocorticoid receptors, also has unique actions on the relative production of IL-2 and IL-4. It would be interesting to determine whether this action has any role in the abortifacient properties of this substance. Thus, in the microenvironment of the feto-placental unit and particularly the chorio-decidual interface, it is likely that interleukin production is significantly different from that observed peripherally, as suggested recently.

Finally, autoimmune disease may eventually be linked to the study of labour (term and preterm). Systemic lupus erythematosus and antiphospholipid syndrome are both associated with utero-placental insufficiency due to decidual vasculopathy and decidual and placental infarction. The spectrum of the resulting clinical problems ranges from early pregnancy loss to fetal growth impairment, fetal distress, and fetal death. It is possible, but unproven, that the associated early pregnancy losses may be due to an immune-mediated initiation of premature uterine activity. But fetal death is of more interest in regard to preterm labour. In sheep, fetal death regardless of cause leads to frank labour within a short time span. However in human, autoimmune fetal death associated with widespread decidual infarction (as well as other categories of fetal death) may precede the onset of labour by several days to several weeks. Is this because the decidua, perhaps a primary initiator of the onset of labour in man, is severely damaged and non-functional? Could such a model of intact pregnancy with significant decidual damage be developed in an experimental animal?

Influence of drug treatments for preterm labour on eicosanoid production

The strong link between increased prostaglandin production and parturition might suggest that treatments for preterm labour should result in reduced prostaglandin output, either directly or indirectly. Hence the effects of three substances that have been used to treat preterm labour (i.e. β-sympathomimetics, magnesium sulfate, and ethanol) will be evaluated. Interestingly, the evidence in favour of an inhibitory action of these substances on prostaglandin production is relatively limited. The preponderance of data suggests that these substances actually enhance prostaglandin production. There is strong evidence that β-adrenergic agonists stimulate prostaglandin production in reproductive tissues both *in vitro* and *in vivo*. Magnesium sulfate has been shown to increase prostacyclin output by human umbilical vein endothelial cells. The urinary output of prostacyclin and thromboxane metabolites is significantly greater in pregnant women that consume alcohol than in those who abstain. Similar results have been reported in rats that have been treated chronically with ethanol. In addition, uterine tissues from pregnant mice treated with ethanol show enhanced production of prostaglandins. These seemingly odd results will require further examination before a clear picture can emerge of the relationship between agents used to suppress preterm labour on their activities on prostaglandin production. It will be necessary to determine specific sites of prostaglandin biosynthesis that are affected as well as the range of prostaglandins secreted. Also it cannot be forgotten that prostacyclin is widely considered to inhibit uterine activity and that ethanol can enhance the activity of prostaglandin catabolizing activities.

Further reading

Garfield, R.E. (ed) (1990). *Uterine contractility.* Serono Symposia, Norwell, Massachusetts.

Hammarstrom, S. (1983). Leukotrienes. *Annual Reviews of Biochemistry*, **52**, 355–77.

McNellis, D., Challis, J.R.G., MacDonald, P.C., Nathanielsz, P.W., and Roberts, J.M. (ed.) (1988). *The onset of labour: cellular and integrative mechanisms.* Perinatology Press, Ithaca, New York.

Mitchell, M.D. (1982). Prostaglandin synthesis, metabolism and function in the fetus. In *Biochemical development of the fetus and neonate* (ed. C.T. Jones), pp. 425–71. Elsevier, Amsterdam.

Mitchell, M.D. (ed.) (1990). *Eicosanoids in reproduction.* CRC Press, Boca Raton, Florida.

Romero, R., Mazor, M., Wu, Y.K., Sirtori, M., Oyarzun, E., Mitchell, M.D., and Hobbins, J.C. (1988). Infection in the pathogenesis of preterm labour. *Seminars in Perinatology*, **12**, 262–79.

Samuelsson, B. and Funk, C.D. (1989). Enzymes involved in the biosynthesis of leukotriene B_4. *Journal of Biological Chemistry*, **264**, 19469–72.

Samuelsson, B., Goldyne, M., Granstrom, E., Hamberg, M., Hammarstrom, S., and Malmsten, C. (1978). Prostaglandins and thromboxanes. *Annual Reviews of Biochemistry*, **47**, 997–1029.

Smith, W.L. (1989). The eicosanoids and their biochemical mechanisms of action. *Biochemical Journal*, **259**, 315–24.

36. The maternal endocrine system during pregnancy and parturition
Graham C. Liggins

Pregnancy

From the point of view of an endocrinologist, pregnancy consists of the addition to the complement of endocrine organs in a non-pregnant woman of a massive new endocrine organ, the placenta. Although the tissues of the placenta that secrete hormones (the syncytiotrophoblast and cytotrophoblast) are the fetal origin, the placental hormones are secreted mainly into the maternal circulation where they have the function of modifying the maternal physiological systems to the advantage of the growing conceptus. It is often not appreciated that, with the exception of the mechanical effects of enlarging uterus, every other familiar maternal change in pregnancy including those in the musculoskeletal system, the gastrointestinal tract, the cardiovascular system, the haemopoietic system, the liver, the breasts, and the kidneys is a response to hormones generated by the conceptus, particularly oestrogen and progesterone. Indeed, a psuedopregnancy that reproduces most of these changes can be induced in the non-pregnant woman by the administration of oestrogen and progesterone in large doses but even the small doses contained in oral contraceptives may cause pregnancy-like side-effects on body weight, breast development, liver function, kidney function, and the central nervous system. In addition to these widespread changes in the maternal physiology induced by pregnancy hormones, there are marked changes in the maternal endocrine system that include every endocrine organ.

Corpus luteum

The first evidence of an endocrine function of the human conceptus is the failure of the corpus luteum to regress 12 days after ovulation and the consequent absence of menstruation. The corpus luteum is maintained by chorionic gonadotrophin (hCG), a glycoprotein secreted by the trophoblast that is detectable in the maternal blood stream within 7 days of conception. The secretion of hCG reaches a peak at about 60 days of pregnancy at a time when the placental production of progesterone has reached a level capable of maintaining pregnancy; removal of the corpus luteum before 7 weeks of pregnancy causes abortion but removal after 7 weeks results in continuing pregnancy. Non-human species vary in the duration of dependence on the corpus luteum, ranging up to dependence to term. In some, a chorionic gonadotrophin has been identified and in others the luteotrophic factors are more complex and include oestrogens, luteinizing hormone (LH), prolactin, and prostaglandin E_2.

Pituitary

The pituitary gland increases progressively in size through pregnancy mainly due to hyperplasia of large, chromophobe prolactin cells. Apart from prolactin synthesis, anterior pituitary function is inhibited by placental hormone production. The high plasma concentrations of oestrogen and progesterone suppress follicle-stimulating hormone (FSH) and LH production. Pituitary growth hormone levels become undetectable as the result of suppression by a steady increase in an immunologically cross-reacting growth hormone variant secreted by the placenta. Plasma concentrations of thyroid-stimulating hormone (TSH) are usually in the low, normal range due to inhibitory feedback by hCG which has intrinsic thyrotrophic activity. The levels of corticotrophin (ACTH) are within the normal non-pregnant range throughout pregnancy despite high circulating concentrations of placental corticotrophin-releasing hormone (CRH), probably because of the presence of a CRH-binding protein. The extent to which placental ACTH or placental pro-opiomelanocortin (POMC) contribute to plasma ACTH levels is uncertain.

The posterior pituitary hormones, oxytocin and arginine vasopressin, show no consistent changes through pregnancy until labour is established, when pulsatile release of both hormones begins; pulse amplitude and frequency reach maximal values during the expulsive phase of labour.

Thyroid

Pregnancy affects most aspects of maternal thyroid function. A relative iodine deficiency is often present in women because of increased renal clearance of iodine. Total thyroxine (T_4) and triiodothyronine (T_3) concentrations in maternal plasma increase within the first month of pregnancy and remain high throughout pregnancy. These increased concentrations are attributable to a rise in thyroxine-binding globulin (TBG) concentration due to the stimulatory effects of oestrogen on synthesis in the liver. Despite the relatively high concentrations of TBG-bound T_4, the concentrations of unbound (free) T_4 and T_3 are within the normal range. Thyroid-stimulating hormone (TSH) is modestly decreased during the first trimester probably because hCG or a fragment thereof has intrinsic thyroid-stimulating activity. The thyroid hormones traverse the placenta in very limited amounts and TSH is completely excluded from the fetal circulation thus allowing autonomous development of fetal thyroid function (see Chapter 29).

Pancreatic β-cells

Pancreatic β-cell function is profoundly changed in pregnancy in ways that enhance the availability of the major metabolic fuel of the fetus (glucose). In early pregnancy, maternal glucose concentrations fall slightly due to β-cell hyperplasia and increased insulin secretion caused by oestrogen, but as pregnancy advances resistance to insulin at a postreceptor site is induced by rising levels of placental lactogen (hPL) and prolactin (hPRL). As a consequence, both glucose concentrations and insulin concentrations rise, simulating a mild non-insulin dependent diabetic state. Glucose levels are increased further both by cortisol which reduces glucose utilization and by glucagon which promotes both glycogenolysis and gluconeogenesis. In addition, pregnant women show an 'accelerated starvation' response to food deprivation in which lipolysis is increased to conserve glucose. The overall effect of these metabolic changes is to give the fetus and placenta priority in the use of maternal carbohydrate stores even when food supplies are limited.

Adrenal cortex

The secretion of cortisol by the maternal adrenal cortex is unchanged in pregnancy. However, oestrogen increases synthesis of cortisol-binding globulin (CBG) leading to a three-to fourfold increase in plasma levels of CBG and reduced clearance of cortisol.

A two–to threefold increase in unbound cortisol and a similar increase in urinary excretion is not associated with depressed levels of plasma ACTH probably due to resetting of the ACTH feedback control mechanism by placental steroid, but a contribution from placental secretion of

ACTH has been suggested. The placenta produces a pro-opiomelanocortin (POMC)-like molecule which gives rise to several hormones including chorionic corticotrophin and α-melanocyte-stimulating hormone (α-MSH) but whether these have physiological significance remains uncertain. The characteristic pigmentation of pregnancy could be related to placental production of α-MSH but this seems unlikely since similar pigmentation is seen in non-pregnant women taking oestrogen-containing oral contraceptives. The secretion of aldosterone increases many-fold through a complex mechanism depending on oestrogen-induced synthesis of angiotensinogen, the precursor of angiotensin I which gives rise to angiotensin II. As with the levels of free cortisol, the increased production of aldosterone does not give rise to overt clinical signs of hypersecretion.

Parathyroid

The high calcium and phosphorus requirements of the fetus as bone mineralization accelerates in the third trimester are met by active placental pumps, which are probably regulated by fetal $1,25\,(OH)_2$ vitamin D. Maternal plasma calcium concentrations tend to fall slightly but are maintained near normal values by increased secretion of parathyroid hormone, reduced secretion of calcitonin, and raised levels of $1,25\,(OH)_2$ vitamin D (see Chapter 30).

Decidua

Under the influence of oestrogen and progesterone the human uterine endometrium forms the decidua, which is characterized by the conversion of the small stromal fibroblasts to large lipid-containing cells. The decidua secretes a number of hormones including prostaglandins (see below), prolactin, relaxin, and renin. Prolactin and renin are produced in relatively large amounts and accumulate in high concentrations in the amniotic fluid. The secretion of prolactin appears to be regulated differently from pituitary prolactin, being stimulated by arachidonic acid but not by thyrotrophin-releasing hormone (TRH) or dopamine antagonists. A decidual prolactin-releasing protein and a decidual prolactin inhibitory protein have been isolated but their regulation is unclear. The physiological role of the protein hormones, prolactin, relaxin, and renin, is not understood, although it has been suggested that they all have paracrine actions; prolactin may control the movement of water and electrolytes across the fetal membranes and relaxin may contribute to the biochemical changes in the cervical connective tissues before labour.

Placental steroids

A detailed description of placental steroidogenesis will be found in Chapter 3. The following section is concerned with the patterns of plasma concentrations of progesterone and oestrogens observed in various species and the factors underlying the differing patterns.

Progesterone

The various patterns of plasma progesterone concentrations can be grouped into two types.

1. Progressive increase through pregnancy with an abrupt fall prepartum;
2. Progressive increase through pregnancy with no fall prepartum.

An abrupt fall in concentration prepartum is much the most common pattern observed in most domestic and experimental animals as well as in wild species. As a result the 'progesterone withdrawal' theory to explain parturition regardless of species was strongly supported until other patterns were recognized. The fall in progesterone concentration depends on one or other of two mechanisms. In the first, which occurs in members of the rodent, canine, feline, caprine, and other families, the maintenance of pregnancy is dependent on the production of progesterone by the corpus luteum throughout pregnancy, and parturition is preceded by luteolysis. In the second, which occurs in members of the ovine and bovine families, the main-

tenance of pregnancy becomes independent of the corpus luteum in the latter half of pregnancy when the placental production of progesterone meets the requirements. The fall in progesterone concentrations results from cortisol-induced metabolism within the placenta (see below).

In primates (including man but excluding the marmoset) and hystricomorph rodents (such as guinea-pigs), the placenta is the major source of progesterone throughout most of pregnancy but it lacks a cortisol-inducible 17α-hydroxylase. Consequently, progesterone secretion continues unchanged up to term and the concentration in plasma does not fall (Fig. 36.1). In general, the concentration of circulating progesterone in late pregnancy differs little from one species to another with a few notable exceptions including man and hystricomorph rodents in which the concentration is an order of magnitude higher. The high levels in hystricomorphs are attributable to the production of a specific, high-affinity, progesterone-binding globulin. Most, if not all, body systems respond to progesterone in ways that enhance the maternal adaptation to pregnancy but none is more important than the uterus in which the response is essential for the maintenance of pregnancy. By acting on the smooth muscle of the myometrium, the fibroblasts of the cervix and the stromal cells of the decidua, progesterone ensures that uterine activity is minimal and ineffective and that the cervix remains indistensible. These actions are mainly inhibitory and include inhibition of gap-junction formation, prostaglandin synthesis, oxytocin receptors, collagenolytic activity, and smooth-muscle cell contractility. In part, these actions are

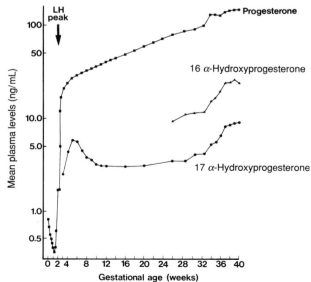

Fig. 36.1 Mean plasma levels of progesterone and metabolites throughout human pregnancy. The pattern of 17 α-hydroxyprogesterone levels during the first 12 weeks of pregnancy reflects mainly the function of the corpus luteum. LH, Luteinizing hormone.

indirect, being mediated by antagonism of stimulatory effects of oestrogen that become evident when the concentration of progesterone falls.

Oestrogen

The sources of oestrogen in pregnancy are the ovaries, the placenta, or both of these. In general, when pregnancy is dependent on the corpus luteum to term and parturition is preceded by luteolysis, the placenta lacks aromatase and the ovary is the sole source of oestrogen. In the dog and many other species, luteolysis is associated with a marked fall in oestrogen levels but the rat differs in that oestrogen levels are unaffected by luteolysis probably because the ovarian stroma uses placental androgen as the substrate for oestrogen synthesis. The goat is an exception in that the placenta is the major source of oestrogen although the corpus luteum remains indispensable to term.

Placental aromatase is invariably present in those species that lose dependence on the corpus luteum but they fall into one or other of two types depending on whether or not the placenta can convert progesterone to oestrogen. Those that lack placental 17α-hydroxylase activity and accordingly are unable to form androgen from preg-

nenolone or progesterone require an alternative source of androgen for placental oestrogen synthesis; in women and higher primates, androgen is supplied by the fetal zone of the fetal adrenal, whereas in the horse the androgen is derived from the fetal gonands. The extraplacental source of androgen also leads to the appearance of large quantities of unusual oestrogens in the maternal circulation. In human pregnancy, the main androgen secreted by the fetal adrenal is dehydroepiandrosterone (DHEA); the fetal liver promptly hydroxylates DHEA to form 16α-OH-DHEA which gives rise to oestriol when aromatized by the placenta. In the horse, the unusual oestrogens are equilin and equilenin, for which the gonadal substrate is unknown. In those species where the ovary or placenta is the source of oestrogen, the major oestrogen in the maternal circulation is usually oestrone with lesser amounts of the more potent oestradiol-17β.

Enormous quantities of oestrogen are produced in human pregnancy for reasons that are not apparent (Fig. 36.2). Much of the oestrogen is either very weakly potent (oestriol) or inactivated by conjugation. It is clear from pregnancies in which oestrogen excretion is less than 10 per cent of normal because of placental sulfatase deficiency that minimal production of oestrogen is compatible with normal pregnancy.

Initiation of parturition

Despite considerable progress in understanding the mechanism in initiating parturition in domestic and experimental animals, the mechanism in humans remains enigmatic. This stems in part from the difficulties, both technical and ethical, in studying the intrauterine environment in pregnant women but, more importantly, because of a fundamental difference in the control of oestrogen and progesterone synthesis and metabolism between the human and other primate species and those species like sheep and cattle (Fig. 36.3). In the latter species, the fetal trophoblast of the placenta contains not only an aromatase (cytochrome P450 aromatase) capable of converting androgen to oestrogen as in the human placenta but also a steroid 17α-hydroxylase (cytochrome P450 17α-hydroxylase) and C17–C20 desmolase (cytochrome P450 sidechain enzyme) that remove the side chain from progesterone thus forming androgen; human and nonhuman primate placentae lack 17α-hydroxylase and accordingly are incapable of converting progesterone to oestrogen. Thus sheep and other placentae can synthesize from cholesterol the full range of primary steroids including progesterone, oestrogens, and androgens, whereas the human placenta can convert cholesterol only to preg-

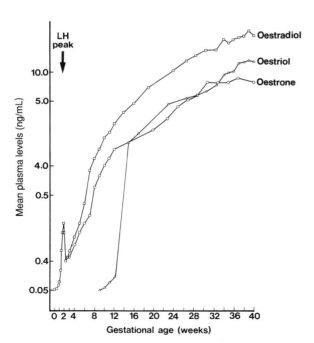

Fig. 36.2 Mean plasma levels of oestrogens throughout human pregnancy. The delay in the rise of oestriol levels reflects the onset of fetal adrenal production of androgen.

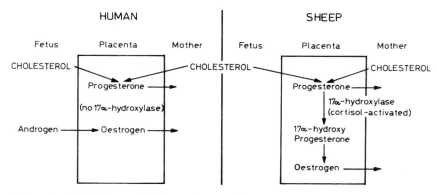

Fig. 36.3 Diagram illustrating the fundamental difference between placental steroid production in human and sheep pregnancies. The presence of the cortisol-inducible enzyme, 17α-hydroxylase, in sheep provides the means for the adrenal of fetal sheep to initiate parturition.

nenolone and progesterone; formation of oestrogen is dependent on a supply of oestrogen precursor from other sources, predominantly the specialized fetal zone of the adrenal, which generates large quantities of dehydro-epiandrosterone (DHEA) which is presented to the placenta mainly as 16-hydroxydehydroepiandrosterone sulfate (16-OH-DHEAS) after 16α-hydroxylation and sulfo-conjugation by the fetal liver. The lack of 17α-hydroxylase activity in the human placenta might be of little fundamental importance in relation to parturition were it not for the fact that the enzyme is inducible by cortisol in those species in which the enzyme is active. The human fetus shares with the fetuses of every species investigated to date an increase in adrenocortical activity leading to a rise in the concentration of cortisol, which has such a vital role in stimulating the maturation of various organs in preparation for birth (see Chapter 11). But the human placenta, lacking a cortisol-inducible 17α-hydroxylase, fails to recognize rising fetal cortisol concentrations, and levels of progesterone and oestrogen in the maternal circulation remain unchanged in the week before term. In sharp contrast, the increasing activity of 17α-hydroxylase in the sheep placenta leads to a rapid fall in levels of progesterone and an equally rapid rise in levels of oestrogen in the circulations of both fetus and mother.

Clearly, in the sheep and similar species, the fetus itself has rigid control of the time of parturition by means of the complex and as yet incompletely elucidated system that determines the rate of secretion of cortisol, thus ensuring close co-ordination of the preparations for birth and the timing of birth. Does the absence of such a system in the human fetus mean that it plays no part in determining the time of birth? The circumstantial evidence derived from experiments of nature provided by various congenital mal-

formations involving the hypothalamus, pituitary, or adrenals points to a contribution by the fetus in refining the timing but it is not essential. For example, in anencephaly in which the hypothalamus is absent and the pituitary is at least hypoplastic if not absent, the mean time of birth is normal term but the range around term is abnormally wide. Bilateral adrenal aplasia is not associated with prolonged pregnancy. Such evidence might be interpreted as an indication that the mother, not the fetus, determines the time of birth but such a conclusion runs counter to the overwhelming evidence implicating the fetus in a wide variety of other species. Furthermore, no disorder of the mother is known that delays timely delivery. If the mechanism initiating term parturition is attributable neither to the fetus itself nor the mother, where does it reside? Most investigators centre their studies on interactions of the placenta and fetal membranes with the contiguous maternal tissues, the endometrial epithelium and stroma. In other words, the mechanism in women is thought to be a paracrine rather than an endocrine system. Various hypotheses based on this assumption are described below.

Prostaglandins in parturition

Although the early links in the chain of events culminating in labour are poorly understood there is general agreement that prostaglandins, particularly prostaglandin $F_{2\alpha}$ ($PGF_{2\alpha}$), form the essential link near the end of the chain, regardless of species. In those species (like sheep) with placental 17α-hydroxylase, the stimulus to prostaglandin synthesis and release is relatively straightforward since the fall in progesterone concentration and the rise in oestrogen concentration at term is a potent stimulus. In women, how-

ever, the ratio of progesterone to oestrogen is unchanged
and the stimulus is more complex and poorly understood. Nevertheless, the evidence is convincing that
prostaglandins have the same essential role in women as in
other species and that an understanding of the control of
prostaglandin release will go far in explaining the mechanism of onset of labour. The evidence falls into four general categories.

1. The onset of labour is associated with a marked
   increase in the concentrations of $PGF_{2\alpha}$ and
   prostaglandin E_2 (PGE_2) in amniotic fluid and the
   maternal circulation. Because most prostaglandins are
   rapidly metabolized in the lungs, the change is most
   evident in assays of the major metabolites, PGFM and
   PGEM. Induction of the labour by amniotomy and
   infusion of oxytocin is likely to fail in the absence of a
   sustained increase in PGFM.
2. Administration of $PGF_{2\alpha}$ or PGE_2 at any stage of pregnancy causes abortion or labour. Provided that the
   prostaglandin is given in small amounts over a period
   of 8–12 h, the ensuing abortion or labour mimics
   exactly the natural process.
3. Physical, chemical, bacteriological, and other agents
   that stimulate the release of endogenous prostaglandin
   cause abortion or labour. For example, digital stretching of the cervix or amniotomy near term are promptly
   followed by release of $PGF_{2\alpha}$ and subsequently by
   labour. Similarly, infection of the amniotic sac is associated with both the release of $PGF_{2\alpha}$ and preterm
   labour.
4. Inhibition of prostaglandin synthesis by cyclooxygenase inhibitors such as indomethacin prevents the
   onset of labour and can temporarily arrest established
   labour.

Sites of synthesis of prostaglandins

For a comprehensive account of the synthesis and
metabolism of eicosanoids, the reader should consult
Chapter 35. The brief description that follows serves only
to place prostaglandins in context with the other factors
that may contribute to the initiation of term labour. The
spectrum of prostaglandins formed *in vitro* by samples of
the various uterine and intrauterine tissues varies considerably from one tissue to another. The uterine smooth
muscle generates predominantly prostacyclin (PGI_2), the
connective tissue of the cervix yields PGE_2, the decidua
forms both $PGF_{2\alpha}$ and PGE_2 with a predominance of
$PGF_{2\alpha}$ and the amnion produces mainly PGE_2. The
chorion has little capacity for synthesis but is active in
metabolism to the extent that PGE_2 generated by the

amnion is unlikely to penetrate to reach the uterine wall in
an unmetabolized and active form. The source of
prostaglandin that activates the uterus is generally thought
to be the decidua rather than the fetal membranes as high
levels of PGFM persist in the maternal circulation after the
placenta and membranes are delivered whereas those of
PGE_2 fall rapidly. An unusual physical property of all
prostanoids is that they are readily soluble in both water
and lipids and this makes them ideally suited to paracrine
functions. There is also no difficulty in accepting that
$PGF_{2\alpha}$ formed in the decidua can penetrate deeply into the
uterine muscle wall or the cervix. The hypothesis that
PGE_2 formed by the amnion is the major prostaglandin
involved in activation of the uterus has waned in popular-
ity since it was shown that 9-ketoreductase, the enzyme
that converts PGE_2 to $PGF_{2\alpha}$, is very low in activity in
intrauterine tissues and that penetration of the chorion is
unlikely.

The decidual cell-type responsible for the synthesis of
$PGF_{2\alpha}$ is uncertain because both the epithelium and the
stromal cells have the capacity for synthesis. However, the
stromal cell is favoured by some researchers because it is
has many properties in common with macrophages and,
indeed, may originate in the bone marrow.

Experiments in animals have identified a number of factors that either inhibit or stimulate prostaglandin production but in human pregnancy the factors responsible for the
release of prostaglandin associated with the onset of labour
remain hypothetical. Most of the studies in women are
necessarily restricted to tissues obtained at Caesarean section or from the placenta and membranes after delivery.
Extrapolation of the results to *in vivo* conditions should be
cautious.

Inhibitors of prostaglandin synthesis

Pregnancy is associated with a large increase in the capacity for prostaglandin production due mainly to specific
enrichment of glycerophospholipids at the *Sn*2 position by
arachidonic acid, the precursor of the prostanoids including
the lipoxygenase products, as well as the prostaglandins.
Clearly, synthesis is firmly inhibited under most conditions
to prevent inadvertent uterine activation. In some species,
production is rate-limited until late pregnancy by low activity of cyclooxygenase but the prompt (within minutes)
release of $PGF_{2\alpha}$ in the human uterus following
amniotomy, for example, is not consistent with a system
restrained by lack of an enzyme that must be synthesized to
increase activity. In general terms, it is more likely that
synthesis is limited by inhibitors of the activity of phospholipase A_2 (PLA_2) or cyclooxygenase (prostaglandin H synthase; PGHS) rather than by lack of enzyme.

Progesterone

In animals in which parturition is preceded by 'withdrawal' of progesterone, the administration of progesterone not only prevents parturition and prolongs pregnancy but also inhibits the release of prostaglandin. In human pregnancy, however, the maintained plasma concentrations of progesterone to term do not suggest 'withdrawal' of progesterone, and administration of large doses of progesterone does not prevent delivery at term. Furthermore, the concentration of myometrial nuclear receptors does not change and they remain fully saturated. Nevertheless, progesterone is probably an inhibitor of prostaglandin synthesis in humans as it is in other species. For example, surgical removal of the corpus luteum before 8 weeks of pregnancy causes abortion; treatment with progesterone prevents abortion and presumably also inhibits prostaglandin release. However, experience with the progesterone receptor blocker, RU486 (mifepristone), suggests that it is likely that inhibitors of prostaglandin synthesis other than progesterone are mainly responsible for suppressed prostaglandin production at least in the second and third trimesters. Treatment with RU486 is unreliable in causing abortion after 8 weeks unless accompanied by treatment with $PGF_{2\alpha}$ or a PGE_2 analogue.

The mechanism by which progesterone inhibits prostaglandin synthesis is uncertain. The release of arachidonic acid from perfused preparations of dispersed human decidual cells is markedly inhibited within 15 min by physiological concentrations of progesterone, indicating that the action depends on a direct physical effect on the activity of one or more phospholipases, probably PLA_2. However, other possible actions such as inhibition of PGHS or prostaglandin synthase have not been studied.

Recent preliminary evidence for an endogenous inhibitor of progesterone action that acts at a postreceptor site suggests that a fresh look at the progesterone withdrawal hypothesis in human pregnancy is needed.

Protein inhibitors

Amniotic fluid from women in late pregnancy inhibits prostaglandin production and various fractions containing uncharacterized proteins such as pregnancy-associated prostaglandin synthase inhibitor (PAPSI) and endogeneous inhibitor of prostaglandin synthase (EIPS) do likewise. Gravidin, an 80kDa protein is present in amniotic fluid in remarkably high concentrations. It is a potent $(10^{-11}M)$ inhibitor of arachidonic acid and prostaglandin release from dispersed human decidual cells, indicating first, that it is a PLA_2 inhibitor and, second, that it acts via a receptor-mediated system. Gravidin has been fully characterized and

sequencing shows it to be homologous if not identical with the secretory component of IgA, a protein bound tightly to IgA which facilitates the passage of IgA through cell membranes. It has not been recognized previously to have PLA_2-inhibitory activity and the rationale is unknown. Of particular interest is the observation that the inhibitory activity of gravidin generated by cultures of dispersed chorionic cells is lost when the cells are obtained after labour has started. Further studies are needed to test the hypothesis that inactivation of gravidin is a major contributor to the onset of labour. The factors controlling the synthesis of gravidin and the nature of its inactivation remain to be investigated.

Stimulators of prostaglandin synthesis

Oestrogen

The effect of oestrogen on prostaglandin production has not been studied in pregnant human tissues but explants of non-pregnant human endometrium respond to oestrogen with increased release of arachidonic acid and prostaglandins. Experimental animals such as sheep in which oestrogen levels are low until near term respond to treatment with oestrogen with substantial production of $PGF_{2\alpha}$. However, it is difficult to postulate a role for oestrogen in human pregnancy in which very high plasma levels of oestrogen are maintained to term. The administration of oestrogen to women at term is ineffective in inducing labour.

Oxytocin

The action of oxytocin is traditionally considered to be mediated by oxytocin receptors on the smooth muscle of the myometrium. Recent work has shown that oxytocin has a second and possibly equally important action on the decidua. Oxytocin in physiological concentrations stimulates arachidonic acid release from cultures of dispersed decidual cells, almost certainly by activating PLA_2 (Fig. 36.4). Oxytocin shares with oestrogen and progesterone a difficulty in being attributed a role in parturition because of unchanging plasma concentrations at the onset of labour. But unlike the receptors for two steroids, oxytocin receptors approximately double in concentration with the start of labour (Fig. 36.5). The increased receptor concentration could then lead to increased action of oxytocin despite the absence of a rise in the concentration of circulating hormone. Support for this hypothesis comes from the clinical observation that intravenous oxytocin infusion following amniotomy for induction of labour at term is associated with sustained release of $PGF_{2\alpha}$ when labour is successfully induced but only transient release when induction is unsuccessful.

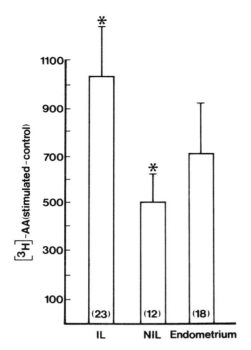

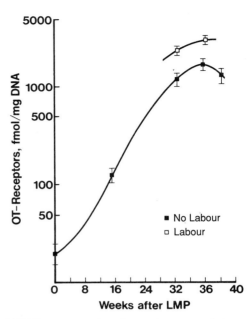

Fig. 36.4 Effect of a pulse of oxytocin on the release of radiolabelled arachidonic acid (AA) from perfused decidual cells from women in labour (IL), before labour (NIL), or not pregnant. Note the increased sensitivity during labour. *, $P<0.05$ for the difference between IL and NIL

Fig. 36.5 The concentration of oxytocin receptors in human myometrium. Note the log scale and that the concentration in labour is twice that before labour (LMP, last menstrual period).

An oxytocin analogue, Atosiban, which acts as a competitive antagonist of oxytocin, is effective in inhibiting the action of oxytocin on human myometrial tissue *in vitro*. Preliminary uncontrolled trials suggest that Atosiban inhibits uterine contractions in preterm labour but assessment of its effectiveness must await rigorously controlled studies. Similar studies to examine the ability of Atosiban to prolong normal pregnancy at term will present ethical problems that may prohibit this critical test of the role of oxytocin in initiating normal term parturition.

Platelet activating factor

Platelet activating factor (PAF), a metabolite of phosphatidylcholine, is known to stimulate prostaglandin synthesis in various tissues by increasing the concentration of cytosolic Ca^{2+}. It stimulates the production of PGE_2 by human amnion *in vitro* but effects on production of $PGF_{2\alpha}$ by decidua have not been reported. PAF is present in the majority of amniotic fluid samples from women in labour although absent from samples taken earlier in pregnancy. The source of amniotic PAF could be either fetal urine or fetal lung fluid, both of which contain PAF in measurable

quantities, and it has been suggested that the fetus triggers prostaglandin synthesis by one or other of these two routes. However, the hypothesis is not supported by the fact that neither the absence of kidneys nor complete atresia of the trachea is associated with prolonged pregnancy.

Cytokines and other factors

The most convincing evidence of a direct relationship between factors stimulating prostaglandin production and the onset of labour comes from studies of preterm labour associated with amniotic infection. The liberation of bacterial toxin (a lipopolysaccharide) into amniotic fluid is associated with the release of at least three cytokines, (interleukin-1α, interleukin 1-β, and tumour necrosis factor) into the amniotic fluid. However, cytokines are not present in normal labour and there is no evidence to suggest they have a role in normal parturition. Interleukin-8, a peptide that both attracts and activates neutrophils, is produced by chorion and decidual cells in normal pregnancy. The chemotactic action of interleukin-8 is greatly enhanced by PGE_2. Neutrophils entering the cervical connective tissue in labour may be a source of the collagenases responsible for the biochemical changes associated with cervical 'ripening'.

Mechanics of parturition

Labour is often thought of as no more than a contracting uterus and the essential contributions of changes in the nature of the myometrial contractions and in the structure of the uterine connective tissue, particularly of the cervix, are overlooked. The necessity for a complete complex of modifications is easily illustrated by describing the response to an intravenous infusion of oxytocin a few weeks before term. The uterus can be made to contract frequently, strongly, and even painfully for many hours but without any signs of progressive labour.

Smooth muscle

The uterine smooth muscle shares with smooth muscle in general a contractile mechanism dependent on the interactions of actin, myosin, and Ca^{2+} but differs in having relatively high concentrations of receptors for oxytocin, ovarian steroid hormones, and prostaglandins. It differs also in having few efferent neural connections as a result of almost complete degeneration of the uterine sympathetic nerves with advancing pregnancy. Thus myometrial activity becomes increasingly dominated by endocrine factors and, in late pregnancy, by paracrine factors.

Lacking a neural network, the myometrium is dependent for the spread of electrical spike potentials from a point of origin to other parts of the uterine wall on direct transmission from cell to cell. Through most of pregnancy, cell-to-cell transmission is hampered by a paucity of gap-junctions resulting in very limited propagation of spike potentials. At the onset of labour, however, gap-junctions appear in large numbers and the muscle behaves as an electrical syncytium so that potentials arising at a pacemaker site pass rapidly over the entire uterus causing a well-co-ordinated contraction. The strength of the contraction is greatest at the fundus because of the greater mass of smooth muscle. Furthermore, the muscle cells are arranged circularly in the inner wall and longitudinally in the outer wall so that a contraction exerts force both circumferentially and along the axis of the uterus.

Endocrine control

Inactivity of the smooth muscle throughout pregnancy is maintained by factors that either have direct inhibitory effects on smooth muscle cells or indirect effects mediated by inhibition of potential stimulatory factors. Of these, progesterone is considered to be the dominant factor; it not only inhibits the development of spike potentials, probably by causing hyperpolarization of the plasma membrane, but it also inhibits the formation of gap-junctions. In addition, progesterone contributes to the suppression of prostaglandin production and oxytocin receptors. The role, if any, of circulating relaxin in maintaining uterine quiescence is uncertain. Removal of the corpus luteum in late pregnancy results in disappearance of relaxin from the circulation but no increase in uterine activity.

The only known oxytocic hormone with a site of origin remote from the uterus is oxytocin which, although present in the maternal and fetal circulations in low concentrations throughout much of pregnancy, shows no tendency to increase until after labour is well established. The identification of oxytocin mRNA (although not oxytocin itself) in membranes raises the possibility that oxytocin has a paracrine as well as an endocrine role.

Paracrine control

While investigators are in general agreement that myometrial activity at the onset of labour is determined by products of the endometrium and fetal membranes, details of the mechanism remain unclear. An attractive hypothesis proposes that the synthesis of progesterone and oestrogen is regulated at a local level although the changes are not reflected in circulating concentrations. Such a system would make human parturition analagous to that in animals in which withdrawal of a 'progesterone block' is firmly established. In support of the hypothesis, it has been shown that dispersed cell preparations of chorion and decidua have the capacity to form progesterone from pregnenolone (which is present in the fetal circulation in high concentrations) and that this capacity is less in tissues obtained after the onset of labour. Likewise, the chorion and decidua contain steroid sulfatase activity which hydrolyses oestrogen sulfates to form free oestrogens; activity of the enzyme is greater after labour has started. Thus it is conceivable that the initiation of labour is associated with an increase in the ratio of oestrogen to progesterone at a local level resulting in both direct effects on the myometrium and indirect effects mediated by the release of prostaglandins. Preliminary evidence of a endogenous progesterone antagonist acting at a postreceptor site may require reassessment of the progesterone-withdrawal hypothesis.

Prostaglandins have multiple actions on the myometrium and undoubtedly are the major factor controlling its activity. Both $PGF_{2\alpha}$ and PGE_2 are oxytocic in that they acutely stimulate uterine contractions but are weak compared to oxytocin itself. Their more important effects have a relatively long latency and probably depend mainly on an increase in myometrial sensitivity to oxytocin and in the concentration of gap junctions.

Connective tissues

The connective tissue of the uterus is typical of that at other sites in being composed of dense bundles of collagen and small bundles of elastin supported in a matrix of proteoglycan in which the predominant glycosaminoglycans are strongly sulfated, particularly dermatan sulfate. However, the uterine connective tissue is unique in undergoing physiological changes prior to labour that are otherwise seen only in pathological states such as inflammation. The changes are most obvious in the human and sheep cervixes which contain approximately 90 per cent connective tissue. The 'unripe' cervix is a firm, rigid structure that is difficult to dilate forcibly without rupturing whereas the 'ripe' cervix is soft and readily distensible. Unless the cervix undergoes 'ripening', even the strongest uterine contractions are incapable of forcing the fetus through the human birth canal. In some species, such as the rabbit and the horse, the cervix is distensible throughout pregnancy and rapid delivery can be achieved before term by single injections of oxytocin. Clearly, the biochemical changes causing increased distensibility of the uterine connective tissues are fundamental to the success of parturition.

The biochemistry of cervical ripening is well understood although the factors initiating the changes are not. Control of the composition of the connective tissue resides in the fibroblasts, which not only synthesize and secrete the fibrous component and the proteoglycan matrix but also secrete a variety of proteases that degrade them. The tightly bound bundles of collagen are attacked by collagenases arising from fibroblasts and possibly also from neutrophils, loosening the bundles and dissolving the old collagen which is replaced by new, weakly cross-linked collagen. The highly charged sulfated glycosaminoglycans that bind the bundles are reduced in concentration and are replaced by the less charged heparan sulfate, allowing the bundles to slip on each other and to be exposed to the action of collagenases. The overall effect of the altered architecture is a softer connective tissue that has lost much of its ability to resist stretch.

Local intravaginal treatment with either prostaglandins or relaxin reproduces all the above biochemical changes. In view of the failure of lutectomy (ablation of the corpus luteum) to prevent normal cervical ripening, it is difficult to invoke relaxin as having a major role as it has in some species such as pigs. However, the decidua secretes relaxin and it is conceivable that it acts locally on the cervix. Prostaglandins, both $PGF_{2\alpha}$ and PGE_2, fit the role more

easily since they are released in increasing amounts in labour. Administration of prostaglandins locally or parenterally in appropriate doses at any stage of pregnancy causes a good replica of spontaneous labour with cervical ripening preceding the onset of painful uterine activity. Prostaglandin inhibitors such as indomethacin arrest not only the uterine activity of preterm labour but also may reverse premature cervical ripening.

Conclusion

Although investigators are generally agreed that human parturition is the outcome of paracrine interactions of the chorioamnion (fetus) and the decidua/myometrium (mother), opinions differ as to the relative importance of prostaglandins, oxytocin, gap-junctions, and progesterone. Ideally, of course, all four would combine their actions. The simplest hypothetical mechanism by which this could be achieved is by inhibiting the action of progesterone, which would have the effect of promoting uterine contractility, stimulating the formation of both oxytocin receptors and gap-junctions, and enhancing the release of prostaglandins. The weight of evidence is against progesterone 'withdrawal', but the recent finding of a protein that antagonizes progesterone at a postreceptor site should encourage further work in this direction. Nevertheless, the failure of the progesterone antagonist, mifepristone, to reliably cause midtrimester abortion unless supported by administration of a prostaglandin suggests that progesterone 'withdrawal' in itself (when it does occur) is insufficient to initiate parturition but requires an additional mechanism such as inactivation of gravidin to promote prostaglandin release. Clearly, the mechanism initiating human parturition is far from resolved.

Further reading

Casey, M.L., MacDonald, P.C., and Mitchell, M.D. (1988). Decidual activation: the role of prostaglandins in labor. In *The onset of labor: cellular and integrative mechanisms* (ed. D. McNellis, J.R.G. Challis, P.C. MacDonald, P. Nathanielsz, and J. Roberts), pp. 141–56. Perinatology Press, Ithaca, New York.

Challis, J.R.G. and Olson, D.M. (1988). Parturition. In *The physiology of reproduction* (ed. E. Knobil and J.D. Neill), pp. 2177–215. Raven Press, New York.

de Zegher, F., Vanderschueren-Lodeweyckx, M., Spitz, B., Faijerson, T., Blomberg, F., Beckers, A., Hennen, G., and Frankenne, F. (1990). Perinatal growth hormone (GH) physiology: effect of GH-releasing factor on maternal and fetal secretion of pituitary and placental GH. *Journal of Clinical Endocrinology and Metabolism*, **71**, 520–2.

Fuchs, A.R., Fuchs, F., Husslein, P., and Soloff, M.S. (1984). Oxytocin receptors in the human uterus during pregnancy and parturition. *American Journal of Obstetrics and Gynecology*, **150**, 734–41.

Garfield, R.E. and Hayashi, R.H. (1981). Appearance of gap junctions in the myometrium of women during labor. *American Journal of Obstetrics and Gynecology*, **140**, 254–9.

Handwerger, S., Markoff, E., and Richards, R. (1991). Regulation of the synthesis and release of decidual prolactin by placental and autocrine/paracrine factors. *Placenta*, **12**, 121–30.

Jenkins, J.S. and Nussey, S.S. (1991). The role of oxytocin: present concepts. *Clinical Endocrinology*, **34**, 515–25.

Liggins, G.C. (1993). The placenta and the control of parturition. In *The human placenta* (ed. C.W.G. Redman, I.L. Sargent, and P.M. Starkey), pp. 273–91. Blackwell Scientific Publications, Oxford.

Nissim, M., Giorda, G., Ballabio, M., D'Alberton, A., Bochicchio, D., Orefice, R., and Faglia, G. (1991). Maternal thyroid function in early and late pregnancy. *Hormone Research*, **36**, 196–202.

Sanborn, B. and Anwer, K. (1990). Hormonal regulation of myometrial intracellular calcium. In *Uterine contractility* (ed. R.E. Garfield), pp. 69–82. Serono Symposia, Norwell, Massachusetts.

Schulte, H.M., Weisner, D., and Allolio, B. (1990). The corticotrophin releasing hormone test in late pregnancy: lack of adrenocorticotrophin and cortisol response. *Clinical Endocrinology*, **33**, 99–106.

Index